Algorithms for Intelligent Systems

This book series publishes research on the analysis and development of algorithms for intelligent systems with their applications to various real world problems. It covers research related to autonomous agents, multi-agent systems, behavioral modeling, reinforcement learning, game theory, mechanism design, machine learning, meta-heuristic search, optimization, planning and scheduling, artificial neural networks, evolutionary computation, swarm intelligence and other algorithms for intelligent systems.

The book series includes recent advancements, modification and applications of the artificial neural networks, evolutionary computation, swarm intelligence, artificial immune systems, fuzzy system, autonomous and multi agent systems, machine learning and other intelligent systems related areas. The material will be beneficial for the graduate students, post-graduate students as well as the researchers who want a broader view of advances in algorithms for intelligent systems. The contents will also be useful to the researchers from other fields who have no knowledge of the power of intelligent systems, e.g. the researchers in the field of bioinformatics, biochemists, mechanical and chemical engineers, economists, musicians and medical practitioners.

The series publishes monographs, edited volumes, advanced textbooks and selected proceedings.

All books published in the series are submitted for consideration in Web of Science.

More information about this series at https://link.springer.com/bookseries/16171

Garima Mathur · Mahesh Bundele ·
Mahendra Lalwani · Marcin Paprzycki
Editors

Proceedings of 2nd International Conference on Artificial Intelligence: Advances and Applications

ICAIAA 2021

Springer

Editors
Garima Mathur
Poornima College of Engineering
Jaipur, India

Mahesh Bundele
Poornima College of Engineering
Jaipur, India

Mahendra Lalwani
Rajasthan Technical University
Kota, Rajasthan, India

Marcin Paprzycki
Polish Academy of Sciences
Warsaw, Poland

ISSN 2524-7565 ISSN 2524-7573 (electronic)
Algorithms for Intelligent Systems
ISBN 978-981-16-6331-4 ISBN 978-981-16-6332-1 (eBook)
https://doi.org/10.1007/978-981-16-6332-1

This Springer imprint is published by the registered company Springer Nature Singapore Pte Ltd.
The registered company address is: 152 Beach Road, #21-01/04 Gateway East, Singapore 189721, Singapore

Preface

The Conference on Artificial Intelligence: Advances and applications-2021 is the Second International Conference that has been organized with publication support of Springer. It is jointly organized by Poornima College of Engineering, Jaipur, RTU (ATU) TEQIP-III sponsored Rajasthan Technical University, Kota, Rajasthan, India, in Association with Soft Computing Research Society. The objective of the conference was to bring together researchers from academia, government agencies, research laboratories and the corporate sector from all over the world to present their research works carried out in artificial intelligence domain. Artificial intelligence research is leading to evolve in the area of machine learning, deep learning applications in health care, agriculture, business and security, etc. It covers research in core concepts of computer networks, intelligent system design and deployment, real-time systems, WSN, sensors and sensor nodes, SDN, NFV, etc. It is useful for students, faculty and industry working on AI applications. AI is an emerging technology that shall be used in almost all applications/technologies/systems in one or other form.

On behalf of RTU Kota, Poornima College of Engineering, I am pleased to welcome all the readers of Proceedings of International Conference on Artificial Intelligence: Advances and Applications-ICAIAA 2021.

This conference has provided an environment to conduct intellectual discussions and exchange ideas that are instrumental in shaping the future of artificial intelligence. It has been planned in two tracks, viz. Track 1: Artificial Intelligence and Track 2: Intelligent Sensor, Devices & Application. The conference could get very good response from thirteen countries such as India, Japan, Singapore, Bangladesh, Brunei Darussalam, Ethiopia, Malaysia, Nigeria, Serbia, Viet Nam, South Africa, Turkey and Spain.

A total of 298 papers were received and were reviewed by 148 reviewers from different reputed institutions across India. A total of 97 papers were accepted for presentation in the conference, out of which 71 papers were registered and presented during March 27–28, 2021.

I am thankful to RTU Kota for giving this opportunity to organize this conference under TEQIP III project.

Jaipur, India Garima Mathur
Jaipur, India Mahesh Bundele
Rajasthan, India Mahendra Lalwani
Warsaw, Poland Marcin Paprzycki

Contents

About the Editors

Garima Mathur is currently working as Head of Department, Poornima College of Engineering, Jaipur since 1st July 2016. Previously she was working as Head of Department in Electronics & Communication Engineering at Jaipur Engineering College, Jaipur, Rajasthan, India since Oct. 2000 to Dec. 2015. She has total 20 years experience in teaching and research. She is working on various research projects sponsored by various agencies. She did his doctoral degree in Performance Evaluation of Modified Sphere Decoding Scheme for MIMO Systems. Dr. Mathur published and presented International, National Journals, Conferences, Symposium and Seminar more than 30 research papers. She has guided more than 15 M.Tech. Dissertation thesis. Her area of interest is Wireless channel, Channel modeling, and Ad hoc Networks. Dr. Mathur is a life member of the Institution of Electronics and Telecommunication Engineers (IETE), The Indian Society for Technical Education (ISTE), India, and many other professional bodies.

Mahesh Bundele is currently working as Principal and Director of Poornima College of Engineering, Jaipur since 1st September 2018. He has total 34 years experience in teaching and research. He has developed many unique research methodology concepts and implemented. He is the mentor and controller of quality research and publications at the Poornima. He is also responsible for inculcation of innovative and critical analysis concepts across the University and across the Poornima Foundation involving three other campuses. He did his doctoral degree in Wearable Computing and guiding research in Pervasive & Ubiquitous Computing, Computer Networks, Software Defined Networking. His areas of interests are also Wireless Sensor Networks, Algorithmic research, Mathematical Modeling, and Smart grids. He has more than 60 publications in reputed journals and conferences. He has been reviewer of few IEEE Transactions. He is actively involved in IEEE activities in Rajasthan Subsection and Delhi Section and holding the responsibility on Standing Committee of IEEE Delhi Section for Technical & Professional activities for controlling quality of conferences and publications in IEEE. He is also member of IEEE Delhi Section Execom. He has organized 5 editions of IEEE ICRAIE in India and Malaysia with conference position as General Co-Chair. He has organized PerCAA

2019 with Elsevier Computer Science Procedia as Program Chair. He is General Chair, Advisor in many other International Conferences including Springer ICAIAA 2019, 2021, IOP publication ICGSSET 2021 and IEEE Delhi Section Oversight committee member of all IEEE International Conferences being organized in IEEE Delhi Section jurisdiction since last 3 years. He is Editorial Manager for Elsevier's Computers and Electronics in Agriculture, Environmental Science and Pollution Research, Renewable and Sustainable Energy, Applied Mathematical Modelling etc. He has received Excellence in Leadership award by ITSR foundation in 2020.

Mahendra Lalwani is recently working as an Associate Professor in the Department of Electrical Engineering, Rajasthan Technical University, Kota. After completion of his M.Tech. in Power System from Malaviya National Institute of Technology, Jaipur, he served as an Assistant Professor in Stani Memorial College of Engineering and Technology, Phagi, Jaipur where he held various administrative and academic positions. He received his Ph.D. from MNIT, Jaipur in 2013 and then joined RTU, Kota. He has more than 17 year experience in teaching and research field. He is life member of IETE and ISTE. He has authored three books in the field of electrical engineering.

Marcin Paprzycki is an associate professor at the Systems Research Institute, Polish Academy of Sciences. He has an MS from Adam Mickiewicz University in Poznan, Poland, a Ph.D. from Southern Methodist University in Dallas, Texas, and a D.Sc. Degree from the Bulgarian Academy of Sciences. He is a senior member of IEEE, a senior member of ACM, a Senior Fulbright Lecturer, and an IEEE CS Distinguished Visitor. He has contributed to more than 450 publications and was invited to the program committees of over 500 international conferences. He is on the editorial boards of 12 journals and a book series.

Chapter 1
Detecting Depressive Online User Behavior During Global Pandemic by Fusing LSTM and CNN Models

Bhuvaneswari Anbalagan

1 Introduction

Various mental health organizations have concluded that because of the stigma attached to mental health issues, people end up venting out on social media about their feelings [1]. In 2020, Covid-19 has instigated momentous distress around the world. Unfortunately, there's still a huge stigma that surrounds mental health problems, and quite a large number of people opinion disorders during COVID lockdowns. Depression affects people from all areas of life, no matter their respective backgrounds. Because of the stigma attached to mental illness, many people suffer in silence and end up venting out on social media. So, the goal of the research is to analyze a user's Twitter post to detect if the person is under any forbidden mental health conditions. In order to assist the online users, a chatbot is developed for depression support. In this research paper, machine learning and Natural Language processing are applied to achieve our goal. The key objective of the research defines the factors to look for depression detection in Twitter posts and comments, in order to extract these factors from Twitter comments. According to various survey conducted, the wide-reaching of depressive conditions are found among the online users. The main motivation is the current pandemic situation [2] where depression and depressive tendencies have increased dramatically due to an uncertain future because of lockdown, unstable job market, health concerns, etc. Psychological disorders such as downheartedness can be lifecycle bullying, with depression as a conceivable result [3]. The lifetime danger of self-destruction among mental illnesses can be identified among 20% of online users with untreated burdensome issues. Various survey conducted for psychological well-being says that individuals are utilizing online media as a substitute for depressive opinion posting podium and alternate approach to vent criticism [4].

B. Anbalagan (✉)
School of Computing Science Engineering, Vellore Institute of Technology, Chennai, Tamil Nadu, India

G. Mathur et al. (eds.), *Proceedings of 2nd International Conference on Artificial Intelligence: Advances and Applications*, Algorithms for Intelligent Systems,
https://doi.org/10.1007/978-981-16-6332-1_1

The rest of the paper is organized as follows. Section 2 presents the related work on user behavior analysis using online social media. Section 3 presents the problem formulation and contributions, and fusion of LSTM (LSTM) and Convolution Neural Networks (CNN) models are applied on Twitter data to detect if the person who tweeted is depressive or not. The chatbot for depressive users are developed is discussed in Sect. 4. The experiments and results demonstrate the fusing model of deep learning techniques and are discussed in Sect. 5. Section 6 concludes the paper with significant future directions.

2 Related Work

Various research works are carried out to attain certain optimization erudition background that exploits technique [5] content information of microblogging messages and the relations between result classes. The limitations are that the accuracy level is moderate. Appreciative online health groups for depression like social network use morphological perceptions technique [6] for modality contribution analysis, feature extraction, and data collection. The performance speed of the model is slow. The depressive dictionary learning technique uses the image social networks technique [7] to find depressive, feature extraction where the precision results are moderate. The mental health-related depressive user detection technique [8] identifies demographic, clinical, and personal variables are detected using machine learning with limited accuracy. The future of mental healthcare technique [9] identifies peer support, emotion prediction using NLP concepts like n-gram and bag of words. The survey that conducted the study by the Roundtree technique [10] uses techniques like carrying out an experimental strategy to test online social network interferences which includes text scraping for online user sentiment examination with performance accuracy degradation.

Social media-conducted systematic review technique [11] uses lexicon-based filtering for scraping expression of desperate thoughts and feelings. It gives an unknown, open, and non-critical gathering for sharing encounters using machine learning. The online based social media in youth'<s method [12] were used search for subject heading using n-gram processing. The chatbot strategy [13] utilizes strategies like recognizing a client's temperament absolutely from cell phone sensor information, customized AI model dependent on client information; chatbot gives health proposals by means of miniature mediations.

Deep learning-enabled machine learning methods and applications, CNN [14] and LSTM [15], utilize strategies like Textual substance investigation, for example, dictionary-based sifting, and Word Cloud perception. Feature designing includes even, printed, and emotional elements, and profound learning-based portrayal learning like LSTM-based content encoders. Its restriction is annotation bias, data imbalance, and data deficiency. Based on the literature survey, a novel CNN and LSTM fusion framework is proposed to resolve using renowned deep learning algorithms. Recently, machine learning [16, 17] and deep learning [18] have been deployed for Covid-19 prediction. Various research works discussed in [19–21] about social network datasets are referred to in the proposed work.

3 Proposed LSTM + CNN Fusion Methodology

For depression detection, a deep learning-enabled classification baseline model is developed by fusing LSTM and CNN. The online users Twitter posts during lockdown is collected for the proposed work implementation. The python package Keras is used to decide to check the social media user posts are depressive or not. Finally, we evaluate each model and compare its accuracies. Deep Learning techniques including 1-layer LSTM, 2-layer LSTM, and CNN + 2-LSTM are carried out in the datasets. LSTM is a recurrent neural network which by default retains the information for a long period of time in memory.

In our training model, LSTM is equipped for learning long haul conditions and function estimation on a number of classifications. While RNNs undergo the issues with long pull conditions, LSTMs are unmistakably intended to evade the drawn-out dependence issue. The following equations govern our training:

$$Z_k = \sigma(W_k[h_{k-1}, x_k]) \tag{1}$$

$$r_k = \sigma(W_r[h_{k-1}, x_k]) \tag{2}$$

$$\tilde{h}_k = \tanh(W[r_k * h_{k-1}, x_k]) \tag{3}$$

$$h_k = (1 - z_k) * h_{k-1} + z_k * \tilde{h}_k \tag{4}$$

where $x_{k-1}, r_{k-1}, z_{k-1}, \tilde{h}_{k-1}$ represents the input, reset gate, update gate, and reset memory, respectively. The following is a real tweet generated by our model and its process is given below.

$$\begin{aligned}\langle \text{begin} \rangle \rightarrow h_1 =& (k_1,\ k_2,\ \ldots,\ k_{|V|}) \rightarrow \text{stress} \rightarrow h_2 = (k_1,\ k_2,\ \ldots,\ k_{|V|}) \\ &\rightarrow \text{anxiety} \rightarrow h_3 = (k_1, k_2,\ \ldots,\ k_{|V|}) \rightarrow \text{depressive mood} \\ &\rightarrow h_4 = (k_1,\ k_2,\ \ldots,\ k_{|V|}) \rightarrow \langle \text{end} \rangle\end{aligned}$$

The LSTM model is used to train the model for epoch times using Eqs. (1)–(4). The LSTM + CNN model takes in an input. Each input sentence is replaced by its embeddings, and then the model runs the new embedding vector through a convolutional layer. The convolutional layer then passes the structure that it learns from the sequential data into an LSTM layer. The output is a single number representing the probability that the tweet indicates depression or not. The output of the LSTM layer is then fed into a dense model for prediction. The training includes fusing LSTM and CNN with epoch times. The K-fold cross validation is applied to the datasets. The fusing of LSTM and CNN models takes in an input and then outputs a single number representing the probability that the tweet indicates depression. The model takes in each input sentence, replaces it with its embedding's, then runs the new embedding vector through a convolutional layer. CNNs are excellent at learning spatial structure from data; the convolutional layer takes advantage of that and learns some structure from the sequential data then passes into a standard LSTM layer. Last but not the least, the output of the LSTM layer is fed into a standard dense model for prediction.

The model is trained for epoch time, and the Early Stopping argument is used to end training if the loss or accuracy doesn't improve within 3 epochs. In order to evaluate the effectiveness of the LSTM + CNN model, a logistic regression model is trained with the same train data and the same number of epochs, and tested with the same test data. The logistic regression model is used for predicting the categorical dependent variable using a given set of independent variables. Logistic regression is used for solving the classification problems, which predicts two maximum values (0 or 1). Logistic Regression is a significant machine learning algorithm because it has the ability to provide probabilities and classify new data using continuous and discrete datasets (Fig. 1).

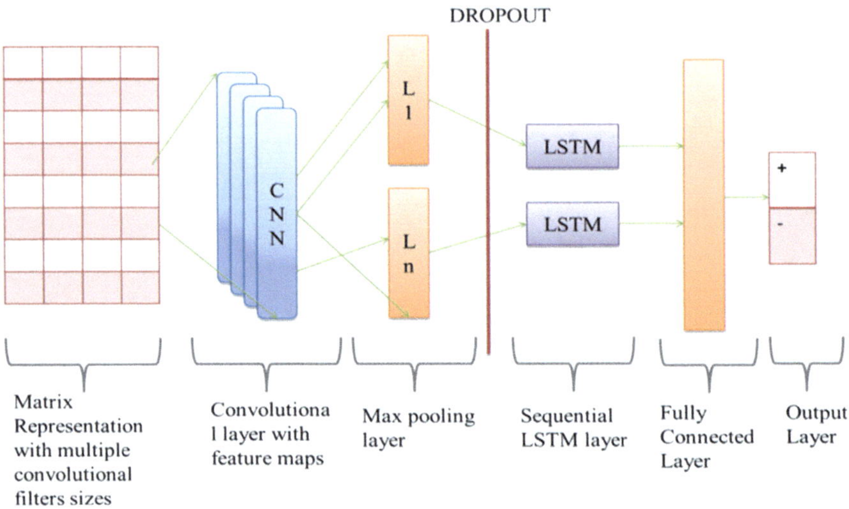

Fig. 1 Proposed fusion LSTM + CNN Model

4 Chatbot Assistance for Depressive Users

The homogenous depressive online users are identified and clustered using the Jaccard Coefficient Index which is used to find the distance between user pairs via K-Means. Initially, the tweets are classified as similar users who retweeted similar tweets. Let $rt_a andrt_b$ be the pair of tweets for which the number of similar users is identified who shared or retweeted the same tweet. Let the cluster of users who retweeted be rt_a be U_A.The overlap in the number of users who retweeted both $rt_a andrt_{b,}$ in clusters U_A and U_B as U_{AB}, is found using Jaccard Coefficient Index which is represented as $J(U_A, U_B)$ of common Twitter users.

$$J(U_A, U_B) = \frac{|A \cap B|}{|A| + |B| - |A \cap B|} \tag{5}$$

If $J(U_A, U_B)$ exceeds the threshold value TS, then link $rt_a andrt_b$ to the same cluster. Similar users with the same depressive behavior are identified and assisted using a chatbot. The chatbot is developed using Multi-Layer Perceptron (MLP). During the execution of the proposed dynamic chatbot, the user interface receives the text the user types as input, applies the MLP model on the chat text, and classifies the text as fear or sadness or joy or anger. Then it will ask the user if they want a suggestion. If the user presses 'Y', then the chatbot will show them suggestions like therapists, tips, self-evaluation quiz, and online group therapy form, if the text is classified as anger or sadness or fear. The computation is carried out from the last layer, backward to the first layer. During forward propagation, ReLU and Softmax activation functions are used. In the later part, the linear part of backward propagation and loss function is found. Finally in backward propagation, the linear part of backward propagation is implemented and weights are adjusted. The number of hidden layers is set as 100 and the number of iterations is set as 50. Given two probability distributions of two texts, Twitter post A and Twitter post B, the KL divergence calculates the expected number of extra data necessary to code phrases from A when using a code based on B. It obtains the entropy of A and B. Given a discrete random variable k over topics, the probability distributions of A and B, KL divergence is defined as follows:

$$D_{KL}(A||B) = H(A, B) - H(A) \tag{6}$$

where H(A, B) is the cross-entropy of A and B, and H(A) is the entropy of A. The KL divergence is asymmetric; the real distance function is calculated as follows:

$$D_T(A, B) = \frac{1}{2}(D_{KL}(A||B) + D_{KL}(B||A)) \tag{7}$$

The D_T is symmetric real distance function, although it reserves the benefits of KL divergence to validate the effectiveness of the proposed deep learning classification model.

5 Experiments and Results

The depressive tweets are scraped from Twitter and random tweets from the real-time dataset from Twitter and loaded for the deep learning algorithm. The dataset is loaded from a stored folder. The training data is stored as a text file and each tweet is accompanied by the magnitude of its sentiment 0–1. The tweets are observed such that a threshold of 0.5 is good enough to classify a tweet according to its sentiment. Tweets with lesser threshold were not definitive to be trained as per their mentioned classification. Using Twitter API, the tweets are crawled during Covid-19 lockdown as 40,000 tweets are collected, out of which 35,000 tweets are considered to be valid tweets suitable for training. In data preprocessing, cleaning the tweet data removes punctuations, name tags, and converts everything into lower case. Tokenization is the process of tokenizing or splitting a string, text into a list of tokens. The input data is pre-trained using tokenization as parts like a word is a token in a sentence, and a sentence is a token in a paragraph. An embedding matrix is a linear mapping from the original space (one-of-k) to a real-valued space where entities can have meaningful relationships using bag of words. Split the data into test and train data, and shuffle the order of data. Load the pre-trained vectors for the Word2Vec model. The sample tweet classification is shown in Table 1. Using a Keras Vectors file, we can get the embedding of any word by calling word_vec(word) to obtain all the words in the model's vocabulary. Use a Tokenizer to assign indices and filter out unfrequented words. Tokenizer creates a map of every unique word and an assigned index to it. The parameter called num_words indicates that we only care about the top 20,000 most frequent words. Lexicon padding processing Fig. 2 sequences all to the word cloves with the same length of 0–140 words.

The embedding matrix is an n x m matrix where n is the number of words and m is the dimension of the embedding. In this case, m = 300 and n = 20,000. The input matrix takes the minimum number of unique words in our tokenizer and maximum words in case there are less unique words than the specified maximum value. The data is chunked and formatted by assigning labels to the depressive tweets and random tweet data, and splitting the arrays into test (80%), validation (20%), and train data (20%). The depressive tweet and random tweet arrays are combined and shuffled to give as input to the training model. The training and validation accuracy using LSTM (1-layer and 2-layer) is shown in Figs. 3 and 4.

The randomized tweets lead to the use of a neural network with LSTM and CNN. The training and validation accuracy using LSTM + CNN (1-layer and 2-layer) is shown in Figs. 5 and 6. The neural networks applied on the tweets give almost

Table 1 Sample Tweet dataset—classification labeling

#	Tweet	Classification
1	I don't know what to do anymore I lost my job	Depressive Post
2	I am so fed up with my life	Depressive Post
3	I watched new web series on Netflix	Not Depressive Post

Fig. 2 1-layer LSTM

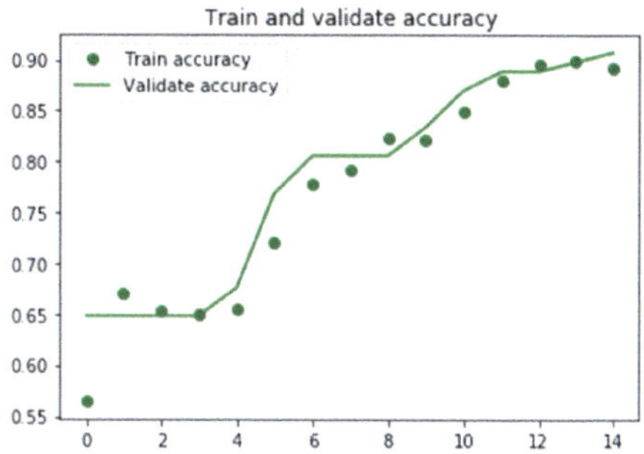

Fig. 3 2-layer LSTM

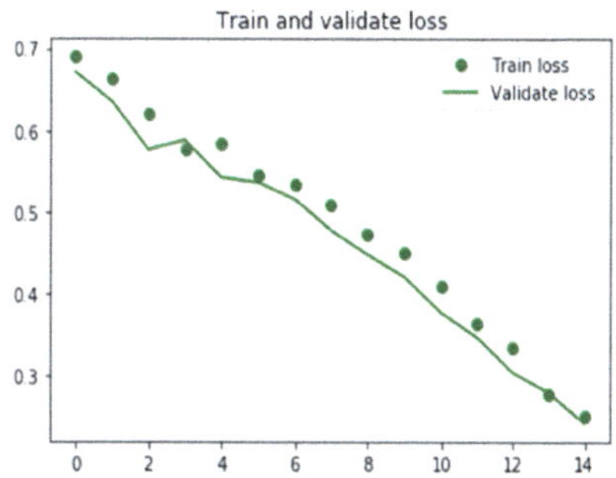

Fig. 4 1-layer LSTM + CNN

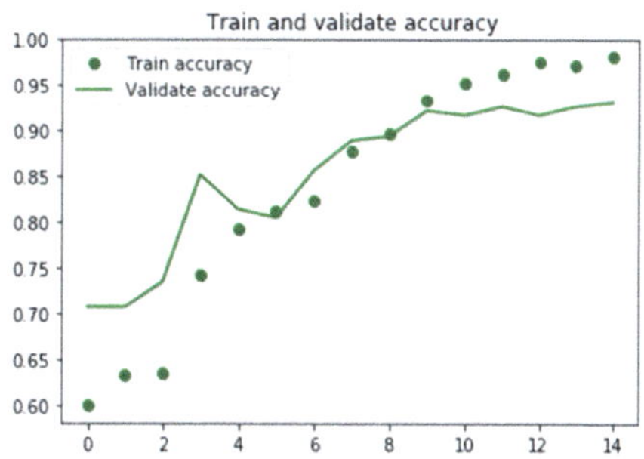

Fig. 5 2-layer LSTM + CNN

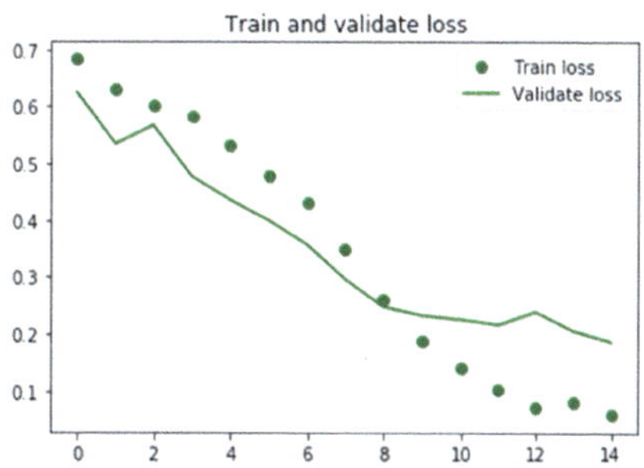

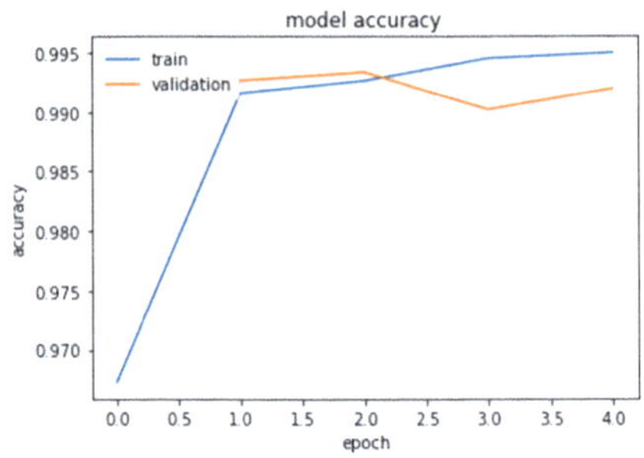

Fig. 6 Fusion model accuracy

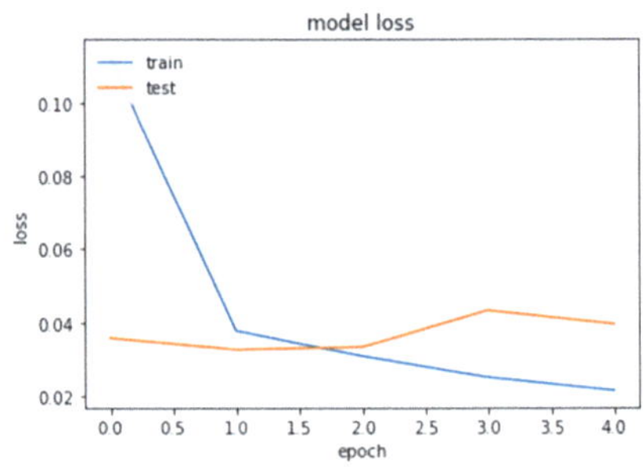

Fig. 7 Fusion model loss

the same accuracy after 5 iterations, although CNN with 2 layers produces higher accuracy when compared to LSTM. The experiments are targeted to determine in empirical LSTM + CNN method, where global information is supportive to classify locally, even though each tweet sample is presumed to be independent of one another. It is observed from Figs. 6 and 7 that there is an acute increase in validation accuracy by increasing the number of iterations for LSTM and CNN networks, using Word Embedding, classifier ensembles, and lexicons.

From the above, we can find the fusion of LSTM + CNN has higher accuracy of 90% than logistic regression and is hence more preferable as given in Table 2. Using Jaccard Coefficient, similar depressive users are identified as 78% among the non-probabilistic sampling method.

The KL-divergence is chosen for the experiments to estimate PMI score which identifies the overlap between original Twitter post and classified depressive Twitter post for discrete distributions. It is chosen to obtain message similarity which

Table 2 Proposed model accuracy

Model	Accuracy
Logistic regression	0.73
LSTM (2 layer) [14]	0.85
CNN [15]	0.79
Proposed Fusion LSTM + CNN	0.90

is invariant under parameter transformations. During the validation process KL-divergence, it shows that the overall probability distribution of KL-divergence between users' Twitter posts leads to a very negligible information loss up to 0.03 which is practically acceptable.

6 Conclusion and Future Work

During Covid-19 pandemic lockdown situations, a huge stigma surrounds depressive health problems and quite a large number of people opinions disorders. In this paper, a technical solution is proposed to identify demographic, clinical, and personal variables associated with depressive health-related via deep learning methods. The proposed fusing of LSTM and CNN detects depressive ideation using a user's Twitter post and assists the online user through a depression chatbot which the user can interact with to obtain suggestions. The experiment is carried out with 1-layer LSTM, 2-layer LSTM, and fusion of LSTM + CNN with a maximum accuracy of 90% which is more preferable when compared with state-of-the-art models. The proposed model is validated using KL-divergence with a negligible information loss of up to 0.03 which is practically acceptable. A dynamic chatbot is deployed to assist 78% of similar depressive online users. As a part of future work, the system can be trained and tested for dynamic datasets which can be handled by multiple social networking platforms and streaming frameworks.

References

1. Ji S, Pan S, Li X, Cambria E, Long G, Huang Z (2020) Suicidal ideation detection: A review of machine learning methods and applications. IEEE Trans. Computer Soc. Syst.
2. Rehman U, Shahnawaz MG, Khan NH, Kharshiing KD, Khursheed M, Gupta K, Kashyap D, Uniyal R (2020) Depression, anxiety and stress among Indians in times of Covid-19 lockdown. Community mental health journal, pp 1–7
3. Chancellor S, De Choudhury M (2020) Methods in predictive techniques for mental health status on social media: a critical review. NPJ digital medicine 3(1):1–11
4. Mdhaffar A, Cherif F, Kessentini Y, Maalej M, Thabet, JB, Maalej M, Freisleben B (2019) DL4DED: Deep Learning for Depressive Episode Detection on Mobile Devices. In: Proceedings of International Conference on Smart Homes and Health Telematics. Springer, Cham, pp 109–121
5. Akbari M, Hu X, Nie L, Chua TS (2016) From Tweets to Wellness: Wellness Event Detection from Twitter Streams. In AAAI, vol 16, pp 87–93
6. Xu R, Zhang Q (2016) Understanding online health groups for depression: social network and linguistic perspectives. J. Med. Internet Res. 18(3): e63
7. Wang X, Jia J, Tang J, Wu, Boya, Cai L, Xie L (2015) Modeling emotion influence in image social networks. IEEE Trans Affect Comput 6(3):286–297
8. Villagonzalo K-A, Arnold C, Farhall J, Rossell SL, Foley F, Thomas N (2019) Predictors of overall and mental health-related internet use in adults with psychosis. Psychiatry Res 278:12–18

9. Naslund JA, Aschbrenner KA, Marsch LA, Bartels SJ (2016) The future of mental health care: peer-to-peer support and social media. Epidemiology and psychiatric sciences 25(2):113–122
10. Roundtree AK (2017) Social health content and activity on Facebook: A survey study. J Tech Writ Commun 47(3):300–329
11. Jo, Robinson, Cox G, Bailey E, Hetrick S, Rodrigues M, Fisher S, Herrman H (2016) Social media and suicide prevention: a systematic review. Early Interv Psychiatry 10(2):103–121
12. Sedgwick R, Epstein S, Dutta R, Ougrin D (2019) Social media, internet use and suicide attempts in adolescents. Curr Opin Psychiatry 32(6):534
13. Ghandeharioun, Asma, Daniel McDuff, Mary Czerwinski, and Kael Rowan (2019) "EMMA: An Emotion-Aware Wellbeing Chatbot." In: Proceedings of 2019 8th International Conference on Affective Computing and Intelligent Interaction (ACII), IEEE, pp 1–7
14. Bhuvaneswari A, Timothy Jones Thomas J, Kesavan P (2019) "Embedded Bi-directional GRU and LSTMLearning Models to Predict Disasterson Twitter Data." Procedia Computer Science 165: 511–516
15. Bhuvaneswari A, Aishwarya K, Bhuvaneshwari S, Sai Chandni C, Sundara Akilesh P (2020) "Detecting New Events from Microblogs Using Convolutional Neural Networks." In Advances in Electrical and Computer Technologies, Springer, Singapore, pp 1–9
16. Singh V, Poonia RC, Kumar S, Dass P, Agarwal P, Bhatnagar V, Raja L (2020) Prediction of COVID-19 corona virus pandemic based on time series data using Support Vector Machine. Journal of Discrete Mathematical Sciences and Cryptography 23(8):1583–1597. https://doi.org/10.1080/09720529.2020.1784535
17. Bhatnagar V, Poonia RC, Nagar P, Kumar S, Singh V, Raja L, Dass P (2020) Descriptive analysis of COVID-19 patients in the context of India. J. Interdiscip. Math. 24(3):489–504. https://doi.org/10.1080/09720502.2020.1761635
18. Kumari R, Kumar S, Poonia RC, Singh V, Raja L, Bhatnagar V, Agarwal P Analysis and Predictions of Spread, Recovery, and Death Caused by COVID-19 in India, Big Data Mining and Analytics, 4(2): 65–75. doi: https://doi.org/10.26599/BDMA.2020.9020013
19. Bhuvaneswari A, Karthikeyan M, Lakshminarayanan T (2012) Improving diversity in video recommender systems and the discovery of long tail. J Theor Appl Inf Technol 37(2):224–233
20. Bhuvaneswari A, Valliyammai C (2019) "Information entropy-based event detection during disaster in cyber-social networks." J. Intell. Fuzzy Syst. IOS Press 36, no. 5
21. Valliyammai. C, Bhuvaneswari A (2018) "Semantics-based sensitive topic diffusion detection framework towards privacy aware Online Social Networks", Cluster Computing, Springer, pp 1–16

Chapter 2
Simplifying and Optimizing the Convolution Encoding Algorithm in Error Control Codes

Constance Izuchukwu Amannah and Jennifer N. Igwe

1 Introduction

The unnecessary addition or subtraction in valid measurement is scientifically accepted as error. The entire activities of humans are error-prone: from science to engineering, from arts to law, from communication to business, and from actions to reactions. Errors are admittedly inevitable and measurably significant in both science and engineering. Their effect is understandably negative and counterproductive. Ironically, we cannot do without errors because what you cannot eliminate lives with you and you work with it. However, we can control the degree of occurrence of error in scientific activities and practice. Errors can be measured, controlled, and their effect adequately manipulated.

In digital communication, some critical error control techniques exist, namely linear block codes (LBC) and convolution code encoder (CCE). Both techniques share some similarities and differences. This study focuses on convolution codes. It closely evaluated the mathematical principles enshrined in convolution codes with a view to making them more practical and comprehensible. The outcome of this study is therefore a simplified and optimized convolution coding algorithm.

Statement of the Problem

Everywhere in the literature, mathematical principles of the CCE algorithm (CCEA) abound. As logical as the principles may appear to be, they are at best only closer to the comprehension and application of the experts. Many though experts in information theory and digital communication struggle to transfer these logics to learners in books

C. I. Amannah (✉)
Ignatius Ajuru University Of Education, P.M.B. 5047, Rumuolumeni Port Harcourt, Rivers State, Nigeria

J. N. Igwe
Rivers State University, P.M.B. 5080, Port Harcourt, Rivers State, Nigeria

G. Mathur et al. (eds.), *Proceedings of 2nd International Conference on Artificial Intelligence: Advances and Applications*, Algorithms for Intelligent Systems,
https://doi.org/10.1007/978-981-16-6332-1_2

and journals as well. As a result, many only allude to convolution codes passively without emphasis on the logics and procedural techniques. Such allusion is full of assumptions and ultimately sketchy pictures of knowledge, understanding that this method is unique and integral in error control coding. This study is therefore poised to advance the procedural logics and art of computing the parameters, associated with the methods. The requisite parameters include the input sequence (IS), output sequence (OS), start state (SS), end state (ES) or final state (FS), leftmost digit (LMD), and rightmost digit (RMD).

Purpose and Specific Objectives

The aim of the study was to simplify and optimize the convolution code encoder algorithm (CCEA) in error control codes. The objectives were to

(i) critically evaluate the mathematical principles in CCEA;
(ii) simplify the encoding procedure and logic in the CCEA;
(iii) optimize the procedural logics in the CCEA;
(iv) evaluate the performance of the CCEA in comparison with simplified and optimized CCEA (SOCCEA).

Scope of the Study

The study is restricted to evaluation, simplification, and optimization of the CCEA. Consequently, the results of the study are restricted to dry test results.

2 Related Literature

Digital Communication

When information is encoded in a format that is electronically acceptable, it is subsequently transferred from the source of encoding to the point of receipt called destination. This phenomenon is called digital communication [1]. This field of study is widely explored in science and engineering. Its computing features include source, medium, encoding, decoding, receiver, error, and so on [2]. Figure 1 highlights these basic features [15, 16, and 17].

Information Theory

This field of study was discovered by Claude Shannon in 1948 [3] in his quest for logical principles in communication. The major components of this field include measurement, storage, and communication of information contents [4]. Figure 2 shows the essential concepts in this subject area [14].

Error Control Coding (ECC)

ECC is the approach designed to check the occurrences of errors. ECC serves two purposes: error detection and error correction [5]. Many ECCs exist. In this study, two key ECCs were considered: the linear black codes (LBC) and the convolution

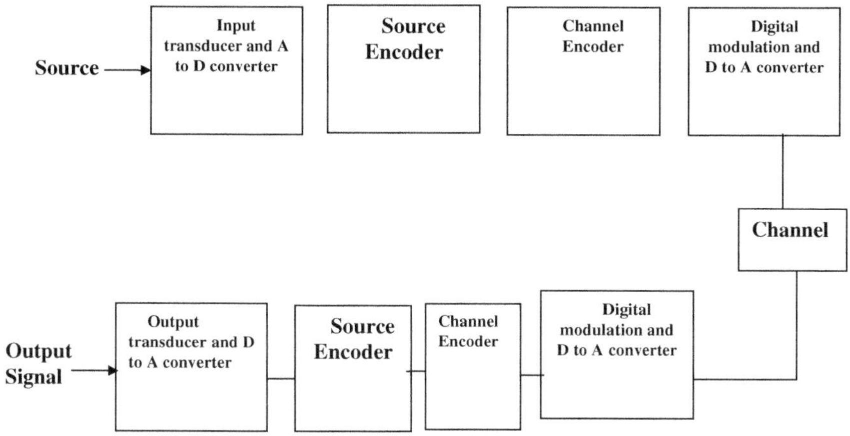

Fig. 1 Features of a Digital Communication System; *Source* [2, 15, 16, 17]

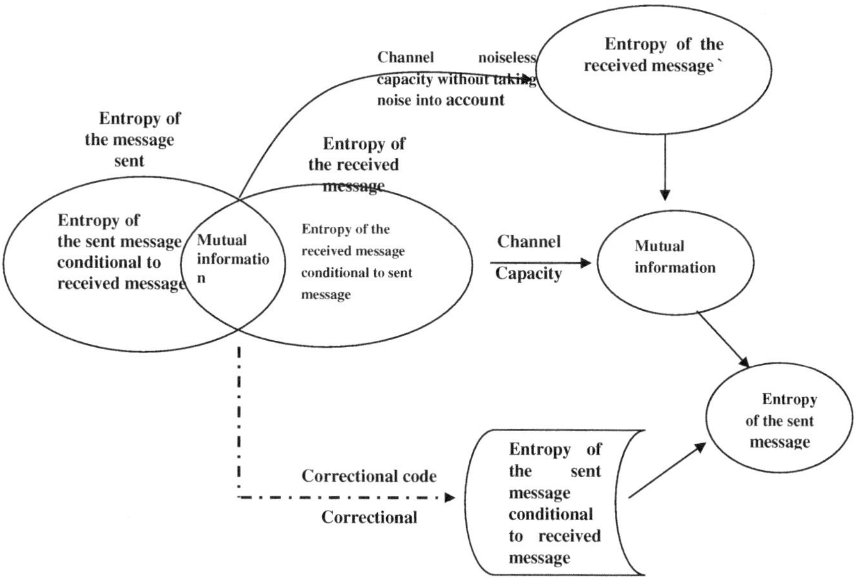

Fig. 2 Key Concepts in Information Theory (adapted from [5, 18])

codes (CC). In LBC, the resultant code word (RCW) is simply the linear combination of any two code words. The output of this method is independent of the preceding inputs in the shift registers. The example below designates the structure of code words in relation to the LBC [11].

The CC's approach differs slightly from the LBC's. It allows the parity code to be subtracted from the entire data while having the message bits encoded. Consequently, the parity bits are further added while the entire data are encoded. Figure 3 gives a succinct picture of the block of data and stream of data as applicable to information transmission. The overview highlighted above is bedeviled with enormous computing bottlenecks. These constraints are, however, resolvable using the CC approach. This is basically the main trust of this study which converges at the convolution code encoder algorithm (CCEA) [12, 13].

Coding Theory

There are various ways of applying information theory; one of the key areas of application is Coding Theory (CT). CT s branched into two major sub-areas: source coding and channel coding [6]. While channel coding deals with error correction or control, source coding handles data compression [7]. The latter underscores the focus of this study.

Conceptual Parameters in Error Control Codes

Some computing parameters are responsible for the effective error control operations [8]; they include

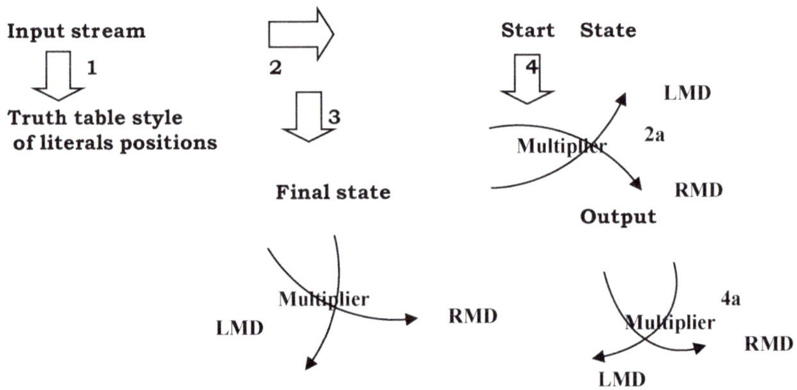

Fig. 3 Incremental Stages of the SOCCEA

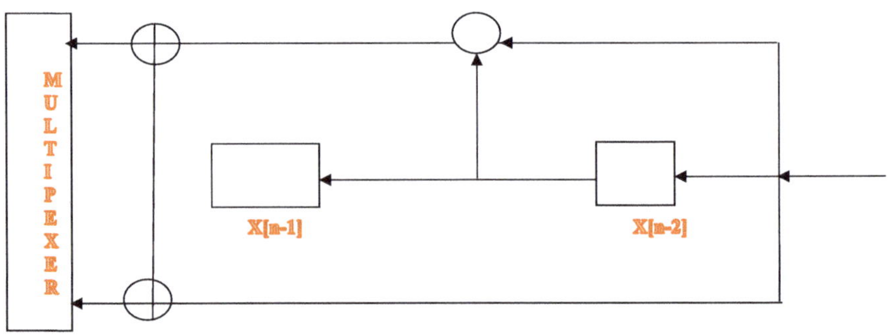

Fig. 4 The Convolution Encoder (Adapted from [10])

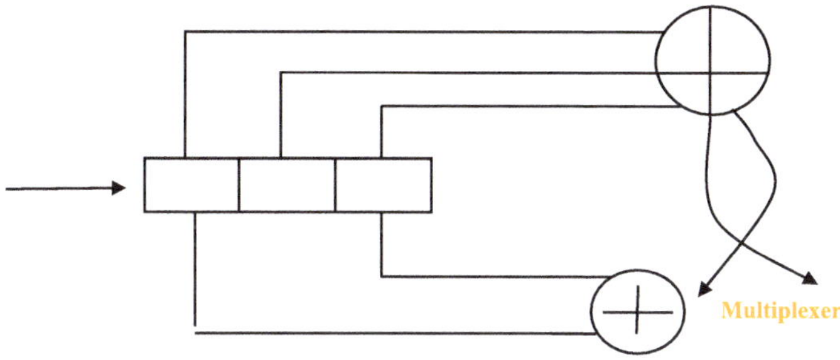

Fig. 5 Adjusted Convolution Code Encoder (Adapted from [11])

Code Word: This is encoded block of bits.
Block Length: It refers to the number of bits 'n' after coding.
Code Rate: The ratio of message bit (k) to bit number (n) in a code word: **r = k/n.**
Constraint Length: The number of shift registers, denoted by k.

3 Methodology

The method adopted in this is the incremental model. It allows the procedure to be handled incrementally until a final output is achieved. This is required in simplifying and optimizing the convolution code encoding algorithm. In a piecemeal manner, the processing in the shift registers is explained and the technique is well illustrated. What happens at the registers is crucial to the outputs that would be multiplexed to generate the two-bit output sequence of the register.

Incrementally, the leftmost digit (LMD) and rightmost digits (RMD) of start state, final state, and output are patterned and associated with a prescribed logic for convenient and efficient generation at all times.

4 Analysis and Design

The convolution code encoder (CCE) is characterized by the input layer, output layer, multiplexer, shift registers, the LMD operator, and RMD operator.

The operational logic of the encoder requires that the input bit follows the n-modulo-2 adder. The bits are moved into the shift registers causing all other bits to shift to the left. This discharges the previously stored bit completely from the registers. At any given input, the encoder starts by clearing the register with state 00 and the encoding process commences as required. Table 1 shows the operation of the convolution encoder. Understudying the operations highlighted in Table 1 offers

Table 1 State table of the convolution encoder (Adapted from [9])

Start state	Input	End state	Output
0 0	0	0 0	0 0
0 0	1	0 1	1 1
0 1	0	1 0	1 0
0 1	1	0 1	0 1
1 0	0	0 0	1 1
1 0	1	0 1	0 1
1 0	0	1 0	0 0
1 1	1	1 1	0 1

Table 2 Truth Table of XOR Logic Gate

Literals A	B	A ⊕ B	B ⊕ A
0	**0**	**0**	**0**
0	**1**	**1**	**1**
1	**0**	**1**	**1**
1	**1**	**0**	**0**

little comprehension of the practicability of the encoding algorithm. Its supporting state increases the complexity of the algorithm.

The complexities embedded in the state diagram can drastically be reduced only when the encoder diagram and its complimentary state table are apt enough in terms of their logics and mathematical principles. The proposed encoding algorithm is both a simplifier and optimizer to the CCEA. Its simplification begins from the encoder and subsequently makes a tabular representation of the encoder's results. Within the table, the operations of the encoder are aptly illustrated. Comparing the encoder and the table's logics breadths simplicity to the state diagram. This makes for smooth reengineering of the SOCCEA.

Simplifying and Optimizing the CCE from design.

The CCE is simplified and optimized by clarifying the mathematical principles in the CCE.

The simplification process is initiated by providing computing explanations to the figure.

Figure 6 above introduces the expected simplicity of logic and clarity of the computing procedure. R_1, R_2, and R_3 are the shift registers with a constraint length of 3.

THE LMD of the output = n-mod-2 adder. That is, given n-digits, LMD = 1 iff the number of 1 is odd, otherwise LMD = 0. Similarly, the RMD is obtained by applying the Boolean XOR logical operation.

The input sequence is buffered into the shift registers from the most significant bit (MSB) as shown in Fig. 6. At the beginning of the operation, R_1 and R_1 are initiated as zero, zero. Movement of the first input into R_1 moves the initial $R_1 \rightarrow R_2$, $R_2 \rightarrow$

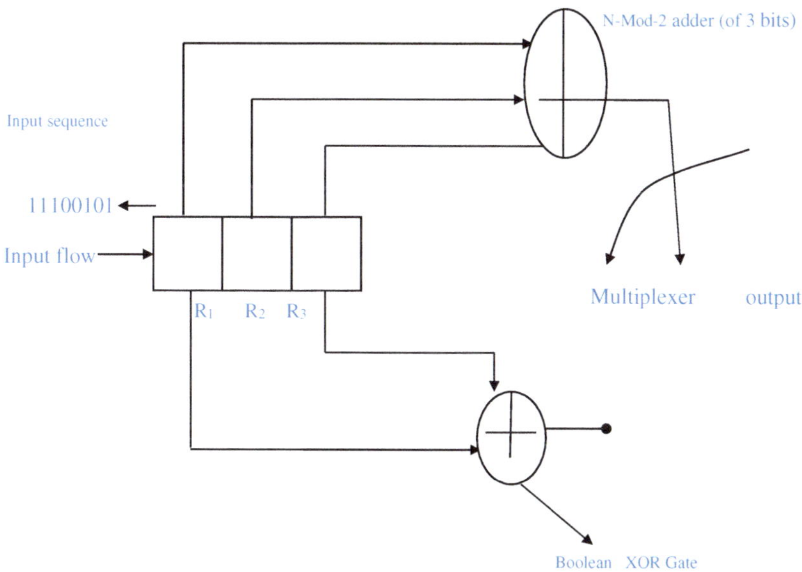

Fig. 6 Elucidated convolution code encoder

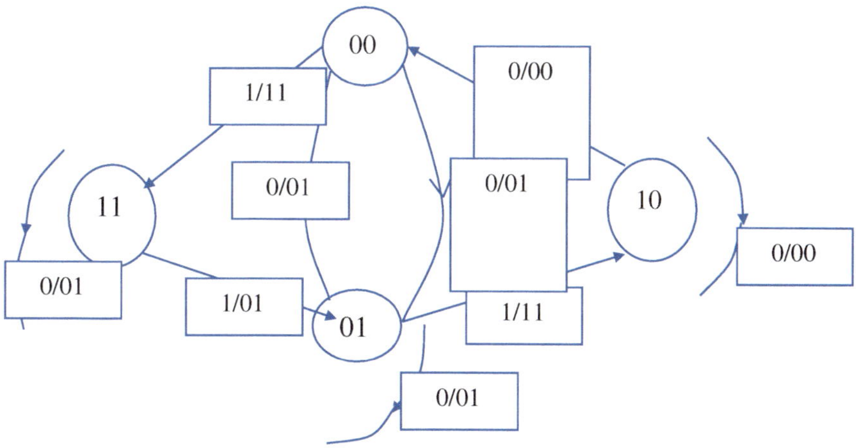

Fig. 7 State diagram of SOCCEA

R_3 flushing out R_3. This trend continues until the last input from the input sequence is moved into the shift registers. The multiplexer simply juxtaposes the LMD with the RMD, i.e. LMD RMD (i.e. 2-digit output).

The SOCCEA

Step 1:Start.

Step 2:Set up input sequence.

Step 3:Set up shift registers.
Step 4:Set up the LMD unit.
Step 5:Set up the RMD unit.
Step 6:Flush the shift registers with zeroes as their initial values.
Step 7:Move input n_1 (from MSB) to shift register R_1.
Step 8:Compute LMD (n-mod-2 adder) and RMD (X0R on R_1 and R_3).
Step 9:Move input n_2 (from MSB) to shift register R_1 and move $R_1 \rightarrow R_2$, $R_2 \rightarrow R_3$ to flush R_3.
Step 10:Compute LMD (n-mod-2 adder) and RMD(XOR on R_1 & R_3).
Step 11:Multiplex resultant LMDs and RMDs.
Step 12:Repeat steps 7–11 until the last input is moved from the input sequence.
Step 13:End.

5 Results and Discussion

With an input sequence of 1,110,001,101, the resultant start states, final states, and corresponding output sequence are computed and results tabulated,
with K = constraint length = 3.
Input sequence: 1,110,001,101.
LSB MSB.
Results 1: Start State and Final State streams of bits.
1,110,001,101 ← input.

0	0	0

Start
Result 2: Output Sequence of Bits
1,110,001,101 ← input.

0	0	0

Start
Tables 3 and 4 highlighted the products of the SOCCEA. The simplicity is seen in the ease with which the parameters are resolved, and the optimality is observed in the limited time and possibly the space required to store the computing instructions together with their data. Indeed, little time is required to achieve results 1 and 2 as against the time required for computing the same results using the CCEA. Having simplified the CCEA, the state diagram becomes nothing but a complimentary procedure devoid of any constraints.

Table 3 Result 1—State Table for SOCCEA

Input	Coded digits	Start State	Input	Coded Digits	Final State
1	1 0 0	1 0	1	1 1 0	1 1
0	0 1 0	0 1	0	0 0 1	0 0
1	1 1 0	1 1	1	1 1 1	1 1
1	1 1 1	1 1	1	1 1 1	1 1
0	0 1 1	0 1	0	0 0 1	0 0
0	0 0 1	0 0	0	0 0 0	0 0
0	0 0 0	0 0	0	0 0 0	0 0
1	1 0 0	1 0	1	1 1 0	1 1
1	1 1 0	1 1	1	1 1 1	1 1
1	1 1 1	1 1	1	1 1 1	1 1

Table 4 Result 2—Output State Table of SOCCEA

Input	Shift Registers	LMD (mod 2 adder)	RMD (XOR on R1 and R3	Multiplexer (of LMD and RMD)
1	1 0 0	1	1	1 1
1	1 1 0	0	1	0 1
1	1 1 1	1	0	1 0
0	0 1 1	0	1	0 1
0	0 0 1	1	1	1 1
0	0 0 0	0	0	0 0
1	0 0 0	1	1	1 1
1	1 1 0	0	1	0 1
0	0 1 1	0	1	0 1
1	1 0 1	0	0	0 0

6 SUMMARY AND CONCLUSION

The study identified the constraints of the CCE as non-predictability of the patterns of the start state, final state, output sequence of bits, and the lack of clarity between the encoder and the complimentary state table and state diagram. In addressing these concerns, the study simplified the CCE from design down to method and results. The encoder was provided with key computing logics and mathematical principles. Each component of the encoder can be easily comprehended and manipulated more efficiently than it could with the CCE. We can now perform convolution code encoding directly in the state table having digested and perfected the logics associated with the encoder.

The study successfully achieved its aim. It realized the SOCCEA. It provides convenient and efficient error control coding in information theory and digital communication and other related fields where the technique applies. The SOCCEA is therefore recommended as a compliment or an alternative to the CCEA. As is the case with CCEA, the SOCCEA is not designed for experts alone but especially for trainees for speedy learning and mastering of the rudiments of ECC.

References

1. Leslie D (2020) What is digital communication and why are skilled professionals in such high demand? Accessed on July 15 2020, from www.northeastern.edu/bachelors-complete/news/what-is-digital-communication
2. Moon TK (2005) Error correction coding. Mathematical methods and algorithms, Jhon wiley and son
3. Graham PC, Claude E (2002) Shannon: founder of information theory. References about Quantum information science in Scientific American
4. Martignon L (2001) In international Encyclopedia of the Social & Behavioral Sciences
5. van Lint JH (2012) Introduction to Coding Theory. GTM, vol 86. 2nd edn. Springer-Verlag, pp 31
6. MacWilliams FJ, Sloane NJA (2014) The Theory of Error-Correcting Codes. North-Holland, pp 35
7. Huffman W, Pless V (2003) Fundamentals of error-correcting codes. Cambridge University Press
8. Lin S, Costello DJ. Jr (2015) Error Control Coding: Fundamentals and Applications. Prentice-Hall
9. Butman SA, Deutsch LJ, Miller RL (1981) "Performance of concatenated codes for deep space missions." The Telecommunications and Data Acquisition Progress Report 42–63, March–April 1981 (1981): 33–39
10. Sadjit PH (2020) Punctured convolution coding. Accessed on May 10, 2020 from https://www.mathworks.com/help/comm/ref/poly2trellis.html
11. Huffman D (1953) "A Method for the Construction of Minimum-Redundancy Codes" (PDF). Proc IRE 40(9):1098–1101
12. Van L (2020) "On the construction of Huffman trees" (PDF). ICALP: 382–410. Retrieved 2020–02–20
13. Huffman K (1991) "Profile: David A. Huffman: Encoding the "Neatness" of Ones and Zeroes". Scientific American: 54–58
14. Gribov A (2012) "Optimal Compression of a Polyline with Segments and Arcs, Pretence Press
15. Gallager RG, van Voorhis DC (2011) Optimal source codes for geometrically distributed integer alphabets. IEEE Trans Inf Theory 21(2):228–230
16. Abrahams J (1997) Written at Arlington, VA, USA. Division of Mathematics, Computer & Information Sciences, Office of Naval Research (ONR). "Code and Parse Trees for Lossless Source Encoding". Compression and Complexity of Sequences 1997 Proceedings. Salerno: IEEE: 145, 171. CiteSeerX 10.1.1.589.4726
17. Hu TC, Tucker AC (2007) Optimal Computer Search Trees and Variable-Length Alphabetical Codes. SIAM J Appl Math 21(4):514
18. Knuth DE (1998) "Algorithm G (Garsia–Wachs algorithm for optimum binary trees)", The Art of Computer Programming, vol 3, Sorting and Searching 2nd edn. Addison–Wesley, pp 451–453. See also History and bibliography, pp 453–454

Chapter 3
An Efficient HIDS System Using Machine Learning Algorithm and Evidence Theory

Surbhi Solanki, Chetan Gupta, Kalpana Rai, and Minal Saxena

1 Introduction

Currently, in the world the number of intruders are rising rapidly with the increasing number of internet technologies. Therefore, it is necessary to have the IDS system to protect the intruder from doing the malicious access of the information. Its main concentration is on reporting the identified attacks to the administrator. It considers misuse and anomaly detection module. The abnormal behavior is first identified by the misuse detection and the left one is classified as normal behavior [1, 2]. And in opposite of this the anomaly detection is used to detect the unknown attacks which identify the left one behavior as abnormal behavior after identifying all the normal behavior mode. Machine learning (ML)-based SVM algorithm is used to deal with these kind of attacks. This is supervised ML algorithm which is used for both regression and classification challenges. Where in the field of IDS there is an algorithm called machine learning is used. This is the procedure of data scanning that robotize analytical mould building [3–5].

Here, we have rehabilitated PHAD, ALAD as anomaly-based statistical algorithm and SNORT as misuse-based intrusion detection system. Packet header anomaly detection (PHAD) is the anomaly detection methods which first model the protocols and afterward current-based model depending on the rapid changes of network. SNORT is an open-source network intrusion that prevents system, which have the capability of logging of packet on IP network and traffic analysis at the real time [6, 7].

S. Solanki (✉) · C. Gupta · K. Rai · M. Saxena
Sagar Institute of Research Technology & Science, Bhopal, Madhya Pradesh, India

G. Mathur et al. (eds.), *Proceedings of 2nd International Conference on Artificial Intelligence: Advances and Applications*, Algorithms for Intelligent Systems,
https://doi.org/10.1007/978-981-16-6332-1_3

2 Literature Review

In [1], this paper researcher proposed a hybrid intrusion detection method which is based on Fuzzy C-Means and SVM. In his paper, the author tries to improve the SVM classifier by reducing the complexity of large-scale datasets and he also directed the use of NSL-KDD datasets to verify the feasibility of the method.

In [2], there are some techniques like SVM, J48 Decision, and Naïve Bayesian tree used for the detection of different kinds of attacks and along with them this also considers the dissimilar types of accuracy with respect to their algorithms. This method takes NSL-KDD dataset for testing all tests.

In [3], the author proposed different classifiers like random forest (RF), deep convolution neural network (DCNN) for attacks detection. The author also differentiates the classifier on the basis of minority and majority which conclude that the minority class will conduct to numerous inconsistencies.

In [8], the researcher proposed the machine learning techniques like supervised learning K-NN and unsupervised learning K-Means for immensity of data packets and for feature selection to make proposed work effectual. The author also describes the corresponding performance for locating all attack categories which is contemplating an objection in intrusion detection.

In [9], the author proposed an algorithm for intrusion detection in communication network on the basis of data mining. Here, the researcher compares terms like accuracy, recall on the base of classification performance for the identification approaches. These decrease the cost and computational time by keeping the distribution execution.

In [10], the researcher proposed a NIDS for feature selection and influence analysis using NSL-KDD dataset. Here, the author describes the discretization feature process for justify the decision-making valued feature. However, his results show the decreasing rate in accuracy but also speeding in testing.

In [11], the researcher proposed a model called oppositional particle swarm optimization–probabilistic neural network (OPSO-PNN) for anomaly based intrusion detection system. Here, the NSL-KDD dataset is compared with the models like PSO-PNN and PSO-RB. They contain the low false positive rate and higher detection rate by having the better classification Fig. 1 capabilities to compare with PSO-PNN.

3 Proposed Work

The improvement of the demonstration of HIDS is done by marking the identification after effects of SNORT and ALAD, PHAD by capitalizing semi-supervised ML technique [12]. The resulting build measures are upgraded by displaying improved proof of hypothesis [13,14]. The upgraded ET and SVM enhanced the identification precision by proposed Fig. 3 HIDS technique.

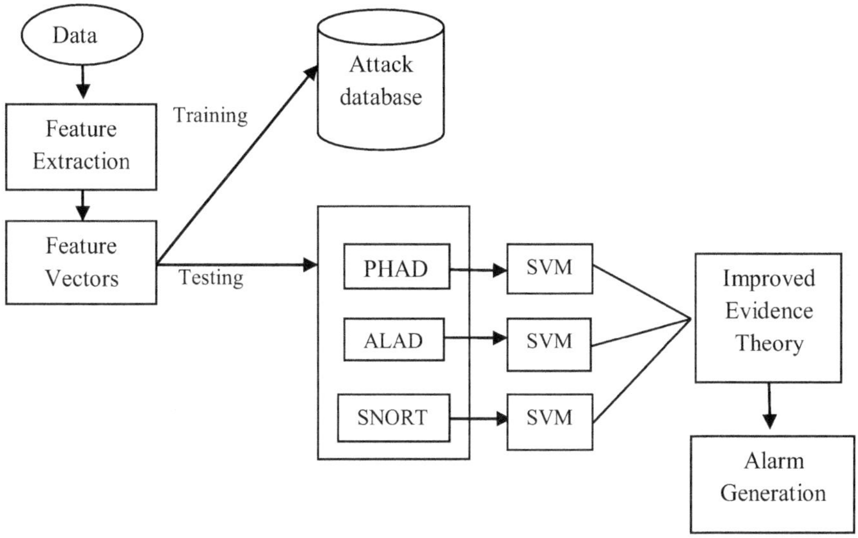

Fig. 1 Overall Process Flow

3.1 Generalized Delta Learning Rule

The algorithms use the Machine learning techniques through which the generalized delta gives us the exact result of the proposed methodology [15]. From the net the total squared error of output is computed and the minimum by gradient accurate method known as back propagation or generalized delta rule [10].

3.1.1 Derivation

Consider the arbitrary activation function f(x). The derivation of the activation function is denoted by F(x).

$$y_{-ink} = \sum_i z_i w_{jk}$$

$$Z_{-inJ} = \sum_i V_{ij} X_i$$

$$Y_k = f(y_{-ink})$$

The error to be minimized is

$$E = 0.5 \sum_k [t_k - y_k]^2$$

$$\frac{\partial E}{\partial w_{jk}} = \frac{\partial}{\partial w_{jk}} 0.5 \sum_k [t_k - y_k]^2$$

$$= \frac{\partial}{\partial w_{jk}} (0.5[t_k - t(y_{ink})]^2)$$

$$= -[t_k - y_k] \frac{\partial}{\partial w_{jk}} f(y_{-ink})$$

$$= -[t_k - y_k] f(y_{-ink}) \frac{\partial}{\partial w_{jk}} (y_{ink})$$

$$= [t_k - y_k] f^1(y_{ink}) Z_j$$

Let us define

$$\delta_k = -[t_k - y_k] f^1(y_{-ink})$$

Weight on connection to the hidden unit Z_j

$$\frac{\partial E}{\partial v_{ij}} = -\sum_k [t_k - y_k] f(y_{ink}) \frac{\partial}{\partial v_{ij}} y_k$$

$$= \sum_k [t_k - y_k] f(y_{ink}) \frac{\partial}{\partial v_{ij}} y_{-ink}$$

$$= \sum_k \delta_k \frac{\partial}{\partial v_{ij}} y_{-ink}$$

3.1.2 HIDS SVM Algorithm

Input: Dataset, Evidence vector E = E_1, $E_{2.......}$ En.

Output: detected attacks.

The steps for the proposed algorithm is described as follows:

1- From the network gather the dataset.

2- The data utilizing the PHAD, ALAD, and SNORT detect the intrusions.

3- By making use of SVM distribute the outcomes.

4- Start.

5- Estimate distinguish matrix utilizing equation.

$$d_{ij}(E_i, E_j) = \begin{cases} 0, \ C_1 = C_2 \\ \left(w_1|A_1 - B_1|^2 + w_2|A_1 - B_1|^2 + \cdots + w_n\ |A_n - B_n|^2\right)^{\frac{1}{2}}, C_1 \neq C_2 \end{cases}$$

Where both Ei and E_j are the evidences instance of the detection outcome if discernment Θ, w_i is the evidences and the preposition values are C1 and C_2. For attack disclosure there are n evidences, and to calculate dissimilarity in between E_i and E_j the above formula is used.

6- The n is numerical of evidences in for i = 1 to n.

7- By using equations calculate the entropy of evidences.

$$Entropy_i = supporting\ degree(mi)_i \ln(supporting\ degree(mi))$$

The dissimilarity of upholding degree is corresponded by collected information entropy of evidence. By preserving the entropy amount of the evidence, the virtue of the evidence is estimated.

8- Calculate credibility of evidence using equation.

$$Credibility_{mi} = \frac{1/Entropy_i}{\sum_{i=1}^{n} \frac{1}{Entropy_i}} \quad i = 1, 2, \ldots n$$

The $credibility_{mi}$ can be contemplated as the measurement of $_{mi}$ which is referred by the sum of the $correctness_{mi}$ which is identical to 1. Afterward the evidence can be enumerated and the absorbing basic identity assignment values of modified E_i by absorbing the weight of evidence which is given in equation.

9- Compute the weight of evidences utilizing.

$$m'(A) = Credibility_{mi}.m(A), m'(\theta) = 1 - \sum_{i=1}^{n} m'(A)$$

10- terminate for.

11- For i = 1 to n.

12- if n > 2(if n is greater than 2).

13- $e_x = e_i$.

14- $e_y = e_{i+1}$.

15- By utilizing earlier equation apply upgraded union rule of evidence theory.

16- if i = = n-1.

17- return result.

18- end if.

19- i = i + 1.

20- end if.

21- End.

4 Result Analysis

For this work NSL-KDD dataset is taken; first we selected 10,000 random dataset samples from a large volume of data, i.e., 1,25,973 datasets. First, we applied PHAD, ALAD, and SNORT algorithms on the selected sample. Then,

SVM is applied to get the final results as shown in Table 1. And we have concluded that we get True positive (TP) is 3898, True negative (TN) is 6134, Recall is 0.54, and precision is 0.32 (Table 2).

The two Bar Graph which is mentioned above shows the comparison value of accuracy and false alarm rate (FAR) with the previous method which clearly shows that the accuracy of proposed work is much higher than the previous work, i.e., 99.31% as shown in Fig. 2 and also the FAR rate is slow compared to previous method which is 0.70% as shown in Fig. 3.

Table 1 Results Comparison

IDS/ Attacks Types	Total Attacked	TP	TN	Recall	Precision
PHAD	10,000	3218	6928	0.26	0.19
ALAD	10,000	3399	6467	0.38	0.26
SNORT	10,000	3608	6299	0.39	0.24
HIDS SVM	10,000	3898	6134	0.54	0.32

Table 2 Comparison with Previous Work

IDS/Attack Types	Accuracy	FAR
Previous Method [1]	99.19%	0.76%
HIDS SVM Algorithm	99.31%	0.70%

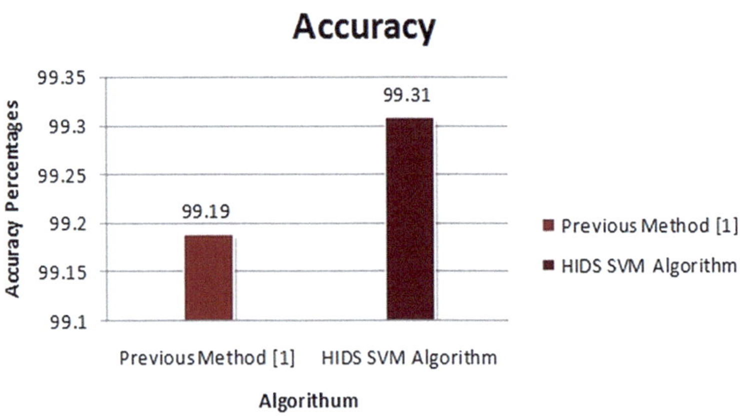

Fig. 2 Bar Graph showing the comparison of Accuracy with previous paper

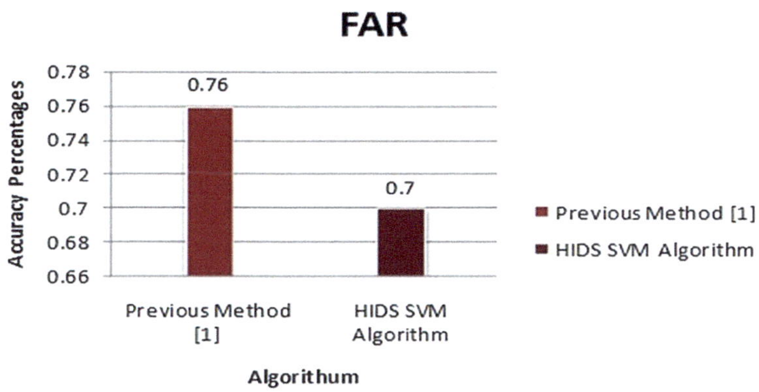

Fig. 3 Bar Graph showing the comparison of FAR with previous work

5 Conclusion

In this paper, a HIDS based on SVM algorithm is proposed. The method first takes training and testing dataset which undergo through the various detection techniques after which SVM is used which classify them in normal or attacked node. An alarm is generated after this whole process when an attack is detected. The proposed method contains 99.31% accuracy and 0.70% FAR. In future, we can focus on more different types of attack for detection and can check few more parameters for better result.

References

1. ZHIYOU ZHANG et al. (2019) "A Hybrid Intrusion Detection Method Based on Improved Fuzzy C-Means and SVM". In: Proceedings of International Conference on Communication Information System and Computer Engineer [CISCE]. https://doi.org/10.1109/CISCE.2019.00056
2. AFREEN BHUMGARA et al. (2019) "Detection of Network Intrusion Using Hybrid Intelligent System". In: Proceedings of International Conferences on Advances in Information Technology
3. RITUMBHIRA UIKEY et al. (2019) "A Survey on Classification Techniques Applied to Intrusion Detection System and its Comparative Analysis". In: Proceedings of 4th International Conference on Communication & Electronics System IEEE Conference (ICCES)
4. SINGH P, VENKATESAN M (2018) "Hybrid Approach for Intrusion Detection System". In: Proceedings of International Conference on Current Trends Towards Converging Technologies(ICCTCT), Coimbatore
5. TAVALLAEE M et al. (2009) "A Detailed Analysis of the KDD CUP 99 dataset". In: Proceedings of IEEE Symposium on Computational Intelligence for Security and Defense Applications, Ottawa, ON.
6. ADITYA PHADKE et al. (2019) "A Review of Machine Learning Methodologies for Network Intrusion Detection". In: Proceedings of 3rd National Conference on Computing Methodologies and Communication, IEEE Xplore part Number: cfp19k25-art; ISBN; 978–1–5386–7807–4

7. SIVANTHAM S, ABIRAMI R, GOWSALYA R (2019) "Comparing in Anomaly Based Intrusion Detection System for Networks". In: Proceedings of International conference on Vision towards Emerging Trends in Communication and Networking (ViTECon)
8. XIAOYAN WANG, HANWEN WANG (2018) "A High Performance Intrusion Detection Method Based on Combining Supervised and Unsupervised Learning". In: Proceedings of IEEE Smart World, Ubiquitous Intelligence & Computing Advanced $ Trusted Computing, Scalable Computing, Internet of People and Smart City Innovations, INSPEC Accession Number: 18342295. https://doi.org/10.1109/SmartWorld.2018.00304
9. KARUNA S.BHOSALE et al. (2018) "Data Mining Based Advanced Algorithm for Intrusion Detection in Communication Networks". In: Proceedings of International Conference on Computational Techniques, Electronics & Mechanical System (CTEMS)
10. LUKMAN HAKIM, RAHILLA FATMA NOVRIANDI (2019) "On Influence Analysis of Feature Selection to Network Intrusion Detection System Performance Using NSL-KDD Dataset", Jember, Indonesia at ICOMITEE
11. SREE KALA T, CHRISTY A (2019) "An Intrusion Detection System Using Opposition Based Particle Swayam Optimization Algorithm and PNN". In: Proceedings of International Conference on Machine Learning, Big Data, Cloud and Parallel Computing India 14th–16th
12. Dr. UMA KUMARI, UMA SONI (2017) "A Review of Intrusion Detection using Anomaly Based Detection", at 2nd International Conference on Communication and Electronics System, IEEE Xplore Compliant—part Number: CFP17AWO-ART, ISBN:978–1–5090–5013–0 " in 2017.
13. HASSAN AZWAR et al. (2018) 'Intrusion Detection in Secure Network for Cyber Security Systems Using Machine Learning and Data Mining" In: Proceedings of IEEE 5th International Conference on Engineering Technologies $ applied Sciences, 22–23 Nov 2018, Bangkok Thailand
14. CHIRAKAR R, HUANG C (2012) "Anomaly Based Intrusion Detection Using Hybrid Learning Approach of Combining K-medoids Clustering and Naïve Bayes Classification". In: Proceedings of 8th International Conference on Wireless Communications, Networking and Mobile Computing, Shangai.
15. AZAR ABID SALIH et al. (2019) "Combining Best Features selection Using Three Classifiers in Intrusion Detection System". In: Proceedings of International Conference on Advanced science and Engineering (ICOASE), University of Zakho, Duhok Polytechnic University, Kurdistan Region, Iraq

Chapter 4
Self-Supervised Learning for COVID 19—An Envision to Salvage Model

Anjali Jivani, Hetal Bhavsar, and Kshitij Gupte

1 Introduction

It has been seen over the history of mankind that unforeseen and dreadful pandemics have shattered the lives of people and brought the world to a situation of disastrous stagnancy. Whether it was the Black Death pandemic of the fourteenth century or the Spanish Flu of the early nineteenth century, to name a few, they brought with them the devastation of death and turmoil among those living [1, 2].

During those periods, technology and medical sciences were not so enhanced, and combating the pandemics was by itself a great challenge. As is already known, a pneumonia-like disease of unknown cause was detected in Wuhan city of China, and was first reported to the WHO Country Office in China on December 31, 2019. Due to its widespread outbreak, by the end of January 2020, it was declared a Public Health Emergency of International Concern. The World Health Organization (WHO) in the second week of February 2020 announced a name for this viral illness as the Corona Virus Disease of 2019: COVID-19 [3].

It would be surprising if right now when the world is under the impact of the COVID-19 pandemic, technocrats and scientists do not think of options to envision the containment of the disease and salvage successfully using the power of AI. The focus of this article is mainly on the AI options with their applicability to be implemented and used at every stage of the COVID-19 pandemic.

A. Jivani (✉) · H. Bhavsar · K. Gupte
Department of CSE, The Maharaja Sayajirao University of Baroda, Vadodara, India
e-mail: anjali.jivani-cse@msubaroda.ac.in

H. Bhavsar
e-mail: hetal.bhavsar-cse@msubaroda.ac.in

K. Gupte
e-mail: kshitij.gupte-cse@msubaroda.ac.in

G. Mathur et al. (eds.), *Proceedings of 2nd International Conference on Artificial Intelligence: Advances and Applications*, Algorithms for Intelligent Systems,
https://doi.org/10.1007/978-981-16-6332-1_4

2 Stages of COVID-19

It is important that before implementing an AI application or technique, the understanding and the right analysis of the data is available to exactly identify the best suitable method to move forward. In Fig. 1, the depiction of the different stages of COVID-19, and at each phase how a learning procedure can be implemented are given.

The different stages of the disease outbreak have been marked and along with them the junctures where AI could assist in the suppression and prediction for the next stage have been identified. The disease itself transports fast, so should the AI handling algorithms start giving restraining and combating options. It is evident that after the initial outbreak of COVID-19, what happened next was identifying those which were confirmed positive cases and those which were suspected to have been affected [4]. Obviously, the confirmed cases were to be handled in hospitals resulting in either cure or death. The degree of intensity of the disease decided the level of treatment.

Those cases where there was suspected infection to be quarantined in hospitals, specially arranged facility centers or at home. These cases need to be closely monitored to see the intensity of the infection or return to normalcy. The very important task of finding all those who were in physical contact with the positive or suspected cases and putting them all in quarantine becomes the next magnanimous mission [5, 6].

Starting from the first stage till the last one, the medical facilities available, the medical facilities to be instantly created, the medical staff and their safety equipment,

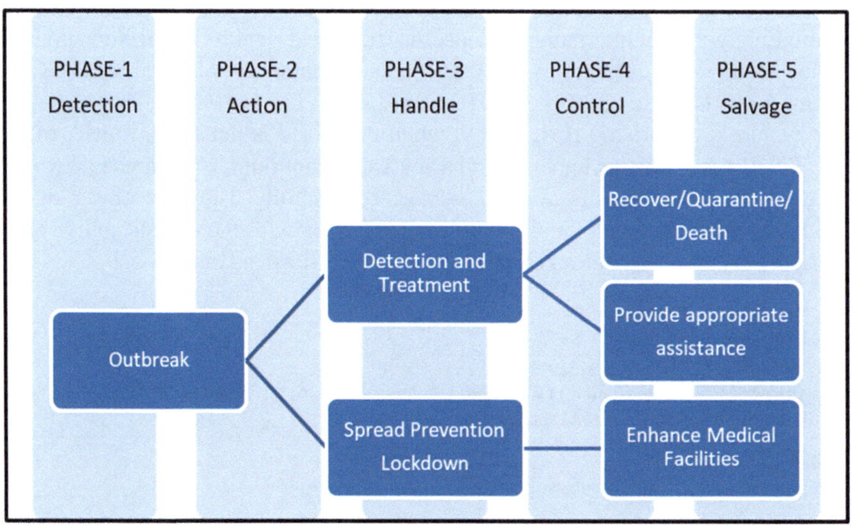

Fig. 1 COVID-19 stages and AI phases

etc. each needs to be addressed appropriately. Moreover, the precautionary lockdown resulting in economic, social and psychological impact is a major area of concern [7].

Before actually discussing the proposed model, it is important to discuss which area of AI would be appropriate in a pandemic. The concepts of supervised, unsupervised and reinforcement learning are assumed to be known. Moreover, the drawback of implementing Deep Learning (DL) when a large amount of data and large computation power is not initially available restricts going in that direction in the initial phases. The planned strategy involves applying a suitable method in the absence of adequate data and trying to generate the best possible output. So the recently proposed and lucrative field of AI—the Self-Supervised Learning—is considered here in this proposal.

3 The Self-Supervised Learning

Recently, Yann LeCun who is the pioneer behind the concept of DL and especially the concept of Convolutional Neural Networks (CNN) discussed the limitations of DL and how a newer, finer concept can be explored to overcome these restrictions. Self-Supervised Learning is one way to create data-efficient artificial intelligence systems [8]. This perception is neither unsupervised learning where the class labels are not known nor supervised where they are known. It is akin to learning and classifying on the go the way babies start learning by observing. It is somewhat similar to how we correlate 'cow' with the sound it makes—'moo' [9].

To go further in this, it is essential to discuss the functionality of the brain which is supposed to be replicated in this novel AI area. Daniel Kahneman, a Nobel Laureate psychologist and economist, has in his book discussed our brain of having two operating systems and he has called them System 1 and System 2. Both have been very aptly contrasted with the way they operate. System 1 being the one which is fast defines characteristics automatically and controls a major part of our thinking. System 2, however, in the one which contains conscious, effortful and rational thinking. It tries to seek new and missing information and also fill in the gaps wherever possible. This is a very small percentage of our thinking [10]. The working of System 1 has been implemented in our deep learning process but it's System 2 which we need to develop now. It is the one which would be efficient in the task of cognitive and abstractive output in areas where appropriate actions are to be taken without previous knowledge about the system.

Self-Supervised Learning is supposed to replicate System 2 functionality and give predictions and conclusions after intense deliberation on the input parameters, and try to plug in appropriate action plans wherever it is unavailable. More about this is discussed along with the phases of the model where it is to be implemented.

The basic comparison among Supervised Learning, Unsupervised Learning, Semi-supervised Learning and Self-Supervised Learning is given below in Table 1. It can be clearly observed that Self-Supervised Learning is the only method which

Table 1 Comparison of different Machine Learning Methods

	Supervised Learning	Unsupervised Learning	Semi-supervised Learning	Self-Supervised Learning
Definition	Previously labeled data used by machine to learn	Machine learns from inherent attributes without previously labeled data	Machine learns using combination of supervised and unsupervised methods	A form of unsupervised learning where the data provides the supervision
Types of problems	Classification, Prediction	Clustering, Association Mining	Classification, Prediction, Clustering	Classification, Prediction
Types of data	Labeled data	Unlabeled data	Labeled and Unlabeled data	Unlabeled data
Popular algorithms	K-NN, Naïve Bayes, SVM, Linear and logistic regression, etc.	K-Means, PAM, DBSCAN, CLARANS, etc.	Semi-supervised SVM	CNN, Computer Vision algorithms, Image Processing Algorithms

uses unlabeled data and learns on the go. The algorithms related to Neural Networks, Computer Vision, Image Processing, etc. are popular in it.

4 The Proposed Envision-Salvage Model

The model proposed in this article is related more to bridging between DL and Machine Learning (ML). Both are subsets of AI wherein DL comes under ML and ML under AI. While ML represents the capability of a machine to learn from the available data without being explicitly programmed, AI reflects the larger picture of creating intelligent machines that can simulate human thinking capability and behavior. A subdivision of ML, DL, is supposed to be the next evolution in AI. It is supposed to automatically discover the features to be used for classification, and in contrast to ML it needs high-end machines and considerably big amounts of training data to deliver accurate results [11].

To overcome the drawback of creating a model which requires high processing power, the proposed model uses Self-Supervised Learning as discussed earlier, which balances between the size and capability of data and machines and delivers better outcomes. This model would be easy to implement and would predict the requirement and prerequisites at each stage of the pandemic.

The model has been named the Envision-Salvage Model (E-S Model) as it is supposed to envisage or predict the pandemic and help to recover from its devastation. Figure 2 depicts the model and its implementation intertwined with the different stages of COVID-19. Each phase of the E-S Model has been named. The phases are Detection, Action, Handle, Control and Salvage. Though the phases look as if they

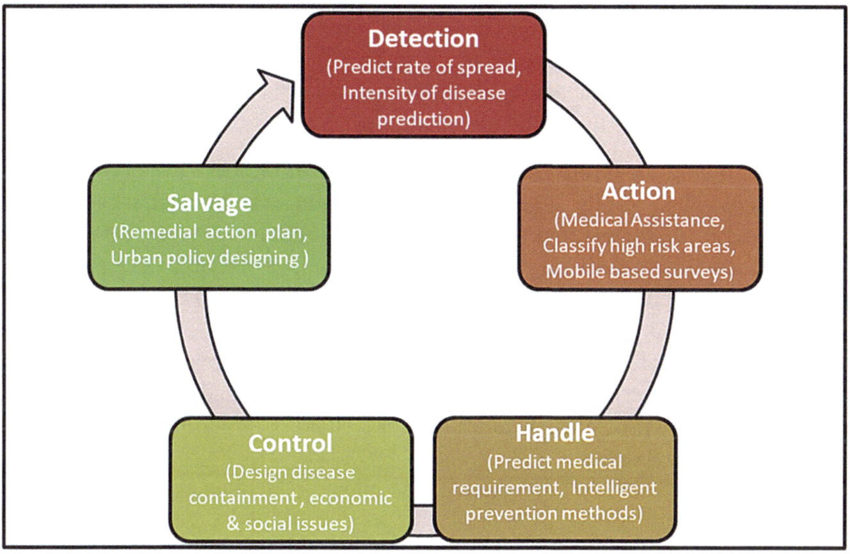

Fig. 2 The Envision-Salvage Model

come one after the other, each phase is somewhat executing in overlapping mode with the other. What has been designed for COVID-19 can be used as a prototype for any pandemic or disaster which could have similar devastating implications on mankind.

4.1 The Detection Phase

This is the initial phase where the pandemic starts. This stage though it has passed for the COVID-19, is of great importance for future disease outbreaks. The AI applications could be used to handle issues related to the following:

The transmission dynamics: Predicting the pattern that the contagion follows.

Predicting the peak of infection in terms of days or weeks.

Envisaging the number of people who would be infected as well as the level of infection.

Assessing and estimating the spread across cities, countries and continents.

Since the time when this phase starts, very little data is available and the need to predict the above is very critical, Self-Supervised Learning would work perfectly in this scenario. It can be a combination of convolutional neural networks and artificial neural networks.

4.2 The Action Phase

This is the phase where the action actually starts after the detection of the disease. This phase can leverage the following conditions:

- Start predicting for positive/negative cases. As of now, these tests take time for confirmation and results. If self-supervised learning is to be used, it can be used from images of the respiratory specimens to construct image representations that are semantically meaningful via pretext tasks that do not require semantic annotations for a large training set of images [12]. So applying AI on the test itself to speed up the results.
- AI applications of mobile apps: This can be used to generate surveys and GPS information of every person infected with his tracking and movement history. This can help in predicting who all have the chances of getting infected due to contact with this person.
- The model can also help in predicting who all are to be quarantined and the level of seclusion required.

4.3 The Handle Phase

This phase is totally overlapped with the Action phase and the Control phase. This is the phase where it is very important to take action to properly handle the disease as well as prepare to control the situation. This phase needs the following AI-based applications:

- Intelligent patient monitoring system—related to treatment, cure and convalescence. The concept of Transfer Learning which is related to Self-Supervised Learning that can be used to develop an intelligent system [13].
- Predicting medical requirements in terms of medical staff, equipment, medical precautionary aides, etc. This system would take into account the predicted cases that would erupt in the near future and estimate the medical requirements accordingly.
- AI can be used to identify areas where there could be a high concentration of people and accordingly monitor those areas.

4.4 The Control Phase

By the time adequate measures have been taken to handle the disease outbreak, the procedures dealing toward the containment and restriction of spread are to be designed. This phase is totally engaged toward those measures. They are listed as follows:

- Designing a system for food and grocery distribution considering the parameters related to the red alert areas or prospective areas which might become sensitive.
- Developing routes which are best for fast garbage collection, transportation, etc. using AI tools and methods.
- Focusing on the urban as well as rural policies to control the pandemic by means of intelligent mining.

4.5 The Salvage Phase

Each affected area would go through all these phases and hence the arrow goes back to the first phase where the implementation improves upon what has been done using heuristic and deep learning approaches also as per the requirement. This is the phase of containment, recovery and salvaging back from the disaster. This phase has

- Smart systems to be developed to recover from the economic, social and psychological impact which has caused havoc and mayhem in the lives of the people and society. This is a very important area of AI as there could be so many different ways in which techniques could be developed depending on the different parameters related to each area mentioned above.
- Rapid drug development techniques for cure and precaution are to be focused on to prevent this pandemic from erupting again. Medical discipline uses a lot of AI practices and there are researches related to Self-Supervised Learning in the Medical Science area too which are becoming popular [13, 14].
- The procedures to deal with this disease as part of our life and learn to move forward considering the fact that it may never completely vanish. What is required are the measures to fight and resist getting infected using precautionary and effective processes.

In the Envision-Salvage Model at each phase, there are a number of AI application areas which have been envisaged. The model is linked to the different stages of the pandemic outbreak which allows for a quick and apt response to contain it and efficiently fight it. The exact data required at each stage could be acquired from the WHO or the official government sites.

5 Conclusion

The whole world is still reeling under the impact of the COVID-19 pandemic, and most of the countries are still struggling with the Handle and Control phase. The Salvage phase is yet to come but definitely, the human race will come out mightier, wiser and definitely more judicious toward future disasters. Such catastrophes have befallen before but this time in spite of all the capabilities of the advanced countries with their superior technologies, mankind has been shaken and traumatized. What is still not clear is why and how it actually started? We need AI to find that too.

References

1. History's Worst Pandamics homepage: https://www.history.com/news/pandemics-end-plague-cholera-black-death-smallpox
2. Yongsen Ruan, Haijun Wen, Xionglei He, Chung-I Wu, A theoretical exploration of the origin and early evolution of a pandemic, Science Bulletin, 2020.
3. WHO homepage, https://www.who.int/emergencies/diseases/novel-coronavirus-2019/
4. W. Clark Lambert, Albert Alhatem, Muriel W. Lambert, Robert A. Schwartz, What we have learned from the COVID-19 pandemic: Time to think outside the box, maybe far outside, Clinics in Dermatology, (2020).
5. Daniel, Sir John, Education and the COVID-19 pandemic, Prospects: Comparative Journal of Curriculum, Learning, and Assessment, 2020, pp 91–96.
6. John Watkins, Preventing a covid-19 pandemic, BMJ (2020).
7. Nicola M, Alsafi Z, Sohrabi C, Kerwan A (2020) The socio-economic implications of the coronavirus pandemic (COVID-19): A review. Int J Surg 78:185–193
8. TNW homepage, https://thenextweb.com/neural/2020/04/05/self-supervised-learning-is-the-future-of-ai-syndication.
9. Virginia R, Learning Classification with Unlabelled Data, NIPS'93: Proceedings of the 6th International Conference on Neural Information Processing Systems, (1993), pp 112–119
10. Kahneman D (2013) Thinking Fast and Slow. Farrar, Straus and Giroux, Paperback
11. Towards Data Science homepage, https://towardsdatascience.com/clearing-the-confusion-ai-vs-machine-learning-vs-deep-learning-differences-fce69b21d5eb.
12. Misra I, Maaten L, Self-Supervised Learning of Pretext-Invariant Representations, https://arxiv.org/abs/1912.01991.
13. Raghu M, Zhang C, Kleinberg J, Bengio S, (2019) Transfusion: Understanding Transfer Learning for Medical Imaging, 33rd Conference on Neural Information Processing Systems (NeurIPS 2019), Vancouver, Canada. pp.1–22.
14. Kaiming H, Haoqi F, Yuxin W, Saining X, Girshick R, Momentum Contrast for Unsupervised Visual Representation Learning, (2019).

Chapter 5
Forecast of Covid Cases Using Deep Learning Algorithm

N. Vijaya Anand and G. Chandra Mouli

1 Introduction

Coronavirus disease (COVID-19) is an infectious disease caused by a newly discovered coronavirus. It is a deadly virus and is especially lethal to those with underlying medical conditions related respiratory, cardiac and cancer diseases. The primary mode of transmission occurs when saliva or nasal discharge form the infected person encounters the eyes, mouth or nose of a healthy individual. The World Health Organization (WHO) advises people to follow respiratory etiquette and to wash hands with alcohol-based sanitizers and handwashers. The WHO suggests that the only way to (at present) stop irrecoverable damage from happening is to slow down the transmission of the disease by 'flattening the curve' [1–3]. So far wearing masks, following basic hygiene and social distancing seems to be most effective when it comes slowing down the transmission [1–3]. In this paper we are focusing on the topic of 'Social distancing', as countries go as far as to apply a full-scale Lockdown to implement this. The reason for focusing on this topic is that, apart from being really effective at flattening the curve, the practice of lockdown tends to take its toll on the country's economy and this is especially dangerous for developing countries like India [4, 5]. So, in this paper we have discussed about a ML algorithm that can aid the governments to keep the economy from collapsing while still maintaining a flatter curve. Some recent studies predicted end of pandemic using machine learning [6–8].

Machine Learning (ML) is one of the subfields of Soft Computing. In this paper we are using Deep Learning algorithms to achieve the desired accuracy as DNN's perform well when compared to shallow networks at complex tasks. ML is really good at pattern recognition which is really helpful when it comes to forecasting. For the implementation of DNN in this paper we have acquired data on Italy about

N. V. Anand (✉) · G. C. Mouli
PVP Siddhartha Institute of Technology, Vijayawada, Andhra Pradesh, India

G. Mathur et al. (eds.), *Proceedings of 2nd International Conference on Artificial Intelligence: Advances and Applications*, Algorithms for Intelligent Systems,
https://doi.org/10.1007/978-981-16-6332-1_5

total number of cases (cumulative) and the time-to-time actions taken by the local government. This data is used to calculate some required factors. This data along with the calculated factors were used to train the DNN.

2 Proposed Materials and Methods

2.1 Activation Function

An activation function is a mathematical equation used in ML algorithms at the end of computation in each layer. In machine learning only nonlinear activation functions are used for various reasons [9].

Rectified linear unit (ReLU):

Rectified linear unit (ReLU) is a popular activation function used in ML. this is due to the fact that it offers better results when compared to other activation functions and gives better performance [9]. Its mathematical relation is given as

$$f(x) = \begin{cases} x\forall x > 0 \\ 0\forall x < 0 \\ not\, defined\, at\, 0 \end{cases} \tag{1}$$

The main purpose of this DNN is to predict the number of upcoming cases. The training set has an output set for each input sample, holding features with the values of total number of covid19 cases on each day for a week. This is to train the AI to tune itself to predict what this week's cases will grow up to. This can later on be extended for a number of weeks. The input data of the training set will have 'n' samples and each sample will have features as "current number of cases, alpha, number of days since the first case and beta" (we will later discuss about what alpha and beta are). After training is finished, it is inevitable for the predictions to have some error, about which We will discuss later in the 'results' section.

Formation of the training set: Table 1 presents the training data set formed using the data from Italy with the total number of cases on each day up to 8 weeks (56 days) starting from February 19 of 2020.

The reason to why only this specific period of 56 days is simply due to the fact that from 19th of February the total number of cases started to rise and before that the cases mostly remained at 3, since our main focus is to let the DNN learn the exponentiation, ignoring the days before February 19 lets the DNN to develop features more biased towards exponential growth and less on the initial incubation periods. It was practically observed that 8 weeks of data was sufficient to get the results at desired accuracy. Though, more data can be used to train Fig. 1 the network, its impact on the network's performance was found to be negligible and thus only 8 weeks' worth of data was collected.

Table 1 Training Data Based on case in Italy (adapted from [10])

Week	Number of cases registered in Italy from 01 Feb to 14 Apr						
	Day 1	Day 2	Day 3	Day 4	Day 5	Day 6	Day 7
Week 1	N/A	N/A	N/A	N/A	N/A	N/A	N/A
Week 2	N/A	N/A	N/A	N/A	N/A	N/A	N/A
Week 3	N/A	N/A	N/A	N/A	3	4	21
Week 4	79	157	229	323	470	655	889
Week 5	1229	1701	2039	2502	3089	3858	4636
Week 6	5883	7375	9172	10,149	12,462	15,113	17,660
Week 7	21,157	24,747	27,980	31,560	35,713	41,035	47,021
Week 8	53,578	59,318	63,927	69,176	74,386	80,589	86,498
Week 9	92,427	97,689	101,739	105,792	110,574	115,242	119,827
Week 10	124,632	128,948	132,547	135,586	139,422	143,626	147,577
Week 11	152,271	156,363	159,516	162,488	N/A	N/A	N/A

The output data: The output data is formed from the above list divided into 8 sub-lists each containing the corresponding data for 7 subsequent days. The dimensions of the resulting data will be (8, 7) i.e., 8 samples with 7 features per sample.

The input data: This also consists of 8 samples each containing the follow values in the same order [first value in the corresponding sample of output dataset, calculated alpha value, number of days since the first case, calculated beta value]. The resulting array dimensions will be (8, 4) i.e., 8 samples with 4 features per sample. Here the beta and alpha values are dynamically calculated according to the lockdown scales applied in Italy between the dates 19th of February and 14th of April (inclusive).

The DNN is then trained using supervised ML algorithm. For training purpose Adam optimization algorithm [11] is used with log-cosh [12] as the loss function.

For this study a standard DNN as shown in Fig. 2 (fully connected network) with one input layer having 4 nodes that takes the number of cases on the present day, alpha, day number and beta. Three hidden layers having 64, 128, 64 nodes respectively and one output layer with 7 nodes that represents the number of cases in each day of the week. All the nodes have the RELU function as activation. The number of hidden layers and the number of neurons in each hidden layer are tuned to achieve the desired accuracy (99%) on training set with minimum network complexity. The above architecture was found to be sufficient with no signs of underfitting or overfitting.

The DNN is trained till 99% of accuracy is reached. From the tests that we've conducted most of the times the error was on the positive side, meaning that the predicted output was a bit more than the actual data. This means that the DNN is showing some extra cases and these extra cases are giving us some room to fiddle with in terms of safety and precaution. The data thus obtained is provided under the 'simulation result' section.

Formulation of 'alpha' and 'beta'

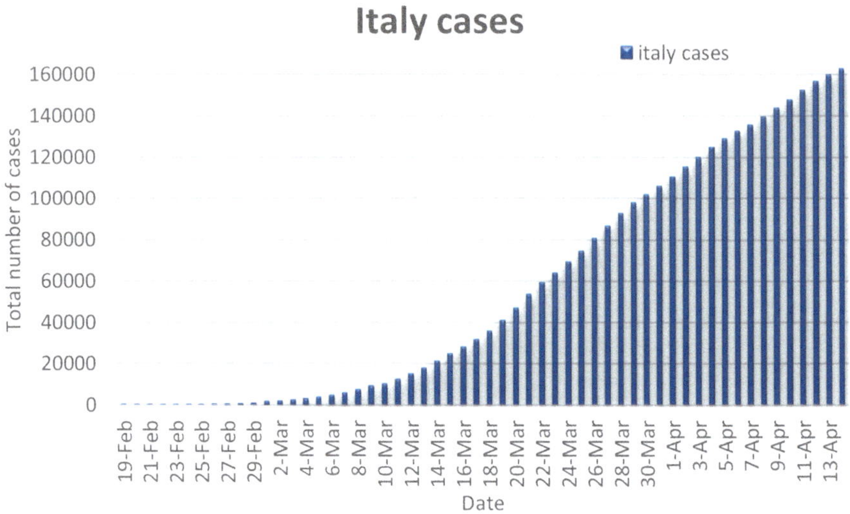

Fig. 1 Number of cases on each day in Italy starting form 19th of FEB till 14 of APR 2020

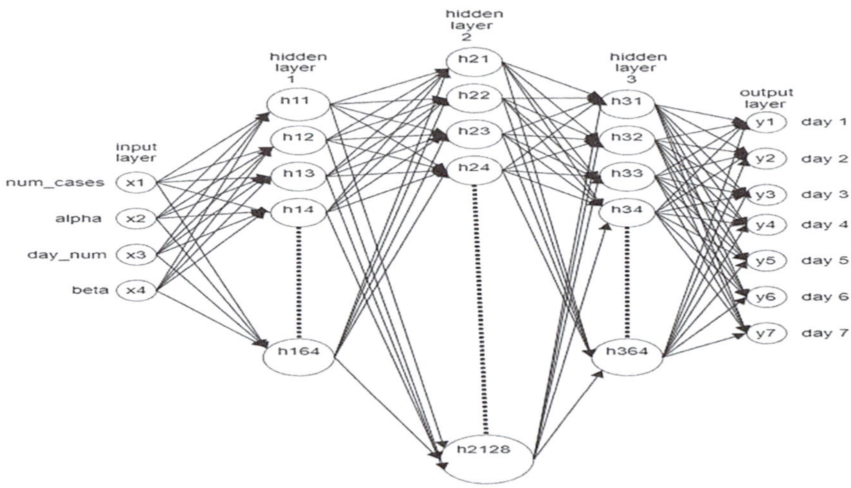

Fig. 2 Architecture *of the DNN*

‘alpha’ and ‘beta’ are based on the two new factors namely ‘transmission factor’ and ‘host factor’. Here the ‘transmission factor’ lies B/W [0.1, 1] and is determined based on how quickly and successfully the disease can be transmitted to others. transmission factor - 0.1 implies host to host jump is nearly impossible (non-infectious disease), 1 implies host to host jump is successful 100% of the time (highly infectious disease). The 'host factor' lies B/W [0.1, 1] and is determined based on the availability for new hosts. host factor - 0.1 implies no new host can be found to infect, 1 implies a new host is readily available (no social hygiene and social distancing). The host factor and the transmission factor together will determine how much of a threat the disease will become. None among the 'host factor' or the 'transmission factor' can be 0 (this is to avoid division with zero in the later part) and the population under lockdown(ppl) value holds a liner relation with 'Host factor' and is given by Eq. (2).

$$\text{Host factor} = 0.1 + 0.9 * \left(1 - {}^{\text{ppl}}\!/_{\text{tp}}\right) \quad (2)$$

Beta is calculated based on the given mathematical equation below.

$$Beta = 1 - (ppl^2)/(pd * LA * tp) \quad (3)$$

Alpha is calculated using the local population density and beta as per the equation below.

$$\text{Alpha} = \text{pd} * \text{Beta} \quad (4)$$

where

tp = total population of the country

ppl = total people under lockdown

*pd = population density of the country*host factor*transmission factor*

LA = total land area of the country.

These two numbers (alpha and beta) define how the growth of cases will be later on. But one thing to be noted is that the value of 'ppl' for India can never be greater than 0.75*tp, because for such large population a lockdown scale over 75% of total population can never be possible. As a matter of fact, for countries with 'tp' as huge as India, the scale strictly lies between [0.1 and 0.4] (this was concluded based on practical observations). Now let us look at what these numbers mean.

Beta can be interpreted as the degree of seriousness and awareness shown towards the global effort to stop the pandemic and alpha can be interpreted as a degree of possible interactions between a certain number of people per day per kilometer. Of course, as the government starts to impose more rules and restrictions this numbers will change as people tend to move out less often. But it is almost certain that not all will stay home or stay away from public places, as a matter of fact the service sector needs to be running (with increase in work hygiene followed). To account for the change in rules and restrictions, the alpha and beta values were calculated dynamically (according to the change in 'ppl') during the preparation of training set by taking 'the changes and actions that the Italy government took at each stage' into account.

3 Simulation Result

After the DNN is trained, it is left to predict the growth of cases in INDIA from day 16 after the first case. At first the DNN seems to be doing well but as it starts to predict the possible cases for the further future, we might observe that the numbers start to deviate from what is possible. This can be because of the fact that the DNN cannot be 100% accurate and so, there is a bit (negligible) of error in the first week's prediction, but remember we need the number of cases of one day to predict the number of cases for that week meaning that, we need the data from the first prediction for the DNN to be able to give the next prediction. The fact that we are feeding the DNN with a predicted value (that which carries some error) gives us an understanding about why the DNN starts to become less accurate when it is forced to predict distant future. The reason is something called 'error amplification' due to exponentiation. But this doesn't mean we can't use it for near future, since It was observed that when the prediction was made for less than 4 months the relative error lies in acceptable error range and 3 months' worth of predictions are more than enough, we can consider the maximum limit distant prediction as 35 to 40 days. Another possible reason for the decrement of accuracy over time is overfitting [13, 14].

If we run the same DNN with different values of 'ppl', it will provide us with different sets of data. This data will give the government a proper understanding about how much working population is permissible to run the services, as we can judge according to the growth of cases.

From different curves obtained by different values of 'ppl' and 'host factor'. We can pick the case where the cases double after some optimum number of days, thus making it possible to keep the economy up and running.

Here are the results after running the trained DNN with different values of 'ppl' and it was asked to predict from 1st of march till 14th week from there (inclusive). Tables 2, 3, 4 shows the output of DNN with ppl is equal to 40%, 30% and 20% respectively. Figures 3, 4, 5 shows the bar charts of predicted cases. It is observed that with 'ppl' of 40% number of cases are less compared with the other two conditions.

One thing to be noted is that, since these predictions are based on static 'ppl' the actual data obtained after the completion of those 14 weeks can be seen to jump in between those above given results. This is due to the fact that the value of 'ppl' won't be static but instead keeps changing (as told in the Italy data section above).

But either way these predictions will help the government to be prepared to what's coming ahead. The preparation may include further improvement in the health standards at the working facilities (this results in direct decrease in 'host factor' because it can also indicate hygiene), increase in capacity of health care facilities such as hospitals, increasing public awareness etc.

Table 2 Cases predicted for 14 weeks with 40%ppl in India

Week	Number of predicted COVID cases per week starting from 01 -March 2020 to 06 June 2020 by DNN						
	Day 1	Day 2	Day 3	Day 4	Day 5	Day 6	Day 7
Week 1	3	6	24	88	221	252	262
Week 2	343	359	373	387	400	407	425
Week 3	555	580	604	626	646	658	687
Week 4	896	935	973	1010	1041	1061	1107
Week 5	1443	1505	1568	1626	1677	1709	1783
Week 6	2323	2423	2523	2617	2698	2749	2869
Week 7	3736	3896	4058	4208	4339	4422	4614
Week 8	6007	6264	6525	6766	6977	7109	7419
Week 9	9657	10,070	10,489	10,877	11,215	11,428	11,925
Week 10	15,523	16,186	16,860	17,483	18,026	18,368	19,168
Week 11	24,949	26,014	27,097	28,098	28,970	29,520	30,806
Week 12	40,096	41,807	43,548	45,156	46,558	47,442	49,508
Week 13	64,437	67,187	69,985	72,569	74,822	76,242	79,562
Week 14	103,552	107,971	112,468	116,620	120,241	122,523	127,859

Table 3 Cases predicted for 14 weeks with 30%ppl in India

Week	Number of predicted COVID cases per week starting from 01 -March 2020 to 06 June 2020 by DNN						
	Day 1	Day 2	Day 3	Day 4	Day 5	Day 6	Day 7
Week 1	4	7	31	115	289	332	344
Week 2	450	470	489	508	524	534	557
Week 3	727	759	790	820	845	861	899
Week 4	1172	1222	1273	1321	1362	1388	1448
Week 5	1887	1968	2050	2126	2192	2233	2331
Week 6	3035	3166	3297	3419	3526	3592	3749
Week 7	4881	5090	5302	5498	5669	5777	6028
Week 8	7848	8183	8524	8839	9114	9287	9691
Week 9	12,615	13,154	13,701	14,208	14,649	14,927	15,577
Week 10	20,275	21,141	22,022	22,835	23,544	23,991	25,036
Week 11	32,586	33,977	35,392	36,699	37,838	38,556	40,235
Week 12	52,368	54,603	56,878	58,978	60,809	61,963	64,661
Week 13	84,159	87,750	91,405	94,779	97,722	99,577	103,913
Week 14	135,245	141,016	146,890	152,312	157,042	160,022	166,991

Table 4 No of cases predicted for 14 weeks by DNN with 20%ppl in India

Week	Number of predicted COVID cases per week starting from 01 -March 2020 to 06 June 2020 by DNN						
	Day 1	Day 2	Day 3	Day 4	Day 5	Day 6	Day 7
Week 1	5	8	37	138	350	401	417
Week 2	544	568	592	614	633	645	673
Week 3	878	917	955	990	1021	1041	10,856
Week 4	1415	1476	1538	1595	1644	1676	1749
Week 5	2278	2376	2475	2566	2646	2696	2814
Week 6	3664	3821	3980	4128	4256	4337	4526
Week 7	5892	6144	6400	6636	6843	6973	7276
Week 8	9472	9877	10,288	10,668	11,000	11,208	11,696
Week 9	15,225	15,875	16,536	17,147	17,680	18,015	18,800
Week 10	24,470	25,515	26,577	27,559	28,414	28,954	30,215
Week 11	39,326	41,005	42,713	44,290	45,665	46,531	48,558
Week 12	63,200	65,897	68,642	71,176	73,386	74,779	78,035
Week 13	101,565	105,899	110,310	114,382	117,934	120,172	125,405
Week 14	163,217	170,181	177,270	183,813	189,521	193,117	201,527

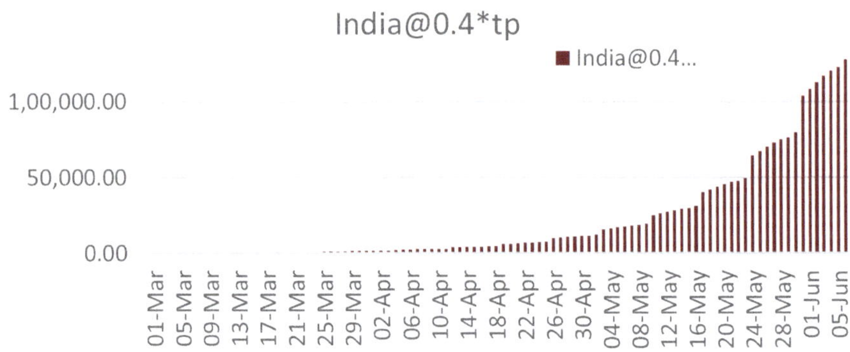

Fig. 3 Number of COVID cases identified with 40% people under lockdown

4 Conclusion

This is a paper on Machine Learning and its potential in aiding governments at times of economic crisis such as pandemics. Before taking proper decisions, it is better to understand about the facts and limitations of ML and to be aware of the possible risks involved. This model can be further developed by taking the suggestions of economists, epidemiologists, engineers and scientists.

limitations:

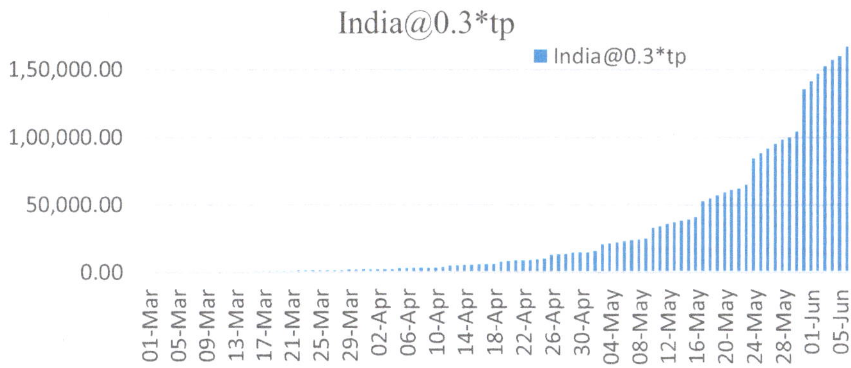

Fig. 4 Number COVID cases identified with 30% people under lockdown

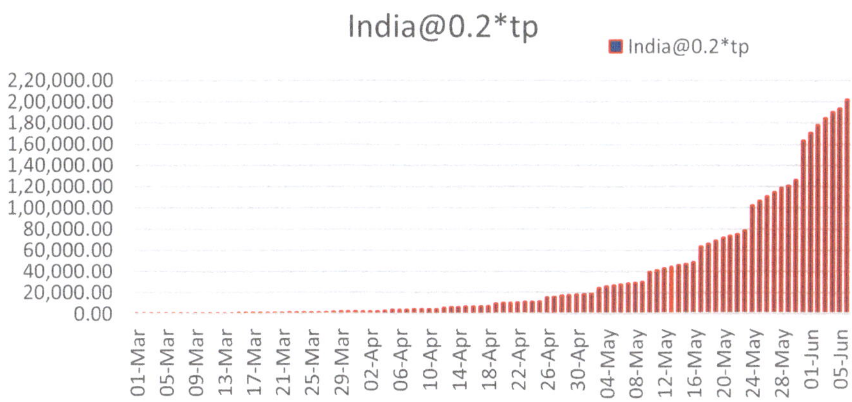

Fig. 5 Number COVID cases identified with 20% people under lockdown

The first limitation is that what we faced with the lack of computational power. This is to convey the fact that better results (when in comparison to this paper) can be achieved with the proposed model given enough time for training, and data acquisition and data processing.

The second limitation is the that there is no way of perfectly knowing the actual number of cases and what prior actions were taken to reach that state. This actually requires some assumptions to be made (assumptions based of available data, which are logically acceptable) to fill in the missing information in order to form a dataset for training. This naturally introduces some noise to the data to which the DNN might show some side effects.

References

1. WHO 2020a Coronavirus Disease 2019 (COVID-19): Situation Report-91 2020. URL. https://www.who.int/emergencies/diseases/novel-coronavirus-2019.
2. WHO 2020b Coronavirus Disease 2019 (COVID-19): Situation Report-51 2020. URL. https://www.who.int/docs/default source/ corona virus/ situationreports/20200311-sitrep-51-covid-19.pdf?sfvrsn=1ba62e57_4
3. WHO 2020c Coronavirus Disease 2019 (COVID-19): Situation Report-30 2020. URL.https://www.who.int/docs/defaultsource/coronaviruse/situationreports/20200219-sitrep-30-covid-19.pdf?sfvrsn=6e50645_2.
4. Monika Chaudhary PR, Sodani1, Shankar Das (2020) Effect of COVID-19 on Economy in India: Some Reflections for Policy and Programme. J. Health Manag 22(2): 169–180
5. Hema S, Gopalana, Anoop Misraa (2020) COVID-19 pandemic and challenges for socio-economic issues, Healthcare and National Health Programs in India. Diabetes Metab. Syndr: Clin. Res. Rev 14(5): 757–759
6. Singh V, Poonia RC, Kumar S, Dass P, Agarwal P, Bhatnagar V, Raja L (2020) Prediction of COVID-19 corona virus pandemic based on time series data using Support Vector Machine. J. Discrete Math. Sci. Cryptogr. 23(8):1583–1597. https://doi.org/10.1080/09720529.2020.1784535
7. Bhatnagar V, Poonia RC, Nagar P, Kumar S, Singh V, Raja L, Dass P (2020) Descriptive analysis of COVID-19 patients in the context of India. J. Interdiscip. Math. 24(3):489–504. https://doi.org/10.1080/09720502.2020.1761635
8. Kumari R, Kumar S, Poonia RC, Singh V, Raja L, Bhatnagar V, Agarwal P Analysis and Predictions of Spread, Recovery, and Death Caused by COVID-19 in India, Big Data Mining and Analytics 4(2), pp 65–75, DOI: https://doi.org/10.26599/BDMA.2020.9020013
9. Ding B, Qian H, Zhou J (2018) "Activation functions and their characteristics in deep neural networks," Chinese Control And Decision Conference (CCDC 2018), Shenyang, pp 1836–1841. doi: https://doi.org/10.1109/CCDC.2018.8407425.
10. The worldo meter COVID Tracking. Retrieved from. https://www.worldometers.info/corona virus/ country/Italy.
11. Diederik P. Kingma*, Jimmy Lei Ba∗ (2015)" adam: a method for stochastic optimization" in Proc. ICRL Conf. 2015.
12. Pengfei Chen, Guangyong Chen, Shengyu Zhang (2019) "Log Hyperbolic Cosine Loss Improves Variational Auto-Encoder" in Proc. ICRL Conf. 2019.
13. Madhiarasan M, Deepa SN (2017) Comparative analysis on hidden neurons estimation in multi layer perceptron neural networks for wind speed forecasting. Artif Intell Rev 48:449–471
14. Gnana Sheela K, Deepa SN (2013) "Review on Methods to Fix Number of Hidden Neurons in Neural Networks", Mathematical Problems in Engineering, vol 2013, ArticleID 425740, 11 pages, 2013. https://doi.org/10.1155/2013/425740.

Chapter 6
Prediction of the Geographical Origin of Soils Using Ultra-Performance Liquid Chromatography (UPLC) Fingerprinting and *K*-Nearest Neighbor (*K*-NN)

Loong Chuen Lee, Hukil Sino, Nor Azman Mohd Noor, Saiful Mohd Ali, and Azhar Abdul Halim

1 Introduction

Forensic soil analysis is one of the most important analyses in forensic investigation. Soil can be used to establish a link between a suspect and a victim or a crime scene, particularly an outdoor scene [1]. Recently, Procter et al. [2] presented a detailed study on the influence of clothing materials on soil transfer. The authors concluded that soil can provide an important lead in assisting a criminal investigation. Typically, a forensic analyst establishes the link between a questioned soil with a known soil by comparing their physicochemical and biological characteristics.

In practice, the physical properties of soils, *e.g.* pH, color and density, are of limited use [3]. Hence, discrimination based on chemical components of soils has attained more attention. Various analytical instruments have been evaluated in acquiring the chemical profile of soil for forensic discrimination and classification purposes [4–8]. For instance, Profumo et al. [5] demonstrated the use of gas-chromatography mass

L. C. Lee (✉) · H. Sino
Forensic Science Program, Faculty of Health Sciences, CODTIS, Universiti Kebangsaan Malaysia, Selangor Kuala Lumpur, Malaysia
e-mail: lc_lee@ukm.edu.my

L. C. Lee
Institute IR4.0, Universiti Kebangasaan Malaysia, 43600 Bangi, Selangor, Malaysia

N. A. M. Noor
Polis Diraja Malaysia, Bukit Aman, Kuala Lumpur, Malaysia

S. M. Ali
Pusat Analisis Sains Forensik, Jabatan Kimia Malaysia, Petaling Jaya, Malaysia

A. A. Halim
School of Environmental and Natural Resources Sciences, Universiti Kebangsaan Malaysia, Selangor Kuala Lumpur, Malaysia

G. Mathur et al. (eds.), *Proceedings of 2nd International Conference on Artificial Intelligence: Advances and Applications*, Algorithms for Intelligent Systems,
https://doi.org/10.1007/978-981-16-6332-1_6

spectrometer (GC-MS) in analyzing volatile organic compounds in soils for forensic purposes. On the other hand, Kikawa et al. [6] employed a semi-automated scanning electron microscopy-energy dispersive X-ray dispersive (SEM–EDX) to evaluate differences between soils based on mineral particles in soils.

In general, data obtained from the analytical instruments are often of high dimensionality. For example, ultra-performance liquid chromatography (UPLC) chromatograms obtained at a particular wavelength via a photodiode array detector can have over hundreds or thousands of descriptor variables (*i.e.* retention time points). Therefore, chemometrics techniques have been employed by forensic analysts in accomplishing different forensic tasks, *e.g.* identification of the origin of an unknown soil. For instance, the factor analysis technique has been reported to be useful in discriminating soils by their origins [9].

Recently, McCulloch et al. [10] have proposed markers for discriminating various soils originated from close proximity sites in the United Kingdom by using a high-performance liquid chromatography (HPLC) technique. The authors first identified a number of markers by retention time points that peaks below the limit of quantification were all excluded. Then, canonical discriminant function analysis was employed to classify the chromatographic profiles of soils by the locations. Despite the discrimination result being encouraging, the authors utilized the condensed chromatographic data that requires the researcher to manually assign the selected peaks by the retention time.

On the other hand, the *K*-nearest neighbor (*K*-NN) is one of the most popular machine learning methods in the field of data mining [11], however, to the best of our knowledge, the performance of *K*-NN in modeling liquid chromatographic data in the context of forensic science is still unavailable in the literature. Therefore, this work aims to evaluate the performance of *K*-NN in modeling ultra-performance liquid chromatography (UPLC) chromatograms of soils. Additionally, the full chromatographic data instead of the condensed chromatographic data were employed herein for constructing the K-NN models.

2 Methodology

2.1 Data

Five proximity locations in Bangi, Selangor, Malaysia, have been identified for collecting soil samples. These included (a) fern garden of Universiti Kebangsaan Malaysia (UKM); (b) forest of UKM; (c) commuter station of UKM; (d) illegal dumping site at Bangi Lama; and (e) commuter station of Bangi Lama. Table 1 shows the GPS coordinates of the five locations. One red and one brown soils were identified from each of the locations and sampled via the grid pattern method [3]. Eventually, ten soil samples were prepared for this study. By injecting each sample

Table 1 GPS coordinates of the five studied locations

Location	Code	GPS Coordinates
Abandoned land nearby Commuter station of Bangi Lama	KB	2° 54′ 02.9″ N 101° 47′ 06.0″ E
Illegal dumping site at Bangi Lama	BL	2° 54′ 05.6″ N 101° 46′ 37.2″ E
Abandoned land nearby Commuter station of UKM	KU	2° 56′ 14.1″ N 101° 47′ 26.7″ E
Forest at UKM	HK	2° 54′ 48.8″ N 101° 47′ 17.1″ E
Fern farm at UKM	PP	2° 55′ 23.6″ N 101° 46′ 57.8″ E

into the UPLC system in triplicate, a total of 30 UPLC chromatograms was obtained. Detailed information and parameters of the UPLC method are reported in [12].

Typically, more than one type of data can be prepared from a chromatographic system. For example, Lee et al. [13] demonstrated merits and pitfalls of full and condensed chromatographic data of pen inks acquired from a UPLC system equipped with a photodiode array detector. First, the researchers have to identify the wavelength at which the sample shows the most number of peaks. Then, the researchers can prepare the condensed and full chromatographic data, respectively, based on the best detection wavelength. The former considers only the components of interest, whereas the latter form of data includes all the data points presented in the chromatogram.

This work is devoted to model the UPLC chromatograms by extracting the full chromatographic data (absorbance values) instead of the condensed chromatographic data (peak areas of selected components). Two types of full chromatographic data were prepared by considering the retention time from (a) 0–15 min and (b) 1–5 min. The former full data have 18 000 variables and the latter was proposed based on the observation that the region in the range 5–15 min showed a limited number of peaks in the chromatogram.

2.2 K-*Nearest Neighbor* (K-*NN*) *Algorithm*

The K-nearest neighbor (K-NN) method is a model-free approach. Strictly speaking, it does not construct any model but involves all the training data whenever predicting a test sample. Two parameters to be considered in constructing a K-NN model are (a) multivariate distance metric and (b) number of neighbor (K). Typically, the best K-NN model is constructed by optimizing only the latter parameter.

Therefore, this work employed only Euclidean distance but four different values in a number of neighbor ($K = 1, 2, 3$ and 4) [11]. The values were selected based on the number of samples available for a class in the UPLC data of soils. As explained above, we have prepared two sets of full chromatographic data, *i.e.* full and truncated chromatograms, thus, a total of eight different K-NN models were prepared in this work.

For prediction purposes, the K-NN algorithm has to evaluate the membership of K training samples showing the shortest distance from the test sample. For instance, assume $K = 3$, and two of the three nearest neighbors are from class 1. In this circumstance, the test sample is assigned to class 1 since the majority class among the nearest neighbors is class 1. In other words, K-NN relies on the majority voting rule to predict a new sample.

Figure 1 illustrates the overall steps employed in constructing the K-NN model using the full and truncated chromatographic data, respectively. Firstly, the chromatogram data were divided into a pair of training and test samples. The former was employed to construct the K-NN model and estimate the internal accuracy rate, whereas the latter was used for computing the external accuracy rate of the model. The internal and external accuracy rates (Acc) are determined by Eq. (1):

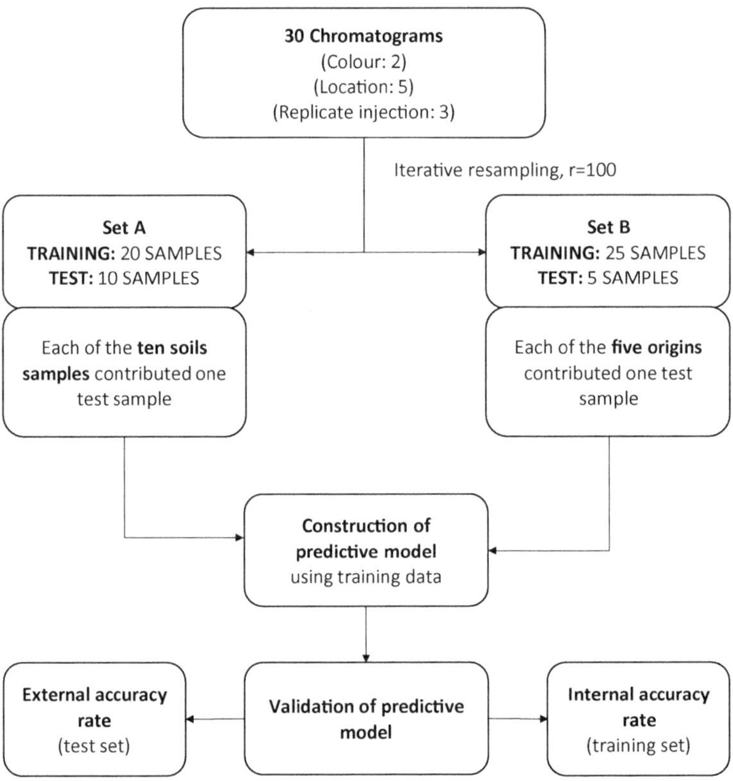

Fig. 1 Schematic flowchart of constructing multiple predictive models using K-nearest neighbor (K-NN) algorithm. Iterative resampling approach has been employed to prepare 100 pairs of training and test samples using strategies A and B, respectively

$$\mathrm{Acc} = \frac{n'}{n} \tag{1}$$

where *n'* and *n* denote the number of correctly predicted sample and total sample of a given test (*i.e.* external accuracy) or training set (*i.e.* internal accuracy).

In order to gain a deeper insight into the potential of K-NN in modeling UPLC chromatographic data, we have prepared 100 pairs of training and test samples by using two different strategies. The first strategy selected only one chromatogram from a particular soil sample which resulted in a test set composing 10 samples. Meanwhile, the second strategy treated the same primary chromatogram data differently that only one chromatogram of a location was selected to form the test set, *i.e.* five test samples.

Eventually, this study has constructed 16 different K-NN models from the same chromatographic data by considering (a) two input regions (*i.e.* full and truncated); (b) four values of *K* (*i.e.* 1, 2, 3 and 4); (c) two resampling strategies (*i.e.* sets A and B).

3 Results and Discussions

3.1 Mean Chromatograms

Figure 2 illustrates the mean chromatograms of the ten soil samples. The red and brown soils were, respectively, represented by thin and thick lines in Fig. 2. In brief, all the ten soil extracts contained multitude numbers of organic compounds as shown in the UPLC chromatograms. The peaks were varied in location (*i.e.* elution time), shape and peak height.

In general, the red and brown soils sharing the same origin demonstrated highly similar profiles, except for the red and brown soil samples from KB. However, the red soil of KB demonstrated a highly similar chromatographic profile to the (red and browns) soils of KU. Meanwhile, the brown soil of KB appeared to be similar to the (red and brown) soils sampled from BL and HK, respectively.

In general, most of the peaks were eluted out from the UPLC column within the first 5 minutes. Meanwhile, peaks that appeared after the first 5 minutes were seen only in a few samples. In order to take a closer look at the most diverse region, the zoomed view of the region is presented in Fig. 2b. Based on the truncated region, we can see that the brown and red soils were quite similar to each other, except that sampled from KU and KB. Additionally, it is also noted from Fig. 2b that most of the peaks were not fully resolved. Therefore, it is infeasible to discriminate the ten soil samples by visual inspection alone.

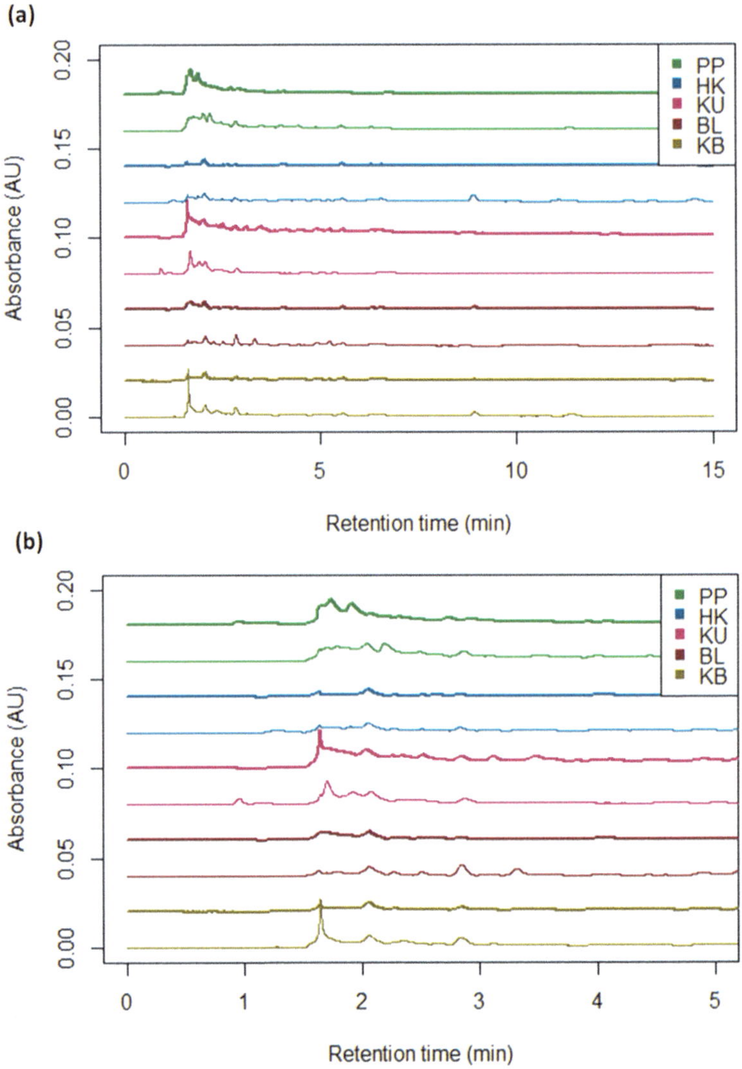

Fig. 2 Mean chromatograms of the ten soils, *i.e.* five (thin line) red and five (thick line) brown soils sampled from five different locations: **a** full chromatogram covering 0–15 min and the respective **b** zoomed view covering 0–5 min

3.2 *Principal Component Analysis (PCA)*

Before modeling via the K-NN method, the data matrix of 30 rows (chromatograms) and 18 000 columns (retention time points) was first explored using the principal component analysis (PCA) technique. The purpose was to evaluate the variability of the 10 soil samples by five locations and two soil colors using the score plot of PCA. After a careful inspection on all the possible score plots, the two most discriminated score plots of the full and truncated chromatograms were, respectively, obtained using PC-1, -4, -5 (Fig. 3a, b), and PC-1, -6, -8 (Fig. 3c, d).

Basically, the first principal component (PC) has explained more than 80% of the total variance in both the chromatographic data. Hence, it is understood why the second and the third principal components were found to be deficient in discriminating the soil samples. These are observed in both the full and truncated chromatographic data. Another possible reason for these could be due to the fact that we have not preprocessed the chromatographic data using any data preprocessing technique.

In general, soils showing similar chromatographic profiles as seen in Fig. 2 were found to be differentiated in the score plots of PCA (Fig. 3). In particular, score plot of PC-1 versus PC-4 of the full chromatograms (Fig. 3a) showed the best discrimination, *i.e.* none of the ten soils is overlapped with each other. However, the other three plots have samples clustered in close distance with other samples. For instance, brown soils of HK and PP were located nearby each other in Fig. 3b, d). In addition, the clusters of red soils of BL and HK were also often located nearby each other (see Fig. 3b, c).

Hence, it seems sound to conclude that the full chromatographic data is much useful than the truncated chromatographic data.

3.3 K-*NN Modeling*

As mentioned in Sect. 2.2, a total of 16 K-NN models have been constructed from primary chromatographic data. The modeling performances of the K-NN models in terms of mean accuracy rates are summarized in Fig. 4. The impact of the number of neighbors (K) on the performances of the K-NN model was presented by the two different resampling strategies: (a) ten samples (Fig. 4a, set A) and (b) five samples (Fig. 4b, set B) as a test set.

In this work, each of the five locations described in Sect. 2.1 was represented by six chromatograms, *i.e.* three replicates from one red and brown soils, respectively. Hence, given the test set consisting of ten samples, a location has only four representatives in the training set (Fig. 4a). On the other hand, a location was represented by five samples in the training data in case the test set consisted of five samples (Fig. 4b).

Basically, K-NN models constructed using only one neighbor ($\#K = 1$) outperformed the other models, in both the internal and external accuracy rates. In brief, the K-NN model performances degrade as the number of neighbors ($\#K$) increases. Moreover, it is worth mentioning that model performance in terms of external accuracy rate degraded faster when strategy B (see Fig. 1) was adopted.

Next, performances of models constructed using the full chromatogram data were found to be slightly better than those built using the truncated chromatograms when $\#K$ is more than 1. This seems to support the interpretation derived from the score

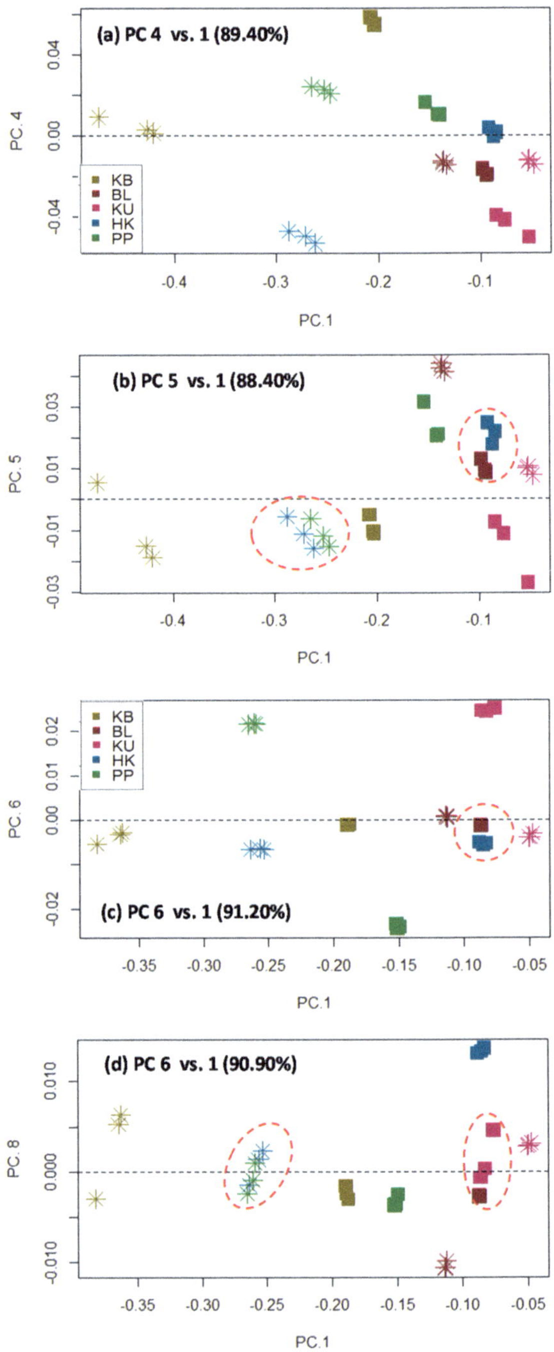

Fig. 3 Score plots of PCA obtained using **a**–**b** full and **c**–**d** truncated chromatograms of five red (square) and five brown (*) soils

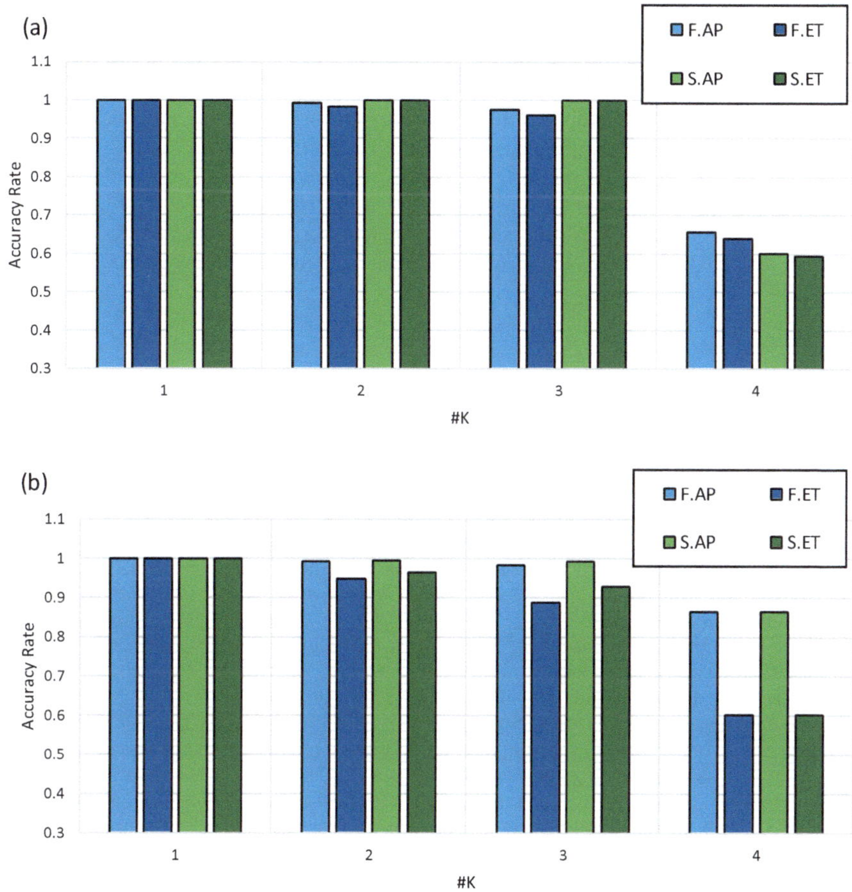

Fig. 4 Mean accuracy rates of *K*-NN models constructed using (F) full and (S) truncated chromatograms which were estimated via (AP) auto-prediction and (ET) external validation procedures. The external validations have been performed on **a** ten chromatograms and **b** five chromatograms as test samples

plot of PCA that the minor peaks eluted after 5 min do play important role in discriminating the ten soil samples.

4 Conclusion

In conclusion, UPLC chromatograms of soils collected from proximity locations could be discriminated using the K-NN method. In this work, we have not attempted to discriminate red and brown soils sampled from the same location. In practice, red and brown soils sampled from the same location can be differentiated from each other based on visual inspection alone. Additionally, the origin of the soil is more important than the color of the soil in a typical forensic investigation. In future, spatial and temporal variations of soils shall be included in evaluating the performance of the K-NN algorithm, and the stability of the K-NN model needs to be addressed by considering more number of chromatograms.

Acknowledgements We would like to thank you Anas Fahmi Zakaria, Ameeta A/P Nai Eyan and Nurul Syahiera Kharudin for performing soil sampling and extraction; and Mr. Aziz Ab Ishak for conducting the UPLC analysis. This research was conducted with the support of CRIM-UKM (GUP-2020-085).

References

1. Bull PA, Parker A, Morgan RM (2006) The forensic analysis of soils and sediment taken from the cast of a footprint. Forensic Sci Int 162:6–12
2. Procter FA, Swindles GT, Barlow NLM (2019) Examining the transfer of soils to clothing materials: Implications for forensic investigations. Forensic Sci. Int. 305, 110030
3. Pye K (2007) Geological and Soil Evidence: Forensic Applications. CRC Press, Taylor & Francis Group, Boca Raton
4. Xu X, Du C, Ma F, Shen Y, Zhou J (2020) Forensic soil analysis using laser-induced breakdown spectroscopy (LIBS) and Fourier transform infrared total attenuated reflectance spectroscopy (FTIR-ATR): Principles and case studies. Forensic Sci. Int. 310, 110222
5. Profumo A, Gorroni A, Guarnieri SA, Mellerio GG, Cucca L, Merli D (2020) GC-MS qualitative analysis of the volatile, semivolatile and volatilizable fractions of soil evidence for forensic application: A chemical fingerprinting. Talanta 219, 121304
6. Kikkawa HS, Naganuma K, Kumisaka K, Sugita R (2019) Semi-automated scanning electron microscopy energy dispersive X-ray spectrometry forensic analysis of soil samples. Forensic Sci. Int. 305, 109947
7. Zeng R, Rossiter DG, Zhao YG, Li DC, Zhang GL (2020) Forensic soil source identification comparing matching by color, vis-NIR spectroscopy and easily-measured physio-chemical properties. Forensic Sci. Int. 317, 110544
8. Woods B, Kirkbride KP, Lennard C, Robertson J (2014) Soil examination for a forensic trace evidence laboratory=Part 2: Elemental analysis. Forensic Sci Int 245:195–201
9. Melo VF, Mazzetto JML, Dieckow J, Bonfleur EJ (2018) Factor analysis of organic soils for site discrimination in a forensic setting. Forensic Sci Int 290:244–250
10. McCulloch G, Dawson LA, Brewer MJ, Morgan RM (2017) The identification of markers for Geoforensic HPLC profiling at close proximity sites. Forensic Sci Int 272:127–141
11. Wu X, Kumar V, Quinlan JR, Ghosh J, Yang Q, Motodo H, McLachlan GJ, Ng A, Liu B, Yu PS, Zhou Z-H, Steinbach M, Hand DJ, Steinberg D (2008) Top 10 algorithms in data mining. Knowl Inf Syst 14:1–37
12. Lee LC, Ishak AA, Ameeta NE, Anas FZ, Nurul SK, Nor Azman MN (2021) Forensic profiling of non-volatile organic compounds in soil using ultra-performance liquid chromatography: a pilot study. In press, Forensic Science Research
13. Lee LC, Ishak AA, Hamid NA, Ravi Y, Ahamd MA, Ali N (2021) A comparison between univariate and multivariate statistical techniques to determine source of pen inks using ultra-performance liquid chromatography (UPLC) chromatograms In press J. Liq. Chromatogr. Relat. Technol

Chapter 7
A Novel Entropy-Based FCM Algorithm Using Inverse Fuzzy Membership Framework and Uncertainty Measure for Segmentation of Brain MR Images

Madhumita Ray, Nabanita Mahata, and Jamuna Kanta Sing

1 Introduction

Segmentation of brain magnetic resonance (MR) image data is a most essential requirement of medical field to analyze and diagnose the complicated and critical diseases as well as different abnormalities and anomalies of human brain. Computer aided magnetic resonance imaging segmentation become crucial task as noise and intensity inhomogeneity affect coupled in the image. The reasons behind this are non-uniform sensitivity of the receiving coil, radio frequency signals related artifacts, movement of the patient at the time of image scanning, and shortfalls of the capturing devices, which make the intersection of the inter organ soft tissue region boundary blurry.

In the recent past, different algorithms proposed. We review some of them related to the present work. The fuzzy clustering algorithms are mostly used in brain MR image segmentation. Fuzzy c-means (FCM) [1] clustering method is very much renowned method for brain image segmentation without considering spatial information and badly affected by high percentage of noise and inhomogeneity. Chuang et al. [2] introduced clustering algorithm with FCM, considering spatial information that can reduce fake and noisy spots. Cai et al. [3] implemented fast generalized FCM (FGFCM) by using original image with spatial coordinate information with gray level local neighborhood window. Kanan et al. [4] proposed entropy tolerant FCM algorithm. Wang et al. [5] incorporated more robust fuzzy clustering algorithm using adaptive spatial information. Iftikhar et al. [6] proposed an enhanced non

M. Ray (✉)
CSE Department, Greater Kolkata College of Engineering and Management, Baruipur, West Bengal, India

N. Mahata · J. K. Sing
CSE Department, Jadavpur University, Jadavpur, Kolkata, West Bengal, India
e-mail: jksing@ieee.org

G. Mathur et al. (eds.), *Proceedings of 2nd International Conference on Artificial Intelligence: Advances and Applications*, Algorithms for Intelligent Systems,
https://doi.org/10.1007/978-981-16-6332-1_7

local mean algorithm to make magnetic resonance imaging noise free and robust. Adhikari et al. [7] introduced an improved version of FCM considering conditional and spatial information involve with each pixel. Bai et al. [8] proposed possibility FCM [PFCM] which is suitable for hypercorrectly distributed data groups. MRI brain image intensity values do not distribute hypercorrectly. So PFCM algorithm has inherent deficiency for MRI brain image segmentation. Kahali et al. [9] proposed modified FCM method in multi-objective framework with two stage of 3D spatial and 3D modified FCM. Askari et al. [10] generalized combination of entropy fuzzy c-means and probabilistic FCM. Gharieb et al. [11] proposed c-means clustering method with membership relative entropy and local membership relative entropy. Kahali et al. [12] proposed non-Euclidian based fuzzy c-means clustering incorporated with entropy and Gaussian function. Some recent research hybridized this clustering approach with spider monkey algorithm [13, 14] and also transformed data sets [15].

Our proposed entropy-based FCM (ebFCM) clustering algorithm introduces an inverse fuzzy membership framework for each pixel in association with Gaussian distribution function. It also uses a local uncertainty parameter along with the fuzzy membership function. The underline information uncertainty is defined by the Shannon entropy function using these uncertainty parameters.

The organization of the paper is as follows. We present the proposed ebFCM algorithm in Sect. 2. While Sect. 3 gives the experimental results with discussions. Finally, we give the concluding remarks in Sect. 4.

2 The EbFCM Clustering Algorithm

In brain MR imaging, uncertainty of soft tissue region classification due to noise and inhomogeneity makes the procedure complex enough. Further, neighboring pixels associated with every other pixels/voxels powerfully and generally dependent on each other. In our approach, we propose an inverse membership function frame work along with Gaussian distribution function. It enforces the membership value to gradually decrease as we move away from the cluster center. The algorithm also incorporates a local uncertainty parameter in conjunction with the fuzzy membership function. Unlike the former one, its value increases with the increasing distance from the cluster center. Its higher value indicates that most unlikely it will be a member of the cluster, indicating its amount of uncertainty. Its value is used in association of a Gaussian distribution function that utilizes local mean intensity value around a center pixel. These two terms are controlled by two complementing regularizing parameters. The framework enables us to utilize the local and global distribution of pixels in the whole image domain. Finally, by using the Shannon entropy function utilizing the uncertainty factor we define the underline total information uncertainty. Here, we exploit the intensity information of brain MR image with size of $(H \times W)$ i.e. $(height \times width)$ with C different regions. In our approach we consider eight features corresponding to each of the pixels to further reduce the effect of noise and intensity inhomogeneity.

The objective function of the proposed algorithm is following:

$$g_{ebFCM} = \sum_{i=1}^{C} \sum_{r=1}^{H} \sum_{c=1}^{W} \alpha\left(1 - \mu_{irc}^{m}\right) A + \sum_{i=1}^{C} \sum_{r=1}^{H} \sum_{c=1}^{W} (1 - \alpha) p_{irc}^{m} \mu_{irc}^{m} B - \sum_{i=1}^{C} \sum_{r=1}^{H} \sum_{c=1}^{W} p_{irc} \ln(p_{irc}), \quad (1)$$

where C = number of segments/cluster, α indicates the control parameter, μ_{irc}= membership value of pixel of row r and column c for the segment i, p_{irc} = uncertainty value, degree of fuzziness $m > 1.0$, σ_i^2 = variance, v_i = ith cluster center, A indicates Gaussian distribution function, and B indicates Gaussian distribution of mean of neighborhood pixels. These are defined as follows:

$$A = e^{\frac{-\|x_{rc} - v_i\|^2}{2\sigma_i^2}}, \quad (2)$$

$$B = e^{\frac{-\|\overline{x_{rc}} - v_i\|^2}{2\sigma_i^2}}, \quad (3)$$

$$\overline{x_{rc}} = \frac{1}{N_{\text{rc}}} \sum_{j=1}^{N_{rc}} x_j, \quad (4)$$

where N_{rc}= number of pixels in the squared neighborhood with centered pixel as x_j. $\overline{x_{rc}}$= mean of the square neighborhood pixels. Here, window size of the neighboring pixels consider as N_{rc}= 3 × 3. Using the partial derivative of the objective function (1) with respect to μ_{irc}, p_{irc}, and v_i we can minimize it with the following constraints:

$$\sum_{i=1}^{C} \mu_{irc} = 1, \quad (5)$$

$$\sum_{i=1}^{C} p_{irc} = 1. \quad (6)$$

The final iterative equations are given below:

$$\mu_{irc} = \frac{1}{\sum_{l=1}^{C} \sum_{r=1}^{H} \sum_{c=1}^{W} \left(\frac{(1-\alpha) p_{irc.}^{m} B - \alpha.A}{(1-\alpha) p_{lrc.}^{m} B - \alpha.A} \right)^{\frac{1}{m-1}}}, \quad (7)$$

$$p_{irc} = \frac{1}{\sum_{l=1}^{C} \sum_{r=1}^{H} \sum_{c=1}^{W} \left(\frac{(1-\alpha)\mu_{irc.}^{m} B - \frac{ln(p_{irc})}{p_{irc}^{m-1}}}{(1-\alpha)\mu_{lrc.}^{m} B - \frac{ln(p_{lrc})}{p_{lrc}^{m-1}}} \right)^{\frac{1}{m-1}}}, \tag{8}$$

$$v_i = \frac{\sum_{i=1}^{C} \sum_{r=1}^{H} \sum_{c=1}^{W} \left[\left(\alpha.\left(1 - \mu_{irc}^{m}\right) A.x_{\mathrm{rc}} \right) + \left((1-\alpha).p_{irc}^{m} \mu_{irc}^{m}.B.\overline{x_{\mathrm{rc}}} \right) \right]}{\sum_{i=1}^{C} \sum_{r=1}^{H} \sum_{c=1}^{W} \left[\left(\alpha.\left(1 - \mu_{irc}^{m}\right) A \right) + \left((1-\alpha).p_{irc}^{m} \mu_{irc}^{m} B \right) \right]}. \tag{9}$$

2.1 Algorithm: EbFCM

User Input: C, m, and the threshold of the error ε.

1. Define the cluster center $v_i^{(0)}$ randomly.
2. Initialize the uncertainty parameter $p_{irc}^{(0)}$.
3. Set iteration number $s = 1$.
4. Repeat.
   a. $s = s + 1$
   b. Calculate the value of the membership functions $\mu_{irc}^{(s)}$ using (7).
   c. Calculate the uncertainty parameters $p_{irc}^{(s)}$ using (8).
   d. Calculate the cluster centers $v_i^{(s)}$ using (9).
5. Until $v_i^{(s)} - v_i^{(s-1)} < \varepsilon$
6. Use the membership values μ_{irc} and the cluster centers v_i to classify a pixel.

3 Experimental Results

We test the ebFCM method on ten volumes of different percentage level of noise and inhomogeneity. These brain MR image volumes contain simulated as well as clinical data. We segment the images into (i) white matters (WM), (ii) gray matters (GM), and (iii) cerebrospinal fluid (CSF) regions. We examine the behavior of the ebFCM algorithm with the following performance indices.

Partition Coefficient (v_{pc}). Partition coefficient v_{pc} is one of the important matrices regarding this purpose. Performance of v_{pc} is very significant when fuzziness is less. The partition coefficient value needs closer to 1, assuring successful clustering. The value of $v_{pc} left$ [7] described as follows:

$$v_{pc} = \frac{\sum_{i=1}^{C} \sum_{r=1}^{H} \sum_{c=1}^{W} \mu_{irc}^{2}}{H \times W}. \tag{10}$$

Partition Entropy (v_{pe}). Minimal value of v_{pe} indicates superiority of the clustering. If result of v_{pe} is 0, means optimal clustering value. The definition of v_{pe} [7] is following:

$$v_{pe} = \frac{-\sum_{i=1}^{C}\sum_{r=1}^{H}\sum_{c=1}^{W}\left[\mu_{irc}\log\mu_{irc}\right]}{H \times W}. \tag{11}$$

Similarity Index (ρ).Similarity index ρ [7] *is* measured by

$$\rho = \frac{1}{C}\sum_{i=1}^{C}\frac{2|M_i \cap R_i|}{|M_i| + |R_i|}, \tag{12}$$

where R_i denotes the set of pixels in the ith region of the ground truth and M_i denotes a set of pixels in the ith region generated by the algorithm. The value of ρ is within [0, 1]. For identical segmented and ground truth regions, its value becomes 1.

Tissue Segmentation Accuracy (TSA). It is very efficient index to measure the accuracy of the inter tissue clustering. The definition of *TSA* [7] is as follows:

$$TSA = \frac{2N_{ak}}{N_{bk} + N_{ck}}, \tag{13}$$

where N_{ak} indicates the number of correctly classified pixels in region k.N_{bk} is the total number of pixels recognized as part of region k by the algorithm, and N_{ck} is the number of ground truth pixels in region k. In an ideal situation, the *TSA* will be 1.

Segmentation Accuracy (SA). The *SA* can be defined by the follow equation:

$$SA = \frac{|M_k \cap R_k|}{|R_k|}, \tag{14}$$

where M_k is the set of kth cluster pixels generated by the algorithm and the set of ground truth pixels of the kth cluster is denoted by R_k. If the segmented region exactly matches with the ground truth region, its value becomes 1 and said to be its optimal value.

3.1 Experimental Results Using Simulated Brain MR Image Data

From BrainWeb [16], we collect the T1-weighted brain MR images. The image resolutions of each simulated image is 181 × 217, and pixel size is 1 mm × 1 mm. We test the ebFCM algorithm with $m = 1.50$ and neighborhood window size = 3 × 3. Table 1 presents the values of the segmentation accuracy (SA) generated by

Table 1 Tabulated values of segmentation accuracy with different percentage levels of noise and IIH

Image volume (noise, IIH)	Segmentation accuracy							
	Region	**ebFCM**	PFCM	ASIFC	sFCM	k-Means	FGFCM	FCM
(5, 20)	WM	**0.953**	0.959	0.946	0.938	0.845	0.941	0.848
	GM	**0.924**	0.879	0.919	0.916	0.817	0.828	0.834
	CSF	**0.923**	0.907	0.911	0.907	0.824	0.861	0.881
(5, 40)	WM	**0.938**	0.925	0.933	0.926	0.843	0.916	0.840
	GM	**0.908**	0.837	0.911	0.909	0.810	0.821	0.825
	CSF	**0.919**	0.875	0.867	0.861	0.806	0.832	0.837
(7, 20)	WM	**0.928**	0.949	0.918	0.912	0.833	0.909	0.829
	GM	**0.923**	0.815	0.907	0.902	0.785	0.801	0.818
	CSF	**0.906**	0.836	0.859	0.852	0.794	0.816	0.819
(7, 40)	WM	**0.933**	0.926	0.908	0.903	0.789	0.906	0.795
	GM	**0.897**	0.772	0.989	0.895	0.738	0.792	0.782
	CSF	**0.901**	0.817	0.852	0.849	0.743	0.795	0.807
(9, 20)	WM	**0.941**	0.932	0.901	0.897	0.783	0.873	0.781
	GM	**0.873**	0.762	0.875	0.871	0.728	0.736	0.755
	CSF	**0.869**	0.777	0.836	0.827	0.731	0.739	0.753
(9, 40)	WM	**0.909**	0.914	0.880	0.873	0.738	0.876	0.765
	GM	**0.880**	0.737	0.868	0.862	0.721	0.725	0.742
	CSF	**0.879**	0.753	0.829	0.824	0.726	0.731	0.742

different algorithms by changing amount of noise and IIH in the image volumes. It shows that the ebFCM algorithm is not only superior to the other methods but also yields consistent results as the levels of noise and IIH increase. Likewise, we present the values of the tissue segmentation accuracy (TSA) yielded by the algorithms in Table 2. Similar to the earlier one, here also the ebFCM algorithm gives better results and outperforms the other algorithms.

To visualize the segmented regions, we present them in Fig. 1. The results presented here use a T1-weighted brain MR image, which is contaminated with 9% noise and 40% IIH. The results clearly indicate that the ebFCM algorithm can efficiently segment the image despite with high quantity of noise and IIH. Further, Figs. 2 and 3 show the comparative study of partition coefficient and partition entropy of the ebFCM algorithm along with other previous algorithms. The comparative graphs indicate the supremacy of the ebFCM algorithms over others. Its partition coefficient and partition entropy values remain consistent in the nearer to its optimal values. These values even get better as the noise and IIH increase. Finally, we present a comparative graph of similarity index in Fig. 4. Again, it indicates that as the noise and IIH increase, the performance of the ebFCM algorithm also increases and

Table 2 Tabulated values of tissue segmentation accuracy with different percentage levels of noise and IIH

Image volume (noise, IIH)	Tissue segmentation accuracy							
	Region	**ebFCM**	PFCM	ASIFC	sFCM	k-Means	FGFCM	FCM
(5, 20)	WM	**0.864**	0.717	0.846	0.839	0.718	0.714	0.716
	GM	**0.822**	0.690	0.829	0.817	0.694	0.689	0.693
	CSF	**0.604**	0.494	0.623	0.615	0.437	0.442	0.439
(5, 40)	WM	**0.849**	0.711	0.846	0.839	0.718	0.714	0.716
	GM	**0.787**	0.690	0.829	0.817	0.694	0.689	0.693
	CSF	**0.849**	0.494	0.623	0.615	0.437	0.442	0.439
(7, 20)	WM	**0.853**	0.690	0.835	0.822	0.695	0.691	0.694
	GM	**0.812**	0.651	0.816	0.795	0.657	0.654	0.660
	CSF	**0.611**	0.418	0.611	0.602	0.425	0.421	0.417
(7, 40)	WM	**0.830**	0.684	0.829	0.815	0.696	0.689	0.693
	GM	**0.782**	0.629	0.792	0.778	0.647	0.637	0.645
	CSF	**0.601**	0.397	0.601	0.592	0.387	0.392	0.390
(9, 20)	WM	**0.831**	0.660	0.824	0.808	0.670	0.664	0.666
	GM	**0.797**	0.605	0.786	0.762	0.624	0.613	0.621
	CSF	**0.587**	0.366	0.579	0.558	0.379	0.381	0.377
(9, 40)	WM	**0.831**	0.662	0.817	0.792	0.673	0.670	0.671
	GM	**0.787**	0.590	0.769	0.751	0.608	0.605	0.610
	CSF	**0.578**	0.358	0.571	0.552	0.359	0.372	0.361

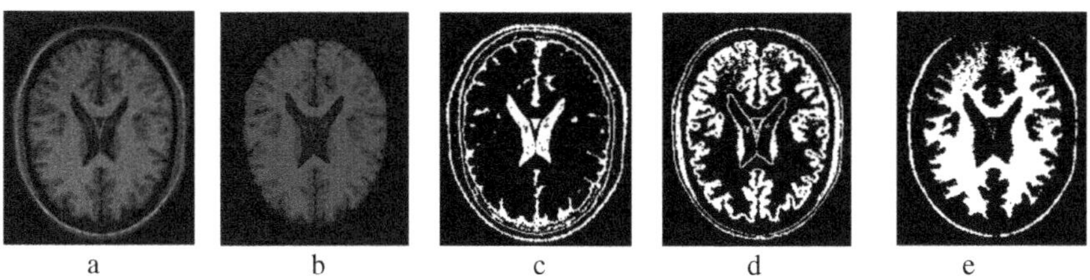

Fig. 1 The segmented output. **a** Input image; **b** Ground truth image; **c** CSF region; **d** GM region; and **e** WM region

performs better than the other algorithms. In summary, by analyzing all the performance indices, it establishes that the proposed ebFCM algorithm is efficient and provides superior results.

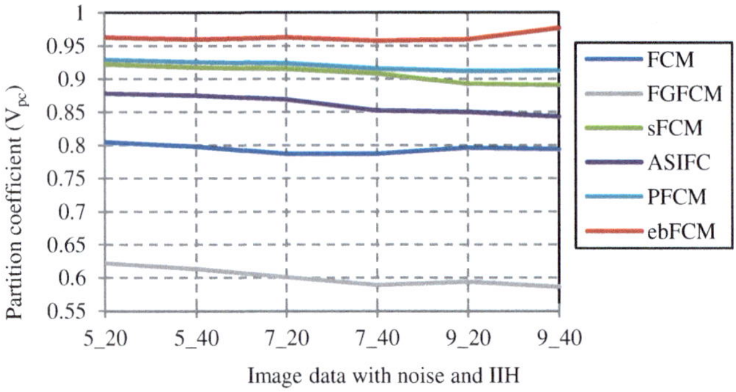

Fig. 2 Comparative graphs of partition coefficient versus image data for different algorithms

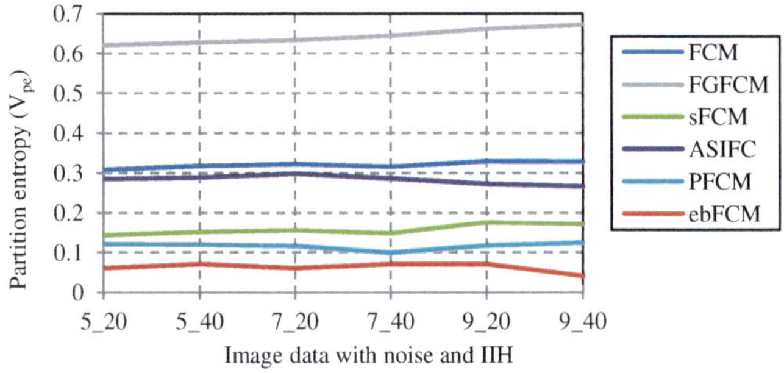

Fig. 3 Comparative graphs of partition entropy versus image data for different algorithms

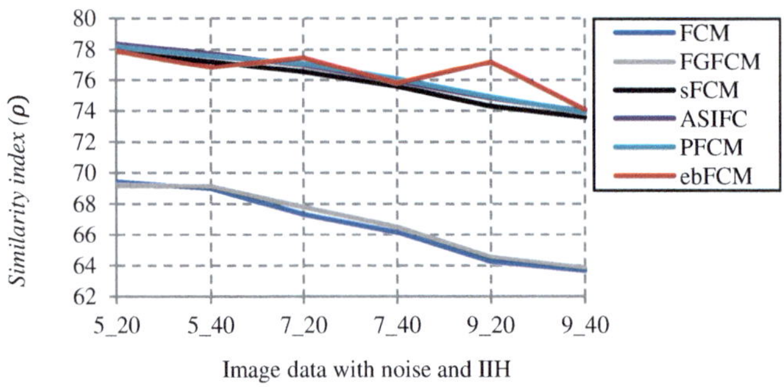

Fig. 4 Comparative graphs of similarity index versus image data for different algorithms

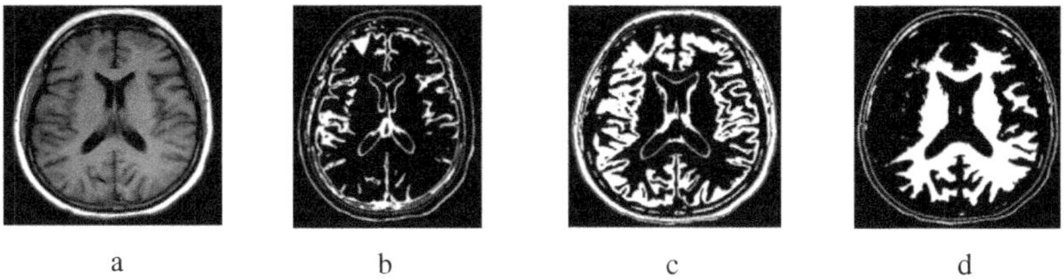

Fig. 5 Outputs using a brain MR image of a real patient. **a** Input image; **b** CSF region; **c** GM region; and **d** WM region

3.2 Experimental Results Using Brain MR Image Data of Real Patients

We also apply our proposed ebFCM method on clinically generated brain MR images of real patient and validate the results.

Here, Fig. 5 shows the segmented outputs. This analysis demonstrates its efficiency and supremacy over others.

Since the ground truth for these images are absent, we present the results in partition coefficient and partition entropy indices, as shown in Table 3. The results show that the values of the partition coefficients generated by the ebFCM algorithm are very nearer to 1 in all four cases. While the values of the partition entropies are also very nearer to 0. These results are also found to be better than the other algorithms.

4 Conclusion

The paper presents a new entropy-based FCM clustering algorithm by introducing an inverse fuzzy membership function framework and uncertainty parameter. This framework is realized in association of a Gaussian density function. It enforces that the fuzzy membership function will increase as the pixel intensity value comes closer to the center of the Gaussian density function. On the other hand, an uncertainty parameter coupled with fuzzy membership function is used with locally influenced Gaussian function. Finally, the underlying total information uncertainty is defined by using the Shannon entropy function with the help of uncertainty parameters. The present framework utilizes the local well as global distribution of pixel intensity across the whole image domain. We test the effectiveness of the ebFCM algorithm on simulated brain MR images by changing the amount of noise and IIH. Further, we also conduct experiments on brain MR images of real patients. The analysis of the results indicates it superiority over some FCM related algorithms.

Table 3 Tabulated results using brain MR images of real patients

Image	Method	Partition coefficient	Partition entropy
Patient 1 (Female)	**ebFCM**	**0.992**	**0.013**
	PFCM	0.924	0.137
	ASIFC	0.911	0.206
	sFCM	0.886	0.294
	FGFCM	0.815	0.329
	FCM	0.705	0.558
Patient 2 (Female)	**ebFCM**	**0.995**	**0.007**
	PFCM	0.921	0.129
	ASIFC	0.912	0.196
	sFCM	0.883	0.228
	FGFCM	0.852	0.287
	FCM	0.741	0.507
Patient 3 (Female)	**ebFCM**	**0.994**	**0.010**
	PFCM	0.935	0.112
	ASIFC	0.922	0.143
	sFCM	0.907	0.178
	FGFCM	0.893	0.193
	FCM	0.870	0.273
Patient4 (Male)	**ebFCM**	**0.993**	**0.010**
	PFCM	0.905	0.053
	ASIFC	0.897	0.056
	sFCM	0.835	0.074
	FGFCM	0.811	0.159
	FCM	0.791	0.253

Acknowledgements Authors are thankful to SERB, Government of India for partially supporting this work (File No.: EEQ/2016/000145). They also very much thankful to the Eko X-ray & Imaging Institute, JN Road, Chowrangee, Kolkata, and the AMRI Hospital at Dhakuria, Kolkata, for sharing clinically generated brain MR image data of real patients.

References

1. Bezdek JC (1981) Pattern recognition with fuzzy objective function algorithms, Kluwer academic publisher, Norwell, MA, USA
2. Chuang KS, Tzeng HL, Chen S, Wu J, Chen TJ (2006) Fuzzy c-means clustering with spatial information for image segmentation. Comput Med Imaging Graph 30(1):9–15
3. Cai W, Chen S, Zhang D (2007) Fast and robust fuzzy C-means clustering algorithms incorporating local information for image segmentation. Pattern Recogn 40(3):835–838

4. Kannon SR, Ramathilagam S, Devi R, Huang Y (2011) Entropy tollerent fuzzy c- means in medical images. Journal of Innovative Optical Health Sciences 4(4):447–462
5. Wang X, Song Q, Soh YC, Sim K (2013) An adaptive spatial information-theoretic fuzzy clustering algorithm for image segmentation. Computer Vis. Image Underst 117(10):1412–1420
6. Iftikhar MA, Jalil A, Rathore S, Hussain M (2014) Robust brain MRI denoising and segmentation using enhanced non-local means algorithm. Intelligent Imaging Systemand Technology 24(1):52–66
7. Adhikari SK, Sing JK, Basu DK, Nasipuri M (2015) Conditional spatial fuzzy c- means clustering algorithm for segmentation of MRI images. Appl Soft Comput 34:758–769
8. Bai X, Chen Z, Lie M, Zhang Y (2015) Center-free PFCM for MRI brain image segmentation. IEEE international conference on Image Processing (ICIP), IEEE, Quebec City, pp 656–660
9. Kahali S, Adhikari SK, Sing JK (2017) A two-stage fuzzy multi-objective framework for segmentation of 3D MRI brain image data. Appl Soft Comput 60:312–327
10. Askari S, Montazerin M, Zarandi MF, Hakimi E (2017) Generalized entropy-based possibilistic fuzzy c-means for clustering noisy data and its convergence proof. Neurocomputing 219:186–202
11. Gharieb RR, Gendy G, Abdelfattah A (2017) C-means clustering fuzzified by two membership relative entropy functions approach incorporating local data information for noisy image segmentation. SIViP 11(3):541–548
12. Kahali S, Sing JK, Saha PK (2018) A new entropy-based approach for fuzzy c-means clustering and its application to brain MR image segmentation. Appl Soft Comput 23:10407–10414
13. Kumar S, Sharma B, Sharma VK, Poonia RC (2018) Automated soil prediction using bag-of-features and chaotic spider monkey optimization algorithm. Evolutionary Intelligence, pp 1–12. Doi:https://doi.org/10.1007/s12065-018-0186-9.
14. Kumar S, Sharma B, Sharma VK, Sharma H, Bansal JC (2018) Plant leaf disease identification using exponential spider monkey optimization. Sustainable Computing: Informatics and Systems 28. https://doi.org/10.1016/j.suscom.2018.10.004
15. Shekhawat SS, Sharma H, Kumar S, Nayyar A, Qureshi B (2021) bSSA: Binary Salp Swarm Algorithm With Hybrid Data Transformation for Feature Selection. IEEE Access 9:14867–14882. https://doi.org/10.1109/ACCESS.2021.3049547
16. Brainweb Simulated brain database, http://www.bic.mni.mcgill.ca/brainweb/, last accessed 2018/06/06.

Chapter 8
A Quick and Single-Ended Scheme for Fault Detection and Classification on Transmission Line

Gaurav Kapoor

1 Introduction

The two significant components of the digital relaying scheme are fault detection and fault classification. The overall performance of the digital relaying scheme is decided by the speed and accuracy of these components. Many techniques have been introduced up to now for the digital relaying of transmission lines. Several of those techniques are Data mining and modal decomposition [1], artificial intelligence [2], Fourier transformation [3], and wavelet and Clarke's transformation [4]. Moreover, extra schemes are optimized ensemble of regression tree [5], S-transform and FFNN (feed forward neural network) [6], PCA (principal component analysis) [7], synchronized PMUs (phasor measurement units) [8], cumulative sum approach [9], and DWT (discrete wavelet transform) with fuzzy logic [10].

In this paper, a scheme is proposed for fault detection in transmission system based on the approximate coefficients index value (ACIV) of wavelet transform.

The application of the scheme is presented in Sect. 2. Presented fault detection scheme is provided in Sect. 3. The effects of simulations are illustrated in Sect. 4, while Sect. 5 concludes the paper.

2 Application of the Scheme

The presented scheme is applied to the system of 500 kV illustrated in Fig. 1. The model has two 50 Hz line sections of 150 km each. Numerous faults are incepted at different locations through MATLAB simulations in SIMULINK.

G. Kapoor (✉)
Department of Electrical Engineering, Modi Institute of Technology, Kota, Rajasthan, India

G. Mathur et al. (eds.), *Proceedings of 2nd International Conference on Artificial Intelligence: Advances and Applications*, Algorithms for Intelligent Systems,
https://doi.org/10.1007/978-981-16-6332-1_8

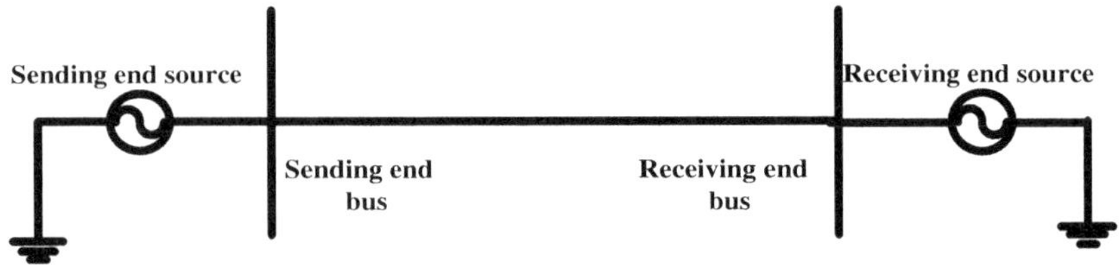

Fig. 1 Simulated model

Fig. 2 Block diagram of FD scheme

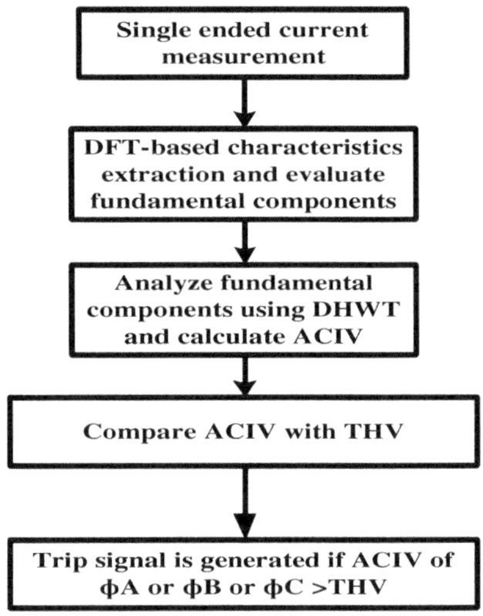

3 Presented Fault Detection (FD) Scheme

The block diagram of FD scheme is presented in Fig. 2.

4 Simulation Effects

The effects of simulations in view of numerous fault scenarios to bear out the implementation of the scheme are provided in this section.

4.1 Fault Resistance (FR) Variation

An AG fault is applied at 0.1 s (FL = 150 km), and the effect is validated in Fig. 3 with FR of 5 Ω. It represents that the ACIV of phase-A crosses the THV (threshold value)

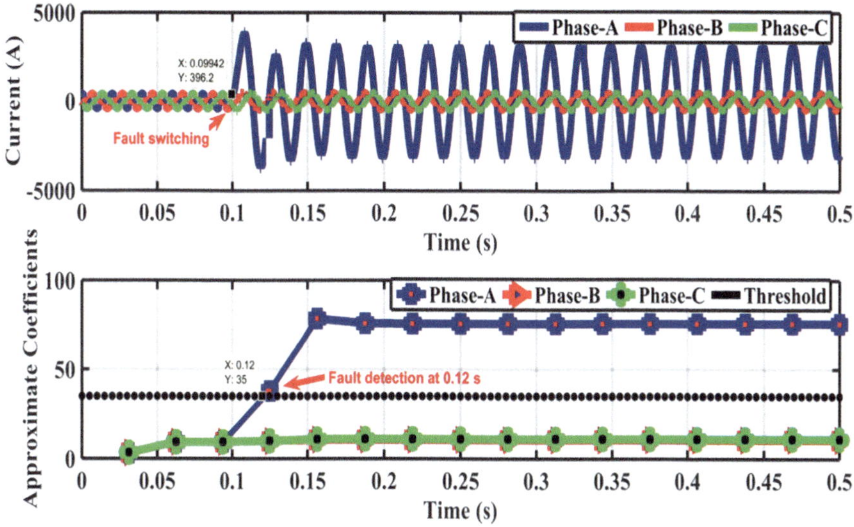

Fig. 3 Performance during AG fault initiated at 0.1 s and detected at 0.12 s with FR = 5 Ω

Table 1 ACIV and FDT for faults with different FRs

Fault Case	FAT (s)	FR (Ω)	ACIV			Fault Detection Time (ms)		
			ACIV-A	ACIV-B	ACIV-C	FDT-A	FDT-B	FDT-C
AG	0.1	5	78.6447	10.6111	11.0559	20	–	–
ABG	0.1	10	167.1964	164.9557	10.8926	5	5	–
BC	0.1	15	9.4717	161.3728	156.7563	–	10	10
ACG	0.1	20	154.8230	10.8893	150.3144	10	–	10
ABCG	0.1	25	161.3452	160.7352	160.6619	10	10	10

at 0.12 s. In AG fault case, the FDT (fault detection time) calculated is 20 ms. The ACIV and FDT results for some extra faults are provided in Table 1. The maximum FDT is 20 ms for AG fault, and minimum FDT is 5 ms for ABG fault.

4.2 *Fault Application Time (FAT) Variation*

Scheme implementation during ABG fault started at 0.05 s (FL = 140 km and FR = 2 Ω) is displayed in Fig. 4. The fault in this case is detected at 0.065 s. The detection time in this case is 15 ms. The ACIV and FDT outcomes varying FAT are presented in Table 2. The maximum FDT for AG and CG faults is 20 ms and minimum FDT is 5 ms for BC fault.

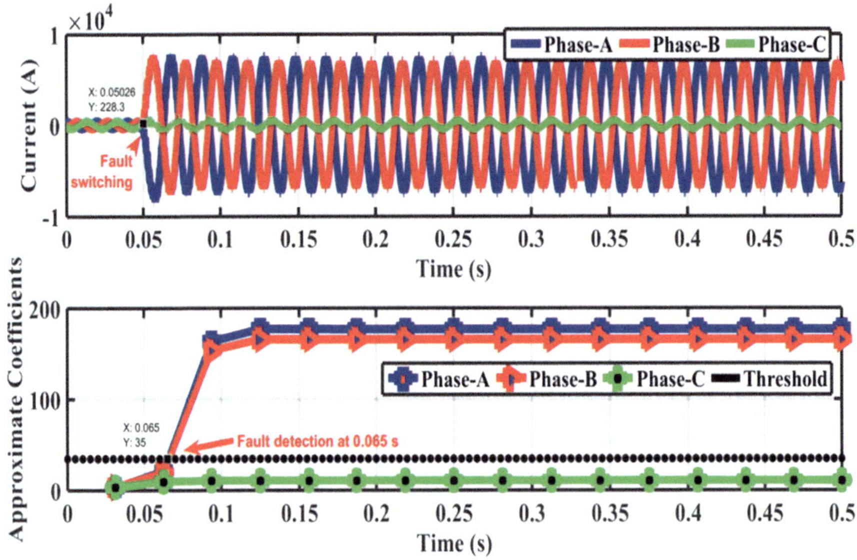

Fig. 4 Performance during ABG fault initiated at 0.05 s and detected at 0.065 s

Table 2 ACIV and FDT for faults at different FATs

Fault Case	FAT (s)	FL (km)	ACIV			Fault Detection Time (ms)		
			ACIV-A	ACIV-B	ACIV-C	FDT-A	FDT-B	FDT-C
ABG	0.05	140	176.5526	165.1473	10.8475	15	15	–
BC	0.1	140	9.4723	173.7232	171.8279	–	5	5
ABCG	0.15	140	190.5280	190.4307	190.4184	10	10	10
AG	0.2	140	79.3583	10.6620	10.9912	20	–	–
CG	0.25	140	10.6475	10.9975	75.1348	–	–	20

4.3 *Fault Location (FL) Variation*

The scheme is corroborated for a BG fault applied at 148 km (FAT = 0.05 s and FR = 7 Ω), and the trip waveform is demonstrated in Fig. 5. The fault is detected in B-phase at 0.075 s, i.e., in 25 ms after the starting of fault. The ACIV and FDT effects are provided in Table 3. The maximum FDT for BG and ACG fault is 25 ms, and for the remaining faults, FDT is 15 ms.

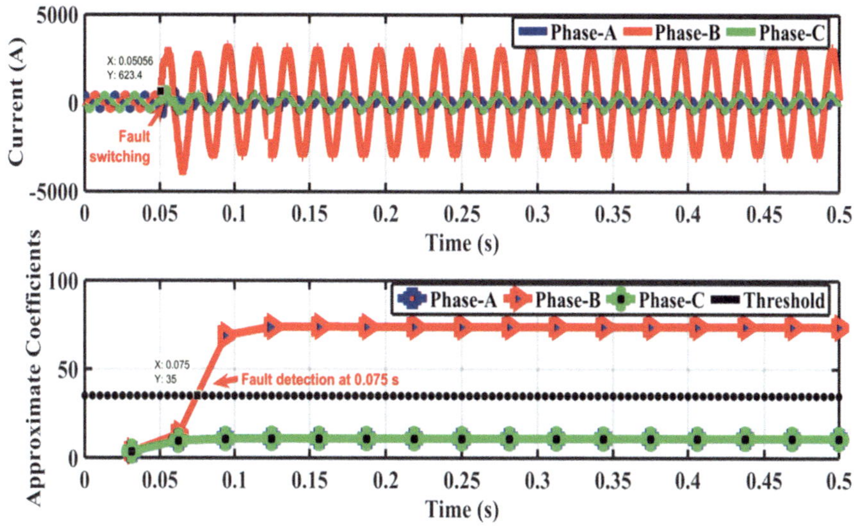

Fig. 5 Performance during BG fault initiated at 0.05 s and detected at 0.075 s at 148 km

Table 3 ACIV and FDT for faults at different FLs

Fault Case	FAT (s)	FL (km)	ACIV			Fault Detection Time (ms)		
			ACIV-A	ACIV-B	ACIV-C	FDT-A	FDT-B	FDT-C
BG	0.05	148	10.8024	73.7874	10.7616	–	25	–
ACG	0.05	159	153.4753	10.7877	163.3830	25	–	25
AB	0.05	167	151.2072	148.8817	9.4709	15	15	–
ABC	0.05	136	210.5460	201.5565	201.5637	15	15	15
ABCG	0.05	172	164.9587	164.9407	165.0222	15	15	15

5 Conclusion

An algorithm with DFT and DWT for fault detection and phase type classification is illustrated in this paper. The scheme does not need distant-end data as it makes use of only measured currents at relaying-end as inputs to reach absolute effects. The minimum FDT is calculated as 5 ms, and the maximum FDT is computed as 25 ms. Results provide evidence that the scheme is accurate and reliable from the general analysis.

References

1. Jamali S, Ranjbar S (2021) Phase selective protection in microgrids using combined data mining and modal decomposition method. Int. J. Electr. Power Energy Syst 128 106727
2. Wang S, Dehghanian P (2020) On the Use of Artificial Intelligence for High Impedance Fault Detection and Electrical Safety. IEEE Trans Ind Appl 56(6):7208–7216
3. Nadeem MH, Zheng X, Tai N, Gul M, Yu M, He Y (2021) Non-communication based protection scheme using transient harmonics for multi-terminal HVDC networks. Int. J. Electr. Power Energy Syst 127 106636
4. Saini M, Mohd Zin AA, Mustafa MW, Sultan AR, Nur R (2018) Algorithm for Fault Location and Classification on Parallel Transmission Line using Wavelet based on Clarke's Transformation. Int J Electr Comput Eng 8(2):699–710
5. Valabhoju A, Yadav A, Pazoki M et al. (2021) Optimized ensemble of regression tree-based location of evolving faults in dual-circuit line. Neural Comput & Applic
6. Shafiullah M, Abido MA (2018) S-Transform Based FFNN Approach for Distribution Grids Fault Detection and Classification. IEEE Access 6:8080–8088
7. Mukherjee A, Kundu PK, Das A (2021) A Differential Signal-Based Fault Classification Scheme Using PCA for Long Transmission Lines. J. Inst. Eng. India Ser. B
8. Saber A, Emam A, Elghazaly H (2018) A Backup Protection Technique for Three-Terminal Multisection Compound Transmission Lines. IEEE Transactions on Smart Grid 9(6):5653–5663
9. Fasihipour H, Seyedtabaii S (2020) Fault detection and faulty phase(s) identification in TCSC compensated transmission lines. IET Gener Transm Distrib 14(6):1042–1050
10. Rambabu K, Naidu KB (2020) Identification of double transmission line shunt faults by using combined DWT and interval type-2 fuzzy logic for digital distance protection. Materials Today: Proceedings

Chapter 9
A Method of Micro Pixel Similarity for Lung Cancer Diagnosis Using Adaboost

G. Kale Vaishnaw

1 Introduction

The Globocan, International Agency for Research on Cancer (2018) [1], has issued a statistical summary about the number of new cancer cases in India which have risen to 1, 15, 7, and 294 (including male and female). The Globocan Observatory figures indicate that the number of cancer deaths is 78, 482 including both the sexes. These published figures give a clear indication of the risk of dying from cancer in India. This paper concentrates on the analysis and diagnosis of lung cancer [2]. The abnormal random behavior of the diseases has made it difficult to take a firm decision on lung cancer using radiological imaging techniques [3]. Doctors prefer lung biopsy [4] as an initial option to detect lung cancer. Anyhow the medical imaging technique is going to be needed step even if the image processing technique is to be applied. The Micro pixel Similarity method is a statistical method implemented on microscopic lung images using image processing technique with MATLAB. The computational histopathological approach improves the knowledge about lung cancer diseases and also improve treatment strategies. In this paper, we have introduced a new statistical method to diagnose lung cancer accurately by applying identified statistical and mathematical parameters on microscopic lung images. The lung images are pre-verified as cancerous and noncancerous by pathological experts.

Atsushi Teramoto [5] et al. introduces the cytology in the diagnosis of lung cancer. Deep CNN automatically classifies the cytological images into benign and Malign with an accuracy of 81%. The performance of the classification accuracy of Deep CNN is improved with the use of actual and synthesized cytological images with GAN (Generative Adversarial Network). With this improvement, the overall classification accuracy of lung cells increased to 85.30%. Another method related to deep

G. K. Vaishnaw (✉)
Department of Electronics &Telecommunication, SVERI's College of Engineering, Pandharpur, Sholapur, Maharashtra, India

G. Mathur et al. (eds.), *Proceedings of 2nd International Conference on Artificial Intelligence: Advances and Applications*, Algorithms for Intelligent Systems,
https://doi.org/10.1007/978-981-16-6332-1_9

learning-based framework is introduced by Lei Cui [6] et al. for lung cancer survival analysis. Here the method is divided into three sections: the first is an end to end cellular feature learning module with global average pooling, the second is robust feature selection and survival analysis, and the third is the biomarker interpretation module that contributes to the survival modules decision. The survival module has excellent predictive power which is verified through the log-rank test (p-value) and concordance index (C value).

Besides deep learning and machine learning, lung cancer analysis and diagnosis is achieved using an image processing technique which is also one of the favorite tools among the researchers. Lal Hussain [7] et al. introduced Refined Fuzzy Entropy Methods. The methods say that the hidden information of Cancer is best analyzed using dynamical measures. The highest significant results are obtained through texture features using Multiscale Fuzzy Entropy with standard deviation, through morphological features with Refined Multiscale Fuzzy Entropy with mean and through elliptic Fourier descriptors features with Multiscale Fuzzy Entropy with Variance. These methods help to extract the hidden information from lung cancer images and distinguish the lung images into NSCLC and SCLC. Another method in this category is introduced by Noor Kehrak [8] et al. His method is about lung cancer detection using statistical and shape-based features. These features are extracted to form feature vectors which are classified using SVM. The paper focuses on the sensitivity parameter which is 93.75%.

Besides the introduction of the work carried out by the other authors, the next section is the introduction of some of the previously implemented statistical methods on lung cancer analysis and diagnosis by the author. The proposed method is new in the series after the successful accomplishment of statistical methods such as.

1.1 Average Information Method [9, 10]

Statistical and mathematical parameters utilized are mean, variance, standard deviation, entropy, and MSE.

Standard Diagnostic Test Results: Specificity 41.66%, Sensitivity 86.84%, and Accuracy 68.32%.

1.2 A Coalition of Photometric and Quality Metric Method [11]

Statistical and mathematical parameters utilized for Photometric Method are Brightness, Contrast, Luminance, and LC index.

Standard Diagnostic Test Results for Photometric Method: Specificity 31.54%, Sensitivity 79.76%, and Accuracy 57.45%.

Statistical and mathematical parameters utilized for Quality Metric Method—Visual Image Fidelity, VSNR, PSNR, and MSE.

Standard Diagnostic Test Results for Quality Metric Method: Specificity 46.03%, Sensitivity 88.77%, and Accuracy 72.04%.

Statistical and mathematical parameters utilized for Coalition Method are Brightness, Contrast, Luminance, LC index, Visual Image Fidelity, VSNR, PSNR, and MSE.

Standard Diagnostic Test Results for Coalition Method: Specificity 68.80%, Sensitivity 93.89%, and Accuracy 85.40%.

This paper is organized in four sections, initially, we have explained the principle working of Micro-pixel Similarity technique with the identification and working of identified parameters with the same principal characteristics of the proposed technique. Then, the proposed methodology with block diagram is explained utilizing Micro-pixel Similarity technique. It also focuses on ANN Adaboost function for lung image classification as cancerous and noncancerous and then the standard diagnostic test is explained. Next section is about results and discussions which is again divided into five subsections. The comparative analysis of the proposed methodology with the other methods is also explained in this section, and the last section is about the conclusion.

2 Micro Pixel Similarity Technique

The statistical analysis [12, 13] is preferred here due to random behavior of the lung cancer diseases. Micro pixel Similarity Technique is a statistical technique which shows how image variables are vigorous in relation with each other. It is used to investigate the principle component at the micro-level in an image which is similar to the reference image. The reference image is slide over the image to find the principle component, where the template overlaps the reference image to get aligned with similar values in the image. This technique is used to determine the grey level of the pixels which are having linear dependency at the particular positions relative to each other that also means a measurement of pixel similarities at micro-level between the two. Various statistical and mathematical parameters identified under this technique working on the same principle are as follows.

2.1 Mean Absolute Error (MAE)

Mean Absolute Error is calculated to find the difference between predicted and true values produced. It is the most common metric used to measure the accuracy of continuous variables. The initial step to calculate MAE is to find SAE that is Sum of Absolute Errors. Repeat the step for each image to forecast the predicted values. Add the absolute values to generate SAE. Once the SAE is calculated, then find the mean or average value of the absolute errors using the formula $MAE = SAE \div n$

$$MAE = \sum_{i=1}^{n} \frac{|y_i - x_i|}{n} = \sum_{i=1}^{n} \frac{|e_i|}{n}. \quad (1)$$

There is no such functional difference between SAE & MAE.y_i is the predicted value, and x_i is the true value; both are the variables of an image. y_i versus x_i include comparisons of predicted values and observed values. When an average of x values does not equal the average of y values, then the difference exists between two images. MAE is interpreted as the average absolute difference between y_i and x_i. It is conceptually more simple and interpretable.

2.2 *Absolute Difference (AD)*

It is the mathematical operator which computes absolute differences between two input images and is given as

$$Q = |P_1(i, j) - P_2(i, j)|, \quad (2)$$

where P1 (i, j) and P2 (i, j) are the pixels of an original image or reference image and a filtered image or a test image.

2.3 *Skewness*

In statistics, Skewness is a measure of an imbalance and asymmetry from a mean of data distribution. The equation is given by

$$Skewness = \sum_{i=1}^{N} (Yi - Y)^3/(N-1)S^3, \quad (3)$$

where Yi is the image data, Y is the mean, S is the standard deviation, and N is the number of images. The Skewness for normal data is zero and near to zero for any symmetric data.

2.4 *Mean Square Error (MSE)*

Mean square value is mathematically given as

$$MSE = \sum_{0}^{m-1} \sum_{0}^{n-1} \| f(i, j) - g(i, j) \|^{2} . \quad (4)$$

Practically, the mean squared error is used to compare the correct pixel values of a reference image to a test image. The MSE is calculated by averaging the squares of the errors between a test and a reference image. The error is the amount by which the two images differ from each other.

2.5 *Structural Similarity Index (SSIM)*

SSIM Index [14, 15] is highly sensitive to degradations in the spatial structure of image luminance. Original image and textured image being compared are properly allied and scaled to compare them point by point. Structural Similarity Indices Matrix is also been used in measuring the quality of an image. The procedure for calculating the SSIM is explained in Fig. 1.

Three similarity functions luminance similarity, contrast similarity, and structural similarity are computed on the window image data, which for two images X and Y are calculated as follows:

$$l(x, y) = \frac{2\mu_X(x, y)\mu_Y(x, y) + C_1}{\mu_X^2(x, y)\mu_Y^2(x, y) + C_1}, \quad (5)$$

$$c(x, y) = \frac{2\sigma_X(x, y)\sigma_Y(x, y) + C_2}{\sigma_X^2(x, y)\sigma_Y^2(x, y) + C_2}, \quad (6)$$

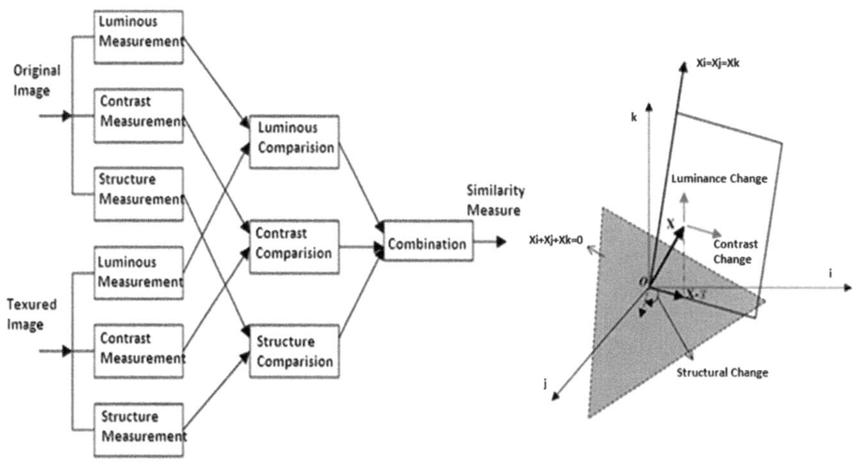

Fig. 1 Flow Diagram to Calculate Texture Index through SSIM

$$s(x, y) = \frac{\sigma_{XY}(x, y) + C_3}{\sigma_X(x, y)\sigma_Y(x, y) + C_3}, \tag{7}$$

$$SSIM = \frac{(2 \times \overline{x} \times \overline{y} \times +C_1)(2 \times \sigma_{xy} + C_2)}{(\sigma_x^2 + \sigma_y^2 + C_1) \times ((\overline{x})^2 + (\overline{y})^2 + C_2)}, \tag{8}$$

where

$$\overline{x} = \frac{1}{N}\sum_{i=1}^{N} x_i^2,\ \overline{y} = \frac{1}{N}\sum_{i=1}^{N} y_i^2,\ \sigma_x^2 = \frac{1}{N-1}\sum_{i=1}^{N} (x_i - \overline{x})^2,\ \mu_y = \frac{1}{N}\sum_{i=1}^{N} y_i,$$

$$\sigma_{xy} = \frac{1}{N-1}\sum_{i=1}^{N} (x_i - \overline{x})(y_i - \overline{x}),\ \sigma_y^2 = \frac{1}{N-1}\sum_{i=1}^{N} (y_i - \overline{y})^2,\ \mu_x = \frac{1}{N}\sum_{i=1}^{N} x_i,$$

μ_x—Average of x, μ_y—Average of y,σ_x^2—Variance of x,σ_y^2—variance of y,σ_{xy}—Covariance of x, and y,$C_1 = (k_1 L)^2$,$C_2 = (k_2 L)^2$ are two variables to stabilize the division with the weak denominator,L —the dynamic range of pixel values given as $(2^{bitsperpixel} - 1)$,$k_1 = 0.01$,$k_2 = 0.03$ are default values.

3 Material and Methodology

3.1 Material

Microscopic lung images are used for evaluating the performance of the proposed methodology. The microscopic images are histopathological images [16]. The database is of 322 lung images which are checked and pre-verified by the radiological experts. The microscopic lung image samples are collected from different hospitals and medical pathologies from Ahmednagar and Aurangabad districts, Maharashtra, India.

3.2 Methodology

The basic aim of the method is to identify some of the statistical and mathematical parameters working based on micro pixel similarity technique as described in section two for lung cancer analysis and diagnosis. The identified parameters are applied to each image. Then the individual parameter statistical range is calculated for the cancerous lung images. If the value of the individual parameter of the new test image lies within the calculated statistical range, then it is cancerous otherwise

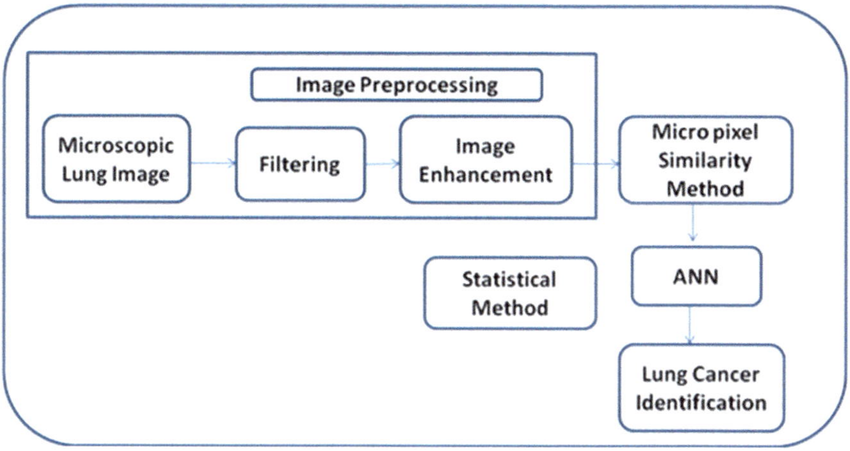

Fig. 2 Block diagram of the proposed methodology

noncancerous. It does not seem to be simple as it looks because the number of values overlaps for both the types of images and hence need a decision-maker. ANN is used as an image classifier method. Adaboost function of ANN [17, 18] as an image classifier is successful in achieving the objectives. The methodology is applied for a specific image database only. The block diagram of the proposed methodology is shown in Fig. 2.

The test image is the microscopic lung image which is normally a color image. The image is normalized and rescaled to standard size. It is converted into grayscale as per the mathematical expression of each parameters. It is preprocessed using median filtering and image enhancement technique for efficient representation of a lung image. Image enhancement is about varying the contrast and brightness of the image to such level where they are ready for the next step.

The processed lung images are passed through the Micro-Pixel Similarity Technique that has identified some of the selected statistical and mathematical parameters like MAE, AD, MSE, SSIM, and Skewness. These parameters are applied one by one on each of the input images. Then a statistical range is achieved for the individual parameter in real-time. The statistical range obtained for the cancerous lung images plays an important role in image classification. As some of the values overlap for cancerous and noncancerous lung images, due to which ANN comes in the picture for exact classification. ANN classifies the images into cancerous and noncancerous based on the statistical ranges by resolving the overlap issues. The performance of the proposed method is analyzed in terms of the involvement of each selected parameter for lung cancer analysis using image processing. The proposed method is implemented in MATLAB.

3.3 Artificial Neural Network (ANN)

The ANN is implemented using AdaBoost function. The AdaBoost ANN [19] is preferred here since it perfectly deals with nonlinear functions. The input parameters are all nonlinear hence adapts to the nonlinearity, correctly minimizing the problem. The features of AdaBoost function includes identifying some important features from huge library of potential features, used in combination with many other learning algorithms to improve their performance, is adaptive, sensitive to noisy data and less vulnerable to over fitting problem as compared to the other algorithms.

3.4 Standard Diagnostic Test

In this section, the proposed methodology is validated using standard diagnostic tests [20]. There are four possible outcomes of the proposed method which are implemented on the test image that is True Positive (TP) means that the cancer is identified correctly, True Negative (TN) means that no cancer is identified correctly, False Positive (FP) means that no cancer is identified in the image but is incorrectly identified by the method, and False Negative (FN) means that the cancer is not identified by the method. To enumerate the performance of the proposed methodology, the statistical measurement parameters which work on above four possible outcomes under standard diagnostic test are Sensitivity means cancerous is tagged as cancerous, Specificity means that noncancerous is tagged as noncancerous, Accuracy means how the classifier correctly identifies or exclude the condition, and Precision means how is the method precise in identifying true positive results.

4 Results and Discussions

This section is divided into five subsections as explained below.

4.1 Calculation of Statistical Parameter Range

In this paper, the arbitrary data of the random patterns of lung cancer cases are statistically analyzed. Measurements include a discrete arrangement of information that is characterized by statistical parameters such as MAE, AD, MSE, Skewness, and SSIM. These parameters are applied to cancer infected microscopic lung images. With these figures, a particular range of each parameter has been distinguished for the investigation and analysis of lung cancer. The statistical range for the identified parameters is shown in Table 1.

Table 1 Statistical Parameter Range for Cancerous Lung Images for Micro pixel Similarity Method

Micro pixel Similarity Parameters	Statistical range	Average Value
SSIM	0.041–0.018	0.0295
MAE	45.359–47.173	42.266
AD	45.359–47.173	46.266
Skewness	0.878–1.507	1.1925
MSE	55–162	108.5

The analysis of each parameter for lung cancer images is described in 4.2

4.2 *Parameter Analysis of Lung Images*

This section includes the analysis of each input parameter for the lung cancer images. The investigation of the significant number of parameters can be seen as measurable from the statistical response of the plot. The actual response of five statistical parameters for cancerous and noncancerous lung images under Micro pixel Similarity strategy is observed in the form of a graphical plot from Figs. 3, 4, 5, 6 and 7. The graph shows the scattered points for cancerous and noncancerous lung images.

Following things are observed from the statistical responses of the parameters.

- If the overlapping of the scattered points for cancerous and non-cancerous lung images are less, then better is the usefulness of the parameter for the lung cancer investigation.
- It is also observed from the individual performances of the parameters that AD and Skewness are showing the maximum overlap of the discrete values for cancerous and noncancerous, whereas SSIM and MSE have shown the least overlapping of the points.

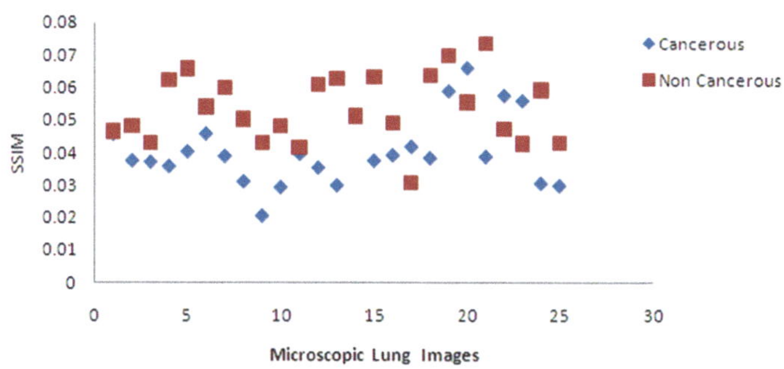

Fig. 3 Statistical Response of SSIM

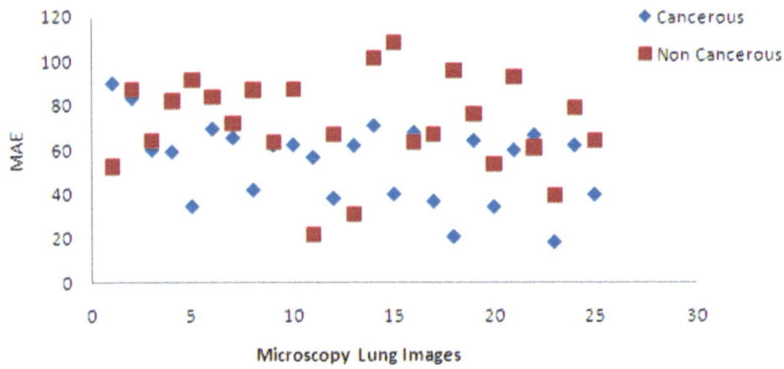

Fig. 4 Statistical Response of MAE

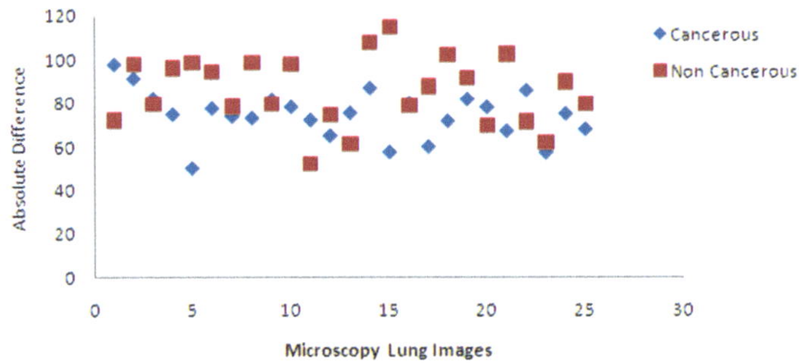

Fig. 5 Statistical Response of AD

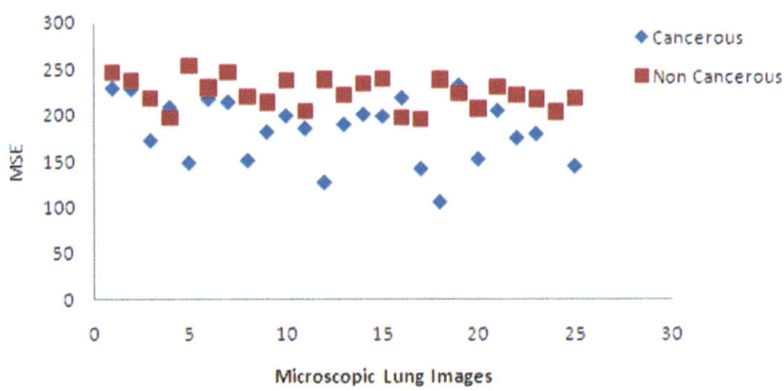

Fig. 6 Statistical Response of MSE

- MAE is the average performer compared to the rest of the parameters, but is significant and has contributed less or more in lung cancer diagnosis.
- Tables 2 and 3 show the calculation of statistical parameter values for the particular of cancerous and noncancerous microscopic lung image. At this point, the

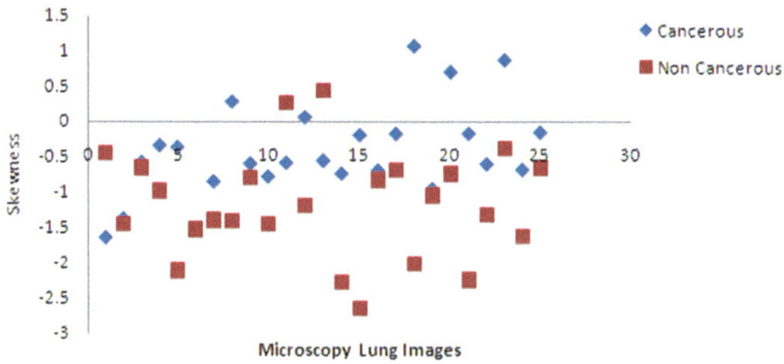

Fig. 7 Statistical Response of Skewness

Table 2 Statistical Parameter Values for Cancerous Lung images

Micro pixel Similarity parameters	Image1	Image2	Image3	Image4	Image5
SSIM	0.0457	0.0375	0.0371	0.0358	0.0453
MAE	89.529	82.628	59.455	55.584	33.969
AD	97.484	90.990	89.695	74.473	77.192
Skewness	1.648	-1.382	-0.581	-0.340	-1.540
MSE	228.958	227.384	171.910	208.312	148.033

Table 3 Statistical Parameter Values for Noncancerous Lung Images

Micro pixel Similarity parameters	Image 1	Image 2	Image 3	Image 4	Image 5
SSIM	0.0465	0.0482	0.0431	0.0622	0.0657
MAE	52.108	87.147	63.776	81.687	91.454
AD	71.756	97.608	79.440	95.573	97.992
Skewness	-0.443	-1.460	-0.658	-0.985	-2.122
MSE	245.856	236.802	217.135	196.607	252.352

computations are applied over the image database and a range for all identified parameters under micro pixel similarity strategy is acquired.

The range is useful in discriminating the lung images. But it is not that simple as it looks as some of the identified parameter values overlaps, and then subsequently, this disarray is overcome using ANN, which classifies the lung images into cancerous and noncancerous.

Table 4 Accuracy of the Parameters for Cancerous Lung

Sr. No	Micro pixel Similarity parameters	Accuracy (%)
1	SSIM	69.65
2	MAE	39.00
3	AD	32.20
4	Skewness	32.20
5	MSE	61.92

4.3 Accuracy of the Parameters

This section explains how the accuracy of each parameter affects decision making. MAE and AD have shown lesser response as compared to SSIM and MSE as observed in Table 4.

SSIM and MSE have shown good accuracy in the range of 61 to 69%, whereas others have shown an average accuracy in the range of 30 to 40%. The accuracy of the statistical and mathematical parameters under the proposed method is calculated on how accurately the cancerous images are differentiated from noncancerous images. The successful working of the proposed method is also validated using a standard diagnostic test.

4.4 ANN-Adaboost an Image Classifier

- Assigning equal or same weights to all observations or to each record in the dataset
- Classify random samples using stumps which are weak classifiers
- Total error is calculated by taking sum of weights of misclassified data
- The performance of the stumps is calculated using total error
- The stump value is used to update the weights and also used for final prediction
- Weights are updated according to the performance of the stumps
- Update weights in iteration
- Final prediction is carried out by obtaining the sign of the weighted sum of final predicted value

4.5 Standard Diagnostic Test Results of the Proposed Method

The True Positive (TP), True Negative (TN), False Positive (FP), and False Negative (FN) values are calculated for the implemented and proposed method as shown in Table 5.

The results of the standard diagnostic test are shown in Table 6. The sensitivity is 94.95%, specificity is 66.34%, Precision is 85.53%, F-score is 90%, and ultimately the accuracy of the proposed method is 85.71%. High the value of Sensitivity

Table 5 Comparative Analysis of Standard Diagnostic Test Results

Comparative Analysis					
Standard Diagnostic Parameter	Average Info Method [13]	Photo Metric Method	Quality Metric Method	Coalition of Method [14]	Micro-Pixel Similarity Method
TP	165	138	174	200	207
TN	55	47	58	75	69
FP	77	102	68	34	35
FN	25	35	22	13	11

Table 6 Standard Diagnostic Test of Micro-pixel Similarity Method

Statistical Method	Sensitivity	Specificity	Accuracy	Precision	F-score
Micro pixel Similarity Method	94.95%	66.34%	85.71%	85.53%	90%

and Precision, higher the value of F-score. The performance test indicates that the proposed method has unanimous results for lung cancer diagnosis. Although the specificity number has been on the lower side but can be enhanced in the future scope.

Looking at the comparative analysis from Table 7 and Fig. 7, it is observed that the proposed method has shown a good rise in Sensitivity, Specificity, and Accuracy as compared to the previously implemented statistical methods.

Table 8 shows the comparative analysis of the number of parameters used, with the proposed method. The coalition method is the combination of two methods Photometric and Quality metric method having 08 parameters. With these 08 parameters, the accuracy of the method achieved is 85.40%. The advantage of the proposed method is that Fig. 8 it has achieved an accuracy of 85.71% with only 05 parameters.

Table 7 Comparative analysis of Proposed Method with previous implemented Methods

Comparative Analysis					
Standard Diagnostic Parameter	Average Info Method	Photo Metric Method	Quality Metric Method	Coalition Method	Micro-Pixel Similarity Method
Accuracy	68.32%	57.45%	72.04%	85.40%	85.71%
Specificity	41.66%	31.54%	46.03%	68.80%	66.34%
Sensitivity	86.84%	79.76%	88.77%	93.89%	94.95%

Table 8 Comparative analysis of Number of Parameters Implemented for all the Statistical Methods

Comparative Analysis					
Statistical and Mathematical Parameters Implemented	Avg. Info Method [13]	Photo Metric Method [14]	Quality Metric Method [14]	Coalition Method [14]	Micro-Pixel Similarity Method
Number of Parameters	05	04	05	08	05
Accurcay	68.32%	57.45%	72.04%	85.40%	85.71%

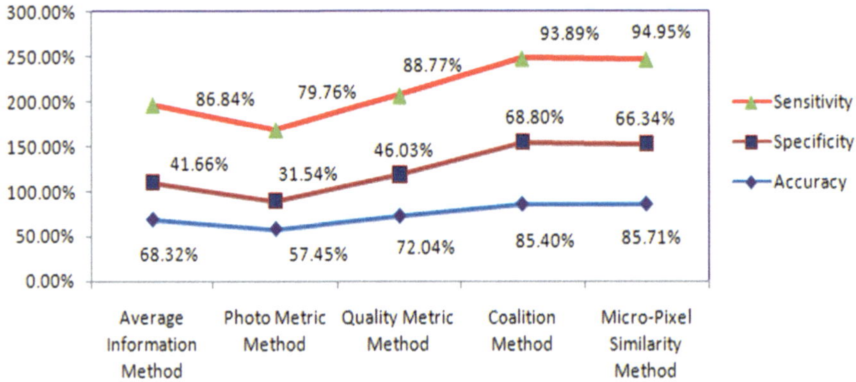

Fig. 8 Comparitive Analysis of Micro-pixel Similarity Method with previous implemented Methods

5 Conclusions

Around different measures, some of the identified specific and selective parameters have been utilized for lung cancer classification using proposed methodology. These parameters under micro pixel similarity technique follow the principle working of finding the micro pixel similarities between the lung images. The micro pixel similarity technique plays a very important role in finding the statistical range over a database of 322 microscopic lung images. Each identified parameters plays a very crucial role in lung cancer diagnosis. The overall range of the accuracy of the identified statistical and mathematical parameters is from 32.20 to 69.65%. The SSIM is the stand out performer, whereas AD and Skewness are lower performers as compared. The proposed method is successfully validated through a standard diagnostic test. The accuracy of the proposed method achieved is 85.71%, Sensitivity is 94.95%, Specificity, 66.34%, Precision is 85.53%, and F score is 90%, which is ascertained based on the number of lung images analyzed and diagnosed accurately and verified through a standard diagnostic test. The Adaboost function of ANN works effectively as an image classifier for this method. The accuracy of each parameter shows that the parameter under this method is working successfully and has its significance for lung cancer analysis and diagnosis. The proposed method successfully has achieved maximum accuracy with only 05 parameters, which is one of the significant achievements in the field of Biomedical Engineering. The application area includes the method is applicable for lung cancer diseases, can be applied test any radiological images including microscopic images, and can be applied to study various other diseases.

Conflict of Interest: There is no relevant financial interest in the manuscript and no other potential conflict of interest to disclose.

References

1. Freddie Bray, Jacques Ferlay sabelle Soerjomataram, Rebecca L. Siegel, Lindsey A. Torre, Ahmedin Jemal," Global Cancer Statistics (2018) GLOBOCAN Estimates of Incidence and Mortality Worldwide for 36 Cancers in 185 Countries". CA CANCER Journal CLIN 2018:394–424. https://doi.org/10.3322/caac.21492
2. Bonnie J Addario (2018)" Navigating Lung Cancer-360 degree of Hope", Lung Cancer Foundation, Fourth Edition—November 2018, pp 1–250
3. Kasban H, El-Bendary MAM, Salama DH (2015)" A Comparative Study of Medical Imaging Techniques" , Int. J. Inf. Sci. Intell. Syst vol 4(2): 37–58
4. Muhammad Qurhanul Rizqie, Nurul Shafiqa Mohd Yusof, Rino Ferdian Surakusumah, Dyah Ekashanti Octorina Dewi, Eko Supriyanto, Khin Wee Lai (2015)"Review on Image Guided Lung Biopsy", IJN-UTM Cardiovascular Engineering Center, Springer Science and Business Media, pp 41–50
5. Atsushi Teramoto,Tetsuya Tsukamoto, Ayumi Yamada, Yuka Kiriyama,Kazuyoshi Imaizumi, Kuniaki Saito, Hiroshi Fujita (2020) " Deep learning approach to classification of lung cytological images: Two-step training using actual and synthesized images by progressive growing of generative adversarial networks", Yuchen Qiu, University of Oklahoma, US, PLoS ONE 15(3):e0229951,March2020,pp.1–12, doi.org/https://doi.org/10.1371/journal.pone.0229951.
6. Lei Cui, Hansheng Li, Wenli Hui, Sitong Chen, Lin Yang, Yuxin Kan1, Qirong Bo, Jun Feng (2020)" A deep learning-based framework for lung cancer survival analysis with biomarker interpretation ",Cui et al. BMC Bioinformatics 21:112, pp 1–14, https://doi.org/10.1186/s12859-020-3431-z.
7. Lal Hussain , Wajid Aziz, Abdulrahman Alshdadi,Malik Sajjad Ahmed Nadeem, Ishtiaq Rasool Khan,and Qurat-ul-ain Chaudhry,"Analyzing the Dynamics of Lung Cancer Imaging Data Using Refined Fuzzy Entropy Methods by Extracting Different Features", IEEE Access, vol 7, pp 64704–64721, D.O.I.https://doi.org/10.1109/ACCESS.2019.2917303.
8. Noor Khehrah , Muhammad Shahid Farid , Saira Bilal, Muhammad Hassan Khan (2020) " Lung Nodule Detection in CT Images Using Statistical and Shape-Based Features ", J. Imaging 2020 MDPI, pp 1–20, doi:https://doi.org/10.3390/jimaging6020006.
9. Kale Vaishnaw G, Vandana B. Malode," Exploring Average Information Parameters over Lung Cancer for Analysis and Diagnosis", Publication by Springer nature (AISC series), Chapter 54, Book ID: 450375_1_En, Book ISBN: 978–981–10–8054–8, doi.org/https://doi.org/10.1007/978-981-10-8055-5_54.
10. Kale VG, Vandana B (2018) Malode," A Novel Approach based on Average Information Parameters for Investigation and Diagnosis of Lung Cancer using ANN", Pleiades Publishing Ltd. Pattern Recognit Image Anal 28(2):301–309

11. Kale Vaishnaw G, Vandana B. Malode (2020) "Lung Cancer Analysis and Diagnosis by Coalition of Photometric and Quality Metric Parameters" IEEE Explorer,ESCI-2020, DOI:https://doi.org/10.1109/ESCI148226.2020.9167532, ISBN: 978–1–7281–5263–9, pp 1–6
12. Kale Vaishnaw G, Vandana B. Malode (2016) "New Approach of Statistical Analysis for Lung Disease Diagnosis using Microscopy images" , ICACDOT—2016 IEEE Conf. Proceed, pp 378–383, DOI: 10.1109@ICACDOT.2016.7877612.
13. Md Rashidul Hasan, Muntasir Al Kabir (2019) Lung Cancer Detection and Classification based on Image Processing and Statistical Learning. Cornell University, 2019, pp 1–6, arXiv.org > stat > arXiv:1911.10654.
14. Sung-Ho Bae, Munchurl Kim (2015) " A Novel SSIM Index for Image Quality Assessment using a New Luminance Adaptation Effect Model in Pixel Intensity Domain", IEEE VCIP 2015, pp 1–4
15. Jana Zujovic,Thrasyvoulos N. Pappas, David L. Neuhoff (2013) " Structural Texture Similarity Metrics for Image Analysis and Retrieval", IEEE Trans. Image Process vol 22(7): 2545–2558
16. Gurcan MN, Boucheron LE, Can A, Madabhushi A, Rajpoot NM (2009) and Bulent Yener," Histopathological Image Analysis: A Review". IEEE Rev Biomed Eng 2:147–171
17. Periyasamy N, Thamilselvan P, Sathiaseelan JGR "A Comparative Study of ANN, K- Means and Adaboost Algorithms for image classification", In: Proceedings of IEEE Sponsored 2nd International Conference on Innovations in Information Embedded and Communication Systems,ICIIECS'15, pp 1–5
18. Shakeel PM, Tolba A, Al-Makhadmeh Z, Jaber MM (2020) Automatic detection of lung cancer from biomedical data set using discrete AdaBoost optimized ensemble learning generalized neural networks. Neural Comput Appl (3):201–212
19. Kim K, Choi HI (2015) Adjusting Initial Weights for Adaboost Learning, pp 1–5
20. Tharwat A (2020) Classification Assessment Methods, Applied Computing and Informatics, Elsevier. J Saudi Comput Soc, 1–13. https://doi.org/10.1016/j.aci.2018.08.003

Chapter 10
Multi-Agent Intrusion Detection System Using Sparse PSO K-Mean Clustering and Deep Learning

Tanushri Jain and Chetan Gupta

1 Introduction

An intrusion detection system (IDS) tracks internet traffic for unusual behavior that may indicate an assault or security breach. Network-based network services (IDS) are resources that enforce these network protection mechanisms against interference from in or out of the network [1]. These messages contain inbound and outbound network traffic, conduct periodic analyses and report on intrusion as shown in Fig. 1.

An integral method in terms of information protection is the intrusion detection system (IDS). An IDS is meant to inform users of a compromise with their programs. An intrusion prevention system (IPS) is usually seen as different from an intrusion system. As the names suggest, IPS can prevent attacks, and IDS can track attacks after they have occurred. An IDS is a helpful method [2]. IDS may work at or combined with a host level or network level. There are several methods for violation identification, whether on the server or the network level. From such a decent quality, the main techniques used by IDS are anomaly detection and signature detection [3, 4].

1.1 Host-Based IDS

A host-based IDS tries to detect invasions of privacy using the host information, without taking into account network behavior. That is, the behavior of host-based IDS monitors malicious behavior on a certain host or set of hosts. This intrusion sensing method depends on data stored in logs, trails of audit, checksum values, user behavior properties, etc. Since logs may mimic the behaviors of individual users, a host-based IDS might be able to identify who is behind malicious activity. Depending

T. Jain (✉) · C. Gupta
Sagar Institute of Research Technology & Science, Bhopal, Madhya Pradesh, India

G. Mathur et al. (eds.), *Proceedings of 2nd International Conference on Artificial Intelligence: Advances and Applications*, Algorithms for Intelligent Systems,
https://doi.org/10.1007/978-981-16-6332-1_10

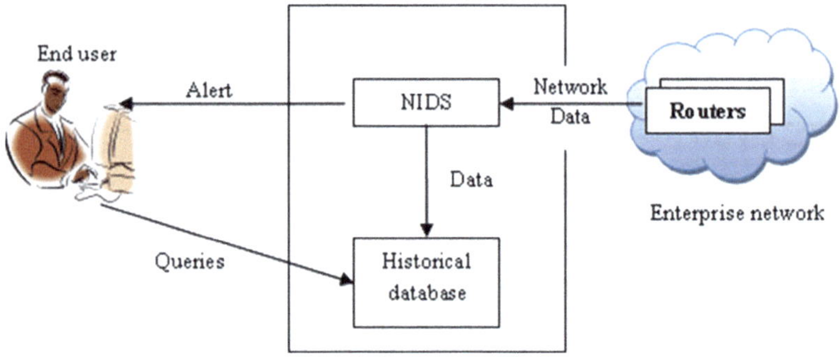

Fig. 1 Network intrusion detection system

on the specific implementation, host-based IDS has some inconvenience. Host-based IDS, for example, can require massive storage to retain the data needed for the IDS. Normally, many hosts tend to have their host-based IDS, which makes it complicated to set up, customize and manage [5].

1.2 Network-Based IDS

IDS-based networks try to detect network intrusions. Such IDS monitor network traffic and normally check the header and other aspects of networked packets. The system topology IDS has the advantage that it can detect scans, DoS attacks and other network-based attacks, for example. In contrast to a host-based method, this form of IDS is often easy to use and manage. One downside of network-based IDS is that the host activity is not transparent. Network-based and host-based intrusion detection systems (IDS) are also widely used in service, at least to some extent. It provides layered protection and comprehensive defense [6, 7].

1.3 Anomaly and Signature Detection

The identification of abnormality and the detection of signatures can be viewed as two broad categories for IDS classification—host- or signature-based. We strive to model system characteristics in anomaly detection and, if the system behavior varies adequately from the model, we use it as a potential attack. While it is difficult to detect abnormalities, it can help us to detect attacks from Zero-day [8, 9]. Signature recognition, by contrast, is a type of pattern. In that method, a pattern or signature is derived from a known attack, such that the subsequent attack may have occurred when this pattern is observed. Signature scans can only detect common threats for which a signature was previously retrieved, though reasonably reliable and precise [10].

2 Literature Review

A dimensionality problem is present with the intrusion detection system, i.e. large data sets simulating real network data increase training and test time in IDS. Huge quantities of information can also impact detection stability by determining resource usage [11, 12]. As a result, an effective extraction and downsizing technique is developed, which not only helps reduce training time but also improves accuracy and protects against malicious activities. Selection of features reduces computational complexity, redundancy of information, improves the accuracy of the learning algorithm, enables comprehension and generalization. Liang et al. [12] have used a multi-agent intrusion detection system hybrid positioning strategy. This framework consisted of a module for data acquisition, an analysis module and an answering module. For this research, the analysis module is carried out using an algorithm for a deep neural network for intruder detection. The findings show that deep learning algorithms are successful at predicting transport layer assaults. Zhao et al. [13] proposed a deep belief network (DBN) and probabilistic neural network (PNN)-based intrusion detection system. The model's success in binary classification and the classification of different groups was investigated by Chuanlong Yin et al. [14]. The simulation result is ideal for modeling a highly precise classification model. Yuan et al. [15] suggested a deep learning-based DDoS attack detection approach. To learn patterns from network traffic sequences and monitor network attack behavior, a recurring deep neural network project is proposed. The model outperforms traditional machine learning models in the larger dataset, with an error rate of 2.103% compared to 7.517% for traditional machine learning approaches. A few of the intrusion detection contributions are discussed hereafter (Table 1).

3 Proposed Methodology

This describes the flow of the proposed method:

- A numerical form for all symbolic attributes of the data set. TCP, UDP and ICMP protocols, for example, are translated to 1, 2, and 3 respectively.
- Feature extraction is used to find IDS features to reduce the dimension of the dataset for further processing. Before feature extraction, the redundant features are removed.
- Decision-making on trained rules.

Figure 2 depicts the proposed algorithm's block diagram, which consists of four

Table 1 Review of various feature selection and reduction approaches

Authors	Remarks	Conclusion and limitations
Bolon-Canedo et al. [1]	Symmetrical uncertainty and correlation are used for feature reduction	For function reduction, symmetrical insecurity and correlation are used. Normal, DoS and U2R attack detection rate is low
Uguz [2]	Feature reduction is performed in two stages	For all classes, result doesn't outperform better
Mukherjee and Sharma [3]	Feature vitality is used for feature reduction	Computational complexity is high
Li et al. [4]	Used 19 features out of 42	Only 71% of normal data packets are identified
Karimi et al. [5]	Information gain and symmetrical uncertainty are merged to get a better result	Detection rate of U2R and R2L is not good
Al-Jarrah et al. [6]	Feature weight is identified and the lowest feature weight is removed out	Only accuracy rate is focused
Akashdeep et al. [7]	Information gain and correlation feature reduction techniques are merged to find out important features	Normal and U2R accuracy is low Computational overhead due to two feature reduction techniques

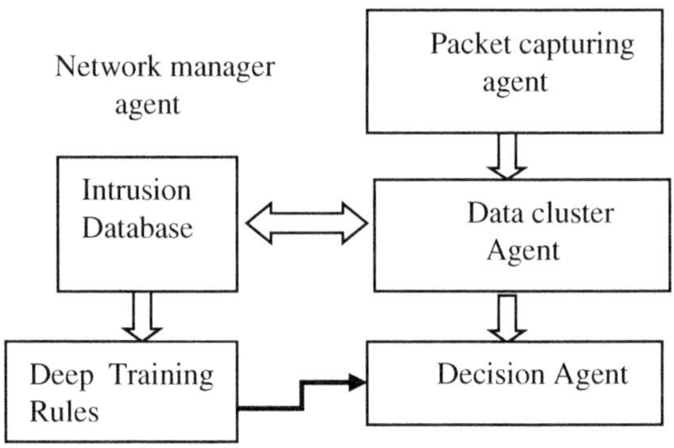

Fig. 2 Proposed multi-agent architecture

agents that interact with one another to help identify network attacks. The following is a list of them:

Network Manager Agent: This agent handles the training process from the existing intrusion database and trains the LSTM network for creating decision rules. The training is performed based on protocol type TCP, UDP and ICMP.

Packet Capturing Agent: The whole advisor carries out the tasks of recording traffic throughout the network through sniffing protocols, by listening to the network.

Intrusion Database: In this unit, pre-processed intrusion dataset is stored. The database taken in raw form is first converted into the numerical format and redundant features are removed. The preprocessed data are then further can be used by a deep learning tool to learn the pattern with the least complexity.

Data Cluster Agent: This agent will cluster data based on sparse PSO K-Mean clustering and make three clusters of data on protocol type, i.e. TCP, ICMP and UDP. The sparse PSO K-Mean clustering algorithm clusters the data for the reduction of training complexity and losses. More accurate and efficient output can be obtained. PSO is used here because it is more likely to find a near-optimal solution.

Decision Agent: This agent is responsible for deciding to analyze and process the uncategorized data packets. The data packets are classified according to protocols, i.e. TCP, UDP and ICMP. It helps network and security administrators to communicate with the intruder database and conduct reporting tasks on all observed intrusions and deep training algorithm (Fig. 3). The algorithm for a multi-agent-based intrusion system is described below:

Algorithm:

1. Begin
2. Input: Intrusion Feature set = {F1, F2,....Fn}
3. Output: Intrusion Report → Data(status)
4. Agent initialization and deployment
5. Feature Extraction
6. Feature clustering based on protocols
7. LSTM training
8. Decision-making agent → Data(status)
9. Generate alert

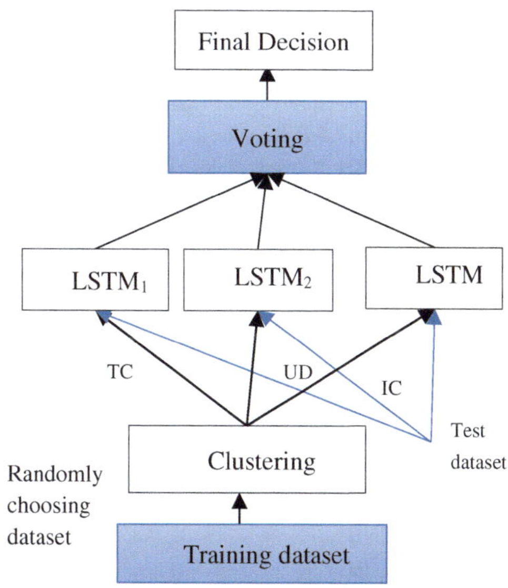

Fig. 3 Proposed training architecture

10. Update learning model
11. End

3.1 Sparse PSO K-Mean Clustering Algorithm

The convergence of clustering of data is considered a problem in high-dimensional data. So, there is a requirement for the design of an accurate converging clustering algorithm. In this paper, a sparse PSO K-mean clustering algorithm is proposed. The data matrix (X, k) is considered with X as the collection of data whereas k is several clusters formed. So, each particle in data collection, X is represented as: $X = (Cent_1, Cent_2, \ldots, Cent_i)$. At each iteration, the cluster is updated according to fitness value, which is stated as Eq. 1:

$$fitness = \frac{\sum_{a=1}^{m}\left\{\frac{\sum_{b=1}^{p} dist(Cent_a, m_{ab})}{X_a}\right\}}{k} \tag{1}$$

$$Clus_{indx} = \frac{1}{X_a}\sum dist(cent) \tag{2}$$

where,

$m_{ab} = a_{th}$data sample in bth cluster.

$dist(cent_a, m_{ab})$ = distance between data value and centroid.

X_a = total number of data in the cluster.

K = number of clusters.

The algorithm is summarized as:

1: Choose N sub-samples from X samples.
2: Apply K-mean for initialization of random samples.
3: Apply the K-means on each sub-sample using random initial centroids.
4: Application of PSO on each sub-sample 'N' using the fitness of centroids.
5: Finally apply K-means on all clusters and identify the number of best fit cluster for further training process.

Algorithm: Modified Sparse K-Mean Clustering
Input: X (Dataset), K (number of cluster), Compression factor C_f (must be < 1)
Output: Cluster index (C_{indx})

1. Start
2. Y ← random(X)
3. For i = 1 to n do
4. $Clus_{indx}$ ← fitness(Y_i) {see Eqs. (1, 2) }
5. For each iteration do
6. Update $Clus_{init}$
7. Update $Clus_{indx}$
8. Return $Clus_{indx}$

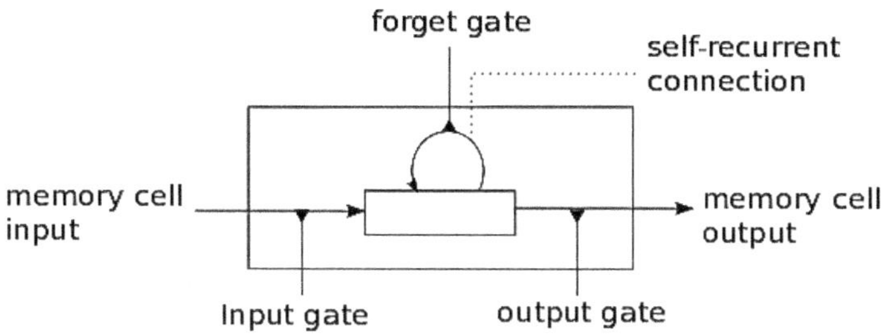

Fig. 4 Long Short-Term Memory (LSTM) Units

3.2 Ensemble LSTM Training

After clustering of data samples, it is fed into a deep classifier (ensemble LSTM) network. As in a traditional LSTM network, the last layer is the softmax layer whose efficiency is less. So, it is replaced by a random forest classifier. This will help to analyze the protocol data type. Figure 3 represents the block diagram of the proposed training architecture.

One type of the recurring neural network, a deep learning technique, is long short-term memory (LSTM). A memory cell, an input unit, an output unit and a forget unit are the four units that make up each node in the LSTM network (Fig. 4). The memory unit's purpose was to keep the internal parameter values for a period of time, while the other three units control the flow of data values for performance evaluation. Sample t, an input gate it, a forget gate ft, an output gate ot and a memory cell Ct are all initialized at the same time.

The output of the hidden layer ht is computed as follows using all of these:

$$f_t = \sigma(W_f * x_t + U_f * h_{t-1} + b_f) \tag{3}$$

$$i_t = \sigma(W_i * x_t + U_i * h_{t-1} + b_i) \tag{4}$$

$$\widehat{C}_t = tanh(W_c * x_t + U_c * h_{t-1} + b_c) \tag{5}$$

$$C_t = i_t * \widehat{C}_t + f_t * C_{t-1}) \tag{6}$$

$$o_t = \sigma(W_o * x_t + U_o * h_{t-1} + b_o) \tag{7}$$

$$h_t = o_t * \tanh(C_t)) \tag{8}$$

Throughout this framework, tanh is the hyperbolic tangent activation function, xt is the input at time t, W_i, W_C, W_f, W_o, U_i, U_C, U_f, U_o are weight matrices to control the input, and bi, bC, bf, bo are bias vectors.

4 Result Analysis

The performance of the proposed methodology is evaluated based on the following parameters:

$$\text{Accuracy} = (\text{TP} + \text{TN})/(\text{TP} + \text{TN} + \text{FP} + \text{FN}) * 100 \tag{9}$$

$$\text{Precision} = \text{TP}/(\text{TP} + \text{FP}) * 100 \tag{10}$$

$$\text{Recall} = \text{TP}/(\text{TP} + \text{FN}) * 100 \tag{11}$$

$$\text{F\_Measure} = 2 * \text{Precision} * \text{Recall}/(\text{Precision} + \text{Recall}) \tag{12}$$

Where,

True Positive (TP) = If actual and predicted data samples are an anomaly in nature, then TP is evaluated.

True Negative (TN) = If actual and predicted data samples are not an anomaly in nature, then TN is evaluated.

False Positive (FP) = If actual and predicted data samples are normal and anomaly in nature respectively, then FP is evaluated (Table 2).

False Negative (FN) = If actual and predicted data samples are an anomaly and normal respectively, then FP is evaluated.

On the NSL-KDD dataset, Tables 3, 4 display the performance evaluation of the proposed multi-agent detection method Figs. 5, 6, 7. The success rate of the proposed method obtained the best result, according to the results report .

5 Conclusion

This paper used deep learning to develop and simulate a multi-agent multilevel intrusion detection system. Because of the simplicity of a multi-agent scheme, this new IDS can be used in a variety of environments. The article is designed to predict intrusions on various protocol types, including UDP, ICMP and TCP. The simulation is run on the NSL-KDD dataset, which indicates an increase of about 2–3% over previous work.

Table 2 Result analysis for TCP protocol

Performance	Accuracy	Precision	Recall	F_Measure
Sample_1	99.59	100	97.5	98.7
Sample_2	100	100	100	100
Sample_3	100	100	100	100
Sample_4	100	100	100	100
Sample_5	100	100	100	100
Sample_6	100	100	100	100
Sample_7	98.77	100	93.54	96.66
Sample_8	100	100	100	100
Sample_9	100	100	100	100
Sample_10	99.59	97.05	100	98.50
Proposed Average	99.79	99.70	99.10	99.39
Liang et al. [12]	99	99	99	99

Table 3 Result analysis for UDP protocol

Performance	Accuracy	Precision	Recall	F_Measure
Sample_1	98.58	96.42	100	98.18
Sample_2	99.20	98.70	99.35	99.02
Sample_3	96.28	91.22	99.52	95.19
Sample_4	96.28	91.97	97.67	94.73
Sample_5	99.73	99.32	100	99.66
Sample_6	100	100	100	100
Sample_7	99.73	99.22	100	99.61
Sample_8	100	100	100	100
Sample_9	99.73	99.20	100	99.6
Sample_10	99.64	99.52	99.52	99.52
Proposed Average	98.92	97.56	99.60	98.55
Liang et al. [12]	98	97	97	97

Table 4 Result analysis for ICMP protocol

Performance	Accuracy	Precision	Recall	F_Measure
Sample_1	97.81	97.43	99.84	98.62
Sample_2	100	100	100	100
Sample_3	99.15	99.70	99.26	99.48
Sample_4	98.90	99.78	98.92	99.35
Sample_5	100	100	100	100
Sample_6	100	100	100	100
Sample_7	100	100	100	100
Sample_8	99.81	99.77	100	99.88
Sample_9	100	100	100	100
Sample_10	100	100	100	100
Proposed Average	99.56	99.66	99.80	99.73
Liang et al. [12]	93	94	94	94

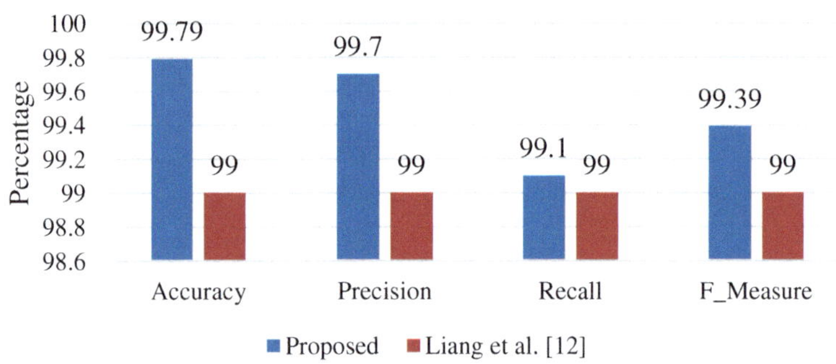

Fig. 5 Performance evaluation on NSL-KDD dataset on TCP protocol

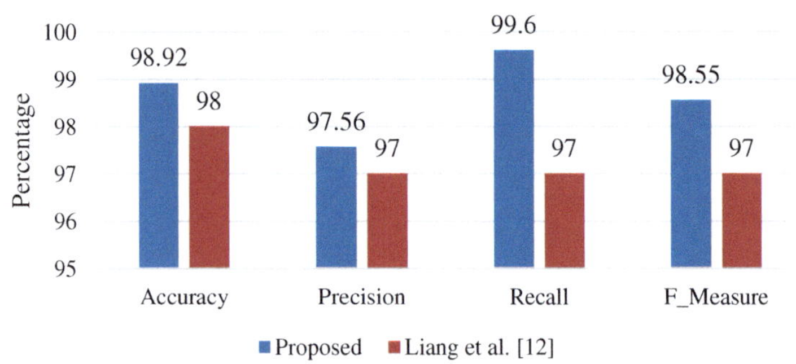

Fig. 6 Performance evaluation on NSL-KDD dataset on UDP protocol

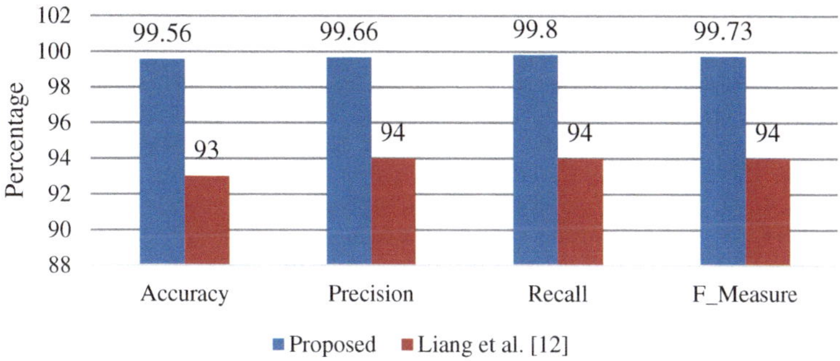

Fig. 7 Performance evaluation on NSL-KDD dataset on ICMP protocol

References

1. Bolon-Canedo V, Sanchnez-Marano N, Alonso- Betanzos A (2011) Feature se-lection and classification in multiple class dataset: An application to KDD cup 99 dataset. Expert Syst Appl 38:5947–5957
2. Uguz H (2011) Two stage feature selection method for text categorization by using information gain, principal component analysis and genetic algorithm. J Knowl Based Syst 24:1024–1032
3. Mukherjee S, Sharma N (2012) Intrusion detection using Naïve Bayes classifier with feature reduction. Procedia Technol 4:119–128
4. Li Y, Xia J, Zhang S, Yan J, Chuan X, Dai K (2012) An efficient intrusion detection system based on support vector machine and gradually features removal method. Expert Syst Appl 39:424–430
5. Karimi Z, Mansour M, Harounabadi A (2013) Feature ranking in intrusion detection dataset using combination of filter methods. Int J Comput Appl 78:21–27
6. Al-Jarrah OY, Siddiqui A, Elsalamouny M, Yoo PD, Muhaidat S, Kim K (2014) Machine learning based feature selection techniques for large scale intrusion detection. IEEE, pp 177–181
7. Manzoor AI, Kumar N (2017) A feature reduced intrusion detection system using ANN classifier. Expert Syst Appl 88:249–257
8. Mohamed S, Gadal AM, Mokhtar RA (2017) Anomaly detection approach using hybrid algorithm of data mining technique. In: Proceedings of the International Conference on Communication, Control, Computing and Electronics Engineering, IEEE
9. Kuang F, Xu W, Zhang S (2014) A novel hybrid KPCA and SVM with GA model for intrusion detection. Appl Soft Comput J 18:178–184
10. Hanifive K, Bank H (2016) Network Intrusion Detection Using Machine Learning Anomaly Detection Algorithms. In: Proceedings of the International Conference on Computer and Information Sciences (ICCOINS), pp 369–373
11. Al-Yaseen WL, Othman ZA, Nazri MA (2017) Multi-Level hybrid support vector machine and extreme learning machine based on modified K-means for intrusion detection system. Int J Expert Syst Appl Elsevier 67:296–303
12. Liang C, Shanmugam B, Azam S, Jonkman M, Boer FD, Narayansamy G (2019) Intrusion Detection System for Internet of Things based on a Machine Learning approach. In: Proceedings of the 2019 International Conference on Vision Towards Emerging Trends in Communication and Networking (ViTECoN), Vellore, India, pp 1–6. https://doi.org/10.1109/ViTECoN.2019.8899448

13. Zhao G, Zhang C, Zheng L (2017) Intrusion detection using deep belief network and probabilistic neural network. In: Proceedings of the 2017 IEEE International Conference on Computational Science and Engineering (CSE) and IEEE International Conference on Embedded and Ubiquitous Computing (EUC), Guangzhou, China, 21–24 July, vol 1, pp 639–642
14. Yin C et al (2017) A deep learning approach for intrusion detection using recurrent neural networks. IEEE Access 5:21954–22196
15. Yuan X, Li C, Li X (2017) DeepDefense: Identifying DDoS attack via deep learning. In: Proceedings of IEEE International Conference on smart computing (SMARTCOMP)

Chapter 11
Application of Hybrid of ACO-BP in Convolution Neural Network for Effective Classification

Suruchi Chawla

1 Introduction

Deep learning techniques are well developed and widely used for various real-world applications such as human action recognition [1, 2] text processing and applications [3, 4], medical image processing [5–8] and computational biology [9]. Deep supervised machine learning generates the classification model through training based on labeled data set. The training of the model based on labeled data set involves adjusting the weights and biases of the neural network in order to minimize the difference between predicted and actual output. Training of neural network plays an important role in classification precision.

The effective learning of neural network for pattern classification depends on the appropriate network architecture and weight configuration of the network. The most popular algorithm used for training Neural Network (NN) is backpropagation for updating the weights. Backpropagation is based on gradient descent and has a drawback of getting trapped in local minimum and low convergence speed [10, 11]. Gradient descent method has been used with swarm intelligence for weight initialization to speed up the training process. The training of NN using BP will be faster using global search ability of ACO and avoid local minima [12, 13].

In [16], a hybrid of the ant colony and backpropagation algorithms was applied to optimize the neural network weights for effective learning. ACO was implemented to find the optimal combination of initial weights. Once the initial weights of NN were optimized using ACO then BP algorithm was applied to obtain the accurate optimal solution quickly. BP was used because it executes fast during the algorithm iterations. The hybrid ACO-BP shows good performance for problems of large size.

S. Chawla (✉)
University of Delhi, Department of Computer Science, Shaheed Rajguru College of Applied Sciences for Women, New Delhi, India

G. Mathur et al. (eds.), *Proceedings of 2nd International Conference on Artificial Intelligence: Advances and Applications*, Algorithms for Intelligent Systems,
https://doi.org/10.1007/978-981-16-6332-1_11

The performance of gradient descent techniques is low for large problem size in comparison to the hybrid ACO-BP algorithms [15].

Backpropagation has been used for training of CNN and tuning of parameters. BP algorithm has a drawback of local minima and unable to find global optima, specifically when search space is large for optimal solution [14]. In this paper, novel method is proposed using hybrid of ACO and BP for CNN weight initialization and optimization using training based on data set of MINST for effective classification. Thus for effective pattern classification and to avoid the local minima problem using BP, hybrid of ACO-BP has been used for weight optimization of the CNN network. Thus combining the global search capability of ACO and local as well as the fast search capability of BP provides the powerful and efficient method for determining the optimal weight configuration of the network. The effectiveness of CNN training using hybrid of ACO-BP was confirmed based on results showing the increase in accuracy of classification using CNN with ACO-BP in comparison to CNN (BP only) with random initialization of weights.

2 Related Work

In [17–19] Artificial Neural Network (ANN), k nearest neighbor, and Bayesian are the classification model used in various domains. The classification model is built in two phases: learning and classification. The training of classification model is done based on training data set and parameters of the model are modified during training. The training of classification model is continued for sufficient number of iterations till the terminating condition is reached. The classification model learnt during training is used to predict the class label of test data set and its performance is evaluated to confirm its effectiveness.

The most commonly used learning algorithm for Artificial Neural Network is Backpropagation. The learning of NN is done based on weight modification using backpropagation based on gradient descent. The gradient descent is implemented based on the difference between the actual output and predicted output. The single objective algorithm was proposed using both PSO and DE algorithms for training multilayer FFNN with one hidden layer [20].

The Hybrid Particle-move Artificial Bee Colony (HPABC) algorithm was proposed to use in Neural Network Training for pattern classification using UCI dataset [21, 22]. In [24], backpropagation was applied after every iteration of ACO to determine the weights for Neural Network initialization. The experimental results showed that ACO select the good initial weight values for optimization using BP. The performance of the hybrid ACO-BP was the best especially on the large size problem cases. In [25], hybridization of genetic algorithm along with local search was used for classification problem.

It is found that nature-inspired algorithm like ACO is best for weight initialization, GA is used for improving the computational speed due to parallel search capability. This optimization algorithm provides the good heuristic values for classification problem. Deep learning integrated with metaheuristic method has become a major research area for improving the performance of classification models [15]. The three metaheuristic techniques were evaluated for its effectiveness in improving the accuracy of CNN. The primary goal is to improve the accuracy of original CNN by using SA, DE, and HS algorithms. The experiment results show that CNNSA presents the best accuracy for all epoch in comparison to other proposed method and the original CNN. This strategy can also be developed for other metaheuristic algorithms such as ACO, PSO, and BCO to optimize CNN [26].

The training of the Neural Network for pattern identification in a given data set is optimized using Backpropagation algorithm. Backpropagation modifies the weights of Neural network with the objective of minimizing error computed using the difference between actual output and predicted output. Backpropagation is local optimization technique and prone to local minima [27].

To overcome local minima using BP, hybrid training algorithms are used. It is found that hybrid training using GA or ACO produces promising results rather than independent standalone metaheuristic or gradient descent algorithm. Metaheuristic techniques are global optimization techniques and less prone to local minima. In [23], ant colony optimization was used for training of neural network to classify patterns using different combinations of weights. The ACO was executed to initialize the NN with promising weights that are further optimized using BP thereby avoiding local minima. The experimental results showed that ACO was a good option to select good BP's values for optimization using BP and gradient descent [16, 24].

In this paper, hybrid of ACO-BP metaheuristic algorithms is applied to improve the training of CNN based on MINST data set and hence therefore improves the classification accuracy. Since BP has a drawback of local optima during weight optimization of Neural Networks for large problem space therefore ACO global optimization method is used for weight initialization of CNN and overcome local optima. Subsequently, BP is applied to rapidly find the accurate solution and improves the classification accuracy.

3 Background

3.1 Deep Learning Model-CNN

Deep learning models like CNN have been widely applied in domains like web Information Retrieval, natural language processing such as word embedding, search, question answering, and image retrieval. CNN is a multilayer supervised learning neural network which consists of convolutional layer and a pooling layer. Learning of NN is done using backpropagation based on gradient descent during each iteration and improves the network accuracy [28].

CNN uses convolution operation and maxpooling layer to extract the feature at abstract level. In convolution layer, the filter weights matrix of a given region size are moved through the input matrix with a stride size and generates the variable length feature map. Thus, the application of multiple filters with a given stride on input will generate the multiple variable length feature map. Maxpooling operation extracts the maximum feature value from each feature map and generates the fixed length global abstract feature vector. This global feature vector is connected to FNN and undergoes nonlinear transformation at hidden layer and is classified at the output layer. The training of CNN is done using Backpropagation based on gradient descent. During training the objective is to minimize the error based on the difference between predicted and actual output. The gradient of the error is backpropagated and weights are modified to minimize the error. The modification of weights based on backpropagation continues till the objective function-based error is minimized [29].

3.2 Ant Colony Optimization

ACO was first introduced in 1990 by Marco Dorigo. ACO is inspired from foraging behavior of Ants. Ants use pheromone to find the shortest path from nest to food source. Ants once reach the food source then during return trip to nest, it deposits the pheromone on path they followed to reach the food source. The quantity of pheromone deposited on the path during return trip depends on the quality and quantity of food found. Thus, the amount of pheromone deposited on the path will guide the other ants to follow the shortest path. This indirect communication between ants using pheromone is called stigmergy [30–32].

In ACO each ant built its solution starting from initial state and select the component of solution according to some probability that is problem specific. The selection of nodes in solution path is guided by local heuristic information and pheromone. Consider a graph having N nodes, representing possible solution to a problem, and a set E of edges fully connecting the nodes N. The decision of an ant at node i to choose node j at timestamp t depends on the value of a_{ij} (t). The a_{ij} (t) depends on local pheromone trail values and the local heuristic values as follows:

$$a_{ij}(t) = \frac{\left[\tau_{ij}(t)\right]^{\alpha} * \left[\eta_{ij}\right]^{\beta}}{\sum_{l\epsilon\in N_i}\left[\tau_{il}(t)\right]^{\alpha} * \left[\eta_{il}\right]^{\beta}} \forall j\epsilon N_i \tag{1}$$

where $\tau_{ij}(t)$ is the quantity of pheromone deposited on arc $(.i; j)$ at time t, η_{ij} is the heuristic value associated with the path from node i to node j, N_i is the set of neighbors of node i, and α as well as β are the weights that control the relative importance of pheromone trail and heuristic value. The values of the parameters α and β are defined as the following. If the value of parameter $\alpha = 0$, then the heuristic values guide the ant for the selection of next node j from a given node i during the exploration of shortest path from nest to food source.

If value of $\beta = 0$, then selection of nodes is guided by pheromone in the construction of shortest path that yields suboptimal solution. A trade-off between heuristic value and pheromone trail intensity is necessary and depends on the selection of value of α and β.

The probability with which an ant *k moves* from node *i* to node *j* at the *t*th algorithm iteration is

$$p_{ij}^{k}(t) = \frac{a_{ij}(t)}{\sum_{l\epsilon N_i^k} a_{il}(t)} \tag{2}$$

where *Ni* is the set of nodes in the neighborhood of node *i* that ant *k* has not visited yet.

After the completion of their respective tour by the ants, the ants on their return trip deposit the pheromone on the best path found, the amount of pheromone deposited depends on the error at the NN output layer. The pheromone evaporation on all arcs is triggered and depends on the evaporation rate. The pheromone evaporation rate is set so as to prevent the local optima during exploration of the shortest path.

Each ant *k* deposits a quantity of pheromone on their return trip,

$$\tau_{ij}(t) = (1-\rho)\tau_{ij}(t) + \Delta\tau_{ij}(t) \tag{3}$$

$$\Delta\tau_{ij}(t) = \sum_{k=1}^{m} \Delta\tau_{ij}^{k}(t) \tag{4}$$

where T is the tour done by ant *k* at iteration *t*, and L is its length where m is the number of ants.

In Eq. (4), it implies that for all those ants k that an edge (i,j) in the path selected by ant k has the value of $\Delta\tau_{ij}^{k}(t) = Q/L_k$ where Q is the constant and value is taken to be 1 and L_k is the length of the path selected by ant k otherwise $\Delta\tau_{ij}^{k}(t) = 0$ if $(i, j) \notin L_k$

The shorter the tour done, the greater the amount of pheromone deposited. $\rho \in (0, 1]$ is the pheromone evaporation rate. The high value of pheromone increases the probability of choosing that path by ants and thus the more pheromone will be deposited on that path [33].

The initial amount of the pheromone is set to a same small positive constant value on all arcs, and the recommended values of α, β, are 1, 5, and 0.5 Dorigo [34].

Ant Algorithms had been used for many optimization problems such as traveling salesman problem, quadratic assignment problem, vehicle routing, sequential ordering, graph coloring, routing in communications networks, etc. [35].

3.3 Training of Neural Network Using Hybrid of ACO-BP

Supervised learning of neural network is done using large collection of labeled data set. The data set is partitioned into training and validation set. The training data set is used as input data during training of neural network. The input data is passed through nodes of hidden layers and undergoes non-linear transformation and generates the output at nodes of output layer. Since the training of NN is based on labeled data set means the actual label of input data is known and compared with the predicted output of NN. The difference between the actual and predicted output is used to adjust the weights and bias parameters using Backpropagation stochastic gradient descent during learning of NN.

The Deep learning model has been widely used for machine learning. Convolution Neural Network model has been widely used in image recognition and Natural Language processing. The performance of CNN depends on filter weights, biases, and weights of FNN for effective machine learning.

These weight configurations determine the performance of the network. Weight is initialized using random real values between 0 and 1 during the early training phase. BP algorithm has been widely used in supervised training of NN but it has the disadvantage to get trapped in local optima. The ACO metaheuristic is the global optimization technique and is useful for optimal weight initialization in the training of NN for pattern classification. The NN initialized with optimal weights is further optimized using BP algorithm for effective learning of NN. The hybrid of ACO-BP combines the advantage of global optimization of ACO and fast convergence of BP on large problem space to avoid the problem of local optima for effective learning of NN for pattern classification.

ACO has been used for determining the global optimum initial weights of Neural Network. Each weight's range of values is split into m discrete points and each point is the possible value of weight parameter. Thus, an ant k for a given weight parameter Wi selects from among the m possible m points according to the probability function as follows:

$$p_i^k(t) = \frac{\tau_i(t)}{\sum_{1 \le j \le m_i^k} \tau_j(t)} \tag{5}$$

Thus, the weights parameter values selected by ants using the probability function are used for weight initialization of NN and Err is computed based on the difference in the actual and predicted output.

The pheromone of points for each selected weight parameter is updated using $\Delta\tau_i(t)$ as follows:

$$\tau_i(t) = (1 - \rho)\tau_i(t) + \Delta\tau_i(t) \tag{6}$$

$\rho \in (0, 1]$ is the pheromone evaporation coefficient. $\Delta\tau_i(t) = \frac{Q1}{E}$ represents the updation of pheromone, *Q1* is a constant. Thus, the ACO is applied to select the weight values based on updated pheromone value and the selected weight values are used to initialize the weights of NN and generate the output. Thus, the weight optimization using ACO continues for certain fixed number of iterations or till the E is less than threshold value.

Thus, the optimal weights generated using ACO are used for weight initialization of NN and are modified during learning of NN using Backpropagation and avoid local minima [16].

BP algorithm was developed by Rumelhart and McClelland [36, 37]. For training of NN using BP, the input data and weights of network are initialized. The input values were non-linearly transformed at the hidden layers as weighted sum and non-linear activation as shown in Eq. (7) and (8) and generate output at the output layer. The learning using BP objective is to minimize the difference between the actual and predicted output as given in Eq. (9)

$$x_j = \sum\nolimits_l u_l w_{jl} \tag{7}$$

$$y_j = \frac{1}{1 + e^{-x_j}} \tag{8}$$

Total Error E is defined as

$$E = \frac{1}{2}\sum\nolimits_c \sum\nolimits_j \left(y_{j,c} - d_{j,c}\right)^2 \tag{9}$$

where c is the index over number of training samples. J is the number of output units in the output layer. The u is the input, y is the actual output, and d is the expected output. The objective is to minimize the error E using gradient descent by taking the partial derivative of E for each weight in the network. The simplest method is to update the weight by an amount proportional to as given in Eq. (10) and (11)

$$\Delta W = -\varepsilon \frac{\partial E}{\partial W} \tag{10}$$

$$W_{t+1} = W_t + \Delta W \tag{11}$$

The training using BP algorithm continues for the specific number of iterations or till the error is minimized. The proper weight initialization of NN will prevent the BP to get trapped in local minima during training. The use of ACO for NN weight initialization in BP training will enhance the learning of network using BP for effective classification.

4 Proposed Method

In this paper hybrid of ACO and BP has been proposed for network weight initialization as well as optimization using BP based on gradient descent for effective training of CNN for classification problems. CNN uses convolution layer and maxpooling operation on input to extract the features at abstract level. These abstract feature values are input to FNN that undergoes non-linear transformation at hidden layer and classified at the output layer.

The training of CNN is done based on minimizing the difference between actual and predicted output and the learning is done through weight adjustment using back-propagation gradient descent based on minimizing the difference at the output layer. BP has a drawback of local optima during weight optimization for large problem space therefore ACO global optimization method is used for weight initialization of CNN and overcomes local optima. Subsequently, BP is applied to rapidly find the accurate solution and improves the classification accuracy.

The deep learning model Convolution Neural Network has been widely used for image recognition using MINST data set. The MINST data set has been widely used for evaluating the performance of machine learning algorithms. The MINST data set contains data of handwritten characters in real number matrix form. The MINST data set is split into 60,000 training samples and 10,000 test samples.

CNN with a given architecture (filter size, number of filters, pooling, activation function, number of hidden layers, and number of output neurons) is selected. The training of CNN using MINST data set involves the learning of filter weights, biases, and FNN weights based on minimizing the error between predicted and actual output. The method is proposed in this paper for training of CNN using Hybrid ACO-BP. The steps used for training of CNN with ACO-BP are given as follows:

Steps:

1. Weights range is split into m discrete points.
2. The ant for each weight selects weight point according to probability function with a maximum value using Eq. (5).
3. The discrete weight point selected by ants for each weight are used for weight initialization of CNN.
4. Input is applied to CNN configured with initial weights and undergoes convolution operation, maxpooling followed by non-linear transformation in FNN and generates the output.
5. The error is computed based on difference between the predicted and actual output.
6. The feedback of error at the output layer is used to update pheromone on selected weight point according to pheromone updation rule using Eq. (6).
7. The initial weight optimization of CNN using ACO continues till the error of the model minimizes.
8. The final optimal weight set obtained using ACO is used for CNN weight initialization and further optimized using BP during training of CNN using Eqs. (7)–(11).

Thus, ACO-based optimized weights used for initialization of CNN thus provide the promising neighborhood in subsequent search using BP for global optimal solution. BP has been used for fast convergence to optimal solution. Therefore trained CNN based on ACO-BP is used subsequently on test dataset. The stepwise execution of offline and online processing of the algorithm for the proposed approach of training of CNN with ACO-BP is mentioned as follows.

Algorithm: Offline processing:

Input:CNN architecture with maxpooling (2*2), number of convolution layer d = 3, number of filter n.

CNN weights, FNN weights, biases.

1. **Initialization**
   a. Divide weights N total range [–1,1] into m points uniformly so that each of weight w have m edges possible coming out of it thus corresponding to m points
   b. Each point is associated with same initial pheromone value τ 0.
   c. Create a pheromone table for each weight parameter.
   d. Initially, the probability of choosing paths by ants are equal so the paths chosen by ants are random otherwise after some time of algorithm execution, the path that has a more pheromone level than others will have a higher probability for ants to follow.
2. for each ant k = 1 to (Nant = 1000) (currently in state iteration t)
   a. Ant k will select one of the m weight values for each weights of the network according to probability using Eq. (5).
   b. Ant k will select one value based on P(i) for each weight parameter i and store the index of selected weight value.
   c. Once ants select values for all the weight parameters. The selected weight values initialize all the parameters of the network.
   d. Network initialized with weights selected by ants are trained using input samples.
   e. Evaluate the Error function E of the network with respect to selected value of weights using Eq. (9)
   f. Update the pheromone on every edge selected by ant k corresponding to the path of weights using Eq. (6)
3. Repeat loop step 2, until the maximum number of iteration = 1000 is reached or there is no change in the path of weights selected by all ants.
4. Use the weight parameters found by ACO algorithm for CNN weights initialization and is used for training using BP based on MINST data set.
5. Calculate the error function, and the gradient is backpropagated from the output layer back to its forward layer one by one. The weight vector is modified according to eqs. (7)–11).
6. The training process continues until the error E minimizes or the number of maximum epochs has been run.
7. The trained network is tested using the test samples and the generalization ability of the well-trained network is confirmed. If the errors satisfied the set criterion, stop. Otherwise, restart step 1.

Thus, the training of CNN using hybrid of ACO-BP overcomes the local optima drawback found in BP and uses global optimization of ACO to initialize the CNN weights. BP is simple and has the ability of quick convergence to optimal solution. So the use of hybrid of the ACO-BP algorithms in training of CNN for classification problems may accelerate the learning speed and improve the precision of the trained neural network in classification problem.

5 Experiment

The performance of the proposed system was evaluated based on MINST data set. MINST data set is collection of handwritten character digital images. The digital image is represented as a matrix of 28*28. with the values ranging between 0 and 255. MINST data has been widely applied for the evaluation of machine learning algorithms. The proposed method was implemented in tensorflow and python. ACO algorithm was implemented in python for the optimization of initial weights of CNN. The hardware used for experimental evaluation were i7 processor, linux operating system, and 8 GB RAM. ACO algorithm was implemented using user-defined functions in python like generate path, spread pheromone, etc.

In order to evaluate the performance of the CNN using hybrid of ACO-BP, MINST data set is split into 60,000 training samples and 10,000 samples. The accuracy of the proposed method was evaluated using k-fold cross validation technique. During each iteration of algorithm execution in the training of CNN model, the training data set samples in a given batch are split into k independent data subsets and k-fold method is executed k times. During every iteration of k-fold method, one data subset of a given batch is used as test data set and remaining subset of data set is used as training data set therefore each subset of dataset will be used as test set after k iterations. At the end of execution of k-fold method for a given batch of training data set, accuracy calculated is average based on k-fold method for a given iteration of the CNN algorithm. The accuracy average is plotted for different number of iterations using CNN with hybrid ACO-BP and simple CNN method.

Initially, the ACO optimization based on pheromone updation is done for initial weight configuration. The weight optimization using ACO continues till the specified number of iterations are done or there is no change in the selection of path of weight values by ants during ACO optimization. Upon completion of ACO optimization of CNN initial weight configuration, the pheromone value corresponding to the weights for the best path is applied to CNN for training using BP on the MINST data set and minimizes the error difference between the predicted and actual output for effective classification.

The accuracy is averaged for k = fivefold method over 25 iterations using both CNN model and proposed method CNN with Hybrid ACO-BP as shown in Table 1. The performance of hybrid ACO-BP for CNN was measured based on average accuracy of CNN classifier and compared with CNN with Hybrid ACO-BP. It was found that classifier generated using CNN-ACO is superior in comparison to CNN-NONACO. Hence use of ACO algorithm with the BP algorithms means combining the advantages of both global and local global exploration abilities for effective

Table 1 Shows the accuracy of CNN (ACO-BP) and Simple CNN over 25 iterations using k- fold method (k = 5)

No of iterations	CNN with ACO-BP					CNN with BP only				
	Fold1	Fold2	Fold3	Fold4	Fold5	Fold1	Fold2	Fold3	Fold4	Fold5
1	0.15	0.47	0.66	0.75	0.78	0.14	0.2	0.39	0.61	0.7
2	0.78	0.83	0.85	0.87	0.89	0.68	0.75	0.75	0.8	0.84
3	0.82	0.85	0.88	0.87	0.88	0.74	0.75	0.79	0.82	0.84
4	0.84	0.85	0.86	0.88	0.91	0.83	0.84	0.86	0.89	0.9
5	0.87	0.88	0.89	0.89	0.91	0.84	0.85	0.87	0.88	0.89
6	0.87	0.87	0.88	0.89	0.9	0.86	0.88	0.88	0.88	0.89
7	0.88	0.89	0.9	0.91	0.92	0.9	0.9	0.92	0.93	0.94
8	0.9	0.92	0.92	0.92	0.93	0.91	0.92	0.91	0.92	0.93
9	0.93	0.93	0.94	0.95	0.96	0.92	0.92	0.93	0.94	0.95
10	0.92	0.93	0.95	0.95	0.96	0.87	0.89	0.9	0.91	0.92
11	0.88	0.9	0.92	0.93	0.93	0.87	0.88	0.89	0.91	0.92
12	0.88	0.88	0.9	0.91	0.91	0.88	0.88	0.89	0.9	0.91
13	0.91	0.9	0.91	0.93	0.93	0.91	0.92	0.92	0.93	0.93
14	0.91	0.91	0.94	0.93	0.93	0.87	0.88	0.89	0.9	0.91
15	0.88	0.89	0.9	0.92	0.94	0.88	0.9	0.91	0.93	0.94
16	0.92	0.93	0.93	0.94	0.95	0.9	0.92	0.92	0.93	0.93
17	0.91	0.91	0.93	0.94	0.95	0.92	0.94	0.93	0.94	0.96
18	0.94	0.94	0.95	0.96	0.97	0.88	0.9	0.91	0.92	0.92
19	0.92	0.93	0.93	0.95	0.96	0.9	0.91	0.92	0.93	0.94
20	0.92	0.93	0.94	0.94	0.96	0.94	0.94	0.95	0.96	0.96
21	0.96	0.96	0.97	0.97	0.97	0.94	0.94	0.95	0.94	0.96
22	0.94	0.96	0.96	0.97	0.97	0.93	0.93	0.95	0.95	0.97
23	0.95	0.96	0.96	0.96	0.97	0.93	0.94	0.94	0.94	0.95
24	0.94	0.95	0.96	0.96	0.97	0.92	0.92	0.93	0.93	0.95
25	0.92	0.93	0.93	0.95	0.95	0.92	0.93	0.94	0.95	0.95
Avg	0.87	0.89	0.91	0.92	0.93	0.85	0.87	0.88	0.9	0.92

training of CNN in pattern classification. The results confirm the effectiveness of ACO-BP algorithm.

The training of CNN involves learning of Wc, biases, and FNN weights. The initial random weights values of CNN are optimized using ACO. During weight optimization using ACO following parameters were selected Wmin $= -1$, Wmax $= -2$, number of split points m was varied in the range 30–50. The number of ants N was selected as 30 tradeoff between running time and exploration of new paths during ACO optimization, pheromone evaporation ρ is varied between 0.8 and 0.95. During weight optimization using ACO, N ants select the weights values according to probability function and generate N paths that correspond to selected weights values in every iteration.

CNN (optimal weight initialization using ACO) is trained using BP. The learning rate parameter is set to 0.001 and number of iterations was set to 25. The values of hyperparameters selected are number of filters = 300, size of filter = 3, maxpooling, and tanh activation function (Fig. 1).

The graph shown in Fig. 1 plot the classification accuracy for five folds for 25

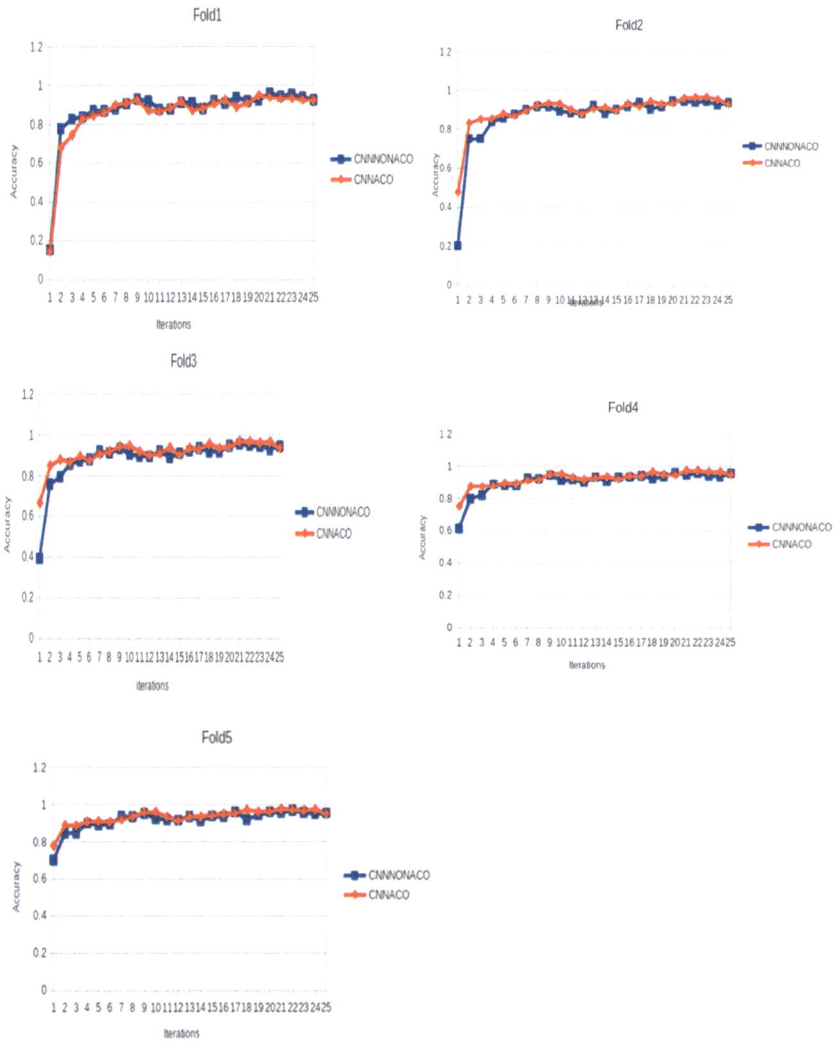

Fig. 1 The classification accuracy values of the CNN training with 25 epochs of the individual classifier using the ACO-BP algorithm and Simple CNN for five runs of fivefold cross validation on the MNIST dataset

iterations using both CNN with ACO-BP and CNN with BP only. At every iteration of CNN training, there is an improvement in average accuracy in comparison to CNN with random weight initialization Thus, the improvement in accuracy of CNN model during training confirms that using hybrid of global optimization technique ACO along with BP will improve the training of CNN for effective classification and converge to optimal solution without trapping in local minima. Therefore, the effective training of CNN using ACO-BP improves the classification accuracy in comparison to CNN with BP only.

6 Conclusion

In this paper, hybrid of ACO-BP has been used for weight initialization of CNN for effective classification. ACO is used for initial weight optimization of CNN and overcomes the local optima of BP. During training of CNN, filter weights Wc, FNN weights, and biases are modified with the objective of increasing the accuracy of classification and minimizing the value of error function based on difference between the actual and predicted output. ACO uses ants to select the path of weights with the maximum pheromone value for CNN weight initialization and input is applied to measure the error in the output. The pheromone value is updated on selected path of weights based on error function value. Once best set of initial weight values are selected and applied to CNN. The CNN initialized with optimal weights is trained using BP for specific number of iteration using k-fold cross validation method. Experiment was conducted on MINST data set for handwritten character recognition. The average classification accuracy is computed using k-fold cross validation during training of CNN using both (ACO-BP/BP). It was found that average classification accuracy is improved using CNN with ACO-BP in comparison to simple CNN.

References

1. Chun S, Lee CS (2016) Human action recognition using histogram of motion intensity and direction from multiple views. IET Comput Vision 10(4):250–257
2. Ji X, Cheng J, Tao D, Wu X, Feng W (2017) The spatial Laplacian and temporal energy pyramid representation for human action recognition using depth sequences. Knowl-Based Syst 122:64–74
3. Chaturvedi I, Ong YS, Tsang IW, Welsch RE, Cambria E (2016) Learning word dependencies in text by means of a deep recurrent belief network. Knowl-Based Syst 108:144–154
4. Poria S, Cambria E, Gelbukh A (2016) Aspect extraction for opinion mining with a deep convolutional neural network. Knowl-Based Syst 108:42–49
5. Ngo TA, Lu Z, Carneiro G (2017) Combining deep learning and level set for the automated segmentation of the left ventricle of the heart from cardiac cine magnetic resonance. Med Image Anal 35:159–171
6. Dhungel N, Carneiro G, Bradley AP (2017) A deep learning approach for the analysis of masses in mammograms with minimal user intervention. Med Image Anal 37:114–128

7. Kamnitsas K, Ledig C, Newcombe VF, Simpson JP, Kane AD, Menon DK, Rueckert D, Glocker B (2017) Efficient multi-scale 3D CNN with fully connected CRF for accurate brain lesion segmentation. Med Image Anal 36:61–78
8. Chen S, Qin J, Ji X, Lei B, Wang T, Ni D, Cheng JZ (2016) Automatic scoring of multiple semantic attributes with multi-task feature leverage: a study on pulmonary nodules in CT images. IEEE Trans Med Imaging 36(3):802–814
9. Romo-Bucheli D, Janowczyk A, Gilmore H, Romero E, Madabhushi A (2017) A deep learning based strategy for identifying and associating mitotic activity with gene expression derived risk categories in estrogen receptor positive breast cancers. Cytometry A 91(6):566–573
10. Sutton R (1986) Two problems with back propagation and other steepest descent learning procedures for networks. Proceedings of the eighth annual conference of the cognitive science society, pp 823–832
11. Whitley D, Starkweather T, Bogart C (1990) Genetic algorithms and neural networks: optimizing connections and connectivity. Parallel Comput 14(3):347–361
12. Hecht-Nielsen R (1992) Theory of the backpropagation neural network. Neural networks for perception. Academic Press, London, pp 65–93
13. Leonard J, Kramer MA (1990) Improvement of the backpropagation algorithm for training neural networks. Comput Chem Eng 14(3):337–341
14. Albeahdili HM, Han T, Islam NE (2015) Hybrid algorithm for the optimization of training convolutional neural network. Int J Adv Comput Sci Appl 1(6):79–85
15. Hemeida AM, Hassan SA, Mohamed AAA, Alkhalaf S, Mahmoud MM, Senjyu T, El-Din AB (2020) Nature-inspired algorithms for feed-forward neural network classifiers: A survey of one decade of research. Ain Shams Eng J 11(3):659–675
16. Liu Y. P., Wu, M. G., & Qian, J. X.. Evolving neural networks using the hybrid of ant colony optimization and BP algorithms. In *International Symposium on Neural Networks* (pp. 714–722). Springer, Berlin, Heidelberg(2006, May).
17. Adem K, Kiliçarslan S, Cömert O (2019) Classification and diagnosis of cervical cancer with stacked autoencoder and softmax classification. Expert Syst Appl 115:557–564
18. Bala R, Kumar D (2017) Classification using ANN: a review. Int J Comput Intell Res 13(7):1811–1820
19. Edla DR, Cheruku R (2017) Diabetes-finder: a bat optimized classification system for type-2 diabetes. Procedia Comput Sci 115:235–242
20. Leema N, Nehemiah HK, Kannan A (2016) Neural network classifier optimization using differential evolution with global information and back propagation algorithm for clinical datasets. Appl Soft Comput 49:834–844
21. Lichman M (2013) UCI machine learning repository
22. Al Nuaimi ZNAM, Abdullah R (2017) Neural network training using hybrid particlemove artificial bee colony algorithm for pattern classification. J Inf Commun Technol 16(2):314–334
23. Mavrovouniotis M, Yang S (2015) Training neural networks with ant colony optimization algorithms for pattern classification. Soft Comput 19(6):1511–1522
24. Mavrovouniotis M, Yang S (2013) Evolving neural networks using ant colony optimization with pheromone trail limits. 2013 13th UK workshop on computational intelligence (UKCI). IEEE, New York, pp 16–23
25. Mane S, Sonawani SS, Sakhare S (2016) Classification problem solving using multi-objective optimization approach and local search. 2016 international conference on electrical, electronics, and optimization techniques (ICEEOT). IEEE, New York, pp 243–247
26. Rere LM, Fanany MI, Arymurthy AM (2016) Metaheuristic algorithms for convolution neural network. Comput Intell Neurosci 2016: 1537325
27. Alba E, Chicano JF (2004) Training neural networks with GA hybrid algorithms. Genetic and evolutionary computation conference. Springer, Berlin, pp 852–863
28. Yang, J., & Li, J. Application of deep convolution neural network. In 2017 14th International Computer Conference on Wavelet Active Media Technology and Information Processing (ICCWAMTIP) (pp. 229–232). IEEE(2017, December)..

29. Li H, Lu Z (2016) Deep learning for information retrieval. Proceedings of the 39th international ACM SIGIR conference on research and development in information retrieval. ACM, New York, pp 1203–1206
30. Dorigo M (1992) Optimization, learning and natural algorithms. PhD thesis, Politecnico di Milano
31. Dorigo M, Maniezzo V, Colorni A (1991) Positive feedback as a search strategy
32. Dorigo M, Maniezzo V, Colorni A (1996) Ant system: optimization by a colony of cooperating agents. IEEE Trans Syst Man Cybernetics, Part B 26(1):29–41
33. Blum C (2005) Ant colony optimization: Introduction and recent trends. Phys Life Rev 2(4):353–373
34. Dorigo M, Caro GD, Gambardella LM (1999) Ant algorithms for discrete optimization. Artif Life 5(2):137–172
35. Yang XS (2014) Swarm intelligence based algorithms: a critical analysis. Evol Intel 7(1):17–28
36. Rumelhart DE, Hinton GE, Williams RJ (1985) Learning internal representations by error propagation (No. ICS-8506). California University San Diego
37. Rumelhart DE, Hinton GE, Williams RJ (1986) Learning representations by back-propagating errors. Nature 323(6088):533–536

Chapter 12
Artificial Intelligence-Enabled IoT-Based Smart Blood Banking System

E. Muthu Kumaran, K. Velmurugan, P. Venkumar, D. Amutha Guka, and V. Divya

1 Introduction

Human blood is very special because the donated blood of one person cannot be used approximately for the needs of another person. Human blood is classified into four main groups: O, A, B, and AB. In addition, human blood is further subdivided into two subtypes depending on the Rh factor, Rh positive and those without Rh factor are called Rh negative. Blood is a perishable substance, so whole blood can always be stored in the blood bank refrigerator at temperatures ranging from +2 °C to + 6 °C for up to 42 days. But platelets that help to treat patients have a lifespan of only 5 days. Of the 118.5 million blood donations collected globally, this is 16% of the world's population, and survey reveals based on a sample of 1000 people, the donation rate is 5.0 in low-income countries, 6.8 donations in lower-middle-income countries, 15.9 donations in upper-middle-income countries, and 31.5 in high-income countries [1]. Even when so many people come forward to donate blood, there are signals to get the right amount of blood, where and when it is needed [18]. The blood bank receives its blood from organizing external blood donation camps. Another source of blood supply is donors who come to the blood bank and donate blood. In some cases, the blood bank asks the patient to arrange for someone to donate blood equivalent to the amount being transferred to the patient [23]. These are called alternative donations. Some blood banks do not collect blood by organizing blood donation camps, but

E. Muthu Kumaran
D.R. B.R. Amedkar Institute of Technology, Port Blair, Andaman & Nicobar Islands 744103, India

K. Velmurugan · P. Venkumar (✉) · D. Amutha Guka
Kalasalingam Academy of Research and Education, Virudhunagar, Tamilnadu 626126, India
e-mail: p.venkumar@klu.ac.in

V. Divya
Pondicherry University, Port Blair Campus, Andaman & Nicobar Islands 744103, India

G. Mathur et al. (eds.), *Proceedings of 2nd International Conference on Artificial Intelligence: Advances and Applications*, Algorithms for Intelligent Systems,
https://doi.org/10.1007/978-981-16-6332-1_12

they do order blood from a large blood bank at regular intervals. According to the WHO, voluntary, unpaid donations are considered safe.

In the current situation, it is difficult for a person who needs blood to receive blood at the right time. There are many problems with donating blood, such as having a donor spend too much time, not having the right equipment to manage blood collected from shelters through blood bank systems, failing to find the patient's compatible blood or platelets in an emergency, and limited days Inability to use blood before use [18, 19]. The primary purpose of a blood bank is to ensure that the appropriate blood and its components are available when a blood transfusion is needed. The functional responsibilities of a blood bank include planning, collecting blood, and responding to patient requests. It is important to control the scarcity of blood products in the blood bank, as surgeries can be delayed and, in some cases, even fatal, due to the lack of blood units available [23]. Waste of blood products is a serious problem because if one unit of blood is wasted in a particular blood bank it can cause deficiency in some other blood bank. Also, today if a person donates one unit of his blood, then he may not be able to donate blood again for the next few months, so wasting one unit of blood is tantamount to creating a unit shortage for the next few months.

One of the new smart tools is the Internet of Things (IoT), which combines the latest IT applications with the ability to integrate the devices and objects themselves into connected networks all over the world [20, 22]. At present it has become an effective and necessary tool in the market with its connected products and services and activities of daily living. Since its introduction, its popularity has grown and is now spoken of as Smart City, Smart Home, and Smart Health [19]. It has become widely used in health and wellness services [18]. Blood is collected and stored in glass bottles or plastic bags. But nowadays glass bottles are rarely used to collect blood because they have to be opened to separate the blood and they can easily break and lead to contamination. Blood to be stored as Whole Blood (WB) is collected in single bags, while double/three/four times the amount of blood to be stored is collected in the bag. The process of unifying whole blood presents three main components: Red Blood Cells (RBCs), platelets, and plasma. Some blood banks separate the fourth component, called cryoprecipitate, from the plasma fraction, which may be useful in patients with high concentrations of clotting factors and bleeding disorders such as hemophilia [21].

Blood transfusions are required for various surgeries, such as liver transplantation and open heart surgery, in cases of accident and leukemia. RBCs are required in cases where a patient has suffered from sickle cell anemia, gastrointestinal bleeding, and kidney failure, as well as severe blood loss conditions due to trauma and surgery. Platelets, essential for blood clotting, are provided to a patient suffering from bleeding disorders. Platelets are often given to patients suffering from leukemia and those undergoing cancer treatment or open heart surgery. Plasma is administered to trauma and burn victims and patients with coagulation disorders. However, plasma changes are less and less popular because the risks associated with plasma changes are high. Plasma changes can cause circulatory overload, allergic responses, and certain infections in a recipient's body. White Blood Cell (WBC) donations can cause adverse reactions in a recipient's body, which is usually not recommended.

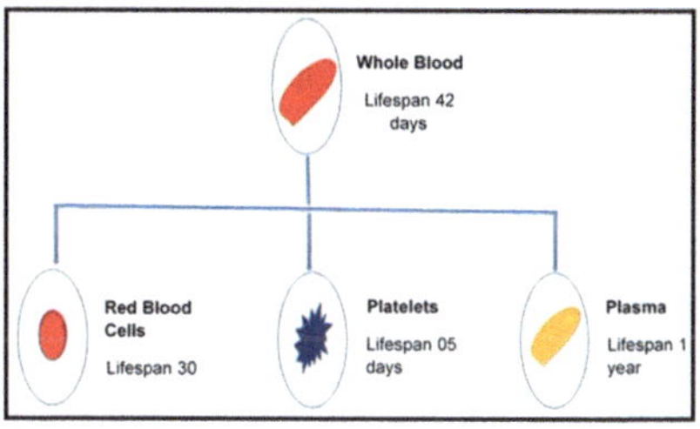

Fig. 1 Components of blood

Tactical decision-making process and operation of the blood banks is complex based on the following factors:

- More numbers of products due to various blood groups (8 groups) in the blood
- Combining whole blood consists of different components—as shown in Fig. 1
- Blood and its components perishability
- Using the standard useful lifetime
- Inconsistency in the supply of blood and its components
- Exchangeability between blood groups.

Due to the durability and complexity of the different blood components, the use of AI/IoT technologies to properly utilize it in a timely manner to avoid wastage is very important. With the IoT system, a specific blood component can be used to indicate their self-lifespan so that connected blood banks can effectively use rare blood. This will be a sign for the Coalition Service Group to make plans for blood collection and storage. This will help to manage the blood inventory system efficiently. Similarly, in the emergency situation, identification and communication to right blood donors is a very important life-saving mechanism achieved by artificial intelligence tools.

2 Literature Review

There is a large number of research articles published in the blood bank inventory and donors management system in the recent world. This section consists of the critical overview of the research articles based on our current research process. The advantages and difficulties of today, as well as approaches to avoiding the difficulties of using and integrating the IoT devices in healthcare settings, and presenting this discussion in the context of the REMOA program aimed at solving home care/telemonitoring for patients with chronic diseases are discussed [1]. A framework for realistically analyzing the inventory problems of blood banks at the individual hospital and regional level, and also the effects of various alternatives to multiple inventory policies is proposed [2]. Many mobile apps already exist for blood donation. With smartphones, blood banks cannot find the right donor for the patient's blood group. An integrated solution that connects donors and those in need with ease is proposed [3]. Then it is translated into a social list as a solution to eliminate the suffering of patients receiving blood in an emergency. They proposed (1) Determine the availability of blood and the nearest blood bank center based on the current location of the patients using GPS tracker. (2) Emergency message notification for registered mobile number. (3) Polio vaccine/dose reminder based on the age of the child [4]. The Short Message Service (SMS) and internet communication platform of the smart blood bank inventory management system are explained. From this research, they proposed the optimal communication framework between the blood bank and blood donor by sending the predefined SMS through the utilization of the Internet Communication Technology [5]

Comparison of the present and future blood bank communication systems is demonstrated. They introduced the challenges, applications, and opportunities to improve the blood bank inventory system and the conversion method of the desktop blood bank system into the mobile communication systems [6]. They discussed the data mining technique for identifying the classification of the blood donor with the utilization of the CART algorithm in weka software. Based on this implemented model predicting the voluntary blood donor behavior for organizing the blood donation camp and controlling the blood bank inventory system [7]. The analysis of real-time case study on implementation of the new blood donation center in Hong Kong city is described. In this research, two different sections are analyzed. The section one focuses on predicting the behavior, relationship among the blood donor's age, gender, and blood type, etc. based on the past donation data and the distance from the donation center. The next section focuses on selecting the best and suitable location for implementing the blood donation center in the city [8]. The new framework of the Blood donation and Blood Transfusion Services (BTS) with the utilization of the social media and smartphone applications is explained. In this research, a new model Blood Donation System (BDS) was implemented through the cutting edge information technologies of cloud and mobile computing. This BDS facilitates the smart communication between the blood donors, blood donation center, and health organizations [9]. The critical overview of the psychology of blood donor's behavior based on the past record of the blood donation is illustrated. In this research, they have addressed current research and future direction with the numerous challenges, and psychological factors in the blood donation [10].

The real-time case study of the BTS between the Hong Kong Red Cross BTS and public hospitals for predicting the major challenges in the anonymizing healthcare data analysis is explained. Based on this research, the new privacy model of BTS with the application of the anonymizing algorithm is proposed [11]. The smart blood bank inventory control system to organize the BTS between the blood bank and the public hospitals is described. Based on this research to share the information of the blood status in inventory, the effective blood utilization of the blood bank through the application of computerized communication system is implemented [12]. The smart blood donor community with recent mobile applications in the health care organizations is explained. Based on this research, the new model BTS with a mobile phone application for organizing the smart communications between the blood donors, blood banks, and local nearest health care organizations is implemented [13]. The past five year's implementations of the blood bank information systems in Macau is demonstrated. This research survey addressed the numerous challenges of blood bank information management in blood donation and BTS processes [14]. The blood donor recruitment, information retrieval, and management system through the mobile applications in the South East Asia Region (SEAR) is discussed. Based on this research, the smart privacy aware blood bank information to the quality management of BTS with blood donor recruitment, ensuring quality of blood and efficiency of the operation management in the health care organizations is proposed [15]. The optimal framework of the blood donor information management system with the utilization of the extended web application Technopedia is explained. From this research, the new model information management system for the better BTS and secure blood donor communication system through the technopedia application in the smart mobile phones is implemented [16]. A review and emphasis on big data technologies and IoT paradigms used in the healthcare industry are provided. The IoT-based framework of blood supply chain management is proposed to conceptualize the interactions between blood products, donors, patients, practitioners, processes, and service functional applications through network connectivity [17].

All real-time conditions associated with the blood bank's available blood flow are shown on the web page, so that the proposed system for the blood seeker to obtain blood from their nearby blood bank is divided into three sections, the first is the temperature sensor, the IR sensor nodes are installed on the blood bank rack, and the GSM module for sending blood requests to donors and blood banks All of this is connected to the Arduino Mega, the second section has a Wi-Fi module for data transfer to the server, and the third section shows the status of available blood cells. Blood is the most important medical substance, so it is important for the blood bank to manage it well and deliver blood to patients with minimal blood transfusion errors. However, because of the number of steps in the Blood Bank Management Guide, it is difficult to provide high levels of accuracy, reliability, and automation in blood banks in the blood storage and transfer process, so the proposed system improved the blood bank management and response time by linking all blood banks to cloud storage.

3 Smart Blood Bank Inventory System Design

The proposed system consists of a customized flexible arrangement in the existing infrastructure for storing whole blood and medical components, which are fabricated with medical grade materials as shown in Fig. 2. The size of the components can be measured using installed load cell sensors or by RFID technology, with special displays incorporated to check the availability of blood components without opening refrigerators that reduce CFC emission. The number and size of the rack are arranged according to their need, one large and the other small according to their need. The preset value can be programmed for an alarm/special indicator to check the availability of a specific blood group. A retrofit sensor module irrespective of the sensor module or by trapping the data through the provided inbuilt interfaces such as RS 232

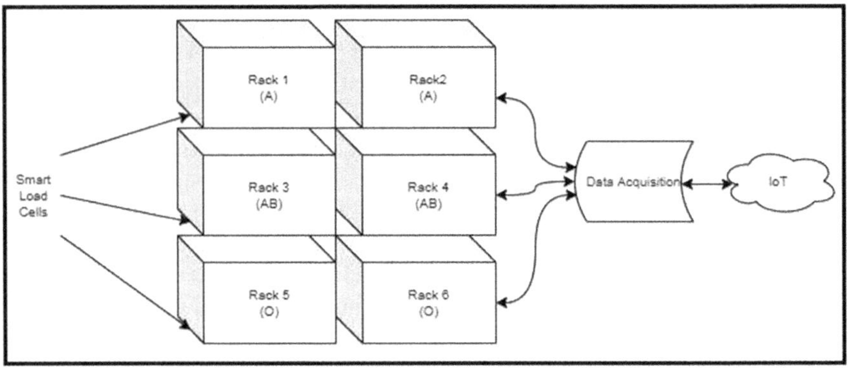

Fig. 2 Customization in existing refrigerators—Sensors/RFID, DAQ, IoT interfaces

or RS 485 in existing refrigerators was employed. All sensor values can be collected in real-time and monitored from any part of the world through IoT.

This is illustrated in Fig. 3. Where the entire system is connected to all government and private blood banks through IoT, the specially designed standalone modules are connected to the refrigerator interfaced through IoT. Based on a database available with NGOs and UIDAI, blood donors are easily identified when there is a need for a specific blood group. With the help of mobile network operators, an intelligent

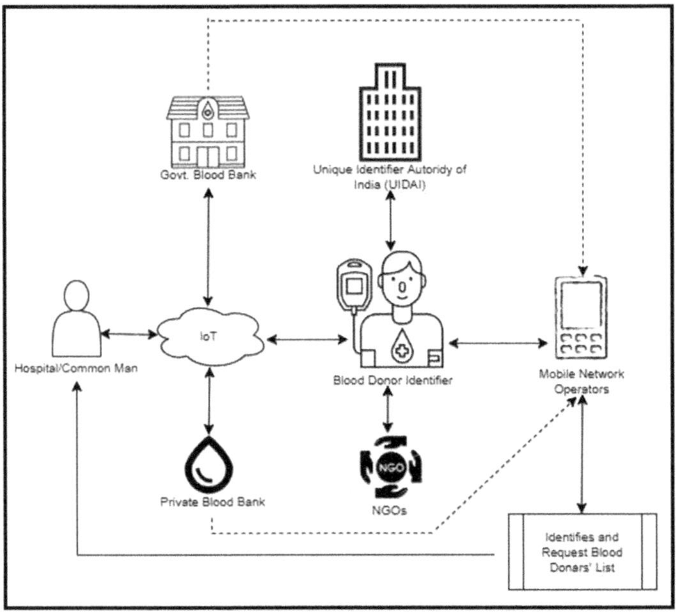

Fig. 3 Workflow of the system

system has been designed that searches for the donor with the closest radius, which reduces time and resources to save lives in the short term.

A technology with slight modifications to the existing systems in the collection and inventory system, and the IoT-based design, sensors, and software could alter the entire system and greatly reduce the mortality rate is proposed. The proposed change in the blood bank refrigerator or inventory system will enable the creation of a networking program for smart blood banks operating in a local region, where entire blood banks' data on the status of various blood products will be available through IoT in real-time. The apps/website will act as an information provider for any authentic person who needs blood, reducing the need to chase the blood banks whenever they want blood. It is also possible to create more donors in an approved manner.

The proposed system possesses three different services

- Blood Data Tracer.
- Database based-Non-Real-Time Donor Request.
- GPS-Based Real-Time Donor Request.

3.1 Blood Data Tracer

The real-time status of availability of each blood component in each blood bank can be obtained through IoT-enabled technology including the date of packing. To ensure the blood supply is as safe as possible, all the donors must meet the specific eligibility criteria outlined by the Food and Drug Administration. As soon as the test results are received, suitable blood units for transfusion are labeled and stored. Whole blood is separated into several components. Red Blood cells are stored in refrigerators at 6 °C for up to 42 days. Platelets are stored at room temperature in agitators for up to five to seven days. Plasma and cryo are frozen and stored in freezers for up to one year. As per AABB standards, the typical set-point of Blood Bank Refrigerator is 4 °C.

The proposed system incorporates NFC sensor technology that possesses low cost flexible Temperature sensor incorporated NFC tag at the blood components packets and IOT connected NFC readers are placed at various stages of blood components packing, storing, and transferring including transportation. As shown in Fig. 4, the details like date of packing, date of expiry, temperature of blood maintained at various stages can be traced by NFC sensor—IOT-enabled real-time data logging mechanism. Since the life-shell is the key factor for usage of blood components and at present these parameters are monitored manually, whereas the proposed technology will pave the way for optimize utilization of the blood components at right time which minimize the loss before the expiry date and monitor all the vital parameters at all stages as well in real-time. Temperature sensor inbuilt NFC tags are reusable and reduce the overall cost of the system. The access to temperature data with Android/IOS devices is also achieved through NFC. Fully passive, battery less mode, NFC-enabled sensor harvest energy from incoming RF emission; low cost, low power RF-based Temperature sensing sticker pasted in the blood component package is employed in

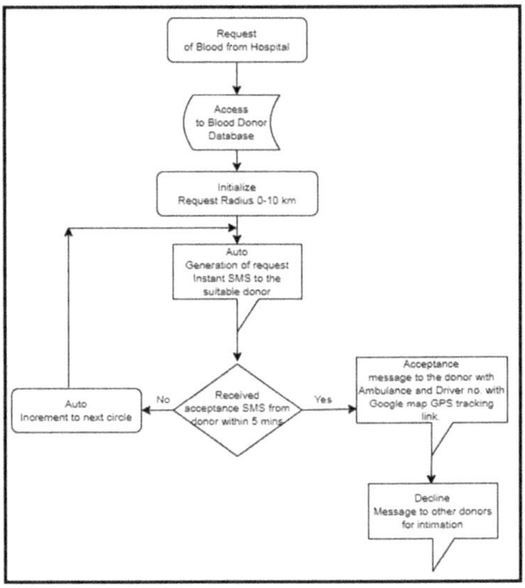

Fig. 4 Flowchart of database-based donor location request for blood donation

this project for blood data tracking and logging including other vital parameters of the refrigerators was monitored through indigenous module or tapping through the existing communication module in the refrigerators.

3.2 Database-based Non-real-time Location Donor Request

Figure 4 illustrates the mechanism to review the contact details of potential blood donors along with their willingness to donate blood that will be available in the database. The auto contact mode will assist in emergency cases to contact the willing donors of the specific blood group from the area close to the demand area from the database as shown in Table 1. If no response is received from any donors, it automatically switches to the next closest area circle for contact. An auto SMS will be generated for the request of donation of blood group with details of blood donation, on acceptance the donor will receive the ambulance number with source and destination details, time, and GPS tracking details of ambulance to the accepted donors. Other all contacted donors will receive an auto SMS with thanks and decline message mentioning already a donor accepted for blood donation

3.3 GPS-Based Real-Time Donor Request

The real-time GPS tracking of the potential donors will assist to locate the availability of the donors around the demand place and AI-enabled algorithm to contact those

Table 1 Database of the potential blood donors

Blood group	Donor ID	Donor name	Mobile No	Address	Distance from blood bank (Km)
B + ve	SMD528	A.Ramesh	94432xxx56	Chernamahadevi	26
O-ve	KIC221	P.Mani	97547xxx73	Kallidaikurichi	32
AB + ve	ASD654	D.Sampath	90434xxx54	Ambasamudram	35
B + ve	PEA334	R.Sami	87879xxx65	Pettai	6
O + ve	TEN733	K.Aravind	67373xxx76	Tvl Junction	4
AB-ve	TEN823	J.Kishore	88342xxx99	Palayamkottai	2
A + ve	SMD433	J.Dawood	97879xxx67	Pattamadai	22
...	...	...	...	...	...
...	...	...	...	...	...

closely available to the blood demanded hospital /place automatically as shown in Fig. 5. On acceptance the donor will receive the ambulance number with source and destination details, time, and GPS tracking details of the ambulance to the accepted donors. Other all contacted donors will receive an auto SMS with thanks and decline message mentioning already a donor accepted for blood donation as shown in Fig. 6.

An effective soft computing technique was applied for two main objectives to (1) predict the suitable and nearby blood donor requirements with the minimum travel distance and (2) analyze the stored blood with specific collect and stored date of blood in the blood bank. The widely used K Nearest Neighbor (KNN) AI algorithm was utilized in this research work to classify the available nearest blood donor's data with reference to the stored information of the blood bank, by measuring the Euclidean distance. From this algorithm, we classified the three different features or groups such as quick reach. Possible to reach and difficult to reach the blood donation center.

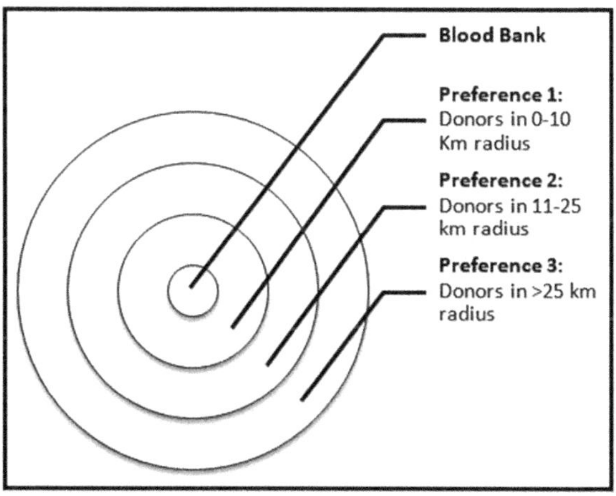

Fig. 5 AI Algorithm-based blood request from donor

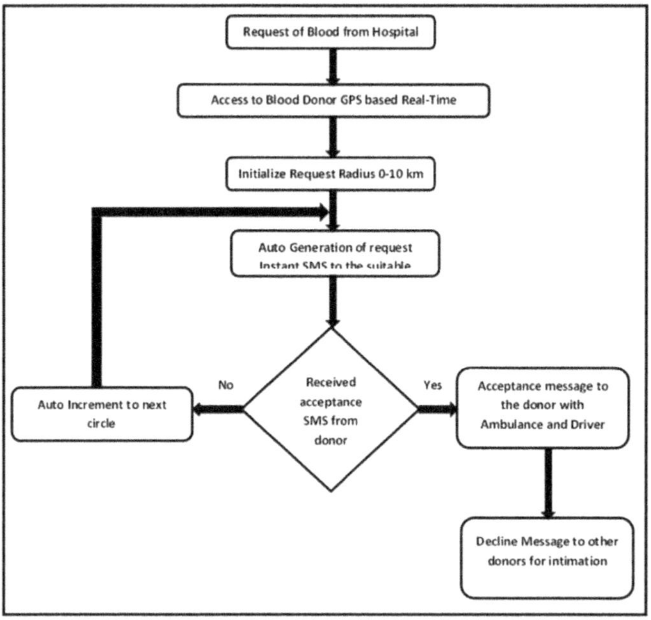

Fig. 6 Flowchart of GPS-based real-time donor location request for blood donation

Step 1: Locate the blood donation center by choosing the random place (K centroid) based on the stored blood donor's information of the blood banking system in the specific region (K = 1,2,3,…n).

Step 2: Measure the Euclidean distance (d) or Manhattan distance or Minkowski distance of the donor's exact locations through the GPS technique based on the stored mobile information.

$$\mathrm{d(X, Y, Z)} = \sqrt{(X_1 - X_2)^2 + (Y_1 - Y_2)^2 + (Z_1 - Z_2)^2} \quad (1)$$

$$\mathrm{SSEx} = \sum_{i=0}^{n} d(X_i - C_x),\ \mathrm{SSEy} = \sum_{i=0}^{n} d\left(Y_i - C_y\right),\ \mathrm{SSEz} = \sum_{i=0}^{n} d(X_i - C_z) \quad (2)$$

$$\mathrm{SSE} = SSEx + SSEy + SSEz \quad (3)$$

Step 3: Predict the nearest neighbor of the blood donor based on the minimum distance to travel to the blood donation center and classify the three different groups.

Step 4: Repeat step 1 to 3 until achieving the maximum solution or data changing the classifications.

4 Conclusion

We have implemented the smart framework with autonomous information sharing activity of the healthcare organization to the blood bank inventory management system. This AI-enabled IoT-based architecture organizes the sustainable and optimal BTS with minimum processing time in the blood banking system and customization of the blood storage refrigerator, ultra-freezer for plasma component storage compatible with IoT application to improve the availability of various blood products in a timely manner and then to reduce the wastage of blood and its components. The optimal and short-timely response of the smart blood bank system can save human lives. In this research, smart communications have been developed to connect donors through smartphone platforms and real-time GPS tracking. The proposed IoT technology makes the blood bank inventory system more efficient. The proposed system will be incorporated into the resource available for efficient management. The same concept can be implemented in home refrigerators also for auto order from supermarkets for grocery items, vegetables, and fruits based on the customer choice and demand.

References

1. Tarouco LMR, Bertholdo LM, Granville LZ, Arbiza LMR, Carbone F, Marotta M, De Santanna JJC (2012) Internet of things in healthcare: interoperability and security issues. In 2012 IEEE international conference on communications (ICC), pp 6121–6125
2. Jennings JB (1973) Blood bank inventory control. Manage Sci 19(6):637–645
3. Siruvoru V, Kumar NV, Kumar YBS (2019) Smart blood bank system using IOT. In International Conference on Computer Networks and Communication Technologies, Springer Singapore, pp 755–765
4. Sharma R, Soni P, Shah K, Panchal B (2016) Health care application for android smartphones using internet of things (iot). In 2016 3rd International Conference on Computing for Sustainable Global Development (INDIACom), IEEE, pp 1430–1433
5. Krishna GM, Nagaraju S (2016, August) Design and implementation of short message service (SMS) based blood bank. In 2016 International Conference on Inventive Computation Technologies (ICICT), IEEE Vol. 2, pp. 1–4
6. Pandit T, Niloor S, Shinde AS (2015) A survey paper on e-blood bank and an idea to use on a smartphone. IJCA 113(6):48–50
7. Santhanam T, Sundaram S (2010) Application of CART algorithm in blood donors classification. J Comput Sci 6(5):548
8. Cheng E, Chan CW, Chau M (2010) Data analysis for healthcare: a case study in blood donation center analysis. In AMCIS, p. 242
9. Mostafa AM, Youssef AE, Alshorbagy G. A framework for a smart social blood donation system based on mobile cloud computing. arXiv preprint arXiv:1412.7276
10. Masser BM, White KM, Hyde MK, Terry DJ (2008) The psychology of blood donation: current research and future directions. Transfus Med Rev 22(3):215–233
11. Mohammed N, Fung BC, Hung PC, Lee CK (2009) Anonymizing healthcare data: a case study on the blood transfusion service. In: Proceedings of the 15th ACM SIGKDD international conference on Knowledge discovery and data mining, pp. 1285–1294
12. Catassi CA, Peterson EL (1967) The blood inventory control system helping blood bank management through computerized inventory control. Transfusion 7(1):60–69

13. Bhowmik A, Nabila NA, Imran MA, Rahman MAU, Karmaker D (2015) An extended research on the blood donor community as a mobile application. IJ Wireless and Microwave Technologies 6:26–34
14. Li BN, Chao S, Dong MC (2007) SIBAS: a blood bank information system and its 5-year implementation at Macau. Comput Biol Med 37(5):588–597
15. Rahman MS, Akter KA, Hossain S, Basak A, Ahmed SI (2011) Smart blood query: a novel mobile phone based privacy-aware blood donor recruitment and management system for developing regions. In: 2011 IEEE Workshops of International Conference on Advanced Information Networking and Applications, IEEE, pp 544–548
16. Priya P, Saranya V, Shabana S, Subramani K (2014) The optimization of blood donor information and management system by Technopedia. Int J Innov Res Sci Eng Technol 3(1)
17. Valan JA, Raj EB (2018) Machine Learning and Big Data Analytics in IoT based Blood Bank Supply Chain Management System 4(12)
18. Fernandez F, Pallis GC (2014) Opportunities and challenges of the Internet of Things for healthcare: systems engineering perspective. In 2014 4th International Conference on Wireless Mobile Communication and Healthcare-Transforming Healthcare Through Innovations in Mobile and Wireless Technologies (MOBIHEALTH), IEEE, pp 263–266
19. Kang Y, Han M, Han K, Kim JB (2015) A study on the internet of things (IoT) applications. Int J Soft Eng Applic 9(9):117–126
20. Rajakumar G, Ananth Kumar T, Arun Samuel TS, Muthu Kumaran E (2018) IoT based milk monitoring system for detection of milk adulteration. Int J Pure Appl Mathem 118(9):21–32
21. Cohen MA, Pierskalla WP (1979) Target inventory levels for a hospital blood bank or a decentralized regional blood banking system. Transfusion 19(4):444–454
22. Pegels CC, Jelmert AE (1970) An evaluation of blood-inventory policies: A Markov chain application. Oper Res 18(6):1087–1098
23. Prastacos GP (1984) Blood inventory management: an overview of theory and practice. Manage Sci 30(7):777–800

Chapter 13
An Attention-Based Medical NER in the Bengali Language

Tanvir Islam, Sakila Mahbin Zinat, Shamima Sukhi, Zakir Hossain Zamil, Aynur Nahar, and M. F. Mridha

1 Introduction

Named Entity Recognition (NER) is a task where it identifies entity type according to pre-training. Many researchers do NER for various purposes. Some tried to identify people's names, locations, objects, and so on, and also some researchers do NER with a particular intention to make efficient any particular task by NER. According to recent years, NER has been done in the medical sector, also in the biomedical sector with machine learning and deep learning. We can collect data for the advancement of our technology and can simplify those data for a better target and the NER method is one of them.

Deep learning methods are the most popular in the NLP sector and especially most of the NLP tasks are done by sequential model. Those models are Long Short-Term Memory (LSTM) [1], Bidirectional LSTM (BiLSTM) [2, 3], Bidirectional Gated Recurrent Unit (BiGRU) [4], Bidirectional Convolutional Neural Network (Bi-CNN) [5], and so on. Many models are being used with the Conditional Random Field (CRF) layer in deep learning also to predict the final output or target label. In medical sectors and telemedical services, who faces patient for the first time they try to understand their problem and then refer to a specialist according to the patient's problem. So, they do this rapidly and need to think and their thinking level depends on their experiences and learning about that which specialist they have to refer. On the other hand, a well-educated person can understand which specialist he or she needs based on their medical problem but in the rural area and the people who are not much educated they couldn't understand which specialist they should go to according to their medical problem.

T. Islam · S. M. Zinat · S. Sukhi · Z. H. Zamil · A. Nahar · M. F. Mridha (✉)
Department of Computer Science and Engineering, Bangladesh University of Business and Technology, Dhaka, Bangladesh
e-mail: firoz@bubt.edu.bd

G. Mathur et al. (eds.), *Proceedings of 2nd International Conference on Artificial Intelligence: Advances and Applications*, Algorithms for Intelligent Systems,
https://doi.org/10.1007/978-981-16-6332-1_13

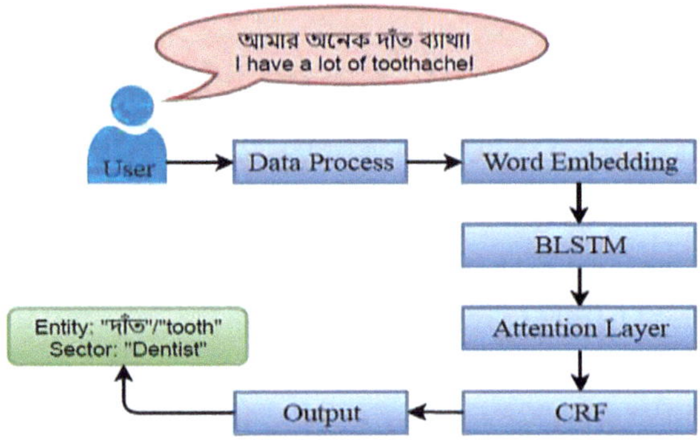

Fig. 1 Applying the attention-based model for telemedicine consultancy

So, we try to solve those kinds of issues and make this system much more efficient with artificial intelligence with the technique of NER where a patient's problem statements automatically identify which kind of specialist the patient needs based on pre-trained symptoms and disease labeled entities in the Bengali language. Figure 1. illustrates better visualization of our task.

In this paper, our contributions are

- patients can easily know about their specialization sector in medical based on their problem.
- this research will make the medical consultancy service and referring service more convenient.

This paper has been partitioned in this way; some previous recent works which are based on attention-based NER and especially for the medical sector's task have alluded shortly in Sect. 2. We have made a telemedicine dataset for this paper described in Sect. 3 when patients' messages will be tokenized for finding specialists. We proposed our attention-based BiLSTM-CRF model in Sect. 4. The experimental setup is described with dataset and hyperparameter in Sect. 5. The result analysis of our proposed model has been described with accuracy level in Sect. 6. And finally, we mention the challenges of our research work including baseline and further plan in Sect. 7.

2 Related Work

Named Entity Recognition (NER) is an information extraction or consolidation technique of automatically identifying entities in text and classifying them into predefined categories. It is also called entity identification. We built a research work that can be useful in every sector where there needs to extract main particular keywords to find a better result. NER is used in various sectors for information extraction tasks.

Weihua Wang et al. developed an attention-based NER for Mongolian News [6]. This NER is done before with the method of the traditional Bi-LSTM-CRF joint model. There the accuracy and sentence tagging inconsistency are so poor because of Mongolian word characters but by adding attention base the newly developed model is so good than before according to accuracy improvement and the way of training is much easier.

Li et al. proposed for eliminating semantic interference and improving the ability of autonomous learning of internal features [7]. They integrated BiLSTM-Att-CRF which is an improved method of named entity recognition for Chinese electronic medical.

Gao et al. proposed a research paper about extracting Chinese Electronic Medical Records which is done in English [8]. A practical annotation model on Resident Admit Notes (RANs) which can specifically extract medical entity is proposed in this research.

Another research work was proposed by Xiaoyu Han et al. that identifies the relationship between two specified entities in a sentence [9]. They generated a large-scale Chinese relation extraction using the attention-based model.

Fan et al. proposed a research work that extracts geological hazard named entities from the important body of geological hazard literature [10]. They built a geological hazard knowledge graph based on the NER model; namely, the deep, multi-branch BiGRU-CRF model that makes a combination between multi-branch Bidirectional Gated Recurrent Unit (BiGRU) layer and a Conditional Random Field (CRF) model.

Another research work has been done by Li et al. where they proposed the BiLSTM-Attended–CRF model for EMR in the Chinese language [11].

Hao Wei et al. established a research work for Biomedical Named Entity Recognition through the BiLSTM-CRF model for JNLPBA corpus with a 73.50 F1 score [12].

Keyword extraction for online medical questions especially in the Chinese language is a challenging part, as well as this task, has been done by Chaochen Wu et al. by establishing a BiLSTM-CRF model [13].

BioCreative corpus represents a vital role for Chemical Named Entity Recognition. Ling Luo et al. proposed a research work for the Chemical Named Entity Recognition task through the Att-BiLSTM-CRF model [14].

Guohua Wu et al. proposed a research work for the Chinese Clinic Named Entity Recognition task through the Att-BiLSTM-CRF model for Chinese CNER task and also for capturing graphical knowledge of Chinese characters [15].

Yang Pei et al. established an attention-based BiLSTM-CRF architecture Drug and Chemical Compound Named Entity Recognition task through BioCreative corpus [16].

We analyzed many related papers and notice that the telemedicine consultancy task in the Bengali language has not been established yet. As we have many regional populations, if we built this task in the Bengali language, it will benefit those who are not educated and have no idea of essential medical issues.

3 Dataset

We collected some telemedical patients' data where the patients say their problems at the first consult on a phone call or message. We gathered around 190 patients' issues. Then, we structured the data manually on the table and tokenized them. There are around more than four thousand words we have gotten. The structured information is labeled manually with some tags.

3.1 *Structuring the Dataset*

Here, we pick only those lines when a patient says their problem or symptoms initially. As like "আমার তিন দিন যাবত অনেক জ্বর"it means in English that " I am suffering from fever for last three days". Those lines are separated according to the patient then each patient problem statement is separated into words.

3.2 *Tokenization and Lemmatization*

After getting words separated, each word lemmatization into a base word like Fig. 2. In the English language, there are so many options for lemmatization. But in the Bengali language, there are only a few ways to tokenize them and lemmatize or

Table 1 Illustrates some sample of raw data before processing that are collected from telemedicine consultancy

নামঃ লিজা, লিঙ্গঃ মহিলা, বয়সঃ ২৩, সমস্যাঃ আমার নাকে আর থুতনিতে সাদা বিচির মত উঠে আর মুখ অনেক বেশি খশখশা হয়ে থাকে। Name: Liza, Gender: Female, Age: 23, Problem: My nose and chin rises like whiteheads and my face is very rough.
নামঃ জুয়েল, লিঙ্গঃ পুরুষ, বয়সঃ ২০, সমস্যাঃ আমার মায়ের অনেক মাথা ব্যথা করে। ঘাড় ব্যথা করে, শ্বাসকষ্ট হয়। Name: Juel, Gender: Male, Age:20, Problem: My mom has a lot of headaches. Neck pain, shortness of breath.
নামঃ সাদিয়া, লিঙ্গঃ মহিলা, বয়সঃ ৩০, সমস্যাঃ আমি অতিরিক্ত টেনশন করলে ঘুম ঠিকমত হয় না, অনেক বেশি অস্থিরতা কাজ করে। Name: Sadia, Gender: Female, Age: 30, Problem: when I have extra tension, I cannot sleep properly and much more restlessness works.
নামঃ মুন্না, লিঙ্গঃ পুরুষ, বয়সঃ ৪৩, সমস্যাঃ আমার ডান গালে ফুলে লাল হয়ে আছে। আর মুখের এই জায়গাটায় হলদে পুঁজ জমে আছে। Name: Munna, Gender: Male, Age: 43, Problem: My right cheek is swollen and red. And there is yellow pus in this part of the mouth.
....

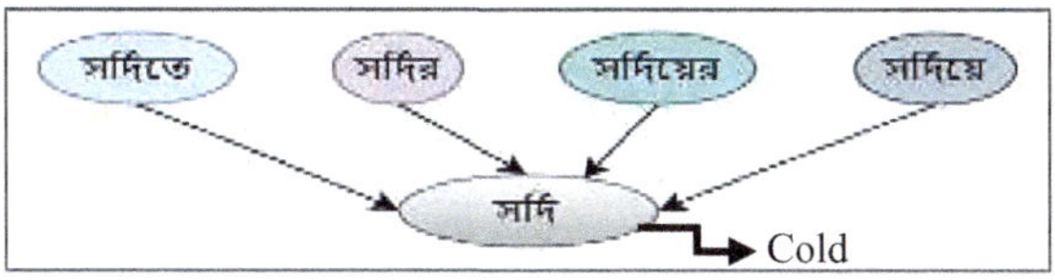

Fig. 2 Lemmatization of words

steam them into the base word. For getting an accurate word matrix or turning the words into a numeric value, we avoid the same word variation form.

4 Methodology

We illustrate an Attention-based BiLSTM-CRF model for detecting entities in this paper. This model is working so well in NLP sectors especially on NER according to our review. When the Attention layer work with BiLSTM-CRF any sequential data prediction accuracy will increase. Our model working procedure is written below stepwise,

1. First, we prepared our annotated data.
2. Make every word into some vector sequentially.
3. We then put our sequential vector to the embedding layer, making it easier to understand the learning layers.
4. We are getting input from the embedding layer to the BiLSTM layer.
5. After getting the signals from the BiLSTM layer, the attention layer combines all signals and makes it easier to focus on one.
6. Then, the output will decide from the CRF layer by normalizing and solve some bias issues.

The whole working procedure has been shown in Fig. 3.

4.1 BiLSTM

This model is called sequence modeling in RNN. Here, we have two LSTM one is for the forward direction and the other is for the backward direction. By iteration, the algorithm will understand the available data or context and learn properly about the dataset that what to predict for one entity. The forward LSTM computes the sequence representation h_p from the left side to the right side for every single word p and the other backward LSTM computes the same sequence representation h_p from right to left side which has been shown in Fig. 4. The two hidden layer vector of LSTM is formed by concatenating as $h_p = [\overrightarrow{h}_p; \overrightarrow{h}_p]$. The basic functions of BiLSTM are

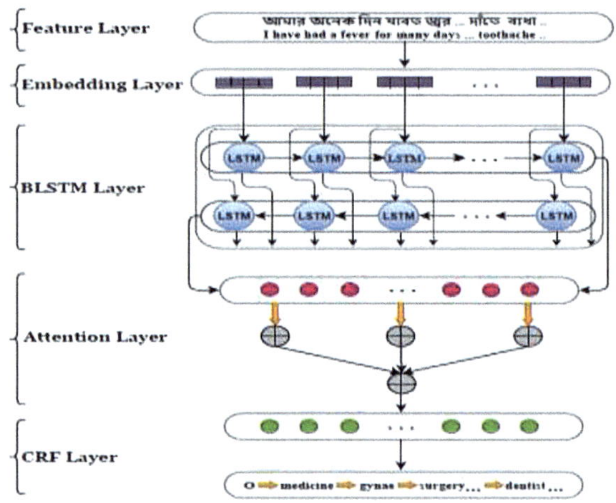

Fig. 3 The attention-based BiLSTM-CRF model for the Bengali language

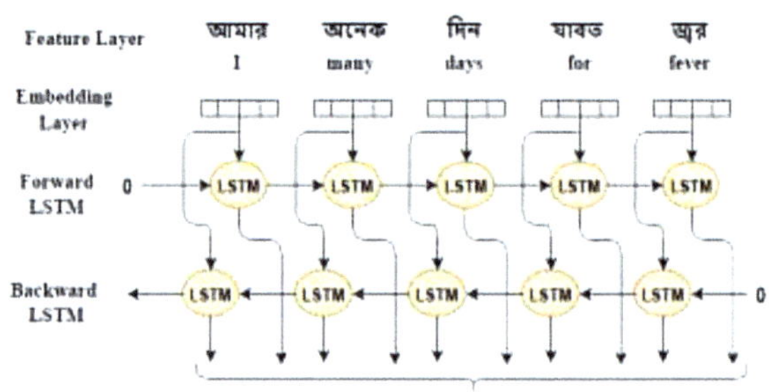

Fig. 4 Bidirectional LSTM

$$\begin{bmatrix} \tilde{c}_p \\ f_p \\ o_p \\ i_p \end{bmatrix} = \begin{bmatrix} \sigma \\ \sigma \\ \sigma \\ tanh \end{bmatrix} \left(W^p \begin{bmatrix} x_p \\ h_{p-1} \end{bmatrix} + b \right) \tag{1}$$

$$C_p = i_p \odot \tilde{c}_p + f_p \odot c_{p-1} \tag{2}$$

$$h = o_p \odot \tanh(c_p) \tag{3}$$

Here, i_p denotes the input gate; f_p denotes the forget gate; o_p denotes the output gate; W^p and b represent the trainable parameters; σ denotes the nonlinear sigmoid function; is the dot product function; x_p represents the input vector of the present time step.

4.2 Attention Layer

The attention mechanism is useful for extracting more features in NLP. It gives more attention to certain words that are relative to entity vocabulary by assigning various attention probability weights. The attention layer is used for connecting the work of classification. h_p represents a matrix that contains output vectors as $[h_1, h_2, \ldots, h_N]$, where N is considered as the sentence length. γ is formed by the representation of a weighted summation of the vectors.

$$M = \tan(h_p) \tag{4}$$

$\alpha = \text{softmax}(\omega^T M)$, Here, ω^T is considered as a transpose.

$$\gamma = h_p \alpha^T \tag{5}$$

So, for classification, the final sentence-pair representation is calculated as

$$h* = \tan h(\gamma) \tag{6}$$

4.3 CRF Layer

This layer measures which predicted output is valid according to the predefined dataset. The primary CRF layer is applied on the top of $h^* = h^*_1, h^*_2, \ldots, h^*_p$. Label sequence $y = l_1, l_2, \ldots, l_p$.

So the probability of

$$y = P(y|s) = \frac{exp(\sum_j (W^{l_j} h_j + b^{(l_{j-1}, l_j)}))}{\sum_y exp(\sum_j (W^{l_j} h_j + b^{(l_{j-1}, l_j)}))} \tag{7}$$

Here, $W^{l_j}{}_{CRF}$ is a model parameter that is specific to l_j; y denotes the arbitrary label sequence; $b^{(l_{j-1}, l_j)}$ is considered as bias specific to l_{j-1} and l_j.

5 Experimental Setup

5.1 Dataset

Our dataset has nine unique labels or tags that we annotated manually. The label quantity for visualization is shown in Fig. 5. except the label "O" which represents

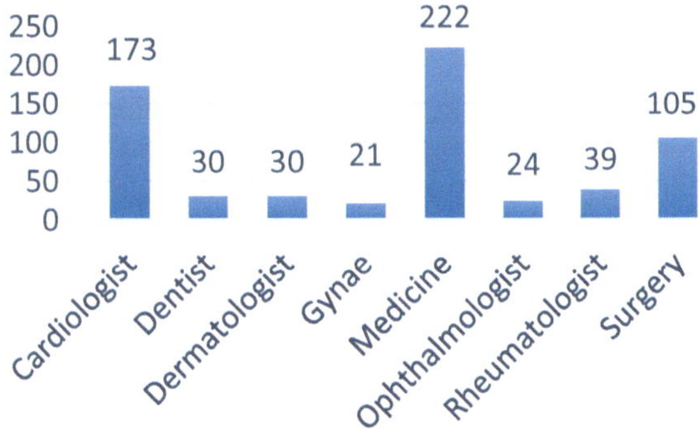

Fig. 5 The dataset's target label

Table 2 Indicates the summary of our dataset

Patient	190
Word	4095
Unique label	9
Label names	O, cardiologists, dentist, dermatologist, gynae, medicine, ophthalmologist, rheumatologist, surgery

non-annotated words. Our overall data are described in Sect. 3 that how we process and filter our dataset. We illustrate Table 2 for a summary of our dataset in detail.

5.2 *Hyperparameter*

In deep learning models for sequential data learning and specifically, we can say the models which are worked on the NLP sector are all known who work for NLP learning tasks. So, we have seen BiLSTM-CRF work so well in the NER sector and we set our hyperparameter by testing severally and use a built-in API "GridSearchCV" from Sklearn for setting up our hyper-parameter is shown in Table 3.

Table 3 Illustrates the hyperparameter of our model

Parameter	Value
Dropout	0.02
Learning rate	0.001
Optimizer	"rmsprop"
Batch size	12

Table 4 Represents the accuracy level of our analysis

Method	F1-score (%)
Conditional random field	90.00
BiLSTM-CRF	94.00
Attention-based BiLSTM-CRF	**95.60**

6 Result Analysis

This section represents our accuracy, data measurement, and some model element that we set for our task. We got an F1 score on Machine learning 90%, and on the deep learning model we got 94% and when we work with the attention layer it rises 95.60%. Though we got lower accuracy than 95.60%, after increasing the number of data in our dataset, the accuracy increases by 95.60%.

We represent our accuracy which has been shown in Table 4. We have nine target categories but our dataset is not so big so it was quite challenging to get better accuracy.

7 Conclusion

There are many variations in the Bengali language that we faced, like regional changes. So, data processing is quite challenging for us, and make the data suitable to feed into the NER model. In the future, this work can be updated using knowledge-based NER that can also predict which medication the patients need according to the patients' problem. There are a few works on the Bengali language in the medical sector. This paper illustrates that an attention-based BiLSTM-CRF NER model through a telemedicine dataset will make the manual service of telemedicine and other medical sectors more efficient.

References

1. Tang B et al (2019) Entity recognition in Chinese clinical text using attention-based CNN-LSTM-CRF. BMC Med Inf Decs Making 19(3):74. https://doi.org/10.1186/s12911-019-0787-y
2. Naseem U et al (2020) Biomedical named-entity recognition by hierarchically fusing biobert representations and deep contextual-level word-embedding. In: 2020 international joint conference on neural networks (IJCNN). IEEE. https://doi.org/10.1109/IJCNN48605.2020.9206808
3. Patel H (2020) BioNerFlair: biomedical named entity recognition using flair embedding and sequence tagger. arXiv preprint arXiv:2011.01504
4. Li Y et al (2019) A joint model of clinical domain classification and slot filling based on RCNN and BiGRU-CRF. In: 2019 IEEE international conference on big data (big data). IEEE. https://doi.org/10.1109/BigData47090.2019.9005449

5. Zheng J, Zheng L (2019) A hybrid bidirectional recurrent convolutional neural network attention-based model for text classification. IEEE Access 7:106673–106685. https://doi.org/10.1109/ACCESS.2019.2932619
6. Tan M et al (2019) An attention-based approach for mongolian news named entity recognition. In: China national conference on Chinese computational linguistics. Springer, Cham. https://doi.org/10.1007/978-3-030-32381-3_35
7. Li L et al (2019) An attention-based deep learning model for clinical named entity recognition of Chinese electronic medical records. BMC Med Inf Decis Making 19(5):235. https://doi.org/10.1186/s12911-019-0933-6
8. Gao Y et al (2019) Constructing a Chinese electronic medical record corpus for named entity recognition on resident admit notes. BMC Med Inf Decis Making 19(2):67–78. https://doi.org/10.1186/s12911-019-0759-2
9. Han X et al (2020) An attention-based model using character composition of entities in Chinese relation extraction. Information 11(2):79. https://doi.org/10.3390/info11020079
10. Fan R et al (2020) Deep learning-based named entity recognition and knowledge graph construction for geological hazards. ISPRS Int J Geo-Inf 9(1):15. https://doi.org/10.3390/ijgi9010015
11. Li M et al (2019) Named entity recognition in Chinese electronic medical record using attention mechanism. In: 2019 international conference on internet of things (iThings) and IEEE green computing and communications (GreenCom) and IEEE cyber, physical and social computing (CPSCom) and IEEE Smart Data (SmartData). IEEE. https://doi.org/10.1186/s12911-019-0767-2
12. Wei H et al (2019) Named entity recognition from biomedical texts using a fusion attention-based BiLSTM-CRF. IEEE Access 7:73627–73636. https://doi.org/10.1109/ACCESS.2019.2920734
13. Wu C et al (2020) An attention-based multi-task model for named entity recognition and intent analysis of Chinese online medical questions. J Biomed Inf 108:103511. https://doi.org/10.1016/j.jbi.2020.103511
14. Luo L et al (2018) An attention-based BiLSTM-CRF approach to document-level chemical named entity recognition. Bioinformatics 34(8):1381–1388. https://doi.org/10.1093/bioinformatics/btx761
15. Wu G et al (2019) An attention-based BiLSTM-CRF model for Chinese clinic named entity recognition. IEEE Access 7:113942–113949. https://doi.org/10.1109/ACCESS.2019.2935223
16. Pei Y et al (2018) An attention-based approach for chemical compound and drug named entity recognition. J Comput Res Dev 55(7):1548. https://doi.org/10.7544/issn1000-1239.2018.20170506

Chapter 14
Malware Classification Based on Various Machine Learning Techniques

Vinay Gautam and Prabhjot Kaur

1 Introduction

Malware is the combination of two words malicious and software [1]. Malware is an executable file which when executed on the computer can harm computer. It is used to perform harmful actions on computer systems [2]. These malwares directly impose security threats to individuals, governments, and financial institutions. One of the fundamental explanations behind this great capacity of tests is broad utilization of confusion procedures by inventors, which implies that malicious software from the same category is always altered to make a new one [3, 4]. The malware is of different types and found in different forms and factors. Many attributes are used to define these forms and factors. These attributes are used to classify malwares. The malware classification is done by using various machine learning techniques such as Decision Tree, Multi-layer, and Multi-SVM. The aim of this paper is to focus on different machine learning and to show how these algorithms performed on a given dataset. Keeping in mind the end goal is to adapt to quick advancement, it is necessary to create malware classification techniques that are tolerant of variations of malware records that have a place within the same family. This paper presents an Xgboost classifier system for malware classification. Recent research in malware classification recommends that malware samples that come under same family have same traits, i.e., most new malware are variants of existing ones. Henceforth, the possibility of building a technique that can proficiently classify malware in view of its family independent of being a variation appears especially profitable and a way

V. Gautam (✉) · P. Kaur
Chitkara University Institute of Engineering and Technology, Chitkara University, Rajpura, Punjab, India
e-mail: vinay.gautam@chitkara.edu.in

P. Kaur
e-mail: prabhjot.k@chitkara.edu.in

G. Mathur et al. (eds.), *Proceedings of 2nd International Conference on Artificial Intelligence: Advances and Applications*, Algorithms for Intelligent Systems,
https://doi.org/10.1007/978-981-16-6332-1_14

to deal with the drastic growth of malware [5]. The classifier is tested on a dataset of Microsoft provided by the Microsoft in a Kaggle competition. The paper is organized as follows. Starting with Sect. 2, deals with previous work in detail, Sect. 3 describes our methodology, including dataset, feature extraction, and classifier tuning. Results are compiled in Sect. 4 and Future work and Conclusive section is described in 5.

2 Related Work

This section explains various machine learning techniques used for malware classification based on their behaviors. Different algorithms are used to classify malware such as Association Rule, Support Vector Machine, Decision Tree, Random Forest, Naive Bayes, and Clustering. Some of these are described in this section.

The concept of data mining and static feature such as Portable Executable (PE), strings, and byte sequences are used in [23] to classify malware. A Naive Bayes algorithm is used in [24] to detect different patterns in DLL file with an accuracy of 97% [25]. n-gram and data mining techniques are used to classify malware in executable files. Image processing technique is used to detect malware using grayscale image and K-NN algorithm is used to classify malware [26]. A novel technique is used in [27] to classify packed and unpacked malwares. A function call graph and discriminate distance metric learning is used to classify malware based on structure [28]. References [29, 30] function length frequency is used to classify Trojan. WEKA machine learning algorithm is used using WEKA library for classification [31]. Semi-supervised learning algorithm is used to classify malwares in [32]. Various techniques are described above which classify malware correctly. Now next part of the paper covers description of techniques.

A. Multiclass classification

More than two classes are used in this type of classification problem. Each training data point in the dataset can only belong to one of the N classes defined. The objective is to formulate a function which, when fed with a new data point, will accurately predict the class to which the new data point belongs. In its most basic form, multiclass classification problem decomposes itself into a set of smaller binary problems, which can be solved using our techniques for binary classification. Henceforth, techniques like K-Nearest Neighbors, Random forest, Logistic regression, and XgBoost can be used to perform the classification [2].

B. Assembly files

Domain knowledge is required to apply feature extraction on assembly files. These files contain the actual instruction that the CPU understands. .asm is the file extension for assembly files. Assembly files have a sequence of human readable instructions that map usually 1:1 to machine instructions. When fed to the assembler it generates machine code [3].

III. K-Nearest Neighbor (K-NN)

The primary capability of K-NN is to store all available instances and distinguish them into new cases primarily based on similarity measurements (distance, weight). When "k" is a very small quantity, K-NN can overfit, it'll classify based totally on nearest acquaintances in preference to gaining knowledge of an awesome separation between classes. However if "okay" is a completely big variety, K-NN will lower, if k = n K-NN in the variety thinks that each point belongs to a class that has greater samples [4]. The CNN-based novel technique is used in [20] to detect both packed and unpacked types of malware. In this CNN is used to detect malware based on sematic-based image data. In [21], a hybrid intelligent technique is used to detect malware of n order non-linear system using Neural Network. In [22], a CNN technique is proposed to identify and classify malware and to convert a malware binary to an image is introduced and then NSGA-II is used to detect it.

IV. Random Forest

Random wooded area tree is a machine learning algorithm based totally on choice bushes. It act as ensemble learning method or ensemble class in ML algorithm that 'Ensemble 'class. via Ensemble we imply (In Random wooded area Context), distinctive selection of various decision timber. In RFT (Random woodland Tree), we make a prediction approximately a class, based now not best on a decision tree, but an (almost) unanimous prediction, made by means of "k" decision timber. In Random forest, the prediction (a group of "okay" decision timber) is honestly ensemble, i.e., for every choice tree, expect the class of instances after which return the magnificence that became predicted maximum regularly.

E. XgBoost

It is basically supposed to increase the peak gradient. Its miles binary is optimized and simply designed to increase the tree set of rules. Its main function is to examine the extremes of the computation limits of the gadget and to provide a correct, portable, and scalable for tree growing. Xgboost version is a linear combination of choice timber. To make a prediction Xgboost computes predictions of person timber and provides them [5].

3 Methodology

The experiment is performed using python and libraries of python like SciKit-Learn, Pandas, NumPy, MatPlotLib, and SeaBorn.

A. Dataset

A Kaggle competition for malware classification is conducted by Microsoft in 2015 [6]. The dataset released in the competition is of 200 GB and it is bifurcated into two parts in which 150 GB are .asm files and rests are .bytes files. Each sample of the malware can be distinguished into one of the nine different malware classes. The malware sample in the training data is not uniform that is the number of malware samples of some families significantly outnumbers the samples of other families. For each malware in the data set we have two files .bytes file (It contains a hexadecimal representation of the binary content of the file) and .asm file (Assembly file, it contains the assembly level features like address, segments, opcodes, registers, function calls). There are a total of 10,868 .bytes file and a total of 10,868 [7] .asm files making a total of 21,736 data points. The y-axis obtained shows the count of files, and the x-axis shows the respective classes [8, 9]. Class 5 has the least number of files while Class 3 has the greatest number of files. We divided dataset randomly into three parts train, cross-validation, and test 64, 16, 20% of data, respectively.

B. Performance Metric

We use multi-class log-loss as performance metrics. Suppose we have dataset having "m" data points and c different classes, then multi-class log-loss is

$$-\frac{1}{n}\sum_{i=1}^{m}\sum_{j=i}^{c} y_{ij} \times \log P_{ij}$$

where Y_{ij} is 1, if x_i belongs to Class j else 0 and is the ground truth value that is given in the training data. P_{ij} is the probability predicted by the model when x_i belongs to Class j [10]. Figure 1 shows Bar Graph and Fig. 2 shows Box Plot.

III. Feature Extraction

This part describes the graphical view of feature extracted under .byte file.

Fig. 1 Bar graph

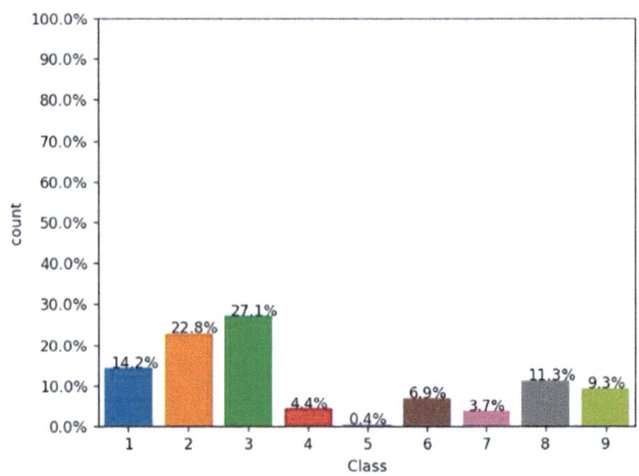

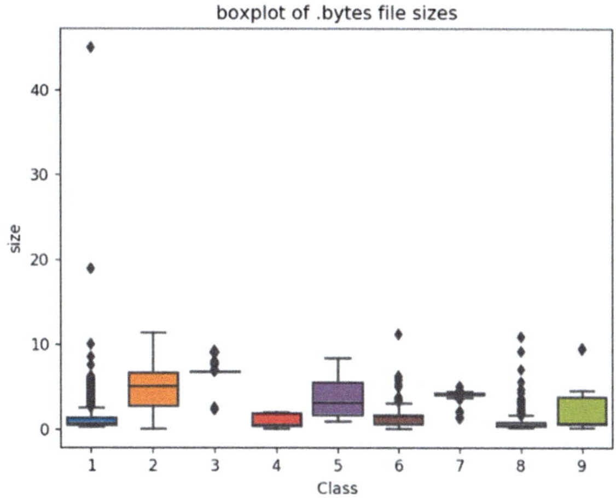

Fig. 2 Box plot

IV. Classifier Tuning

As know that multi-class log-loss can vary between 0 and infinity so to find the worst possible log-loss a random model is used here. In random model, first is to generate the random values of the Pij and then normalize the values. The log-loss is 2.485039 and the number of misclassified points is 82.5%. It should be taken care that the worst case should not exceed.

In this experiment, first applied classifiers on .bytes files only. First K-Nearest Neighbor (K-NN) is used to classify and calibrate the classifier (sigmoid function) to get the exact probability values [11]. After hyperparameter tuning, it has noted that K = 1 is the best hyperparameter. The y-axis shows the size of files and the x-axis shows the class. The box-plot shows that the class 6 and class 5 are completely separated, whereas there is some overlapping in some classes. Therefore, the file size can be used as a feature because some of the classes can be separated based on size [12]. Two folders are created, i.e., BYTES and ASM. The respective files are stored after prior categorization in .bytes files, removed the starting address and encoded each hexadecimal pair as a word. Each hexadecimal pair is used as a unigram which has two 256 different pairs making a total of 257 features [13]. A custom code is used to calculate the bag-of-words. From asm files, 52 features are extracted by acquiring domain knowledge and also size of .asm files as per derived making a total of 53 features. Here, multithreading is used and the dataset is sub-divided to count the bag-of-words [14]. Thus, a total of 310 features were in hand and used t-SNE with perplexity 30 and 50 for multi-variant analysis of the dataset.

After t-SNE all the 310 features can be used for classification. Each color represents different class ranging from 1 to 9. Cross-validation log-loss came out to be 0.225386237304, training log-loss is 0.0782947669247, and testing log-loss is 0.241508604195 [15]. The difference between the training and testing is large which suggests that K- NN is over fitting. After applied logistic regression and with alpha 100 it has found that log-loss for cross-validation is 0.549929846589, log-loss for training data is 0.498923428696, log-loss for test data is 0.528347316704. In case of very few features (257), the random forest classifier can be used. In this, all the hyperparameters [6] are kept as default and just tweaked with the number of trees [16]. In this scenario, when the number of trees is 1000, the train log-loss is 0.0266476291801, cross-validation log-loss is 0.0879849524621, and test log-loss is 0.0858346961407. It is quite an improvement so Xgboost in order is best and check whether to improve log-loss or not. In this, the XgBoost is used with random search for hyperparameters [7] and different hypermeters are set as follows "subsample: 1, 'n_estimators':500,'max_depth':5,'learning_rate':0.05,'colsample_bytree':0.5". The training log-loss is 0.022540976086, cross validation log-loss is 0.0928710624158, and test log-loss is 0.0782688587098.

Then various classifiers are applied only on .asm files. During feature extraction, it has found that there are 53 features. Using box-plots, it has found that each feature has its own unique importance for the classification [17]. Then K-Nearest Neighbor is applied with best K = 3 and found that log-loss for training data is 0.0476773462198, log- loss for cross validation data is 0.0958800580948, and log-loss for test data is 0.0894810720832. Logistic Regression with Stochastic gradient descent and found out that for alpha 100 the log-loss for training data is 0.396219394701, log-loss for cross validation data is 0.424423536526, and log-loss for test data is 0.41568559251 7. The training log-loss is 0.0102661325822, cross-validation log-loss is 0.0501201796687, and test log-loss is 0.0483908764397. Random forest and Xgboost are applied on both .bytes and .asm files together [18, 19]. There are a total of 310 features and applied Random forest classifier. The best value for number of trees is 3000 and the training log-loss is 0.0166267614753, cross-validation log-loss is 0.0355909487962 and the test log-loss is 0.0401141303589. Xgboost with random search for parameters is set as follows "'subsample': 1, 'n_estimators': 1000, 'max_depth':-10,'learning_rate': 0.15, 'colsample_bytree': 0.3". The training log-loss is 0.0121922832297, cross-validation log-loss is 0.0344955487471, test log-loss is 0.0317041132442. It can be concluded that Xgboost with random search for parameters classifies the malwares in the best possible way. The logistic regression is the worst performing classifier.

4 Result and Discussion

In this paper, the multi-class log-loss and percentage misclassified points are used as the performance metric. The Table 1 is the compilation of the results of the experiment. The Random model had the maximum percentage of misclassified points of 88.5%. K-NN, logistic regression with SGD classifier, Random Forest, and XgBoost are implemented with random search for parameters. The K-NN used only on .bytes files produced 4.5078196 percentage misclassified points which got increased to 12.327507 when logistic regression is used. Logistic regression clearly is the worst of all the classifiers. When only .asm files are used for classification and applied K-Nearest Neighbor, the percentage misclassified points are 2.0239191 which got

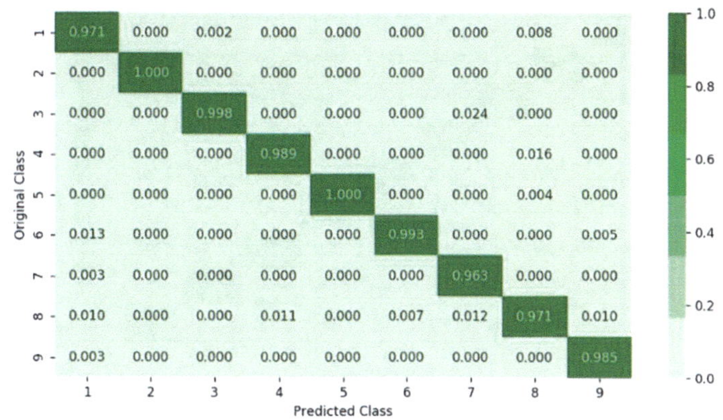

Fig. 3 Precision matrix

increased to 9.6136545 when logistic regression is used. Random forest decreased percentage misclassified points to 1.1499541 which is a good improvement but again XgBoost produced the best which is 0.8739651 percentage misclassified points. The K-NN and logistic regression are not producing good results so when the whole dataset is used with 310 features only applied random forest and XgBoost with Random search for parameters. The Random forest produced 0.7365425 percentage misclassified points but XgBoost with Random search for parameters outperformed it by reducing the percentage misclassified to 0.5769342. Therefore, Xgboost with random search gave us the best results.

The precision matrix is shown in Fig. 3 and recall matrix of the XgBoost with random search is used to get a better understanding of the results.

From the above precision matrix (sum of the values in a column is 1) can derive that:

- Out of expected all points to be in orbit 1, only 97.1% actually belonged to class 1.
- Out of expected all points to be in orbit class 2, 100% points belonged to class 1.
- Out of expected all points to be in orbit class 3, 99.8% actually belonged to class 1.
- Out of expected all points to be in orbit class 4, 98.9% actually belonged to class 4.
- Out of expected all points to be in orbit class 5, 100% belonged to class 5.
- Out of expected all points to be in orbit in class 6, 99.3% actually belonged to class 6.
- Out of expected all points to be in orbit class 7,96.3% actually belonged to class 7.
- Out of expected all points to be in orbit class 8, 97.1% actually belonged to class 8.

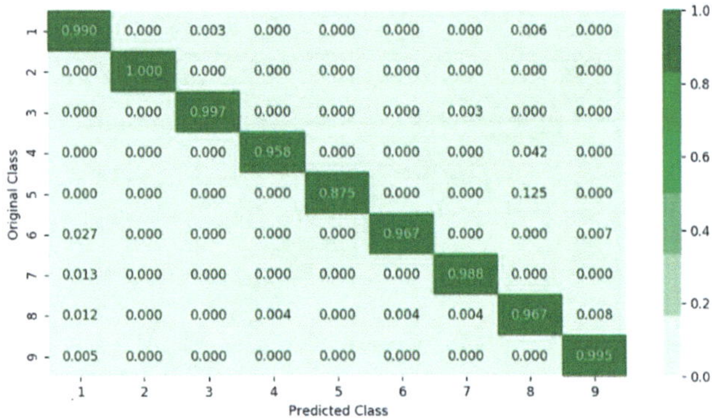

Fig. 4 Recall matrix

From the above recall matrix shown in Fig. 4 (sum of elements in a row is 1) can derive that:

- Out of all the points that belonged to class 1, 99% points are predicted to be in class 1.
- Out of all the points that belonged to class 2, 100% points are predicted to be in class 2.
- Out of all the points that belonged to class 3, 99.7% points are predicted to be in class 3.
- Out of all the points that belonged to class 4, 95.8% points are predicted to be in class 4.
- Out of expected all points to be in orbit class 5, 87.5% points are predicted to be in class 5.
- Out of all the points that belonged to class 6, 96.7% points are predicted to be in class 6.
- Out of all the points that belonged to class 7, 98.8% points are predicted to be in class 7.
- Out of all the points that belonged to class 8, 96.7% points are predicted to be in class 8.
- Out of all the points that belonged to class 9, 99.5% points are predicted to be in class 9.

5 Conclusion and Future Work

The XgBoost produced the best results. XgBoost with number of trees 1000, maximum depth 10, and learning rate 0.15 are implemented. This Xgboost produced a multi-class classification log-loss of 0.0317041132442 which is the best result. System Specification is a major challenge during the experiment. Due to the large size of the dataset, the multithreading techniques are used to fit our data in RAM and

Table 1 Values of all classifiers

Classifier	Files used	Log-loss	Percentage misclassified
Random	.bytes + .asm	2.485039	88.500459
K-NN	.bytes	0.241508	4.5078196
Logistic regression	.bytes	0.528347	12.327507
Random forest	.bytes	0.085834	2.0239190
XgBoost + random search	.bytes	0.078268	1.2419503
K-NN	.asm	0.089481	2.0239191
Logistic regression	.asm	0.415685	9.6136545
Random forest	.asm	0.571234	1.1499541
XgBoost + random search	.asm	0.048391	0.8739651
Random forest	.bytes + .asm	0.040114	0.7365425
XgBoost + random search	.bytes + .asm	0.031704	0.5769342

user has to define functions to perform multithreading. As for the future work, plan is to add bi-grams and n-grams features on .bytes files and in .asm files to explore additional features by acquiring more domain knowledge. To decrease our log-loss below 0.01%. In future, deep learning technique will be used to identify and classify packed and unpacked malware based on images.

References

1. Yuan B, Wang J, Liu D, Guo W, Wu P, Bao X (2020) Byte-level malware classification based on markov images and deep learning. Comput Sec 92, 101740
2. Anderson, H.S., Roth, P. Ember: An open dataset for training static pe malware machine learning models. arXiv preprint arXiv:1804.04637 (2018).
3. Yue S (2017) Imbalanced malware images classification: a cnn based approach. Cornell University Library, pp 1–6
4. Shalaginov A, Banin S, Dehghantanha A, Franke K (2018) Machine learning aided static malware analysis: a survey and tutorial. In Cyber Threat Intelligence. Springer, pp. 7–45
5. Lim H, Yamaguchi Y, Shimada H, Takakura H (2015). Malware classification method based on sequence of traffic flow. International Conference on Information Systems Security and Privacy (ICISSP), IEEE, pp. 1–8
6. Ni S, Qian Q, Zhang R (2018) Malware identification using visualization images and deep learning. Comput Secur 77:871–885
7. Pekta A, Acarman T (2017) Classification of malware families based on runtime behaviors. J Inform Sec Applic 37:91–100

8. Chen CW, Su CH, Lee KW, Bair PH (2020) Malware family classification using active learning by learning. In 2020 22nd International Conference on Advanced Communication Technology (ICACT), pp. 590–595. IEEE
9. Park S, Gondal I, Kamruzzaman J, Oliver J (2019) Generative malware outbreak detection
10. Torres JLG, Catania CA, Veas E (2019) Active learning approach to label network traffic datasets. J Inform Sec Applic 49:102388
11. Nari S, Ghorbani AA (2013) Automated malware classification based on network behavior. In 2013 International Conference on Computing, Networking and Communications (ICNC), pp. 642–647. IEEE
12. Kolosnjaji B, Zarras A, Webster G, Eckert C (2016) Deep learning for classification of malware system call sequences. In Australasian Joint Conference on Artificial Intelligence, pp. 137–149. Springer, Cham
13. Vasan D, Alazab M, Issan S, Safaei B, Zheng Q (2020) Image-based malware classification using ensemble of CNN architectures (IMCEC). Computers & Security, p 101748
14. Vinayakumar R, Alazab M, Soman KP, Poornachandran P, Al-Nemrat A, Venkatraman S (2019) Deep learning approach for intelligent intrusion detection system. IEEE Access. https://doi.org/10.1109/ACCESS.2019.2895334
15. Vasan D, Alazab M, Issan S, Naeem H, Safaei B, Zheng Q (2020) IMCFN: Image-based malware classification using fine-tuned convolutional neural network architecture. Comput Netw 171:107138
16. Han KS, Lim JH, Kang B, Im EG (2014) Malware analysis using visualized images and entropy graphs. Int J Inf Secur 14(1)
17. Venkatraman S, Alazab M (2018) Use of data visualisation for zero-day malware detection. Secur Commun Networks
18. Milosevic N, Dehghantanha A, Choo KKR (2017) Machine learning aided Android malware classification. Comput Electr Eng 61:266–274
19. Narayanan BN, Djaneye-Boundjou O, Kebede TM (2017) Performance analysis of machine learning and pattern recognition algorithms for Malware classification. In Proceedings of the IEEE National Aerospace Electronics Conference, NAECON
20. Danish Vasan AB, Mamoun Alazab E, Sobia Issan CD, Babak Safaei F, Qin Zheng A (2020) Image-based malware classification using ensemble of CNN architectures (IMCEC). J Comp Sec
21. Faezeh Farivar, Alireza Jolfaei (2020) Artificial intelligence for detection, estimation, and compensation of malicious attacks in nonlinear cyber-physical systems and industrial IoT. IEEE transactions on industrial informatics, 16(4)
22. Zhihua Cui A, Lei Du A, Penghong Wanga, Xingjuan Cai A, Wensheng Zhang (2019) Malicious code detection based on CNNs and multi-objective algorithm. J Parall Distrib Comput, pp 50–58
23. Schultz M, Eskin E, Zadok F, Stolfo S (2001, May 14–16) Data mining methods for detection of new malicious executables. Proceedings of 2001 IEEE Symposium on Security and Privacy, Oakland, pp 38–49
24. Cohen W. Fast effective rule induction. Proceedings of 12th International Conference on Machine Learning, San Francisco, pp 115–123
25. Kolter J, Maloof M. Learning to detect malicious executables in the wild. Proceedings of the 10th ACM SIGKDD International Conference on Knowledge Discovery and Data Mining, pp 470–478
26. Nataraj L, Karthikeyan S, Jacob G, Manjunath B. Malware images: visualization and automatic classification. Proceedings of the 8th International Symposium on Visualization for Cyber Security, article no 4
27. Nataraj L, Yegneswaran V, Porras P, Zhang J (2011) A comparative assessment of malware classification using binary texture analysis and dynamic analysis. Proceedings of the 4th ACM Workshop on Security and Artificial Intelligence, pp 21–30
28. Kong D, Yan G (2013) Discriminant malware distance learning on structural information for automated malware classification. Proceedings of the ACM SIGMETRICS/International Conference on Measurement and Modeling of Computer Systems, pp 347–348

29. Tian R, Batten L, Versteeg S (2008) Function length as a tool for malware classification. Proceedings of the 3rd International Conference on Malicious and Unwanted Software, Fairfax, 7–8 October 2008, pp 57–64
30. Tian R, Batten L, Islam R, Versteeg S (2009, October) An automated classification system based on the strings of Trojan and virus families. Proceedings of the 4th International Conference on Malicious and Unwanted Software, Montréal, pp 13–14, 23–30
31. Hall M, Frank E, Holmes G, Pfahringer B, Reutemann P, Witten I (2009) The WEKA data mining software: an update. ACM SIGKDD Explorations Newsletter, pp 10–18
32. Santos I, Nieves J, Bringas PG (2011) Semi-supervised learning for unknown malware detection. Int Symp Distrib Comput Artif Intell Adv Intell Soft Comput 91:415–422

Chapter 15
Predicting Survivability in Oral Cancer Patients

Diksha Sharma, Neelam Goel, and Vivek Kumar Garg

1 Introduction

Oral cancer (OC) is among the most-deadly disease in today's era. A malignant tumor enlarges in any part of the mouth. The main risk factors for oral cancer are the intake of alcohol and tobacco. According to the statistics proposed by Globocane, oral cancer ranked as the second most common cancer in India with an annual estimation of 72,616 responsible deaths and 1,19,992 new incident cases that are reported only, out of those cases approximately 2,65,255 patients accomplished a 5-year survival rate [1]. Even among those who survive, many factors affect the survivability of patients. Surgical treatment can improve the survival rate of OC patients, but it affects the patient's life in terms of their ability to perform their daily activities, facial appearance, and expensive treatments [2].

Recent research introduced tumor biomarkers in the field of molecular pathology, which is related to various cancer's prognosis. Although the usage of biomarkers for OC has not been acquitted in regular practice, however biomarkers are conferred as reliable predictors that provide knowledge about the stage, grade, and depth of invasion [3]. Machine learning technology also advances the research environment for the diagnosis and treatment of the patient suffering from oral cancer in the field of medicine [4–8]. Many techniques like decision trees, logistic regression, and artificial neural network (ANN) were used to establish a five-year survival rate (SR) predictor. The study of these models will give references

D. Sharma (✉) · N. Goel
University Institute of Engineering & Technology, Panjab University, Chandigarh, India

N. Goel
e-mail: erneelam@pu.ac.in

V. K. Garg
Punjab Biotechnology Incubator, Mohali, India

G. Mathur et al. (eds.), *Proceedings of 2nd International Conference on Artificial Intelligence: Advances and Applications*, Algorithms for Intelligent Systems,
https://doi.org/10.1007/978-981-16-6332-1_15

for future work and will help in decision-making for physicians and will provide feedback for diagnosis [2].

It would be useful for physicians to avoid unnecessary treatment procedures by selecting the right treatment options using survival prediction methods. This is because the patient does not get a quick diagnosis as the disease is not recognized accurately which is the main reason behind the increasing rate of mortality and morbidity. The main concern is about developing the method which helps in categorizing patients suffering from the same type of oral cancer by using the information of pre-treated cancer patients and improving their overall treatment rate [2]. Also, the early prediction of cancer is necessary to increase the survival time of cancer patients so that multiple strategies can be developed for predicting the outcomes of cancer treatment, by collecting a large amount of cancer data and the data can be made available to the research community. However, this is the most challenging task for the researchers to get accurate results [9]. The main aim of this paper is to review the research work related to survivability prediction for OC.

2 Oral Tumorigenesis

Oral Tumorigenesis is an extremely complex mechanism that happens when the squamous epithelium is affected by various genetic modifications. Diagnosis of oral premalignant lesions using different molecular biological approaches will surely enhance the identification of those variations that are not detected by the microscope. It would be helpful to identify those patients who are at high risk of developing OC [10]. Figure 1 illustrates the sequence of genetic alterations of oral squamous cell carcinoma (SCC) [11]. The microarray technology, proteomics, array comparative genomic hybridization, methylation microarrays, and micro-RNA arrays are few

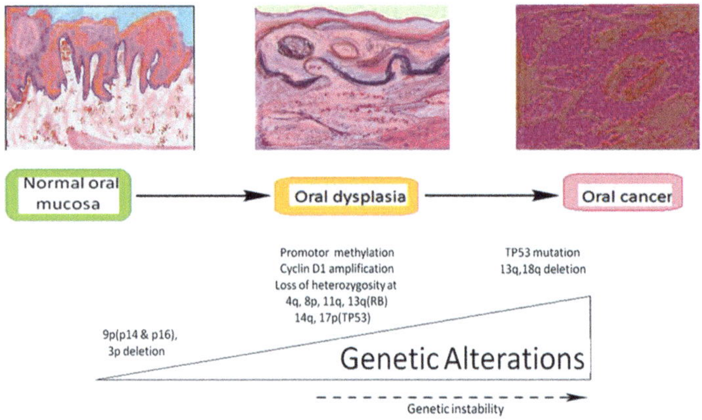

Fig. 1 The sequence of genetic Alteration [11]

approaches that can be used for better clarification of oral cancer's molecular basis [12–14]. Up to now, high-throughput methods have been used to find OC biomarkers in saliva and serum [15].

It is generally known that all cancerous cells occur due to the mutations in genes that control the behavior of cells. Multiple alterations of genes should be there to make cells cancerous. The cells that grow and increase rapidly in an uncontrolled manner are not capable to fix their DNA damage by themselves. To create neoplastic cells, there are specific types of genes that have to be altered several times, resulting in uncontrolled growth of cells. Genetic errors occur randomly every day in the process of replacing multiple cells in our body. In addition to these random modifications, genetic changes can also be developed by radiation or chemical exposure, or influenced by virus infections. Normally, our bodies have mechanisms to destruct these abnormal growing cells [16].

3 Data Sources

There are several data repositories and analysis tools that are available for data analyzing, sharing, and reporting with the cancer research community, and researchers are not required to run their experiments again or to create and resubmit a new dataset. Moreover, this information is also helpful for analysts who don't have high computational abilities. The largest genomic data collection repository is TCGA, which made data available from different molecular profiling platforms like RNA, miRNA, DNA sequencing, protein expressions, and somatic mutations for more than 10,000 cancer types.

TCGA's data assortment is jointly initiated by the National Human Genome Research Institute (NHGRI), National Cancer Institute (NCI), Department of Health and Human Services, and the National Institutes of Health (NIH) [17].

Genomic datasets can be created on a large scale from testing model frameworks or tumor samples in cancer research. Figure 2a illustrates the number of total cases having six different cancer types, whereby it is analyzed that more cases are being found in people with Floor of mouth and Tonsils in those who consume tobacco and alcohol in excess. Oral cancer is categorized by the different types of cells in which cancer begins to proliferate. A total of 9 out of 10 OC patients were having squamous cell (SC) which shows SC is the most prominent form in OC patients [18]. Therefore, in the present study, a detailed analysis of data is done by using custom cases, gene, and mutation filters with 177 cases of patients having Squamous Cell Neoplasm from the years 1998–2019. Oral Squamous Cell Neoplasm means cancer that appears in between lips lines and 1/3rd of the tongue. Over 95% of people with this disease consume alcohol, smoke tobacco, or both. Surgical procedure and Radiotherapy and chemotherapy are the best treatment strategies, although surgical treatment is the best way to treat the squamous cell [19]. Figure 2b shows the cancer survivor's patients and a forecast of their survival conditions, which helps physicians to interpret the results. This is the main parameter to help physicians in determining

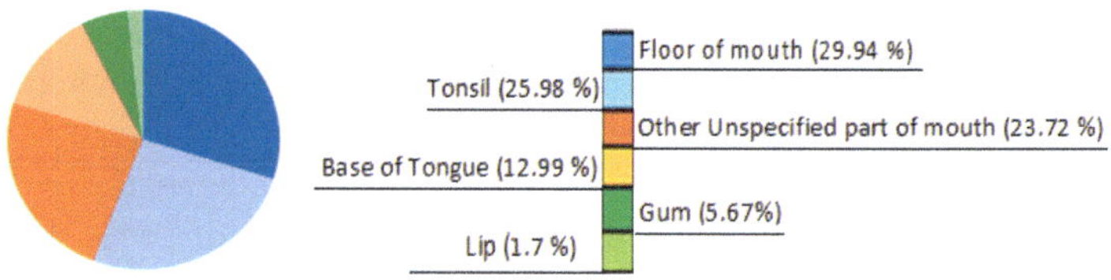

(a) Primary Site

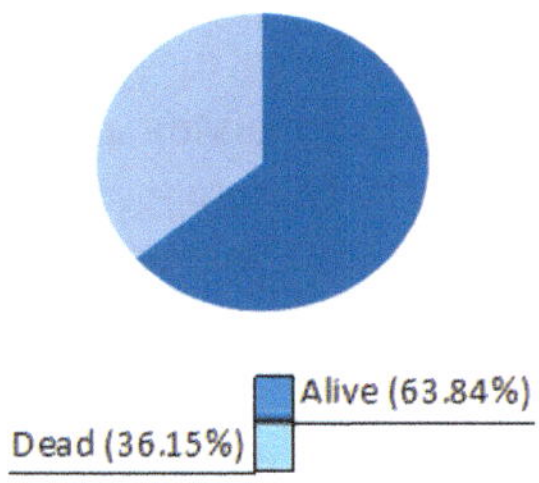

(b) Vital Status

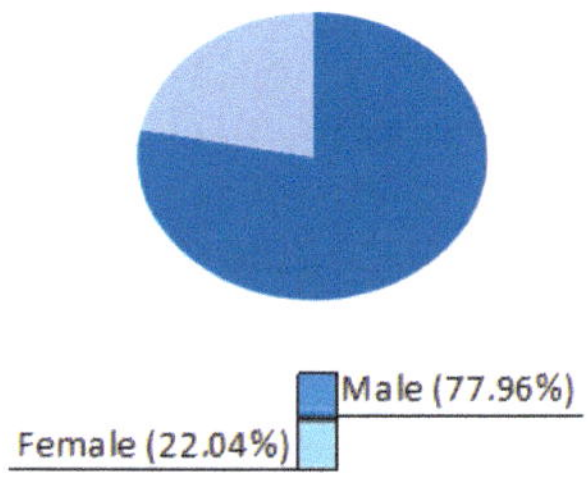

(c) Gender

Fig. 2 The number of cases that meet the filter criteria having types of oral cancer, disease type, vital status, and gender. **a** Primary site. **b** Vital status. **c** Gender

the health outcomes of cancer patients and prioritize treatment options for similar cases. The result shows that the ratio of male patients among 177 cases is higher than that of female patients who have a habit of chewing betel nut, smoking, and other tobacco as shown in Fig. 2c.

Figure 3 illustrates the distribution of mutated genes. TP53 is the most commonly mutated gene in patients having OC. The irreversible modification of the DNA sequence that makes up a gene is known as gene mutation. These alterations can be classified as hereditary or somatic, which are present in certain cells and causes genetic changes that lead to disease. This can be occurred by environmental factors like UV Rays, smoking, excessive consumption of alcohol due to which an error

Fig. 3 An illustration of a bar graph of the most frequently mutated genes

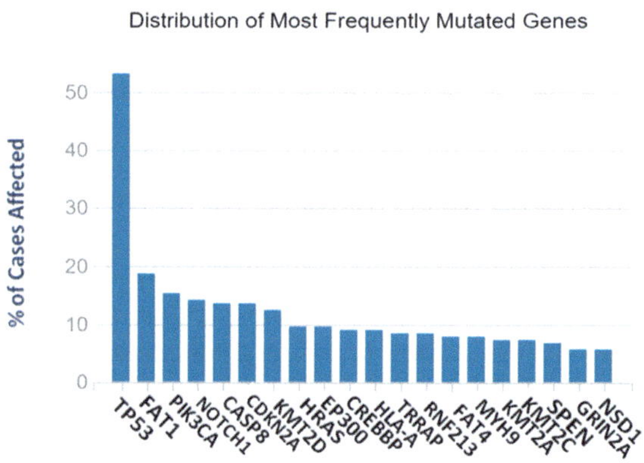

occurs in DNA copies by itself during the cell division. Different techniques have been applied to identify the genetic mutations which are leading to diseases and compare them for the genetic diagnosis.

The purpose is to make physicians familiar with the techniques and tools used for genetic prognosis so that they can easily choose treatment alternatives for better survival of patients according to mutated genes and avoid unnecessary treatment.

4 Survival Prediction in Oral Cancer

Several studies have addressed the survivability of OC. The results of these studies give information on clinical management and health policies, the characteristics of various data types used, the medical oncology of patients, and enhanced awareness for predicting the survival time and survival rates of patients with OC. Survival rates are often used by physicians as a standard way of discussing a person's prognosis.

Using clinical and Microarray data, an ANN model predicts the survival time of cancer patients. The prediction made by the model shows a higher level of correctness [20]. For predicting the accurate survivability of oral squamous cell carcinoma (OSCC) patients, an attempt has been made with the data of 96 cases using molecular biomarkers and clinical variables. The approach is constructed, as a nomogram that can accurately predict the 5 and 10-year survival after treatment of an individual patient of OSCC in terms of c-index [21].

In 2011, the survival of oral cancer patients was predicted using several classification algorithms, the Random Forest technique precisely classified the survival of cancer and also led to less absolute relative error as compared to other algorithms [22]. In 2013, three models were proposed as Single Tree, Decision Tree Forest, and Tree Boost model. These proposed models were employed to predict the survival of patients as one of the effective models. Based on various standards and observations, a comparison was made among these models and found that all models have similar performance. However, the tree boost model (TBM) was slightly better than others in identifying the survival rate of patients with OC [23].

Two techniques, ANN and decision tree were used in 2015, to examine previous cases of oral cancer, and their functionings were compared with logistic regression. The result of the study indicated that the decision tree was relatively more suitable than ANN for practical implementation. Clustering and classification analysis were used to find the significant difference between the symptoms among the patients where they died or survived from OC [2]. In 2016, machine learning techniques were used to improve the prediction of survivability using gene expressions. ANN is used to identify distinctive signs of gene expression in cancer patients and improves the efficiency and accuracy of gene expression analysis to predict survival [24].

A model called multi-layer feed-forward (MLFF) neural network for survival time prediction of OSCC is described in the year 2018. Data from OSCC patients who consume Alcohol and Betel Quid, their tumor size, and tumor site are important features to develop the best MLFF model. Significant factors are used to test the performance of the model for different combinations. Mean error training (MAT) and mean absolute error (MAE) of the analyzed samples were used to evaluate the model's overall performance. From this method, researchers can predict a better forecasting result in the future for decision-making [25]. In 2019, data on 255 patients of oral SCC who underwent the surgery were used to validate the performance of the survival prediction model, namely DeepSurv. The performance of survival prediction models namely DeepSurv, Cox Proportional Hazard Model, and random survival forest (RSF) were compared, and the results indicated that DeepSurv has higher performance among the mentioned models. This can help physicians and patients by classifying risk factors and directing treatment options to save even more lives and time by eliminating inefficient treatment [26].

A 5-year overall survival-based machine learning model was implemented among patients diagnosed with OSCC and the variables used in the model for predicting the survival were mainly the collection of pathology, demographic, and treatment variables. Then, the comparative analysis was made with another prediction model based on the pathological and clinical stage of tumor, node, metastasis (TNM) and it was found that the model predicts better overall survival [27].

In 2019, a study was conducted to identify multi-transcription factors (TF) signatures expression pattern that could predict outcomes for patients with head and neck squamous cell carcinoma (HNSCC). Based on 6 TF expression patterns, researchers developed a prognostic signature by using patient outcomes with RNA sequence data and their analysis indicates that the 6TF signature is a strong forecasting outcome approach for HNSCC patients. This could also assist with the patient's disease level based on expected clinical responses [28].

To predict the survival of patients with tongue squamous cell carcinoma (TSCC) in 2020, researchers assess the prognostic significance of systemic immune-inflammation index (SII) and established a prognosis nomogram model using 120 patients diagnosed with TSCC. The observation showed that the SII-based prognostic nomogram model was suitable for determining disease-free survival (DFS) and survival rates in TSCC patients and also helped in enhancing the precision of clinical prognosis in patients undergoing cervical artery dissection [29].

A 6-gene signature prognostic model was established for predicting the survival of HNSCC patients described in [30]. This 6-gene signature prognostic model will lead to improved AUC results for the training and testing sets, and prediction of survival risk will also be improved by gene classifiers as compared to clinical features. Researchers are therefore advised to use these gene classifiers as a diagnostic test of molecules for analyzing the predictive risk in patients with HNSCC (Table 1).

5 Conclusion

The purpose of this review is to give an overview of published studies regarding the prediction of survivability of Oral Cancer patients. The most recent works relevant to cancer prediction are done using machine learning. Classification techniques are used

Table1 Summary of work done for survival prediction of oral cancer

References	Data source	Data type	Technique	Evaluation parameters	Prediction	Number of patients
Chen et al. [20]	Public domain	Microarray and clinical data	ANN	Probability of Survival, comparison of observed, and predicted results	Survival time	368
Kim et al. [21]	NCC in Korea	Clinical variables and Molecular markers	Nomogram	Effect of p53 and c-index	Survival	96
Kaladhar et al. [22]	Oxford Dataset	Clinical data and Genes expression	Classification algorithms	Accuracy	Survivability	-
Sharma and Om [23]	Oral healthcare industry	Clinical and Genes expression	Data Mining	Sensitivity and Specificity	Survivability	1024
Tseng et al. [2]	Medical center in Southern Taiwan	Clinical	Data Mining	Efficiency	Survival analysis	673
Bashiri et al. [24]	-	Genes expression	Machine Learning	Accuracy, Effectiveness, and Comparison	Survival	-
Shafiq et al. [25]	UNIVERSITI SAINS MALAYSIA	Clinical Pathological	MLFF Neural Network	Mean Error Training and Mean Absolute Error	Survival Time in months	16
Kim et al. [26]	AJCC	Clinical data	Deep Learning	Validation, comparison	Survival	255
Karadaghy et al. [27]	NCDB	Clinical data	Machine Learning	Accuracy and area under the curve (AUC)	Overall survival	33,065
Zhang et al. [28]	TCGA	Genomic data	Data Mining	Sensitivity and Specificity	overall survival	502
Lu et al. [29]	Hospital of Sun Yat-sen University	clinical data, pathological data	Nomogram	Significance of SII, Accuracy of the model	Survival rate	120

(continued)

Table1 (continued)

References	Data source	Data type	Technique	Evaluation parameters	Prediction	Number of patients
Wang et al. [30]	TCGA	Genomic data	6-gene signature model	Classification performance, AUC, risk regression analysis	Survival	771

to suspect cancer, its recurrence, and survivability. These techniques were correctly classified and had a relatively low error as compared to other methods. SVM and ANN classifiers are extensively used among applied machine learning algorithms. The prediction of survival time of oral cancer is achieved by evaluating the probability of survival and comparing the predicted results concerning specific predictive outcomes. A decision support system for clinical data is build up using an ANN algorithm that can increase the predictability and survivability of cancer patients.

In the last decade, a large amount of ML studies have been published that provide accurate results but work on survival prediction has been done more than survival time. As the work done on the survival time of oral cancer patients is less, it becomes difficult for clinicians to estimate the quality of life of a patient, decision-making, and cancer risk. Also, by exploring the dataset it is analyzed that the survival rate decreased over the years because the disease could not detect early. So, predicting survival time is the need of the hour to decrease the mortality rate.

References

1. Oral Cancer-India. http://cancerindia.org.in/oral-cancer/
2. Tseng WT, Chiang WF, Liu SY, Roan J, Lin CN (2015) The application of data mining techniques to oral cancer prognosis. J Med Syst 39. https://doi.org/10.1007/s10916-015-0241-3
3. Almangush A, Heikkinen I, Mäkitie AA, Coletta RD, Läärä E, Leivo I, Salo T (2017) Prognostic biomarkers for oral tongue squamous cell carcinoma: a systematic review and meta-analysis. Br J Cancer 117:856–866. https://doi.org/10.1038/bjc.2017.244
4. Pande P, Soni S, Kaur J, Agarwal S, Mathur M, Shukla NK, Ralhan R (2002) Prognostic factors in betel and tobacco related oral cancer. Oral Oncol 38:491–499. https://doi.org/10.1016/S1368-8375(01)00090-2
5. Lu HY, Li TC, Tu YK, Tsai JC, Lai HS, Kuo LT (2015) Predicting long-term outcome after traumatic brain injury using repeated measurements of glasgow coma scale and data mining methods. J Med Syst 39. https://doi.org/10.1007/s10916-014-0187-x
6. Nahar J, Tickle KS, Ali ABMS, Chen YPP (2011) Significant cancer prevention factor extraction: an association rule discovery approach. J Med Syst 35:353–367. https://doi.org/10.1007/s10916-009-9372-8
7. Chao CM, Yu YW, Cheng BW, Kuo YL (2014) Construction the model on the breast cancer survival analysis use support vector machine, logistic regression and decision tree. J Med Syst 38:1–7. https://doi.org/10.1007/s10916-014-0106-1

8. Yilmaz N, Inan O, Uzer MS (2014) A new data preparation method based on clustering algorithms for diagnosis systems of heart and diabetes diseases. J Med Syst 38. https://doi.org/10.1007/s10916-014-0048-7
9. Kourou K, Exarchos TP, Exarchos KP, Karamouzis MV, Fotiadis DI (2015) Machine learning applications in cancer prognosis and prediction. Comput Struct Biotechnol J 13:8–17. https://doi.org/10.1016/j.csbj.2014.11.005
10. Joseph BK (2002) Oral cancer: prevention and detection. Med Princ Pract 11:32–35. https://doi.org/10.1159/000057776
11. Sapkota D (2011) S100 gene family members in oral squamous cell carcinomas (OSCCs)
12. Campo-Trapero J, Cano-Sánchez J, Palacios-Sánchez B, Sánchez-Gutierrez J, González-Moles MA, Bascones-Martínez A (2008) Update on molecular pathology in oral cancer and precancer. Anticancer Res 28:1197–1205
13. Patel V, Leethanakul C, Gutkind JS (2001) New approaches to the understanding of the molecular basis of oral cancer. Crit Rev Oral Biol Med 12:55–63. https://doi.org/10.1177/10454411010120010401
14. Tsantoulis PK, Kastrinakis NG, Tourvas AD, Laskaris G, Gorgoulis VG (2007) Advances in the biology of oral cancer. Oral Oncol 43:523–534. https://doi.org/10.1016/j.oraloncology.2006.11.010
15. Leoncini E, Ricciardi W, Cadoni G, Arzani D, Petrelli L, Paludetti G, Brennan P, Luce D, Stucker I, Matsuo K, Talamini R, La Vecchia C, Olshan AF, Winn DM, Herrero R, Franceschi S, Castellsague X, Muscat J, Morgenstern H, Zhang ZF, Levi F, Dal Maso L, Kelsey K, McClean M, Vaughan TL, Lazarus P, Purdue MP, Hayes RB, Chen C, Schwartz SM, Shangina O, Koifman S, Ahrens W, Matos E, Lagiou P, Lissowska J, Szeszenia-Dabrowska N, Fernandez L, Menezes A, Agudo A, Daudt AW, Richiardi L, Kjaerheim K, Mates D, Betka J, Yu GP, Schantz S, Simonato L, Brenner H, Conway DI, Macfarlane TV, Thomson P, Fabianova E, Znaor A, Rudnai P, Healy C, Boffetta P, Chuang SC, Lee YC, Hashibe M, Boccia S (2014) Adult height and head and neck cancer: a pooled analysis within the INHANCE consortium. Head Neck 36:1391. https://doi.org/10.1002/HED
16. OC-facts. https://oralcancerfoundation.org/facts/
17. Creighton CJ (2019) HHS public access, pp 1–19. https://doi.org/10.1002/cpmb.49.Making
18. NHS. https://www.nhs.uk/conditions/mouth-cancer/
19. Bradley A, Schiff MD. Montefiore Medical Center, T.U.H. of A.E.C. of M.: OSCC. https://www.msdmanuals.com/professional/ear,-nose,-and-throat-disorders/tumors-of-the-head-and-neck/oral-squamous-cell-carcinoma
20. Chen YC, Yang WW, Chiu HW (2009) Artificial neural network prediction for cancer survival time by gene expression data. 3rd Int. Conf. Bioinforma. Biomed. Eng. iCBBE 2009, pp 1–4. https://doi.org/10.1109/ICBBE.2009.5162409
21. Kim K-Y, Li S-J, Cha I-H (2010) Nomogram for predicting survival for oral squamous cell carcinoma. Genom Inform 8:212–218. https://doi.org/10.5808/gi.2010.8.4.212
22. Kaladhar D, Chandana B, Kumar P (2011) Predicting cancer survivability using classification algorithms. Int J Res Rev Comput Sci 2:340–343
23. Sharma N, Om H (2013) Data mining models for predicting oral cancer survivability. Netw Model Anal Heal Inform Bioinf 2:285–295. https://doi.org/10.1007/s13721-013-0045-7
24. Bashiri A, Ghazisaeedi M, Safdari R, Shahmoradi L, Ehtesham H (2017) Improving the prediction of survival in cancer patients by using machine learning techniques: experience of gene expression data: a narrative review. Iran J Public Health 46:165–172
25. Shafiq M, Ibrahim M, Ali Z, Aleng NORA, Husein A, Halim NA (2018) Modeling of survival time of oral squamous cell carcinomas (oscc) in hospital universiti sains malaysia using multilayer feedforward neural network. 4, 1045–1050
26. Kim DW, Lee S, Kwon S, Nam W, Cha IH, Kim HJ (2019) Deep learning-based survival prediction of oral cancer patients. Sci Rep 9:1–10. https://doi.org/10.1038/s41598-019-43372-7

27. Karadaghy OA, Shew M, New J, Bur AM (2019) Development and assessment of a machine learning model to help predict survival among patients with oral squamous cell carcinoma. JAMA Otolaryngol. - Head Neck Surg. 145:1115–1120. https://doi.org/10.1001/jamaoto.2019.0981
28. Zhang B, Wang H, Guo Z, Zhang X (2019) A panel of transcription factors identified by data mining can predict the prognosis of head and neck squamous cell carcinoma. Cancer Cell Int 19:1–10. https://doi.org/10.1186/s12935-019-1024-6
29. Lu Z, Yan W, Liang J, Yu M, Liu J, Hao J, Wan Q, Liu J, Luo C, Chen Y (2020) Nomogram based on systemic immune-inflammation index to predict survival of tongue cancer patients who underwent cervical dissection. Front Oncol 10:1–11. https://doi.org/10.3389/fonc.2020.00341
30. Wang J, Chen X, Tian Y, Zhu G, Qin Y, Chen X, Pi L, Wei M, Liu G, Li Z, Chen C, Lv Y, Cai G (2020) Six-gene signature for predicting survival in patients with head and neck squamous cell carcinoma. Aging (Albany. NY) 12, 767–783. https://doi.org/10.18632/aging.102655

Chapter 16
Semantic Similarity Extraction on Corpora Using Natural Language Processing Techniques and Text Analytics Algorithms

Nisha Varghese and M. Punithavalli

1 Introduction

The advent of the digital era leads to information overloading or information explosion, consequently, the extraction of the relevant information is an arduous task. The application of Text Mining (TM) is one of the solutions to alleviate this problem. It can efficiently retrieve meaningful insights from the corpus by using prominent algorithms and techniques. This study has been started with the hypothesis that there is some similarity between the two tropes, the Bible and the Quran. The Middle East is the origin of these sacred texts, as a result, the comparative analysis of these holy books may reveal some fascinating insights. So, the Bible and the Quran have been taken as the dataset for this study of Semantic similarity extraction. The Bible is a multi-lingual corpus with 66 books including 39 old and 27 new testament books. The Christians, Jews, and others are believed that the compilation of this religious book of various formats that are finally assembled as per the revelation of God. Many kinds of people from various places are involved in the compilation of this sacred text. The Bible is authored by the peoples including kings, fishermen, shepherds, and others, so the religious book contains laws, proverbs, songs and other concealed information in the form of revelations, parables and dreams. The book illuminates the archaeological details and history of medieval culture. The Quran is compiled in the eastern Arabian dialect with 30 Juz or 114 chapters (Surah) and the chapters are chronologically arranged according to the length, not by historical order. These semi-structured corpora do not adhere to any specific patterns. Semantic similarity

N. Varghese (✉) · M. Punithavalli
Department of Computer Applications, Bharathiar University, Coimbatore 641046, India
e-mail: nisha.dca@buc.edu.in

M. Punithavalli
e-mail: punithavalli@buc.edu.in

G. Mathur et al. (eds.), *Proceedings of 2nd International Conference on Artificial Intelligence: Advances and Applications*, Algorithms for Intelligent Systems,
https://doi.org/10.1007/978-981-16-6332-1_16

extraction measured based on the syntactic and semantic structure of corpus using the similarity metrics and measures.

2 Literature Review

This section listed out some relevant studies based on semantic similarity and information retrieval related to sacred texts. Salha Hassan Muhammed et al. [1] investigated through various similarity measures and metrics to reveal the semantic similarity between the Bible and Quran using some of the popular families are Lp family or Lp norm and Lp norm is subdivided into Lp Minkowski, L1 norm and L2 norm. Intersection, Inner product family, Matching based family, Fidelity and squared chord family and entropy family are some other families for similarity. Daniel Merrill McDonald [2] article explained automatic knowledge retrieval and semantic analysis of nine Holy Books using Self Organizing Maps. MayuriVerma [3] presented the lexical information retrieval from ten sacred texts. A. Ta'a et al. [4] implemented an ontology modeling for the extraction of the hidden concepts in the Holy Quran using the RDF schema/OWL (web ontology language) and semantic search approach. The extracted information can be machine-readable structure. Mohamed Osman Hegazi [5] presented a work to extract Arabic texts and that can be transformed into a structured format, that can be applicable for various semantic similarity and machine learning models. Ramona Popa et al. [6, 7] proposed a study using knowledge-based ontological modeling of the Bible. Both of these studies unveiled the concepts and relationships using the Text2onto tool, the knowledge extraction ontology modeling tool. But the problem occurred due to the proper noun elimination some important central concepts have been eliminated from the corpus, which may affect the proper semantic similarity extraction.

3 Word and Meaning Relationships

A word is an unbreakable-smallest syntactic unit. Words are arranged in sequence to form sentences, which are governed by the grammar of a language. The main focus of semantic similarity analysis lies in meaning. According to J R Firth, the semantic information of a word depends on the context words surrounded by it [8]. The semantics are lexical semantics (word meanings) and compositional semantics (Meaning of sentences). Words can be classified into eight standard parts of speech (POS) and are subdivided into content (vocabulary) and function (grammatical) words. The proper combination of a string of words, phrases, and punctuations makes sentences. Every element in a sentence or corpus is considered as the tokens. Each token holds meaning, even if it a punctuation mark and that may or may not be related to the neighboring words. Lexical semantics concerning the sense of the word, that lies in the arrangement of the context word and the surrounding words concerning the window size of the context word. According to wordnet, a word may have several meaning [9]. The challenges in the word and meaning relationship lie in multiple resolutions of words such as synonyms, polysemes and their sub-categories. The major polysemes are unidirectional, bilingual, double pivotal and contronyms.

4 Pre-Processing or Text Normalization

Text Wrangling or Text preprocessing or Text Cleaning is the process of converting the text into an analyzable format. The raw data may contain various noisy or irrelevant information, consequently, it compromises the accuracy of the extraction of information. Lower case conversion, special character removal, tokenization, stop words removal, Parts of speech tagging, stemming or lemmatization, shallow parsing or chunking, constituency parsing and dependency parsing are the various Pre-processing steps. The process of text cleaning may include all or some of these steps depending upon our text mining task. After the proper fine-tuning of these steps, the input data turns to a predictable structured format. The resultant data is normalized data, then it can be used for any downstream tasks of text mining. Tokenization is the process of splitting the corpus into segments of constituent words (word tokenization) or sentences (sentence tokenization). Chunk is the resultant words after the partial parsing or Chunk extraction. Chinks are the words obtained after the chunk extraction.

The first column 'Text Tokens' of Fig. 1, shows the number of tokens from a fragment of text from the first book of the Bible, Genesis chapter one. The second column is the list of the most frequent tokens and the 'Normalized Text' column is the words or tokens after the text normalization. Text Normalization or wrangling is the cleaning process to make standardized data from raw textual data. Figure 2 shows the Parts of speech (POS) extraction of the verses from Genesis 1:1, the data set is the Bible, King James Version (KJV) has taken directly from the python Gutenberg natural language tool kit ('bible-kjv.txt') and the Chunks tagging. The tagging types are listed in the POS tag list, which includes more than 30 types. The chunks tagging includes various parts of speech tagging listed in Fig. 2.

Fig. 1 Tokenization

Text Tokens	Frequent items	Normalized Text
['In',	[('the', 60),	['In',
'the',	('and', 46),	'the',
'beginning',	(':', 26),	'beginning',
'God',	(',', 23),	'God',
'created',	('.', 18),	'created',
'the',	('god', 17),	'the',
'heaven',	('1', 16),	'heaven',
'and',	('was', 11),	'and',
'the',	('let', 9),	'the',
'earth',	('light', 9)]	'earth',
'.',		'And',
'1',		'the',
':',		'earth',
'2',		
'And',		

Fig. 2 POS and chunk tagging

Parts of Speech(PoS)	Chunks Tagging
[('In', 'IN')]	(S
[('the', 'DT')]	In/IN
[('beginning', 'VBG')]	the/DT
[('God', 'NNP')]	beginning/NN
[('created', 'VBN')]	(PERSON God/NNP)
[('the', 'DT')]	created/VBD
[('heaven', 'NN')]	the/DT
[('and', 'CC')]	heaven/NN
[('the', 'DT')]	and/CC
[('earth', 'NN')]	the/DT
	earth/NN
	And/CC
	the/DT
	earth/NN

5 Word Statistics, Lexical Analysis, and Comparison of Corpora

Word statistics contain some parameters related to the corpus such as the total number of words, the average length of a word; the total number of unique or identical words in the corpus. The vocabulary or number of word types will be higher with respect to the increase in the number of words. Entropy is another parameter that is computed from the letter frequencies for each text. The size of the vocabulary can predict by using Heap's law, the law states that the size of the vocabulary, $v = Kn^{\beta}$, Where n is the number of words, $0 < \beta < 1$ and $0 < K < 100$. Most perfect approximations to predict the vocabulary size with $0.4 < \beta < 0.6$ [10].

Zipf's law claims some interesting insights with the frequency, rank and meaning of the words in a text [11]. By this law, the product of the word frequency and the rank of the word will be approximately equal to the product of the frequency and rank of another word. Zipf's law again claims that if the words are ranked by its frequency of occurrence in the text, then the count of the compositional semantics is inversely proportional to the square root of its rank.

Figure 3 and Fig. 4 listed below show the relative and absolute frequencies of the

Fig. 3 Bible Words frequency

Most common words	Absolute frequency	Relative Frequency
lord	7830.0	0.024094
god	4443.0	0.013672
man	2613.0	0.008041
israel	2565.0	0.007893
son	2370.0	0.007293
king	2256.0	0.006942
people	2139.0	0.006582
house	2024.0	0.006228
come	1971.0	0.006065
one	1967.0	0.006053

Fig. 4 Quran Words frequency

Most common words	Absolute frequency	Relative Frequency
allah	2799.0	0.040557
lord	988.0	0.014316
people	792.0	0.011476
say	570.0	0.008259
surely	554.0	0.008027
day	490.0	0.007100
one	438.0	0.006347
earth	404.0	0.005854
may	399.0	0.005781
believe	373.0	0.005405

Fig. 5 Differences in relative frequencies

Most Common Words	Bible RelFreq	Quran RelFreq	Difference in RF
allah	0.000000	0.040214	0.040214
lord	0.023355	0.014195	0.009160
people	0.006380	0.011379	0.004999
say	0.003150	0.008189	0.005040
surely	0.000847	0.007960	0.007112
day	0.005172	0.007040	0.001868
upon	0.008196	0.006623	0.001573
one	0.005867	0.006293	0.000426
earth	0.002944	0.005804	0.002860
may	0.003063	0.005733	0.002669

most common words in the Bible and the Quran and Fig. 5, shows the comparison between the Bible and the Quran relative frequencies and the differences in relative frequencies of both corpora.

Lexical Density (LD) and Lexical Variation (LV) are the popular measures to calculate the lexical richness of a corpus. LD is the percentage of lexical words in the text and LV is the ratio in percentage between different words and the total number of tokens. Lexical Density and Lexical Variation measures of the Bible and the Quran are included in Table 1.

$$LD = \frac{\text{No. of Lexical Tokens}}{\text{Total no. of Tokens}} \times 100 \quad LV = \frac{\text{No. of various words}}{\text{Total no. of Tokens}} \times 100$$

Table 1 Lexical parameters

Sacred texts	# lexical tokens	# types of tokens	# tokens	Lexical density (%)	Lexical variation (%)
Bible	442,053	16,167	936,682	47.19	1.73
Quran	68,200	7925	149,402	45.5	5.3

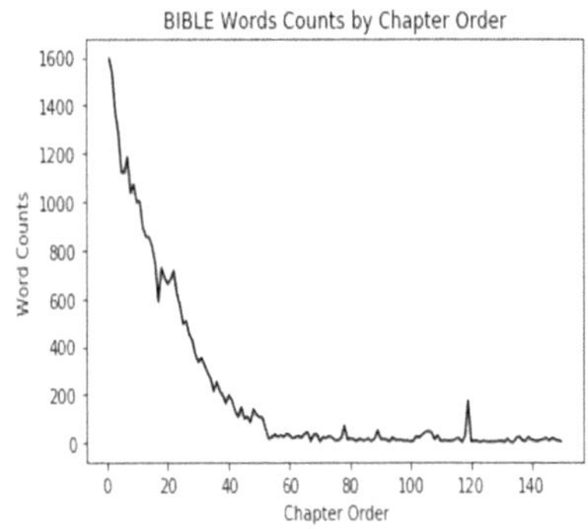

Fig. 6 Bible chapter word counts

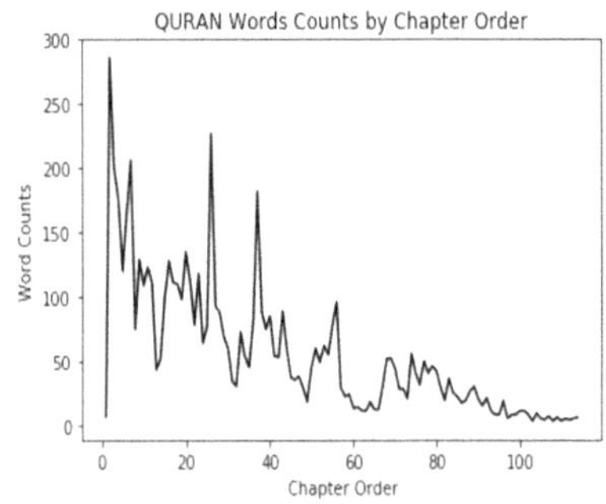

Fig. 7 Quran chapter word counts

Figures 6 and 7 show the word count of both corpora with respect to the chapter order. In the Bible the first book, Genesis shows the highest word counts because it contains the most number of proper nouns and other words. In the Quran Surah 2 is Al-Baqarah means the cow, is the longest chapter with 286 verses, so the graph shows a high peak from the first chapter Al-Fatihah with 7 verses.

Figures 8 and 9 show the dispersion graph of the Bible and the Quran, both shows some interesting insights about some personalities in the Bible as well as in the Quran, the names listed 'Jesus', 'Mary', 'Moses', 'David' and 'Abel'. In the dispersion graph, the black lines indicate that the presence of the word and the white shows the absence. In the Bible, the word 'Jesus' starts only from the New Testament and there is only one white line indicate the absence of the word in the book of 3rd John. Similar to the word 'Jesus', 'Mary' is also present only in the New Testament. In the Quran, due to the differences in chronological order of the chapters, the words are spread in the corpus. The religious text Quran is not according to the subject matter or the order of revelations to the prophet. Then 'Moses' and 'David' also present in both corpora according to the narrative time and finally the word 'Abel' is only present in the Bible.

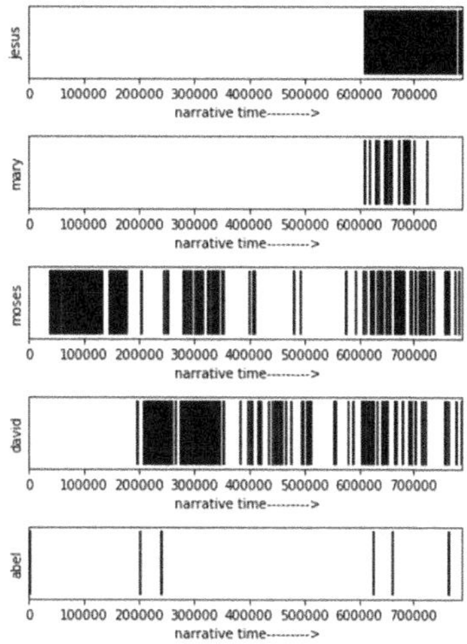

Fig. 8 Bible dispersion graph

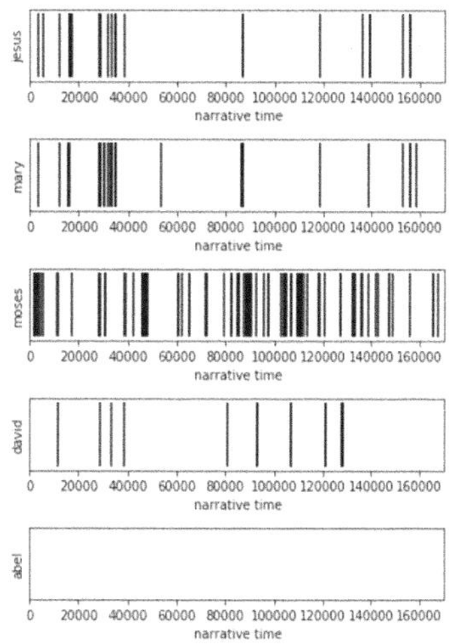

Fig. 9 Quran dispersion graph

6 Semantic Similarity Analysis

Semantic similarity is one of the challenging problems in the area of text data mining or text analytics. The semantic similarity can be measured by using various distance measures and metrics. These measures and metrics calculated the numerical difference and the closeness between two entities, and that returns a whole number [12]. The similarity and dissimilarity represent the term proximity, normally the proximity measures are represented in the interval [0, 1]. A distance or similarity metric is used to calculate the proximity between texts, but all similarity measures are not metrics. The degree of similarity can be represented by the d(u, v), where d is the distance measure, u and v are the text document entities. If the d is said to be the distance metric if it satisfies the conditions listed below: [13, 14].

1. The between two words must be always non-negative.
2. If the two words are identical then the proximity of the words must be zero.
3. The proximity measure is symmetric in all cases.
4. The measure should agree with the triangle inequality property, $d(u, w) \leq d(u, v) + d(v, w)$.

If the distance measures violate any of these conditions, it will be non-metric or also known as divergence. The distance or similarity measures can be classified into different families based on the concepts and changing the value of the parameters [15–17]. Some of the popular families are Lp family or Lp norm and Lp norm is subdivided into Lp Minkowski, L1 norm and L2 norm. Intersection, Inner product family, Matching based family, Fidelity and squared chord family and entropy family are some other families for similarity. The types of similarities are mainly classified into Term or token, sentence and Document or corpus similarity. The term similarity calculates the proximity between two entities, which means it measures the similarities between the root vector and term vectors. Some of the popular term similarity measures are Hamming, Levenshtein, Euclidean, Cosine, Manhattan distances and cosine similarity. The proximity between sentences and documents or corpora are termed as sentence similarity and Document similarity respectively. There may be two or more sentences or corpus for the estimation of the proximity measures. Some of the popular document similarity measures are Cosine similarity, Hellinger- Bhattacharyya distance and Okapi Best Match 25 Ranking and its various versions; all these methods can estimate the similarity measure irrespective of the size of corpora.

The cosine similarity of two tokens is estimated based on the value of the cosine angle and it is ranging from 1 (cos 0) to −1 (cos 180). Cosine similarity measure ignores the magnitude and it retrieves the similarity between the texts with variant sizes of texts, but invariant to a linear transformation. The complement of cosine similarity is cosine distance and it is derived from Cosine similarity. The Cosine similarity is the ratio between the dot product of the term vectors u and v and the Product Euclidean norm [18]. $u.v = ||u||||v||\cos\theta$, Where θ is the angle between the term vectors. The Cosine similarity is

$$\cos_{sim}(u, v) = \cos(\theta) = \frac{u.v}{||u||||v||} = \frac{\sum_{i=1}^{n} u_i v_i}{\sqrt{\sum_{i=1}^{n} {u_i}^2}\sqrt{\sum_{i=1}^{n} {v_i}^2}} \tag{1}$$

Cosine distance is estimated Euclidean distance like comparison but estimated in a normalized base and u and v are the term vectors

$$\cos_{dist}(u, v) = 1 - \cos_{sim}(u, v) = 1 - \cos(\theta) = 1 - \frac{\sum_{i=1}^{n} u_i v_i}{\sqrt{\sum_{i=1}^{n} {u_i}^2}\sqrt{\sum_{i=1}^{n} {v_i}^2}} \tag{2}$$

Hellinger-Bhattacharyya is an f-divergence function for document similarity measurement, which can calculate the variation between two probability distributions

U and V.

$$\mathrm{hbd}(u, v) = \frac{1}{\sqrt{2}}\sqrt{\sum_{i=1}^{n}(\sqrt{u_i} - \sqrt{v_i})^2} \tag{3}$$

where u = (u1, u2, u3, u4......) and v = (v1, v2, v3, v4......) are the document vectors with length n. The advantage of Hellinger- Bhattacharyya distance is efficient in the estimation of both discrete and continuous probability distributions.

Okapi Best Matching 25 [19] Ranking technique is a similarity measure based on probabilistic relevance, which is normally used for document ranking. The function uses the bag of words model for extracting matching corpus with respect to the input statement. Query Document (QD) input document with 'n' number of terms, such as q1, q2, q3, q4......., qn). The Best Matching 25 returns the score between Query Document (QD) and Corpus Document (CD) can be represented as in (4). The Okapi best matching 25 ranking estimates the rank of CD with respect to CD and it results from the inverse term frequency of the tokens in CD. The higher score shows more similarity between the corpus and query documents. The f(qi,CD) returns the term frequency of the Corpus Document and |CD| is the total length of CD. The avgdl parameter shows the average length of the document in the corpus.

$$\mathrm{BM25}(CD, QD) = \sum_{i=1}^{n} \mathrm{idf}(q_i).\frac{f(q_i, CD).(k_1 + 1)}{f(q_i, CD) + k_1.\left(1 - b + b.\frac{|CD|}{\mathrm{avgdl}}\right)} \tag{4}$$

7 Vectorization and Weighing Schemes

The text data needs to be transformed into a standard word vectorization format from an unstructured or semi-structured format. Here comes the role of Word embeddings, these are the representation of document vocabulary as vectors. Any statistical techniques or Machine Learning—deep learning algorithms are executed only after the conversion to the vectors [20]. There are various types of Sparse or Frequency Based vectors and Dense or Prediction Based vectors models. The semantic information of a word depends on the context words associated with it. The sparse vector representation holds limited semantic information and it also exploits huge memory, but the dense vector or prediction-based representation models can preserve more semantic information efficiently without sacrificing huge memory. The methods in Word2Vec word vectors like Skip-gram and CBoW [21] can perform algebraic operations on word vectors by adding and subtracting and by distance calculation and they can also predict the word from the existing context word. The process of conversion corpus to the word vectorization using any vectorization models [22]. The vectorization models are mainly classified into sparse and dense vectorization models [23] based on the representation of tokens into numerical values in a vector space. One of the examples for the sparse vectorization models is One hot encoding. and it encodes the features in the form of 0 and 1. But these binary lists in this model cause the loss of memory in the case of higher dimensionalities. Then the mathematical or machine learning algorithms or statistical methods can be implemented corpus.

Feature Extraction or feature engineering is one of the core concepts in Machine Learning, which extracts the most relevant attributes from the data for prediction, classification and clustering based on categorical or numerical values. Bag of Words (BoW) and tf-idf (Term Frequency and Inverse Document Frequency) are the major feature-extraction techniques [24]. BoW contains a multi-set of words, but it will not consider the grammar and word order in the text. Bag of words only keeps the word frequency. BoW disregards the semantic or synonyms of the words and considers word frequency as a classification feature. The BoW gives flexibility and identifies and implements efficiently. Due to the sparse model, it is inadequate for the higher dimensionality corpus and consequently, this sparsity nature leads *t* memory leakage. This model provides more attention to the frequent words even if do not carry any semantics and are not much focused on the infrequent words even if it carries any vital information. This model does not capture any semantics in the sentences with the same word but different word order. For example, it does not return any similarity changes between "*cat jumped at the dog*" and "*dog jumped at the cat*". This model is hardly detecting acronyms, abbreviations and chat texts and other misspelled words.

TF (Term Frequency) IDF (Inverse Document Frequency) weighing scheme is a numerical statistic used for the ranking of search engines and document information retrieval. The *tf* measures the term vector frequency of a corpus. The *idf* eliminates the drawback of the BoW model, idf provides attention to the most significant infrequent words. The *idf* is the logarithmic scaling of the ratio between the total number of documents and the number of documents with term vector in the document. The *tf-idf* keeps the relevant word score and lowers the frequent word score that is easy to get the document similarity. The *tf-idf* can be calculated using $tfidf = tf \times idf$ and the *idf* can be measured by Eq. (5).

$$idf(w) = 1 + log\frac{N}{(1 + df(w))} \tag{5}$$

where df(w) is the number of document frequency with the term vector. The *tf-idf* weighing scheme hardly retrieves the semantic information. The *tf-idf* model weak to handle the multiple resolutions of the words like synonyms, Homonyms, polysemes and much more. This weighing scheme does not any closeness between paraphrased sentences. For example, consider the following sentences "*the plane leaves at 12 pm*" and "*the flight departs at noon*". The model does not show any similarity between these sentences. Cosine similarity and Hellinger- Bhattacharyya distance uses the *tf-idf* and OKAPI Best Matching uses BoW model.

8 Perfect Matching Algorithm

The novel method, Perfect Matching Algorithm (PMA), developed to reduce the over penalization on context corpus with taken into account the length of both Query and Context Documents by the length normalization. The first phase is the Text Normalization, which includes Tokenization, special characters and stops words removal Lemmatization, and Named Entity Recognition. The second phase is the information retrieval with lexical analysis and finally the semantic similarity extraction. In addition to PMA, a contraction map is included in the implementation. The experimental results exhibit PMA achieves accuracy and improved efficiency in semantic extraction by pivot length normalization. The PMA can be calculated using the Eq. (6). Where CD is the Context Document, QD is the Query Document, *idf* is the vectorization of inverse document frequency and k_1 is the constant value.

$$PMA(CD, QD) = \sum_{i=1}^{n} idf(QD)\left(\frac{f(CD, QD).(k_1 + 1)}{f(QD, CD) + k_1.\left(\frac{|CD|}{|QD|.k_2}\right)}\right) \quad (6)$$

9 Results and Discussion

The dataset has been selected as King James Version of the Bible with its 66 books for chapter-wise similarity analysis. For the semantic similarity, analyses use the libraries in jupyter notebook sklearn and also import the cosine_similarity, for the feature extraction uses the *tfidf* vectorizer. After the pre-processing steps such as removal of stop words, unnecessary symbols and numeric values from the corpus. Lemmatization was used instead of stemming for capturing the semantic content of the text. Unlike stemming, lemmatization preserves the root word with meaning. For the analysis of the Quran, it has been divided into 30 Juz and not considered by chapters, because of the uneven nature of the chapter verses that may compromise the proper analysis and visualization of the results. Figure 10 shows the similarity analysis of 66 Bible books with 30 Quran Juz and Fig. 11 shows the similarity analysis of 30 Quran Juz with 66 Bible books. The graph shows the similarity results according to the chapter-wise total and the curvature of the graph shows the accuracy of similarity measures.

Table 2 shows the results of the most similar Quran Juz with the Bible chapters with a similarity percentage is greater than 40 extracted using Cosine Similarity. The results illuminate some interesting insights such that the Juz 16 named Qāla'alam shows the most similarity with these Bible Books Genesis, Exodus, Deuteronomy, Job, Psalms, Isaiah, Jeremiah, Ezekiel, Micah, Mathew, and Luke. Juz 16 starts from surah 18 ayat 75 to surah 20 ayat 135. Surah 19 named Maryam (named after 'Mary', the mother of Jesus) mentioned the birth of John, the son of Zachariah, the birth of

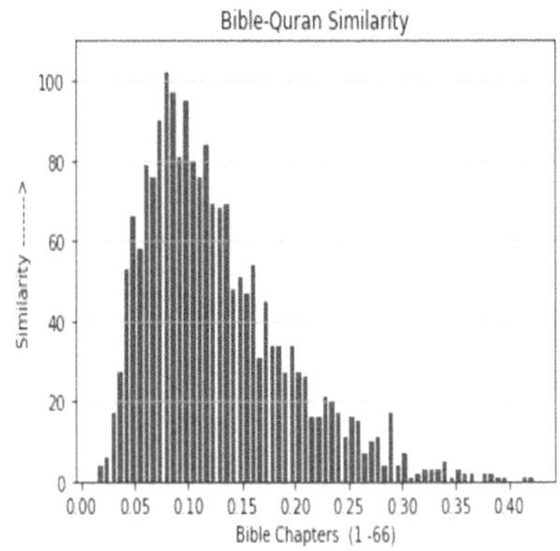

Fig.10 Bible–Quran Similarity

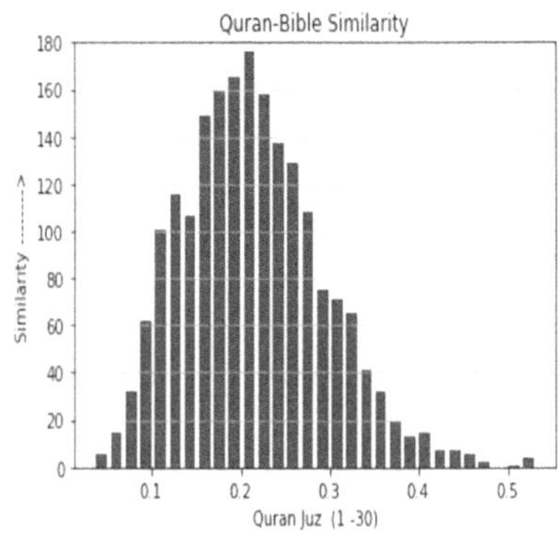

Fig.11 Quran–Bible Similarity

Jesus and also states about the generations of Abraham, Isaac, and Jacob. Surah 20 recites about the birth of Moses is the same as in the second book of the Bible Exodus chapter 2 and also mentioned the missions together with Moses and Aaron against Pharaoh for the rescue of Israel people.

The similarity measure comparison of the methods in Fig. 12 shows the accuracy rate in semantic capturing concerning the length Normalization and less penalization. Cosine Similarity and BM25 Plus show higher accuracy. PMA, the proposed method can achieve the accuracy with Contraction Map than all other methods.

10 Conclusion

The information retrieval based on semantic similarity is an arduous task because of the internal elucidation, aeon, allegorical pattern, semi-structured nature, and the phrases, polysemes, and paraphrases. Traditional word-based retrieval techniques and methods are not sufficient to extract semantic information contents from the corpus. Instead of the traditional vectorization models, the prediction and space-based

Table 2 Similarity percentage of the Bible chapters and the Quran Juz

Bible books	Quran Juz with similarity percentage							
Genesis	12 (46)	16 (41)	–	–	–	–	–	–
Exodus	15 (40)	16 (45)	–	–	–	–	–	–
Deuteronomy	9 (42)	12 (45)	15 (50)	16 (53)	19 (43)	20 (40)	30 (44)	–
Job	15 (45)	16 (44)	30 (40)	–	–	–	–	–
Psalms	9 (43)	15 (52)	16 (53)	19 (44)	20 (40)	30 (46)	–	–
Isaiah	8 (41)	12 (43)	15 (47)	16 (52)	19 (42)	20 (40)	29 (41)	30 (48)
Jeremiah	16 (41)	–	–	–	–	–	–	–
Ezekiel	16 (42)	17(46)	–	–	–	–	–	–
Micah	16 (44)	30 (40)	–	–	–	–	–	–
Mathew	16 (40)	–	–	–	–	–	–	–
Luke	8 (40)	12 (46)	15 (42)	16 (46)	19 (42)	–	–	–

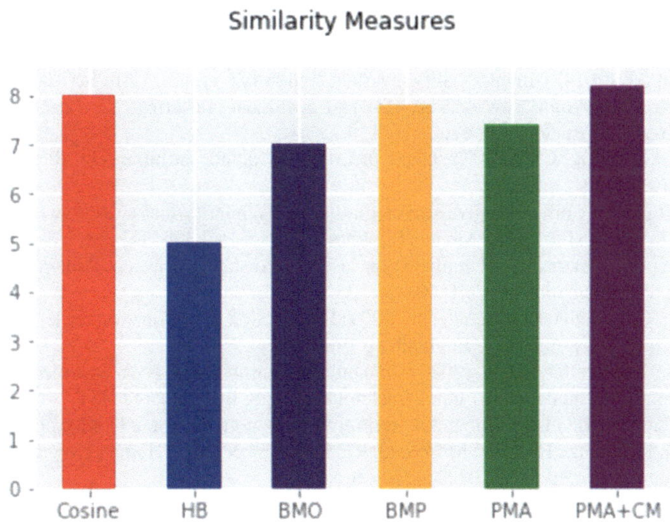

Fig. 12 Similarity measures

models can extract more semantic information and also able to capture polysemic words by using improved semantic approaches for word representations such as Word Vectors, GloVe (Global Vectors), Fast text, and Transformer models. Knowledge-based models and deep learning concepts can extract more semantically relevant data efficiently than the traditional mathematical and statistical methods.

References

1. Muhammed SH (2014) An automatic similarity detection engine between sacred texts using text mining and similarity measures. Rochester Institute of Technology
2. McDonald DM (2014) A text mining analysis of religious texts. J Bus Inq 13(1):27–47
3. Verma M (2017) Lexical analysis of religious texts using text mining and machine learning tools. Int J Comput Appl 168(8):39–45
4. Taa A, Abed QA, Ahmad M (2018) Al-Quran ontology based on knowledge themes. J Fundamental Appl Sci 9(5):800–810
5. Hegazi MOA, Hilal A, Alhawarat M (2015) Fine-grained Quran dataset. Int J Adv Comput Sci Appl 6(12):308–313
6. Popa RC, Goga N, Goga M (2016) Ontology learning applied in education: a case of the new testament. The European proceedings of social & behavioural sciences EpSBS, Edu World 2016 7th ınternational conference, pp 1032–1039
7. Popa RC, Goga N, Goga M (2019) Extracting knowledge from the Bible: a comparison between the old and the new testament. International conference on automation, computational and technology management (ICACTM). IEEE, New York, pp 505–510
8. Firth JR (1957) The technique of semantics. Papers Linguistics 37(2):191–200
9. Miller K (1993) Five papers on wordnet. Technical report. Prinston University, Prinston.
10. Baeza-Yates R, Robeiro-Neto B (1999) Modern information retrieval. ACM Press Books
11. Zipf GK (1949) Human behaviour and the principal of least effort. Addison-Wesley
12. Hassanat AB (2014) Dimensionality invariant similarity measure. J Am Sci 10(8):221–226
13. Huang A (2008) Similarity measures for text document clustering. New Zealand computer science research student conference, pp 1–8
14. Kocher M, Savoy J (2017) Distance measures in author profiling. Inf Process Manage 53(1):1103–1119
15. Cha S-H (2007) Comprehensive survey on distance/similarity measures between probability density functions. Int J Math Models Methods Appl Sci 1(4):300–307
16. Cha S-H (2008) Taxonomy of nominal type histogram distance measures. American conference on applied mathematics (MATH ‘08), pp 325–330
17. Manning CD, Raghavan P, Schütze H (2009) Introduction to information retrieval. Cambridge University Press, Cambridge, England, pp 192–195
18. Varghese N, Punithavalli M (2020) Semantic similarity analysis on knowledge-based and prediction based models. Int J Innov Technol Exploring Eng 9(6):447–481
19. Lv Y, Zhai C (2011) Lower-bounding term frequency normalization. CIKM, pp 7–16
20. Cotterell R, Schütze H (2019) MorphologicalWord-embeddings. Human Language Technologies. Preprint at arXiv: 1907.02423, pp 1–6
21. Mikolov T, Chen K, Corrado G, Dean J (2013) Efficient estimation of word representations in vector space. Preprint at arXiv: 1301.3781, pp 1–12
22. Mikolov T, Sutskever I, Chen K, Corrado G, Dean J (2013) Distributed representations of words and phrases and their compositionality. Preprintat arXiv: 1310.4546, pp 1–9
23. Varghese N, Punithavalli M (2020) Word vector representations: sparse versus dense vectors. Working Papers Linguistic Lit 12(1): 360–367
24. Kwon YM, Jun SH, Gal WM, Lim MJ (2018) The performance comparison of the classifiers according to binary bow, count bow and Tf-Idf feature vectors for malware detection. Int J Eng Technol 7(3):15–22

Chapter 17
Reliability Enhancement in Harmony with Prudent Coding for Flight Critical Embedded Automatic Control Software

Shobha S. Prabhu and H. L. Shashirekha

1 Introduction

Critical automatic control systems are built with intelligence and majorly composed of application oriented embedded computer using requirement specific software with deterministic timing schedule. These critical systems with real-time behavior are considered to be functioning precisely when the system produces logically correct output with strict timelines, irrespective of system load, or adverse environmental conditions. Accurate system predictability is expected from such critical systems irrespective of unpredictable external or internal stimulus. Examples of automatic critical control systems are in the field of medical and aircraft applications, where defects directly cause loss of life/mission. Core quality of critical control systems is achieved through its intelligent fault tolerant embedded software with high reliability and safety. Reliability is a dynamic system characteristic, which is depicted through software failures and efficiency. Appropriate methodologies are being identified and studied in order to check their effect on enhanced reliability of critical automatic embedded control software.

Software reliability is the probability of failure-free operation of software over a specified time within a specific environment for its intended purpose. This indicates that reliability depends on the availability of continued performance of the software functionality during the complete operational period irrespective of external environmental conditions. Working in tandem with software life cycle processes, the reliability is considered as basic non-functional characteristic to be complied in safety critical control software. Hence, while building highly reliable software, prime importance is given to error prevention through assessment of static software quality

S. S. Prabhu (✉)
Gas Turbine Research Establishment, Defense R&D Organization, Bangalore, India

H. L. Shashirekha
Department of Computer Science, Mangalore University, Mangalore, India

G. Mathur et al. (eds.), *Proceedings of 2nd International Conference on Artificial Intelligence: Advances and Applications*, Algorithms for Intelligent Systems,
https://doi.org/10.1007/978-981-16-6332-1_17

attributes prior to testing. It is expected to keep strict vigil on the quality/reliability through its proactive fault tolerance and removal of coding errors identified through stringent static analysis process. Analyzing the static software attributes in order to improve the quality of software is called static analysis [1]. This analysis is very useful in exposing errors that may escape during dynamic execution of the software through various testing process. Detection and mitigation of such software errors/faults at early stages of software implementation prior to testing process results in improved software reliability or availability [2]. Using formal method-based tools to carryout static analysis ensures reducing the turnaround time and cost to fix the defects at a later stage.

The organization of this paper is as given here: Section 2 explains the relevance of reliability and the software failures attributed to code for the case study. Section 3 talks about the survey carried out and the motivation to identify the coding attributes to improve the reliability. Section 4 presents the methodologies adopted to implement the relevant concepts during the software implementation phase. This section highlights the relationship between the prudent coding practices and the reliability enhancement in critical airborne automatic control software. Section 5 concludes the study with some scope for improvement in future.

2 Background and Current Work

The case study of **Sa**fety **C**ritical **S**oftware for **E**mbedded **R**eal-time **C**ontroller (SACSERC) for an airborne system along with the processes carried out to build better reliability is discussed here. This intelligent airborne system provides fault tolerant engine control and monitoring of the unmanned aerial vehicle propulsion over the complete flight envelope with its software featuring functionality, performance, reliability, and safety aspects embedded in it [3]. This automatic control software enables the system to work efficiently in terms of functionality and safely in terms of reliable performance [3]. SACSERC is expected to have Mean Time Between Failures (MTBF) of 1 lakh hours (continuous running for more than 10 years without failures), i.e., with a reliability score of more than 0.9 [4]. This flight critical embedded control software is developed with approximately 1 lakh Lines of Code (LOC), and its reliability is highly depending on the various static software attributes. This work includes the study of such attributes for their moderation, highlighting reliability enhancement through prudent coding methodologies. Coding is said to be prudent if it is written wisely and cautiously to build intelligence, which under any circumstances will lead the system to a safe state. Planned and careful judgment is required while implementing certain critical logics to avoid risks and uncertainties that may be executed during unexpected events. Prudent coder is always vigilant of potential threats and attentively takes care of such consequences during the software implementation.

Reliability, maintenance, and efficiency are basic non-functional quality characteristics which are highly dependent on the best coding practices for any critical system. Since software failures are one of the most dominant causes of failures in today's critical systems, reliability directly depends on the failure free functioning of the software. MTBF and Mean Time to Recover (MTTR) are basic metrics which are directly linked with the failure density/recovery time and at the same time indirectly dependent on the quality of coding attributes to a great extent. Even though it is impossible to eliminate all the faults in the software, usage of well-defined development process, adoption of relevant standards, and software engineering best practices will always help in producing better reliable intelligent systems in repeatable and predictable way.

Since airborne DO-178B/C process with Level A (highest level of safety) guidelines is followed for developing the SACSERC, achieving better reliability through minimizing faults/failures [5] is embedded in the SDLC itself. In addition, reliability depends much on the software code and its identified quality attributes with respect to its traceability, compliance, and invisible run-time errors. These attributes could be improved during coding phase of SDLC to obtain a better performance, functionality, and safety.

Experiments carried out are described here to prove the importance of prudent coding for the reliability enhancement of airborne embedded control software. The baseline static software attributes namely number of lines of dead code, compliance to coding standards, and extant of run-time errors hidden in the source code are considered along with the standardized process.

3 Related Work

Every researcher applies unique techniques for improvement which is suitable for specific software [6]. Literatures were surveyed in this regard to find and derive a suitable methodology for enhancing the reliability of SACSERC. The essence of limitations, improvements, and future scope of work is taken as foundation and utilized for yielding improved reliability.

Rosenberg et al. [7] discussed the improvement of quality and reliability of software products in NASA projects through identification of metrics for critical failures. Lockhart et al. [8] highlighted the fact that additional formal techniques in the software development process are crucial for developing better reliable software systems for safety critical applications. These techniques may include static analysis process for removal of development errors and reporting of run-time errors before the actual launching of the software in order to build better reliability into the embedded system. Debasmita et al. [9] discussed the system level reliability being a combination of reliabilities of various components, strict safety constraints need to be envisaged for complex software. The authors insist on the usage of formal method based tools for software analysis to identify faults and eliminate them for reliability enhancement. Nelson [10] concluded through studies that disciplined structured software development with programming standards could contribute to improved reliability.

Krasich [2] highlighted the fact that faults which have a high impact on reliability, if discovered early during design, analysis, or testing, will help in reliability improvement with correction of potential faults. Moy et al. [11] brought out the importance of certification standard, which provides guidance for verification based software life-cycle processes for avionics software. Groce et al. [12] talked about the process of producing a good code through usage of indispensable and cost-effective method for catching unrecognizable coding errors using static analysis tools. The capabilities of traditional static analysis tools are even though limited; they achieve a higher level of confidence compared to standard software testing. Formal method based static analysis tools are better choice when it comes to safety critical systems. Dvorak [6] discussed that residual software defects will occur with near certainty in complex software systems and hence there must be a strategy for dealing with them to prevent system failures.

Literature survey highlights the impact of best practices in code in reduction of failures and enhancement of reliability during the development of flight critical automatic control software. In the case study of SACSERC, prudent coding was focused in terms of increasing MTBF for evaluating the improvement on reliability. Several exercises were carried out to eliminate remotely occurring implementation errors in order to enhance the reliability of the said critical software.

This paper accentuates the study conducted while identifying various strategies towards obtaining prudent and error free software code for SACSERC. This study also proves that the code thus obtained is complete, consistent, and traceable to the requirements through which enhanced reliability of flight critical embedded control software is ensured.

4 Methodology

Following paragraphs detail the experimental study carried out on the concepts discussed above in the context of SACSERC for an airborne application.

The basic purpose of carrying out the static analysis is to identify areas of the code potentially causing software failures through untraceable portion of code, hard coding, uncalled functions, unsafe code, probable occurrence of run-time errors, and their elimination to enhance the software reliability. During the developmental phase, mandatorily the software was exposed to rigorous static analysis, which in turn helped in unearthing obscure errors/faults (which otherwise could have escaped) prior to releasing the object code for dynamic testing.

As per the DO-178B/C (Level A) process, mandatory guidelines toward the reliability enhancement were strictly adhered to deliver the safe and reliable SACSERC for the airborne system [3]. Examples of such guidelines include complete absence of dead code, 100% compliance to suitable coding standard, accurate, and consistent algorithms to produce object code with no run-time errors.

Further, usage of formal methods is recommended for the static analysis of flight critical control software as they remove ambiguities and imprecision, enabling early detection of faults and better maintainability. Formal methods carry out the analysis based on the exhaustive set of mathematically derived test cases which increase confidence while embedding reliability in the software [3]. Since complete formalization is tough to achieve, Dino Mandrioli [13] adds his modest personal view that attempting the formal method during any stage of SDLC would bring a partial victory to the developers [3]. Daniel Jackson et al. [14] suggested the application of formal method based tools with approaches emphasizing practicality with partial modeling and analysis to overcome this obstacle. Formal method based static analysis tools were used to carry out the experiments in a disciplined manner with their consistent, repeatable, precise, quick, and cost effective way.

Following paragraphs explain the exercises carried out on the aspects stated above to prove their capability in enhancement of reliability through prudent coding. These are evaluated by relevant measurements, and effectiveness is described subsequently.

4.1 Non-Existence of Dead Code

System requirements which are performed through software possess bidirectional traceability with the embedded object code to ensure producing accurate, complete, and consistent behavior throughout the operational phase of the system. 100% traceability confirms that there does not exist any portion of code with—unimplemented requirements, incomplete design, dead code, and untested functionality/performance characteristics. Dead code means either the code will never be executed at run-time or executed but results are never used. This is considered unsafe and unreliable in airborne automatic control software as dead code may be invoked unintentionally during unexpected events and may cause hazardous effect to the system. Such instances are avoided completely through removal of dead code, thereby ensuring consistent behavior, MTBF increase, and reliability improvement of the SACSERC.

In the embedded controller of this case study, 109 system requirements are designed into moderately good software architecture for feasibility of implementation. It gave birth to baseline version of software with 590 modules/functions spread across 369 files. When the code was analyzed for existence of dead code with formal method based static analysis tool, there were 8 of them initially, covering function not called, variable not used, initialization value not used, and empty function. Even though the program compiles without errors, it is risky to have such code in automatic control software because such errors cause system failure during least expected events. This enforced the developer to rewrite the code to eliminate all existing dead code. Sample dead codes which were present in SACSERC are given in Table 1.

It is evidenced from Fig. 1 that the dead codes were removed over few iterations without compromising on any of the quality attributes. The updated version (Ver. 4) achieved 100% traceability with no dead code and increased MTBF of SACSERC. 100% absence of dead code ensured complete removal of code attributing to unexpected failures. This produced a software with MTBF of large value, and hence the enhancement in reliability of SACSERC is obtained.

Table 1 Some cases of dead code incidents in SACSERC

Sl No	Type of dead code	Sample code from SACSERC	Reason and resolution

1	Function not called	void Speed_valid (void)
		{
		Speed = validate(Sp1, Sp2);
		}
		void Temperature_valid (void)
		{
		Temp = validate(Temp1, Temp2);
		}
		void Pressure_valid (void)
		{
		Pres = validate(Pres1, Pres2);
		}
		void P3_valid (void)
		{
		}
		Void main ()
		{
		Speed_valid ();
		Temperature_valid ();
		P3_valid ();
		}
			Pressure_valid is considered dead code because, the function is not called
			P3_valid is considered dead code because, the function is empty
			The developer discussed the purpose of such requirements with system team and found that the code is built for futuristic requirements. This dead code was resolved by removing P3_valid function (futuristic) and commenting Pressure_valid (missing). It is advisory to remove them as dictated by prudent coding

2	Empty function		

3	Variables value not used	float ControlFunc(void)
		{
		float ULimit = 20.0f;
		float LLimit = -1.0f;
		float gain = 3.0f;
		float intermed = 0.0f;
		float inputSig = 13.0f;
		float outputSig = 0.0f;
		float modSig = 8.23f;
		intermed = ULimit – LLimit;
		outputSig = inputSig/modSig;
		outputSig = outputSig * gain;
		return (outputSig)
		}
			outputSig is considered dead code at the initialization stage as the variable is assigned new value without using its initialized value
			intermed is considered dead code because, it is not used both at initialization and at main execution place
			The warning is given by the tool to indicate "developer's intention may be to use the value prior to new assignment. This may result in unintended function output"
			Resolved this dead code by removing the code for intermed computation and removing the initialization of local variables: outputSig, intermed, ULimit, LLimit

4	Variables updated without using first value		

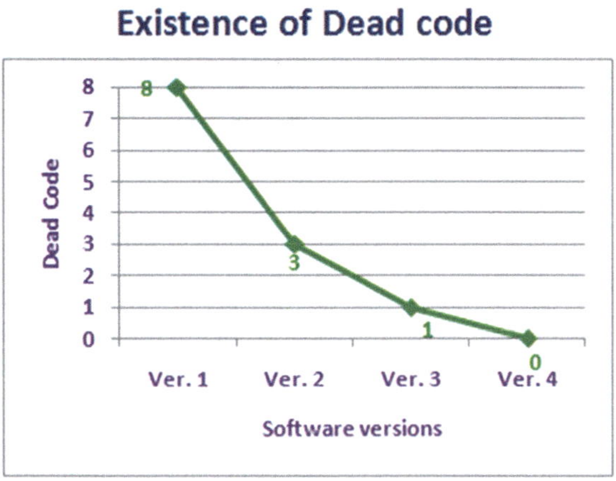

Fig. 1 Dead code in the software

4.2 Compliance to Coding Standards

Safety critical control software, in particular, needs to have a mechanism to generate the efficient code. The code thus generated should not have any grey areas which create threat to safety resulting in unexpected software failures. In case if it exists, identify and eliminate such hidden errors prior to integration, object code generation, and testing. Compliance to safe coding standard exposes software errors that may escape other verification techniques prior to running the program or dynamic execution of the software. This ensures the enhanced safety which gives rise to lesser failures in the control software resulting in higher MTBF and enhancement in reliability.

SACSERC is developed in C programming language, which is inherently flexible, mature, easily portable, and generating small and efficient code. Since 'C' language has several insecurities, software may result in failures when most unexpected and may experience system disasters during the execution of the object code. Despite its popularity and the excellent tool support available, C is suitable for use in critical systems only within firm constraints and guidelines [15]. Motor Industry Research Association (MIRA) has provided a safe subset of C language for the development of safe and reliable software for critical systems of automotive/aerospace industry which is called as MISRA C (Motor Industry Software Reliability Association C) [16]. SACSERC Coding Standard (SCS) is evolved with a combination of MISRA C 2012 mandatory guidelines along with other internally required coding rules, coding styles, and conditions/constrains/conventions suitable for seamless software development. Coding compliance with SCS helps in identifying and removing unintended faults which help in producing reliable, maintainable, efficient, and safe code.

It is not practical to manually check the compliance with the comprehensive SCS. As an easier, consistent, time saving and cost-effective option, a partial auto code generator was developed in-house to deliver software code (from design models) which is compliant to SCS during the iterative development of software [17]. For checking the manually coded portion of software, suitable formal method based static analysis compliance tool was used with appropriate configuration to capture the non-compliances with the said coding standard. When the first version of the SACSERC was checked, few non-compliances were present and they are given in Table 2.

Initially, basic version of the SACSERC was developed with the standard process and it was compiled without errors. Subsequently, the software was checked for code compliance with respect to SCS to ensure 100%. It is observed from Fig. 2 that first version of the software was having a compliance of 93%. Though non-compliances were not affecting the system performance visibly, they may affect the system unexpectedly during operational execution, reducing the MTBF. During the process of this static analysis, the non-compliances were studied thoroughly for their respective disadvantages, analyzed in terms of reliability and eliminated them subsequently by redesigning the algorithms appropriately to comply with the coding standard. Automatic control software got the complete facelift while making it compliant to coding standard, thereby ensuring removal of hidden errors, increasing MTBF, and enhancing software reliability.

4.3 Absence of Run-Time Errors

Main target while enhancing reliability is to detect errors and eliminate them during critical software development. As discussed above, confidence on code quality increases through disciplines adopted for coding and avoiding all threats during traceability and compliance related analysis. Further, run-time error detection is a static activity carried out to evaluate errors occurring during run-time prior to the testing/operational phase. This will benefit the end user in identifying the hidden run-time errors and aids in eliminating them to produce the higher availability of software. In fact, this activity contributes immensely to reduce critical errors while producing enhanced reliability through increased MTBF.

Formal method based sound static analyzer tool is utilized for detecting run-time errors and data race conditions. The tool raises notifications called alarms about the presence of potential run-time errors or invalid concurrent behavior or any such discrepancies in the program. First version of the SACSERC was exposed to this tool, and errors/alarms are observed as given in Fig. 3. There were no errors, no alarms due to memory locations, no flow anomalies in the code, no MISRA C rule violations but there were 25 run-time errors. This means, even though there are no direct faults/alarms, there exist portion of code which may create run-time errors during execution. Important among these were the following: Invalid Function Calls, Invalid Ranges and Overflows, Division or Modulo by Zero. Basically, during the execution of the object code during the run-time, these errors cause unpredictable results, data races, predictable over approximation outcome, and output undesired behavior. Such kinds of errors are completely unaccepted for safety critical automatic

Table 2 Some cases of non-compliances in SACSERC

Sl No.	Type of non-compliance	Sample code from SACSERC	Reason and resolution

1	Mixed mode arithmetic/Data type mismatch	Unsigned int T1average (unsigned int T1_1, unsigned int T1_2)
		{
		unsigned int ret_value;
		ret_value = (T1_1 + T1_2) /2;
		return ret_value;
		}
			In this case, the ret_value computed is a float and the output of the function is unsigned int. The function output will lose on the data value due to data type mismatch. Casting needs to be carried out consciously, when the developer agrees with the data loss. To remove the non-compliance, the statement was rewritten as ret_value = (unsigned int) (T1_1 + T1_2) /2;
			Operands on both sides of operator should be of same type to avoid non-compliance
			Alternatively, function output and ret_value both can be made as float to resolve this

2	Unreachable code	unsigned int status (unsigned int Input)
		{
		unsigned int ret_value;
		float divide;
		divide = Input/5;
		if (divide > Input)
		ret_value = 1;
		else ret_value = 2;
		return ret_value;
		}
			In this case, only the 'else' part is executed irrespective of the input. The 'if' case is never executed and hence it is unreachable. The function is defined wrongly or the code is inconsistent with the requirement
			To avoid non-compliance, the requirement was revisited to revise the function definition as unsigned int status (int Input)

3	Declaration compatibility with static during external linkage	variable declaration with same names in different files	A variable or function that is of file scope and not used in another file shall be declared with the static qualifier. This is mandatory rule to avoid any inconsistencies due to similar names of parameters in different files

(continued)

Table 2 (continued)

Sl No.	Type of non-compliance	Sample code from SACSERC	Reason and resolution

4	Right syntax but wrong and unintended statement and output	unsigned int PreStatus (int Limit1, int Limit2)
		{
		unsigned int ret_value = 0;
		if (Limit1 = Limit2);
		{
		ret_value = 1;
		}
		return ret_value;
		}
			Even though the compiler doesn't give warning or error, the ret_value was always giving wrong value as 1
			There were non-compliances as follows:
			1. Statements under if loop shall always be enclosed in braces
			2. Assignment operators shall not be used in expressions
			Existence of non-compliances showed that the code is implemented wrongly. The compliance is obtained as follows:
			1. if (Limit1 = Limit2) Wrongly implemented semicolon at the end of 'if'statement is removed
			2. Assignment is converted to intended equality. Statement is rewritten as if (Limit1 = = Limit2)

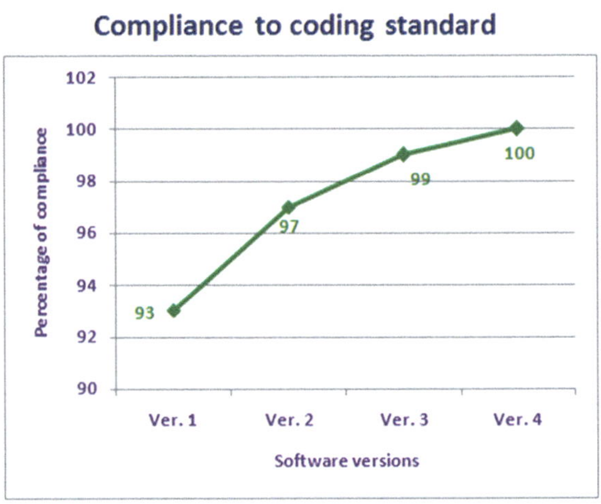

Fig. 2 Growth of code compliance

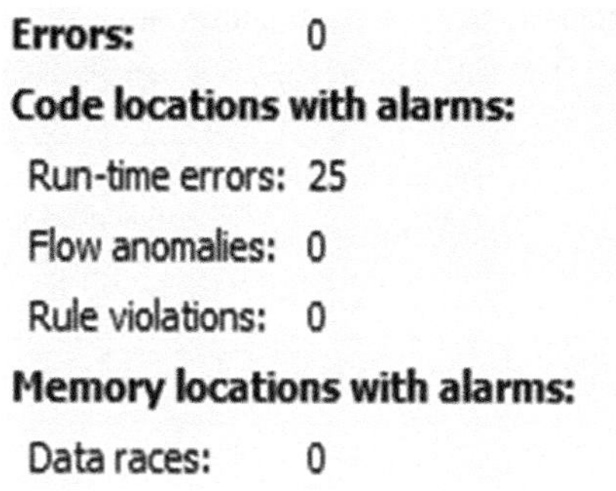

Fig. 3 Details of errors and alarms

control software and need to be resolved by removing their traces. Details of the errors found in first version of SACSERC are described in Table 3.

Figure 4 represents the improvements carried out through prudent coding over various versions of software and the reduction in number of run-time errors. It is also observed that final version did not contain either run-time errors or invalid concurrent behavior. This gives better confidence to the end user during the system acceptance process. The improved version of the control software SACSERC ensured removal of hidden run-time errors and threats, ensuring enhanced MTBF and software reliability.

5 Conclusions and Future Work

Reliability, as an indicator of quality and performance of embedded automatic control software, required to be enhanced through improvement in software attributes. In this paper, the important coding attributes which are affecting the reliability of automatic control software are highlighted. Static analyses are carried out on the software to identify the fault creating areas and eliminated them proactively by adopting prudent coding methodologies before they create emergency in the flight critical automatic control systems. This process of removing static defects thus enhanced the reliability through the increased MTBF and decreased MTTR of the system significantly.

Airborne control software SACSERC follows DO-178B based process keeping reliability in view. During the static analysis, it is observed that the coding best practices adopted while implementing the software reduced the code related alarms, warnings, errors, and faults, not only during the generation of executable object code but also during the run-time. The process of building this kind of safe and reliable code is achieved in automatic control software through prudent coding techniques. These techniques include non-existence of dead code, 100% coding compliance and absence of all run-time errors/alarms. This emerged with the fault-free software to give rise to enhanced reliability to the intelligent embedded control system.

The contribution of improvement in software reliability through moderation of metrics and elimination of static defects need to be studied in detail in many similar projects to evolve a common model which suits all safety critical control systems.

Table 3 Some cases of run-time errors in SACSERC

Sl No	Type of run-time error	Sample code from SACSERC	Reason and resolution
1	Invalid Function Calls	Usage of *math.h* library functions *sqrt* and *pow* in the code to obtain square root and power of a variable	Since the values are obtained through calls to external C library functions, their declarations are not available in the code. This is considered as a serious threat since unknown and undeclared functions are not allowed in safety critical automatic software. As per the safety norms, these library functions are re-written during coding phase to replace the *math.h* library functions and used appropriately by calling valid declared functions to avoid any run-time errors

2	Invalid Ranges and Overflows	float params[20];
		void initData(void);
		{
		for (i = 0;i < 35;i + +)
		params[i] =
		readData(i);
		}
			In this case, params array is of size 20 and we are filling the array with readData. The software is compiled without errors. But while checking for run-time error, tool throws 'Invalid range'
			The reason for this is: The params array index is from 0 to 34 while assigning output of readData function (which is declared with size 20)
			The solution to this is to increase the params array to the required number of readData function outputs

3	Division by Zero	Usage of operator division in a mathematical operation	This was a warning and in SACSERC, removal of this alarm situation was already taken care during the coding phase by including various exception handlers such as divide by zero. Before every instance of division module, divisor was checked for non-zero value and appropriate safe action was taken

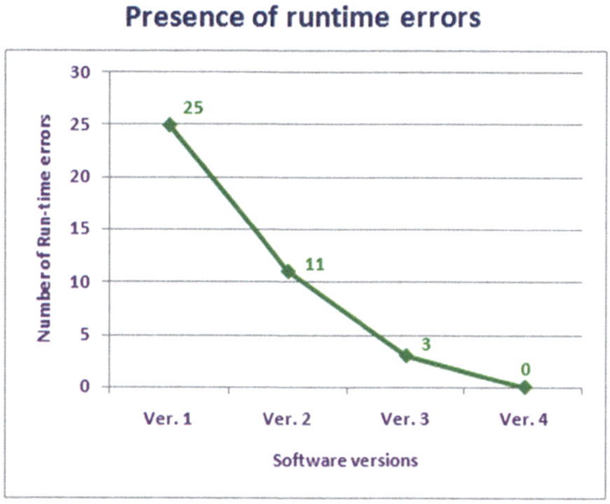

Fig. 4 Trend line for run-time errors

This is an area of research which is yet to be evolved for deriving a model for reliability enhancement and is a mammoth task for any critical system. It is necessary to study various software attributes, metrics, and factors affecting MTBF to understand their influence on the reliability. Airborne embedded automatic control software development where reliability stands strong requires this area of indagation to mature.

Acknowledgements The authors are thankful to Director, Gas Turbine Research Establishment, DRDO, for permitting to publish this paper. The authors also thank all those personnel who supported the detailed study presented here.

References

1. Thakur D http://ecomputernotes.com/software-engineering/what-is-static-analysis-how-is-it-performed-what-are-its-uses
2. Krasich M (2015) Modeling of SW reliability in early design with planning and measurement of its reliability growth. In: IEEE conference on reliability and maintainability symposium (RAMS). IEEE, New York
3. Shobha P, Hem K, Shashirekha HL (2018) Safety critical embedded software: significance and approach to reliability. 7th international conference on advances in computing, communication and informatics, ICACCI-2018. IEEE, New York, pp 449–455
4. Shobha PS, Shashirekha HL (2020) Effectiveness of software metrics on reliability for safety critical real-time software. In: Congress on intelligent systems (CIS2020), soft computing research society, world conference
5. Radio Technical Commission for Aeronautics (2011) Guidelines, standard for: software considerations in airborne systems and equipment certification. RTCA DO-178B/C
6. Dvorak DL. NASA study on flight software complexity. Commissioned by the NASA Office of Chief Engineer. https://www.nasa.gov/pdf/418878main_FSWC_Final_Report.pdf
7. Rosenberg L, Hammer T, Shaw J (1998) Software metrics and reliability. In: IEEE international symposium on software reliability engineering. IEEE, New York

8. Lockhart J, Purdy C, Wilsey P (2014) Formal methods for safety critical system specification. IEEE 57th international midwest symposium on circuits and systems (MWSCAS). IEEE, New York, pp 201–204
9. Debasmita L, Dey S (2015) Integrating formal methods with testing for reliability estimation of component based systems. IEEE international symposium on software reliability engineering workshops (ISSREW). IEEE, New York, pp 33–36
10. Nelson E (1978) Estimating software reliability from test data. Micro Electron Reliab 17(1):67–73
11. Moy Y, Ledinot E, Delseny H, Wiels V, Monate B (2013) Testing or formal verification: DO-178C alternatives and industrial experience. IEEE Softw 30:50–57
12. Groce A, Havelund K, Holzmann G, Joshi R, Xu R-G (2014) Establishing flight software reliability: testing, model checking, constraint-solving, monitoring and learning. Ann Math Artif Intell 70(4):315–349
13. Mandrioli D (2015) On the heroism of really pursuing formal methods. In: IEEE/ACM 3rd FME workshop on formal methods in software engineering, FORMALISE'15, pp. 1–5.
14. Jackson D, Wing J (1996) Lightweight formal methods. IEEE. Computer 29(4):21–22
15. Hatton L (1995) Safer C: developing software for high-integrity and safety-critical systems. McGraw Hill Book, Maidenhead, UK
16. Misra C (2013) Guidelines for: the use of the C language in critical systems. Motor Industry Research Association Limited, Nuneaton, UK.
17. Vinod L, Mahesh R, Hem K, Shobha P, Rao ANV (2016) An automated change process for embedded controller software of a full authority digital engine control system. Int J Comput Appl 975:5–10

Chapter 18
Multi-Location Faults in Transmission Lines: Detection and Classification

Gaurav Kapoor

1 Introduction

The two significant components of the numerical relay are fault detection and phase type classification. The overall performance of the numerical relay is decided by the speed and accuracy of these components. Many techniques have been introduced up to now for the numerical relaying of transmission lines. Several of those techniques are Wavelet packet transform (WPT) [1], discrete wavelet transforms (DWT), traveling waves (TWs) [2], DWT, and artificial neural network (ANN) [3]. Furthermore, a technique based on Bergeron model equations is reported for fault detection in [4]. Moreover, extra schemes are differential admittance [5], ANN [6], redundancy parameter estimation [7], ANN [8], fuzzy inference system [9], and probabilistic neural network (PNN) [10].

In this paper, a single-ended scheme is proposed for multi-location faults detection in transmission line based on the approximate coefficients index value (ACIV) of wavelet transform.

The application of the scheme is presented in Sect. 2. Presented FD and PTC scheme are provided in Sect. 3. The effects of simulations are illustrated in Sect. 4, while Sect. 5 concludes the paper.

G. Kapoor (✉)
Department of Electrical Engineering, Modi Institute of Technology, Kota, India

G. Mathur et al. (eds.), *Proceedings of 2nd International Conference on Artificial Intelligence: Advances and Applications*, Algorithms for Intelligent Systems,
https://doi.org/10.1007/978-981-16-6332-1_18

Fig. 1 Illustrated model

2 Application of the Scheme

The presented scheme is applied to the system of 500 kV as illustrated in Fig. 1. The model has two 50 Hz line sections of 150 km each. Numerous faults are incepted at different locations (FIL) through MATLAB simulations in SIMULINK.

3 Presented Scheme

The block diagram of FD and PTC scheme is presented in Fig. 2.

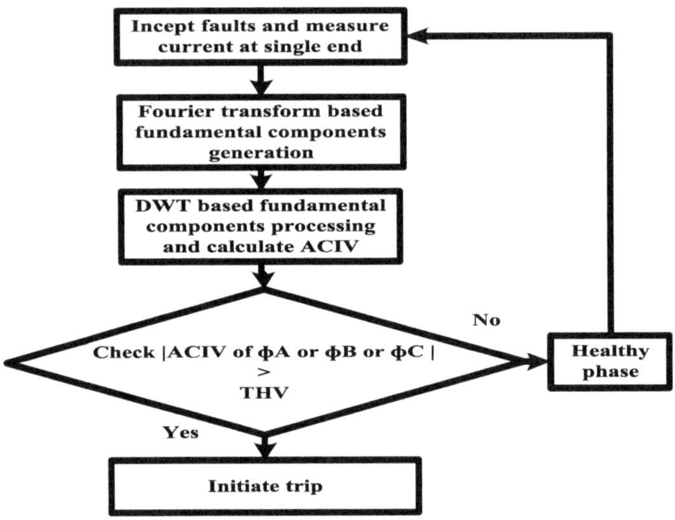

Fig. 2 Block diagram of FD and PTC scheme

4 Simulation Effects

The effects of simulations in view of numerous fault scenarios to bear out the implementation of the scheme are provided in this section.

4.1 Faults When FAT, FT, and FIL Are Varied

The operation of scheme during AG fault (FIL = 145 km) and BG fault (FIL = 155 km) is represented in Fig. 3. Both faults are incepted at 0.05 s. Trip signals are initiated when the ACIV of both phase-A and phase-B crosses the THV (threshold value) at 0.065 s and 0.075 s respectively. Thus, the FDT (fault detection time) calculated for AG fault is 15 ms and for BG fault is 25 ms. The ACIVs for some more faults are displayed in Table 1. It can be observed that the scheme operates correctly and promptly for the detection of faults under simultaneous variation of FAT (fault application time), FT (fault type), and FIL (fault inception location).

4.2 Faults When FR, FT, and FIL Are Varied

The scheme is validated for a BG fault (FIL = 159 km, FR = 5 Ω) and CG fault (FIL = 141 km, FR = 5 Ω). The current and trip waveforms are presented in Fig. 4. For

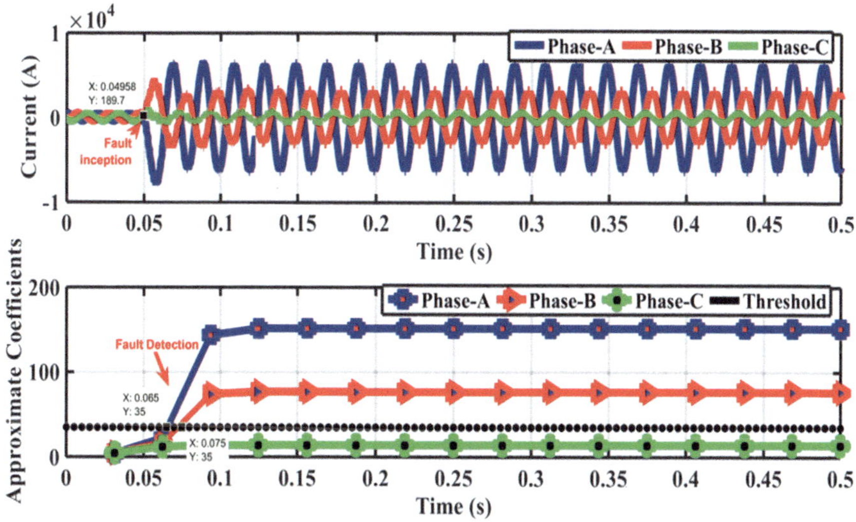

Fig. 3 Performance during AG fault (145 km) incepted at 0.05 s and detected in 15 ms and BG fault (155 km) incepted at 0.05 s and detected in 25 ms

Table 1 ACIV for faults varying FAT, FT, and FIL

FT-1 (km)	FT-2 (km)	FAT (s)	ACIV			Fault Detection Time (ms)		
			ACIV-A	ACIV-B	ACIV-C	FDT-A	FDT-B	FDT-C
AG (145)	BG (155)	0.05	151.2297	76.2724	13.6483	15	25	–
BG (157)	CG (143)	0.1	13.6232	144.5484	79.1235	–	5	25
CG (142)	AG (158)	0.2	79.6417	13.6484	154.7416	10	–	25
AG (161)	CG (139)	0.25	131.7515	13.4466	84.6377	10	–	15
CG (147)	BG (153)	0.3	13.5237	83.5969	141.6144	–	15	20

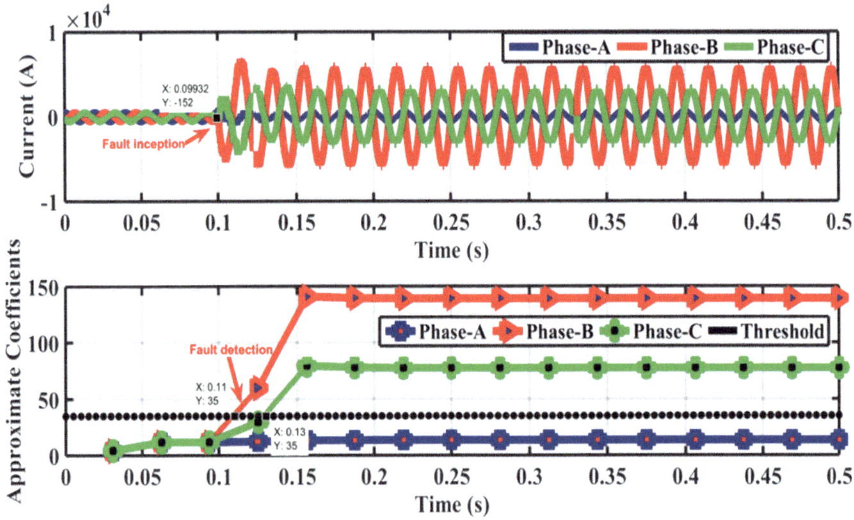

Fig. 4 Performance of scheme during BG fault (159 km) incepted at 0.1 s and detected in 10 ms and CG fault (141 km) incepted at 0.1 s and detected in 30 ms

this case study, both faults are incepted at 0.1 s, and the trip signals for BG and CG faults are initiated at 0.11 s and 0.13 s respectively. The time taken by the scheme to detect these faults is 10 ms and 30 ms respectively. Table 2 presents the performance evaluation for different faults. It is monitored that even with the inception of faults under FR (fault resistance), FT, and FIL variation; the scheme detected the faults correctly and quickly.

4.3 Faults When GR, FT, and FIL Are Varied

The scheme is verified for an AG fault (FIL = 153 km) and CG fault (FIL = 147 km). The performance is demonstrated in Fig. 5. For AG and CG faults, the faults are

Table 2 ACIV for faults incepted at 0.1 s and varying FR, FT, and FIL

FT-1 (km)	FT-2 (km)	FR (Ω)	ACIV			Fault Detection Time (ms)		
			ACIV-A	ACIV-B	ACIV-C	FDT-A	FDT-B	FDT-C
BG (159)	CG (141)	5	13.5199	141.0123	79.5917	–	10	30
AG (138)	BG (162)	15	147.8551	76.3823	13.3316	5	20	–
CG (164)	BG (136)	25	13.7073	74.9530	121.8786	–	20	25
BG (144)	AG (156)	35	67.6927	123.1385	13.8713	10	20	–
AG (160)	CG (140)	45	109.8951	13.7599	65.7908	15	–	30

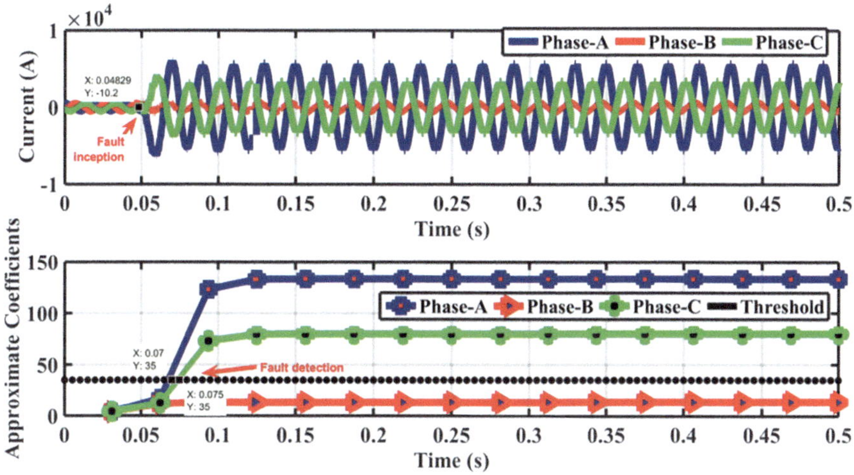

Fig. 5 Performance of scheme during AG fault (153 km) incepted at 0.05 s and detected in 20 ms and CG fault (147 km) incepted at 0.05 s and detected in 25 ms

incepted at 0.05 s and at 0.07 s, the ACIV of phase-A crosses the THV and at 0.075 s, and the ACIV of phase-C crosses the THV, authenticating the AG and CG faults. To detect these faults, time taken by the scheme is 20 ms and 25 ms respectively. Response time taken by the scheme for various faults is given in Table 3. The maximum FDT is 30 ms for one case. With the beginning of faults under GR (ground resistance), FT, and FIL variation, the scheme detected the faults accurately.

5 Conclusion

For transmission line protection, this paper has presented a quick scheme for FD and PTC of multi-location faults. The illustrated scheme use currents measured only from a sending end of the system. FT (Fourier transform) was used as a feature extraction method, and the FD and PTC assignments were accomplished with WT (wavelet

Table 3 ACIV for faults incepted at 0.05 s and varying GR, FT, and FIL

FT-1 (km)	FT-2 (km)	GR (Ω)	ACIV			Fault Detection Time (ms)		
			ACIV-A	ACIV-B	ACIV-C	FDT-A	FDT-B	FDT-C
AG (153)	CG (147)	10	133.4354	13.6310	80.0672	20	-	25
BG (149)	AG (151)	20	74.9204	130.3118	13.7636	15	25	–
CG (168)	AG (132)	30	74.8826	13.1095	117.1530	20	–	25
BG (139)	CG (161)	40	12.9291	120.1742	69.3341	–	20	30
AG (157)	BG (143)	50	104.6614	68.0542	12.8883	20	25	–

transform). The presented scheme is able to detect fault within 30 ms and realized classification exactness above 99% for all cases as indicated by the achieved effects.

References

1. Patel B, Bera P (2019) Fast fault detection during power swing on a hybrid transmission line using WPT. IET Gener Transm Distrib 13(10):1811–1820
2. Jnaneswar K, Mallikarjuna B, Devaraj S, et al (2020) A real-time DWT and traveling waves-based multi-functional scheme for transmission line protection reinforcement. Electr Eng
3. Mishra SK, Tripathy LN (2020) A novel relaying approach for performance enhancement in a STATCOM integrated wind-fed transmission line using single-terminal measurement. Iran J Sci Technol Trans Electr Eng 44:897–910
4. Elgamasy MM, Taalab AMI, Kawady TA, Izzularab MA, Elkalashy NI (2020) Virtual difference voltage scheme for fault detection in VSC-HVDC transmission systems. J Modern Power Systems Clean Energy 8(5):991–1004
5. Khadke P, Patne N, Singh A et al (2020) Pilot Relaying Scheme Based on Differential Admittance Concept for FSC-Compensated EHV Transmission Line. Iran J Sci Technol Trans Electr Eng 44:949–961
6. Aslan Y, Yağan YE (2017) Artificial neural-network-based fault location for power distribution lines using the frequency spectra of fault data. Electr Eng 99:301–311
7. Zhang K, Zhu Y, Liu X (2019) A fault locating method for multi-branch hybrid transmission lines in wind farm based on redundancy parameter estimation. J. Mod. Power Syst. Clean Energy 7:1033–1043
8. Satapathy, A. K., Pati, S., Sahu, A. K., Nanda, R. P., Panigrahi, B. K.: Intelligent Technique for Fault Classification in Transmission Line. In: 3rd International Conference on Computing Methodologies and Communication (ICCMC), pp. 508–511. Erode (India) (2019).
9. Agarwal, S., Swetapadma, A., panigrahi, C. *et al.*: Fault analysis method of integrated high voltage direct current transmission lines for onshore wind farm. J. Mod. Power Syst. Clean Energy, 7, 621–632 (2019).
10. Ngaopitakkul A, Leelajindakrairerk M (2018) Application of probabilistic neural network with transmission and distribution protection schemes for classification of fault types on radial, loop, and underground structures. Electr Eng 100:461–479

Chapter 19
Early Epilepsy Seizure Prediction Using CNN

Aditya Karmokar, Chris David, Shaun Jacob, and Rupali Deshmukh

1 Introduction

Epilepsy is the disease in which the seizure affects the patient, which is caused by a disorder in the functions of the brain. While quite 50 million people around the world are diagnosed with epilepsy, within us, about 3 million patients are suffering from epilepsy. Epilepsy is that the commonest encephalopathy. There are several possible causes of epilepsy. In one case, it can be a molecular mutation due to irregularity in neuronal behaviour or even due to the migration of neurons. Early prediction is usually useful for treating epilepsy. Epilepsy patients are treated with drugs and surgical procedures, but these methods are not effective. Seizures cannot be treated medically, and it will limit the active lifetime of the patient. Patients cannot independently do their work properly. This leads to the social isolation and depression. Prediction of epileptic seizures ensures time before it actually occurs; it is because the attack is often avoided by using drugs. Detecting the looks of the preictal state predicts the seizure. Therefore, our aim is to detect the preictal state for epileptic seizures. Machine learning and deep learning models are used to predict epileptic seizures. The target of the prediction model is detecting the preictal state's sufficient time before seizure onset starts. However, time for the predictive preictal state of the brain and maximum sensitivity is vital, which are the important factors for the performance within the prediction of epileptic seizures (Fig. 1).

Present Address:
A. Karmokar (✉) · C. David · S. Jacob · R. Deshmukh
Fr. Conceicao Rodrigues Institute of Technology, Navi Mumbai 400703, India

G. Mathur et al. (eds.), *Proceedings of 2nd International Conference on Artificial Intelligence: Advances and Applications*, Algorithms for Intelligent Systems,
https://doi.org/10.1007/978-981-16-6332-1_19

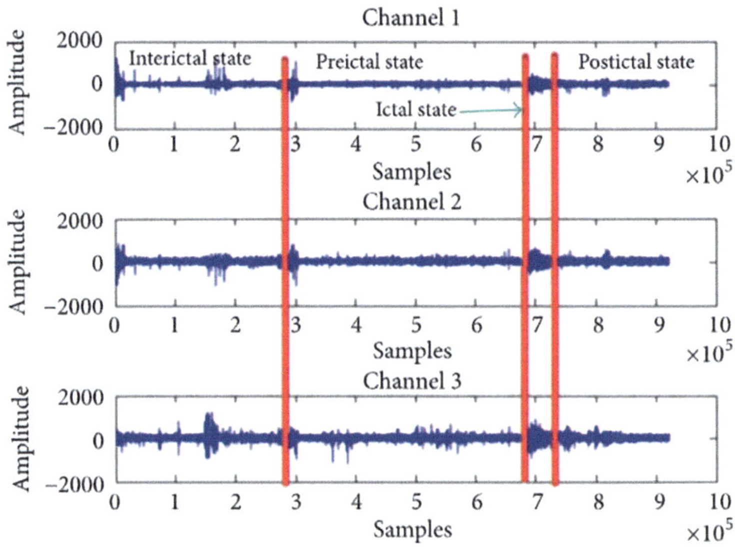

Fig. 1 States of epileptic seizure [2]

2 Related Work

Much work has been wiped out the world of seizure prediction supported EEG signals.

Usman et al. [1] have proposed a model that focuses on both preprocessing and have extraction techniques. To implement the model, the authors have used a publicly accessible dataset of CHB-MIT [10]. After applying preprocessing methods, features were extracted in both the time and frequency domain. Naïve Bayes classifiers, K-Nearest Neighbour, and Support Vector Machine are evaluated on the model from which the authors have selected SVM to differentiate test data between preictal and interictal status because sensitivity performs better. Seizure prediction was made sufficient time before the onset of the seizure by the model with a far better true positive rate, a better true positive rate of 92.23%, maximum anticipation time of 33 min, and therefore the average prediction time of 23.6 min was exhibited by the model. The authors suggested that preprocessing of the EEG signal is often further improved by hybrid preprocessing methods with adaptive window size to urge an increased sensitivity of seizure prediction. Also, the web system for the prediction of epileptic seizures is often implemented.

By incorporating cloud computing and deep learning technology, Hosseini et al. [2] developed a BCI seizure prediction model supported both scalp EEG and intracranial EEG signal data. The tactic uses principle component analysis (PCA), independent component analysis (ICA), and differential search algorithm along with a side optimization approach on existing deep learning structures. For the storage of a giant amount of knowledge, real-time processing, and automatic computing a cloud computing approach has been adapted Patel et al. [7].

It is very tough to accumulate the scalp EEG signals reception for seizure detection and prediction system. To deal with this problem, Vandecasteele et al. [3] have developed a model that uses wearable devices supported electrocardiography (ECG) and photoplethysmography (PPG). The model does seizure classification supported pulse features extracted from the guts rate increase. For classification, vector machine help with the Gaussian kernel was used. Parameters like Sensitivity, False Positive per Hour (FP/h), and Positive Predictive Value (PPV) are used because of the evaluation criteria.

Also, [12] a huge amount of knowledge is being generated on regular bases for iEEG and sEEG which may be a big data problem and requires special attention in terms of computational resources to store and process. To deal with these challenges, Hosseini et al. [4] have developed a cloud computing BCI system. Also, the authors have proposed a seizure prediction technique which is deep learning-based unsupervised feature extraction methods. To enhance the performance in terms of increased accuracy and decreased computation time, the dimensionality of EEG data has been reduced by principle component analysis.

The architecture proposed by Pineau et al. [5] is predicated on a recurrent convolutional neural network (RCNN) (Patel et al. [7]). The model is developed to simultaneously capture the spectral, temporal, and spatial features while learning a general spatially invariant representation of a convulsion (Patel et al. [7]). Most of the previous EEG-based seizure classification work centred on the identification of seizures using patient-specific data.

Fergus et al. [6] have focused on to generalize the seizure prediction model on various subjects of CHB-MIT dataset (Patel et al. [7]). An intense approach for preprocessing EEG signals was done by the authors and have ranking techniques were wont to extract features from pre-processed EEG signals. KNN and SVM classifiers are to evaluate the model in terms of sensitivity, specificity, and area under the curve. Patel et al. [7] directed that the evaluation of various parameter adjustment settings is often checked to ascertain the performance of the seizure prediction model (Patel et al. [7]). Also, advanced artificial neural network architectures like higher-order and spiking neural networks are often incorporated to enhance the general system. Some recent research suggested data transformation [8] and use of nature inspired algorithms for improving performance [9].

3 Proposed System

The ultimate aim of treatments for epilepsy is to provide all patients with seizure control while eliminating side effects. Improved intervention specificity by on-demand interventions can overcome many of the constraints of existing intervention strategies.

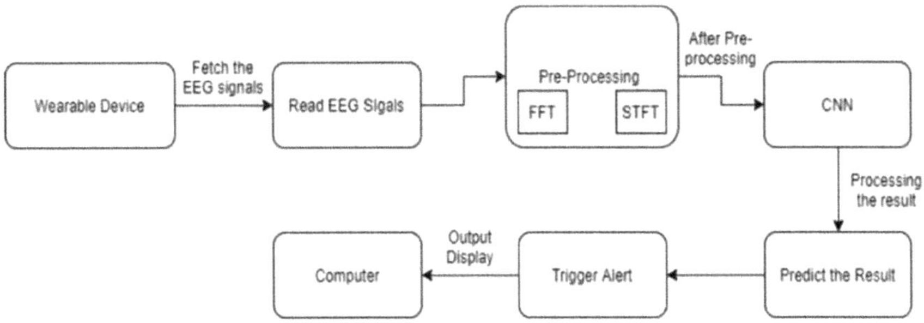

Fig. 2 Architecture design

This paper explores advances in the prediction and diagnosis of seizures, possible new treatments to provide enhanced specificity, and devices for realizing these ends. Specifically, we address (1) possible signal modalities and seizure detection and prediction algorithms, (2) approaches to closed-loop intervention, and (3) hardware to incorporate these algorithms and interventions. The longer-term of epilepsy treatments may reflect seizure prediction and therapies optimizing efficacy while minimizing side effects through improved specificity. It is an exciting time to engage in the conceptualization and advancement of medical device technology, with easy access to resources such as cheap hardware, as well as open-sourced machine learning algorithms and data (Fig. 2).

The hardware part is the most integral part of the project. It main process is to process the software and give the desired output with the given input. Our hardware is a low cost device which was made from scratch. Its main process is to fetch the EEG signals form the brain and give the desired output, with which we can treat the patient successfully (Fig. 3).

4 Implementation Details

4.1 View Data

Intracranial EEG (iEEG) data clips are arranged for any human or canine subject in folders containing training and testing data. Training data is divided into ten-minute EEG clips labelled 'Preictal' for segments of pre-seizure data, or 'Interictal' for segments of non-seizure data. American Epilepsy Society Seizure Prediction Challenge [10] Training data segments are numbered sequentially, while testing data is in random order. Data segments inside directories are stored as follows: in.mat files:

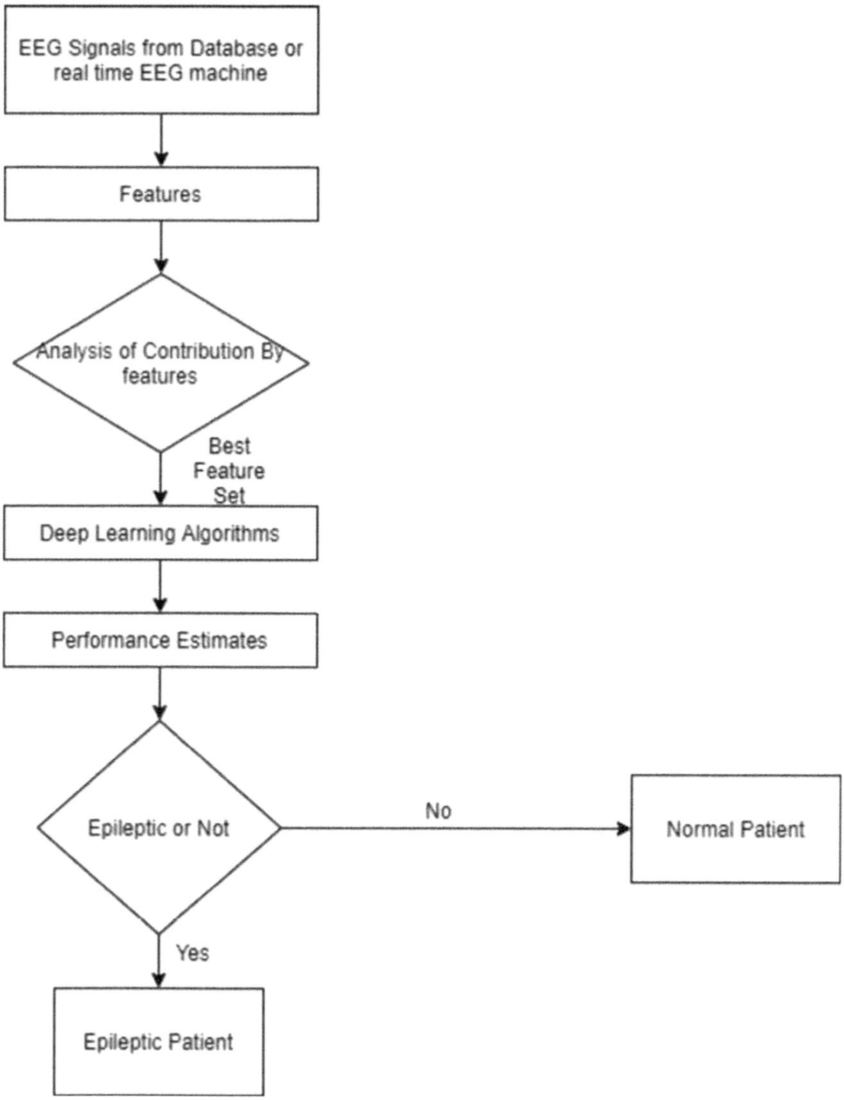

Fig. 3 Flowchart of working

American Epilepsy Society Seizure Prediction Challenge [10] preictal_segment_N.mat—the Nth preictal training data segment interictal_segment_N.mat—the Nth non-seizure training data segment test_segment_N.mat—the Nth testing data segment Each.mat file contains a knowledge structure with fields as follow:

American Epilepsy Society Seizure Prediction Challenge [10] Data: matrix of EEG sample values arranged as electrode × time for row × column. data_length_sec:

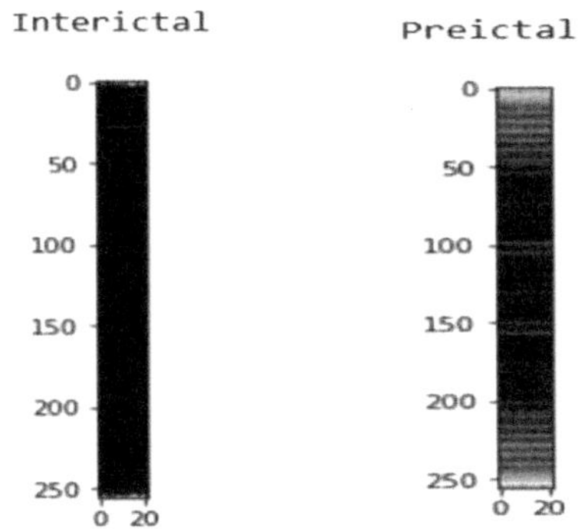

Fig. 4 Spectrogram Image of interictal and preictal states

the time duration of every data row sampling_frequency: the amount of knowledge samples representing 1 s of EEG data. channels: an inventory of electrode names like the rows within the data field sequence: the index of the info segment within the one-hour series of clips. For instance, preictal segment 6.mat has a sequence number of 6 and represents the iEEG data into the preictal data from 50 to an hour [10]. Preictal data segments are presented for training and research covering one hour before seizure with a five-minute seizure horizon (i.e. 1:05–0:05 before seizure onset).

This pre-seizure horizon guarantees that (1) seizures can be predicted with adequate notice to enable fast-acting drugs to be given, and (2) any seizure activity that the epileptologist ignores before the annotated onset will not impact the competition outcome. The original dataset is made up of.mat files, each of which comprises 10 min of 15 EEG channel readings [10]. We train on the data of one patient for convenience and presume that each channel is equally well represented by the signatures that we intend to learn in distinguishing whether or not the segment precedes a seizure occurrence. Thus ultimately want to deploy this model on a resource-limited wearable so the sample 1 s snippets from a single electrode before computing the spectrogram using scipy signal. After Fourier soothes the noisy and non-stationary signal over 1 s time windows, this makes it possible to exploit strong and effective CNNs on the vision problem to work on devices as small as an Arduino (Fig. 4).

4.2 *Algorithm Used*

1. **FFT Algorithm**

   A fast Fourier transformation (FFT) is an algorithm that calculates the discrete Fourier transformation (DFT) of a sequence or its inverse (IDFT). Fourier analysis converts a sign from its original domain (often time or space) to a representation within the frequency domain and the other way around. The DFT is obtained by decomposing a sequence of values into components of various frequencies.

The advantage of regularization is achieved through Fourier Transform [9]; the FFT algorithm is extremely fast and may be performed on resource-limited hardware.

2. **Convolutional Neural Network**

   Regularized versions of multilayer perceptrons are CNNs. Usually, multilayer perceptrons mean completely connected networks, that is, each neuron in one layer in the next layer is connected to all neurons. The "fully-connectedness" of these networks makes them vulnerable to data overfitting. Typical ways of regularization include adding to the loss function some form of magnitude calculation of weights.

   It has been a common approach to speech and signals processing tasks to apply convolutionary neural networks (CNNs) to spectrograms [11]. This borrows from the contributions of computer vision by reframing the question of the time–frequency representation of a signal to classify visual signatures called formants.

3. **Simple Keras MNIST ConvNet**

   From the official example of Keras MNIST Convnet, our model architecture is modelled. The number of filters has been reduced as our spectrogram images are smaller. A simple convnet that on MNIST achieves ~ 99% test precision.

4. **Hardware Details**

   The hardware part is the most integral part of the project. Its main process is to process the software and give the desired output with the given input. Our hardware is a low-cost device that was made from scratch. The device consists of an important component known as AD620. The AD620 may be a low cost, high accuracy instrumentation amplifier that needs just one external resistor to line gains of 1–10,000.

The AD620 features 8-lead SOIC and DIP packaging that is smaller than discrete designs and provides less power (maximum supply current of just 1.3 mA), making it an honest fit EEG reading battery-powered, portable (or remote) devices. The AD620 is then fixed into the PCB board along with the LED light and a capacitor. The capacitor maintains the power supplying through the circuit provided by the 4.5v battery and the red LED light makes sure that the circuit is getting power supply and the AD620 device is functioning. The PCB board also has resistors for reducing the current flow and equally distributing the power throughout the device. The two wires i.e. the green and yellow, are the active electrodes that will give the input and are connected to the patient forehead. The white wire will act as the neutral electrode (ground) and is clipped to the ear. The red wire is connected to the positive side of the battery, and the black is connected to the negative side of the battery for the battery connection. The next sets of wires i.e. the purple and black wires are connected to the positive and negative points of the microphone plug for connecting to the laptop or computer to display the output. All the listed wires and the components are connected to the headers which are soldered to the PCB board (Figs. 5, 6 and 7).

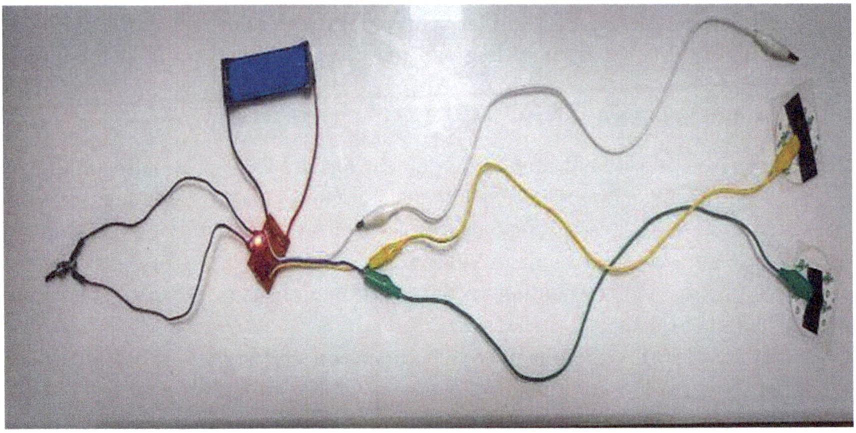

Fig. 5 Wearable device

Fig. 6 AD620 chip connected to PCB Board with other components

The AD620 chip is then integrated with the software which we have collected and tested with the help of TensorFlow, which will help us get the maximum accuracy and gives us the prediction of when will the patient will suffer from epilepsy so treatment can take place immediately.

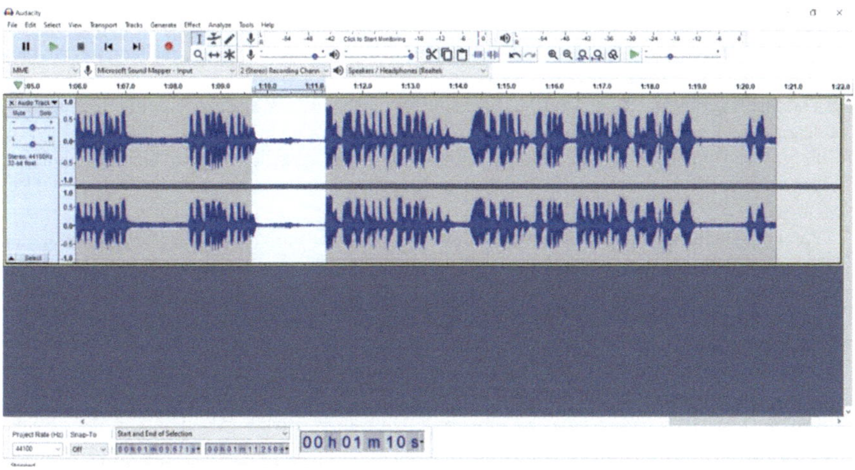

Fig. 7 EEG signals generated in Audacity App

5 Conclusion

Epilepsy is a complex public health problem that requires integrated multidisciplinary approach. A system which predicts the seizure before it occurs could highly improve the quality of life of person with epilepsy. To address the issue, epileptic seizure prediction models were studied. Though wide research has already been done in this area, no methods have achieved acceptable results which could be used in real time applications for seizure prediction.

With rapidly evolving computing power, machine learning paradigms, and big data technology, new methods can be proposed to predict the epileptic seizure with increased sensitivity (due to hardware device) and decreased false positive rate. Simulation and comparative analysis of existing techniques of epileptic seizure prediction can be performed to finalize the techniques that contribute best to improve the performance of epileptic seizure prediction.

References

1. Usman SM, Usman M, Fong S (2017) Epileptic seizures prediction using machine learning methods. Comput Math Methods Med
2. Hosseini M-P, Pompili D, Elisevich K, Soltanian-Zadeh H (2017) Optimized deep learning for EEG big data and seizure prediction BCI via internet of things. IEEE Trans Big Data 3(4):392–404
3. Vandecasteele K, De Cooman T, Gu Y, Cleeren E, Claes K, Van Paesschen W, Van Huffel S, Hunyadi B (2017) Automated epileptic seizure detection based on wearable ECG and PPG in a hospital environment. Sensors 17(10):2338

4. Hosseini M-P, Soltanian-Zadeh H, Elisevich K, Pompili D (2016) Cloud-based deep learning of big EEG data for epileptic seizure prediction. In: Global conference on signal and information processing (GlobalSIP). IEEE, New York, pp 1151–1155
5. Thodoroff P, Pineau J, Lim A (2016) Learning robust features using deep learning for automatic seizure detection. In: Machine Learning for healthcare conference, pp 178–190
6. Fergus P, Hignett D, Hussain A, Al-Jumeily D, Abdel-Aziz K (2015) Automatic epileptic seizure detection using scalp EEG and advanced artificial intelligence techniques. BioMed Res Int
7. Patel V, Buch S, Ganatra A (2020) A review on EEG based epileptic seizure prediction using machine learning techniques. In: Pandian A, Ntalianis K, Palanisamy R (eds) Intelligent computing, information and control systems: ICICCS 2019: advances in intelligent systems and computing, vol 1039. Springer, Cham
8. Shekhawat SS, Sharma H, Kumar S, Nayyar A, Qureshi B (2021) bSSA: binary salp swarm algorithm with hybrid data transformation for feature selection. IEEE Access 9:14867–14882. https://doi.org/10.1109/ACCESS.2021.3049547
9. Chugh A, Sharma VK, Kumar S, Nayyar A, Qureshi B, Bhatia MK, Jain C (2021) Spider monkey crow optimization algorithm with deep learning for sentiment classification and information retrieval. IEEE Access 9:24249–24262. https://doi.org/10.1109/ACCESS.2021.3055507
10. American Epilepsy Society Seizure Prediction Challenge (2014) https://www.kaggle.com/c/seizure-prediction/data. Accessed 17 Nov 2014
11. Deep Learning for Seizure Prediction Wearable (2019) https://www.hackster.io/cnns4eegs/deep-learning-for-seizure-prediction-wearable-5ad2d3. Accessed 8 Feb 2019
12. Pandian A, Ntalianis K, Palanisamy R (2019) Intelligent computing, information and control systems, p 7

Chapter 20
Particle Swarm Optimization with Weighted Extreme Learning Machine for Software Change Prediction

Ruchika Malhotra, Deepti Aggarwal, and Priya Garg

1 Introduction

Software maintenance has become an essential part of today's industry standards. Constant changes and upgrades take place in software due to various reasons like bug-fixes, technology upgrades, functionality addition, etc. This systematic process can be time-consuming and difficult to manage [1]. Here comes the role of Software Change Prediction (SCP), which helps in identifying the change-prone classes, which are the likely source of error and modifications in the upcoming releases. SCP helps the project managers properly plan and effectively allocate limited resources [1, 2].

Various metrics have been used in SCP as independent variables to predict the change-proneness [1, 3, 4]. Various feature selection techniques combined with classification techniques have been used to choose the most effective combination of metrics and build effective models using these metrics for SCP [1, 2, 4]. While a large number of SCP models have been proposed, two open issues still exist in SCP. The first issue is to choose a subset of metrics that is effective in predicting change-proneness of classes. Due to the large variety of available metrics, it is difficult to judge which of them are redundant and non-correlated with the outcome variable. The second issue is regarding the imbalanced nature of the SCP datasets. Since the number of change-prone classes is significantly less than the not change-prone classes, SCP models show biased results leading to low prediction performance [2].

In this study, we propose a hybrid model PSOWE to solve the two key-issues of SCP discussed above. PSOWE uses PSO(4–2) [6], a variant of Particle Swarm Optimization (PSO) to search for an effective feature subset that will be used for predicting change-proneness. It then uses Weighted Extreme Learning Machine (WELM) as

R. Malhotra · D. Aggarwal · P. Garg (✉)
Delhi Technological University, Bawana Road, Delhi, India

R. Malhotra
e-mail: ruchikamalhotra@dtu.ac.in

G. Mathur et al. (eds.), *Proceedings of 2nd International Conference on Artificial Intelligence: Advances and Applications*, Algorithms for Intelligent Systems,
https://doi.org/10.1007/978-981-16-6332-1_20

the classification technique. WELM [7] is a randomized neural network architecture (NNRW) that uses cost-sensitive learning to deal with class-imbalance. PSO(4–2) is chosen for feature selection in PSOWE since it is an evolutionary technique. Evolutionary techniques offer various benefits over conventional filter-based or wrapper-based techniques like less computation time and the ability to search for globally optimal solutions [6]. WELM is used for classification in our model since it has a short training time due to its randomized nature and has good predictive capability in imbalance learning scenarios. In this study, PSOWE is applied on six Java-based SCP datasets that contain OO metrics and graph-based metrics as independent variables. The model is compared with ten state-of-the-art techniques.

The rest of the paper is divided into five sections. Section 2 presents the related work of the study. Section 3 describes the components of the proposed hybrid model in detail. Section 4 describes the experimental setup used in the study. Section 5 presents the experimental results, and Sect. 6 concludes the study by presenting the necessary inferences of the study.

2 Related Work

This section presents the SCP studies that have been previously done by the researchers. Various researchers have encountered issues with selecting a single feature set from a large variety of available feature sets for SCP datasets including Object-Oriented (OO) metrics, graph-based metrics, and developer-related factors [1, 3]. Some studies have been proposed which have concluded that a combination of features is much more efficient than one set of features, especially a combination of OO metrics and graph metrics [3]. Due to the availability of many features, the effect of more important features is reduced over the dependent variable due to the presence of less important features. Due to which the main issue that occurs is selecting the essential features that increase predictive efficiency. Malhotra and Khanna [1] conducted a systematic review of studies in SCP, and more than half studies used feature selection strategy along with the ML technique. Tang et al. [8] proposed a study highlighting the effects and results of feature selection in classification studies. PSO has been used as an efficient feature selection technique in various fields [9, 10], but it has been never used in SCP. Motivated from these studies, we use an evolutionary technique PSO for feature selection for the proposed model. The factor of class imbalance also has adverse effects on SCP datasets and predictive efficiency. Malhotra and Khanna [2] explore the studies in change prediction using imbalanced data and handling techniques. Other fields of studies have highly recommended the use of cost-sensitive imbalance handling techniques [5, 17]. These studies have also proven that cost-sensitive techniques like WELM are better than resampling techniques and ensemble techniques like AdaBoost and Bagging. Hence, in this study, we use WELM as the class imbalance and classification technique.

3 PSOWE—The Proposed Model for Software Change Prediction

3.1 *Feature Selection Using Particle Swarm Optimization*

PSO is an evolutionary metaheuristic technique based on the biological behavior and social interactions depicted by bird flocks searching for corn. This study uses a PSO variant, i.e., PSO(4–2) [6] that has unique initialization and update mechanisms. PSO(4–2) initializes the particle positions such that 2/3rd of the total particles have small-sized feature subsets and 1/3rd of the total particles have large-sized feature subsets. This ultimately results in a medium and optimal size feature subset as the solution. The update strategy of PSO(4–2) is focused on the maximization of the fitness function, i.e., AUC-ROC while minimizing the feature subset size. The pseudo-code of PSO(4–2) is given in Fig. 1.

3.2 *Classification Using Weighted Extreme Learning Machine*

WELM [7] is a cost-sensitive variant of ELM that handles class-imbalance by giving a weight to every sample in the dataset according to the class label. WELM is a randomized neural network architecture similar to an iterative neural network architecture with one input layer, one or more hidden layers, and one output layer.

Consider a dataset (X, Y) where $X = \{x_1, x_2, ... x_N\}$, $x_i = \{x_{i1}, x_{i2}, ..., x_{im}\}$, $Y = \{y_1, y_2, ..., y_N\}$, N = number of samples in the dataset, and m = number of independent features in the dataset. The output of the WELM architecture with x_i as an input is as follows:

$$o_i = \sum_{k=1}^{p} \beta_k t(w_k \cdot x_i + b_k), \quad (1)$$

where 'o_i' represents the output of ith sample where i = 1, 2, ..., N, function 't' represents the activation function that converts linear data to non-linear. 'p' represents the number of hidden layer nodes, 'w_k' represents the weight from the input layer nodes to the kth hidden layer node, and 'b_k' represents the bias of the hidden layer node.

Equation (1) can also be optimized as follows:

$$H\beta = O, \quad (2)$$

```
Algorithm 1: Pseudo Code of PSO(4-2)
Initialization:
1. Number of particles: P
2. Threshold: Θ
3. Initial position of each particle:
     Pos = {pos_i}, i= 1,2,3...P and pos_i is an m-dimensional vector.
4. Initial Velocity of each particle:
     V = {v_i} i= 1,2,3...P and v_i is an m-dimensional vector
5. pbest = [pbest_i], i= 1,2,3...P and pbest_i is a m-dimensional vector
6. gbest = {0} m-dimensional vector
7. maxiteration = it_max

Functions:
 nof(pos): Calculate number of features according to threshold Θ an pos
   vector.
 f(pos): Calculate the fitness function i.e., AUC-ROC value from the
   selected features in pos vector according to threshold Θ.

Algorithm:
for each particle i in (0, 1..., P):
     pbest_i = pos_i
     if(f(pbest_i)>f(gbest))
     gbest = pbest_i

Repeat for it_max iterations:
     for each particle i in (0, 1..., P):
        Calculate f(pbest_i), f(pos_i), f(gbest), nof(pbest_i), nof(pos_i),
        nof(gbest)
        if(f(pos_i)>f(pbest_i))
           pbest_i = pos_i
        else if (f(pos_i)=f(pbest_i) and nof(pos_i)<noff(pbest_i))
           pbest_i = pos_i

        if(f(pbest_i)>f(gbest))
           gbest = pbest_i
        else if (f(gbest)=f(pbest_i) and nof(pbest_i)<noff(gbest))
           gbest = pbest_i
     for each particle i in (0,1,.., P):
        Update v_i to v_i' and pos_i to pos_i' according to equation (1) & (2)
        respectively
        v_j^(t+1) = v_j^t*w + c_1*r_1j*(pbest_j - x_j^t) + c_2*r_2j*(gbest - x_j^t)     (1)
        x_j^(t+1) = x_j^t + v_j^(t+1)          (2)
 return gbest
```

Fig. 1 Algorithm for PSO(4–2)

$$\text{where} H = \begin{bmatrix} t(w_1x_1 + b_1) & \cdots & t(w_px_1 + b_p) \\ \vdots & \vdots & \vdots \\ t(w_1x_N + b_1) & \cdots & t(w_px_N + b_p) \end{bmatrix}_{nXp}. \tag{3}$$

The optimized solution for output is as follows:

$$\|\mathrm{Y} - \mathrm{O}\| = 0 \tag{4}$$

From Eq. (2) and Eq. (4), the optimized solution for β is as follows: $||H\beta$ - $Y|| = 0$, where Y is the target variable.

$$\beta = H^{-1}Y. \tag{5}$$

But for weighted ELM, a weight matrix M is also multiplied to the input before it is sent to the hidden layer. This weight matrix is a diagonal matrix of size NXN where N is the number of samples in the dataset. There are multiple ways to define weight matrix M. In this study golden ratio [7] is used for initializing the matrix which implies that majority samples get reduced weights and minority samples get higher weights to balance out the results overall. The weight matrix is initialized as follows:

$$\mathrm{M} = \begin{bmatrix} \mathrm{M}_{0,0} & \cdots & 0 \\ \vdots & \ddots & \vdots \\ 0 & \cdots & \mathrm{M}_{\mathrm{N,N}} \end{bmatrix} \tag{6}$$

where $\mathrm{Mi,j} = \begin{cases} 0.618/n_{\min}, \; i \text{ is minority class sample} \\ 1/n_{\mathrm{maj}}, \; i \text{ is majority class sample} \end{cases}.$

The final optimized β is now written as after the use of weight matrix M,

$$\beta = H^T\left(\frac{I}{C} + MHH^T\right)^{-1} MY, \tag{7}$$

where I represents the identity matrix, C represents the regularization parameter, and Y represents the target output, and M is the weight matrix that can be calculated by Eq. (6).

4 Experimental Setup

4.1 Dataset Collection and Description

In this study, we have used six datasets which are collected using open-source Java projects. Each dataset consists of 60 OO metrics and 26 graph-based metrics as independent variables. To collect the datasets, we first collect the source code of two consecutive releases of a project using GitHub or SourceForge. This source code is input to the UNDERSTAND tool (http://www.scitools.com/) which calculates the class-level OO metrics [11–13]. For each class common among both releases, we calculate the number of lines added, deleted, and modified. If no modifications are present, the class is labeled 0 else it is labeled as 1. For the 26-graph based metrics, we merge our collected dataset with the open-source dataset [14] class wise.

The OO metrics [11–13] describe the class structure and characteristics like functions, variables, size, coupling, cohesion, etc. The graph-based metrics measure the inter-class dependency relationships like inheritance, containership, etc. derived

Table 1 Description of the dataset used in the study

Project	Release 1	Release 2	No. of classes	% change-prone classes
Neuroph	2.1.1	2.2	112	26.5
JavaClient	2.2.0.0	2.2.0.1	211	4.24
Hibernate	5.2.11	5.2.12	2022	6.77
Wro4j	1.6.2	1.6.3	169	24.7
Tomcat	9.0.3	9.0.4	354	1.97
Ant	1.6.0	1.6.1	319	10.3

from the dependency graph. They also include centrality measures like K-Coreness, Betweenness, hub value, etc. The change-proneness of the class is the binary dependent variable where 1 means a change-prone class and 0 means a not change-prone class. The six datasets are summarized in Table 1. These datasets are available at our online repository [15].

4.2 Performance Measures

In this study, we use AUC-ROC and F-measure for evaluating the proposed model and other techniques. These performance measures have been widely used in previous studies [1, 3]. AUC-ROC represents the area under the Receiver Operating Characteristics Curve represented with a true positive rate (TPR) at the y-axis and false positive rate (FPR) at the x-axis. TPR or Recall represent the ratio of samples that are predicted change-prone accurately to the samples that are change-prone. FPR represents the ratio of samples that are predicted change-prone inaccurately to the samples that are not change-prone. F-Measure represents the harmonic mean of Recall and Precision. Precision represents the samples that are predicted change-prone accurately to the total samples that are predicted change-prone by the ML technique.

4.3 ML Techniques

This study compares PSOWE with ten state-of-the-art techniques described in Table 2.

4.4 Parameter Settings

In this study, each ML technique has been tuned with the best hyper-parameter settings. For the proposed model PSOWE, PSO has been initialized with c1, c2 =

Table 2 ML Techniques compared with PSOWE and their descriptions

ML technique	Description
Resampling techniques	Random Oversampling (RO), Random Undersampling (RO), and SMOTE (SM) are used for imbalance handling [16] followed by Random Forest (RF) for classification
AdaBoost	AdaBoost (Ada) [17] combines the weak learners to produce a strong learner using a weighing technique for training a classifier using results of previous classifiers
Bagging	Bagging [17] trains multiple learners on bootstrap samples generated from random sampling with replacement from the original dataset
CEL	CEL [18] converts a two-class imbalanced dataset to a multi-class binary dataset. It then uses a coding scheme like one-against-all or one-against-one to form diverse binary training datasets for the ensemble
HEELM	HEELM [19] combines representation learning and ensemble learning. It forms new features from predictions of multiple ELMs and use them for training an ensemble of ELMs
KPWE	KPWE [20] is a hybrid technique similar to PSOWE that uses Kernel Principal Component Analysis (KPCA) for feature extraction followed by WELM for cost-sensitive learning
FFWE	FFWE is a hybrid technique that uses Firefly algorithm (FF) [21] for feature selection and WELM for classification. FF is a metaheuristic evolutionary algorithm that models the luminescent behavior of the fireflies
ABCWE	ABCWE is a hybrid technique that uses Artificial Bee Colony Optimization (ABC) [21] for feature selection and WELM for classification. ABC is a metaheuristic swarm-based technique that models the food search behavior of honey bees

1.496, w = 0.7298, number of particles = 30, number of iterations = 50, maximum value of velocity = 2, and threshold for feature selection = 0.55. WELM has been initialized with golden ratio technique for the weight matrix with the initial randomization range for weights and biases to be (-1,1), activation function = sigmoid, and regularization parameter C = 0.3125. For Firefly (FF), the population size = 100, beta = 0.5, gamma = 1, and number of function evaluations = 1 were taken. For ABC, the number of employed and onlooker bees = 20, maximum iterations = 50, the limit for abandonment = 10, and the threshold for feature selection to be 0.6 has been taken. Techniques like RO, RU, SM, Ada, and Bagging were implemented using python Sklearn library. For techniques that were used with RF, the RF parameter settings were taken according to the previous study [22]. CEL is implemented using the one-against-one coding scheme using RF classification technique, as suggested in [18]. For HEELM, the number of component ELMs is chosen to be 40, sparsity ratio to be 0.6, the number of hidden layer nodes to be 75, and the number of subsamples to be 10. For KPWE, the parameter settings selected by its original paper are selected [20].

4.5 Validation Technique and Statistical Tests

This study uses tenfold cross-validation for training and testing the SCP models that has been widely used validation procedure [1, 3].

This study uses the Friedman test [23] and Wilcoxon signed-rank test [24] for statistical validation. Consider the scenario where the performance of m techniques is compared over d datasets. In the Friedman test, we form the null hypothesis (H_0) that all the m techniques have similar prediction performance and calculate the average Friedman rank (AFR) of every technique based on the performance measure. Next, we calculate the Friedman test statistic (FTS) using Eq. (8). FTS is based on the chi-square distribution having m-1 degrees of freedom. If the FTS value is greater than chi-square distribution value, we reject H_0 at 0.05 significance level (α) else we accept H_0,

$$\text{FTS} = \frac{12D}{m(m+1)}\left(\sum\nolimits_{i=1}^{m} AFR_i^2 - \frac{m(m+1)^2}{4}\right). \tag{8}$$

If we reject H_0, we compare the performance of m techniques pairwise using the Wilcoxon-signed rank test.

5 Experimental Results

This section compares the prediction performance of PSOWE with the 10 ML techniques described in Sect. 4.3 using performance measures and statistical tests. Figure 2a is a boxplot showing the variation of AUC-ROC values for PSOWE and the other 10 ML techniques. PSOWE achieves the highest median AUC-ROC value of 0.9337. It shows remarkable improvement in the median AUC-ROC over the other the techniques—60.3% (for HEELM), 78.19% (for Ada), and 86.74% (for Bagging).

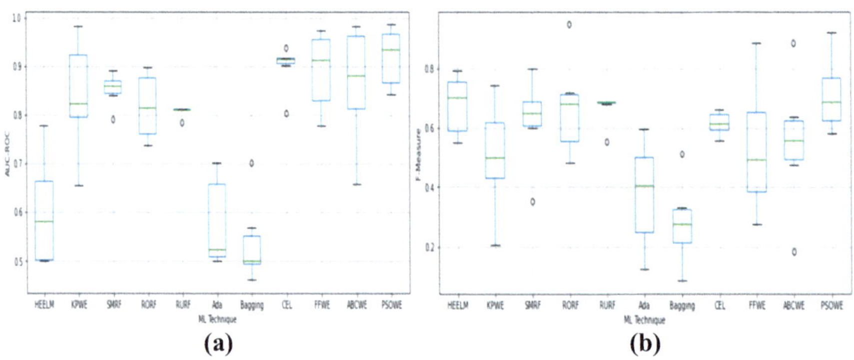

Fig. 2 Box plot for 11 ML-techniques. **a** AUC. **b** F-Measure

Figure 2b shows the F-Measure values for PSOWE and the other 10 ML techniques. PSOWE achieves the second-highest median F-Measure (0.6883) after HEELM (0.7). The median F-Measure value of PSOWE is greater than the highest F-Measure achieved by techniques like ABCWE, CEL, Bagging, and Ada, which shows remarkable improvement in prediction performance. Table 3 shows the average Friedman rankings for PSOWE and the other ten techniques. The rankings are calculated based on both AUC-ROC and F-Measure values. The FTS was found to be 42.529 and 29.024 for AUC-ROC and F-Measure, respectively. Hence, H_0 was rejected for both the cases at a 0.05 significance level. According to Table 3, PSOWE achieves the best ranking according to both AUC-ROC (1.667) and F-Measure (2.417). Table 4 shows the Wilcoxon signed-rank test results. In the Wilcoxon test, PSOWE is compared individually with the other ten techniques based on AUC-ROC and F-Measure values. Here, " ← *" symbol indicates that PSOWE is significantly better than the compared technique. " ← " symbol represents that performance of PSOWE is better, but no significant difference is observed. For AUC-ROC, PSOWE is statistically different from all the ML techniques except CEL and FFWE. For F-Measure, PSOWE is statistically significant from all the ML techniques exempt HEELM, RORF, RURF, and SMRF.

The experimental results indicate that PSOWE shows remarkable improvement over resampling based imbalance-learning techniques like RORF, RURF, and SMRF in terms of AUC-ROC values. One possible reason could be that ROS, RUS, and SMOTE disturb the original data distribution, which can cause overfitting leading

Table 3 Average Friedman Ranking of PSOWE and other techniques in terms of AUC-ROC and F-Measure

ML technique	AFR(AUC-ROC)	AFR(F-measure)	ML Technique	AFR(AUC-ROC)	AFR(F-measure)
PSOWE	1.667	2.417	RURF	6.000	4.167
ABCWE	4.167	6.417	Ada	9.000	9.333
FFWE	3.917	6.667	Bagging	10.250	10.000
KPWE	5.750	7.333	CEL	3.500	6.000
RORF	6.667	4.833	HEELM	10.083	3.667
SMRF	5.000	5.167	–	–	–

Table 4 Significance of PSOWE with other techniques in terms of AUC-ROC and F-Measure

ML Technique	AUC-ROC	F-Measure	ML Technique	AUC-ROC	F-Measure
SMRF	← *	←	RURF	← *	←
ABCWE	← *	← *	Ada	← *	← *
FFWE	←	← *	Bagging	← *	← *
KPWE	← *	← *	CEL	←	← *
RORF	← *	←	HEELM	← *	←

to bad prediction performance [25]. However, PSOWE uses WELM for imbalance-learning, where the weight matrix is calculated using the original data distribution. PSOWE shows superior performance than the feature extraction technique KPWE. The possible explanation for the superiority of PSOWE over KPWE can be that PSO is supervised while KPWE is unsupervised.

6 Conclusions

In this study a novel SCP model, i.e., PSOWE, is proposed which uses PSO(4–2) for feature selection and WELM for class-imbalance handling and classification of both of which have never been applied to SCP studies before for the tasks described above. PSOWE has been compared with 10 state-of-the-art technique including various imbalance learning and ensemble techniques. The major conclusions drawn from this study are as follows:

1. PSOWE is an efficient technique that has outperformed all other techniques in terms of the median of AUC-ROC values and average Friedman rankings for both AUC and F-Measure. The results collected indicate the high predictive capability of the proposed model in software change prediction.
2. The superiority of PSOWE over state-of-the-art feature extraction and imbalance learning techniques indicate its ability to solve the crucial issues in SCP.
3. The small difference between the metaheuristic techniques (PSOWE, ABCWE, and FFWE) according to the median values of AUC-ROC indicate that all of these techniques are comparable to each other and can be applied in SCP studies for more datasets.

References

1. Malhotra R, Khanna M (2019) Software change prediction: a systematic review and future guidelines. e-Inf Software Eng J 13(1):227–259
2. Malhotra R, Khanna M (2017) An empirical study for software change prediction using imbalanced data. Empir Softw Eng 22(6):2806–2851
3. Giger E, Pinzger M, Gall HC (2012) Can we predict types of code changes? An empirical analysis. In: 9th IEEE working conference on mining software repositories (MSR). IEEE, New York, pp 217–226
4. Malhotra R, Bansal AJ (2016) Software change prediction: a literature review. Int J Comput Appl Technol 54(4):240–256
5. McCarthy K, Zabar B, Weiss G (2005) Does cost-sensitive learning beat sampling for classifying rare classes? In: Proceedings of the 1st international workshop on utility-based data mining—UBDM'05
6. Xue B, Zhang M, Browne WN (2014) Particle swarm optimisation for feature selection in classification: Novel initialisation and updating mechanisms. Appl Soft Comput 18:261–276
7. Zong W, Huang GB, Chen Y (2013) Weighted extreme learning machine for imbalance learning. Neurocomputing 101:229–242

8. Tang J, Alelyani S, Liu H (2014) Feature selection for classification: a review. Data classification, algorithms and applications. CRC Press, London, pp 37–47
9. Sakri SB, Abdul Rashid NB, Muhammad Zain Z (2018) Particle swarm optimization feature selection for breast cancer recurrence prediction. IEEE Access 6:29637–29647
10. Sharkawy R, Ibrahim K, Salama MMA, Bartnikas R (2011) Particle swarm optimization feature selection for the classification of conducting particles in transformer oil. IEEE Trans Dielectr Electr Insul 18(6):1897–1907
11. Chidamber SR, Kemerer CF (1994) A metrics suite for object oriented design. IEEE Trans Software Eng 20(6):476–493
12. Tang MH, Kao MH, Chen MH (1999) An empirical study on object-oriented metrics. In: Proceedings of metrics, pp 242–249
13. Thomas J, McCabe J (1976) A complexity measure. IEEE Trans Software Eng SE-2(4):308–320
14. Şora I, Chirila C-B (2019) Data for: Finding key classes in object-oriented software systems by techniques based on static analysis. https://doi.org/10.17632/7h57xwtrcb.1
15. https://figshare.com/s/df5529168b9641bdd96e
16. Basgall MJ, Hasperué W, Naiouf M, Fernández A, Herrera F (2019) An analysis of local and global solutions to address big data imbalanced classification: a case study with SMOTE preprocessing. In: Naiouf M, Chichizola F, Rucci E (eds) Cloud computing and big data: JCC&BD communications in computer and information science, vol 1050. Springer, Cham
17. Catolino G, Ferrucci F (2019) An extensive evaluation of ensemble techniques for software change prediction. J Software Evol Process 31(5):e2156
18. Sun Z, Song Q, Zhu X (2012) Using coding-based ensemble learning to improve software defect prediction. IEEE Trans Syst Man Cybernet Part C (Appli Rev) 42(6):1806–1817
19. Cai Y, Liu X, Zhang Y, Cai Z (2018) Hierarchical ensemble of extreme learning machine. Pattern Recogn Lett 116:101–106
20. Xu Z, Liu J, Luo X, Yang Z, Zhang Y, Yuan P, Zhang T (2018) Software defect prediction based on kernel PCA and weighted extreme learning machine. Inf Softw Technol 106:182–200
21. Agarwal S, Singh AP, Anand N (2013) Evaluation performance study of Firefly algorithm, particle swarm optimization and artificial bee colony algorithm for non-linear mathematical optimization functions. In: 2013 fourth international conference on computing, communications and networking technologies (ICCCNT)
22. Elish KO, Elish MO (2008) Predicting defect-prone software modules using support vector machines. J Syst Softw 81(5):649–660
23. Friedman M (1940) A comparison of alternative tests of significance for the problem of m rankings. J Ann Math Stat 11(1):86–92
24. Wilcoxon F (1992) Individual comparisons by ranking methods. In: Kotz S, Johnson NL (eds) Breakthroughs in statistics. Springer series in statistics (perspectives in statistics). Springer, New York
25. Rodriguez D, Herraiz I, Harrison R, Dolado J, Riquelme JC (2014) Preliminary comparison of techniques for dealing with imbalance in software defect prediction. In: Proceedings of the 18th international conference on evaluation and assessment in software engineering—EASE'14

Chapter 21
Advanced Identification of Alzheimer's Disease from Brain MRI Images Using Convolution Neural Network

Neha Rajawat, Bharat Singh Hada, Mayank Meghawat, Soniya Lalwani, and Rajesh Kumar

1 Introduction

Advancement in the health care facility has given a better and healthier life and increased the average years of human life. On the other side, this increased the risk of non-communicable diseases like diabetes, hypertension, dementia, cardiovascular diseases, etc [1]. Dementia is a chronic and progressive disease that is generally found in older people and deteriorates the brain tissues, cognitive ability, and general human behavior. With the progression of dementia, one can become unable to perform day-to-day activities. Alzheimer's disease (AD) is the most common form of dementia which affects 60–70% of the total cases found around the world. In 2008, it was declared as a primary health condition by the World Health Organization (WHO) [2] being a major cause of disability and dependency among elderly people. AD affects the patients, as well as the people who care about them, by degrading the mental, physical, social, psychological, and economic status of a human being. The progression of AD starts from unnoticeable brain tissue changes to the disability of cognitive and physical skills where Mild cognitive impairment (MCI) is a stage from where a person ends up having AD. Once the MCI turns into AD, it starts causing the symptoms of dementia which ranges from mild to severe [3, 4] mental disability. To avoid the risk involved in the MCI stage, early detection of AD is very necessary. It helps the person to be aware of this condition and necessary precautionary steps

N. Rajawat
Department of Mathematics, Career Point University, Kota, India

B. S. Hada · M. Meghawat
Samsung R&D Institute, Noida, India

S. Lalwani (✉)
Department of Mathematics, Bal Krishna Institute of Technology, Kota, India

R. Kumar
Department of Electrical Engineering, MNIT, Jaipur, India

G. Mathur et al. (eds.), *Proceedings of 2nd International Conference on Artificial Intelligence: Advances and Applications*, Algorithms for Intelligent Systems,
https://doi.org/10.1007/978-981-16-6332-1_21

can be taken. Before developing the noticeable symptoms of AD, patients develop measurable changes in brain structure which work as biomarkers to help in the early diagnosis of AD. Using neuroimaging such as MRI scan, these biomarkers can be identified and necessary treatment can be provided.

AD detection using neuroimaging has become a widely discussed topic among researchers. A considerable amount of prior work exists in this domain which applies CNN and image processing. Salehi et al. [5] implemented CNN-based deep learning method using Alzheimer's disease Neuroimaging Initiative (ADNI) dataset containing brain MRI images. Hussain et al. [6] developed a 12 layered CNN binary classification model for AD and compared the performance with preexisting CNN architectures (InceptionV3, Xception, VGG19, and MobileNetV2). Niu et al. [7] has also used the ADNI dataset for classification between AD and MCI. Ebrahimi et al. [8] implemented a transfer learning-based approach by using a 3D ResNet-18 deep learning model for AD on a dataset of 3D brain MRI scans. Data transformation [9] and nature-inspired algorithms [10] can improve the performance of these methods.

The combined capability of image processing algorithms and Convolution Neural Network (CNN) has made early detection of AD by neuroimaging analysis to be a feasible task. Inspired by this thought, this work proposes an efficient and lightweight CNN model to detect the presence of AD using brain MRI scan with high accuracy. The model contains a total of 8 layers including 5 convolutions and 3 fully connected layers. It is trained on publicly available dataset, namely, Alzheimer's Dataset (4 class of Images) [11] which contains brain MRI scans arranged into four classes: NonDemented, VeryMildDemented, MildDemented, and ModerateDemented. Raw images were preprocessed through an image preprocessing pipeline which extracts Region of Interest (ROI) to highlight the useful information present as biomarkers in the brain structure. Preprocessed ROI was combined with the input image and resultant 2-dimensional (2D) input was fed to the model. This paper contributes toward an efficient and lightweight CNN model with 653,000 trainable parameters, the training accuracy and F1-score of this model is 99.3% and 99.5%, respectively. An intuitive image preprocessing pipeline which to highlight the ROI where the biomarkers of AD are present. Rest of the organization of this paper contains related work in Sect. 1. Section 2 describes the methodology and the proposed model. Results and key findings are discussed in Sect. 3, and the paper is concluded in Sect. 4 with future remarks.

2 Methodology and Proposed Model

The performance and effectiveness of a machine learning model significantly depends on the quality of data, efficient preprocessing methods, and choice of input features and architecture of the model. After choosing the training dataset carefully, the major focus of this work was to implement an efficient and intuitive image preprocessing

Table 1 Dataset overview

Directory	Non demented	Very mild demented	Mild demented	Moderate demented	Total directory sample
Train	2560	1792	717	52	5121
Test	640	448	179	12	1279
Total	3200	2240	896	64	6400

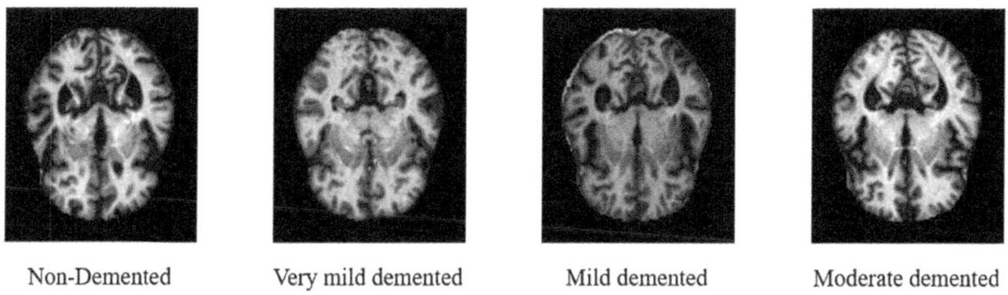

Fig. 1 Preview images of four classes of AD

module. As the current research deals with image dataset, choosing the CNN-based approach was preferred from other deep learning techniques.

2.1 Dataset

The brain MRI scan dataset used in this work is Alzheimer's Dataset (4 classes of Images) [11] publicly available on the Kaggle machine learning platform. As the name suggests, this dataset contains brain MRI images arranged in 4 target classes for both training and testing purpose. Table 1 provides the details about this dataset.

The images presented in Fig. 1 depict some visible changes in the human brain structure in various stages of AD. The image processing module discussed in the next section tries to capture these biomarkers.

2.2 Image Processing Module

Early diagnosis of AD using brain MRI images requires identifying the presence of biomarkers which are the results of structural changes in the human brain. Taking inspiration from this thought, an efficient image preprocessing module was designed to segregate the most informative ROI from the noise background. This ROI mask is used to highlight the input area where possible biomarkers are present for the successful detection of AD. Figure 2 illustrates the steps involved in the image

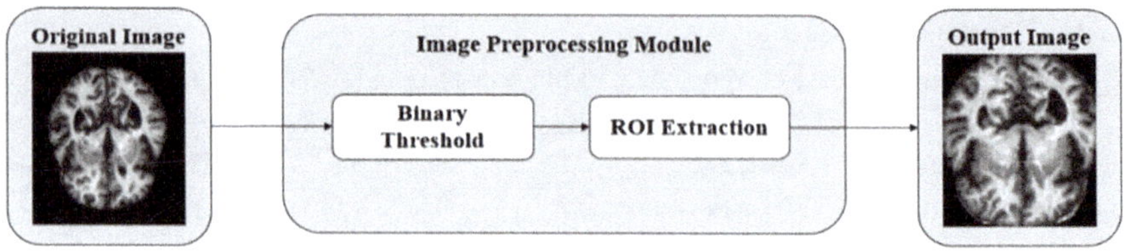

Fig. 2 Image processing module

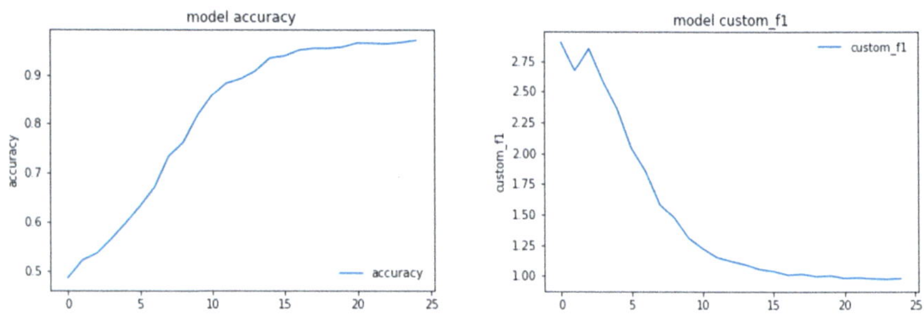

Fig. 3 **a** Accuracy versus epochs, **b** F1-score versus epochs

processing module.

Binary Threshold. Image thresholding is the simplest method for image segmentation. The binary threshold is used to convert a grayscale image to a binary format. It is useful in highlighting the interesting region of the image from the rest of the background [12]. Pixel values below a predefined threshold value are assigned zero intensity or black and other assigned maximum intensity as 255.

ROI Extraction. The thresholding step is required for foreground highlighting and is used for ROI extraction. In this step, a rectangular binary is drawn around the foreground and the ROI mask is generated. After superimposing this mask on the input image, the required output can be generated. The images shown in Fig. 3 are example of ROI images generated from all four types of target classes.

2.3 Convolution Neural Network (CNN)

CNN is a derivative of the deep learning methods which are proved to be efficient in image classification tasks. While other machine learning algorithms require considerable efforts to be put in the preprocessing stage, CNN needs much lower efforts. The working principle of CNNs is inspired by human vision where an object or entities are identified or learnt as a whole [13]. In a similar fashion, the CNN network tries to learn an input volume after strategically dividing it into many smaller sub-volumes. These sub-volumes are termed as receptive fields of CNN nodes. As the layers begin to grow deeper, the network manages to learn from low-level features to high and complex level features. CNN algorithms use the technique of kernel trick or filters to learn the hidden features. These kernels or filters are designed to identify or learn a specific feature associated with it, i.e., there will be two different kernels to learn horizontal edges and vertical edges in the images.

Basic CNN Architecture. In its basic form, the CNN model is composed of the following three building blocks.

Convolution Layers: These layers are used to perform the kernel trick by convolving the input image with a predefined set of kernels. In this step, the presence of specific features is detected in a sub-volume and input feature vectors are designed in the form of the resultant convolved image [14]. This feature vector will either be used by another convolution layer to identify next-level features or by pooling layers to downsample the input volume.

Pooling Layers: Pooling technique is used to consolidate the feature map constructed in convolution operation by downsampling in input volume. This operation reduces the input size while maintaining the same amount of information identified in prior operations [15]. It can identify the most activated presence of feature vector located in various sub-volumes of the image input.

Fully Connected or Dense Layers: Once the feature vector construction is finished using the convolution and pooling techniques into a reduced input volume shape, CNN starts to use conventional deep learning connections to further optimize the learning process. These layers try to optimize the feature vector space using complex activation functions.

2.4 Proposed Model

Proposed CNN model works for the disease diagnosis using the medical imaging analysis to the current problem of AD. Table 2 format explains the design of each layer.

Architecture. The model is composed of a total of 8 layers which include 5 convolutions and 3 densely connected layers. Each of these layers is followed by a dropout layer so that model overfitting can be removed. Both convolution and dense layers use the RELU activation function except the last output layer which has SOFTMAX activation to provide probabilistic distribution over the four target labels. This CNN architecture contains a total of 653,000 trainable parameters which is not a big number for a deep learning algorithm.

Model Training. Proposed model accepts two-dimensional image input of shape 128×128. Before feeding this input to the network, feature scaling was done at the pixel level. Each input image has to pass through the image processing module to generate the ROI. To make model training efficient and fast, a custom image data generator was designed to generate a batch of training images on demand basis. Each batch is a randomized selection of images from the training set, chosen for

Table 2 Proposed model architecture

#	Layer type	# of kernel/probability	Kernel size	Activation function	Input shape	Output shape	Parameter (including bias term)
1	Convolution	24	5 × 5	RELU	128 × 128 × 1	62 × 62 × 24	624
	Dropout	0.2	–	–	62 × 62 × 24	62 × 62 × 24	0
2	Convolution	36	5 × 5	RELU	62 × 62 × 24	29 × 29 × 36	21,636
	Dropout	0.2	–	–	29 × 29 × 36	29 × 29 × 36	0
3	Convolution	48	5 × 5	RELU	29 × 29 × 36	13 × 13 × 48	43,248
	Dropout	0.1	–	–	13 × 13 × 48	13 × 13 × 48	0
4	Convolution	64	3 × 3	RELU	13 × 13 × 48	11 × 11 × 64	27,712
	Dropout	0.1	–	–	11 × 11 × 64	11 × 11 × 64	0
5	Convolution	64	3 × 3	RELU	11 × 11 × 64	9 × 9 × 64	36,928
	Dropout	0.1	–	–	9 × 9 × 64	9 × 9 × 64	0
Flatten the input: number of nodes after flatten = 5184							
6	Dense	100	–	RELU	5184	100	518,500
	Dropout	0.5	–	–	100	100	0
7	Dense	50	–	RELU	100	50	5050
	Dropout	0.3	–	–	50	50	0
	Dense	4	–	SOFTMAX	50	4	204

preprocessing and model training. Pseudocode, shown by algorithm 1, explains the batch image generation.

Algorithm 1: Pseudo code for Image data generator

```
imageDataGenerator (features, labels, batch_size):
batch_features ← init_zeroes (batch_size, 128, 128, 1)
batch_labels ← init_zeroes (batch_size)
FOR (i = 0 to batch_size):
index ← generate_random_number (0, length (features))
processed_image ← preprocessImage (features[index])
processed_image ← processed_image.reshape (128, 128, 1)
batch_features[i] ← processed_image
batch_labels[i] ← labels[index]
RETURN batch_features, batch_labels
```

Image data generator produces a batch of training images of batch size, randomly sampled from the set of input features and labels. For every randomly selected index, the feature image is preprocessed using the image processing module explained in Sect. 2.2 and image dimensions are reshaped to 128 × 128 in 2D. Finally, the processed images and respective labels are stored in the batch array. Such batches are generated every time before starting the new epoch of model training process which is explained in pseudocode 2.

Algorithm 2: Pseudo code for model training process

```
D: brain MRI image dataset.
Xtrain, ytrain, Xtest, ytest: Train and testing split of
dataset D
FOR each_epoch:
batch_features, batch_labels ← imageDataGenerator (Xtrain,
ytrain, batch_size)
normalized_features = normalizeInputData (batch_features)
fit_model (normalized_features, batch_labels)
ypred = predict_model (Xtest)
RETURN ypred
```

For each epoch, a randomized set of batch_features is generated using the training features and training labels denoted by X_{train} and y_{train}, respectively. Before feeding to the model, generated features are normalized on a scale from 0 to 1.

2.5 Evaluation Metrics

While solving the problems related to disease diagnosis, it is highly required to eliminate false prediction cases. The real-life cost of false prediction can be life-threatening. Therefore, the proposed model tried to achieve the highest possible F1-score to ensure robustness and effectiveness. Evaluation metrics [16] used in model performance optimization are explained in the coming subsections:

Confusion Matrix: It is a tabular format to show the classification results of the model. It indicates where the model got confused and happened to predict the wrong class labels. Using this representation, one can improve the model performance using various techniques like hyperparameter tuning, collecting more data, increasing the model training time, etc.

Accuracy: It is the ratio of correctly classified data samples with the total samples. Accuracy is a very basic and straightforward measurement of performance, so one should investigate some more metrics for better understanding.

$$\text{Accuracy} = (TP + TN)/(TP + TN + FP + FN) \tag{1}$$

where *TP*, *TN*, *FP*, and *FN* represents True Positive, True Negative, False Positive, and False Negative, respectively.

F1-Score: This parameter is the way to satisfy the precision–recall tradeoff. F1-score is calculated as the harmonic mean of precision (ratio of correctly predicted positives out of total positive predictions made) and recall (ratio of correctly predicted positives out of total all the positives samples of the dataset). F1-score represents robustness and exactness of the model.

$$\text{F1 - score} = 2(\text{Precision} * \text{Recall})/(\text{Precision} + \text{Recall}) \tag{2}$$

where Precision = $TP/(TP + FP)$ and Recall = $TP/(TP + FN)$.

3 Experimental Results

The proposed model is trained for multiclass classification of the various stages of Alzheimer's disease where 80% of data samples were used for training. After training for 25 epochs, the model optimizes well and achieved a satisfactory level of accuracy and F1-score while trained using Adaptive Moment Estimation (ADAM) technique. Figure 3a, b represents and convergence of accuracy and F1-score as the number of epochs increases. The resultant accuracy and F1-score for training data is 99.3% and 99.5%, respectively and for validation data, it is 97.5% and 98.2%, respectively. As presented in Table 3, performance did not show any considerable difference for training and validation dataset which proved that model did not overfit.

As discussed in the previous section, it is recommended to look into the model confusion scenario. It provides a clear idea that exactly where one should look to

Table 3 Performance results of proposed model

Dataset type	Size	Accuracy (%)	F1-score (%)
Validation	1281	97.5	98.2
Training	5121	99.3	99.5

improve the performance further. The confusion matrix also helps in identifying if the model has biasness toward any particular target label. Table 4 represents the combined confusion matrix for both training and validation dataset.

Comparative Analysis of the proposed model. Table 5 represents the overall performance compared with other related work done in the detection of Alzheimer's disease. The proposed model outperforms all of them with a simple and lightweight CNN model.

Table 4 Confusion matrix for training and validation phase

Evaluation phase	Target class label	Mild demented	Moderate demented	Non demented	Very mild demented
Train	Mild demented	167	0	0	1
	Moderate demented	0	11	0	0
	Non demented	3	0	632	17
	Very mild demented	2	0	8	440
Test	Mild demented	716	0	0	1
	Moderate demented	0	52	0	0
	Non demented	3	0	2540	17
	Very mild demented	2	0	8	1782

Table 5 Comparative analysis

Author	Model	Accuracy (%)	F1-score (%)
Niu et al. [7]	CNN + RDB	89.9	89.9
Embrahimi et al. [8]	3D ResNet-18	96.8	–
Hussain et al. [6]	CNN	97.75	98
Salehi et al. [5]	CNN	99	–
Proposed model	CNN	99.3	99.5

4 Conclusion and Future Remarks

Current work is contributing toward the early diagnosis of Dementia caused by Alzheimer's disease. It proposes a lightweight and efficient CNN model trained over brain MRI scan images. An image preprocessing module was designed to enhance the input images by extracting the ROI. After normalizing the processed images, model training was done for optimal performance. The current approach tried to maximize the F1-score so that misclassification cost can be minimized. This methodology can be extended in the diagnosis of other diseases where neuroimaging analysis is used, e.g., cancer, tuberculosis, etc. The exact approach of image processing and design of deep learning model can be changed as per the disease, but the knowledge gathered in this work may help significantly.

References

1. Duthey B (2013) Alzheimer disease and other dementias. WHO.int. https://www.who.int/medicines/areas/priority_medicines/BP6_11Alzheimer.pdf
2. World Health Organization (2020) Dementia. https://www.who.int/news-room/fact-sheets/detail/dementia
3. Liu S et al (2014) Early diagnosis of Alzheimer's disease with deep learning. In: 2014 IEEE 11th international symposium on biomedical imaging (ISBI), IEEE
4. Alzheimer's Association (2018) 2018 Alzheimer's disease facts and figures. Alzheimers Dement 14(3):367–429
5. Salehi AW, Baglat P, Sharma BB, Gupta G, Upadhya A (2020) A CNN model: earlier diagnosis and classification of Alzheimer disease using MRI. In: IEEE international conference on smart electronics and communication (ICOSEC), Sept 10 2020, pp 156–161
6. Hussain E et al (2020) Deep learning based binary classification for Alzheimer's disease detection using brain MRI images. In: 15th IEEE conference on industrial electronics and applications (ICIEA)
7. Niu J, Tang X (2020) 3D residual dense convolutional network for diagnosis of Alzheimer's disease and mild cognitive impairment. In: IEEE international conference on mechatronics and automation (ICMA)
8. Ebrahimi A, Luo S, Chiong R (2020) Introducing transfer leaming to 3D ResNet-18 for Alzheimer's disease detection on MRI images. In: 2020 35th IEEE international conference on image and vision computing New Zealand (IVCNZ) 2020 Nov 25, pp 1–6
9. Shekhawat SS, Sharma H, Kumar S, Nayyar A, Qureshi B (2021) bSSA: binary salp swarm algorithm with hybrid data transformation for feature selection. IEEE Access 9:14867–14882. https://doi.org/10.1109/ACCESS.2021.3049547
10. Chugh A, Sharma VK, Kumar S, Nayyar A, Qureshi B, Bhatia MK, Jain C (2021) Spider monkey crow optimization algorithm with deep learning for sentiment classification and information retrieval. IEEE Access 9:24249–24262. https://doi.org/10.1109/ACCESS.2021.3055507
11. Dubey S (2019) Alzheimer's dataset (4 class of images). Kaggle.com. https://www.kaggle.com/tourist55/alzheimers-dataset-4-class-of-images
12. Opencv: image thresholding (2021). https://docs.opencv.org/master/d7/d4d/tutorial_py_thresholding.html
13. Albawi S, Mohammed TA, Al-Zawi S (2017) Understanding of a convolutional neural network. In: IEEE international conference on engineering and technology (ICET), 2017 Aug 21, pp 1–6

14. Guo T, Dong J, Li H, Gao Y (2017) Simple convolutional neural network on image classification. In: 2nd IEEE international conference on big data analysis (ICBDA), 2017 Mar 10, pp 721–724
15. Sainath TN, Mohamed AR, Kingsbury B, Ramabhadran B (2013) Deep convolutional neural networks for LVCSR. In: IEEE international conference on acoustics, speech and signal processing, 2013 May 26, IEEE, pp 8614–8618
16. Harikrishnan NB (2021) Confusion matrix, accuracy, precision, recall, F1 score. https://medium.com/analytics-vidhya/confusion-matrix-accuracy-precision-recall-f1-score-ade299cf63cd

Chapter 22
Flower Classification in Videos: A HOG-PCA-NN Method

K. N. Chaitra, V. K. Jyothi, M. Chandrajit, and D. S. Guru

1 Introduction

One of the prominent aspects of organic life is its remarkable diversity. The world inherits a very large number of flower species and the estimation of flower species ranges from 2,20,000 to 4,20,000. It demands specialized knowledge to understand the taxonomic information of flowers. As a consequence, taxonomic knowledge and plant identification skill are restricted to a limited number of individuals [1]. To address the taxonomist's flower species identification requirement, a significant amount of research work has been carried out in the field of Artificial Intelligence and Image Processing for automatic flower recognition. Further, the applications of flower recognition systems have been extended to medicinal use, cosmetics, decoration, floriculture, etc. [2].

Development of a classification model for flower species is a difficult task, due to many challenges such as dynamic environment, illumination variations, scale and viewpoint variations, and occlusions. Further, the large intra-class variability and less inter-class similarity of flowers make the task even more challenging. The segmentation plays a vital role to eliminate the complex background in flower classification [3]; after segmentation, it demands a large number of features to be extracted from the flowers for accurate classification. Contrarily, the large number of features poses a threat to processing due to increased dimensions. Therefore, the segmentation of the flower region is a challenging task.

K. N. Chaitra (✉) · M. Chandrajit
Department of Computer Science, MIT First Grade College, Manandavadi Road, Industrial Suburb, Mysore 570008, India

V. K. Jyothi · D. S. Guru
Department of Studies in Computer Science, University of Mysore, Manasagangotri, Mysore 570006, India

G. Mathur et al. (eds.), *Proceedings of 2nd International Conference on Artificial Intelligence: Advances and Applications*, Algorithms for Intelligent Systems,
https://doi.org/10.1007/978-981-16-6332-1_22

2 Related Works

In the literature, we can find numerous articles on flower classification due to its potential applications and inherent challenges. In [4], the authors proposed a Support Vector Machine (SVM)-based classification model which uses Deep Convolution Neural Network (DCNN) for feature extraction and the Minimum Redundancy Maximum Relevance method for efficient feature selection. The authors in [5] classified flowers using K-Nearest Neighbor (KNN) by extracting texture features such as Gray-Level Co-Occurrence Matrix (GLCM) and Gabor responses. The Threshold segmentation method is applied to segment flowers from the dynamic backgrounds and achieved an accuracy of 98.88%.

In [6], the authors present a novel method for attribute reduction by using the genetic algorithm on extracted features such as Scale-Invariant Feature Transform (SIFT), Pyramid Histogram of Oriented Gradients (PHOG), and Local Color Histogram. Further, an SVM classifier is applied for classification. The authors in [7] used Visual Vocabulary for the classification of flowers. Also, they combined the extracted vocabularies with scale-invariant feature transform (SIFT), hue, saturation, value (HSV), and maximal response 8 (MR8 filter) after Grabcut segmentation and achieved 85.98% accuracy.

In [8], Hu's seven-moment algorithm is applied to acquire edge characteristics, and RGB and HSV components were derived from histogram analysis with KNN classification for flower classification. In [9], the author indicated that the Colored Scale-Invariant Feature Transform (CSIFT) is robust and stable than the scale-invariant feature transform (SIFT) and Dense Scale-Invariant Feature Transform (DSIFT) for flower representation.

A novel approach based on Sum-Product Networks (SPN) for flower classification is proposed in [10]. The authors in [11] reported that the combination of PCA-SIFT with SURF performed significantly better in comparison with lone SIFT. In [12], flower classification was performed using color and GIST features. In [13], linear discrimination analysis (LDA) is used for feature extractor and Principal Component Analysis (PCA) for dimensionality reduction. In [14], Convolution Neural Network (CNN) is used to extract features and Support Vector Machine (SVM) for flower classification. In [15], segmentation using whorl-based region merging was performed and a symbolic classifier for classification was explored.

It can be noted from the literature review that a variety of techniques for flower classification are proposed which investigates the usage of classifiers and prominent features. Further, it can be noted that the Gray-Level Co-Occurrence Matrix (GLCM) has been used in the majority of the methods, which indicate the prominence of texture features for flower classification. Also, it is evident that the majority of the work is carried out on images, and only one work [16] is reported on video flower classification. Therefore, this research focuses on the development of a model for the classification of flowers in videos. We propose a histogram analysis-based auto-threshold segmentation. Next, the texture features from the segmented flower region are extracted which is subjected to dimensionality reduction for handling the curse of dimensionality. Finally, the segmented flower is classified using the Nearest Neighbor (NN) classification technique.

The rest of the paper is organized as follows. In Sect. 3, the model of the proposed flower classification is described. In Sect. 4, the experimental setup, results with reasoning along with a comparative analysis are presented and the article is concluded in Sect. 5.

3 Proposed Method

The proposed model consists of the following stages: segmentation, feature extraction, dimensionality reduction, and classification. The segmentation process is described in Sect. 3.1. Feature extraction phase is shown in Sect. 3.2. In Sect. 3.3, the dimensionality reduction method is depicted and finally, the classification stage is depicted briefly in Sect. 3.4. The proposed model of segmentation and classification is schematically shown in Fig. 1.

3.1 Segmentation

Segmentation in image processing is a typical example of a two-class classification problem which classifies each pixel in an image into the foreground or background region. Otsu thresholding method has been adopted to segment flowers in our proposed model. Otsu method is a global thresholding technique which uses the histogram of the image for automatic thresholding. In [3], histogram on each component of Lab color space is computed and considered for flower segmentation.

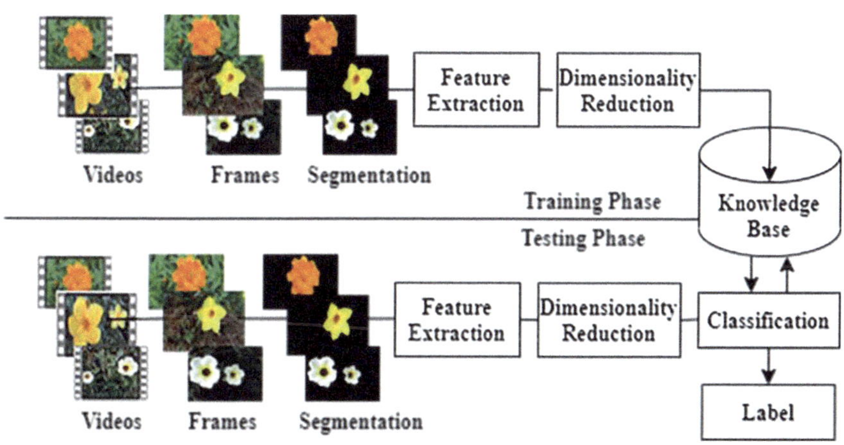

Fig. 1 Block diagram of proposed method

Consequently, the computational time is significantly higher due to color conversion from RGB to Lab. In the proposed method, we consider only the R component and input it for histogram analysis. This helps in reducing the computation time and improved segmentation result as per experimental evidence. Global thresholding using an appropriate threshold T on R component is given in (1). The formulation of Otsu's method is given in (2), (3), (4), (5), (6), and (7):

$$g(x, y) = \begin{cases} 1 & \text{if } f(x, y) > T \\ 0 & \text{if } f(x, y) \leq T \end{cases} \tag{1}$$

where $g(x, y)$ is the thresholded image and $f(x, y)$ is the intensity value of red scale image. The probability (P_i) for each intensity level is given by (2):

$$P_i = \frac{\sum_{i=1}^{L} n_i}{N} \tag{2}$$

where L is the red component level of the image in the range [1, 2 … L], i is the number of pixels at each level and it is denoted by n_i, and N is the total number of pixels.

The probability of foreground pixel class w_0 is given by (3):

$$w_0 = \sum_{i=1}^{k} P_i \tag{3}$$

where k is in the range [1, 2 … k] of foreground pixels.

Similarly, the probability of background pixel class w_1 is given by (4):

$$w_1 = \sum_{i=k+1}^{L} P_i \tag{4}$$

where $k + 1$ is in the range [$k + 1$ … L] of background pixels.

The mean μ_0 of class w_0 is given by (5):

$$\mu_0 = \frac{\sum_{i=1}^{k} \mu_k}{w_0} \tag{5}$$

where $\mu_k = \sum_{i=1}^{k} i P_i$ is the increasing moments of the histogram up to kth level.

Similarly, the mean μ_1 of class w_1 is given by (6):

$$\mu_1 = \frac{\sum_{i=k+1}^{L} \mu_T - \mu_k}{w_1} \tag{6}$$

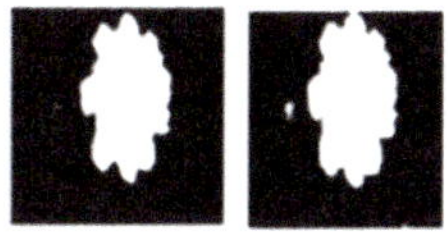

Fig. 2 Before and after hole filling

Fig. 3 Segmentation results

where $\mu_T = \sum_{i=1}^{L} i P_i$ is the total mean level of the original image.

The variance σ between the foreground and background pixel classes is given by (7):

$$\sigma = w_0 x w_1 (\mu_0 - \mu_1)^2 \tag{7}$$

It is observed that some foreground pixels are removed after the segmentation. Therefore, a hole filling operation has been performed. Figure 2 shows the representative images of before and after the hole filling process.

The segmentation results after the hole filling process are shown in Fig. 3. It can be observed that the proposed segmentation method effectively extracts the flower from the background.

3.2 Texture Feature

As the flowers exhibit the same color and shape information and retain their textural characteristics, we have considered texture features in the proposed work. Generally, HOG is applied to color images to extract color feature information. In the proposed method, for the grayscale frames the HOG texture feature is extracted. From the flower region, to calculate HOG features, orientation histogram of edge intensity is considered in the local region using (6) and (7).

The first step of the HOG descriptor is to resize the image into 64 × 128 pixels. The authors in [17] indicated that an ideal image size for HOG descriptor should be 64 × 128. However, we argue that any image which maintains a 1 : 2 ratio is ideal for processing without any hindrances. The next step is to choose the color scales and color normalization. In [17], it is mentioned that the results of using RGB and LAB are identical and there is no much differences. Also, it was noted that the use of grayscale images affected the performance for HOG processing. Contrarily, we argue that the HOG feature performs best on grayscale images as per the experiments.

$$m(x, y) = \sqrt{\mathrm{d}x(x, y)^2 + \mathrm{d}y(x, y)^2} \tag{8}$$

$$\Theta(x, y) = \begin{cases} \tan^{-1}\left(\frac{dy(x,y)}{dx(x,y)}\right) - 180^\circ & \text{if } \mathrm{d}x(x, y) \text{ and } \mathrm{d}y(x, y) < 0 \\ \tan^{-1}\left(\frac{dy(x,y)}{dx(x,y)}\right) + 180^\circ & \text{if } \mathrm{d}x(x, y) \text{ and } \mathrm{d}y(x, y) < 0 \\ \tan^{-1}\left(\frac{dy(x,y)}{dx(x,y)}\right) & \text{Otherwise} \end{cases} \tag{9}$$

where $\mathrm{d}x(x, y)$ is the gradient of x, $\mathrm{d}y(x, y)$ is the gradient of y, $m(x, y)$ is the magnitude, and $\Theta(x, y)$ is the orientation. Further, Kernel to calculate in x-direction is $[-1, 0, 1]$ and

$$\text{Kernel to calculate in } y \text{ - direction: } \begin{bmatrix} -1 \\ 0 \\ 1 \end{bmatrix}$$

After computing the magnitude and orientation, the image is divided into 8 × 8 cells for computing HOG for each of the cell. Therefore, there will be 8 cells and 16 cells in horizontal and vertical directions, respectively, with each cell having 8 × 8 × 3 pixels as represented in Fig. 4. Furthermore, each cell comprises 8 × 8 × 2 values pertaining to gradient and direction information and is represented using nine orientation bins generated using a 2 × 2 filter.

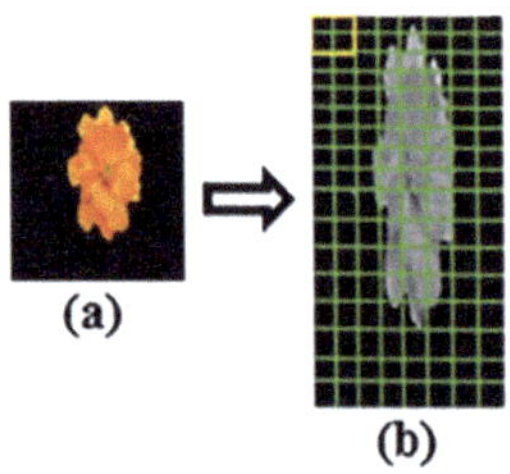

Fig. 4 **a** Input image, **b** 64 × 128 dimension image, 2 × 2 block over the 8 × 8 pixel cell grids

Further, we concatenate all the bins into one vector and obtain a 45,360-dimension vector for the considered dataset, whereas in [18] they have obtained ten thousand dimensions to detect a pedestrian. It is clearly indicated that the dimension of the HOG feature is domain-dependent as per the experimentation.

3.3 Dimensionality Reduction

Principal Component Analysis (PCA) is a well-known dimensional reduction technique. It is based on the composition of variances and covariances through several linear combinations of the original variables, without missing an important part of the original information.

PCA is applied to reduce the 45,360 features extracted from HOG feature applied on segmented frame as shown in Fig. 4b. The computed new variables are called principal component scores. The obtained principle component scores are the compressed data without losing any information with more relevant information of the HOG feature. So, in the proposed system 999 dimension is the reduced feature vector consists of minimum intra variation with maximum inter-class variation attributes by calculating variance and co-variance of an extracted feature.

3.4 Classification

The proposed method uses the Nearest Neighbor (NN) classifier for classification. NN is a supervised learning algorithm used to predict the target label. In the NN classifier, the reduced features of referenced flower images are compared against the reduced features of the test flower image using Manhattan distance. If the distance between the reference flower image and the testing flower image is minimum, then the assigned class label of the reference image is assign to tested flower image. The same procedure is followed for classifying all the remaining test flower images.

4 Experimentation

4.1 Dataset

Experimentation has been performed using the dataset created by the authors in [16]. The dataset consists of ten different classes; each class consists of ten samples of flower videos, with multiple flower objects within each sample video. It is clearly understood that the dataset has less inter-class and large intra-class variation. Further, for processing the flower videos, 1000 images were considered from ten flower classes.

4.2 Experimental Setup

The input video is converted to frames, then stored in JPEG image format. In order to balance the dataset, a hundred frames are selected from each class. Therefore, the balanced dataset consists of 1000 flower images from all ten classes. Subsequently, the images are resized into 250×350 to maintain uniformity among classes. Experiments were performed by dividing the dataset into training and testing samples as per the ratio $20:80$, $30:70$, $40:60$, $50:50$, $60:40$, $70:30$ 30 and $80:20$ of randomly selected samples. Experiments were repeatedly conducted to tabulate the maximum, minimum, and average value for ascertaining the classification accuracy.

4.3 Experimental Result

To evaluate the performance of the proposed classification system, the measures, namely Precision, Recall, and F-Measures, are utilized. The observed result is tabulated in Table 1.

Table 1 reveals the obtained experimental results for varied training and testing sets. It can be observed from the last row of the table that the best result is obtained during the experimentation when the input data for recognition of a flower is trained with 80% of the sample. Further, Fig. 5 shows the graphical representation of the proposed quantitative result where X-axis indicates the accuracy measures and Y-axis indicates the score.

It was observed during experimental analysis that some flowers were misclassified as shown in Fig. 6 (rows 5 and 6). The image in row 5 is expected to classify for the fifth class of the flower, but it is classified as the fourth class. Further, the image in row 6 is expected to classify for the first class of the flower, but it has been to the fourth class. The reason for the misclassification is due to the inaccurate segmentation of some samples having less red intensity levels. Further, the representative of correctly classified flower images has been presented from rows 1 to 4. The result clearly indicates the qualitative effectiveness of the proposed method.

4.4 Comparative Analysis

A comparative study with a contemporary method is performed to showcase the effectiveness of the proposed model. The results of the study are tabulated in Table 2.

Table 1 Quantitative results of the proposed method

	Precision			Recall			Accuracy			F-measure		
Train-test%	Min	Max	Avg	Min	Max	Avg	Min	Max	Avg	Min	Max	Avg
20–80	88.26	92.81	90.52	86.88	92.50	89.87	86.87	92.50	89.87	87.56	92.66	90.20
30–70	91.74	95.08	92.87	91.43	94.57	92.29	91.42	94.57	92.28	91.58	94.83	92.58
40–60	95.73	96.19	96.17	95.50	96.17	96.00	95.50	96.16	96.00	95.61	96.18	96.08
50–50	93.60	96.27	95.57	93.00	96.20	95.20	95.20	96.20	93.00	93.30	96.24	95.39
60–40	97.09	98.07	97.92	97.00	98.00	97.75	97.00	98.00	97.75	97.05	98.03	97.83
70–30	96.47	97.77	97.35	96.33	97.67	97.33	96.33	97.66	97.33	96.40	97.72	97.34
80–20	**98.10**	**99.09**	**99.05**	**98.00**	**99.00**	**99.00**	**98.00**	**99.00**	**99.00**	**98.05**	**99.05**	**99.02**

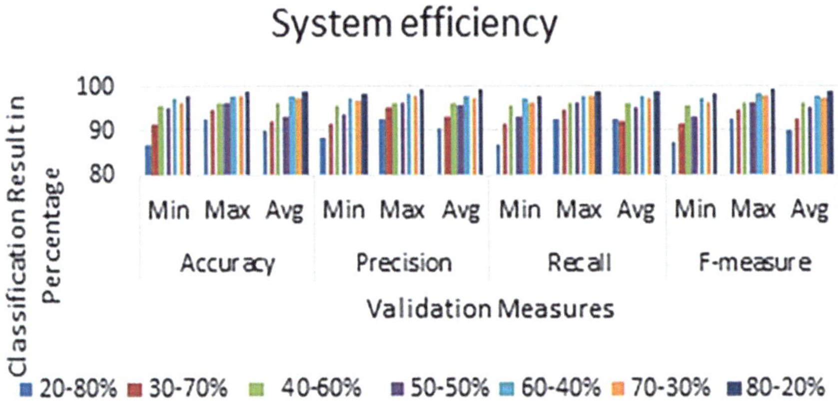

Fig. 5 Classification result in terms of minimum, maximum, and average for various training and testing samples

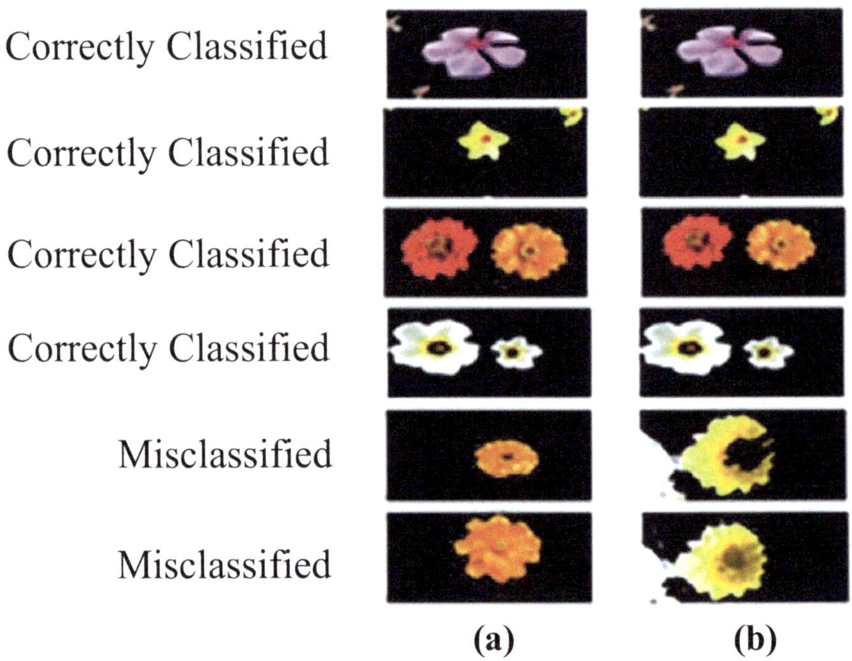

Fig. 6 **a** Test flower **b** classified flower corresponding to test flower

Table 2 Comparative analysis with well-known existing method

Method	No. of classes	Feature type	Classifier used	Accuracy in %
[16]	20	SIFT-DCT	DCNN-MSVM	98.16
Proposed	10	PCA +HOG	NN	99.00

Table 2 indicates that the work reported in [16] uses SIFT and DCT texture features with DCNN and MSVM classifiers and achieved a 98.88% recognition rate. Further, due to the usage of deep network, the method is computationally intensive. Contrarily, in the proposed work an accuracy of 99.00% is achieved using only the HOG feature with NN classifier.

5 Conclusion

In this paper, flower classification in videos using texture features is presented. The modified version of Otsu's method is presented for segmentation which yields better segmentation results after segmentation texture features are extracted. Further, the Nearest Neighbor (NN) classifier is applied for classification. The proposed model gives the accuracy 99.00% and F-measure 99.05%. Further, comparative analysis with existing method is carried to show the stability of the proposed method. The proposed method is robust even with less inter-class and large intra-class variations. In the future, investigation with respect to the increased number of flower classes will be taken into consideration.

References

1. Wäldchen J, Rzanny M, Seeland M, Mäder P (2018) Automated plant species identification-trends and future directions. PLoS Comput Biol e1005993
2. Das M, Manmatha R, Riesman E (1999) Indexing flower patent images using domain knowledge. IEEE Intell Syst 24–33
3. Patil AB, Shaikh JA (2016) OTSU thresholding method for flower image segmentation. Int J Comput Eng Res 2250–3005
4. Cıbuk M, Budak U, Guo Y, Ince MC, Sengur A (2019) Efficient deep features selections and classification for flower species recognition. Measurement 7–13
5. Guru DS, Sharath YH, Manjunath S (2010) Texture features and KNN in classification of flower images. In: IJCA special issue on recent trends in image processing and pattern recognition, pp 21–29
6. Cheng K, Tan X (2014) Sparse representations based attribute learning for flower classification. Neurocomputing 416–426
7. Kulkarni T, Uke NJ (2014) Implementation of image based flower classification system. Int J Comput Sci Bus Inform 35–44
8. Tiay T, Benyaphaichit P, Riyamongkol P (2014) Flower recognition system based on image processing. In: Third ICT international student project conference, pp 99–102
9. Qi W, Liu X, Zhao J (2012) Flower classification based on local and spatial visual cues. In: IEEE international conference on computer science and automation engineering, pp 670–674
10. Shi X (2017) Research on flower image classification using sum-product networks. In: 13th international conference on computational intelligence and security, pp 107–110
11. Rani R, Kumar R, Singh AP (2018) An empirical evaluation of translational and rotational invariance of descriptors and the classification of flower dataset. Pattern Anal Appl 1–18
12. Lodh A, Parekh R (2017) Flower recognition system based on color and GIST feature. In: Device for integrated circuit, pp 790–794

13. Soleimanipour A, Chegini GR, Massah J (2018) Classification of Anthurium flowers using combination of PCA, LDA and support vector machine. Agric Eng Int CIGR J 219–228
14. Togacar M, Ergen B, Comert Z (2020) Classification of flower species by using features extracted from the intersection of feature selection methods in convolutional neural network models. Measurement 107703
15. Sharath Kumar YH, Guru DS (2013) Classification of flowers: a symbolic approach. In: Proc. of Int. Conf. on Multimedia Processing, Communication and Info. Tech., pp 229–235
16. Jyothi VK, Guru DS, Sharath Kumar YH (2019) Classification of natural flower videos through sequential keyframe selection using SIFT and DCNN. In: recent trends in image processing and pattern recognition—second international conference, pp 305–318
17. Bangare SL, Dubal A, Bangare PS, Patil ST (2015) Reviewing Otsu's method for image thresholding. Int J Appl Eng Res 21777–21783
18. Kobayashi T, Hidaka A, Kurita T (2008) Selection of histograms of oriented gradients features for pedestrian detection. Neural Inf Process 598–607

Chapter 23
Impact of Overall Service Quality and Technology Factors on the Intention to Use the Internet of Things (IoT) at Bescom

Kavitha Desai, S. Mahalakshmi, R. Sivaretinamohan, and Macherla Bhagyalakshmi

1 Introduction

The Bangalore Electricity Supply Company Limited (BESCOM) is one among the four companies supplying electricity to consumers, in the distribution sector of Karnataka state, India. It distribute power to 4 operating Zones which covers eight districts of Karnataka state. An area of 41,092 square kilometers is covered with a population of over Rs. 2.07 crore people and consuming around 12,000 MW of power. For ease of operations, 453 section offices under 119 subdivisions of 28 divisions of the nine circles in Karnataka run the company. To advance power distribution management by improving operational efficiency and enhanced service quality, BESCOM has come up with implementing Internet of Things (IoT) enabled smart grid and automating their distribution functionalities through their distribution automation system. To understand the consumer's perception of their quality of service, this proposed research targets the management of the power distribution through the infusion of a dynamic concept called the Internet of Things, which when pervaded with power distribution aims at providing quality and reliable power supply through improved operational efficiency and service quality thereby increasing energy sales, improving quality of service, customer satisfaction, accurate pinpointing of the fault position [1]. It also integrates IT-related activities to monitor and control the entire distribution network efficiently and reliably thereby enabling real-time energy audit [2]. Furthermore, it introduces a regression model for analyzing BESCOM customers' intentions to adopt IoT in the form of AMI smart

K. Desai (✉) · R. Sivaretinamohan (✉) · M. Bhagyalakshmi
Department of Professional Studies, CHRIST (Deemed to be University), Bengaluru, India
e-mail: kavitha.d@christuniversity.in

S. Mahalakshmi
CMS Business School, Jain Deemed To Be University, Bengaluru, India

G. Mathur et al. (eds.), *Proceedings of 2nd International Conference on Artificial Intelligence: Advances and Applications*, Algorithms for Intelligent Systems,
https://doi.org/10.1007/978-981-16-6332-1_23

grid features, taking into account Service Quality Factors (SQF) and Technology Adoption Factors (TAF) [3].

2 Review of Literature

The Internet of Energy (IoE) stipulates a state-of-the-art concept for distribution of power, energy stowing, grid monitoring, and communication [4]. It permits the transmission of energy units as and when required. Power consumption will be monitored on specific devices at the national and international level. Vermesan et al. and Sah [4, 5] conducted a modest IoT install on two appliances, namely: lights and desert coolers. The study unveiled that IoT applications lessen the electricity charges and carbon footprints by very nearly 36%. The PIR sensor in the setup will routinely sense if the infrared radiation intensities are fixed/constant for a given instant of time or not and if they are, then automatically lights will be switched off, and is turned ON when somebody gets inside as it alters the infrared levels of the room. Coolers ensure the continuous flow of air throughout the day, but at night, they change according to the outside temperature, thereby reducing the need for a cooler [6]. Therefore, with the IoT setup, at whatever time the temperature drops under the pre-set value, the cooler is turned OFF, and whenever the temperature gets above the pre-set value, it turns ON spontaneously. Rob Young et al., from Deloitte [7], says that IoT applications concerning the electric utility has diligently kept an eye on the curve of technology availability. Occasionally on the depleting side, the utilities have leveraged technology to enhance and monitor properties, bring in safety measures, regulate the grid, and hold the lights on [8]. IoT finds its application in the power industry at (a) supervisory control and data acquisition (SCADA) which is comprised of sensors and actuators interacting with and regulated by a central master unit, and make available a user interface through a human–machine interface and (b) advanced metering infrastructure (AMI). Hyun Gi Hong, in his research paper, develops a framework for IoT acceptance factors. In this regard, data was collected from 321 IoT users in Korea, and it was found that four factors like cost, environment, personal, and technical factors affected the users on IoT adoption [9].

Earlier studies have typically engrossed on the technical issues of using IoT like architecture [10], trait-based design [11], and WSN [12]. Users of IoT, on the other hand, are unfamiliar with the technology, and there is no concrete evidence that they see it favourably. The research in the technical aspects of IoT outstrips the behavioral and attitudinal aspects. Study on the acceptance of IoT by consumers is still scarce [2], While most of the existing IoT studies have investigated the IoT business model application from the industry and government viewpoints. The research on IoT acceptance from a customer viewpoint is still in its infancy [11, 13, 14]. Furthermore, exploration of user recognition and adoption of information technology (IT) has always been a significant subject of research [15, 16].

3 Materials and Methodology

The study intends to analyze the service quality factors (as external variables) and technology adoption factors (as mediating variables) on consumers' intention to use IoT. The framework analyses the impact of these factors in BESCOM consumers' intention to use IoT. The study identifies the most relevant variables of Service Quality model and TAM (Technology Acceptance Model) model through literature review. TAM (Technology Acceptance Model) is a behavioral model used for analyzing the behavioral aspects of technology adoption. This model consists of twelve constructs of service quality factors, six each for perception and expectation, namely: reliability, tangibles, pricing, customer service and problem-solving, convenience, and information communication. Three technology adoption factors were used in this study: perceived usefulness, perceived ease of use, and attitude toward technology. Figure 1 displays the conceptual framework of behavioral intention toward the adoption of IoT for the present study.

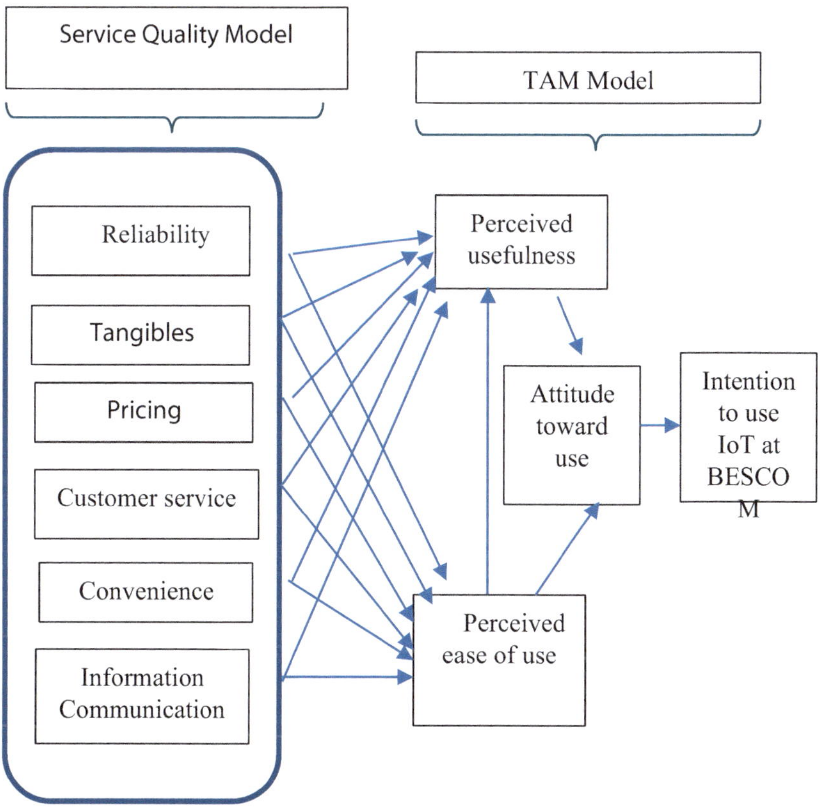

Source: Author

Fig. 1 Conceptual model

Table 1 Outcome of Cronbach's Alpha reliability test

Variable	Number of statements	Reliability value
Service quality factors	28	0.904
Technology factors	3	0.937

Sampling Techniques. According to the thumb rule, as stated by [17], the minimum number of observations would be 4–5 times more than the total number of statements assessing the constructs in the research work. Thus, the minimum sample size was expected to be $96 \times 5 = 480$ respondents. Similarly, Roscoe [18] proposed that a sample size of >30 and <500 are appropriate for most of the research. Accordingly, data was collected from 600 respondents, who are residing in Bangalore and are availing the services of BESCOM were significant respondents.

Cronbach's alpha was employed to measure the reliability of the research instrument. Its coefficient ranges from 0 to 1, with a threshold limit of 0.6, while other researches propose beyond 0.7 for high internal reliability [2]. Since Cronbach's alpha of Service Quality factors, Technology factors are greater than 0.7, and research instruments ensure more reliability to study customer intention to use IoT (Table 1).

4 Results and Discussion

This section unfolds the statistical techniques, viz., Inferential and Descriptive statistics. The researcher has used SPSS 21 for analysis. First, collected raw data was coded into an excel sheet and later exported to Statistical Software SPSS for added investigation. Tools chosen for study are Descriptive statistics, Friedman test, on Correlation Coefficient, and Multiple Regression. Further, the Multiple Regression Analysis was performed using SPSS version 21 for the entire individual constructs such as perceived service quality factors, expected service quality factors, Technology adoption factors, and overall intent to use IoT at BESCOM.

4.1 Mean and Standard Deviation of Expected Service Quality

It is found from Table 2 that the highest mean score of Information Communication (29.0) is the essential factor on expectation in service quality. The service quality factor 'convenience' (12.13) is the least factor. The results found that the respondents expect customized news and updates based on their residential/utility location through a proper communication channel so that it helps them in making proper planning and decision-making in case of outages. Likewise, the respondents have

Table 2 Mean and standard deviation of expected service quality

Expected service quality	Mean	SD	Technology factors	Mean	SD
Reliability	16.27	3.42	Perceived usefulness	71.04	11.40
Tangibles	16.21	3.61	Perceived ease of use	20.57	3.10
Pricing	20.27	4.63	Attitude toward use	12.46	1.87
Customer service	20.55	4.39			
Convenience	12.13	3.10			
Information communication	29.00	6.29			

lesser expectations w.r.t convenience from the service offered by BESCOM, which means they are satisfied with the comfort that BESCOM offers.

In Table 2, based on the mean score, Perceived Usefulness (71.04) is the most essential factor, and attitude toward use (12.46) is the least important factor. Table 3 said that the potential benefits that IoT offers make it the most crucial factor and the respondent's attitude toward factors of technology acceptance makes it the least important factor.

4.2 Exploration of Factors Impacting the BESCOM consumer's Intention to Use IoT by Means of Friedman Test

The present research work intended further to study whether there is any significant variance between mean ranks toward the Service Quality and Technology adoption factors that impact the intention to use IoT at BESCOM.

H_0: There is no major variance among mean ranks toward the service quality and technology adoption factors that impact the intention to use IoT at BESCOM.

As seen in Table 3, above since all the P values for different factors <0.01, the alternative hypothesis is accepted at 1% level of significance. Therefore, there is a substantial variance among mean ranks toward service quality and technology adoption factors on consumers' intention to use IoT at BESCOM. The factors comprised of perceived usefulness, perceived ease of use, attitude, and Intention to use IoT at BESCOM.

Table 3 Test for equality of mean ranks of study variable

Perceived Usefulness	Mean Rank	Chi-square value	P-value
Factors that impact the IoT adoption at BESCOM			
Personal notification about power failure and outages	11.71	773.866	<0.001**
Pay bill online	11.32		
Customized updates on services and energy usage	9.23		
Enhanced information reach by personal notification	8.88		
Time-based pricing	7.66		
Access to real-time energy consumption	8.24		
Net metering	8.80		
Remote monitoring of electrical meters	7.42		
Power quality measurement	8.49		
Mechanism to detect electricity theft, mal-functioning and equipment theft	8.11		
Cost-effective by real-time power monitoring	9.39		
Quicker and automatic restoration of power	8.81		
Improving reliability and trustworthiness to promote IoT over manual system	9.10		
Increased reliability, safety and security	9.04		
Easier to see updates from BESCOM on my gadget	9.13		
Easy access to personalized information	8.09		
Learn and understandable personalized information	9.57		
Perceived ease of use			
Easy bill payment on my gadget	3.25	84.566	<0.001**
Specialized skill to access information from BESCOM	2.78		
It is secured, reliable, safe, energy-saving and cost effective to use IoT to access information from BESCOM	3.08		
Resources, knowledge and ability to use IoT for BESCOM related information	2.78		
Overall using IoT is easier	3.10		
Attitude			
Relying on IoT is a good idea	2.08	14.569	<0.001**
Using IoT is a positive and progressive idea	1.98		
The idea of using IoT in the electric utility sector makes job simple and secured	1.94		
Intent to use IoT at BESCOM			
I will use IoT in near future	2.91	220.268	<0.001**

(continued)

Table 3 (continued)

Perceived Usefulness	Mean Rank	Chi-square value	P-value
I intend to use IoT	2.36		
I will be comfortable to use IoT more frequently	2.21		
Recommend others to use IoT and make an effective management system contributing to the nation's economy	2.52		

Table 4 Expected service quality and factors of technology acceptance

	Perceived usefulness	Perceived ease of use	Attitude toward use	Intention to use IoT at BESCOM
Reliability – E	0.691(**)	0.526(**)	0.570(**)	0.658(**)
Tangibles – E	0.688(**)	0.538(**)	0.561(**)	0.644(**)
Pricing – E	0.725(**)	0.502(**)	0.574(**)	0.646(**)
Customer service – E	0.661(**)	0.482(**)	0.572(**)	0.636(**)
Convenience – E	0.717(**)	0.501(**)	0.540(**)	0.665(**)
Info. comm-E	0.654(**)	0.488(**)	0.502(**)	0.603(**)
Overall expected service quality	0.716(**)	0.526(**)	0.574(**)	0.666(**)

** Denotes correlation is significant at the 1% level (2-tailed)

4.3 *Pearson Correlation Coefficient Between Factors of Expected Service Quality and Factors of Technology Acceptance*

Pearson Correlation is concerned with identifying the association between factors. As seen in Table 4, the Service quality expectations and factors of technology acceptance are positively correlated with each other at 1% level of significance.

4.4 *Multiple Regression Analysis*

Once the degree of relationship between variables has been established using correlation, it is natural to use regression analysis to find out the nature of the relationship and the cause and effect of the relationship between variables. It is framed in this way to examine the relationship between a dependent variable Y and a set of other independent variables X_1, X_2, X_3 … X_n are the coefficients of the variables. b_1, b_2 … b_n are the coefficients of the variables.

The most commonly used linear equation is $Y = b_1X_1 + b_2X_2 + \ldots + b_nX_n + b_0$.

The following null hypothesis is developed to find whether there is any significant effect of the Overall Service Quality factor and technology adoption factors on intention to use IoT at BESCOM.

> H_0: There is no significant impact of overall Service quality factor and technology adoption factors on BESCOM customers' intention to use IoT.

The dependent variable identified in this research is IoT acceptance, while the independent variables are Overall Service Quality, Perceived Ease of use, Perceived Usefulness, Attitude toward use. The multiple correlation coefficient 0.879, which is close to 1, indicates the degree of high positive association between the actual values and the predicted values on the intention to use IoT at BESCOM. The predicted values are acquired as a linear combination of Overall Service quality (X_1), Perceived Usefulness (X_2), Perceived Ease of use (X_3), and Attitude toward use (X_4). Overall service quality factor the coefficient value of 0.879 designates the quite strong and positive correlation between IoT acceptance and the four independent variables (Overall Service Quality, Perceived Usefulness, Perceived Ease of Use, and Attitude toward Use).

The above Tables 5, 6, and 7 portrays the results of Regression analysis and reveal the effect of Overall Service quality, perceived usefulness, perceived ease of use, and attitude concerning the use of Intention to use IoT at BESCOM. The results of the regression analysis is indicating that the model was able to predict 77.3% of the variance in Intention to use IoT at BESCOM. The outcomes of the ANOVA F value $= 506.7953$, $p < 0.001$ point out that the regression model was adequate in envisaging the Intention to use IoT at BESCOM.

Further, when Overall Service quality, Perceived Usefulness, Perceived Ease of use, and Attitude toward use were used as predictors of the outcome; Overall Service quality can predict the impact of Intention to use IoT at BESCOM $b = 0.033$, $p <$

Table 5 Summary of regression model

R	R square	Adjusted R square	Standard Error of the estimate
0.879 (a)	0.7734	0.7721	1.8221

Table 6 Results of ANOVA

	Sum of squares	Degrees of freedom	Mean square	Frequency	Significance
Regression	6732.0641	4	1683.0161	506.7953	0.000
Residual	1975.9364	595	3.3212		
Total	8708.0004	599			

Predictors: (Constant), attitude toward use, Overall Service quality, Perceived Ease of use, Perceived Usefulness

Dependent Variable: Intention to use IoT at BESCOM

Table 7 Variables in multiple regression analysis

Dependent variable	Independent variables	Unstandardized coefficients		Standardized coefficients	T	Sig
		B	Standard error	Beta		
Intention to use IoT at BESCOM	(Constant)	– 5.710	0.646		−8.839	<0.001**
	Overall Service Quality	0.033	0.004	0.195	7.528	<0.001**
	Perceived Usefulness	0.106	0.010	0.318	10.213	<0.001**
	Perceived Ease of use	0.109	0.035	0.089	3.106	<0.001**
	Attitude toward use	0.846	0.061	0.415	13.783	<0.001**

** Denotes significance at 1% level

0.001; while Perceived Usefulness can predict the effect of Intention to use IoT at BESCOM $b = 0.106$, $p < 0.001$; Perceived Ease of Use was able to significantly predict the impact of Intention to use IoT at BESCOM $b = 0.109$, $p < 0.001$ and attitude toward use were able to significantly predict the impact of Intention to use IoT at BESCOM by $b = 0.846$, $p < 0.001$. Hence, it is clinched that all factors such as Overall Service Quality, perceived Usefulness, perceived Ease of use, and Attitude toward the use have a positive influence on the customer's Intention to use IoT at BESCOM.

Estimated model coefficients

The general form of the equation to predict IoT acceptance from Overall Service Quality, Perceived Usefulness, Perceived Ease of use, and Attitude toward use is

$$\begin{aligned}\text{Predicted IoT acceptance} = &-5.710 + 0.33(\text{overall service quality})\\ &+ 0.106(\text{perceived usefulness})\\ &+ 0.109(\text{perceived ease of use})\\ &+ 0.846(\text{attitude towards use})\end{aligned}$$

This is obtained from the Coefficients Table 7, as shown above.

The coefficient of $X_1 = 0.33$ represents that the IoT acceptance value will increase by 0.33 for every unit rise in the overall service quality and the coefficient value is significant at 1% level. The coefficient of $X_2 = 0.106$ represents that the IoT acceptance score would increase by 0.106 for every unit increase in overall service quality, and this coefficient value is significant at 1% level. The coefficient of $X_{3=}$ 0.109 represents that the IoT acceptance score would increase by 0.109 for every unit increase in overall perceived ease of use, and this coefficient value is significant at

1% level. Similarly, the coefficient of $X_4 = 0.846$ represents that intention to use IoT at BESCOM score would increase by 0.846 for every unit increase in overall perceived ease of use and this coefficient value is significant at 1% level. The essential factors for "intention to use IoT at BESCOM" are Overall attitude toward use (0.846) followed by overall service quality (0.33), perceived ease of use (0.109), and perceived usefulness (0.106). Overall, the model accounted for 77.3% of the total variance on Overall Service quality, Perceived Usefulness, Perceived Ease of use, and Attitude toward use and was able to positively forecast the impact of Intention to use IoT at BESCOM.

Outcomes of hypothesis testing: Since p-value < 0.000; the alternative hypothesis is accepted, and it can be concluded that Overall Service Quality and technology adoption factors have a positive influence on the intention to apply IoT at BESCOM. Moreover, Overall Service quality, Perceived Usefulness, Perceived Ease of use, and Attitude toward use were able to predict the influence of service factors and technology factors on intention to make use of IoT at BESCOM positively.

5 Conclusion

This research made an attempt to understand Power Distribution management using the Internet of Things at BESCOM. This study proposes to develop a regression model that intends to assess BESCOM customers' intention to make use of IoT. This research is aspired to discover and evaluate the significant elements that have affected the intention to use IoT by different customer bases of BESCOM. For carrying out this research, 600 users of BESCOM that including residential, commercial, and industrial users have been surveyed. While adopting newer technologies, it is reflected as intricate, this research makes use of the conceptual framework for determining factors that determine the intention to apply IoT at BESCOM. The study result revealed that BESCOM as a service provider would overcome the drawbacks of the existing legacy system and reap benefits from IoT and the customers have an intent to use IoT when given.

References

1. Jaafreh AB (2018) The effect factors in the adoption of internet of things (IoT) technology in the SME in KSA: an empirical study. IMBR
2. Gao L, Bai X (2014) A unified perspective on the factors influencing consumer acceptance of internet of things technology. Asia Pac J Mark Logist 26(2):211–231. https://doi.org/10.1108/APJML-06-2013-0061
3. Waltzman H, Shen L (2015) The internet of things. Intellect Prop Technol Law J 27(7):19–21
4. Vermesan O, Friess P, Guillemin P, Gusmeroli S, Sundmaeker H, Bassi A, Jubert IS, Mazura M, Harrison M, Eisenhauer M (2011) Internet of things strategic research roadmap. https://www.scienceopen.com/document?vid=c891f8b4-f51e-489b-8e5c-4aeacabcc4f9

5. Sah P (2016) Saving environment using internet of things: challenges and the possibilities. Adv Internet Things 6:55–64. https://doi.org/10.4236/ait.2016.64004
6. Deutsche Welle, Cisco Internet Business Solutions Group, Cisco (2012) Internet of things. http://www.cisco.com/c/en/us/solutions/internet-ofthings/overview.html
7. Rob Y, John M, Christian G (2016) The power is on: how IoT technology is driving energy innovation. Deloitte Insights
8. DuBravac S, Ratti C (2017) The internet of things:evolution or revolution? AIG White Paper. http://www.aig.com/Chartis/internet/US/en/AIG%20White%20Paper%20-%20IoT%20English%20DIGITAL_tcm3171-677828_tcm3171-698578.pdf
9. Hong H-G (2016) Analysis of business framework for internet of things. Indian J Sci Technol 9(37):1–15. https://doi.org/10.17485/ijst/2016/v9i37/102549
10. Gubbi J, Buyya R, Marusic S, Palaniswami M (2013) Internet of things (IoT): a vision, architectural elements, and future directions. Futur Gener Comput Syst 29(7):1645–1660
11. Su J, Cao D, Zhao B, Wang X, You I (2014) ePASS: an expressive attribute-based signature scheme with privacy and an unforgeability guarantee for the internet of things. Futur Gener Comput Syst 33:11–18
12. Turkanović M, Brumen B, Hölbl M (2014) A novel user authentication and key agreement scheme for heterogeneous ad hoc wireless sensor networks, based on the internet of things notion. Ad Hoc Netw 20:96–112
13. Haller S, Karnouskos S, Schroth C (2009) The internet of things in an enterprise context. In: Domingue J, Fensel D, Traverso P (eds) Future internet—FIS. Springer, Berlin, pp 14–28
14. Peoples C, Parr G, Mcclean S, Scotney B, Morrow P (2013) Performance evaluation of green data centre management supporting sustainable growth of the internet of things. Simul Model Pract Theory 34:221–242
15. Bandyopadhyay K, Bandyopadhyay S (2010) User acceptance of information technology across cultures. Int J Intercult Inf Manag 2(3):218–231
16. Li X, Lu RX, Liang XH, Shen XM, Chen JM, Lin XD (2011) Smart community: an internet of things application. IEEE Commun Mag 49:68–75
17. Mitchell V (1994) How to identify psychographic segments: part 2. Mark Intell Plan 12(7):11–17
18. Roscoe JT (1975) Fundamental research statistics for the behavioral sciences, 2nd edn. Holt, Rinehart and Winston, New York

Chapter 24
Transformer Deep Learning Model for Bangla–English Machine Translation

Argha Chandra Dhar, Arna Roy, Md. Ahsan Habib, M. A. H. Akhand, and N. Siddique

1 Introduction

Machine translation (MT) refers to translating texts or documents from the source language into the target language without human intervention. On a fundamental basis, MT was used to conduct direct substitution of words from the source language to the target language [1]. However, it is clear that only the replacement of words rarely provides a meaningful translation in real life. Later, different methods have been developed in the last several decades to improve the quality of translation. Among the different MT approaches, neural MT (NMT) is the most recent as well as a promising method. NMT is a data-driven approach that consists of special neural networks (NN) [2]. Due to the drastic advancement of deep neural network, NMT emerges as a prolific research approach in the MT domain in recent years [3].

The task of MT is inherently language dependent because data preparation (e.g., training data for NMT) is a vital task. Moreover, a particular MT system is required to adjust a specific language pair along with their distinct grammar and phrase rules. A number of remarkable researches are available with valuable resources for English–French, English–German language pairs. On the contrary, there has been very little reported on MT of Bangla language although it is one of the fifth most widely spoken languages with approximately 228 million native speakers.

A number of studies with different methods on Bangla–English language pair have been reported in the last several years. The pioneering studies of Bangla–English are

A. C. Dhar · A. Roy · Md. A. Habib · M. A. H. Akhand (✉)
Department of Computer Science and Engineering, Khulna University of Engineering & Technology, Khulna, Bangladesh
e-mail: akhand@cse.kuet.ac.bd

N. Siddique
School of Computing, Engineering and Intelligent Systems, Ulster University, Newtownabbey, Northern Ireland, UK

G. Mathur et al. (eds.), *Proceedings of 2nd International Conference on Artificial Intelligence: Advances and Applications*, Algorithms for Intelligent Systems,
https://doi.org/10.1007/978-981-16-6332-1_24

rule-based MT (RBMT) methods which are basically based on linguistic information and produce translation through rules generated by human experts incorporating verb, phrase, preposition, and other things [4–8]. Recently, Mumin et al. [9] presented a phrase-based statistical MT (SMT) model (called shu-torjoma), for both Bangla to English and English to Bangla. The proposed system outperforms other developed systems significantly.

A number of researches have reported deep learning-based NMT on Bangla. Mumin et al. [10] have investigated attention-based model and Byte Pair Encoding (BPE) in their NMT model. They have separately examined the basic attention-based model and attention-based model with BPE for both Bangla to English and English to Bangla. It is shown that the attention-based model with BPE gives comparatively better results than other approaches. Hasan et al. [11] investigated an NMT approach for Bangla to English with BiLSTM and their model is found better than SMT-based system. BiLSTM consists of two LSTMs which can take input from the forward direction, as well as backward direction, by which the output layer becomes susceptible to receive information from backward and forward [12]. Hasan et al. [13] also investigated transformer-based deep learning method for Bangla–English language pair and the model gives fairly good result when compared with BiLSTM-based model. In BiLSTM and transformer-based models [11, 13], the model parameters have been adjusted with word-embeddings such as Glove model [14] and Word2vec model [15] for English and Bangla languages, respectively. It is worth mentioning that Glove and Word2vec are algorithms that can distinguish similar words from a large corpus.

The aim of this study is to develop an NMT system to achieve well-performing Bangla–English MT. Modern deep learning-based transformer model is considered for this purpose. A transformer model consists of NN-based encoders and decoders and the model is adapted by tuning different parameters to identify the best performing model for Bangla–English MT.

The rest of the paper is organized as follows. Section 2 describes the proposed methodology. Section 3 presents experimental results of the proposed model and compares performance with prominent existing methods. And finally, Sect. 4 concludes the paper with few remarks.

2 Transformer-Based Bangla–English MT

The purpose of this study is to develop a transformer-based method for Bangla–English MT. The method is divided into two major phases: data preprocessing; and preparing and training a transformer model. The following subsections briefly explain the steps.

2.1 Data Preprocessing

For NMT system, data preprocessing is an important task that includes tokenization, true-casing, normalizing punctuation, removing non-printable characters from the data. Long sentences and empty sentences may cause a problem. Therefore, fixed-length sentence is used in any NMT system; sentence length is limited to 40 in this study. Byte Pair Encoding (BPE) [16, 17] algorithm is applied on the corpus for sub-word segmentation to handle rare words. Preprocessing depends on the data used and will be explained for the selected data in the experimental studies section.

2.2 Transformer Architecture and Its Adaptation

Transformer is a deep learning model proposed in 2017 [18]. The significant property of the model for the natural language processing (NLP) tasks is that it does not require feeding the data into the model in a sequential manner which enables more parallelism and faster training. The model uses semi-supervised learning approach in training. It generally utilizes large amount of unlabeled data (unsupervised) with the combination of smaller amount of labeled data (supervised) [19]. Transformer is widely used in MT, time series prediction [20], document generation [21], named entity recognition [22], biological sequence analysis [23, 24]. Another important issue to choose the model is the open-source model availability and customization facility for a particular task in OpenNMT toolkit [25].

Figure 1 shows the basic building blocks of the transformer model which has three main operating units: Embedding, Encoder–Decoder, and Output Generation.

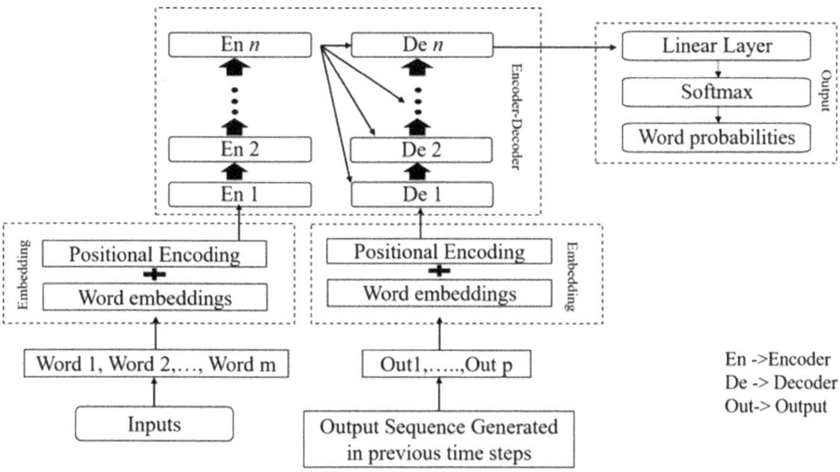

Fig. 1 The building block of a transformer deep learning model

In the embedding layer, each word in a sentence is transferred into word embedding and then each embedding is added to the positional encoding vector (in the range of −1 to 1) of the same dimension. The resultant vector holds all necessary information about the sequence of each word in the input sentence as well as the distance of different words, and the vector of numeral values is fed into the encoder. Figure 2 illustrates layers of an encoder and a decoder. An encoder can be divided into 2 sub-layers: Self-Attention and Feed Forward Neural Network (FFNN). A decoder can be divided into 3 sub-layers: Self-Attention, Encoder–Decoder Attention, FFNN. Layer normalization is common in each layer to reduce training time.

The encoder–decoder module is the core of the transformer model which handles the attention mechanism and is the element to improve the performance of NMT. The numbers of encoders and decoders should be the same although they are slacked separately. The encoder receives a list of vectors of fixed size as input and processes through self-attention layer and FNN layer. Output of a lower-level encoder is sent to its immediate upper-level encoder as an input in a cascading manner and this procedure continues up to the topmost level of the encoder. The output of the top encoder is converted into attention vectors. These vectors are provided to each decoder to generate the first word of the output. These are all done in timestep 1. This output works as an additional input to the decoder in the next time step to predict the upcoming word. The process of word generation continues until an end of the sentence character is encountered; thus, the output sentence is generated. The output generation simply consists of Linear and Softmax layer which converts the decoder output vector into some probabilistic values, which help the model generate the next token.

In order to achieve good performance, the hyperparameters of the transformer model should be adjusted. There are a number of hyperparameters in the transformer model such as batch size, dropout, learning rate, number of encoder–decoder layers, number of heads, and so on. Multiple number of heads help the self-attention to make the attention layer work better in a way where it increases the model's ability to guess what the other words refer to a particular word that is currently being processed.

3 Experimental Studies

This section describes the experimental outcomes of the proposed NMT system on the benchmark dataset. The performance of the proposed method is also compared with the existing methods.

3.1 Benchmark Data and Preprocessing

A few parallel corpora are available for Bangla–English MT. In this study, SUPara [26] dataset is used as a number of recent studies have used this corpus [9–11, 13].

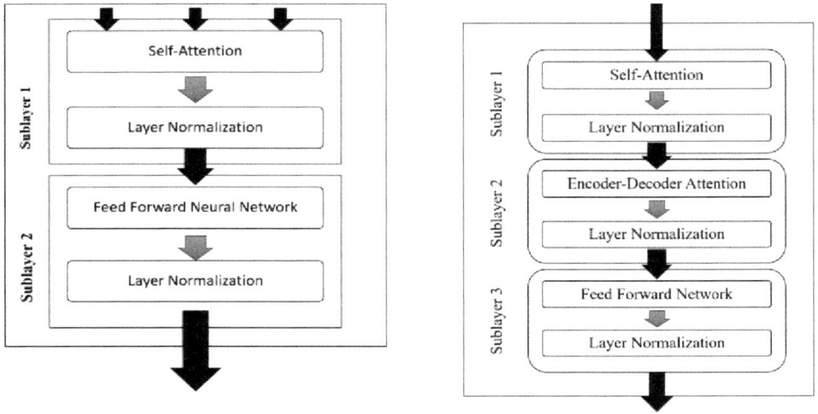

(a) Design of an encoder (b) Design of a decoder

Fig. 2 Design of encoder and decoder of the transformer model

Table 1 Examples of preprocessing effect on sample sentences

	Available in dataset	Sentence after preprocessing
Bangla	সে লাল গোলাপ পছন্দ করে।	সে লাল গোলা@@ প পছন ্ দ করে ।
	মানব জীবন অত্যন্ত সংক্ষিপ্ত।	মানব জীবন অত ্ যন ্ ত সংক ্ ষিপ ্ ত ।
	আমরা বাংলাদেশে বাস করি।	আমরা বাংলাদেশে বাস করি ।
English	She likes red roses	She likes red ro@@ ses
	Human life is very short	Human life is very short
	We live in Bangladesh	we live in Bangladesh

The dataset contains 70,861, 500, and 500 parallel sentences for training, validation, and test sets, respectively. In the processing step, data tokenization, true-casing, normalizing punctuation, and removing non-printable characters are performed on the data using Moses [27]. It changes the raw sentences into number of tokens where words and punctuation marks are separated by a space. Long sentences and empty sentences may cause problem, so the sentence length is kept as 40 in our NMT model. Having a small corpus size results in a poor dictionary which might cause large number of unknown words in the test case. To handle such a scenario, sub-word segmentation is employed using BPE algorithm. The algorithm counts the frequency of each word in a corpus and a special stop symbol </w> is added at the end of each token. Characters are then separated. After that, the algorithm finds out the most frequent two consecutive byte pairs and merges the two-byte pairs into one token. As an example, the BPE algorithm can recognize 'r' and 'o' as a consecutive frequent token pair and thus merge them into one token 'ro'. The same explanation applies to 'ses' token. Then the BPE algorithm divides the token 'roses' into two sub-words: 'ro' and 'ses'; adding '@@' in between them. After all these preprocessing operations, training, validation, and test sets had 65,855, 366, 361 parallel sentences, respectively. Table 1 shows a few sentences having some '@@' after preprocessing.

3.2 *Performance Evaluation and Experimental Setup*

For evaluating the performance of the model, the Bilingual Evaluation Understudy (BLEU) score is measured which is currently the most popular evaluation metric in the MT field [28]. It is a precision-oriented measurement and evaluates the correctness of system output. BLEU score is measured roughly in three steps. At first, n-gram matches are calculated in the candidate sentences (system output) and in the reference sentences. Then the candidate counts are clipped by their corresponding maximum reference value. Next, the clipped n-grams are summed and divided by the number of candidate n-grams [28]. Through these steps, the modified precision score (p_n) is found.

$$p_n = \frac{\sum_{C\in\{Candidates\}} \sum_{n-gram\in C} Countclip(n-gram)}{\sum_{C'\in\{Candidates\}} \sum_{n-gram'\in C'} Count(n-gram')} \tag{1}$$

The second step is the BLEU Brevity Penalty factor (BP) calculation:

$$BP = \begin{cases} 1, \text{ if } c > r \\ e^{1-\frac{r}{c}}, \text{ if } c \le r \end{cases}, \tag{2}$$

here c is the length of the candidate translation and r is the length of reference translation. Finally, the BLEU score is the geometric mean of the precision scores and calculated using Eq. (3).

$$BLEU = BP \cdot \exp\left(\sum_{n=1}^{N} w_n \log p_n\right) \tag{3}$$

Here, N is typically 4 and w_n is the positive weight which is typically set to $^1/_N$.

The proposed NMT model is implemented using OpenNMT toolkit [25]. To train the model, we have used a batch size of 4096 and neurons in FFNN of 2048. The word embedding size was 512. The encoder–decoder layer size was kept as 6. Adam optimizer [29] is used for training the model with a dropout of 0.1. The values of alpha, beta1, and beta2 are 0.00031, 0.9, and 0.998, respectively. The PC in which the experiments were conducted has the following configuration: CPU 7th Generation Intel® Core™ i5-7400 3.50 GHz, RAM 8 GB, and NVIDIA GPU Ge-Force GTX 1070Ti 8 GB.

3.3 Experimental Results and Analysis

A number of experiments have been carried out to improve the performance of the proposed transformer model. Since the number of heads is an important issue, experiments have been performed on varying head numbers to identify the appropriate number. Table 2 represents BLEU scores on the test set and required time, while varying heads from 1 to 32 for fixed 20,000 training steps. From the table, it is observed that the model incurs larger training time for the higher number of heads. As an example, for 20,000 training steps, the required training time for a single head and 32 heads are 51,869 and 63,024 s, respectively. The time continues to increase with the head numbers because the model has to perform the same matrix calculation as many times as the number of heads. From the table, it is also detected that BLEU scores are different, while the number of heads varied but the scores are not correlated with numbers. Therefore, it is a matter of empirical study to identify the best-suited head number for a dataset. The best BLEU score 19.09 is achieved at 20,000 steps when the number of heads is equal to two. It is also observed that the BLEU score

Table 2 Effect of number of heads on Bilingual Evaluation Understudy (BLEU) score on the test set and training time in transformer model

Number of heads	BLEU score		Required time (s)	
	At 10,000 steps	At 20,000 steps	First 10,000 steps	Total 20,000 steps
1	16.42	18.03	25,807	51,869
2	17.24	19.09	26,399	53,098
4	17.06	17.55	26,843	54,001
8	16.35	18.99	27,581	55,491
16	15.41	17.53	28,989	58,379
32	13.89	17.43	31,253	63,024

at 20,000 steps is always better than that of 10,000 steps which indicates that higher training steps may be required for an improved score.

Figure 3 shows the BLEU scores for training, validation, and test sets for two heads while training steps (i.e., iteration) continued for 80,000 steps. From the figure, it is noticed that initially BLEU score improved rapidly but did not improve much after a certain number of steps. As an example, after 10,000 training steps, the BLEU scores for training, validation, and test set are 96.74 (point not visible in the figure), 19.98, and 17.24, respectively. On the other hand, at 30,000 training steps, the scores for the three sets are 99.56 (point not visible in the figure), 21.12, and 20.29, respectively. It is notable that training set BLEU is much better than validation and test sets. For any MT system, test set BLEU is always important which recognizes the generalized ability of the system. The transformer model with two heads achieved the test BLEU of 21.33 at 80,000 training steps.

Table 3 compares the performance of the proposed transformer model with other

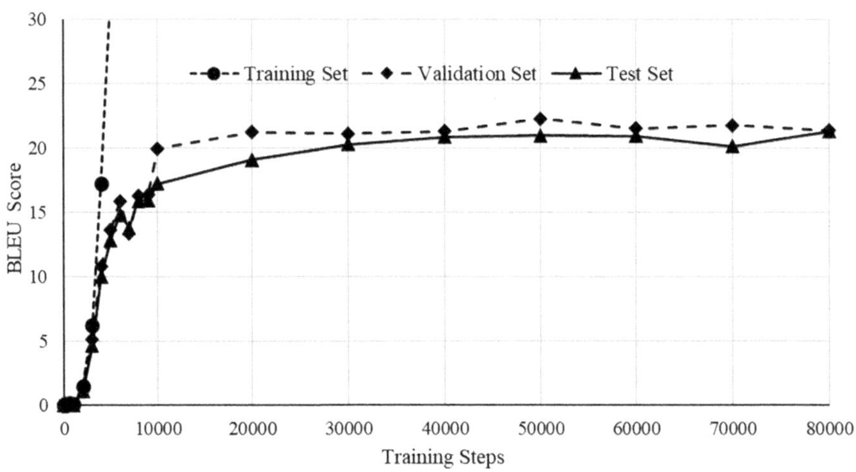

Fig. 3 BLEU score on training, validation, and test sets for number of heads equals two

Table 3 Comparison of the achieved BLEU scores on the test set of the benchmark data

Work Ref., year	Dataset	Train./Val./test set size	Model used	BLEU score	Remarks
Mumin et al. [9], 2019	SUPara, GlobalVoices	197,388/500/500	Phrase based SMT	17.43	Aggregation of SUPara and GlobalVoices used as training set
Hasan et al. [11], 2019	SUPara	70,861/500/500	BiLSTM + Embeddings	19.76	
Hasan et al. [13], 2019	ILMPC, SIPC, PTB, SUPara, AmaderCAT	419,109/500/500	BiLSTM + Embeddings	19.24	Aggregation of the data sets used as training set
	ILMPC, SIPC, PTB, SUPara, AmaderCAT	419,109/500/500	Transformer with base configuration	18.99	Aggregation of the data sets used as training set
	SUPara	70,861/500/500	BiLSTM + Embeddings	19.98	
Our model	SUPara	70,861/500/500	Transformer with optimal head and BPE	21.33	

N.B.: Methods considered the same SUPara validation and test sets while several methods considered a larger training set combining different datasets with SUPara training set

prominent Bangla MT models on the basis of achieved BLEU score on SUPara test set. From the table, it is observed that any NMT method outperformed the SMT method. The SMT method [9] achieved a 17.43 BLEU score. Among the existing NMT methods, BiLSTM [11, 13]- based method with word embedding is the best with a BELU score of 19.98. The existing transformer model [13] achieved a BLEU score of 18.99 with base configuration which includes word embedding size 512, number of heads 8. On the other hand, our transformer model excels with the BLEU score of 21.33 because sub-word segmentation helped the model to guess rare words. In addition, the appropriate number of heads for the proposed model is identified through empirical study which is two. The two heads enhance the ability of the model to put appropriate attention on different positions of words in a sentence which helps the model to perform better in combination with sub-word segmentation. From the table, it is amply clear that the proficiency of the proposed transformer model for Bangla–English MT is better than existing methods.

Moreover, most of the existing methods employed a larger training set combining different datasets with SUPara training set. As an example, Hasan et al. [13] trained the transformer model with a base configuration of 419,109 sentences combining Indic Languages Multilingual Parallel Corpus (ILMPC), Six Indian Parallel Corpus (SIPC), Penn Treebank Bangla–English parallel corpus (PTB), SUPara and AmaderCAT. On the other hand, our transformer model having optimal heads with BPE is trained with SUPara training set only (i.e., 70,861 sentences) which indicates the computational efficiency of the proposed model over the other models.

4 Conclusions

Recently, NMT has drawn attention with different deep learning methods for MT, and the transformer model seems promising to perform well for different language pairs. In this study, the standard transformer model is tuned to achieve improved performance for Bangla–English MT. It is identified that two heads in the model performed better than larger number of heads while testing on the benchmark data set. The proposed model considered Byte Pair Encoding in preprocessing. The proposed model outperformed other leading MT methods on the basis of BLEU score achievement on the benchmark dataset.

Acknowledgements This research is partially supported by REVE Systems, Bangladesh.

References

1. Hutchins WJ (1995) Machine translation: a brief history. In: Concise history of the language sciences, Elsevier Science, Amsterdam. https://doi.org/10.1016/b978-0-08-042580-1.50066-0.
2. Tan Z, Wang S, Yang Z, Chen G, Huang X, Sun M, Liu Y (2020) Neural machine translation: a review of methods, resources, and tools
3. Waibel A, Jain AN, McNair AE, Saito H, Hauptmann AG, Tebelskis J (1991) JANUS: a speech-to-speech translation system using connectionist and symbolic processing strategies. In: [Proceedings] ICASSP 91: 1991 international conference on acoustics, speech, and signal processing, vol 2. IEEE, pp 793–796. https://doi.org/10.1109/ICASSP.1991.150456.
4. Ashrafi SS, Kabir H, Anwar M, Noman AKM (2013) English to Bangla machine translation system using context-free grammars. IJCSI Int J Comput Sci 10:144–153
5. Muntarina K, Moazzam G, Bhuiyan A (2013) Tense based English to Bangla translation using mt system. Int J Eng Sci Invent 2(30):38
6. Alamgir T, Arefin MS, Hoque MM (2016) An empirical machine translation framework for translating bangla imperative, optative and exclamatory sentences into English. In: 2016 5th international conference on informatics, electronics and vision (ICIEV). IEEE, pp 932–937. https://doi.org/10.1109/ICIEV.2016.7760137
7. Anwar M (2018) Bangla to English machine translation using fuzzy logic. Int J Comput Sci Inf Secur 16(156):165
8. Mukta AP, Mamun A, Basak C, Nahar S, Arif FH (2019) A phrase-based machine translation from English to Bangla using rule-based approach. In: 2019 international conference on electrical, computer and communication engineering (ECCE). IEEE, pp 1–5
9. Al Mumin MA, Seddiqui MH, Iqbal MZ, Islam MJ (2019) Shu-torjoma: an English↔Bangla statistical machine translation system. J Comput Sci 15:1022–1039. https://doi.org/10.3844/jcssp.2019.1022.1039
10. Al Mumin MA, Seddiqui MH, Iqbal MZ, Islam MJ (2019) Neural machine translation for low-resource English–Bangla. J Comput Sci 15:1627–1637. https://doi.org/10.3844/jcssp.2019.1627.1637

11. Hasan MA, Alam F, Chowdhury SA, Khan N (2019) Neural versus statistical machine translation: revisiting the Bangla–English language pair. In: 2019 international conference on Bangla speech and language processing (ICBSLP). IEEE, pp 1–5
12. Schuster M, Paliwal KK (1997) Bidirectional recurrent neural networks. IEEE Trans Signal Process. https://doi.org/10.1109/78.650093
13. Hasan MA, Alam F, Chowdhury SA, Khan N (2019) Neural machine translation for the Bangla–English language pair. In: 2019 22nd international conference on computer and information technology (ICCIT). IEEE, pp 1–6. https://doi.org/10.1109/ICCIT48885.2019.9038381
14. Pennington J, Socher R, Manning CD (2014) GloVe: global vectors for word representation. In: EMNLP 2014—2014 conference on empirical methods in natural language processing, proceedings of the conference. https://doi.org/10.3115/v1/d14-1162
15. Alam F, Chowdhury SA, Noori SRH (2017) Bidirectional LSTMs—CRFs networks for bangla POS tagging. In: 19th international conference on computer and information technology, ICCIT 2016. https://doi.org/10.1109/ICCITECHN.2016.7860227
16. Sennrich R, Haddow B, Birch A (2016) Neural machine translation of rare words with subword units. In: 54th annual meeting of the association for computational linguistics, ACL 2016—long papers. https://doi.org/10.18653/v1/p16-1162
17. Gage P (1994) A new algorithm for data compression. C Users J
18. Vaswani A, Shazeer N, Parmar N, Uszkoreit J, Jones L, Gomez AN, Kaiser Ł, Polosukhin I (2017) Attention is all you need
19. Engelen van JE, Hoos HH (2020) A survey on semi-supervised learning. Mach Learn. https://doi.org/10.1007/s10994-019-05855-6
20. Maxime: what is a transformer? An introduction to transformers and… | by maxime | inside machine learning | medium
21. Askell A, Clark J, Brundage M, Hernandez D, Lansky D, Luan D, Radford A, Wu J Better language models and their implications
22. Davydova O 10 applications of artificial neural networks in natural language processing
23. Rives A, Goyal S, Meier J, Guo D, Ott M, Zitnick CL, Ma J, Fergus R (2019) Biological structure and function emerge from scaling unsupervised learning to 250 million protein sequences. https://doi.org/10.1101/622803
24. Nambiar A, Heflin M, Liu S, Maslov S, Hopkins M, Ritz A (2020) Transforming the language of life: transformer neural networks for protein prediction tasks. In: Proceedings of the 11th ACM international conference on bioinformatics, computational biology and health informatics, BCB 2020. https://doi.org/10.1145/3388440.3412467
25. Klein G, Kim Y, Deng Y, Crego J, Senellart J, Rush AM (2017) OpenNMT: open-source toolkit for neural machine translation. In: 20th Annu. Conf. Eur. Assoc. Mach. Transl. EAMT 2017
26. Iqbal MZ (2012) SUPara: a balanced English–Bengali parallel corpus SUST. J Sci Technol 16(46):51
27. Koehn P, Zens R, Dyer C, Bojar O, Constantin A, Herbst E, Hoang H, Birch A, Callison-Burch C, Federico M, Bertoldi N, Cowan B, Shen W, Moran C (2007) Moses: open source toolkit for statistical machine translation. In: Proc. 45th Annu. Meet. ACL Interact. Poster Demonstr. Sess.—ACL '07. https://doi.org/10.3115/1557769.1557821
28. Papineni K, Roukos S, Ward T, Zhu W-J (2001) BLEU: a method for automatic evaluation of machine translation. ACL. https://doi.org/10.3115/1073083.1073135
29. Kingma DP, Ba JL (2015) Adam: a method for stochastic optimization. In: 3rd international conference on learning representations, ICLR 2015—conference track proceedings

Chapter 25
Application of Machine Learning for Heart Disease Prediction

Mohsin Qureshi and Nilima Warke

1 Introduction

Heart is one of the vital organs in the human body. Any abrupt changes in its working or any disturbance to the normal functioning can cause serious damage and can be fatal in some cases. Global organizations like the WHO have reported that nearly 17.9 million people die every year out of which 80% deaths are caused due to Heart Diseases.

Realizing that cardiovascular diseases are the leading causes of increase in mortality rate, measures are being taken globally to bring this situation under control [1]. Since heart diseases are affected due to various factors like age, cholesterol level, blood pressure, smoking, drinking, and unhealthy lifestyles. It is difficult for a regular doctor to predict the disease before it reaches higher stages.

Automation of these problems can be very beneficial as it reduces the doctor's workload in dedicating crucial time for report analysis. To enable the doctors to fully utilize the potential of the latest healthcare instruments, it became necessary to deploy concepts like ML to handle the enormous data from high precision and high-resolution devices [2]. With the implementation of different ML algorithms supportive decision-making systems can be developed to analyze the disease in very short time frame and give highly precise reports for the doctor's final verdicts.

In this particular heart disease prediction model, algorithms that can deal with numeric representation of real-world data are required [3].

In the proposed work, comparative analysis of different ML algorithms like DT, NB, NN, and RF are carried out. Data set is collected from UCI repository. For

M. Qureshi (✉)
Instrumentation and Control, V. E. S. Institute of Technology, Chembur, Mumbai, India

N. Warke
Instrumentation Department, V. E. S. Institute of Technology, Chembur, Mumbai, India
e-mail: nilima.warke@ves.ac.in

G. Mathur et al. (eds.), *Proceedings of 2nd International Conference on Artificial Intelligence: Advances and Applications*, Algorithms for Intelligent Systems,
https://doi.org/10.1007/978-981-16-6332-1_25

programming online Integrated Development Environment (IDE), Kaggle is used. With a larger dataset, accuracy of RF will be further improved keeping the time complexity minimum.

This paper discusses machine learning algorithms as well as other researchers' work on using machine learning to predict heart disease. With a conclusion, the methodology used to develop the system and the results of the proposed work are discussed.

2 Machine Learning

Automation in HealthCare sectors can be very beneficial, and the best way to do so is by implementing concepts like Artificial Intelligence (AI) and ML [4]. Unlike the AI where algorithms are deployed to get shallow information on a wide spectrum of products, the concept of ML itself means developing an in-depth model for a particular problem. ML is generally termed as the "Vertical Product", whereas AI is used for "Horizontal Products". Also, the locus of concentration will be entirely focused on developing models based on ML algorithms that can precisely handle data related to heart disease and then to iterate it for better results with minimum time complexity. Performance of ML techniques can be improved by hybridizing with nature inspired algorithms [5, 6] and by the means of data transformation [7].

A. Working of Machine Learning

It is important to understand the work flow of ML before shortlisting algorithms and selecting procedures for the implementation of any specific method. An overview of the workflow is as shown in Fig. 1.

High emphasis should be given to each blocks of the process and the methodologies used so that there are less errors during the later stages of development.

The data required for implementing the system is collected from various sources like the UCI repository, etc. The selected data may not always be perfect. There may be errors like missing data, repetition of values, obsolete values, etc. that needs to be addressed apart from cleaning the data and normalizing it.

Once pre-processing of the data is completed, it is important to list out the details of each attribute, i.e., the mean, range, counts, minimum/maximum, etc. Plotting graphs helps to have visuals of details of each attribute.

The data set gets divided into two parts, i.e., 80% Training and 20% Testing (depending upon various factors). The model is then trained using a particular algorithm and then tested to check the accuracy.

B. Types of Learning

Algorithms used in ML systems are broadly classified in three major types, i.e., Supervised Learning, Unsupervised Learning, and Reinforced Learning.

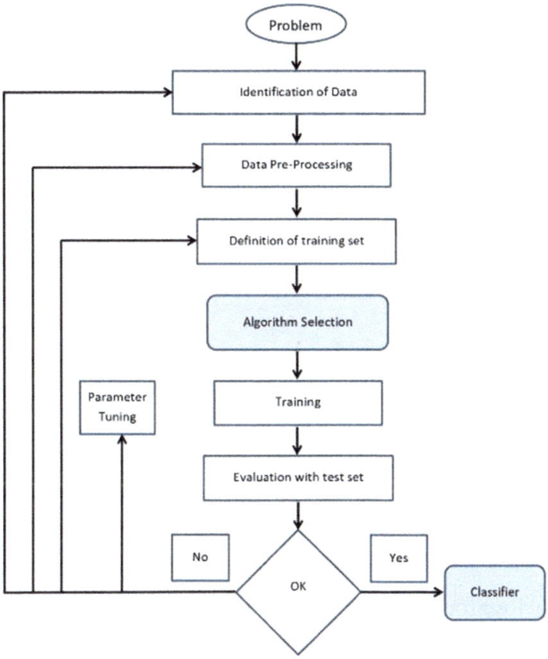

Fig. 1 Work flow of machine learning process

(1) Supervised Learning

In this type of learning, the system maps the inputs with the outputs based on the pre-given input output pairs. This helps the system in learning how to associate the problems with its solution.

After the system gets familiar with the association, the system gives solution to a new problem on its own.

This type of learning is basically referred as a learning in the presence of a teacher. Supervised Learning has the following algorithms.

- Support Vector Machine (SVM)
- Naïve Bayes
- Decision Tree
- Logic Regression
- Linear Regression
- Random Forest

(2) Unsupervised Learning

This type of learning includes input data sets with no solutions associated with the data set.

For the solution to a particular problem, the system has to figure out on its own. The system tries to generate patterns to associate the input data set and the outputs.

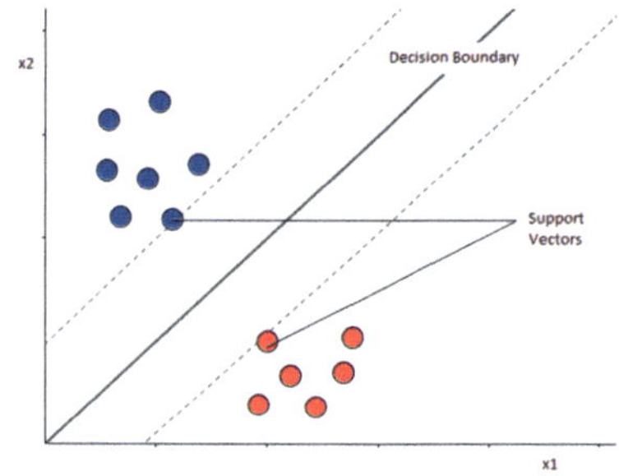

Fig. 2 SVM algorithm showing hyperplane

This type of learning is basically referred as a learning without guidance. Unsupervised Learning has the following algorithms.

- k-means clustering
- Random Projection
- ANN

(3) Reinforced Learning

In Reinforced Learning the system is concerned with how the program interacts to take actions in an environment. The system earns positive points if it performs well and the system establishes trends accordingly in order to optimize the notion of cumulative compensation.

C. Algorithms

(1) Support Vector Machine (SVM)

These models implement associated learning algorithms that analyses data used to perform classification and regression analysis.

In SVM model, examples are represented as points in space as shown in Fig. 2. These points are then mapped so that they fall into separate categories. The concept of hyper plane is used to separate these points. Hyperplane is basically a plane in higher dimensions. The decision boundary is considered as the hyperplane.

(2) Naïve Bayes (NB)

Naïve Bayes classifier algorithms are based on the Bayes theorem as given in Eq. (1). The independence between the attributes of the dataset is the main assumption. Since in Naïve Bayes algorithms, every variable is treated independently, it helps to predict even if variables do not have proper relation [8],

$$P(A|B) = \frac{P(B|A)P(A)}{P(B)}. \tag{1}$$

Using Bayesian probability terminology, Eq. (1) can be written as Eq. (2),

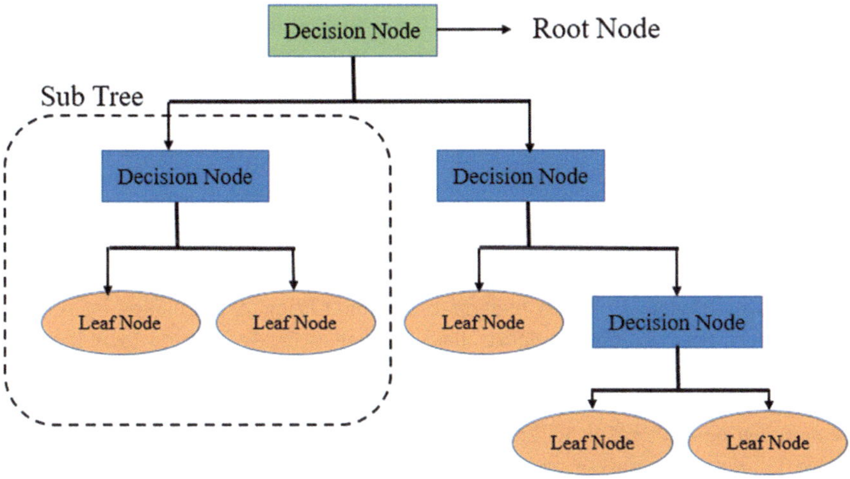

Fig. 3 Decision tree flowchart

$$\text{Posterior} = \frac{\text{priori} \times \text{likelihood}}{\text{evidence}}. \tag{2}$$

Bayes' theorem calculates the posterior probability of an event (A) given some prior probability of event B.

Naïve Bayes algorithms are fast and highly scalable. Even the simplest Bayesian network models when coupled with Kernel density estimation, gives higher accuracy levels with minimum time for execution.

(3) Decision Tree (DT)

DT uses a graphical representation of data and starts from a node known as the Root node and gets split off into various branches or nodes as shown in Fig. 3. Based on the probabilities, the root node starts splitting. The point of splitting is termed as the root node followed by the decision node. The end of the tree consists of end node or leaf node.

Some information about the features are extracted from each node, and each link represents decision rule taken on the nodes. The Tree is generally mapped/drawn on entropy rule which is given by Eq. (3),

$$I_H = -\sum_{j=1}^{c} p_j \log_2(p_j), \tag{3}$$

where p_j: proportion of the samples that belongs to class c for a particular node.

Also, this is the definition for all non-empty classes ($p \neq 0$). The entropy is zero if all samples at a node belongs to the same class.

The node consisting the highest entropy calculation is selected as the root node, and this process is repeated until all the nodes of the tree are calculated or until the tree is constructed. The problem of overfitting may arise if there is an imbalance in the number of nodes. Decision tree has less accuracy as compared to linear regression. To avoid missing values issues, ID3 is used for building decision trees, but the hurdle of overfitting is still not coverable.

(4) Neural Networks (NN)

Replicating the functioning of a human brain, neural networks perform a series of algorithms that attempts to recognize the underlying relationships in a set of data. Neural network also supports back propagation algorithm which helps in classification, prediction, and inference sequential reasoning.

Neural networks can adapt to changing input, and hence the network generates the simplest possible result without having to revamp the output criteria.

Neural Networks have a multi-layer network which contains input, hidden, and output neurons as depicted in Fig. 4. The collection of nodes connected in a network transmits a signal. Point to keep in mind is that not every neuron is connected to every other neuron. Activation node contains weight values. Training data is used for teaching a neural network various problem-solving technique.

During training, the neural network is tested to ensure that it does not over fit the network and to assess its performance. This helps to avoid over fitting issues and make it more reliable than Decision Tree.

(5) Random Forest (RF)

RF is a classification algorithm consisting of many decisions' trees. This algorithm can be used for classification as well as regression. Manifold of trees are created for the data and makes prediction based on that. Large data sets can be used in Random

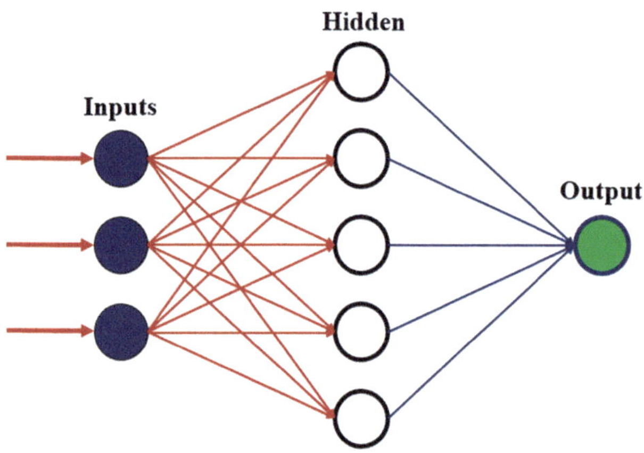

Fig. 4 Neural network layers

Forest algorithm with higher efficiency. Time for execution of the algorithms is highly affected by dataset volume and the number of trees incorporated.

Even if the values in the data set are missing, RF will work smoothly. This algorithm works in two phases, i.e., to create a random forest then make a prediction using a random forest classifier created in the first phase. It uses bagging, where various learning models are combined to improve the overall results. RF when combined with k fold cross validation gives highly accurate results.

3 Related Work

Some researchers have been working on predicting heart diseases by using different algorithms and different datasets. Some of the fine research done by other researchers is reported in Table 1. Algorithms used are mainly RF and DT. Data set used in developing the systems was taken from UCI Repository. Cleveland dataset is one of the most refined dataset available on the online repository with over 300 patient's data set whereas Framingham heart study data set consists of 4240 patient's data.

Majority of the work done in predicting heart diseases is carried out using the Cleveland dataset as there are less preprocessing steps.

4 Methodology

A. Data Set

Cleveland data set from the UCI repository is utilized for the proposed work. There are totally 74 attributes that form the heart disease data set. Only 14 important attributes were used for the proposed work.

The attributes collected to develop a ML-based heart disease prediction system with its description are listed in Table 2 [9]. Data of five patients selected randomly for this work is as given in Table 3.

B. Pre-processing

Imbalances in data can affect the quality of the overall prediction system. Predefined libraries helps in handling the dataset. Since the program is being built using python language, libraries like NumPy, Pandas, and Matplotlib are implemented. For any missing values, mean of the particular row/column is calculated and then the missing values are replaced with the mean values. Rows containing zero attributes or more than 75% of the attributes are deleted. Data duplication is also avoided as it can affect the accuracy and give biased results.

Now to have visuals of the meaningful data, plots were used to get a glance of details of each attribute. Parameter like the mean, median, mode, range, variance, and counts have been plotted using Matplotlib as shown in Fig. 5. Any inconsistency

Table 1 Performance analysis of heart disease prediction using machine learning

Authors	Algorithms/techniques used	Results
Singh and Kumar (2020) [1]	Algorithms used: k-nearest neighbor (kNN), Decision Tree (DT), Linear Regression (LR) and Support Vector Machine (SVM) IDE used: Anaconda (Jupyter Notebook)	Accuracy (in percentage): kNN: 87, DT: 79, LR: 78, SVM: 83
Rajdhan et al. (2020) [2]	Algorithms used: Random Forest (RF), Naïve Bayes (NB), Decision Tree (DT), Linear Regression (LR) Data set: UCI Cleveland data set	Accuracy: Highest accuracy of 90.16% achieved in RF
Krishnani et al. (2019) [3]	Algorithms used: Random Forest (RF), Decision Tree (DT), and kNN Data set: Framingham Heart Study data set (4240 records)	Accuracy (in percentage): RF: 96.8, DT: 92.7, kNN: 92.87
Rajamhoana et al. (2018) [4]	Algorithms used: Multi-Layer Perceptron with Back Propagation Learning and Convolutional Neural Network (CNN)	Accuracy (in percentage): Multi-Layer Perceptron: 94, CNN: 90
Nichenametla et al. (2018) [8]	Algorithms used: Decision Tree (DT), Naïve Bayes (NB) Data set: UCI Cleveland data set	Accuracy: For small data set, Naïve Bayes. For large data set, Decision Tree
Chitra and Seenivasagam (2013) [9]	Algorithms used: Cascaded Neural Network and SVM Data set: UCI Cleveland data set	Accuracy: Cascaded Neural Network gives 3% more accurate results as compared to SVM

in the data can be easily identified using graphs. Appropriate measures can be taken according to the inconsistency.

5 Results and Discussion

Performance of the algorithms is inferred from two important parameter, i.e., Accuracy and Time Complexity. In the proposed work, both accuracy and time complexity are calculated. Table 4 explains the time complexity and the accuracy of the implemented algorithms.

After performing training and testing, it is found that accuracy of RF to be much efficient as compared to other implemented algorithms.

To verify the same, tenfold cross validation is also performed for the given algorithms. Tenfold cross validation helped in getting an unbiased result as the iterations

Table 2 Attributes of the data set

Sr. No	Attribute	Description
1	Age	Age in years
2	Gender	Value 1 for Male, Value 0 for Female
3	Cp	Chest pain 0, 1, 2, 3
4	Tresh Bp	Resting blood pressure
5	Chol	Cholesterol serum in mg/dl
6	Fbs	Value 1 for fasting blood sugar > 120 mg/dl, else Value 0
7	RestECG	Resting electrocardiographic results Value 0 for normal else Value 1
8	OldPeak	ST depression induced by exercise relative to rest
9	Slope	The slope of peak exercise of ST segment Value 1 for up sloping Value 2 for flat Value 3 for down sloping
10	Cal	Number of major vessels coloured by fluoroscopy: 0, 1, 2, 3
11	Thal	Value 3 for normal; value 6 for fixed defect value 7 for reversible defect
12	Exang	Exercise included Angina Value 1 for Yes, Value 0 for No
13	thalach	Maximum heart rate achieved
14	target	5 or 1

Table 3 Overview of data of five randomly selected patients

Attributes	Data				
	197	121	90	173	29
Age	67	59	48	58	53
Gender	1	1	1	0	1
Cp	0	0	2	2	2
Tresh Bp	125	138	124	132	130
Chol	254	271	255	224	197
Fbs	1	0	1	0	1
RestECG	1	0	1	0	1
OldPeak	0.2	0.0	0.0	3.2	1.2
Slope	1	2	2	2	0
Cal	2	0	2	2	0
Thal	3	2	2	3	2
Exang	0	0	0	0	0
Thalach	163	182	175	173	152
Target	0	1	1	0	1

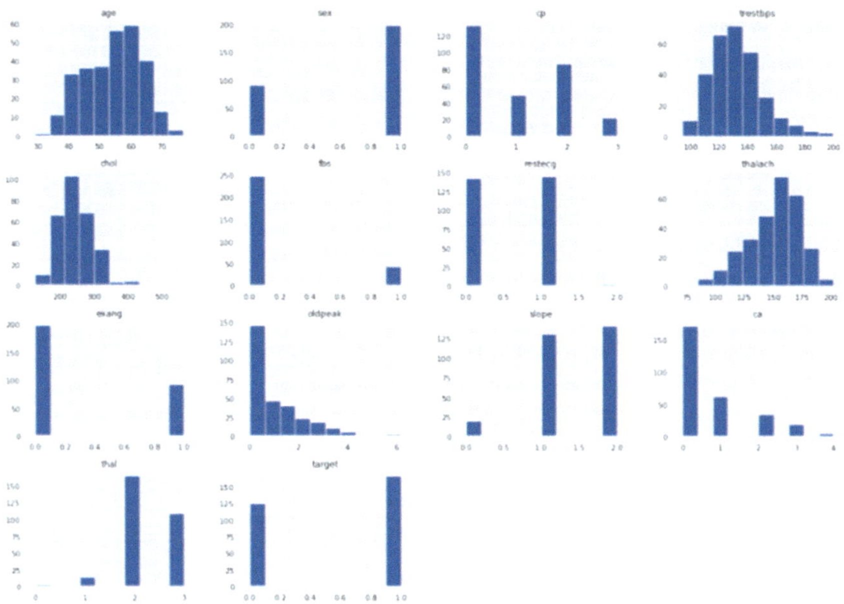

Fig. 5 Data set visualization

Table 4 Accuracy and time complexity of the algorithms

Algorithms	Accuracy (%)	Execution time (s)
RF	82.25	0.73
NN	82.06	2.17
NB	81.72	2.06
DT	74.83	0.0769

are increased 10 times, and the analysis is carried out accordingly. It is observed that RF gives the highest accuracy in this work. The output of tenfold cross validation is as reported in Table 5.

Figure 6 shows accuracy of each iterations. Plotting the iterations helps in understanding the inconsistency in the given data set as for each fold the output of the system changes with some margin.

The final accuracy of the 10 Fold Cross Validation method is calculated by taking the mean of the 10 folds.

To support the argument, confusion matrix is used as shown in Fig. 7.

Table 5 10 fold cross validation

Iteration fold	Accuracy (%)			
	RF	NN	NB	DT
1	86.2	89.65	82.75	82.75
2	82.75	82.75	79.31	82.75
3	75.86	79.31	82.75	79.31
4	82.75	79.31	89.65	72.51
5	98	96.55	86.20	89.65
6	79.31	68.96	72.41	65.51
7	75.86	82.75	93.10	65.51
‘8	89.65	82.75	86.20	65.51
9	86.2	89.65	79.31	79.31
10	68.96	68.96	65.51	65.51
Avg	82.25	82.06	81.72	74.83

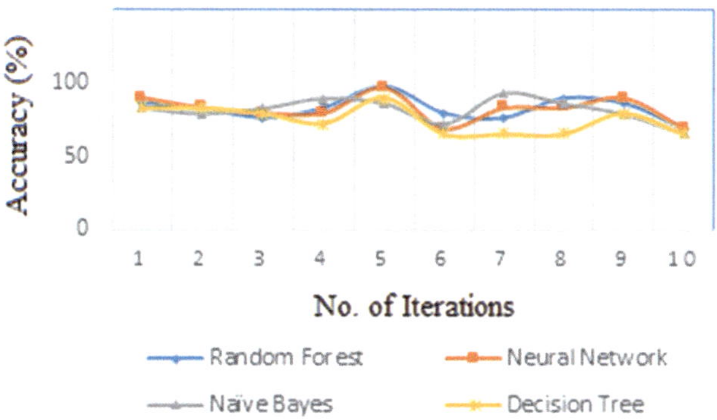

Fig. 6 10 fold cross validation results

6 Conclusion

Based on the data set available with the attributes shown in Table 1, multiple ML algorithms like DT, NB, NN, and RF are implemented.

Experimental results show that RF gives the best accuracy of 82.25% in 0.73 s. Accuracy of other algorithms like NN and NB turns out to be 82.06 and 81.07%, but the time taken is 2.17 and 2.06 s. DT takes more time than others and results in poor output. A larger dataset is required to increase the overall accuracy of the system. With a refined and well pre-processed real time data set, the time for implementing the algorithms can be reduced. The health-care industry will improve treatment by a factor of ten by using machine learning to predict heart disease. A more robust treatment layout can be drafted, and lives can be saved in remote areas as well.

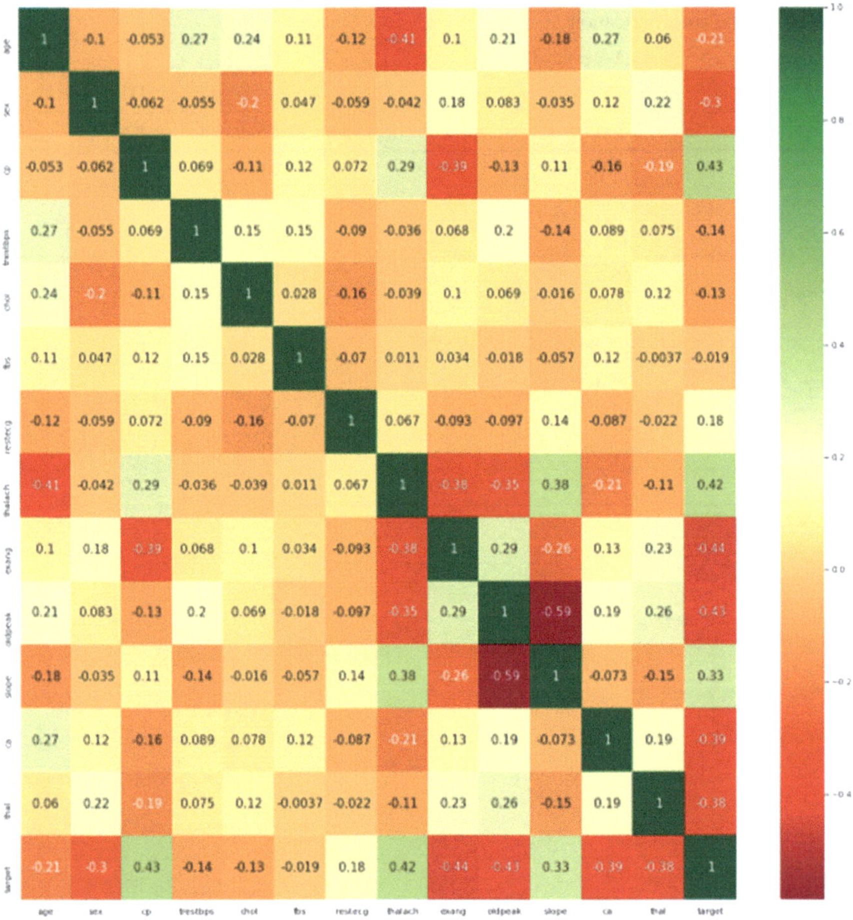

Fig. 7 Confusion matrix

References

1. Singh A, Kumar R (2020) Heart disease prediction using machine learning algorithms. In: 2020 international conference on electrical and electronics engineering (ICE3). Gorakhpur, India, pp 452–457. https://doi.org/10.1109/ICE348803.2020.9122958
2. Rajdhan A, Agarwal A, Sai M, Ravi D, Ghuli P (2020) Heart disease prediction using machine learning. Int J Eng Res Technol (IJERT) 9(4)
3. Krishnani D, Kumari A, Dewangan A, Singh A, Naik NS (2019) Prediction of coronary heart disease using supervised machine learning algorithms. In: TENCON 2019—2019 IEEE region 10 conference (TENCON). Kochi, India, pp 367–372. https://doi.org/10.1109/TENCON.2019.8929434
4. Rajamhoana SP, Devi CA, Umamaheswari K, Kiruba R, Karunya K, Deepika R (2018) Analysis of neural networks based heart disease prediction system. In: 2018 11th international conference on human system interaction (HSI). Gdansk, pp 233–239. https://doi.org/10.1109/HSI.2018.8431153

5. Kumar S, Sharma B, Sharma VK, Poonia RC (2018) Automated soil prediction using bag-of-features and chaotic spider monkey optimization algorithm. Evol Intell 1–12. https://doi.org/10.1007/s12065-018-0186-9
6. Chugh A, Sharma VK, Kumar S, Nayyar A, Qureshi B, Bhatia MK, Jain C (2021) Spider monkey crow optimization algorithm with deep learning for sentiment classification and information retrieval. IEEE Access 9:24249–24262. https://doi.org/10.1109/ACCESS.2021.3055507
7. Shekhawat SS, Sharma H, Kumar S, Nayyar A, Qureshi B (2021) bSSA: binary salp swarm algorithm with hybrid data transformation for feature selection. IEEE Access 9:14867–14882. https://doi.org/10.1109/ACCESS.2021.3049547
8. Nichenametla R, Maneesha T, Hafeez S, Krishna H (2018) Prediction of heart disease using machine learning algorithms. Int J Eng Technol (UAE) 7:363–366. https://doi.org/10.14419/ijet.v7i2.32.15714
9. Chitra R, Seenivasagam V (2013) Heart disease prediction system using supervised learning classifier. In: SOCO 2013

Chapter 26
Time Series Analysis and Forecasting on Crime Data

J. Vimala Devi and K. S. Kavitha

1 Data Analytics

1.1 An Introduction

Data Science is an interdisciplinary field that requires knowledge of Probability and Statistics, domain understanding, and analytical skills. It uses scientific methods, processes, algorithm, and models to extract knowledge and insights from data. Technological advancement of this era and availability of public datasets have encouraged us to devise a systematic analytical approach in every field of science. The contemporary crime datasets are mostly Spatio-temporal and an intelligent compilation of attributes of interest in contrast to traditional datasets.

Data analysis and Data visualization are one of specific steps in the life cycle of data science. Data analysis is often intended to uncover useful patterns and trends of data in the form of visualization such as bar plot, scatter plot, and Histograms. This analytical approach helps us to understand the origin, structure, and dynamics of crime.

J. V. Devi (✉)
Cambridge Institute of Technology, Bengaluru, India

K. S. Kavitha
Global Academy of Technology, Bengaluru, India

G. Mathur et al. (eds.), *Proceedings of 2nd International Conference on Artificial Intelligence: Advances and Applications*, Algorithms for Intelligent Systems,
https://doi.org/10.1007/978-981-16-6332-1_26

1.2 Kinds of Data Analytics

There are three types of data analytics:

Descriptive Analytics. It speaks about the history and explains what has happened from the given data set.

Predictive analytics. It is built on descriptive analytics and predicts what may happen in the future based on past data. Descriptive analytics and exploratory data analysis are always the tools for predictive analytics.

Prescriptive Analytics. It always suggests few course of actions and recommendations based on the past data and predictions.

2 Existing System

The authors of paper [1] use auto regression techniques to locate offender site efficiently with minimum error for time series data. They have explored ARIMA model to predict crimes in advance by identifying relationship among attributes of crime data using R tool.

In paper [2], the authors have discussed the usefulness of analytical algorithms and auto regression techniques on spatio-temporal data in predicting crimes based on crime profiles, crime hotspots, and crime trends [3].

The authors of paper [4] have applied machine learning algorithms such as KNN and boosted Decision tree algorithms to predict crime on 15 years Vancouver dataset and obtained accuracy between 39 and 44%.

The authors of paper [5] have proposed a fresh approach to discover related crime patterns and developments on Indian crime records of different crime locations and to subsequently use those crime trends for future prediction based on dynamic time wrapping and Minkowski model.

In paper [6], the authors have explored dynamic features instead of static features to predict crime events. Correlation analysis has been used to verify the effectiveness of dynamic features. The authors have also considered factors such as social diversity and population distribution that are related to area's safety [7].

In paper [8], ARIMA, a statistical model, was employed to make short-term prediction on property crimes that happened in one of towns of China. Using this model, prediction of crime occurrence for one week's time has been foreseen provided the dataset of property crime for the duration of fifty weeks. The results of the model were compared with the SES. The ARIMA model has appropriate fitting and more prediction accuracy than Holt Winters or exponential smoothing.

In Paper [9], there is a discussion on crime trends and forecasting based on autoregressive models in metropolitan areas. A prophetic model had been employed to forecast the quantity of crimes with respect to rolling time prospects. A case study on an area of Chicago, using open data sources has been conducted to quantify the rate of crime for future [10]. Experimental analysis proved that the proposed methodology predicted the crime rate with an accuracy of 84% and 80% on one-year-ahead and two-year-ahead forecasts respectively (Table 1).

Table 1 Summary and critical analysis of proposed work

References	Objective of proposed work	Methodology/parameters	Accuracy measures	Limitations
Manal Almuammar [11]	Forecasting real time available parking spaces based on time series data	Deep Gated Recurrent units and shallow MLP models	MSE, RMSE and MAE	Simple LP models outperformed deep GRUmodels
Khan Mohammad Habibullah [12]	Finding the most probable crime category	Random forest, Decision tree and Ensemble methods	Accuracy	Ensemble methods give more accuracy than other classifier models
G. Borowik [2]	This paper analyses the effectiveness of algorithms in identifying hot crime spots, crime trends and to create potential criminal profiles	FB Prophet, Yearly seasonality and frequency characteristics of burglary and theft, hooliganism, road traffic offenses etc., have been recorded	MAPE is recommended to measure the accuracy of the model	Usefulness of time series analysis algorithms has been understood compared to ML algorithms
Muhammad Yousaf [13]	Forecasting of COVID-19 cases for the next one month is obtained	ARIMA model is used to forecast COVID-19 cases, no of deaths and recoveries for the next one month	It has been identified that COVID cases would be increased three fold	Number of cases, deaths and Recoveries have been predicted in 95% interval No metrics applied to validate the model

(continued)

Table 1 (continued)

References	Objective of proposed work	Methodology/parameters	Accuracy measures	Limitations
Suhong Kim [14] 2018	Crime analysis and crime prediction has been carried out using Machine Learning algorithms	Two approaches adopted for data pre-processing: In the first approach, category variables such as neighbourhoods and crime category were converted into numerical value by assigning UID (Unique Identification Number) In the second approach, a binary value either 1 or 0 has been given to the crime happened neighbourhood based on the day of week. KNN and Boosted decision tree algorithms have been used to train model	Crime prediction accuracy of 39% to 44%	Low prediction accuracy
Romika Yadav [15]	The objective of this work is to predict crime trends and offender site in advance with more accuracy	Generalised Linear regression Model has been used to understand and draw inferences about time series data set. ARIMA model is used to predict	Predicts the relationship between crime attributes and the impacted crime sites	ARIMA model doesn't suit for non-linear data sets No measures to validate the model
Projected Work	The objective of this work is to forecast crime growth rate for the next 30 days	EDA has been applied to draw usable inferences from the data set and also to understand the geographical divisions of the city Sacremento. After EDA, FB Prophet algorithm has been deployed to forecast the growth of crime for the next 30 days	Crime growth rate is forecast with accuracy of 72%	Hyperparameter such as change points may be tuned further to build a better model While capturing seasonality trends, hyperparameters such as period and order of Fourier transform may be tuned better to model the data

Table 2 List of attributes in the crime dataset

Sl. No	Attribute name	Type of attribute
1	Record_ID	Unique ID of each incident recorded by Sacremento Police
2	Offense_Code	More than one Offense_code maps to offense_category
3	Offense_Ext	SubID of Offense_Code that explains further, type of Offense_Category
4	Offense_Category	66 Unique categories of offense exist
5	Description	Short description about the incident
6	Police District	City divided into six districts
7	Beat	Each district further divided in to beats
8	Grid	Each beat in turn divided into grids
9	Occurrence_Date	Date and Time of incident, reflected in UTC time

3 Proposed System

3.1 About Dataset

Sacremento is the capital city of the U.S. state of California according to https://en.wikipedia.org/wiki/Sacramento,_California. Crime data in the city of Sacremento is publicly available and accessible for all given in the link http://data.cityofsacramento.org/datasets/84e4483fc0624d678d7608a4fa12aae1_0. This dataset is published by Sacremento police department.

Crime data from the city of Sacremento is a Time series and multivariate dataset. The data set contains record of crime incidents on each day along with time. So time is the primary axes for this data set. Each data point in the data set is an incident report of crime. This data set contains 94,810 rows and 9 columns.

There are 66 unique categories of offense, and a total of 114 offense_code exists. More than one offense_code is assigned to the same offense_category. The description of attributes are given in the below table (Table 2).

3.2 Proposed Methodology

The city of Sacremento data set needs to be combined as it is available as three different files. The data set is to be cleaned first and then analyzed statistically to understand the inter relationships among the attributes. Later FB Prophet Model is to be deployed on the dataset (Fig. 1).

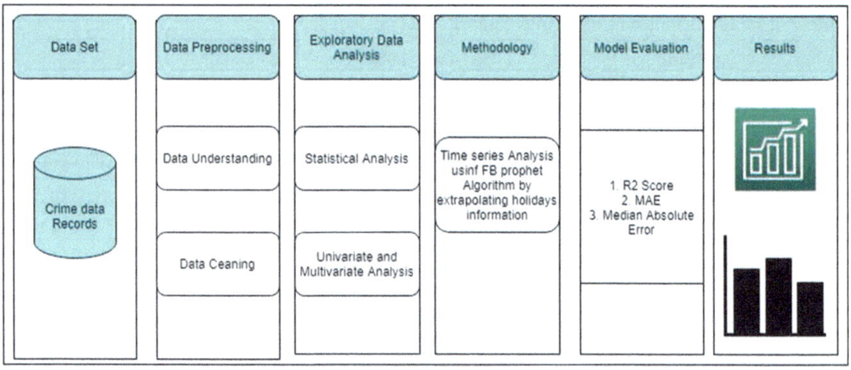

Fig. 1 Architecture of proposed methodology

4 Exploratory Data Analysis on Crime Data of City Sacremento

Exploratory Data Analysis (EDA) and data visualization help us to understand the attributes of the dataset and underlying interactions among them. EDA always brings out the deep insights of the dataset. It also leads to preparedness of the dataset to deploy a suitable model or algorithm on it.

Analysis begins with an attempt to remove the null values, and as the dataset does not contain much null values, there is no notable change in the shape of data set. After removing null values, the shape of dataset became (94791rows, 09cols).

The statistical analysis of dataset draws usable inferences and interesting patterns. The following statistical techniques have been applied for thorough understanding of the dataset, and the following graphs have been generated. These statistical methods allow us to explore data set exhaustively.

Let d = Occurrence_date

C = Total Count of crimes

C_i = Crime, Category wise, where 'i' = Offense_category

C_D = Crime, District wise, where 'D' = District

C_B = Crime, Beat wise, where 'B' = Beat

$X = \text{Count}(C_D) = \text{Count}(d)_{G(D)}$

$Y = \text{Count}(C_B) = \text{Count}(d)_{G(B)}$

$P_{(Ci)}$—Percentage of crime distribution, category wise

$P_{(D)}$—Percentage of crime distribution, District wise

$P_{(B)}$—Percentage of crime distribution, Beat wise

$\tilde{C}$ Mean distribution of crime, District wise

G: Groupby() (Table 3).

To understand the attributes such as Offense_Category, Offense_Code, and Offense_Ext of the dataset, a plot depicting the hierarchical representation of these three attributes is given below (Fig. 2).

Table 3 Statistical formula for analysis

1.	$\text{Count}(C_i) = \text{Count}(d)_{G(i)}$
2.	$P_{(Ci)} = \forall_i \frac{\text{Count(Ci)}}{\text{Count(C)}} * 100$
3.	$P_{(D)} = \left(\frac{X}{\text{Count(C)}}\right) * 100$
4.	$P_{(B)} = \left(\frac{Y}{\text{Count(C)}}\right) * 100$
5.	$\tilde{C} = \frac{X}{\text{Count(B)}}$

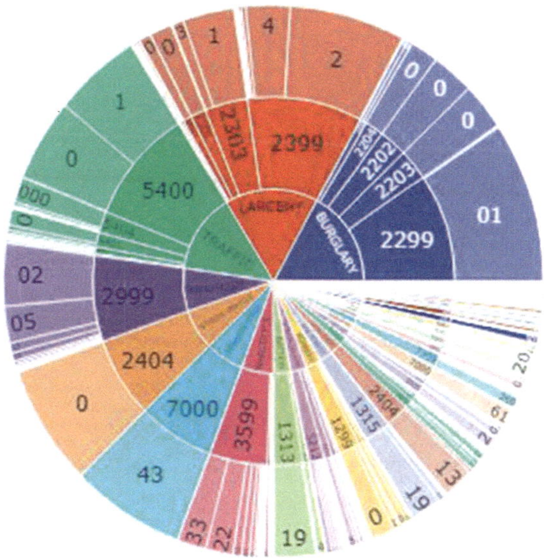

Fig. 2 A donut chart that depicts offense types: Offense_Category, Offense_Code, and Offense_Ext

The city of Sacramento is divided into six districts, districts into Beats and Beats further into Grids. Districts are the largest of the areas. Beats are assigned to Patrolling officers, and Grids are used for reporting purposes. Districts 1, 2, 4, and 5 have three Beats each, District 3 has two Beats, and District 6 has five Beats. Below is given the plot that displays Hierarchical representation of the Patrol division of City Sacramento (Fig. 3).

The dataset is grouped by attribute 'Offense_Category' and aggregated on Occurrence_Date, thereby finding the count of occurrences of that particular crime type. A list of Top ten Crime type has been identified, and the rest of the count of crimes is labeled as 'Others' that counts to 15,496 (Table 4).

Below is the graphical representation of proportion of each crime on the whole crime picture. It is observed that 'others' category occupies the third largest proportion on the whole, even though its occurrences are sparse.

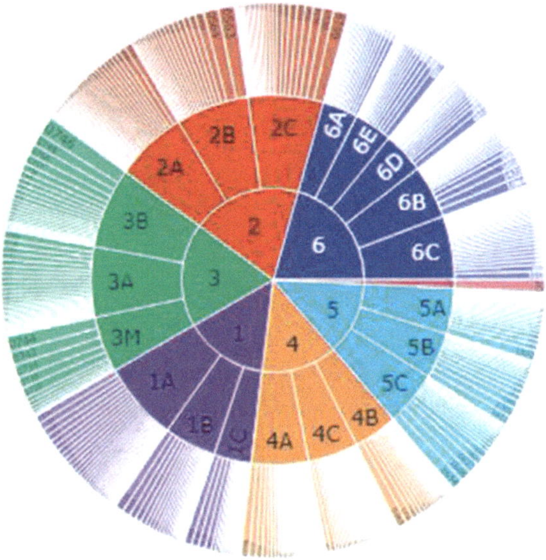

Fig. 3 A plot displaying patrol divisions of the city Sacremento, with the innermost circle representing districts, then beats and the outermost circle representing grids

Table 4 List of top 10 crimes and its count for the period (Jan 2018–July 2020)

Sl. No	Offense_Category	Count of occurrence based on Occurrence_Date
1	Burglary	16,262
2	Larceny	15,827
3	Traffic	13,211
4	Vandalism	7184
5	Stolen vehicle	6731
6	Towed/stored	6491
7	Narcotics	5468
8	Battery	2952
9	Weapon offense	2662
10	Robbery	2526

Now the count of occurrences of particular crime is calculated in every district by aggregating 'Occurrence_Date' district wise. Below is given the plot for distribution of crime district wise.

Pertinent to previous plot, distribution of crime, Beat wise is plotted as given below.

It is concluded from the plot Fig. 4 that District 6 have recorded highest rate of

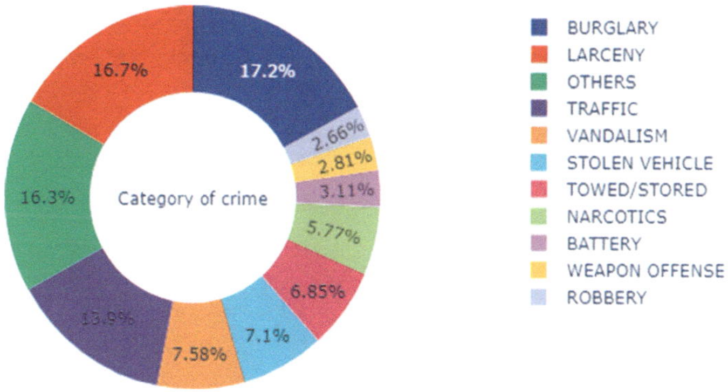

Fig. 4 Pie chart for top 10 crimes of city Sacremento

crimes for the past three years. Instead of count, a statistical score, Mean is considered to be the best as it is measuring the centrality of the data. Hence mean distribution of crime, district wise based on the number of Beats is calculated as given below (Fig. 5).

By comparing Figs. 4 and 6, it is understood that District 2 with 3 Beats (Smaller area) displays higher average crime record compared to District 6 with 5 Beats (Larger area).

In order to understand the pattern of occurrences of crime in District 2, a line plot is plotted with 'Occurrence_Date' on X-axis and 'No of crimes Beatwise' on Y axis (Fig. 8).

From EDA, it is recommended that more policing, monitoring, and stringent law enforcing are required for District 2.

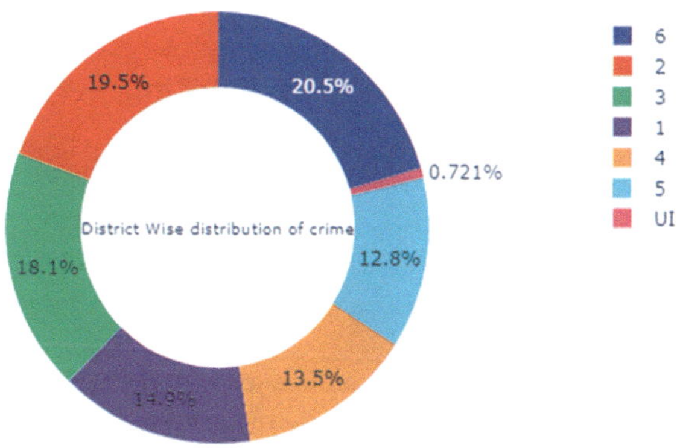

Fig. 5 A plot for distribution of crime district wise

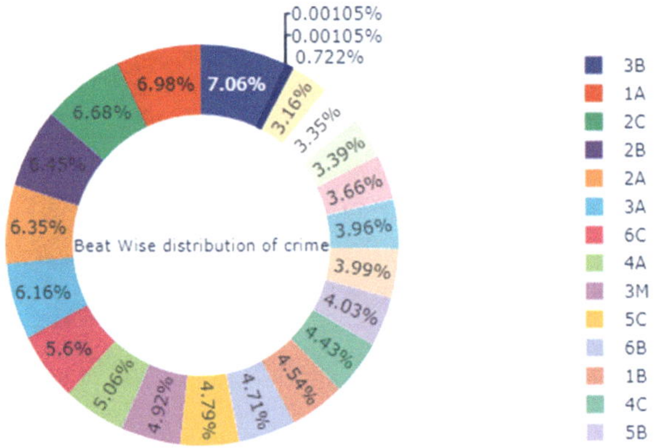

Fig. 6 Beat wise distribution of crime

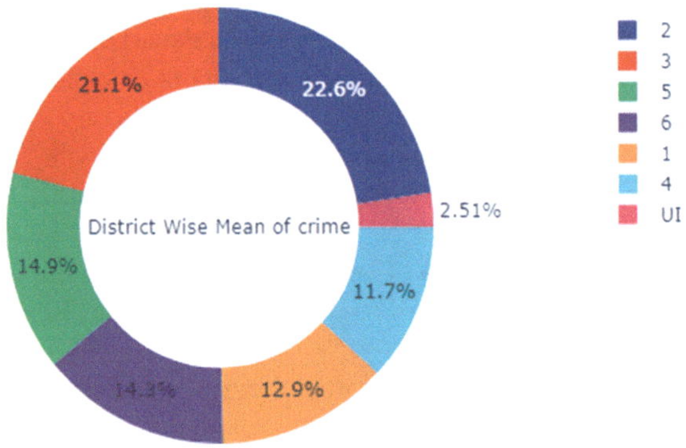

Fig. 7 A Plot that depicts mean crime rate in every district based on the number of Beats

5 Predictive Analysis Using FB Prophet Time Series Algorithm

Exploratory and Descriptive data analysis are the tools for Predictive analytics. It is crucial to understand from the data set what happened in the past in order to predict what will happen in future. As the given dataset is Time series data, it is a perfect candidate for forecasting.

Any time series model has quite a lot of aspects while modelling the data and needs to handle autoregressive and moving average part of the algorithm. Any time series model needs stationary data and to model the non-periodic changes, periodic changes and extra generous things of data like holiday effects. We have chosen FB Prophet Algorithm to perform forecasting of crime data for the following reasons [16–18]:

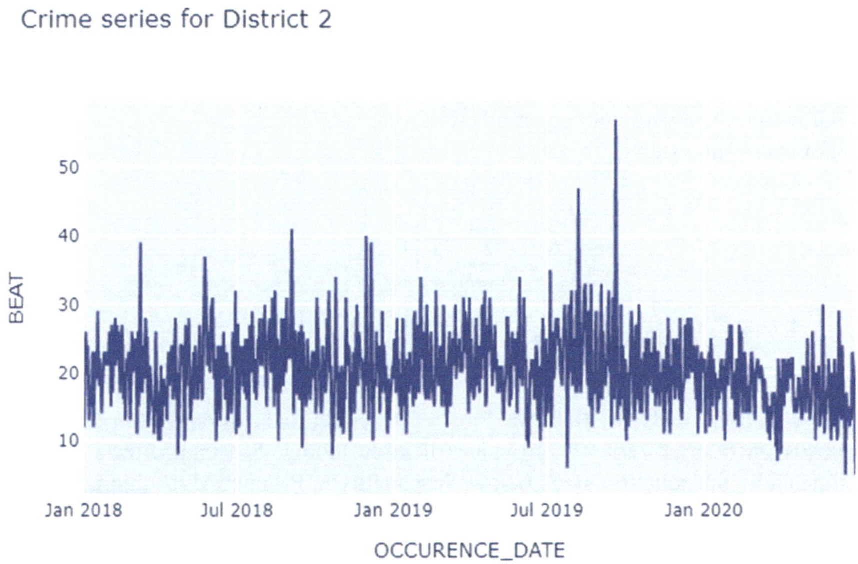

Fig. 8 Line plot in displaying crime pattern in District 2

- Fast and Accurate,
- Tunable parameters for forecasting,
- Fully automatic,
- Robust to outliers and missing data,
- Flexibility to record trends during different period.

This Prophet model is an additive regression model based on four components [19],

$$\mathrm{Y} = \mathrm{P(t)} + \mathrm{S(t)}_{\mathrm{yearly}} + \mathrm{S(t)}_{\mathrm{weekly}} + \mathrm{H(t)} + \mathrm{noise}. \tag{1}$$

P(t) = piece-wise trend model, i.e., growth of crime rate over time
S(t) = Seasonality changes of crime rate over an year
H(t) = holiday effects on crime rate, where t = 1 day
and it does bivariate time series analysis on 911 days of crime data, day wise.
Noise is neglected and cannot be accommodated by the model.

Change points p(t): Change points are the instances in time at which the statistical properties start differing and remain same till next change point occurs. This is one of the hyper parameters that can be tuned. By default, Prophet Algorithm records growth over time for 25 change points. The graph is showing the linear trend with change points [19],

$$P(t) = (k + A\delta)t + (m + A(-s\delta)). \tag{2}$$

$K_{\text{base growth}}$
$\delta_{\text{growth rate adjustment}}$
Number of change points, S = 25
Growth rate adjustment vector: $\delta \, \mathcal{E} \, R^S$
For every time step t,

$$\left(K + \sum_{j:t>\delta j} \delta j\right)t$$

$$K + a(t)T\delta \text{ where } a(t) = \begin{Bmatrix} 1, & \text{if } t \geq \delta j \\ 0, & \text{otherwise} \end{Bmatrix} \text{ and } a(t)\varepsilon\{0, 1\}^S$$

Seasonality Trend S(t): To fit and forecast the effects of seasonality, this algorithm depends on Fourier series to provide a flexible model. Seasonal effects S(t) are estimated by the equation given below. Seasonality in Prophet Algorithm is defined in terms of Fourier series which is a summation of cosine and sine functions [19],

$$\text{S(t)} = \sum_{n=1}^{N} (a_n \cos(2\pi nt/p) + b_n \sin(2\pi nt/p)), \tag{3}$$

where P and N are the hyper parameters that can be tuned to find best fit of the model,
N: Order of Fourier transform, N = 10
P: Period, P = 365 days or 7 days
a_n, b_n: Parameters to be learned
Vectorization of Fourier series is presented as follows:

$$\text{Seasonality Trend S(t)} = x(t).\beta,$$

where $x(t) = \left[\cos\left(\frac{2\pi 1t}{7}\right) \ldots \sin\left(\frac{2\pi nt}{7}\right)\right]$,

$\beta = a_1, b_1, \ \ldots \ a_n b_n.$

Holiday Trend H(t): US default holidays are taken into consideration to predict the behavior of crime growth on those days. For each holiday j, let S_j be the set of holiday dates in the past and future.

If 't' is the time instance during holiday 'j', then k_j is the corresponding forecast change. This can be modeled as follows:

$$\text{H(t)} = \text{R(t)k} \quad \text{and} \quad \text{R(t)} = [1(\text{t} \in \text{S}_1) \ \ldots \ 1(\text{t} \in \text{S}_\text{L})].$$

Forecasting the Trend and Uncertainty: Now the model is concluded past the history to make a forecast trend. The uncertainty in the forecast trend is estimated as follows by extending the generative model [19]:

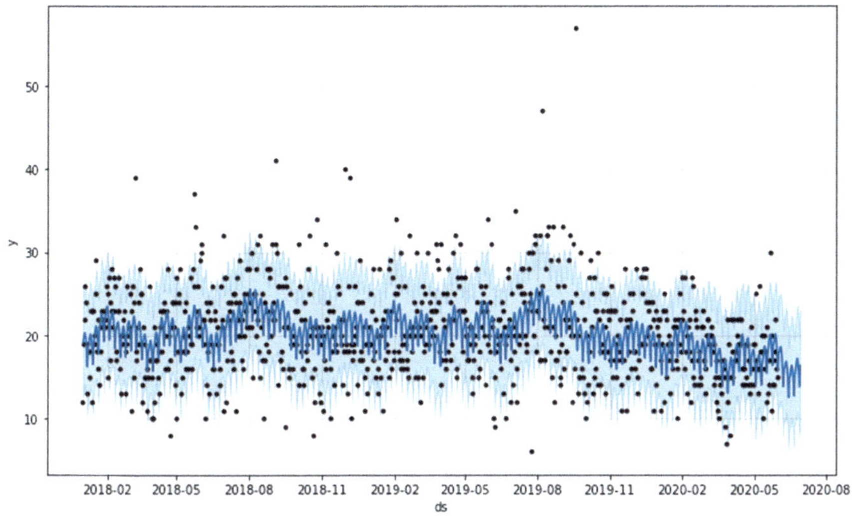

Fig. 9 A plot displaying forecasting of crime data of city Sacremento

$$\forall j > T, \delta j = 0, w.p. \frac{T - S}{T}, \tag{4}$$

where $\delta j\ Laplace(0, \lambda) w.p \frac{S}{T}$ and $\lambda = 1/S \sum_{j=1}^{S} |\delta_j|$.

Using FB Prophet Process [19], forecasting of crime data is captured for the next 30 days, as the prediction is supposed to be more accurate if it is calculated for near future. In the below plot, scattered points are the actual data. On x-axis, time is displayed; and on y axis, forecasting is recorded. The forecast says that the situation is not only reassuring the situation but also shows decrease in crime activity (Fig. 9).

6 Results and Discussion

Below plot is the outcome of FB Prophet Algorithm which displays trend and seasonality of crime data. From the trend plot, the seasonal fluctuations remain constant over time and it is also inferred that crime rate is declining as the year increases. Trend of crime was at the peak in the May month of 2019 and after that declining begins. Both weekly and yearly seasonality are also recorded.

In the weekly seasonality plot, it is observed that crime rate reaches its peak in the middle of week days such as Wednesdays and Thursdays. Yearly seasonality plot shows that there is notable rise in the crime rate from the month of July till September every year (Fig. 10).

US Holidays for a year have been considered: New Year's Day, Martin Luther King Jr. Day, Washington's Birthday, Memorial Day, Independence Day, Labor Day,

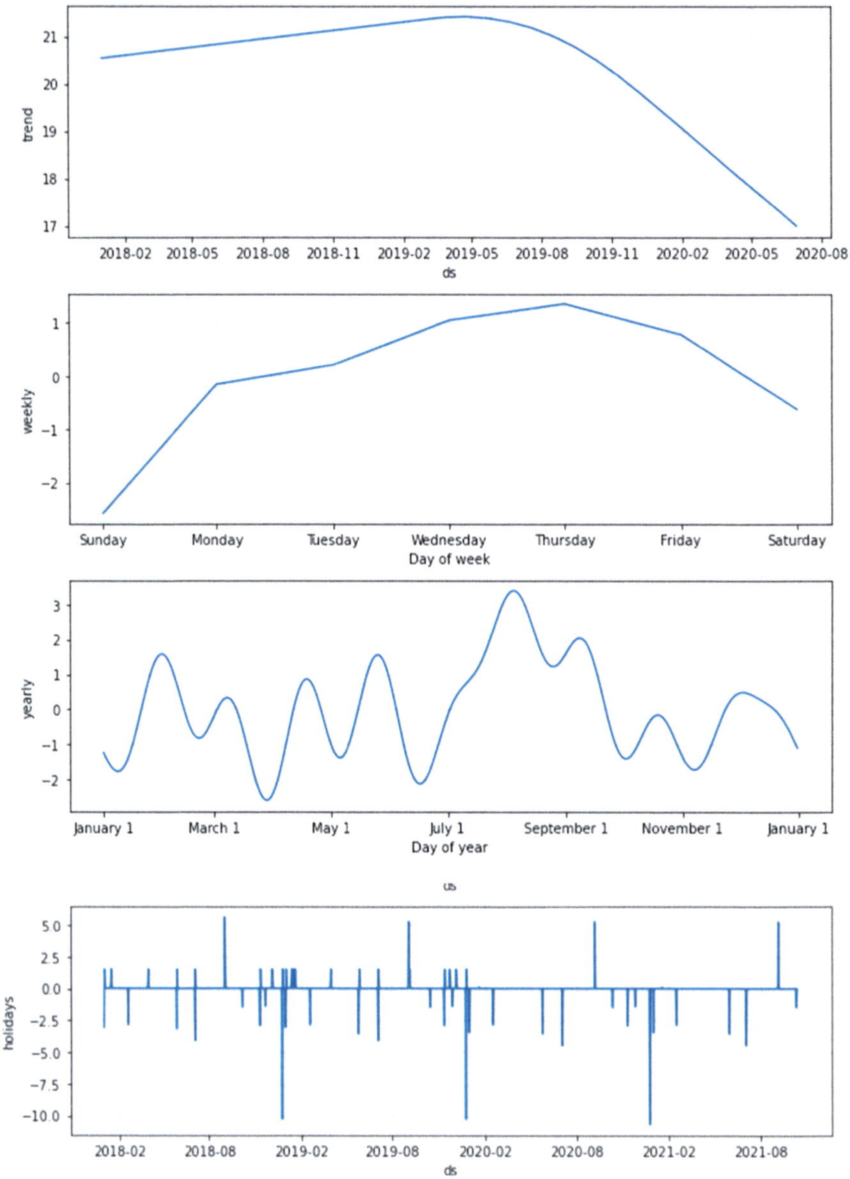

Fig. 10 A plot displaying trend and seasonality of crime data

Table 5 Error measures of the model

R2 Score	0.014019
MAE	4.411915
MedAE	3.843552
MAPE	38.09533

Columbus Day, Veterans Day, Veterans Day (Observed), Thanks giving, Christmas, and Independence Day (Observed) and the pattern of crime occurrences has been recorded. The trend on Holidays shows both positive and negative effect but could not substantiate the fact that crime rate either increases or decreases.

The prediction accuracy of the model has been measured by the following measures and are tabulated in the table given below.

i. R^2 Score represents the percentage of variation between the calculated and the predicted values of the model. It indicates goodness of fit of the model, and it is a measure of how well hidden samples are expected to be foretold by the model.
ii. Mean Absolute Error (MAE) is to find the sum of absolute errors (predicted y_i—calculated y) and then divided by the number of test samples.
iii. The Median absolute error (MedAE) is one of good measures as it is strong enough to respond to outliers and is calculated by finding the median of all absolute differences between the target and the predicted values.
iv. The Mean Absolute Percentage Error (MAPE): It is defined as given below.

$$MAPE = \frac{1}{n}\sum_{i=1}^{n}\left|\frac{Ai - Pi}{Pi}\right|,$$

where At = Actual data and Pt = Predicted value (Table 5).

From the measured accuracy values, the performance of model is found to be appreciable and when compared to the other time series models, the results of the model may be further improved if the following drawbacks are taken care of:

i. Size of data is not good enough to make better predictions.
ii. The data set lacks demographic features about Sacremento.
iii. The algorithm provides a room for tuning of hyper parameter that may lead to a better model fit for this dataset.

7 Conclusions

From the resulting plots, this paper identified crime prone areas and the top crime types that are prevalent in the city. Based on the count measure, District 6 was identified as the highest crime recording location. But the statistical measure, Mean gave a big twist in identifying crime-prone prime location. This paper recorded the trend and seasonality of crime activity in the city of Sacremento.

This work may be improved by tuning the further parameters on which Forecast model is to be built. The results may be further improved by enriching the dataset with features about population such as employability, literacy rate, male–female ratio, etc. There may be a possibility of capturing trend and seasonality of a particular crime, provided the dataset is larger. This paper is an attempt to show preliminary framework for further prediction and analysis.

References

1. Yadav R, Kumari Sheoran S (2018) Crime prediction using auto regression techniques for time series data. In: 2018 3rd international conference and workshops on recent advances and innovations in engineering (ICRAIE), Jaipur, India, 2018, pp 1–5. https://doi.org/10.1109/ICRAIE.2018
2. Borowik G, Wawrzyniak ZM, Cichosz P (2018) Time series analysis for crime forecasting. In: 2018 26th international conference on systems engineering (ICSEng), Sydney, Australia, 2018, pp 1–10. https://doi.org/10.1109/ICSENG.2018.8638179
3. Prophet Python package. https://pypi.python.org/pypi/fbprophet/
4. Kim S, Joshi P, Kalsi PS, Taheri P (2018) Crime analysis through machine learning. In: 2018 IEEE 9th annual information technology, electronics and mobile communication conference (IEMCON), Vancouver, BC, 2018, pp 415–420
5. Chandra B, Gupta M, Gupta MP (2008) A multivariate time series clustering approach for crime trends prediction. In: 2008 IEEE international conference on systems, man and cybernetics, Singapore, 2008, pp 892–896. https://doi.org/10.1109/ICSMC.2008.4811393
6. Rumi SK, Deng K, Salim FD (2018) Crime event prediction with dynamic features. EPJ Data Sci 7(1):43
7. https://www.analyticsvidhya.com/blog/2018/05/generate-accurate-forecasts-facebook-prophet-python-r/
8. Chen P, Yuan H, Shu X (2008) Forecasting crime using the ARIMA model. In: IEEE fifth international conference on fuzzy systems and knowledge discovery, vol 5, pp 627–630
9. Cesario E, Catlett C, Talia D (2016) Forecasting crimes using autoregressive models. In: IEEE 14th Intl Conf on dependable, autonomic and secure computing, 14th Intl Conf on pervasive intelligence and computing, 2nd Intl Conf on big data intelligence and computing and cyber science and technology congress (DASC/PiCom/DataCom/CyberSciTech), Aug 2016, pp 795–802
10. Palaniappan S, Rajinikanth TV, Govardhan A (2016) Enhancement of effective spatial data analysis using R. Indian J Sci Technol 9(21). https://doi.org/10.17485/ijst/2016/v9i21/95149
11. Almuammar M, Fasli M (2019) Deep learning for non-stationery multivariate time series forecasting. In: 2019 IEEE international conference on big data (big data)
12. Yuki JQ, Sakib MQ, Zamal Z, Habibullah KM (2019) Predicting crime using time and location data. In: ICCCM 2019, July 27–29, 2019, Bangkok, Thailand, Association for Computing Machinery (ACM). ISBN 978-1-4503-7195-7/19/07
13. Yousaf M, Zahir S, Riaz M, Hussain SM, Shah K (2020) Statistical analysis of forecasting COVID-19 for upcoming month in Pakistan. Chaos Solitons Fractals. https://doi.org/10.1016/j.chaos.2020.109926
14. Kim S, Joshi P, Kalsi PS, Taheri P (2018) Crime analysis through machine learning. 978-1-5386-7266-2/18/$31.00, IEEE

15. Yadav R, Sheoran SK (2018) Crime prediction using auto regression techniques for time series data. In: 3rd international conference and workshops on recent advances and innovations in engineering, 22–25 Nov 2018
16. Prophet R package. https://cran.rproject.org/package=prophet
17. Prophet homepage. https://facebook.github.io/prophet/
18. Wang S, Yuan K (2019) Spatiotemporal analysis and prediction of crime events in atlanta using deep learning. 2019 IEEE 4th international conference on image, vision and computing (ICIVC), IEEE, 2019
19. Forecsting at Scale. https://peerj.com/preprints/3190v2/

Chapter 27
Linguistic Data Analysis Using Nagel Point-Based Ranking Fuzzy Numbers for Financial Risk Management

Lazim Abdullah, Ahmad Termimi Ab Ghani, and Nurnadiah Zamri

1 Introduction

Ranking fuzzy numbers (RFNs) play a vital role in many applications especially in the ranking of alternatives in decision-making [1]. It was stemming from the brainchild of Zadeh [2] where the new knowledge of fuzzy sets was presented while Jain [3] was the first to introduce RFNs. Jain [4] introduced a method by applying the RFN concept to rank fuzzy numbers with the presence of fuzzy variables. Since then, there are numerous researchers who have made efforts to introduce the new RFNs particularly for the case where human perceptions or opinions are used as the subjects of computing. For example, Yu et al. [5] considered left–right triangular fuzzy numbers, and areas to rank generalized fuzzy numbers in a decision-making application. In their research, the left and right areas and the decision maker's opinions toward risks are used to obtain a final ranking result. Wang et al. [6] proposed an evaluation in knowledge management with a combination of subjective and objective indicators. In their analysis, evaluations are made by transforming linguistic variables into triangular fuzzy numbers. To add further literature in transforming linguistic variables into fuzzy numbers, Singh [7] introduced a novel method in which linguistic variables had been transformed into generalized fuzzy numbers. In another research based on human intuition, Chi and Yu [8] presented the centroids and rank index of ranking generalized fuzzy numbers after considering objective and subjective factors. These objective and subjective factors were represented by the centroid point and height of

L. Abdullah (✉) · A. T. Ab Ghani
Management Science Research Group, Faculty of Ocean Engineering Technology and Informatics, Universiti Malaysia Terengganu, 20130 Kuala Nerus, Malaysia
e-mail: lazim_m@umt.edu.my

N. Zamri
Faculty of Informatics and Computing, Universiti Sultan Zainal Abidin, 21030 Kuala Nerus, Malaysia

G. Mathur et al. (eds.), *Proceedings of 2nd International Conference on Artificial Intelligence: Advances and Applications*, Algorithms for Intelligent Systems,
https://doi.org/10.1007/978-981-16-6332-1_27

a fuzzy number which directly measure humans' optimism. The rank value was used to demonstrate the level of optimism of human decision-making. It is believed that the final ranking would be more realistic, reasonable, and mathematically correct. More importantly, the final ranking is established as a result of the computation of human intuition or famously known as computing with words.

It can be seen that the above literature considers left–right spread, divergence, and centroid of fuzzy numbers to represent human thinking. Apart from the centroid of generalized fuzzy numbers, past research also acknowledged other types of centers of triangular fuzzy numbers such as circumcenter [9, 10] and incenter [11–13]. Circumcenter is the point of concurrence of the triangle's three perpendicular bisectors. It is located at the center of the triangle. The center of circumcircle is another name of the circumcenter fuzzy number. Rao and Shankar [14] used the area method based on the circumcenter of centroids to rank fuzzy numbers. Index of optimism is used to reflect the optimistic attitude of decision makers. On the other hand, Incenter is the center of incircle inscribed in the triangle and is also defined as the point of concurrence of a triangle's angle bisectors. Incenter is the point on the interior of a triangle which is equidistant from the triangle's three sides [15]. Some of the researchers such as [12, 15] made a reformation by using incenter as their centroid method to rank fuzzy numbers. Besides circumcenter and incenter, there is another center in a triangle. This center is called Nagel point. This point is actually a point of intersection of line segments from the vertices to the points of tangency of the opposite excircles [16]. Akyar et al. [15] used this method by applying the concept of the centroid point in the pursuit to rank triangular fuzzy numbers and applied it to an illustrative example of fuzzy risk analysis. In contrast to this illustrative example, this paper analyzes the eight risks of financial management where linguistic data from an empirical research are employed. These linguistic data are analyzed using Nagel point-based RFN to obtain the Nagel points, thereby, the highest risk in financial management can be determined.

2 Preliminaries

The following section briefly describes the related notions that are used in this research. Specifically, theoretical definitions of Nagel point, triangular fuzzy number, and linguistic variables are explained as follows.

2.1 Definition of Nagel Point

As explained in the previous chapters, Nagel point is the point of intersection enclosed by circles [16]. Figure 1 shows the point, and the presence of this point was proved by Nagel in 1836.

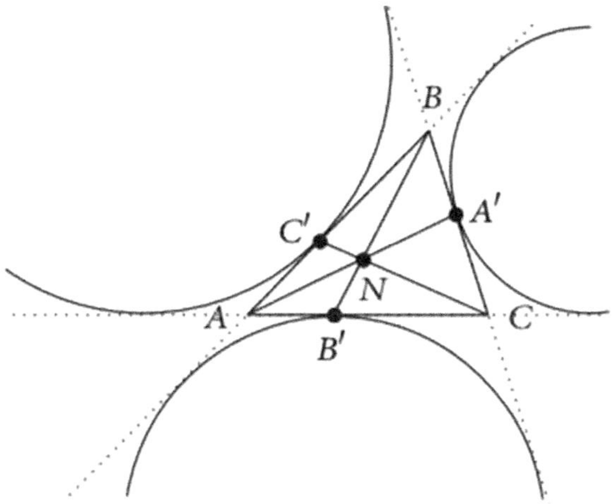

Fig. 1 Location of Nagel point *N* in a triangle ABC [16]

2.1.1 Theorem of Nagel Point

This theorem states that the cevians joining the vertices of a triangle to the points of tangency of the opposite sides with the corresponding excircles are simultaneous [17]. This theorem is also known as the Halfway-Around theorem because points A', B', and C′ are halfway around the perimeter of the triangle from vertices A, B, and C, respectively (refer to Fig. 2).

The trilinear coordinates of a point *P*, with respect to triangle *ABC*, are an ordered triple of numbers, each of which is proportionate to the directed distance from *P* to one of the sidelines. The trilinear coordinates are generally represented by $\alpha : \beta : \gamma$ as shown in Fig. 2.

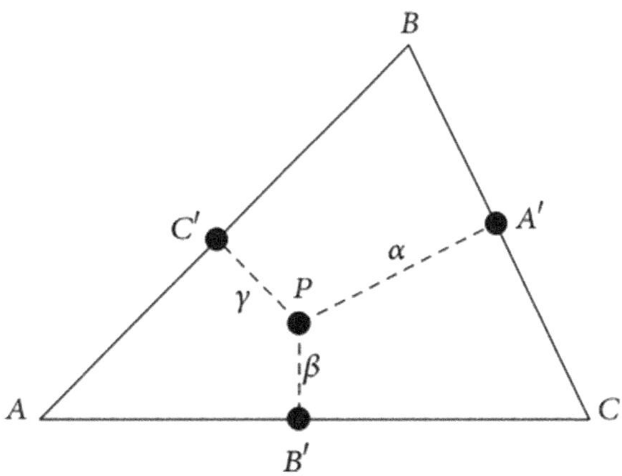

Fig. 2 Trilinear coordinates $\alpha : \beta : \gamma$ of point P [17]

2.2 Triangular Fuzzy Numbers

A fuzzy set $\tilde{A}$ on the universe of discourse U is defined by a mapping $\tilde{A} : U \to [0, 1]$, where $\tilde{A}(x)$ is the membership degree of x [18].

A triangular fuzzy number is a fuzzy set on the real line R with the membership function given by

$$f_{\tilde{A}}(x) = \begin{cases} 0 & x < a \\ \frac{x-a}{b-a}, & a \leq x \leq b \\ \frac{c-x}{c-b}, & \mathrm{b} \leq x \leq c \\ 0, & x > c \end{cases}$$

where $a, b, c \in$ R, $a < b < c$.

For simplicity, it is normally written as a triplet number (a, b, c).

2.3 Linguistic Variables

Zadeh [2] introduced the concept that the value of a linguistic variable is words or sentences instead of a number. For example, let P be a set of linguistic terms where P = {Very High (VH), High (H), Fairly High (FH), Medium (M), Fairly Low (FL), Low (L), Very Low (VL)}.

These three notions are prevalently utilized in the proposed method which is later being applied in an evaluation of financial risk management.

3 Proposed Method for Evaluation of Financial Risks

The RFNs proposed by Abdullah and Azman [9, 19] are the basis for proposing the method. The linguistic scales used in the proposed work are adopted from [19] where seven linguistic scales are predominantly used as subjects of computation. The proposed work is divided into Phase I and Phase II. Prerequisite computations such as defining the linguistic scales and arithmetic mean of fuzzy numbers are made in Phase I. Phase II entails the combination of Nagel point and RFNs. In short, the arithmetic means of fuzzy numbers obtained in Phase I are combined with the Nagel point method in Phase II to establish RFNs. The computational procedures of RFNs based on the Nagel point method for decision-making are proposed as below.

Phase I

There are n number of experts that have been confirmed and are requested to give qualitative evaluation using linguistic variables regarding financial risk management.

The experts are interviewed to provide an evaluation of the risks for financial risk management. In this phase, the computational procedures are presented as follows.

Step 1: Construct a decision matrix.

This decision matrix is constructed based on the evaluation made by the decision maker of the selected risks using the linguistic scales. These evaluations and the matrix are represented by linguistic terms.

Step 2: Transform linguistic terms to trapezoidal fuzzy numbers.

Convert the linguistic term data that have been provided by experts to triangular fuzzy numbers.

Step 3: Calculate the arithmetic mean of triangular fuzzy numbers.

An arithmetic mean of a triangular fuzzy number for each risk is calculated using Eq. (1):

$$TFN_{D_{\tilde{A}n}} = \frac{D_1 + D_2 + D_3 + \ldots + D_n}{n} \tag{1}$$

where $D_{A\overline{n}}$ are the fuzzy numbers for each expert.

Phase II

Consider a triangular fuzzy number $\tilde{a} = \left(a_l, a_p, a_r\right)_T$. The Cartesian coordinates of triangle ABC are $A(a_l, 0)$, $B\left(a_p, 1\right)$, and $C(a_r, 0)$. The algorithm of ranking triangular fuzzy numbers is implemented according to the steps below:

Step 1: Find the length of the corresponding triangle ABC with Cartesian coordinates $A(a_l, 0), B\left(a_p, 1\right)$, and $C(a_r, 0)$ in the Euclidean plane as follows:

$a = |BC| = \sqrt{\left(a_p - a_r\right)^2 + 1}, b = |AC| = (a_r - a_l)$, and

$$c = |AB| = \sqrt{\left(a_p - a_l\right)^2 + 1}. \tag{2}$$

Step 2: Obtain trilinear coordinates $\alpha : \beta : \gamma$ of Nagel point N in triangle ABC (refer to Fig. 2) as follows:

$$\alpha = \frac{\alpha_r - \alpha_l + \sqrt{\left(a_p - a_l\right)^2 + 1} - \sqrt{\left(a_p - a_r\right)^2 + 1}}{\sqrt{\left(a_p - a_r\right)^2 + 1}} \tag{3}$$

$$\beta = \frac{\sqrt{\left(a_p - a_l\right)^2 + 1} + \sqrt{\left(a_p - a_r\right)^2 + 1} - a_r + a_l}{a_r - a_l} \tag{4}$$

$$\gamma = \frac{\sqrt{(a_p - a_r)^2 + 1} + a_r - a_l - \sqrt{(a_p - a_l)^2 + 1}}{\sqrt{(a_p - a_l)^2 + 1}}. \quad (5)$$

Therefore, for all $\tilde{a} \in T$ are points in the Euclidean plane where T is Euclidean space.

Step 3: Calculate the Cartesian coordinates of Nagel point $\mathrm{N}_{\tilde{a}} = (x_{\tilde{a}}, y_{\tilde{a}})$ for $\tilde{a} \in T$ which can be calculated as follows:

$$\mathrm{N}_{\tilde{a}} = (x_{\tilde{a}}, y_{\tilde{a}}) = \left(\frac{a\alpha a_1 + b\beta a_p + c\gamma a_r}{a\alpha + b\beta + c\gamma}, \frac{b\beta}{a\alpha + b\beta + c\gamma} \right) \quad (6)$$

Step 4: Transform $\mathrm{N}_{\tilde{a}} = (x_{\tilde{a}}, y_{\tilde{a}})$ to the function N defined by

$$N(.) : T \to \{(x, y, z)R^3 : 0 < y < 1\}, N(\tilde{a}) = (x_{\tilde{a}}, y_{\tilde{a}}, a_p). \quad (7)$$

These are triangular fuzzy numbers $\tilde{a} = (a_l, a_p, a_r)_T$ and $\tilde{b} = (b_l, b_p, b_r)_T$.

Step 5: Determine the order of triangular fuzzy numbers as

$$\tilde{a} \prec \tilde{b} \text{ if } N(\tilde{a}) <_L N(\tilde{b}). \quad (8)$$

based on the function N where the symbol $<_L$ symbolizes the lexicographical order.

For a triangular fuzzy number, the order is defined as

$$(x_1, x_2, x_3) <_L (y_1, y_2, y_3) \iff (\exists p = 1, 2, 3), (\forall i < p)(x_i = y_i) \wedge (x_p < y_p)$$

Obviously, $(x_1, x_2, x_3) <_L (y_1, y_2, y_3)$ if and only if $x_1 < y_1$ or ($x_1 = y_1$ and $x_2 < y_2$) or ($x_1 = y_1$ and $x_2 = y_2$ and $x_3 < y_3$).

Figure 3 shows that the schematic representation of the computational flows of RFNs based on Nagel point.

The above-proposed method is used to compute the magnitudes or weights of financial risks.

4 Implementation

This section presents the implementation of RFNs based on Nagel point for the case of financial risk analysis. This implementation begins with linguistic data collection. Table 1 demonstrates the representation of linguistic terms and their equivalent fuzzy numbers.

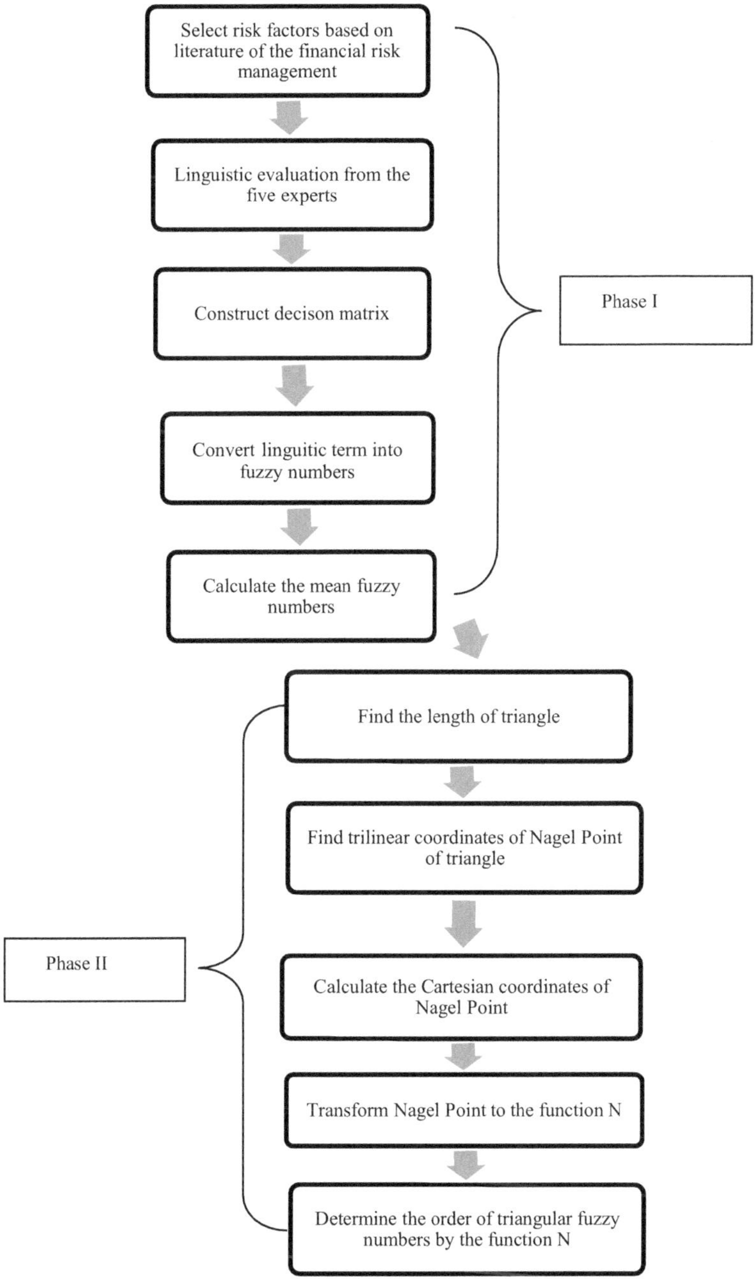

Fig. 3 Flowchart of RFNs based on Nagel point

Table 1 Representation of linguistic terms and their respective fuzzy numbers [19]

Linguistic term	Triangular fuzzy numbers
Very Low (VL)	$(0, 0, 0.1)_T$
Low (L)	$(0, 0.1, 0.3)_T$
Fairly Low (FL)	$(0.1, 0.3, 0.5)_T$
Medium (M)	$(0.3, 0.5, 0.7)_T$
Fairly High (FH)	$(0.5, 0.7, 0.9)_T$
High (H)	$(0.7, 0.9, 1.0)_T$
Very High (VH)	$(0.9, 1.0, 1.0)_T$

The linguistic data using the above scales was collected from five experts in banking and financial institutions. The aim of this computational implementation is to rank the risks of financial management. In this analysis, consideration is given to eight risks in financial management. The eight risks are credit risk (A_1), liquidity risk (A_2), interest rate risk (A_3) equity risk (A_4), commodity risk (A_5), foreign exchange risk (A_6), solvency risk (A_7), and operational risk (A_8).

The implementation is made into two phases in accordance with computational procedures proposed in Sect. 3. All the computations are made based on the equations in Sect. 3. To ease the computations, a spreadsheet software and a computer algebra system are utilized.

Phase I

In this study, RFN based on Nagel point is used to compute the magnitude of each risk and then rank the risk based on the function N. The steps of RFNs based on Nagel point are shown as follows:

Step 1: Suppose that the eight risks are members of set A,

$A = \{A_1, A_2, A_3, A_4, A_5, A_6, A_7, A_8\}$.

Step 2: Suppose that there are five decision makers, denoted by the set $D = \{D_1, D_2, D_3, D_4, D_5\}$.

Step 3: The individual decision opinions are fused into a group of opinions to construct a linguistic decision matrix. The linguistic decision matrix X is constructed as follows:

$$X = \begin{array}{c} \\ A_1 \\ A_2 \\ A_3 \\ A_4 \\ A_5 \\ A_6 \\ A_7 \\ A_8 \end{array} \begin{array}{c} \begin{array}{ccccc} D_1 & D_2 & D_3 & D_4 & D_5 \end{array} \\ \begin{bmatrix} VH & VH & FH & H & VH \\ FH & VH & M & FH & M \\ VH & H & VH & H & M \\ H & FH & M & H & FH \\ M & M & FH & M & FH \\ H & FH & H & H & M \\ H & H & FH & M & M \\ FH & FH & M & M & FH \end{bmatrix} \end{array}$$

Table 2 Risks and their respective triangular fuzzy numbers

Risks	Triangular fuzzy numbers
A_1	$(0.9, 1.0, 1.0)_T$
A_2	$(0.5, 0.7, 0.9)_T$
A_3	$(0.9, 1.0, 1.0)_T$
A_4	$(0.7, 0.9, 1.0)_T$
A_5	$(0.3, 0.5, 0.7)_T$
A_6	$(0.7, 0.9, 1.0)_T$
A_7	$(0.7, 0.9, 1.0)_T$
A_8	$(0.5, 0.7, 0.9)_T$

Step 4: The linguistic data provided by the experts are transformed into triangular fuzzy numbers. Table 2 shows the evaluation given by expert 1 for which the risks are mapped with the respective triangular fuzzy numbers. Similar evaluations are also given by four other experts.

Step 5: By using the arithmetic equation (Eq. (1)), a triangular fuzzy number for each risk is computed. For example, the arithmetic mean of the risk A_1 of expert 1 is obtained as

$$\begin{aligned} D_{\bar{A}_1} &= \frac{D_1 + D_2 + D_3 + D_4 + D_5}{5} \\ &= \left(\frac{0.9+0.9+0.5+0.7+0.9}{5}, \frac{1+1+0.7+0.9+1}{5}, \frac{1+1+0.9+1+1}{5}\right)_T \\ &= \left(\frac{3.9}{5}, \frac{4.6}{5}, \frac{4.9}{5}\right)_T \\ &= (0.78, 0.92, 0.98)_T. \end{aligned}$$

All the evaluations from other experts for each risk are computed similarly. Table 3 shows the arithmetic mean of triangular fuzzy numbers for each risk.

Table 3 Arithmetic mean of triangular fuzzy numbers for risks

Risks	Arithmetic means of triangular fuzzy numbers
A_1	$(0.78, 0.92, 0.98)_T$
A_2	$(0.5, 0.68, 0.84)_T$
A_3	$(0.7, 0.86, 0.94)_T$
A_4	$(0.54, 0.74, 0.9)_T$
A_5	$(0.38, 0.58, 0.78)_T$
A_6	$(0.58, 0.78, 0.92)_T$
A_7	$(0.5, 0.7, 0.86)_T$
A_8	$(0.42, 0.62, 0.82)_T$

Table 4 Length of corresponding triangle ABC for each risk

Risks	a	b	c
A_1	1.0018	0.2	1.0098
A_2	1.0127	0.34	1.0161
A_3	1.0032	0.24	1.0127
A_4	1.0127	0.36	1.0198
A_5	1.0198	0.4	1.0198
A_6	1.0098	0.34	1.0198
A_7	1.0127	0.36	1.0198
A_8	1.0198	0.4	1.0198

These arithmetic means of triangular fuzzy numbers are combined with Nagel point to establish the highest financial risk. Phase II describes details of computation prior to suggesting the highest risk.

Phase II

Step 1: Lengths of the corresponding triangle ABC with Cartesian coordinates $A(a_l, 0)$,$B(a_m, 1)$, and $C(a_r, 0)$ in the Euclidean plane are obtained using Eq. (2).

Table 4 indicates the result of the length of the corresponding triangle ABC for each risk.

Step 2: Trilinear coordinates $\alpha : \beta : \gamma$ of Nagel point N of triangle ABC are found by Eqs. (3), (4), and (5).

For example, the risk A_1

$$\begin{aligned}\alpha &= \frac{\alpha_r - \alpha_l + \sqrt{(a_p - a_l)^2 + 1} - \sqrt{(a_p - a_r)^2 + 1}}{\sqrt{(a_p - a_r)^2 + 1}} \\ &= \frac{0.98 - 0.78 + 1.0098 - 1.0018}{1.0018} \\ &= 0.2076\end{aligned}$$

$$\begin{aligned}\beta &= \frac{\sqrt{(a_p - a_l)^2 + 1} + \sqrt{(a_p - a_r)^2 + 1} - a_r + a_l}{a_r - a_l} \\ &= \frac{1.0098 + 1.0018 - 0.98 + 0.78}{0.98 - 0.78} \\ &= 9.0578\end{aligned}$$

$$\gamma = \frac{\sqrt{(a_p - a_r)^2 + 1} + a_r - a_l - \sqrt{(a_p - a_l)^2 + 1}}{\sqrt{(a_p - a_l)^2 + 1}}$$

Table 5 Trilinear coordinates $\alpha : \beta : \gamma$ of Nagel point N of triangle ABC

Risks	α	β	γ
A_1	0.2076	9.0578	0.1902
A_2	0.339	4.967	0.3313
A_3	0.2487	7.3996	0.2276
A_4	0.3625	4.6459	0.3461
A_5	0.3922	4.099	0.3922
A_6	0.3467	4.9693	0.3235
A_7	0.3625	4.6459	0.3461
A_8	0.3922	4.099	0.3922

$$= \frac{1.0018 + 0.98 - 0.78 - 1.0098}{1.0098}$$
$$= 0.1902$$

Similar computations are executed for other risks. Table 5 summarizes the trilinear coordinates of Nagel points.

Step 3: Cartesian coordinates of Nagel point, $\mathrm{N}_{\tilde{a}} = (x_{\tilde{a}}, y_{\tilde{a}})$ can be obtained using Eq. (7).

For example, the risk A_1

$$\begin{aligned}
\mathrm{N}_{\tilde{a}} &= (x_{\tilde{a}}, y_{\tilde{a}}) \\
&= \left(\frac{a\alpha a_1 + b\beta a_p + c\gamma a_r}{a\alpha + b\beta + c\gamma}, \frac{b\beta}{a\alpha + b\beta + c\gamma}\right) \\
&= \left(\frac{1.0018\times0.2076\times0.78+0.2\times9.0578\times0.92+1.0152\times0.1902\times0.98}{1.0018\times0.2076+0.2\times9.0578+1.0152\times0.1902}, \frac{0.2\times9.0578}{1.0018\times0.2076+0.2\times9.0578+1.0152\times0.1902}\right) \\
&= (0.912, 0.8191)
\end{aligned}$$

Similar computations are iterated for other risks. Table 6 summarizes the Nagel

Table 6 Cartesian coordinates of Nagel point, $\mathrm{N}_{\tilde{a}} = (x_{\tilde{a}}, y_{\tilde{a}})$

Risks	Nagel point, $\mathrm{N}_{\tilde{a}} = (x_{\tilde{a}}, y_{\tilde{a}})$
A_1	(0.912, 0.8191)
A_2	(0.6766, 0.7129)
A_3	(0.8505, 0.7872)
A_4	(0.7329, 0.6991)
A_5	(0.58, 0.6721)
A_6	(0.7699, 0.713)
A_7	(0.6929, 0.6991)
A_8	(0.62, 0.6721)

Table 7 Function $N(\tilde{a}) = \left(x_{\tilde{a}}, y_{\tilde{a}}, a_p\right)$ for each risk

Risks	$N(\tilde{a}) = \left(x_{\tilde{a}}, y_{\tilde{a}}, a_p\right)$
A_1	(0.912, 0.8191, 0.92)
A_2	(0.6766, 0.7129, 0.68)
A_3	(0.8505, 0.7872, 0.86)
A_4	(0.7329, 0.6991, 0.74)
A_5	(0.58, 0.6721, 0.58)
A_6	(0.7699, 0.713, 0.78)
A_7	(0.6929, 0.6991, 0.7)
A_8	(0.62, 0.6721, 0.62)

points for respective risks.

Step 4: Transform Cartesian coordinates of $\mathrm{N}_{\tilde{a}} = (x_{\tilde{a}}, y_{\tilde{a}})$ into the function N using Eq. (7). The transformation results are presented in Table 7.

Step 5: The order of triangular fuzzy numbers by the function N is obtained (see Eq. 8)).

Therefore, $(0.58, 0.6721, 0.58) <_L (0.62, 0.6721, 0.62) <_L (0.6766, 0.7129, 0.68) <_L (0.6929, 0.6991, 0.7) <_L (0.7329, 0.6991, 0.74) <_L (0.7699, 0.713, 0.78) <_L (0.8505, 0.7872, 0.86) <_L (0.912, 0.8191, 0.92)$.

And we find

$$N\left(\tilde{A}_5\right) <_L N\left(\tilde{A}_8\right) <_L N\left(\tilde{A}_2\right) <_L N\left(\tilde{A}_7\right) <_L N\left(\tilde{A}_4\right) <_L N\left(\tilde{A}_6\right) <_L N\left(\tilde{A}_3\right) <_L N\left(\tilde{A}_1\right)$$

Then, the ultimate ordering is

$$A_5 \prec A_8 \prec A_2 \prec A_7 \prec A_4 \prec A_6 \prec A_3 \prec A_1$$

The results of the ordering of triangular fuzzy numbers by the function N in financial risk management using the Nagel point method can be visualized in Fig. 4.

The results of the selection by using ranking fuzzy numbers based on Nagel point methods are summarized in Table 8.

It can be seen that the ranking fuzzy numbers are based on Nagel point which concludes that the highest risk in financial risk management is credit risk, A_1, followed by interest rate risk, foreign exchange risk, equity risk, solvency risk, liquidity risk, and operational risk. It is also good to note that the lowest risk in financial risk management is commodity risk, A_5.

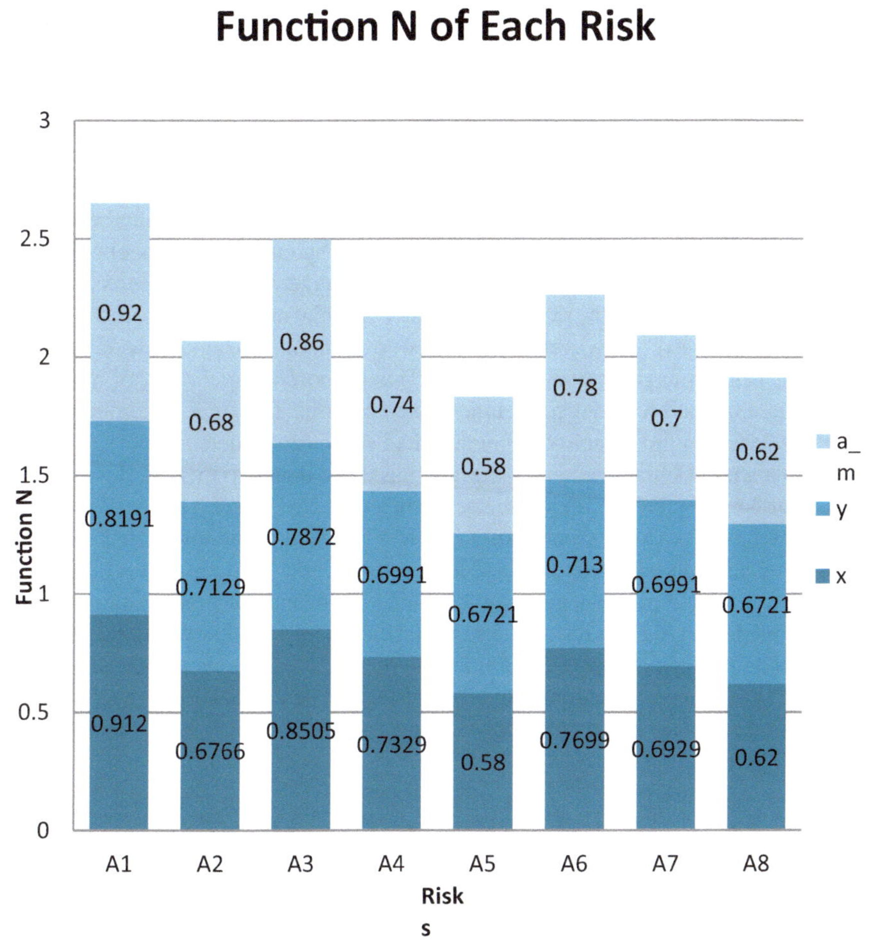

Fig. 4 Result of function N and risks

Table 8 Summary of result by ranking fuzzy numbers based on Nagel point methods

Risks	RFNs based on Nagel point	
	$N(\tilde{a}) = (x_{\tilde{a}}, y_{\tilde{a}}, a_p)$	Ranking
A_1	(0.912, 0.8191, 0.92)	1
A_2	(0.6766, 0.7129, 0.68)	6
A_3	(0.8505, 0.7872, 0.86)	2
A_4	(0.7329, 0.6991, 0.74)	4
A_5	(0.58, 0.6721, 0.58)	8
A_6	(0.7699, 0.713, 0.78)	3
A_7	(0.6929, 0.6991, 0.7)	5
A_8	(0.62, 0.6721, 0.62)	7

5 Conclusion

Financial risk management has been practiced by the banks as the uncovering of various types of risks is seriously affecting the performance of banks. The identification of the risks in financial risk management is extremely important in the banking sector. The purpose of this paper was to identify the highest risk in financial risk management using the proposed method of ranking fuzzy numbers using the Nagel point. Research in ranking fuzzy numbers by introducing Nagel point was used to show the ranking of risks in financial risk management. Five experts provided linguistic evaluation over the ranking of risks in financial management. The linguistic evaluation of the experts was then converted to triangular fuzzy numbers by using linguistic value triangular fuzzy numbers. This proposed Nagel point of a triangle method is a method to order triangular fuzzy numbers. The steps of ranking fuzzy numbers based on the Nagel point method had successfully determined the highest risk in financial risk management in the banking sector that is credit risk. Besides that, the use of Nagel point in ranking fuzzy numbers showed interest rate risk to be the second highest risk value. Commodity risk was revealed as the lowest risk in financial risk management in the banking sector for this proposed method. It is hoped that this research would provide some light on the significance of processing linguistic data for suggesting the degree of risks in financial sectors. Acknowledging the risks in financial management would minimize the financial loss. However, the results are limited to the scope of research. This research could be extended in future research by adopting other natural linguistic processing approaches which may generate more interesting results.

References

1. Modarres M, Sadi-Nezhad S (2001) Ranking fuzzy numbers by preference ratio. Fuzzy Sets Syst 118(3):429–436
2. Zadeh LA (1965) Fuzzy sets. J Inf Control 8(3):338–353
3. Jain R (1976) Decision making in the presence of fuzzy variables. IEEE Trans Syst, Man Cybern 6:698–703
4. Jain R (1978) A procedure for multi aspect decision making using fuzzy sets. Int J Syst Sci 8:1–7
5. Yu FV, Chi HTX, Dat LQ, Phuc PNK, Shen CW (2013) Ranking generalized fuzzy numbers in fuzzy decision making based on the left and right transfer coefficients and areas. Appl Math Model 37(16–17):8106–8117
6. Wang J, Ding D, Liu O, Li M (2016) A synthetic method for knowledge management performance evaluation based on triangular fuzzy number and group support systems. Appl Soft Comput 39:11–20
7. Singh PA (2015) Novel method for ranking generalized fuzzy numbers. J Inf Sci Eng 31:1373–1385
8. Chi HTX, Yu FV (2018) Ranking generalized fuzzy numbers based on centroid and rank index. Appl Soft Comput 68:283–292
9. Abdullah L, Azman FN (2015) Circumcenter of centroid of fuzzy number for identifying risks of obesity: a qualitative evaluation. J Qual Quant 50(6):2433–2449

10. Nesseri SH, Taghi-Nezhad N, Ebrahimnejad A (2017) Note on ranking fuzzy numbers with an area method using circumcenter of centroids, Fuzzy Inf Eng 9(2):259–3268
11. Akyar E, Akyar H, Duzce SA (2013) Fuzzy risk analysis based on a geometric ranking method for generalized trapezoidal fuzzy numbers. J Intell Fuzzy Syst 25(1):209–217
12. Thorani YLP, Rao PPB, Shankar NR (2012) Ordering generalized trapezoidal fuzzy numbers using orthocentre of centroids. Int J Algebra 6(22):1069–1085
13. Maheswari P, Vijaya M (2019) On solving two-person zero-sum matrix games with payoff to generalised trapezoidal fuzzy numbers based on fuzzy ranking values using in center of centroids. Adv Appl Math Sci 18(10):1235–1085
14. Rao PPB, Ravi Shankar N (2013) Ranking fuzzy numbers with an area method using circumcenter of centroids. Fuzzy Inf Eng 5(1):3–18
15. Akyar E, Akyar H, Duzce SA (2012) A new method for ranking triangular fuzzy numbers. Int J Uncertainty, Fuzziness Knowl-Based Syst 20(5):729–740
16. Kimberling C (1994) Central points and central lines in the plane of a triangle. Math Mag 67(3):163–187
17. Martin GE (1982) Transformation geometry: an introduction to symmetry. Springer, New York, NY, USA
18. Ban OI (2011) Fuzzy multicriteria decision making method applied to selection of the best touristic destinations. Int J Math Models Methods Appl Sci 2(5):264–271
19. Akyar H (2016) Fuzzy risk analysis for a production system based on the Nagel point of a triangle. Math Probl Eng 1–9

Chapter 28
Design of AMC-Based Metasurface Loaded Slot Antenna for Wideband RCS Reduction and Gain Improvement

Ankit Sharma, Deepak Kumar, Animesh Chandra, Himanshu Prajapat, Aniket Chauhan, Madan Kumar Sharma, and Hridesh Kumar Verma

1 Introduction

With the evolution of stealth technology, radar cross-section (RCS) of the antennas plays a key role as it has a notable influence on the overall radar cross-section of the object [1]. There are several techniques used to reduce the radar cross-section of the objects such as the Salisbury screen, electromagnetic gap (EBG), and radar absorbing material (RAM) [2, 3]. However, these methods can effectively reduce the RCS of an object but in most cases, it tends to increase the overall material thickness, heavyweight, and maintenance cost of the object. In addition to this, these methods are inherently narrowband [4].

Besides the above methods of RCS reduction, diverting the incident electromagnetic (EM) waves into other directions is an upright idea. The widespread research on Artificial Magnetic Conductor (AMC) and Electromagnetic Band Gap structures have opened new ways to improve the scattering performance of antenna [5, 6]. The principle approach is to redirect the scattered waves away from the radar direction [7]. This approach uses the phase cancellation technique which employs a 180° phase difference of the reflected waves between the different scattering regions [8]. The backscatter cancellation is achieved in the boresight direction by the metasurface which is in the form of a checkerboard which is made up of an Artificial Magnetic Conductor (AMC) and a Perfect Electric Conductor (PEC) [9, 10]. The main limitation of this arrangement is that the AMC structure has narrower RCS reduction bandwidth [11]. Outside this bandwidth, the AMC behaves like a PEC, and the condition for destructive interference is not achieved. To prevail over this limitation, researchers have come up with 2 different methods, i.e. designing AMCs of different dimensions and other methods is implementing different AMC configurations. The

A. Sharma (✉) · D. Kumar · A. Chandra · H. Prajapat · A. Chauhan · M. K. Sharma · H. K. Verma
Galgotias College of Engineering and Technology, Greater Noida, India

Rajasthan Technical University, Kota, India

G. Mathur et al. (eds.), *Proceedings of 2nd International Conference on Artificial Intelligence: Advances and Applications*, Algorithms for Intelligent Systems,
https://doi.org/10.1007/978-981-16-6332-1_28

principle behind employing two dissimilar AMCs is to widen the bandwidth of 180° ± 30° phase difference between the two-unit cells [12]. The backscattering suppression over a wide range can be achieved by using this phase difference criterion. Numerous AMC metasurfaces with modest shapes such as a circular patch, a square patch, and a Jerusalem cross are selected for designing the metasurface to obtain maximum RCS reduction bandwidth [13, 14]. Further, traditional methods such as forming a checkerboard metasurface using two-unit cells are not adopted, instead some unit cells with different phases of reflection are proposed, and at that point by an irregular establishment of the cells a low RCS metasurface is attained [15]. Nowadays, researchers' focus is on developing metasurfaces such that both the scattering and radiation characteristics of the antenna can be improved. In [16, 17], metasurface loaded slot antennas are proposed to achieve gain enhancement and wideband RCS reduction. Further, a holographic metasurface is designed to obtain low scattering and improved gain [18]. Recently, a metamaterial absorber is proposed for in-band RCS reduction of the antenna [19]. A 7 × 7 polarization reconfigurable microstrip array is designed to reduce the in-band RCS of the array by 7 dB [20]. Then, Tiwari et. al. have designed a metasurface using phase cancellation techniques to reduce the RCS in the frequency band 7.7–13.7 GHz [21].

In this work, we have presented an AMC metasurface loaded low-scattering low-profile antenna with improved gain. The unique properties of the research work are mentioned as follows: 1. The proposed AMC metasurface used two different AMC unit cells for designing. 2. Different AMC metasurface configuration loaded antennas are designed and their performances are compared. This manuscript is arranged as follows: in Sect. 2, we elaborate on the designing criteria of AMC metasurfaces and we have also studied the performance of two different AMC metasurface configurations. In Sect. 3, the reference slot antenna is proposed and the metasurface is integrated with the slot antenna. The radiation and scattering characteristics of the slot antenna and metasurface loaded antennas are investigated in Sect. 4. In Sect. 5, the prototypes of the slot antenna and proposed antenna are fabricated to validate the simulated results. Finally, conclusions are drawn in Sect. 6.

2 Design Principle and Analysis of AMC-Based Metasurfaces

2.1 Design Principle of AMC Unit Cell

The AMC-based metasurface employs the principle of reflection phase cancellation for RCS reduction. To realize this, two different AMC unit cells are designed, AMC#1 and AMC#2 as shown in Fig. 1a, b, respectively. Both the AMCs have different reflection phases when illuminated with impinging EM waves. The AMC unit cells are designed on an FR-4 substrate with a thickness of 3.2 mm and the dimensions of the designed unit cells are given in Table 1. When a plane wave illuminates the

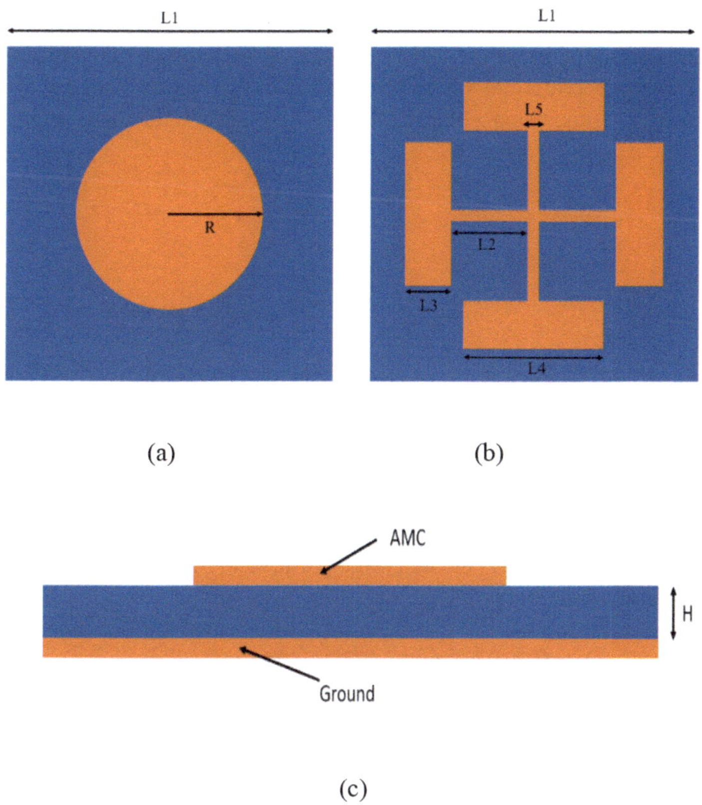

Fig. 1 Geometry of AMC unit cells: **a** AMC unit cell-1, **b** AMC Unit Cell-2, and **c** side view

Table 1 Dimensions of the unit cell

Parameter	Values (mm)	Parameter	Values (mm)
L1	10	L10	11
L2	2.30	L11	80
L3	1.50	L12	41
L4	4	L13	10
L5	0.4	H	3.2
L6	80	R	3.35
L7	7	W1	1
L8	7.5	W2	0.5
L9	9	W3	1.5

metasurface, the total reflected energy is a summation of the reflection from both AMC arrays arranged in orthogonal orientations [19]. Consider the designed AMC unit cells have reflection magnitude near to 1 as the ground is placed at the bottom of the unit cells. Thus, when a plane EM wave impinges on the metasurface [10, 12, 22], then

$$E_{AMC\#1} = A_1 \exp(j \times \alpha_1) \tag{1}$$

$$E_{AMC\#2} = A_1 \exp(j \times \alpha_2) \tag{2}$$

where $E_{AMC\#1}$ and $E_{AMC\#2}$ are the reflected fields of the AMC unit cells. A_1 is the reflected field amplitude, α_1 and α_2 are the phases of reflection for AMC#1 and AMC#2. Then, the total reflected field is calculated as

$$E = (E_{AMC\#1} \times AF\#_1) + (E_{AMC\#2} \times AF\#_2) \tag{3}$$

When a plane EM wave is incident on metasurface, then $AF\#_1 = AF\#_2 = 2$, and the total reflected field is mentioned as

$$E = 2A_1\left[\exp(j \times \alpha_1) + \exp(j \times \alpha_2)\right] \tag{4}$$

A 10-dB scattering suppression criterion is taken into consideration as we need to suppress the reflected EM wave at least 90% in the boresight direction. Thus, to achieve a 10-dB scattering suppression in the boresight direction with respect to a same-sized metallic plate, the ratio of the reflected field of the metasurface and metal plate is given as [19]

$$|E|^2 \big/ |E0|^2 < -10\,\mathrm{dB} \tag{5}$$

E_0 is the electric field of the metal surface under the impinging plane wave. Thus, the effective reflection phase difference between α_1 and α_2 can be expressed as

$$143^\circ \le |\alpha_1 - \alpha_2| \le 217^\circ \tag{6}$$

Thus, to achieve a 10-dB RCS reduction, we need to assume $180^\circ \pm 30^\circ$ as an effective phase difference criterion for the construction of the AMC unit cells. For the analysis of reflection parameters of the AMC unit cells, simulations are done through CST Microwave Studio. The periodic conditions are employed to the x- and y-axes, and a wave-port is employed along the z-axis to simulate the AMC cell. The reflections phase of AMC#1 and AMC#2 are shown in Fig. 2a; it is clear that the resonating frequency of both the unit cells is different. Further, AMC#1 exhibits a zero-degree phase at 7.3 GHz whereas the zero-degree reflection phase of AMC#2 is at 4.9 and phase reversal at 6.3 GHz as shown in Fig. 2a. The phase difference of both the AMC unit cells is shown in Fig. 2b, and it is observed that the phase

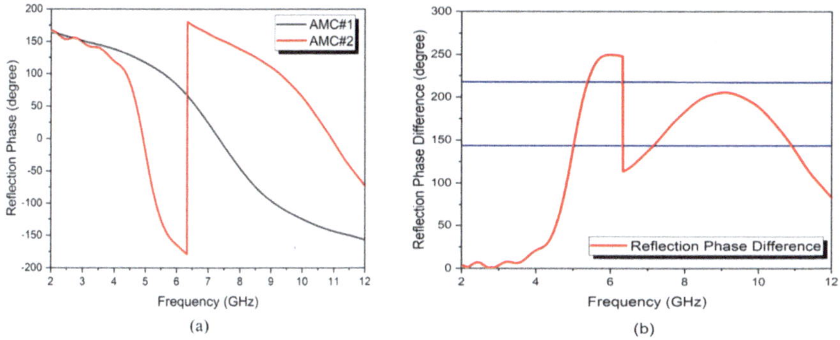

Fig. 2 **a** Reflection Phase of AMC#1 and AMC#2; **b** phase difference between AMC#1 and AMC#2

difference of 180 ± 30^0 is seen in the frequency range of 7.1–10.9 GHz while the phase difference slightly moves up and down in the given frequency range.

2.2 *Design of AMC Metasurface Configurations*

The two different types of metasurface configurations have been proposed which are formed by symmetrically arranging the AMC unit cells to form Meta-1 and Meta-2 as depicted in Fig. 3a, b. Figure 3a which consists of Meta-1 is a checkerboard combination of a 4×4 array of the AMC cells. Further, in Fig. 3b, we have created a triangular-shaped array of the AMC cells which are arranged alternatively to apply the

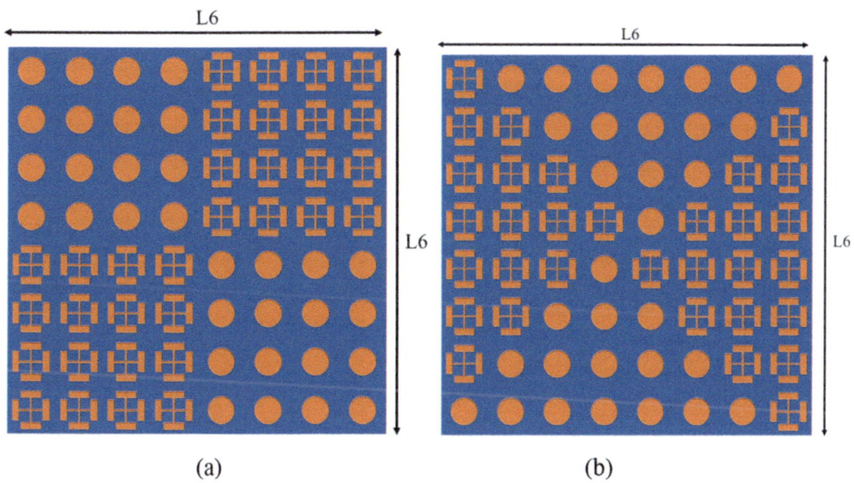

Fig. 3 Metasurface configuration: **a** Meta-1 (top view); **b** Meta-2 (top view)

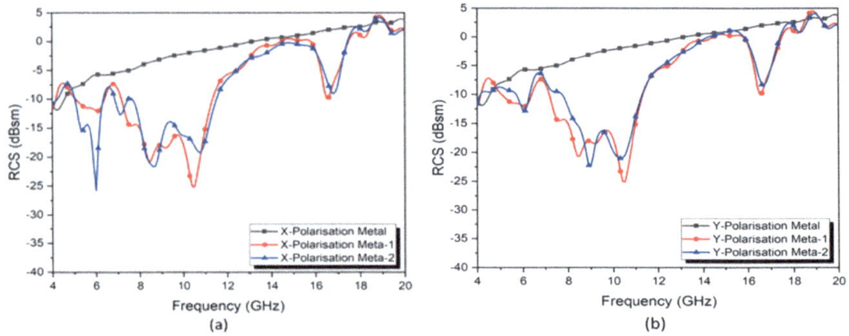

Fig. 4 Simulated Monostatic RCS of Meta-1, Meta-2, and metallic plate: **a** h-polarization; **b** v-polarization

phase cancellation concept. The overall dimension of both the metasurfaces is 80 × 80 mm. The comparison of RCS of the Meta-1, Meta-2, and equal-sized metallic plate are shown in Fig. 4a, b for x- and y-polarized incident plane waves, respectively. It has been observed that the Meta-1 configuration gets significant RCS reduction related to the metallic plate from 7 to 13 GHz and above 10-dB scattering, suppression is achieved in the frequency range of 7.9–11.2 GHz. The designed metasurfaces achieve peak RCS curtailment of −25.19 dB at 10.5 GHz for x-polarized impinging wave and −25.15 dB at 10.5 for y-polarized impinging wave. Further, Meta-2 configuration gets significant RCS curtailment as compared to the metallic plate more than 10 dB RCS reduction from 8 to 11.4 GHz for x-polarized (28.3% relative bandwidth) and 8.2–11.1 for y-polarized wave (24.16% relative bandwidth). The peak reduction in RCS for h- and v-polarized wave value reaches −25.75 dB at 6 GHz and −22.2 dB at 8.9 GHz. To further explore the scattering characteristic of the proposed metasurface, the 3-D scattering pattern of the Meta-1, Meta-2, and metallic plate at 10 GHz are shown in Fig. 5. Figure 5a describes the scattering pattern of the metallic plate at 10.6 GHz; it can be seen that the maximum value of the scattering energy appears in the boresight direction. However, Fig. 5b, c depicts the 3-D scattering pattern of Meta-1 and Meta-2 at 10 GHz. It can be seen that for Meta-1 and Meta-2, the scattered energy gets splits in four different directions causing minimum reflection in the boresight direction as expected, thus implying RCS reduction. Further, by analysing Figs. 4 and 5, we can also conclude that both the configurations of the metasurface, i.e. Meta-1 and Meta-2 are working effectively to suppress the scattering in the band of 7–11 GHz.

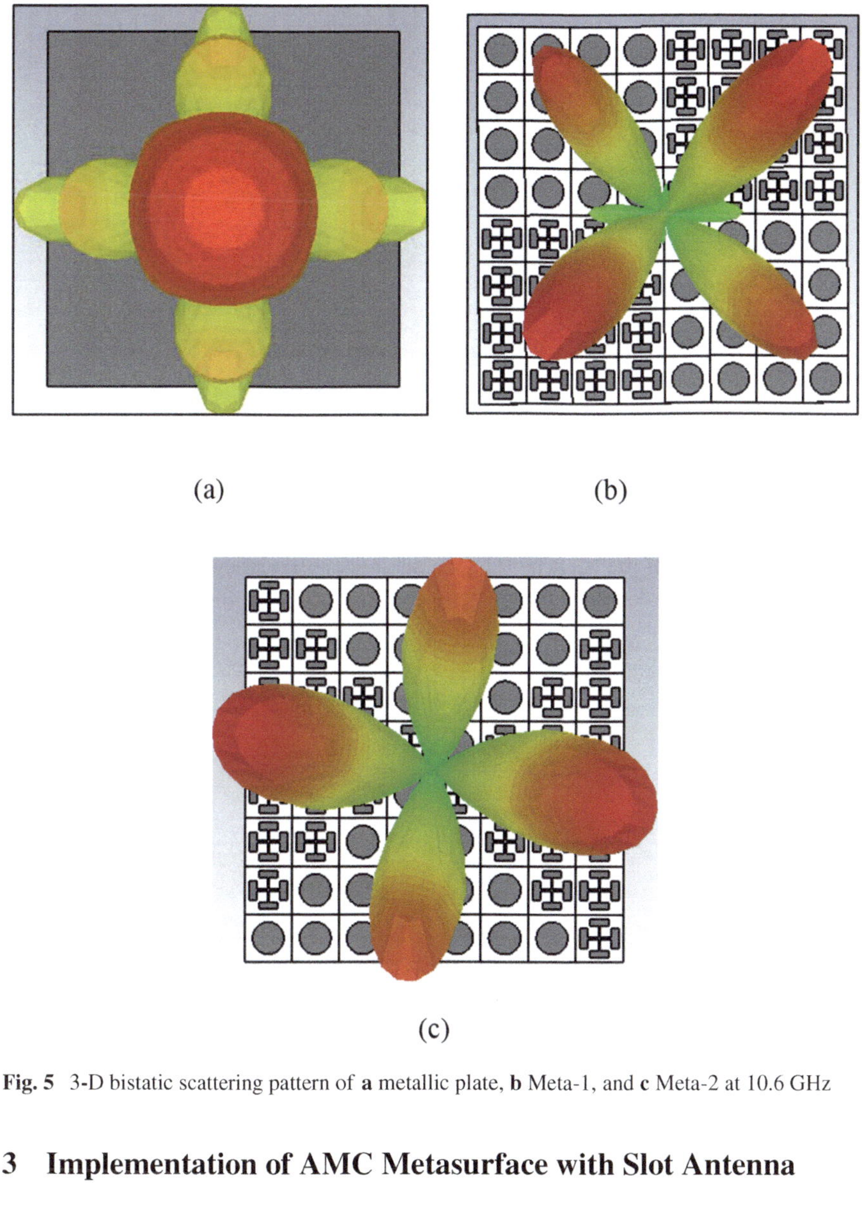

Fig. 5 3-D bistatic scattering pattern of **a** metallic plate, **b** Meta-1, and **c** Meta-2 at 10.6 GHz

3 Implementation of AMC Metasurface with Slot Antenna

3.1 Construction of SRR Inspired Slot Antenna

The projected antenna is constructed on the FR-4 substrate; the dimension of the antenna is 80 × 80 × 0.75 mm. A rectangular-shaped slot is made at the antenna ground and the substrate bottom layer consists of a microstrip feed line and split ring resonator (SRR) as depicted in Fig. 6a, b. The two SRRs placed at the bottom layer

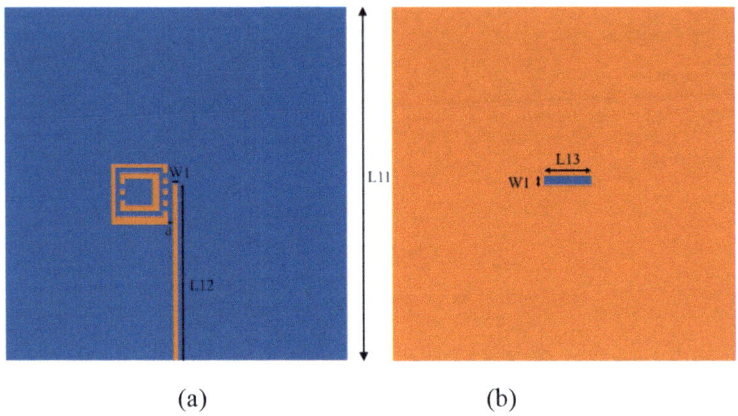

Fig. 6 Geometry of the slot antenna **a** upper view and **b** lower view

of the substrate are employed to achieve wide impedance bandwidth by keeping the resonant frequency of the SRR near to the resonance frequency of the slot antenna. Configuration of the designed SRR is depicted in Fig. 7 and the design parameters are mentioned in Table 1.

Fig. 7 Configuration of the designed SRR

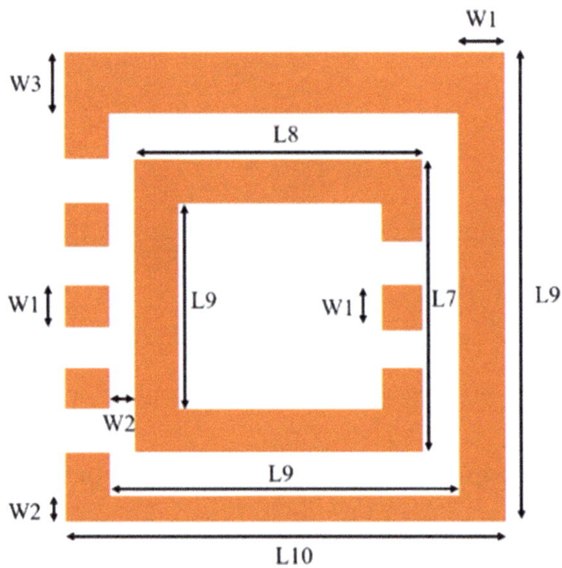

3.2 Implementation of Antenna with Two Different Metasurface Configurations

The SRR inspired slot antenna is integrated with the AMC metasurface arrangements discussed in Section II. The layout of our antennas based on metasurfaces is named Ante#Meta-1 and Ante#Meta-2, and depicted in Fig. 8a, b. To minimize the effect of AMC over the radiation performance of the slot antenna, the centre AMC unit cells are removed.

4 Analysis of Metasurface Loaded Antennas

This section includes the discussion on the performance of the slot, Ante#Meta-1, and Ante#Meta-2 antennas. First, we have discussed the radiation performance of the slot antenna, Ante#Meta-1, and Ante#Meta-2 through simulated results, and then we will take a look at the scattering properties of slot antenna, Ante#Meta-1 and Ante#Meta-2 based on metasurfaces and will choose the better configuration for fabrication.

4.1 Radiation Performance Analysis

We have investigated the radiation properties of Ante#Meta-1, Ante#Meta-2, and slot antenna. The reflection coefficients of Ante#Meta-1, Ante#Meta-2, and slot antenna are shown in Fig. 9. It is observed that the impedance bandwidth of the slot antenna is 9.7–10.59 GHz, Ante#Meta-1 is 9.9–10.6 GHz while that of Ante#Meta-2 is 9.8–10.62 GHz. However, it is clear that the return loss bandwidth of Ante#Meta-1 and Ante#Meta-2 is slightly shifted towards the right side when compared with the slot antenna. The gain values of the slot antenna, Ante#Meta-1, and Ante#Meta-2 are shown in Fig. 10, and it is observed that the gain of Ante#Meta-2 is more than Ante#Meta-1, and slot antenna in the range of 9–10.8 GHz. It is also seen that the peak gain of the Ante#Meta-2 is 5.6 dBi which is 1.61 dB and 2.18 dB higher than the slot antenna and Ante#Meta-1, respectively.

4.2 Scattering Characteristics Analysis

In this section, we are analysing monostatic RCS of the Ante#Meta-1, Ante#Meta-2, and slot antenna as shown in Fig. 11a for the x-polarized and y-polarized incident plane waves in Fig. 11b. Remarkable scattering suppression is achieved in the frequency band of 6–15 GHz for the metasurface-based antennas Ante#Meta-1

Fig. 8 Top view of **a** Ante#Meta-1, **b** Ante#Meta-2

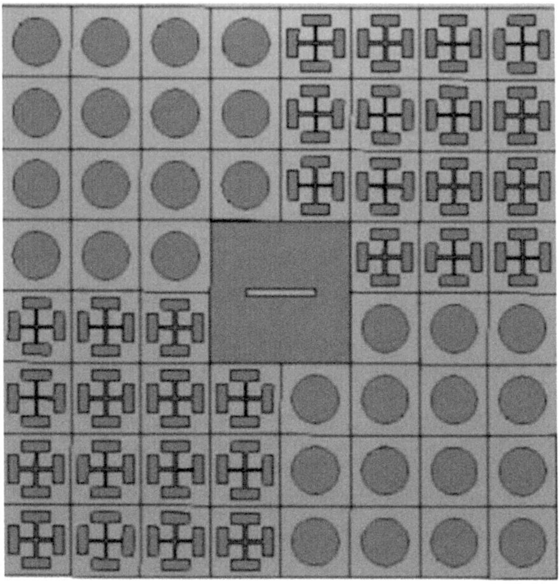

(a)

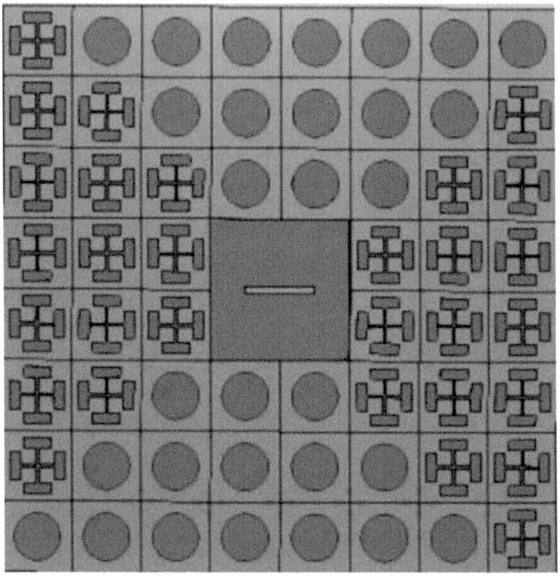

(b)

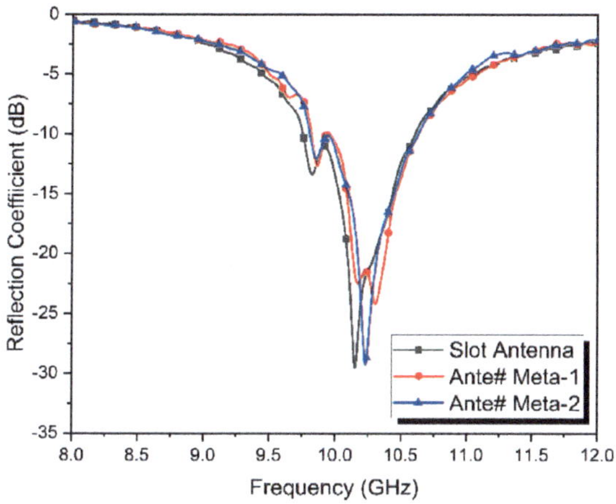

Fig. 9 Reflection coefficients of slot antenna, Ante#Meta-1, and Ante#Meta-2

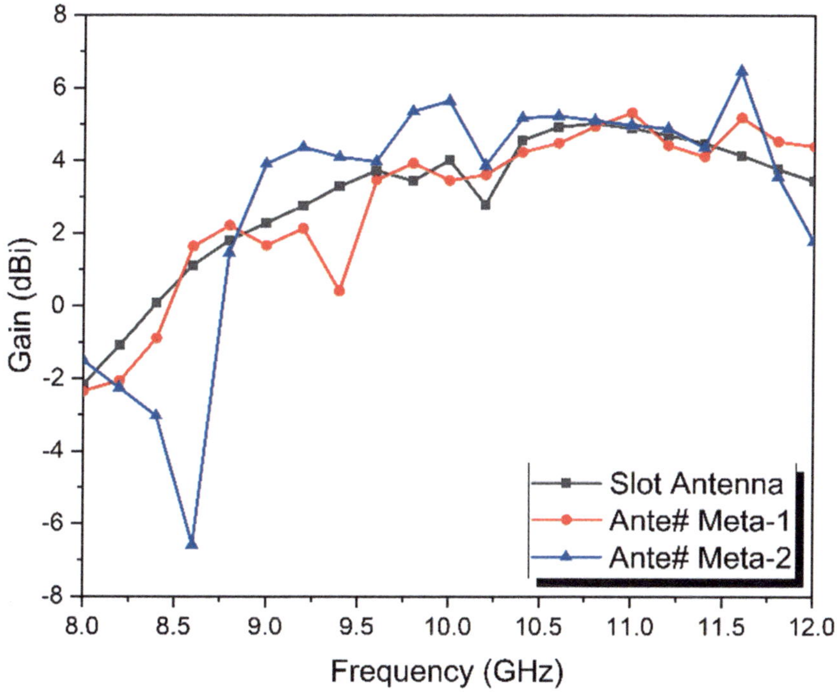

Fig. 10 Gain of slot antenna, Ante#Meta-1, and Ante#Meta-2

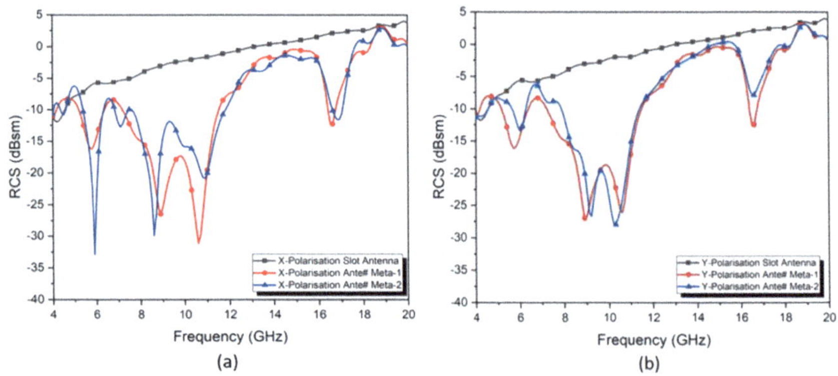

Fig. 11 Simulated monostatic RCS versus frequency with **a** x-polarized incident wave and **b** y-polarized incident wave

and Ante#Meta-2 as compared to slot antenna. The peak in-band RCS reduction of Ante#Meta-1 is 31.12 dB and 26.12 dB at 10.6 GHz for x- and y-polarized waves, respectively, while Ante#Meta-1 achieves peak RCS reduction of 26.2 dB and 26.9 dB at 8.9 GHz for x- and y-polarized waves, respectively. The peak in-band RCS reduction achieved by Ante#Meta-2 is 18.5 dB and 27.9 dB at 10.6 GHz for both polarizations. Further, the peak out of band RCS reduction of Ante#Meta-2 is 32.8 dB and 26.6 dB at 5.9, and 9.2 GHz for x- and y-polarized waves, respectively. It is seen that both the metasurface loaded antennas have almost similar backscattering suppression performance in comparison to the slot antenna. However, Ante#Meta-2 shows better radiation characteristics as compared to Ante#Meta-1. Thus, considering the antenna radiation and scattering characteristics, Ante#Meta-2 is a better choice than Ante#Meta-1.

5 Results and Discussion

We have considered the Ante#Meta-2 as our proposed antenna. To verify the simulated radiation and scattering results, the proposed (Ante#Meta-2) and slot antennas are fabricated and tested in the anechoic chamber. The photograph of both the antennas is depicted in the inset of Fig. 12. The simulated and measured results of the reflection coefficients (S-Parameter) of the proposed antenna and slot antenna are given in Fig. 12. It is observed that the measured resonance frequency of the slot antenna and the proposed antenna slightly shift towards the right side as compared to simulated results. The measured impedance bandwidth of the slot antenna is 9.8–10.48 GHz and that of the proposed antenna is 9.9–10.48 GHz. The measured and simulated gain of the slot antenna and the proposed antenna is shown in Fig. 13. The peak measured gain of Ante#Meta-2 is 5.01 dB at the frequency of 10.2 GHz which

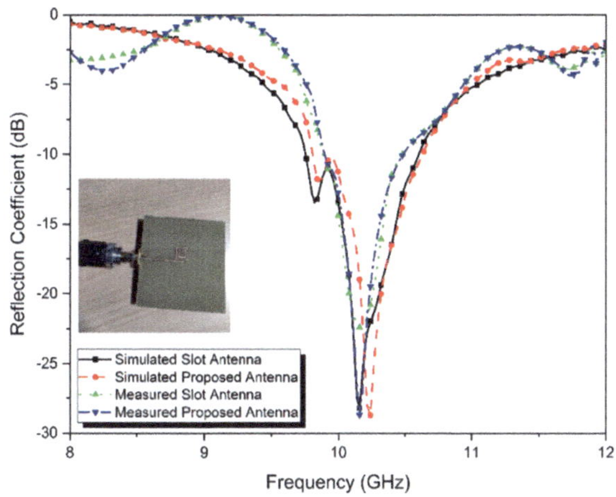

Fig. 12 Reflection coefficients of Slot antenna and Proposed antenna (inset photographs measurement using VNA)

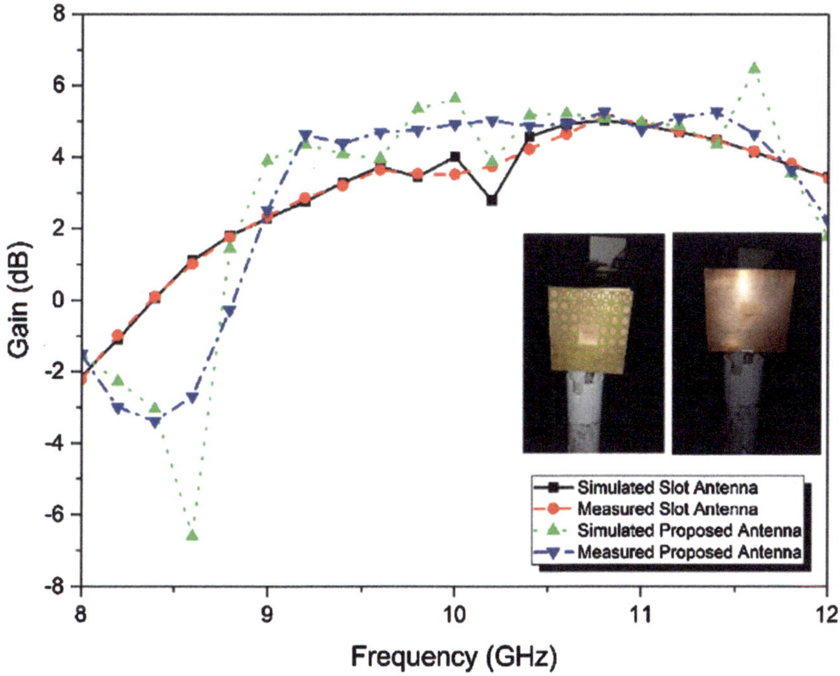

Fig. 13 Gain of slot antenna and designed antenna (inset photograph measurement in anechoic chamber)

is 0.81 dB greater than that of the peak measured gain of the slot antenna. Further, the radiation pattern of slot antenna and proposed antenna at 10 GHz is depicted in Fig. 14. The half-power beamwidth (HPBW) of the slot antenna is much wider than the proposed antenna. The HPBW is 50.9° and 31.1° for slot antenna, whereas for the proposed antenna it is 35.1° and 22.2° for E- and H-planes, respectively. Thus, it is clear that the implementation of the proposed metasurface improves the overall radiation characteristics of the proposed antenna.

Further for the investigation of the scattering performance of the designed antenna, the measurements are done in an anechoic chamber. Two standard antennas covering the range of 1–18 GHz operating as transmitter and receiver are placed at a far-field distance from the antenna under test. From Fig. 15a, b, it is seen that the proposed antenna (Ante#Meta-2) achieves significant scattering suppression as compared to the slot antenna. The designed antenna achieves 10-dB RCS curtailment in the band of 7.7–11.25 GHz and 8.3–11 GHz for x- and y-polarized waves, respectively. Figure 16a, b depicts the simulated and measured RCS versus theta (−90° to + 90°)

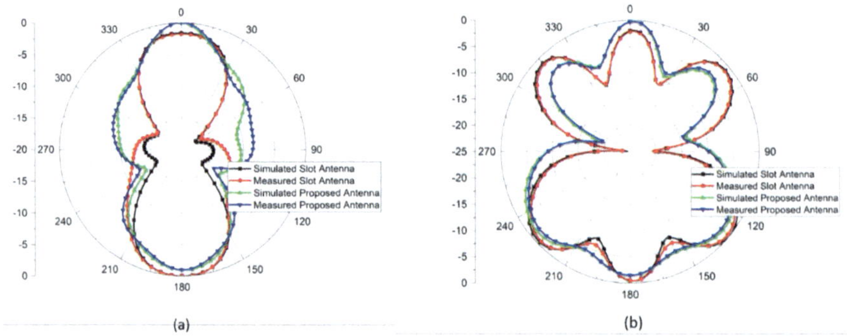

Fig. 14 Normalized radiation patterns of slot antenna and Ante#Meta-2 at 10 GHz **a** E-plane and **b** H-plane

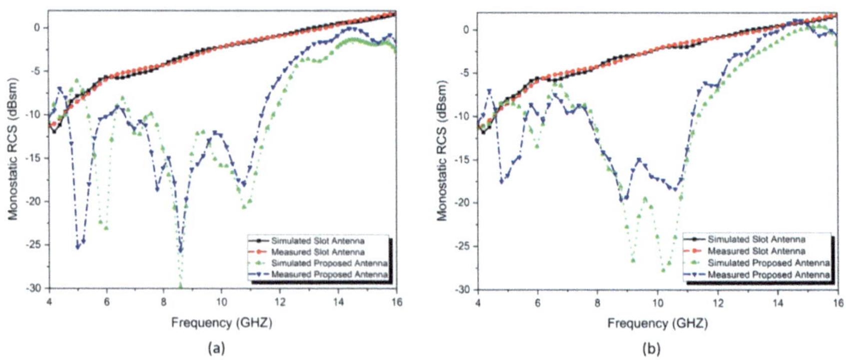

Fig. 15 RCS of slot antenna and proposed antenna **a** x-polarization **b** y-polarization

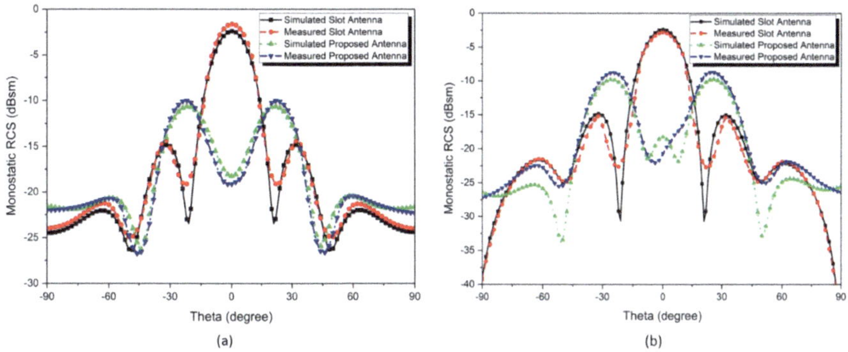

Fig. 16 Monostatic RCS pattern of slot antenna and Ante#Meta-2 **a** x-polarization **b** y-polarization at 10 GHz

graph for x- and y-polarized waves for proposed and slot antennas at 10 GHz. It is seen that the RCS of the proposed antenna shows a significant RCS reduction in the angle range of $\pm 20^\circ$ related to the slot antenna for both polarizations. An assessment between the designed antenna and the antennas mentioned in the previous work is done for comparison with the designed work. Different parameters such as the operating band, maximum gain enhancement, and RCS reduction band are listed in Table 2. The proposed antenna has a lower volumetric dimension as compared to the structure present in the literature with good RCS reduction bandwidth as well as high gain enhancement.

6 Conclusion

This work presents an AMC metasurface-based slot antenna for low scattering and gain enhancement. The proposed AMC cells are improved to implement the 180 $\pm$ 30° phase difference criterion to achieve wideband RCS reduction. Further, two AMC unit cell arrays are arranged in two different checkerboard configurations of the metasurface to obtain wideband RCS reduction. The proposed antenna shows more than 10 dB RCS reduction of 8.1–11.6 GHz for x-polarization and 8.2–11.2 GHz for x-polarization. The maximum gain of the proposed antenna is 5.6 dBi at 10 GHz which is 1.61 dB higher than the slot antenna.

Table 2 Comparison between the proposed antenna and other antennas in reference

References	Operation band (GHz)	Peak gain enhancement	RCS reduction band	Volumetric dimensions	Structural
Proposed	9.8–10.62	1.61 dB	6–15	80 mm × 80 mm × 3.95 mm	Slot antenna loaded with AMC-based metasurface
Ref. [14]	9.6–11.4	1.0 dB	8–20	81 mm × 81 mm × 3 mm	Microstrip antenna loaded with AMC structure
Ref. [17]	5.48–5.84	2 dB	4.95–7.18	120 mm × 120 mm × 3 mm	Slot antenna with low RCS and high gain loaded with thin metasurface
Ref. [18]	9.7–10.7	2.6 dB	5.5–2.15	100 mm × 100 mm × 3 mm	HM as parasitic structure
Ref. [19]	6.6–6.9	N/A	6.6–7	100 mm × 100 mm × 1 mm	Patch antenna with ultra-thin metamaterial absorber

References

1. Wang F, Jiang W, Hong T, Xue H, Gong S, Zhang Y (2014) Radar cross section reduction of wideband antenna with a novel wideband radar absorbing materials. IET Microwaves Antennas Propag 8(7):491–497
2. Fante RL, McCormack MT (1988) Reflection properties of the Salisbury screen. IEEE Trans Antennas Propag 36(10):1443–1454
3. Li YQ, Zhang H, Fu YQ, Yuan NC (2008) RCS reduction of ridged waveguide slot antenna array using EBG radar absorbing material. IEEE Antennas Wirel Propag Lett 27(7):473–476
4. Haji-Ahmadi MJ, Nayyeri V, Soleimani M, Ramahi OM (2017) Pixelated checkerboard metasurface for ultra-wideband radar cross section reduction. Sci Rep 7(1):1–2
5. Yang F, Rahmat-Samii Y (2003) Microstrip antennas integrated with electromagnetic band-gap (EBG) structures: a low mutual coupling design for array applications. IEEE Trans Antennas Propag 51(10):2936–2946
6. Paquay M, Iriarte JC, Ederra I, Gonzalo R, de Maagt P (2007) Thin AMC structure for radar cross-section reduction. IEEE Trans Antennas Propag 55(12):3630–3638
7. Wang Y, Su J, Li Z, Guo Q, Song J (2020) A prismatic conformal metasurface for radar cross-sectional reduction. IEEE Antennas Wirel Propag Lett 19(4):631–635
8. Zhuang Y, Wang G, Liang J, Cai T, Tang XL, Guo T, Zhang Q (2017) Random combinatorial gradient metasurface for broadband, wide-angle and polarization-independent diffusion scattering. Sci Rep 7(1):1
9. Iriarte JC, Paquay M, Ederra I, Gonzalo R, de Maagt P (2007) Combination of AMC and PEC cells for RCS applications. In: 2007 IEEE antennas and propagation society international symposium 2007 Jun 9. IEEE, pp 865–868

10. Galarregui JC, Pereda AT, De Falcon JL, Ederra I, Gonzalo R, de Maagt P (2013) Broadband radar cross-section reduction using AMC technology. IEEE Trans Antennas Propag 61(12):6136–6143
11. de Cos ME, Alvarez-Lopez Y, Las Heras Andres F (2010) A novel approach for RCS reduction using a combination of artificial magnetic conductors. Prog Electromagnet Res 107:147–159
12. Modi AY, Balanis CA, Birtcher CR, Shaman HN (2017) Novel design of ultrabroadband radar cross section reduction surfaces using artificial magnetic conductors. IEEE Trans Antennas Propag 65(10):5406–5417
13. Sharma A, Gangwar D, Kanaujia BK, Dwari S (2018) RCS reduction and gain enhancement of SRR inspired circularly polarized slot antenna using metasurface. AEU-Int J Electron Commun 1(91):132–142
14. Zheng Y, Gao J, Cao X, Yuan Z, Yang H (2015) Wideband RCS reduction of a microstrip antenna using artificial magnetic conductor structures. IEEE Antennas Wirel Propag Lett 24(14):1582–1585
15. Su P, Zhao Y, Jia S, Shi W, Wang H (2016) An ultra-wideband and polarization-independent metasurface for RCS reduction. Sci Rep 11(6):20387
16. Zhao Y, Yu C, Gao J, Yao X, Liu T, Li W, Li S (2015) Broadband metamaterial surface for antenna RCS reduction and gain enhancement. IEEE Trans Antennas Propag
17. Zhao Y, Cao X, Gao J, Yao X, Liu X (2015) A low-RCS and high-gain slot antenna using broadband metasurface. IEEE Antennas Wirel Propag Lett 5(15):290–293
18. Liu Y, Li N, Jia Y, Zhang W, Zhou Z (2019) Low RCS and high-gain patch antenna based on a holographic metasurface. IEEE Antennas Wirel Propag Lett 18(3):492–496
19. Sharma A, Gangwar D, Kanaujia BK, Dwari S (2019) Analysis and design of an ultra-thin metamaterial absorber and its application for in-band RCS reduction of antenna. J Electromagn Waves Appl 33(5):654–667
20. Yin L, Yang P, Gan YY (2020) Microstrip array antenna in-band RCS reduction using polarization reconfigurable elements. J Electromagn Waves App 34(5):655–667
21. Tiwari P, Pathak SK, Anitha VP, Siju V, Sinha A (2020) X-band Γ-shaped anisotropic metasurface-based perfect cross-polarizer for RCS reduction. J Electromagn Waves Appl
22. Zhang C, Cao XY, Gao J, Li SJ, Zheng YJ (2017) Low RCS and broadband ME dipole antenna loading artificial magnetic conductor structures. Radio Eng 26(1):38–44

Chapter 29
A Novel Hybrid ASO-NM Algorithm and Its Application to Automobile Cruise Control System

Davut Izci and Serdar Ekinci

1 Introduction

Optimization problems can be observed in various fields such as engineering, science, economics, business and so on [1]. The diversity of the applications has led to the development of optimization algorithms. The latter case has been gaining incredible attention from researchers of different backgrounds due to its unique contribution to the stability and cost of the systems [2]. Conventional deterministic approaches are available to take care of such problems, however, they are not able to provide efficient solutions since they perform an exhaustive search within limited resources [3]. Therefore, alternative methods are required for efficient solutions with lower computational costs. Metaheuristic algorithms have attracted great attention, as non-deterministic techniques, because of their flexible/problem-independent structure and superior capability of local optima avoidance [4]. The nature of the problem is not a concern for metaheuristic algorithms as they treat the problem as a black box, thus, they are easily implemented to different fields including real-world engineering problems.

Despite many existing metaheuristic algorithms, there is still a continuous effort to develop new ones or enhance currently available approaches. No free-lunch theorem [5] provides a logical answer to this case. As stated by this theorem, an overall best performing optimization approach for all potential problems cannot be found such that some of them may be quite efficient for some issues but not effective at all or competitive for the rest. As part of the requirement stated by this theorem, the atom search optimization (ASO) algorithm [6] has been developed as one of the metaheuristic algorithms to deal with optimization problems. It is a physics-inspired

D. Izci (✉) · S. Ekinci
Batman University, Batman 72060, Turkey
e-mail: davut.izci@batman.edu.tr

G. Mathur et al. (eds.), *Proceedings of 2nd International Conference on Artificial Intelligence: Advances and Applications*, Algorithms for Intelligent Systems,
https://doi.org/10.1007/978-981-16-6332-1_29

and swarm-based intelligent optimization algorithm which mathematically mimics the basic molecular dynamics for atomic motion.

In order to retain interest for any metaheuristic algorithm, the trade-off balance between phases of exploration and exploitation poses a critical role [3]. However, as is the case for most of the metaheuristic algorithms, local optima stagnation and premature convergence are issues that ASO suffers from due to an imbalance between exploration and exploitation phases [7]. Enhancement of the exploitation capability of ASO can help to deal with the balance issue. To do so, Nelder-Mead (NM) algorithm can be adopted. The latter method was proposed to deal with the minimization problems and proved to be a good approach for local search [8].

Considering the abovementioned perspective, this paper attempts to construct a new hybrid algorithm that has a balanced exploration and exploitation that can be used as an effective tool for optimization problems. Therefore, a novel hybrid ASO-NM algorithm has been developed to achieve the objective set by above discussion. In the constructed algorithm, ASO performs the exploration stage whereas NM is used to achieve good exploitation.

Four well-known classical benchmark functions were used to observe the statistical performance of ASO-NM. The results have shown its better capability. In addition, a proportional-integral-derivative (PID) controller-based automobile cruise control system (ACCS) was also considered as a real-life engineering problem to assess the performance. Statistical tests and transient response analysis were performed in addition to comparisons with other existing approaches. The assessments demonstrated the highly competitive capability of the proposed hybrid approach.

2 Atom Search Optimization

The following steps provide a brief mathematical explanation of this algorithm. The interaction force acting on atom i due to atom j at time t is expressed as

$$F'_{ij}(t) = -\eta(t)\left[2\left(h_{ij}(t)\right)^{13} - \left(h_{ij}(t)\right)^{7}\right] \tag{1}$$

$\eta(t)$ is used for the arrangement of the repulsion or attraction and defined as follows: α represents the weight and T is the maximum number of iterations:

$$\eta(t) = \alpha\left(1 - \frac{t-1}{T}\right)^{3} e^{-\frac{20t}{T}} \tag{2}$$

$h_{ij}(t)$ is given as follows: r is the distance between atoms, and $h_{\min}$ and $h_{\max}$ are the lower and upper limits:

$$h_{ij}(t) = \begin{cases} h_{\min}, \frac{r_{ij}(t)}{\sigma(t)} < h_{\min} \\ \frac{r_{ij}(t)}{\sigma(t)}, h_{\min} \leq \frac{r_{ij}(t)}{\sigma(t)} \leq h_{\max} \\ h_{\max}, \frac{r_{ij}(t)}{\sigma(t)} > h_{\max} \end{cases} \tag{3}$$

$\sigma(t)$ denotes the length scale. The total force acting on atom i is given as follows:

$$F_i^d(t) = \sum_{j \in K\text{best}} \text{rand}_j F_{ij}^d(t) \tag{4}$$

where rand_j is a random number in [0, 1]. The constraint force is acquired as

$$G_i^d(t) = \lambda(t)\left(x_{\text{best}}^d(t) - x_i^d(t)\right) \tag{5}$$

where $x_{\text{best}}(t)$ is the best atom position and $\lambda(t)$ is the Lagrangian multiplier. The acceleration of atom i at time t can be written as follows:

$$a_i^d(t) = \frac{F_i^d(t)}{m_i^d(t)} + \frac{G_i^d(t)}{m_i^d(t)} \tag{6}$$

where $m_i(t)$ is the mass of atom i at time t and can be computed as follows:

$$m_i(t) = \frac{M_i(t)}{\sum_{j=1}^{N} M_j(t)} \tag{7}$$

$$M_i(t) = e^{-\frac{Fit_i(t) - Fit_{\text{best}}(t)}{Fit_{\text{worst}}(t) - Fit_{\text{best}}(t)}} \tag{8}$$

where $\text{Fit}_{\text{best}}(t)$, $\text{Fit}_{\text{worst}}(t)$ and $\text{Fit}_i(t)$ are minimum, maximum and function fitness values of atom i at iteration t. The velocity and the position of atom i can be expressed as follows (for iteration $(t + 1)$):

$$v_i^d(t+1) = \text{rand}_i^d v_i^d(t) + a_i^d(t) \tag{9}$$

$$x_i^d(t+1) = x_i^d(t) + v_i^d(t+1) \tag{10}$$

3 Nelder-Mead Method

This algorithm can be explained briefly as follows. An optimal vertex (x_1) is identified by generating vertices of x_1, x_2, ... x_{n+1} and evaluating the respective fitness function values (fit(x_1), fit(x_2), ..., fit(x_{n+1})) which are sorted in ascending order. Four different operations (reflection, expansion, contraction and shrinkage) are used to replace the worst vertex (x_{n+1}). The reflection point (x_r) is identified as follows where x_{cnt} and ρ are the centroid (excluding the x_{n+1} vertex) and the reflection coefficient, respectively:

$$x_r = x_{cnt} + \rho(x_{cnt} - x_{n+1}) \tag{11}$$

The reflection point is then expanded across the search space as follows:

$$x_e = x_{cnt} + \gamma(x_r - x_{cnt}) \tag{12}$$

where x_e is the expansion point and γ is the expansion coefficient. The following equations are used for contraction (x_c) and shrinkage (x_s), respectively:

$$x_c = x_{cnt} + \beta(x_{n+1} - x_{cnt}) \tag{13}$$

$$x_s = x_1 + \delta(x_i - x_1),\ i = 2, 3, \ldots, n + 1 \tag{14}$$

In here, β is the contraction coefficient whereas δ is the shrinkage.

4 Proposed Hybrid ASO-NM Algorithm

In the proposed ASO-NM algorithm, the aim is to provide a well-balanced structure between stages of exploration and exploitation.

To achieve such an objective, the NM algorithm was run for the total number of iterations after each iteration of the ASO algorithm. In this way, a good balance between exploration and exploitation phases was provided. The parameters of the proposed hybrid ASO-NM algorithm were chosen to be 50 for depth weigh (α), 0.20 for multiplier weigh (β), 1 for reflection factor (ρ), 2 for expansion factor (γ), 0.5 for contraction factor (β) and finally 0.5 for shrinkage factor (δ). Figure 1 illustrates the flowchart of the proposed ASO-NM algorithm. As can be seen, it starts with a definition of the above-listed parameters and then ASO is performed for global search. The obtained best solution is then used as an initial starting point by NM which is run for local search. This cycle continues until the termination condition is met.

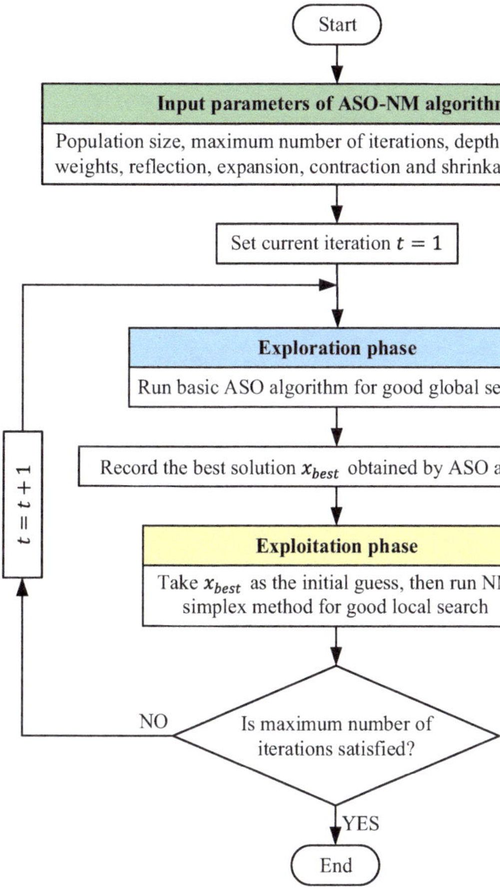

Fig. 1 Flowchart for the proposed ASO-NM algorithm

5 Experimental Results on Benchmark Functions

The performance of ASO-NM was tested using benchmark functions listed in Table 1. In this study, performance comparison with the original version of ASO was evaluated to be sufficient as it has already been compared with simulated annealing algorithm, genetic algorithm, gravitational search algorithm, particle swarm optimization and wind-driven optimization using 37 benchmark functions [6].

The comparisons have already demonstrated the highly competitive performance of ASO. Therefore, Table 2 presents the comparative results of the test functions achieved by ASO and ASO-NM only by running both algorithms for 30 times and adopting a population size of 50 along with a maximum iteration number of 1000. As demonstrated in the table, ASO-NM provides far better results which indicates its highly competitive performance of finding the best optimal solutions.

Table 1 Details of benchmark functions used in experiments

Name	Function	d	$[X_{\min}, X_{max}]$	$F_{\min}$
Sphere	$F_1(x) = \sum_{i=1}^{d} x_i^2$	30	$[-100, 100]$	0
Rosenbrock	$F_2(x) = \sum_{i=1}^{d-1}\left(100(x_{i+1} - x_i^2)^2 + (x_i - 1)^2\right)$	30	$[-30, 30]$	0
Schwefel	$F_3(x) = -\sum_{i=1}^{d}\left(x_i \sin\left(\sqrt{\lvert x_i \rvert}\right)\right)$	30	$[-500, 500]$	$-418.9829 \times d$
Ackley	$F_4(x) = -20\exp\left(-0.2\sqrt{\frac{1}{d}\sum_{i=1}^{d} x_i^2}\right) - \exp\left(\frac{1}{d}\sum_{i=1}^{d}\cos(2\pi x_i)\right) + 20 + e$	30	$[-32, 32]$	0

Table 2 Statistical results of relevant benchmark functions for ASO and ASO-NM algorithms

Function	Statistical metric	ASO	ASO-NM (proposed)
$F_1(x)$	Mean	2.7691E − 21	7.6923E − 240
	Standard deviation	3.5505E − 21	0
	Best (minimum)	6.4831E − 22	3.7974E − 252
$F_2(x)$	Mean	24.9762	6.3526
	Standard deviation	0.3787	2.2386
	Best (minimum)	23.6763	1.2485
$F_3(x)$	Mean	−7.1917E + 03	−8.3021E + 03
	Standard deviation	558.1582	458.6754
	Best (minimum)	−7.9695E + 03	−9.2981E + 03
$F_4(x)$	Mean	2.7911E − 11	1.3509E − 12
	Standard deviation	1.0495E − 11	3.1857E − 13
	Best (minimum)	1.1700E − 11	8.4643E − 13

6 ASO-NM for PID-Based Automobile Cruise Control System

6.1 Modeling of Automobile Cruise Control System

Considering the wind gust speed (V_w) and road inclination (θ) to be zero; the linearized model that provides the relationship between output velocity (ΔV) and control input (ΔU) can be obtained as follows [9]:

$$\frac{\Delta V(s)}{\Delta U(s)} = \frac{\frac{C_1 e^{-\tau s}}{MT}}{(s + \frac{2C_a v}{M})(s + \frac{1}{T})} \tag{15}$$

where C_1 and C_a denote the actuator constant and aerodynamic drag coefficient, respectively, whereas τ and T are the driver's reaction and observation times, respectively. The vehicle mass is represented by M. The block diagram of an ACCS and the model parameters used for this study can be found in [9].

6.2 Proposed PID Tuning Method Using ASO-NM Algorithm

The following provides the transfer function of a PID controller where proportional, integral and derivative gains are represented by K_p, K_i and K_d, respectively [10]

$$G_{PID}(s) = K_p + \frac{K_i}{s} + K_d s \tag{16}$$

The upper and lower limits for PID controller parameters were chosen to be $3 < K_p < 4$, $0.10 < K_i < 0.25$ and $3 < K_d < 4$ [9]. To achieve a system with optimum performance, the integral of squared error (ISE) [11] was employed as an objective function for this study. The stated objective function is given as follows:

$$J_{ISE} = \int_0^T e^2(t) \cdot dt \tag{17}$$

where $e(t)$ corresponds to the difference between reference speed and the actual speed ($v_{\text{ref}}(t) - v(t)$). The integration time (T) was taken to be $10s$ for this study. Figure 2 illustrates the implementation of the proposed hybrid ASO-NM algorithm to PID controlled automobile cruise control system.

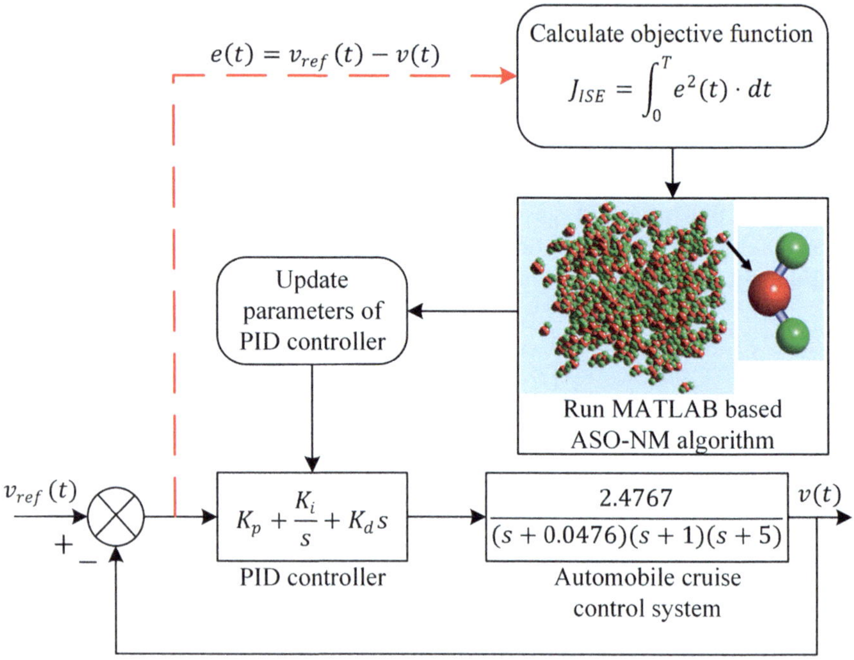

Fig. 2 Block diagram of ASO-NM implementation to optimize the performance of automobile cruise control system

6.3 *Statistical Test Analysis*

The statistical values of mean, standard deviation and best (minimum) using the ISE objective function were obtained for the ACCS system via 30 runs using a population size of 50 and a maximum iteration number of 100 for both ASO and ASO-NM algorithms. The respective values are provided in Table 3. As can be seen, the best values are achieved by the proposed ASO-NM algorithm which demonstrates the better performance of the proposed hybrid approach for ACCS.

Table 3 Statistical results of ISE objective function for ASO and ASO-NM algorithms

Statistical metric	ASO	ASO-NM (proposed)
Mean	0.4035	0.3623
Standard deviation	0.0146	0.0056
Best (minimum)	0.3872	0.3556

6.4 Transient Response Analysis

The best PID controller parameters obtained with the ASO algorithm are $K_p = 3.6629$, $K_i = 0.1520$ and $K_d = 3.5233$. The following closed-loop transfer function is obtained for ASO-based PID-controlled ACCS by substituting the respective parameters:

$$T_{ASO-PID}(s) = \frac{v(s)}{v_{\text{ref}}(s)} = \frac{8.726s^2 + 9.072s + 0.3765}{s^4 + 6.048s^3 + 14.01s^2 + 9.31s + 0.3765} \tag{18}$$

Likewise, the best PID controller parameters obtained with the proposed hybrid ASO-NM algorithm are $K_p = 3.9749$, $K_i = 0.1774$ and $K_d = 3.9956$. The following closed-loop transfer function is obtained for ASO-NM-based PID-controlled ACCS by substituting the respective parameters:

$$T_{ASO-NM-PID}(s) = \frac{v(s)}{v_{\text{ref}}(s)} = \frac{9.896s^2 + 9.845s + 0.4394}{s^4 + 6.048s^3 + 15.18s^2 + 10.08s + 0.4394} \tag{19}$$

The provided transfer functions were used to observe the transient stability of the controlled system. The comparative step responses regarding the velocity change of the ACCS controlled by ASO-based PID and ASO-NM-based PID are provided in Fig. 3. The figure clearly indicates the better time response profile of the proposed ASO-NM-based PID controller, therefore, it is superior to the original ASO-based PID controller for ACCS.

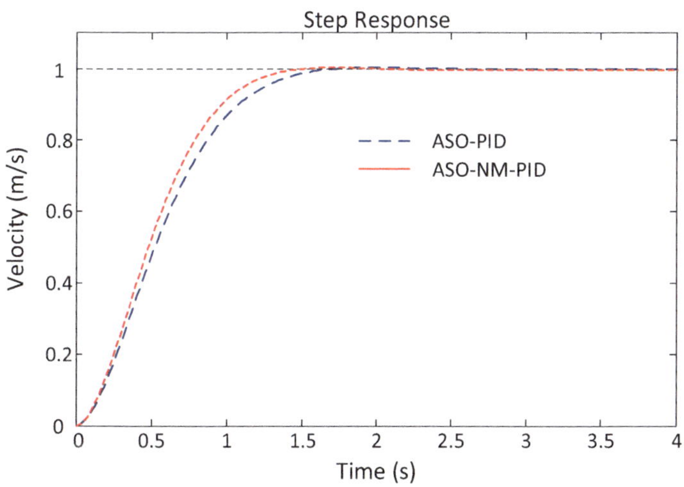

Fig. 3 Velocity step response of the cruise control system

Table 4 Comparison of transient response results with various tuning methods

Control methods	$M_p(\%)$	$T_r(s)$	$T_s(s)$	$T_p(s)$	$E_{ss}(\%)$
PID [12]	10.2	1.7	5.5	3.54	0.01
State space [12]	10	1.38	5	2.97	0.01
Fuzzy logic [12]	1.91	2.21	3.37	3.16	0.01
GA-PID [13]	1.1456	0.9394	1.4559	2.1166	0
ALO-BODE-PID [9]	0.4793	0.8336	1.3093	1.8296	0
ASO-PID (proposed)	0.3678	0.9042	1.4306	2.0034	0
ASO-NM-PID (proposed)	**0.3541**	**0.8052**	**1.2622**	**1.7028**	**0**

6.5 Comparison with the Existing Approaches

To demonstrate the superiority of the proposed hybrid ASO-NM algorithm for the optimization problem of ACCS, other existing approaches (PID, fuzzy logic, state space, genetic algorithm-based PID (GA-PID) and antlion optimizer based on Bode ideal model for PID (ALO-BODE-PID) along with the original ASO-based PID (ASO-PID)) listed in Table 4 were used. In the respective table, corresponding transient response parameters of maximum overshoot (M_P), rise time (T_r), settling time (T_s) and peak time (T_p), along with steady-state error (E_{ss}) of the terminal velocity change of ACCS were used which showed the best performance characteristic (bold values) of the proposed approach.

7 Conclusion

A novel hybrid ASO-NM algorithm was developed which combines both the original ASO algorithm and the NM method. The amalgam of those two approaches has led to the development of a hybrid algorithm that has a good balance between exploration and exploitation stages. The evaluation started by testing the proposed approach against four well-known benchmark functions. The obtained statistical results on those test functions were compared with only the ASO algorithm as the latter algorithm has already been demonstrated to be highly competitive with respect to other competitive algorithms. The best values have been achieved via ASO-NM. Further assessment of the ASO-NM was performed by designing a PID-controlled ACCS as a real-world engineering optimization problem. Statistical tests and transient response analysis have shown the better capability of the proposed ASO-NM for such a real-life problem. Moreover, a comparative analysis was carried out using other existing controller structures including the original ASO-based PID. The comparisons have also shown the better capability of the proposed approach. To sum up, the assessments have proved the highly competitive structure of the ASO-NM for optimization problems.

References

1. Izci D, Ekinci S, Orenc S, Demiroren A (2020) Improved artificial electric field algorithm using nelder-mead simplex method for optimization problems. In: 2020 4th international symposium on multidisciplinary studies and innovative technologies (ISMSIT). IEEE, pp 1–5
2. Hashim FA, Houssein EH, Mabrouk MS et al (2019) Henry gas solubility optimization: a novel physics-based algorithm. Futur Gener Comput Syst 101:646–667. https://doi.org/10.1016/j.future.2019.07.015
3. Hashim FA, Hussain K, Houssein EH et al (2020) Archimedes optimization algorithm: a new metaheuristic algorithm for solving optimization problems. Appl Intell. https://doi.org/10.1007/s10489-020-01893-z
4. Zhao W, Wang L, Zhang Z (2019) Artificial ecosystem-based optimization: a novel nature-inspired meta-heuristic algorithm. Neural Comput Appl 1–43
5. Wolpert DH, Macready WG (1997) No free lunch theorems for optimization. IEEE Trans Evol Comput 1:67–82. https://doi.org/10.1109/4235.585893
6. Zhao W, Wang L, Zhang Z (2019) Atom search optimization and its application to solve a hydrogeologic parameter estimation problem. Knowl-Based Syst 163:283–304. https://doi.org/10.1016/j.knosys.2018.08.030
7. Eker E, Kayri M, Ekinci S, Izci D (2021) A new fusion of ASO with SA algorithm and its applications to MLP training and DC motor speed control. Arab J Sci Eng. https://doi.org/10.1007/s13369-020-05228-5
8. Nelder JA, Mead R (1965) A simplex method for function minimization. Comput J 7:308–313. https://doi.org/10.1093/comjnl/7.4.308
9. Pradhan R, Majhi SK, Pradhan JK, Pati BB (2018) Antlion optimizer tuned PID controller based on Bode ideal transfer function for automobile cruise control system. J Ind Inf Integr 9:45–52. https://doi.org/10.1016/j.jii.2018.01.002
10. Izci D, Ekinci S (2021) Comparative performance analysis of slime mould algorithm for efficient design of proportional–integral–derivative controller. Electrica 21:151–159. https://doi.org/10.5152/electrica.2021.20077
11. Dorf RC (2011) Modern control systems, 12th edn. Pearson, Boston, London
12. Osman K, Rahmat MF, Ahmad MA (2009) Modelling and controller design for a cruise control system. In: 2009 5th international colloquium on signal processing & its applications, pp 254–258
13. Rout MK, Sain D, Swain SK, Mishra SK (2016) PID controller design for cruise control system using genetic algorithm. In: 2016 international conference on electrical, electronics, and optimization techniques (ICEEOT). pp 4170–4174

Chapter 30
Radar Target Recognition and Classification Using Supervised Machine Learning Approaches

Jagan Mohana Rao Pathina and P. Rajesh Kumar

1 Introduction

Classification of targets is one of the challenging tasks in radar signal processing. In military radars, there is a great need to classify the target once detection and tracking target is done. Radars use short-duration pulses and result in a wideband in the frequency domain, and this acts as an advantage for target identification. The target identification is done based on the scattered signals received by radar from the target [1]. It plays an important role in obtaining the radar cross-section (RCS) parameters of a target. The target recognition is done by feature extraction and classification techniques [2].

Many algorithms are developed like Root-MUltiple SIgnal Classification (Root-MUSIC) and Matrix Pencil Method (MPM) to distinguish the targets [3]. MUSIC uses the input covariance matrix and returns the estimated discrete frequency spectrum along with corresponding estimates of signal power.

MPM uses an input pencil matrix rather than covariance matrix and maps noise components to null space thereby distinguishing the signal components. Compared to the MUSIC algorithm, MPM provides accurate results. In order to avoid the effects of false alarm of signal components due to noise, a multi-scale analysis technique, namely MODWPT, is used. MODWPT divides the signal into different frequency bands of equal length and is a time-invariant method and doesn't require any down-sampling like in Discrete Wavelet Transform (DWT).

The performance of the proposed method is applied to various developed models. The paper is organized as follows: Sect. 2 covers the data generation including RCS of different targets and their properties. The proposed feature extraction MODWPT

J. M. R. Pathina
ECE Department, AUCE(A), Visakhapatnam, India

P. R. Kumar (✉)
Professor and Head of the ECE Department, AUCE(A), Visakhapatnam, India

G. Mathur et al. (eds.), *Proceedings of 2nd International Conference on Artificial Intelligence: Advances and Applications*, Algorithms for Intelligent Systems,
https://doi.org/10.1007/978-981-16-6332-1_30

is described in Sect. 3 along with the different Entropy-based feature sets. Section 4 explains the different classifiers we used and the results obtained from the simulation are presented in Sect. 5 and conclusions in Sect. 6.

2 Data Generation

It is a well-known fact that data plays a very important role in the field of machine learning and artificial intelligence. In order to realize machine learning models for automatic RADAR target recognition, we require huge sets of I/Q data samples corresponding to RADAR echo returns from the structures of the chosen target shapes.

Obtaining real-time data for conducting the experimental study in this research domain is cumbersome. It requires a well-established RADAR setup, accompanied by required targets. Moreover, a real-time testbed incurs a lot of expenditure. As a result, in order to overcome the above-mentioned limitations, we propose a novel mechanism to simulate the task of RADAR target recognition by developing the desired recognition models with the help of synthesized I/Q samples.

Synthesized data is defined as the data being created in the software domain using wide varieties of mathematical models which replicate the conditions of real-time scenarios. Initially, the RADAR operating parameters like the operating frequency, set of elevation and azimuthal scan angles, etc., are chosen. Secondly, the desired simple geometrical shape is considered, and its geometrical parameters (radius, height, etc.,) are fixed.

Later, these parameters are passed to the theoretical mathematical equations corresponding to the RCS patterns of the chosen geometrical shape to compute the theoretical RCS pattern. The obtained RCS pattern is then applied to a backscatter RADAR target model to simulate a finite number of echo returns from different aspect angles, which changes from sample to sample. Radar Cross-Section is the measure of the ability of the target to backscatter radar's electromagnetic signals in the radar receiver direction.

$$\sigma is \frac{\text{Power reradiated toward source per unit solid angle}}{\frac{\text{incident power density}}{4\pi}} \mathrm{m}^2 \tag{1}$$

Classification of the target requires the measurement of the aspect angle along with the target properties (like geometry, material type, and azimuth and elevation angles) which includes RCS. For simple targets, RCS depends on the physical size and wavelength of the signal and for complex targets, it depends on aspect angle, frequency, and polarization [4]. To test the effectiveness of the proposed method, four metallic canonical targets of geometry with circular cylinder, sphere, frustum, and circular disk are analyzed.

2.1 Sphere

A sphere is one of the difficult targets for recognition due to its complex exterior and interior resonating modes. Target set data is taken by varying the radii of the sphere, its dielectric properties, and its orientation. Due to the symmetry of the sphere, the scattered waves are co-polarized with the incident waves (Table 1).

RCS of a circular cylinder, frustum, and circular flat plate are mentioned [5] in Table 2 and their corresponding shapes in Fig. 1.

Here the cone angle, $\tan\alpha = \frac{r_2 - r_1}{H}$

RCS of complex targets is computed by combining the RCS of simple targets coherently. The change in aspect angle produces rapid RCS fluctuations. To capture these RCS fluctuations in mathematical models, Swerling RCS models are proposed [6]. For each geometrical shape, 10 different structures with varied dimensions are considered, and their corresponding RCS patterns are synthesized. This results in obtaining 40 (4×10) RCS patterns for 4 shapes. For each geometrical structure/object/target, 250 motion profiles are simulated.

Table 1 RCS of sphere in different regions

Region	RCS of sphere
Optical	$\sigma = \pi r^2 r \gg \lambda$
Rayleigh	$\sigma = 9\pi r^2 (kr)^4 r \ll \lambda$
Mie	$\frac{\sigma}{\pi r^2} = (\frac{j}{kr}) \sum_{n=1}^{\infty} (-1)^n (2n+1)[\left(\frac{kr J_{n-1}(kr) - n J_n(kr)}{kr H_{n-1}^{(1)}(kr) - n H_n^{(1)}(kr)}\right) - (\frac{J_n(kr)}{H_n^{(1)}(kr)})]$

where r is the radius of the sphere, $k = \frac{2\pi}{\lambda}$, λ is the wavelength, and σ is the radar cross-section.
J_n is the spherical Bessel of the first kind of order n,
Y_n is the spherical Bessel of the second kind of order n,
H_n is the Hankel function of order n,
and $H_n^{(1)}$ is the Hankel function of order n, $H_n^{(1)} = J_n(kr) + jY_n(kr)$

Table 2 RCS for circular cylinder, frustum, and disk

Shape	RCS for normal backscattered incidence	RCS for non-normal backscattered incidence
Circular cylinder	$\sigma_{\theta_n} = \frac{2\pi H^2 r}{\lambda}$	$\sigma = \frac{\lambda r \sin\theta}{8\pi (\cos\theta)^2}$
Frustum	$\sigma_{\theta_n} = \frac{8\pi \left(z_2{}^{\frac{3}{2}} - z_1{}^{\frac{3}{2}}\right)^2}{9\lambda}$. $\frac{\sin\alpha}{(\cos\alpha)^4}$	$\sigma = \frac{\lambda z \tan\alpha}{8\pi \sin\theta} (\tan(\theta - \alpha))^2$
Disk	$\sigma_{\theta_n} = \frac{4\pi^3 r^4}{\lambda^2} (\theta = 0°)$	$\sigma = \frac{\lambda r}{8\pi \sin\theta (\tan\theta)^2}$

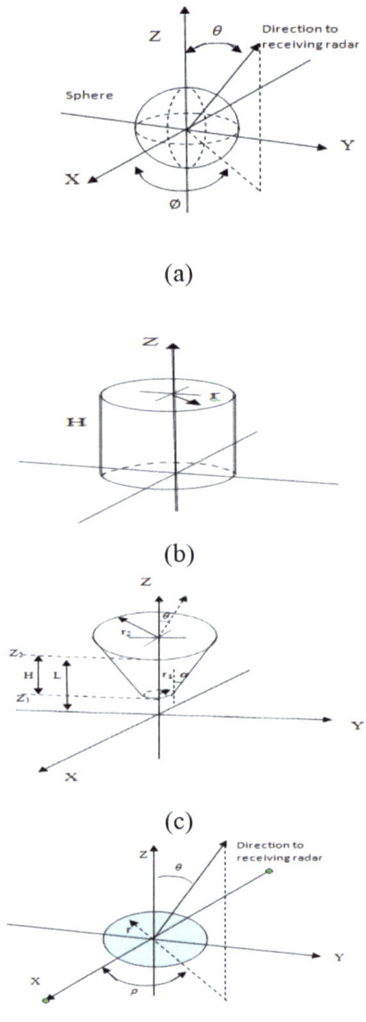

Fig. 1 **a** Sphere, **b** Circular cylinder, **c** Frustum, and **d** Circular flat plate (disk)

In other words, for each geometrical shape, 2500 (250×10) motion profiles are simulated. In total, 20 scans are performed and the corresponding echoes are collected. In each scan, 50 different aspect angles are considered. This results in obtaining 50 echoes per scan and 1000 (20×50) per target.

Therefore, 1 motion profile of a target contains 1000 echo returns. In total, 10,000 (2500×4) motion profiles are obtained for all 40 targets corresponding to 4 geometrical shapes. The above-stated process is initially carried for 4 simple geometrical shapes which are regarded as simple point targets. Later, the same procedure is followed for the case of complex targets. For the targets of complex shapes, reflections are not considered the same across all directions.

The RCS varies with the incident angles (which are also known as aspect angles). RCS patterns which are dependent on Aspect angle can be measured or modeled similar to antenna radiation patterns. The result of such measurements is a table of RCS values as a function of azimuth and elevation angles in the target local coordinate system. For more complicated shapes and materials, computational electromagnetic approaches such as finite element analysis (FEM) or method of moments (MoM) can be used to accurately predict the RCS patterns.

The motion profiles with return echoes are considered to be raw features. Using these raw features directly to realize machine learning models does not help in obtaining deterministic models. This is due to the fact that the return echo patterns (which are basically in time domain) look like a randomly fluctuating signal with respect to time, and it is very hard to infer any crucial information. As a result, the obtained motion profiles are subjected to some frequency (wavelet) domain-based analysis to derive significant information which could help in realizing deterministic classification models. The frequency (wavelet) domain-based analysis considers the use of a novel feature extraction technique, namely maximal overlap discrete wavelet packet transform (MODWPT).

3 Wavelet Packet Transform

Wavelet Packet Transform is a linear combination of wavelets in which level-by-level decomposition of the signal is done [7]. It indicates temporary and permanent features of the signal with desired time–frequency separation.

The wavelet packet $\Psi^i_{j,k}(t)$ contains three indices with integers i, j, and k, namely modulation, scale, and translation parameters, respectively. It inherits the properties of time–frequency localization and orthonormality from corresponding wavelets.

$$\Psi^i_{j,k}(t) = 2^{\frac{j}{2}} \Psi^j \left(2^j t - k\right) \tag{2}$$

The wavelet function is derived from the following rebound functions:

$$\Psi^{2j}(t) = \sqrt{2} \sum_{k=-\infty}^{\infty} h(k) \Psi^j (2t - k) \tag{3}$$

$$\Psi^{2j+1}(t) = \sqrt{2} \sum_{k=-\infty}^{\infty} g(k) \Psi^j (2t - k) \tag{4}$$

$\Psi^1 = \Psi(t)$ is the first wavelet function called the mother function, where $h(k)$ and $g(k)$ are known as quadrature mirror filters. Usually, Shannon entropy is used to select the mother wavelet. These filters have the following properties:

$$\sum_{n=-\infty}^{\infty} g(n) = \sqrt{2}, \sum_{n=-\infty}^{\infty} g^2(n) = 1, \sum_{n=-\infty}^{\infty} g(n)h(n) = 0 \quad (5)$$

$$\sum_{n=-\infty}^{\infty} h(n) = \sqrt{2}, \sum_{n=-\infty}^{\infty} h^2(n) = 1, \sum_{n=-\infty}^{\infty} h(n)g(n) = 0 \quad (6)$$

The discrete wavelet transforms (DWT) down-samples signals at each level (by a factor of 2) so alternate sampling is computed. The quadrature mirror filter splits the signal into two sub-bands. The approximate and detailed coefficients are obtained by the low pass and high pass filters, respectively.

3.1 MODWPT

Wave packets are used to induce a finite frequency partition. MODWPT is a non-decimated version of DWT where the numbers of scaling and wavelet coefficients are equal and is equal to the number of observation sets [8]. When these coefficients are subjected to low pass and high pass filters, the next level scaling and wavelet coefficients are computed.

In MODWPT, the partition of frequency bands is equal and depends on a number of levels. For the *j*th level, it does 2^j equal frequency band partitions. The *n*th node in the *j*th level is associated with the frequency $\left[\frac{f_s n}{2^{j+1}}, \frac{f_s(n+1)}{2^{j+1}}\right]$. $W_{j,n}$ denotes the wavelet coefficients associated with the (*j*, *n*) nodes where *N* is the length of the signal. In synthesis, the coefficients are convolved to reverse low pass and high pass filters for original signal reconstruction.

The signal decomposition using MODWPT is given by

$$s_j^{2z}(k) = \frac{1}{\sqrt{2}} \sum_{n=-\infty}^{\infty} g(n) s_{j-1}^{z}(k-n) \quad (7)$$

$$s_j^{2z+1}(k) = \frac{1}{\sqrt{2}} \sum_{n=-\infty}^{\infty} h(n) s_{j-1}^{z}(k-n) \quad (8)$$

The reconstructed coefficients at any jth level in MODWPT is given by

$$a_j^{2z}(k) = \frac{1}{\sqrt{2}} \sum_{n=-\infty}^{\infty} \tilde{g}(n) s_{j-1}^{2z}(k-n) \quad (9)$$

$$a_j^{2z+1}(k) = \frac{1}{\sqrt{2}} \sum_{n=-\infty}^{\infty} \tilde{h}(n) s_{j-1}^{2z+1}(k-n) \quad (10)$$

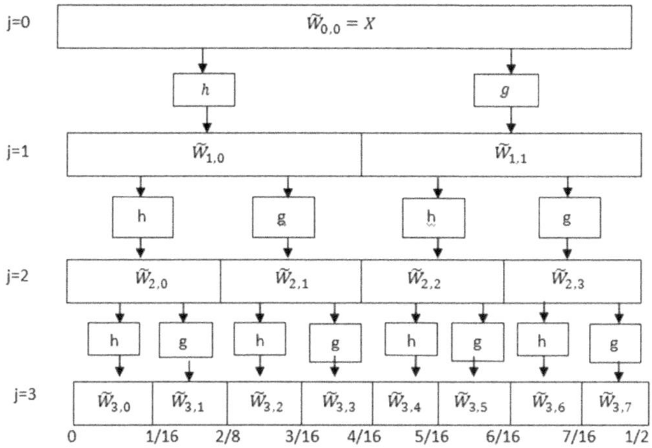

Fig. 2 Decomposition of signal into MODWPT coefficients $\tilde{W}_{j,n}$ with levels $j = 1, 2, 3$ and frequency index n ranging from 0 to $2^j - 1$

The MODWPT at level J partitions the signal energy into 2^J equal-width sub-bands and does this in a way that the total signal energy is preserved. Preserving the signal energy is an important property that is very difficult to achieve with the conventional band-pass filtering. This property of preserving signal energy significantly improves the ability of a classifier to accurately distinguish the classes having frequency separation (Fig. 2).

Using wavelet transform, the extracted features consist of either 8, 16, or 32 predictors per target return. Compared to the original time-domain signal of 1000 points, it is a significant reduction in the data. The number of levels for the wavelet transform can be tuned to improve the performance of the classification algorithms.

3.2 *Entropy-Based Features*

Feature selection plays a crucial role in wavelet coefficient extraction and also in the reconstruction of original signal. Entropy is the measure of uncertainty of random events or amount of information or a system disorder.

Entropy-based features are acquired from the wavelet coefficients of the MODWPT [9]. The energy of individual segments at each resolution level is obtained with the wavelet coefficients. Relative energy is used to extract features accurately and these can be used as an input to the classifiers.

Different entropy-based features, namely Log energy, Sure entropy, Kozachenko-Leonenko entropy, Renyi entropy, and Threshold entropy, are given below [10].

Log energy entropy

$$E = \sum_{n} \log\left(w_{j,k}^{n2}\right) = \sum_{n} \log\left(x_j^2\right) \tag{11}$$

where x_j is the jth sample of signal with length n.

Sure entropy

$$E = N - \#\{j \text{suchthat} \left|x_j\right| \leq \epsilon\} + \sum_{j} \min\left(x_j^2, \epsilon^2\right) \tag{12}$$

ϵ is the positive threshold value.

Renyi entropy. First order (q = 1)

$$S_q = \frac{1}{q-1} \log_2\left(\sum_{j=1}^{N} P_j^q\right) \tag{13}$$

q is a real number and p_i is the probability of random variables $x_1, x_2, \ldots, x_N$.

Kozachenko-Leonenko entropy. It is based on the calculation of the entropy of n_s samples with an interval of l2 between a sample and its nearest neighbor.

Threshold entropy

$$E_T(s_i) = 1 \text{iflsil} > \text{p} \tag{14}$$

where s_i is signal function and p is threshold value.

Novel Extracted non-linear entropy-based feature. It uses MODWPT coefficients. If X is the wavelet coefficient with length $1 \times N$, then Feature F_x = entropy (abs($\frac{x-x^q}{q-1}$)) where q is given by the cost function based on mutual information which is optimized by a grid.

Norm Energy entropy

$$E = \sum_{n} \left|w_{j,k}^{n}\right|^{p} \tag{15}$$

Relative Wavelet Energy Entropy

To extract useful information from the signal, Wavelet energy is used. The relative energy concept is used to get information about the relative energy related to different frequency bands [11]. It extracts features from each level and detects the degree of resemblance between segments of a signal.

$$RWE = \frac{E_j}{E_{\text{total}}} \quad (16)$$

The energy of the last decomposition level, individual wavelet energy, and relative energy of each packet are calculated. Based on the degree of similarity between the segments of the signal, original feature extraction is reduced. Thus, it forms an accurate and compact feature vector which can be used as an input to the classifier.

4 Supervised Learning Methods

Machine Learning is computational statistics or predictive analytics performed automatically by a machine without human intervention. There is a wide range of applications of machine learning approaches. Of these approaches, supervised learning is a category which is subdivided into classification and regression. In supervised learning, the machine is fed with some data called *trained data*. The trained data is used by the machine to classify or predict the test data. Both SVM and k-NN techniques come under classification and ANN under regression techniques.

The classification tasks are implemented by using Statistics and Machine Learning toolbox for SVM and k-NN and Neural Network Toolbox for ANN.

4.1 Support Vector Machine

SVM is a discriminating classifier which maps input vectors to a higher dimensional space. A maximal separating hyper-plane is constructed in the higher dimensional space. In a two-dimensional view, this hyper-plane acts as a line separating data on each side [12]. It divides data like two parallel hyper-planes separating data, and the distance between them is called margin. SVM maximizes margin and increases the distance of separation between hyper-planes. Thus, it creates examples of the training set on either side of it.

For a training set of (x_i, c_i), with x features, c class labels, and $i = 1, 2, \ldots, K$, SVM is constructed from the following mathematical optimization procedure:

minimize $[\frac{1}{2}\left(w^T w\right) + C\sum_{j=1}^{k} \in_j]$

subject to $C_j\left(w^T \varnothing(x_i) + b\right) \geq 1 - \in_j, \in_j \geq 0$

where ω is the decision plane orientation vector,

b is bias,

ε_i is marginal slack variable,

$\varnothing$ is mapping function, and

C is the penalty parameter of the error term.

For unseen test samples, the expected classification error is minimized by finding an optimum linear hyper-plane by SVM.

4.2 k-Nearest Neighbor

k-NN classifier is an intuitive and simple method for the classification of signals [13]. It compares the test data with the available training data and gives results accordingly. k-NN finds k which is the nearest neighborhood in training data from the given values of the data set. Then a class is assigned which appears frequently in its neighborhood. The models are developed by setting the value of k to 5 and 7. The Euclidean distance between training set data $x_1 = \left(x_{1,1}, \ldots..x_{1,k}\right)$ and test data $x_2 = \left(x_{2,1}, \ldots..x_{2,k}\right)$ at any instant is given by

$$d(x_1, x_2) = \sqrt{\sum_{j=1}^{k} (x_{1,j} - x_{2,j})^2} \quad (17)$$

4.3 Artificial Neural Network

In ANN, multi-layer perceptron is a regression method, consisting of the input layer, hidden layer, and an output layer [14]. The number of features used in the input layer represents the number of features used while the output layer nodes are equal to the number of classes to be classified. The number of hidden layers and nodes in the hidden layer depends on the needs. In this paper, we used a single layer.

5 Results

The experimental study is carried out in three different phases, namely Data generation, Feature extraction, and Classifier models. The machine learning algorithms often work on extracted features and make it easier for the classification algorithm to discriminate between returns from different targets/shapes. Here, we have considered 4 geometrical shapes each with 10 different dimensions, and their corresponding RCS pattern is synthesized. The RCS pattern of different shapes is given below.

Figures 3, 4, 5, and 6 are 3D RCS patterns of a circular cylinder, cone, frustum, and sphere, respectively, whose azimuthal angle is varied from 0 to 360° and elevation angle is varied from 0 to 180°. In addition, the features are often smaller in size compared to the original signal. So it takes less computational resources to learn the recognition models. To obtain the right features, it is often useful to look at a time–frequency view of data where the frequency due to motion is varying across RADAR pulses. The time–frequency signature of a signal can be derived by either Fourier transform (the spectrogram) or wavelet transforms.

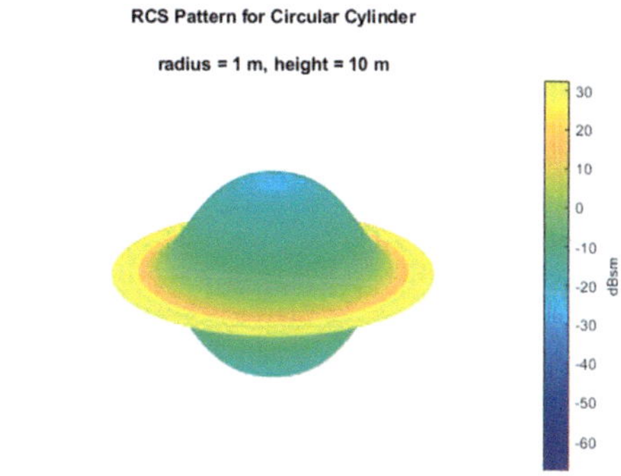

Fig. 3 RCS pattern for circular cylinder

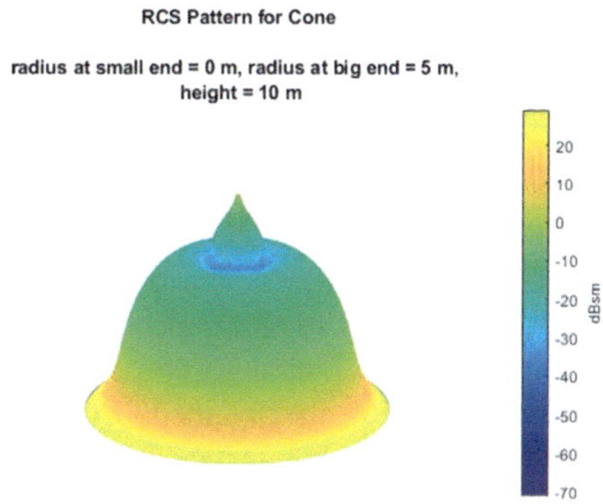

Fig. 4 RCS pattern for cone

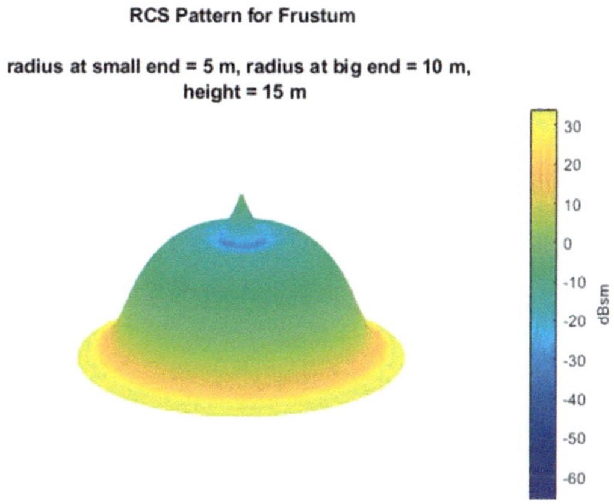

Fig. 5 RCS pattern for frustum

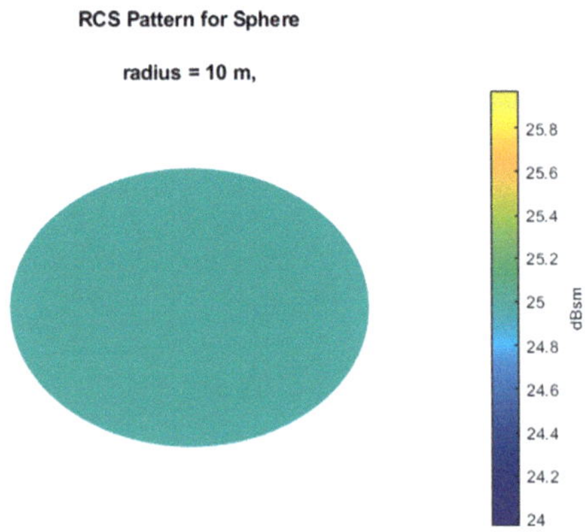

Fig. 6 RCS pattern for sphere

Specifically, in this paper, wavelet packet representation of the signal is rigorously used. The obtained signatures provide some insights that the learning algorithms will be able to distinguish between the target shapes. Specifically, it is observed that there exists a separation between the frequency content over time between the 4 geometrical shape targets. The apparent frequency separation between 4 geometrical shape-based targets suggests using a frequency domain measure to classify signals. So, the MODWPT method is chosen to compare relative sub-band energies.

Upon extraction, the obtained feature data from the radar synthesis tool is divided into two different data sets, i.e., training and testing data sets in the ratios of 70% and 30%, respectively. This results in obtaining 7000 and 3000 feature vector instances for training and testing data sets, respectively. With respect to 4 geometrical shapes, out of 2500 feature vector instances per shape, 1750 are part of the training data set, while the remaining 750 are part of the testing data set. The training data set is used to train machine learning classification models, while the testing data set is used to evaluate the performance of the trained models. The developed machine learning classification model used k-NN, SVM, and ANN, and target recognition performance is compared.

To assess the performance of the proposed method, the Accuracy parameter for each classifier is verified which is obtained from the confusion matrix:

$$\text{Accuracy} = \frac{TP + TN}{(TP + FN) + (TN + FP)} \tag{18}$$

From the obtained confusion matrix during analysis, it is observed that SVM gives an accuracy of 83.7% (Fig. 7).

Similarly, the confusion matrix of the ANN classifier model is given below which shows an accuracy of 73.3% (Fig. 8).

The confusion matrix of the k-NN classifier model is given below which shows an accuracy of 92.5% (Fig. 9).

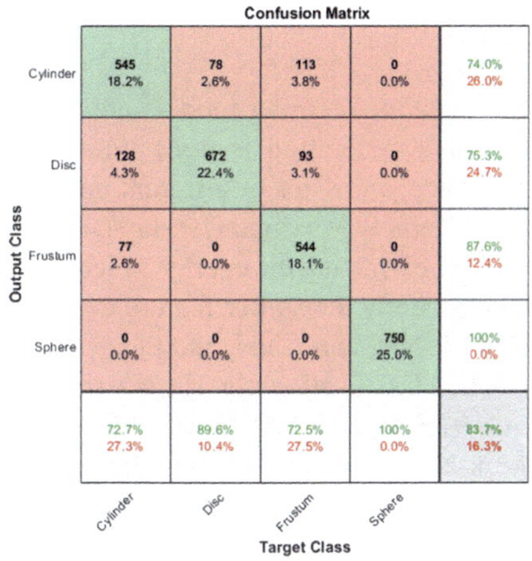

Fig. 7 Confusion matrix of SVM classifier

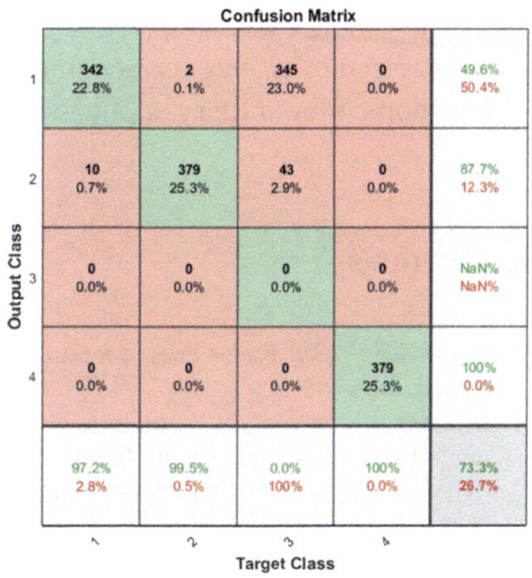

Fig. 8 Confusion matrix of ANN classifier

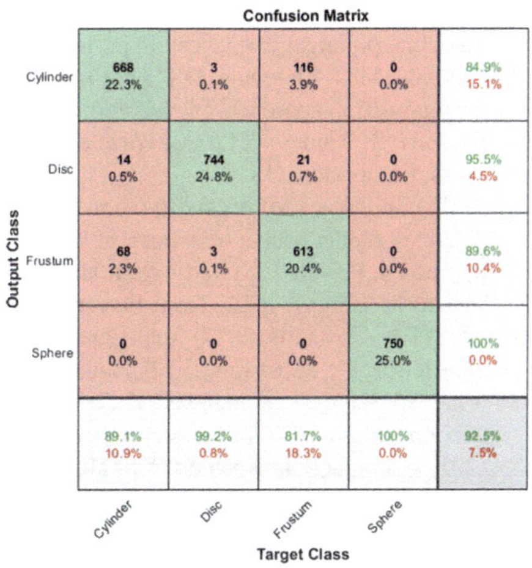

Fig. 9 Confusion matrix of k-NN classifier

6 Conclusions

The machine learning models are created for target recognition of simple targets. The data generation is done on the model by varying aspect angles, operating frequency, and sizes of different geometric shapes. A novel feature extraction technique, namely Maximal Overlap Discrete Wavelet Packet Transform (MODWPT), is used for extracting the energy-based features from the data synthesized. The obtained energy coefficients after feature extraction are given to different classifiers, namely SVM, k-NN, and ANN, and their performance is compared.

The classification algorithm using SVM provided an accuracy of **83.7%**, k-NN with **92.5%,** and ANN with **73.3%**. From the obtained results, it is concluded that the k-NN classifier provides the best results using the MODWPT feature extraction for the target classification. This can be applied to a combination of simple targets which leads to the complex targets and the classification performance can be observed and this can be further extended to classify the moving targets automatically. The feature extraction techniques can be further developed to reduce the noise and clutter in moving targets recognition. This experimental study finds a huge scope of development in military applications for accurate target classification.

References

1. Knott EF (1988) Radar cross section. In: Brooker E (ed) Aspects of modern radar. Artech House, MA
2. Khodjet-Kesba M (2014) Automatic target classification based on radar backscattered ultra wide-band signals. Université Blaise Pascal - Clermont-Ferrand II
3. Khodjet-Kesba M, Lee S (2014) Comparison of matrix pencil extracted features in time domain and in frequency domain for radar target classification. Int J Antennas Propag
4. Skolnik MI (2001) Introduction to radar systems, 3rd edn. McGraw-Hill
5. Mahafza BR, Radar systems analysis and design using Matlab, 2nd edn
6. Alper Selver M, Secmem M (2017) The effects of geometric scattered signal waveform modeling on target identification performance. IEEE
7. Gokhale MY, Khanduja DK (2010) Time domain signal analysis using wavelet packet decomposition approach. Int J Commun Network Syst Sci
8. Ercival DB, Walden AT (2000) Wavelet methods for time series analysis. Cambridge University Press, Cambridge, UK
9. (2017) Epileptic seizure classification using novel entropy features applied on maximal overlap discrete wavelet packet transform of EEG signals. 7th international conference, ICCKE
10. Santoso S, Powers EJ, Grady WM, Hofmann P (1996) Power quality assessment via wavelet transform analysis. IEEE Trans Power Del 11:924–930
11. Alves DK, Costa FB (2017) Real-time power measurement using the maximal overlap discrete wavelet-packet transform. IEEE Trans Industr Electron 64(4)
12. Yang W, Jiaguo L, Changyao Z (2007) Algorithm of target classification based on target decomposition and support vector machine. IEEE
13. (2017) Comparison: KNN & SVM algorithm. Int J Res Appl Sci Eng Technol (IJRASET), vol 5
14. Elgammal MA, Mostafa H (2019) a comparison of artificial neural network (ANN) and support vector machine (SVM) classifiers for neural seizure detection. IEEE

Chapter 31
Nacelle: Knowledge Graph-Based Conversational AI for Skills Gap Analysis to Achieve Sustainable Learning at Workplace

Chao Hong Loh, Sophia Wei, Chee Teck Phua, Azizah Mohd, Albert Tan, and Boon Khoon Seow

1 Introduction

Talent developments are shifting focus to deliver a bigger impact on business. The attention areas in strategic workforce planning include attracting and retaining talent, and ensuring their people have the right skills for today and tomorrow. Survey findings in the global context [1–3] (LinkedIn Learning, Workplace Learning Report) indicate increased budgets in companies to develop Learning and Development (L&D) programs to tackle day-to-day challenges, and marketing L&D programs to learners.

In Singapore, 99% of enterprises are SMEs, which employ 65% of the workforce (Statistics Singapore, Singapore Economy 2017). Through our engagements with local SMEs, there is a general interest to provide L&D opportunities for their employees to develop competencies and retain staff, when necessary. This is validated based on data provided by SkillsFuture Singapore (SSG), which shows in 2016, 418,000 individuals took up 950,000 training places funded by the Ministry of Education (MOE) and SSG. There is an increase from 379,000 individuals in 2015 who took up 835,000 places (SkillsFuture, Singapore State of the Training Industry, 2017). The increase in the number of individuals taking up training places in 2016 seemed to be correlated to the increase in Year-on-Year (YOY) Labour Productivity as shown in Fig. 1 (CEIC, Singapore Labour Productivity Growth). In 2017, the report (Feb 2017) by Committee on the Future Economy mentioned the need "To strengthen the nexus between skills acquisition and utilisation.", which will need employers to expand their role and ownership of employers to utilize skills acquired through training.

C. H. Loh (✉) · S. Wei · C. T. Phua · A. Mohd · A. Tan · B. K. Seow
Nanyang Polytechnic, Ang Mo Kio, Singapore
e-mail: loh_chao_hong@nyp.edu.sg

G. Mathur et al. (eds.), *Proceedings of 2nd International Conference on Artificial Intelligence: Advances and Applications*, Algorithms for Intelligent Systems,
https://doi.org/10.1007/978-981-16-6332-1_31

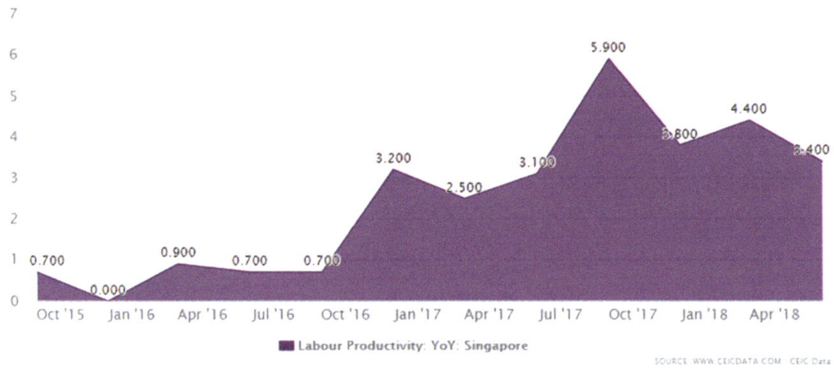

Fig. 1 Singapore's Labour Productivity Growth (YoY) from October 15 to April 18

Singapore set up the National Centre of Excellence for Workplace Learning (NACE) in July 2018 to help companies retain and build competencies to support business growth through Workplace Learning (WpL). One of the key elements of the WpL system is Skills Gap Analysis (SGA), which references the Skills Framework (SFw). The Skills Framework (SFw) provides key information on sector, career pathways, occupations/job roles, as well as existing and emerging skills required for the respective job roles. Using the SFw, SGA can be conducted by trained personnel for companies to identify areas of training required for their staff.

The current methodology to conduct SGA is done through interviews, survey and focus group discussions. This process is time-consuming; requires companies to commit manpower; and outcomes are often dependent on the interviewer, facilitator and scribe. There is a lack of digital tools with built-in intelligence to help the consultant or companies conduct SGA.

To achieve consistent outcomes, conversational AI technologies can be used. The knowledge graphs are constructed using the SGA workflow and skills map to dynamically support the needs of the respective industry sector. Conversational AI—Nacelle—is developed, leveraging the Natural Language Processing to learn through collections of SGA created with companies and using data from the SFw to help companies in SGA.

To the best of our knowledge, it is the first work to develop such a conversational AI to an SGA consultant in the conduct of interviews through online chat. This proposed conversational AI defines an explicit goal to find the skills gap and is achieved through guiding the conversation over the knowledge graph combining both SGA logic and dynamic skills' requirements for individual industries.

2 Related Work

Conversational AI has been successfully applied into different applications and can be generally categorized into mainly three types [4]. The first type serves as Question and Answering system, and it allows users to query a large-scale Knowledge Base or a document collection in Natural Language. The second category belongs to task-oriented dialogue systems. Different from the first type in which the user seeks an answer or certain information, this type of system assists users to complete a task, such as making a hotel reservation or booking movie tickets. The third category belongs to social chatbots. In this category, fully data-driven and end-to-end approaches are explored to conversational response generation. These models are trained entirely from data without resorting to any expert knowledge. However, these models often only reply on utterances and responses as training data, and this leads to uninformative and inappropriate responses. Recently, there is work [5] on proactive conversational AI for social chatbots with explicit conversation goals. It is achieved through explicitly exploiting knowledge associated with a knowledge graph and has shown that the dialogue models that plan over the knowledge graph can make full use of related knowledge to generate more diverse multi-turn conversations.

For SGA purposes, the conversational AI required is with a clear goal, and task-oriented. The conversational AI for SGA is more challenging than the traditional task-oriented dialogue systems in which the task can be completed by asking explicit questions to fill out slots required for the task. For example, for hotel booking, the dates and types of the rooms are the required information to fill out slots. The conversational AI for SGA, on the other hand, requires the combination of explicit and implicit questions leading to multi-turn dialogues for different sub-topics. At the same time, it needs to be proactive to engage users over multiple sub-topics. To solve the challenges, the proposed Nacelle embeds the knowledge related to SGA into a graph to guide multi-turn dialogues to proactively ask both types of questions to realize the goal of identifying the skills gap.

3 Approach

3.1 System Architecture of Nacelle

The traditional way of SGA is done manually. It is a tedious and repetitive process. The outcome also depends on the facilitator or interviewers. Furthermore, the human facilitator/interviewer also may not be able to leverage historical statistics due to the limited memory of humans. Nacelle is thus presented to assist or replace the human facilitator to conduct interviews/surveys and generate the corresponding SGA report automatically.

Figure 2 shows the architecture of Nacelle for SGA. Compared with the traditional conversational AI assistants, Nacelle for SGA has a guided SGA logic captured by a

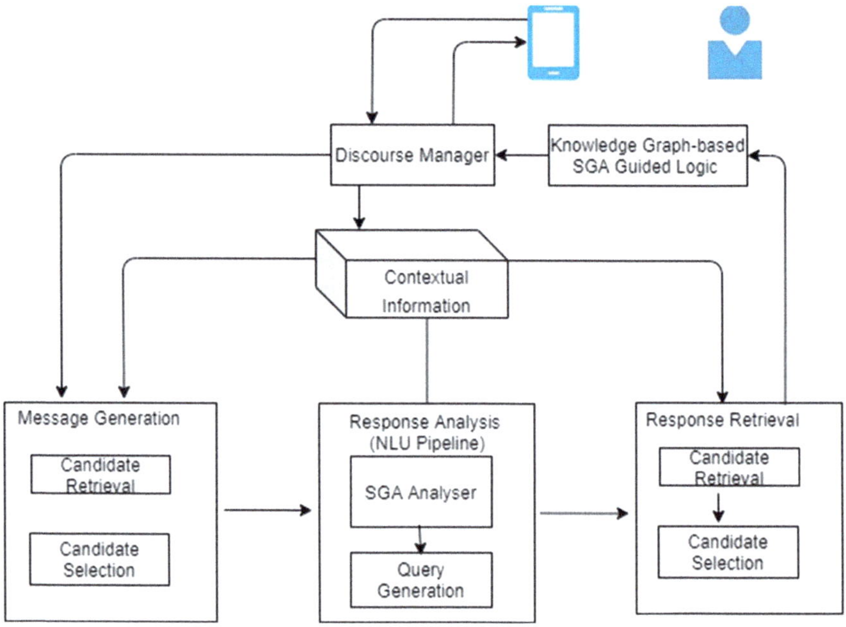

Fig. 2 Architecture of Nacelle

graph with the aim of discovering missing skills and identifying hidden issues. The unit-Discourse Manager oversees the conversation flow with the SGA-Guided Logic. The Guided Logic is decided by the process of SGA combined with dynamically loaded information on skills for specific jobs and industry, and it will be detailed in Sect. 3.2.

There are mainly three processing units in Nacelle. Normally, the facilitator will be initiating the conversation, thus the first processing unit—Message Generation—is responsible to generate guided questions to identify missing skills or hidden issues. Once the user replies, the message is generated. The response is passed to the second processing unit—Response Analysis. Inside Response Analysis, the response is passed through SGA analysis to extract any skills/issue detection, and the extracted information will be provided for the generation of the SGA reports. The detailed process of SGA analysis will be elaborated in Sect. 3.3. As the process for particular skills or topic will last for a few message exchanges, Query generation is required after the SGA analysis so that the relevant query extracted from our graph is generated. The generated query is passed to the third processing unit to retrieve an appropriate response. The response could be a survey form or further interview questions. Contextual information is to store any contextual information collected during the conversation. This will be utilized to retrieve the most appropriate response.

3.2 Knowledge Graph-Based SGA-Guided Logic

Our SGA logic takes reference from Business, Performance, Learning and Learner levels (BPLL), a BPLL model by Tobey [6], and we re-organize them into 5 topics as shown in Fig. 3. In Stakeholder Management, the conversation was used to understand the skills related to handling stakeholder management, while at the Skills Assessment and Cross Check and Performance and Issue, they were used to identify the gaps between the current and desired performance of the learners. In addition, at the learning level, data was used to identify the gaps in the learning conditions, while at the learner level, it was used to assess the gaps in knowledge and skills between the current and desired conditions. Lastly, at the business level, data was used to understand the strategic and business focus of the organization [6].

Currently, our SGA is targeting the Human Resource sector, and our questions are cover topics: Skills Assessment and Cross Check, Learning, Learner and Performance and Issue which shown in Fig. 4a.

We have created knowledge graphs based on SGA conversation logic provided by SGA specialists. Networks, a widely used Python package is implemented as the graph-base network structure to manipulate and analyze the structure, dynamics, and functions of complex conversation Chatbot. This provides a snippet of the visualization of the conversation flow drawn by Neo4j in Fig. 4.

The information of the graph is stored using "node" and "edge": A graph (network) is a collection of nodes together with a collection of edges that are pairs of nodes.

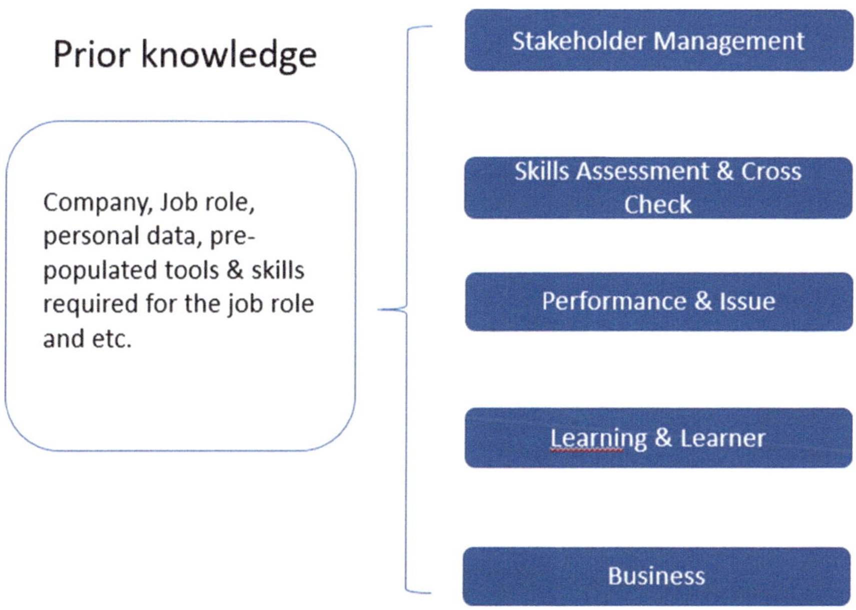

Fig. 3 SGA logic

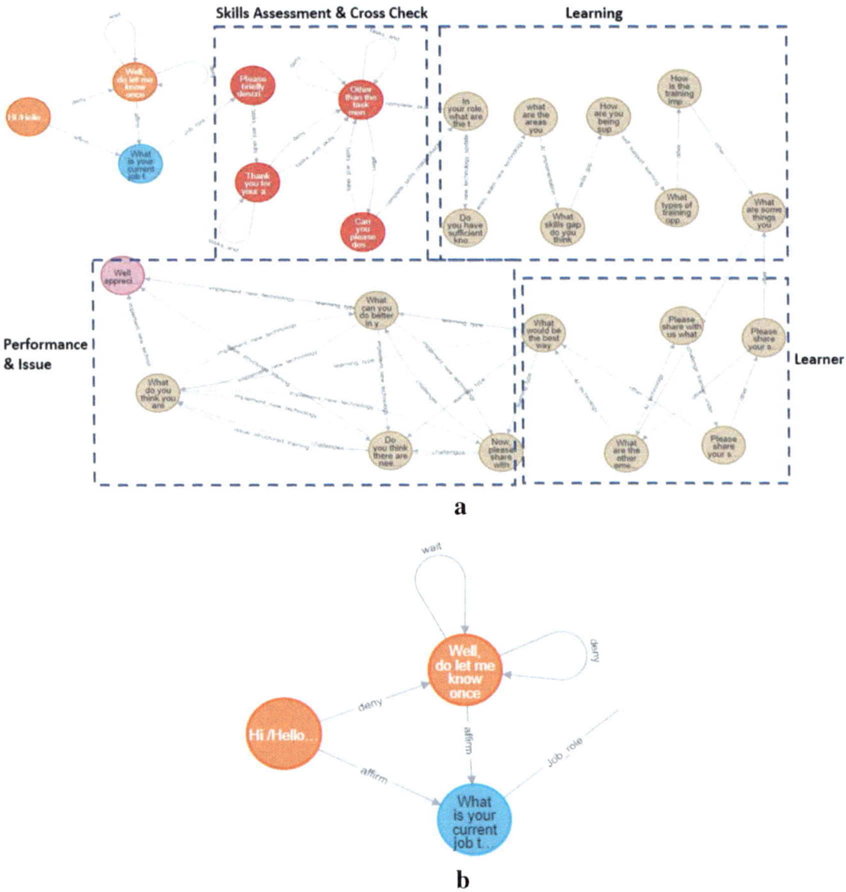

Fig. 4 Conversation flow's network architecture

Attributes are often associated with nodes and/or edges [7]. Below is how we assign the information to the nodes and edges.

Node: Each question is considered one node.

Edge: Arbitrary edge attributes such as probability and intents (the classification result of the response using Dual Intent and Entity Transformer (DIET) architecture which will be described in Sect. 4.1) are associated with an edge. Multiple edges are allowed between each pair of nodes.

Figure 4b shows the potential intents that are stored in edges such as "deny" and "affirm". Other than the potential intents, the probability stored in the edges helps to decide the next question (node) to be asked by Nacelle.

Based on the classification of the response, we have a collection of potential follow-up questions represented by a new node. Among all these potential follow-up questions, we choose the one with the highest probability. Initial probability, which is decided by a domain expert, will be updated after more data has been collected and trained.

3.3 SGA Analyzer

Along the guided conversation process, the SGA analyzer will get the response from the user, conduct the skills gap analysis and generate the SGA report automatically along with the interviews/survey/focus group discussions. It is possible that a temporary report will be generated first so that follow-up interviews/surveys/focus group discussions as the follow-up actions may require other parties such as the peers or supervisors to participate.

Figure 5 shows the flowchart of the SGA Analyzer. Once the response is received, a sectionizer/tokenizer will be applied first to segment the sentence into different phrases. Named Entity Recognition (NER) needs to be conducted to locate and classify named entities in text into pre-defined categories such as the names of persons and organizations. The Pre-trained NER model can be obtained from popular NLP libraries such as NLTK or SsaCy. The pre-trained model needs to be further trained to include new categories such as skills. This is because, for an SGA analyzer, it is important to locate and classify entities into those linked to skills. To achieve that, the skills need to be represented in terms of vectors. One of the most important NLP techniques—word embeddings—are required for vectorization. Word embeddings are distributed representations of text in an n-dimensional space. There exist two popular word embedded models: one is Word2Vector from Google and GloVe from Standford, both having pre-trained models. Unfortunately, the pre-trained models are

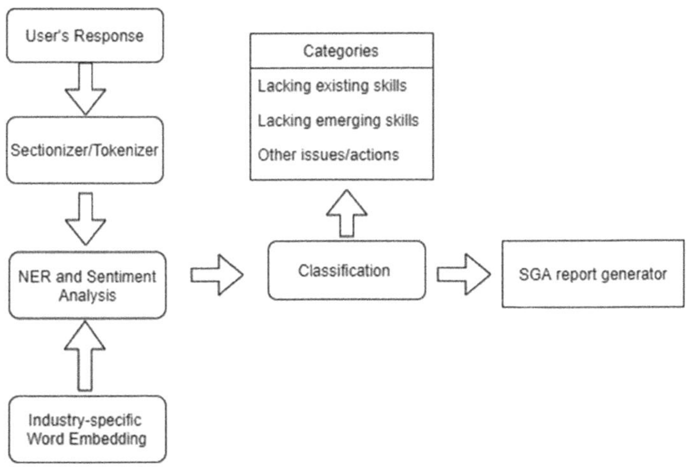

Fig. 5 The architecture of SGA analyzer

based on mainly text gathered from public domains such as Wikipedia and Twitter. Domain adaptation is very important for good performances in targeted applications. For example, python in biology most likely refers to a type of snake, but a programming language in computer engineering.

Thus, for Nacelle, industry-specific embedding is developed by retrofitting the available pre-trained word vectors from BERT trained models and adapting them to be aware of skills in specific job roles or sectors. It is important that sufficient data are available to train the proposed proprietary model. To cater to this purpose, industry-specific data collection has been conducted which will be illustrated in Sect. 4.2. In this way, we could continuously build up our database in terms of skills required for each industry. When necessary, sentiment analysis will also be conducted to detect any frustration or dissatisfaction to identify potential missing skills/skills gap/issues. This process is iterative and as the conversation goes on, the SGA analyzer will be able to finally classify into three categories—Lacking existing skills, Lacking emerging skills and Other issues/actions. This information is fed to the SGA report generator to generate an SGA report.

4 Implementation

Nacelle is built with Rasa open-source framework, a machine learning framework to automate text- and voice-based assistants. The architecture in Fig. 6 shows the overview of the Rasa open-source architecture [8].

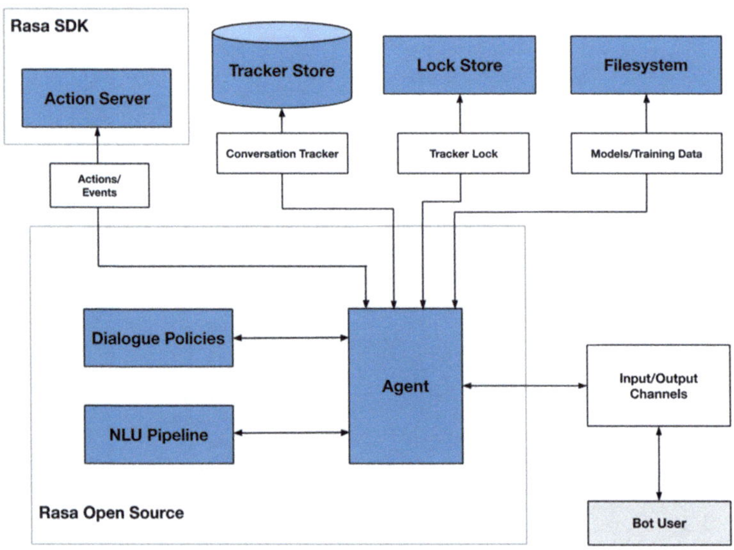

Fig. 6 Rasa open- source architecture

NLU Pipeline in Rasa shown in Fig. 6 is used for Response Analysis shown in Fig. 2 for Nacelle to handle intent classification, entity extraction and sentiment analysis. To be more specific, in Nacelle, the response depends on the contextual information, and the corresponding intent classifier is applied. In general, we developed three types of intent classifiers including skills classifier, issues/challenges classifier and training type classifier. Response Retrieval of Nacelle shown in Fig. 2 is realized through Dialogue Policies in Fig. 6. The knowledge-graph-based SGA-guided logic is implemented through Rasa SDK supported by the Action Server.

4.1 Text Classification in Nacelle

We are using Dual Intent and Entity Transformer (DIET) architecture [9] for the model training of conversational AI.

DIET is a multi-task transformer architecture Fig. 7 that can perform both intent classification and entity recognition together. It is made of multiple components that allow us to have the flexibility to exchange different components. For example, we could experiment with using different word embeddings such as BERT and GloVe.

A lot of the pre-trained language models are very heavy in the sense they require large computing power, and inference time is long and so despite their strong performance, they aren't designed for conversational AI applications. DIET is different as it

- Is a modular architecture that allows software developers to have more flexibility in experimentation.

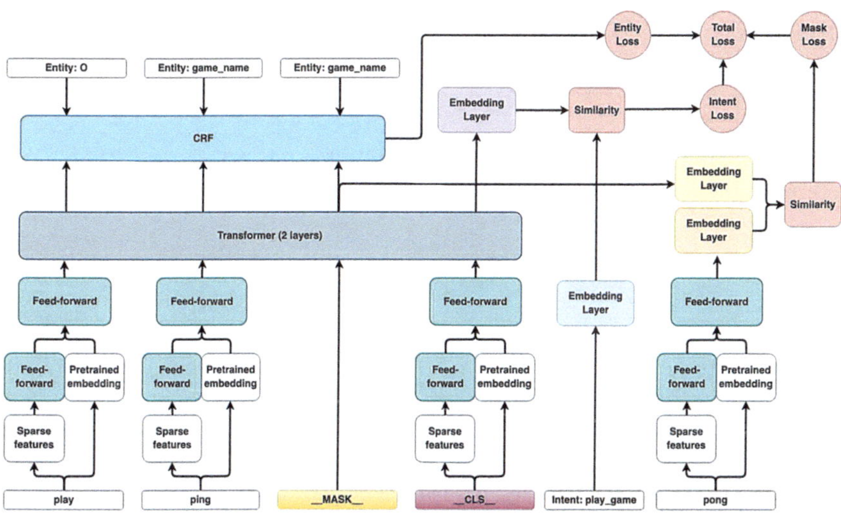

Fig. 7 Dual Intent and Entity Transformer (DIET) architecture

- Matches pre-trained language models in terms of accuracy.
- Outperforms current SOTA and it's 6X faster to train.

a. Intent loss

The transformer output for __CLS__ token α_{CLS} and intent labels y_{intent} are embedded into a single semantic vector space $h_{CLS}= E(\alpha_{CLS})$, $h_{intent}= E(y_{intent})$, where $h \in \mathrm{IR}^{20}$.

$$\mathrm{L_I} = -\langle \mathrm{S_I^+} - \log\left(\mathrm{e}^{\mathrm{S_I^+}} + \sum_{\Omega_\mathrm{I}^-} \mathrm{e}^{\mathrm{S_I^-}}\right)\rangle, \tag{1}$$

where similarity $S_\mathrm{I}^+ = h_{CLS}^T h_{\mathrm{intent}}^+$ is maximized with the target label y_{intent}^+ and similarities $S_\mathrm{I}^- = h_{CLS}^T h_{\mathrm{intent}}^-$ are minimized with negative samples y_{intent}^- using dot-product loss. The sum is taken over the set of negative samples Ω_I^- and the average $\langle . \rangle$ is taken over all examples.

The dot-product similarity serves as a ranker over all possible intent labels during inference time [10].

b. Entity loss

$$L_E = L_{CRF}(\alpha\,,\ y_{\mathrm{entity}}) \tag{2}$$

where $L_{CRF}(.)$ denotes negative log-likelihood for a Conditional Random Field (CRF). A sequence of entity labels y_{entity} is predicted through a CRF tagging layer on top of the transformer output sequence corresponding to an input sequence of tokens [10].

c. Mask loss

The transformer's output α_{MASK} for each selected token y_{token} is fed through a dot-product loss like the intent loss.

$$\mathrm{L_M} = -\langle S_M^+ - \log\left(e^{S_M^+} + \sum_{\Omega_M^-} e^{S_M^-}\right)\rangle, \tag{3}$$

where $h_{\mathrm{MASK}}= \mathrm{E}(\alpha\mathrm{MASK})$ and $h_{\mathrm{token}}= \mathrm{E}(y_{\mathrm{token}})$ are corresponding vectors $h \in \mathrm{IR}^{20}$. The similarity with the target label y_{token}^+ is $S_M^+ = h_{\mathrm{MASK}}^T h_{\mathrm{token}}^+$ and similarities with negative samples y_{token}^- are$S_M^- = h_{\mathrm{MASK}}^T h_{\mathrm{token}}^-$. The sum is taken over the set of negative samples Ω_M^- and the average $\langle . \rangle$ is taken over all examples [10].

d. Total loss

The model was trained in a multi-task fashion by minimizing the total loss L_{total}

$$L_{\text{total}} = L_I + L_E + L_M \tag{4}$$

The architecture can be configured to turn off any of the losses in the sum above [10].

Nacelle was trained using pre-trained language model BERT, 2 layer transformer with a 256-dimension vector. The architecture is designed to allow the model to learn a more general representation of our input sentences. During training, all the weights have to be optimized based on three different losses: intent (L_I), entity (L_E) and mask (L_M) loss and as such, the model can't just learn a representation that heavily minimizes one of the losses. In addition, the architecture is designed in such a way that you can switch on or off multiple components. The architecture is designed to handle intent and entity classification, but if we just want the model to do intent classification, we can "switch" off the entity and mask loss and just focus on optimizing the intent loss during training [9].

4.2 Data Collection

Our data collection is conducted using the following methods:

(1) 290,934 job descriptions through crawling all the largest job boards in Singapore as well as direct company pages of large global corporations for one year.
(2) 300 phone surveys conducted by School of Business Management (SBM), Nanyang Polytechnic, with our alumni whose positions are Professional, Managers, Executives and Technicians (PMET).
(3) 9 sets of online chat surveys using Mattermost [11] (a self-hosted chat service platform) targeting PMET in industry.
(4) 15 sets of SGA conversation data collected by professional SGA companies targeting employees from the HR sector.

We make use of the job description from method 1 to classify them to match with the technical skills and competencies and the generic skills and competencies from Skills Framework.

For 2~4, each conversation is created by a pre-assigned specialist and one crowd-sourced participant, where the pre-assigned specialist plays the role of an SGA specialist with the specific goal of finding the skills gap of the other participant who is the conversation follower. These conversation data play a key role in helping us to design our "Knowledge Graph-based SGA-Guided Logic" that is described in Sect. 3.2.

It is worth mentioning that, among the above-mentioned types of data collection, online chat is the most realistic type of data required by our conversational AI.

5 Result

To evaluate the accuracy of the NLU prediction, we have labeled 3610 data and split the data into "test data" and "train data" in a ratio of 80% train/20% test data. We first train the model using train data, then the test data is used to verify how accurate the trained model is. Below is the outcome of 722 data from the test data file with detailed intent prediction confidence distribution shown in Fig. 8.

accuracy: 0.9903047091412742,
precision: 0.994172932330827,
recall: 0.9917136591478697,
f1-score: 0.9918972638760994.

The above accuracy is mainly based on the classified response from the data we collected. We are in the progress to collect data from the private sector to find out the accuracy of our AI engine during trial.

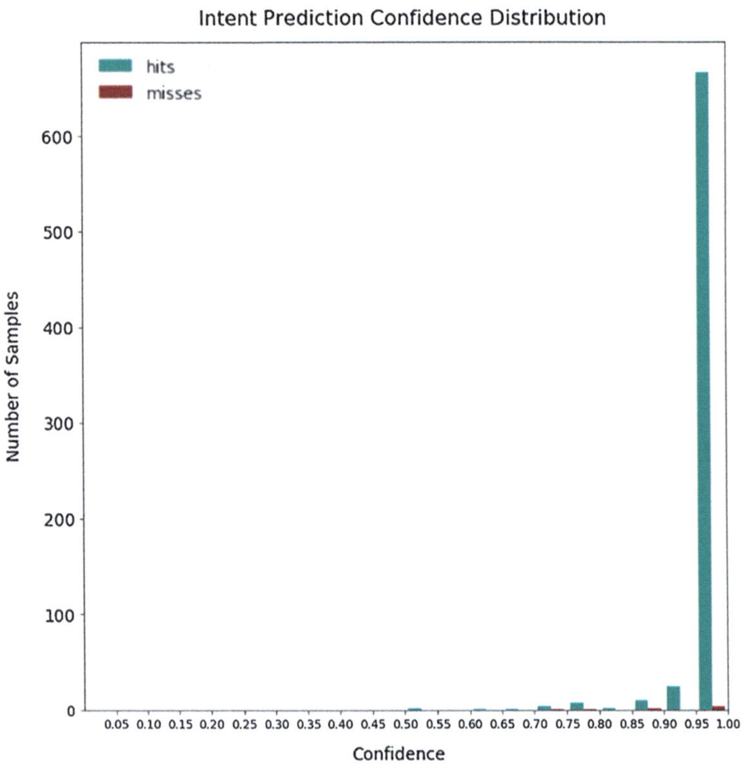

Fig. 8 Intent prediction confidence distribution

6 Conclusion

We have successfully developed Nacelle. The first industry under validation is the Human Resource sector. The model achieves an accuracy of 0.99 with an F1 score of 0.9917. Validation for Nacelle will continue to be carried out at local companies. Two primary research needs were identified through this development effort. The first is the need for additional conversation data that potentially improve the accuracy of the intent classification when deployed to different industries. With larger and comprehensive datasets, Nacelle will be able to find out more information and skills gaps of the interviewee during the chat session. Second, guidance from the SGA expert is needed to streamline the questions and key information needed for each industry. With the framework designed, new industries' requirements for the SGA can be easily trained and brought to life as we have enough quality data.

References

1. 3rd Annual 2019 Workplace Learning Report, LinkedIn Learning
2. Statistics Singapore, Singapore Economy 2017
3. OECD and the World Bank, Indicators of skills for employment and productivity: a conceptual framework and approach for low-income countries
4. Gao J, Galley M, Li L (2018) Neural approaches to conversational AI. Foundations and Trends in Information Retrieval
5. Wu W, Guo Z, Zhou X, Wu H, Zhang X, Lian R, Wang H (2019) Proactive human-machine conversation with explicit conversation goals. ACL
6. Tobey D (2005) Needs analysis basics. American Society for Training & Development Staff, Massachusetts
7. https://networkx.org/documentation/stable//reference/introduction.html
8. RASA docs. https://rasa.com/docs/rasa/arch-overview/. Accessed 20 Jan 2021
9. https://towardsdatascience.com/day-101-of-nlp365-in-depth-study-of-rasas-diet-architecture-3cdc10601599. Accessed 20 Jan 2021
10. Bunk T, Varshneya D, Vlasov V, Nichol A, DIET: lightweight language understanding for dialogue systems. arXiv:2004.09936v3
11. Mattermost. https://mattermost.com/

Chapter 32
On the Use of Machine Learning for Soil Condition Monitoring

Vikash Rameshar, Wesley Doorsamy, and B. S. Paul

1 Introduction

Many of our food resources have been threatened for many years due to the land degradation of the agroforestry industry [1]. Globally this has imposed vast food security issues. Primarily this is due to soil erosion and nutrient loss that has changed the soil's composition altogether. A single factor like soil erosion has affected 83% of the global degradation itself. Other contributing factors such as nutrient loss within the soil, effect on average 80% of African countries [2]. Nutrient loss within the soil is mainly due to synthetic inorganic fertilizers and underprivileged farmers not utilizing sufficient organic fertilizers, such as green biomass or animal manure. This is just one of the factors that have contributed to lower nutrition in crops and have posed an issue in securing food for millions of people [3].

With poverty being high in Africa small scale farming has increased over the years and has overtaken large scale farming which is reinforced with a collaboration of the African Development Bank, the Organisation for Economic Co-operation and Development (OECD), the World Bank Group, and the World Economic Forum [4]. Although assistance and encouragement are reflected from both government and the organizations mentioned, many challenges exist within the small-scale farming industry. One major challenge is the ability of the farmers to replenish the soil with adequate fertilizers and/or the time for the soil to heal with the mulching of fertilizer trees. Farmers within rural locations also lack basic tools that are important for the

V. Rameshar (✉)
Department of Electrical and Electronic Engineering Technology, University of Johannesburg, Johannesburg, South Africa
e-mail: vikashr@uj.ac.za

W. Doorsamy · B. S. Paul
Institute of Intelligent Systems, University of Johannesburg, Johannesburg, South Africa

G. Mathur et al. (eds.), *Proceedings of 2nd International Conference on Artificial Intelligence: Advances and Applications*, Algorithms for Intelligent Systems,
https://doi.org/10.1007/978-981-16-6332-1_32

production of good healthy crops. Tools such as disease control methods of plants and soil management against parasitic elements within the soil in remote locations.

Many areas in the small-scale farming industry have faced plant infestation of soil-inhabiting parasitic elements and degradation of soil composure. With little to no assistance on how to treat these issues, smallholder farmers have to rely on laboratory results to adequately replenish their soil for good yield in crops. The research investigates soil parameters to maintain soil conditioning with the utilization of machine learning algorithms.

In this paper, a review of soil characteristics currently utilized in agriculture by various researchers is within Sect. 2 (current methods). Section 3 demonstrates a case study of K-Means Clustering by utilizing soil data and Sect. 4 discusses the method utilized with further consideration.

2 Current Methods

Within our ecosystem soils provide vital services such as flood prevention, carbon sequestration, and most importantly to humans is the production of food [5]. Their unique ability to partially offsetting fossil fuel emission by intensified carbon uptake is one of the most researched areas. This not only alters the fight against climate change but also assists in the sustenance of degraded soil that leads to increased crop yield and contributes towards food security [6]. Soil organic carbon (SOC) has played an important role in the rejuvenation of topsoil and remote sensing datasets of topsoil spectral clustering and k-means have been utilized to aid in the improvement of current SOC models. Various clustering methods had also been taken into consideration toward the furtherance prediction of SOC models although the clustering methods showed similar results with the reduced spectral library data as a low-cost solution [6].

Soil is predominant in all countries and due to the climate difference around the world, many seek to explore a way of enhancing the quality of the soil. Quality is based typically on the soil type and the nutrient factor within the soil. These nutrient factors depend on features such as Potassium, Phosphorus, Organic Carbon, and Boron as well as parameter soil reaction (pH) according to [7]. Features and parameters listed above have been utilized within the Extreme Learning Machine model to provide enhanced soil fertility and save on chemical costs. The models' accuracy that is determined by different activation functions also aids with soil health and environmental quality, although the pH classifier did not perform adequately [7].

Many of the reviewed articles have worked with soil features and one that stands out is [8]. The use of Electrical Conductivity (EC) in acquiring soil properties method to fertilize the soil in return to generate a better yield of crops had been investigated within this research. Although the research span of 7 years was a bit long, the soil had to be given time to settle in. EC values taken from soil samples are commonly utilized to determine the amount of soil salinity, although the nutrient content can also be determined, the mineralization properties can be measured as well [9]. While the output of the research proved that the higher EC levels that the cattle manure produced, it is still a preferred fertilizer over chemical fertilizers. It also proved that the high EC levels would be obtained from soils that had high nutrients [8]. Although this investigation had positive outcomes with soil conditions using organic fertilizers, it has been exhibited to be very expensive and is very time-consuming.

Soil texture within its many classifications (Sand, Silt, Clay, and Loam), plays a significant role in soil fertility and the type of crop that would thrive within the soil type. Within the research of [10], large datasets were obtained from a soil testing laboratory. This agricultural data mining technique institutes the texture of the soil automatically with the use of a Texture Triangle-Fuzzy C-Mean classifier. The research of [10] also improves the time consumed in processing large data sets and is capable of high predictability. The positive outcome of the research is large datasets have a quicker response time in classifying the soil texture but the accuracy has been reduced [10]. The conclusion of the outcome to [10] is the basic Fuzzy C-Mean classifier has performed better than the Texture Triangle-Fuzzy C-Mean classifier and has proposed an IoT-based system as further research.

As data becomes demanding in the agricultural industry, methods of analyzing data are prominent. In some cases, the data being analyzed require a method that would be user-friendly. Hierarchical Clustering together with other clustering algorithms has been utilized in [11], which was combined with a modified distance matrix to accomplish a soil profile distinct variation. The distance matrix algorithm and clustering method were applied to large datasets obtained from a soil database and soil surveys. The algorithm identified different soil profiles and can assist in large datasets on the soil profile properties produced can improve small to large scale crop modeling applications, soil moisture models, soil surface, and environmental models [11].

Most research categorizes the soil into its physical and chemical properties, which is utilized in [17]. Although the chemical evaluation of soil provides a more definite advantage over the physical analysis, techniques utilized in [17, 18] involve image processing that demines the fraction of sand, silt, and clay that is present within the soil. The use of image analysis is one of the more common methods that has been employed in the agricultural industry, mostly with the identification of crops and predicting the health of crops by images. Within this research [18] the Support Vector Machine (SVM) classifier has been employed with specific feature vectors within 50 samples of the soil from various locations. The main features from the soil sample images included Hue, Saturation and Value (HSV), Gabor feature, and Colour moment amongst others. The results from the SVM were compared with the hydrometer test and the USDA triangle method [18]. The outcomes of the research with the use of a multi-class SVM classifier were successful except for loamy soil and silty clay sand.

Surface Soil moisture is recognized by many small-scale farmers, which utilize the physical data in setting out the plantation of certain crops [19, 20]. Scientifically this means that the energy, carbon flow, and water dynamics between the soil surface and the atmosphere are influenced by these parameters. Factors further influencing the surface soil moisture are texture which has a direct impact on drainage patterns, topography, precipitation which is influenced by temperature, and the types of vegetation. The research within [20] utilized soil texture and slope data by clustering them to find a method of real-time continuous soil moisture monitoring system. The system involves data samples of over 2000 and also considered the Heronian average and the FX model in its data analysis.

As mentioned previously that soil is imperative to all countries and some researchers have taken a resolute method in providing soil knowledge. In [22] soil degradation has been investigated with satellite images that took into consideration features such as type of terrain, soil attributes, land use, and climate variables. When soil is unable to service its eco-system by providing essential elements, this labels soil as degraded [21]. Land use alteration [23], extreme application of fertilizers in agricultural areas [24], overgrazing [25], climate and topography factors are the main drivers of soil degradation. This investigation [22] with the utilization of satellite imagery and organic matter map assisted in forming the bases of the soil degradation index. The satellite imagery mapping was validated with the organic matter (OM) which concluded that the higher the OM the lower the soil degradation would be. The research also takes into consideration the coverage of soil as one of the contributing factors to soil degradation. Digital soil mapping and Remote sensing techniques have been fast and feasible towards the soil degradation index prediction [22]. Spider monkey optimization algorithm successfully applied for soil classification by Kumar et al. [26].

Knowledge areas from the above researchers have shown growth in the use of Machine learning techniques within the agricultural field. Particularly the methods in the analysis of data and cost-effective innovative data collection. As the target of sustainable agriculture toward food security is on the increase the various methods in machine learning research provide a positive background in making low-cost technology accessible to small-scale farmers. The application of low-cost technology would also assist in broadening the communication barrier between innovative technology and farmers, who are the source of between 65 to 80% of the world's food source [16]. An aspect that lacked within the reviews is to make data and results remotely and easily accessible to researchers and small-scale farmers. Whilst the project will be based on the latest technology, a slight challenge to small-scale farmers might occur. Therefore, careful consideration must be placed on certain aspects of the project, e.g. the display of the results. Results should be displayed in the simplest form to not cause any ambiguity to the actual user. The project would be as cost-efficient as possible, but not compromising on the authenticity of the aim.

Many researcher's machine learning techniques utilized unsupervised learning algorithms. The benefactor of unsupervised learning is that it does not require labeled data, and this is significant in the case of small datasets or the presence of class imbalance within data. This may occur with seasonal and climate variation during the capturing of data.

3 Case Study

The study investigates the use of an unsupervised learning approach in soil condition monitoring, with the use of basic features, by utilizing clustering. Unsupervised learning is utilizing datasets that have only independent variables and no target. Unsupervised learning methods are challenging, as the output is more subjective and there is no simple goal for the algorithm, such as predicting the class or continuous variable [15]. Clustering is the segregating of data points into groups or subsets of similar features and assigning them into clusters [12]. These clusters are formed surrounding the closeness of a centroid data point and each cluster can be significantly differentiated from the other. There a few types of clustering algorithms i.e.: Connectivity models, Centroid models, Distribution models, and Density models [13]. The two most popular clustering methods within machine learning are Hierarchical clustering and K-Means clustering [12]. Both of these methods are intriguing whilst utilizing different approaches in data clustering. Within the case study, an approach toward K-Means clustering has been utilized.

Sample data for the case study had been obtained by individual instruments to provide at least a two-digit accuracy for the data to be logged. The data had been logged manually for this study and to avoid any variance, data had been taken at the same time and from the same soil area. The different meters have been calibrated after every sample that had been taken and distilled water was utilized to wash the probes thereafter. This was a precaution to eliminate contamination of samples. Figure 1 below provides a brief flow diagram of the approach to the case study.

The K-Means clustering algorithm is a process of moving the centroids of sub-clusters or k-centers to the significant position of their principal points [14, 15].

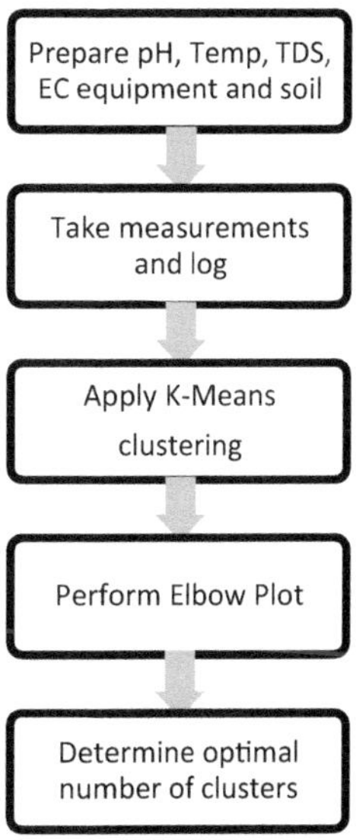

Fig. 1 Flow diagram of Case study approach

By redistributing occurrences to their closest clusters, which continuously iterates until there is no significant change in the number of cluster centroids or the number of iterations has reached its maximum [15]. Table 1 shows the description of the data utilized with the K-Means clustering. These parameters were considered due to their controlling ability with the soil nutrient level of different soil types. This approach would allow one to achieve a method in controlling nutrient content within the soil and would promote healthy crop growth and sustainability of soil concerning degradation [27]. The soil nutrient content is influenced by different temperatures which are also considered under Table 1.

Figures 2 and 3 below show the k-means clustering scatter plot of the soil pH and temperature which are two significant components of healthy soil condition. Value for K = 3 and K = 4 respectively for Figs. 2 and 3. Information from the

Table 1 Statistical description of data

	pH	Temp	TDS	EC
Count	28.00000	28.000000	8.000000	8.000000
Mean	6.194286	18.350714	5.900000	11.795000
SD	0.485680	0.800296	0.014142	0.176777
min	5.120000	17.000000	5.890000	11.670000
25% 50%	6.002500 6.265000	17.500000 18.720000	5.895000 5.900000	11.732500 11.795000
75%	6.485000	18.947500	5.905000	11.857500
max	7.000000	19.300000	5.910000	11.920000

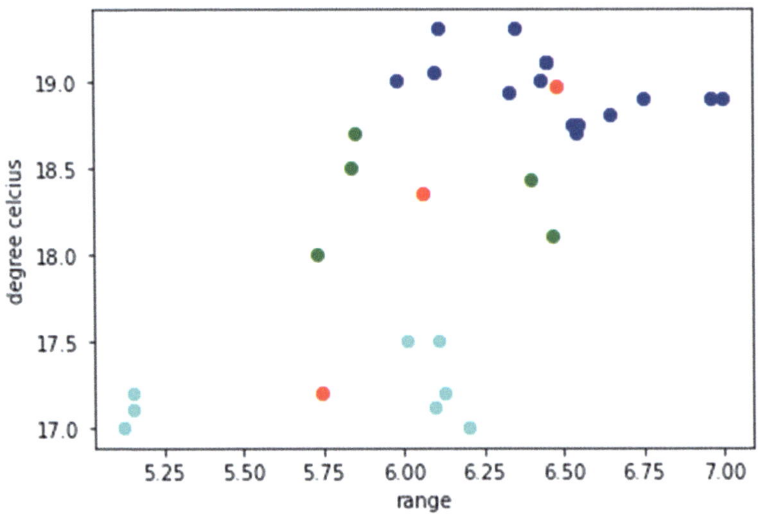

Fig. 2 Three centroids k = 3

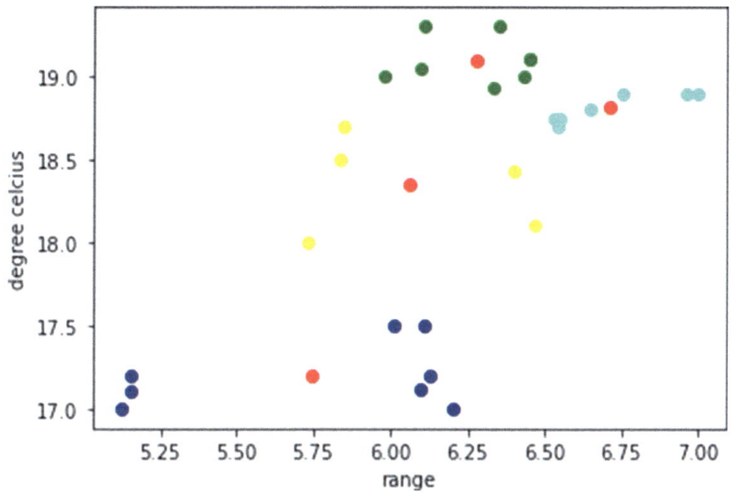

Fig. 3 Four centroids k = 4

plots indicated a higher temperature value as the pH tends to normalize. The pH and temperature are key elements to successful crop germination environments and allows for an improved rooting system that provides plants and vegetation to produce high yields. Scatter plots below demonstrate the level of nutrient value with the different temperatures of the soil directly. The air temperature would also be a factor to consider in future research.

The above Elbow plots (Figs. 4 and 5) have been utilized to determine the optimum k value and indicates a k value between 2 and 2.5 for 20 clusters (Fig. 4). With respect

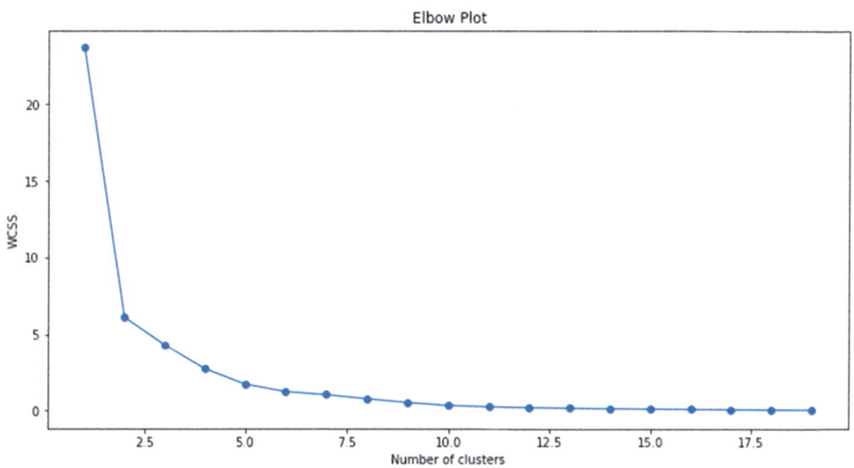

Fig. 4 Elbow plot of 20 clusters

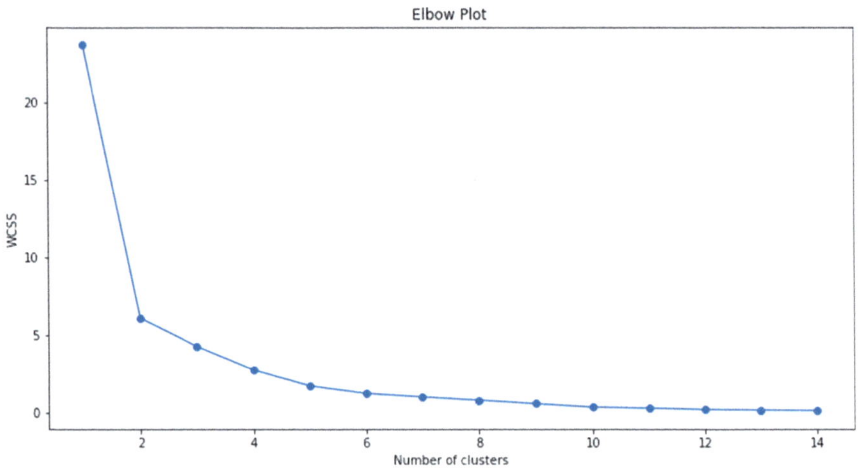

Fig. 5 Elbow plot of 15 clusters

to Fig. 5, the plots indicate a distinct k value of 2. This is prenominal due to the fact that soil samples have been obtained from various locations. The moisture content had also been given sufficient time which allowed the hydrogen ion to react with soil nutrients i.e. settling time after rain or irrigation. Distance between sampling of soil area had been adhered to as per laboratory standards only taken in real-time. The soil sampling locations differed from open area crops that are exposed to all-natural elements and crops that are shade thriving i.e. underneath trees. This insight if predicted by one would diagnose that the shaded crops are exposed to more nutrient value than crops that are fully exposed to all-natural elements, which is for future investigation, but the results show that this is not the scenario.

The method of clustering performed within this study is "unsupervised" which had been able to differentiate between soil conditions without providing labels. The features associated with this study have not been extreme cases of crop exposure which in any way might be misinterpreted by the clustering algorithm. The result indicates that unsupervised learning is suitably sensitive to the difference in soil conditions, even with few features, to assist with soil condition monitoring. The elbow plots in Figs. 5 and 6 indicate more than one cluster is present meaning that the unsupervised approach shows that the method is certainly sensitive to any minor differences in soil condition.

4 Conclusion and Future Work

Collection of soil data and analysis of the data, within the mentioned parameters, stem back to the mid-1980s [28]. It has a great impact on research towards the sustainability of soil as the climate and crop growth have changed over the years.

This impact relates to the quality of soil and as we try and grow crops in different extreme climates, we strive to produce the best harvest. Technology has also ingressed into agriculture at a fast pace to a point where agriculture can now use technology to its full potential, especially in cases such as soil condition monitoring [28]. This research utilizes an aspect of technology, as mentioned below, and allows much more to be unfolded within the agricultural industry.

The K-Means clustering algorithm has been utilized with datasets for pH and temperature as a means of utilizing machine learning techniques in soil conditions. Outcomes from this modest case study indicate that the successful implementation in monitoring soil conditions can be achieved with the use of clustering methods within the machine learning arena. This study provides a broader scope to explore arenas in machine learning that would be beneficial to small-holder farmers and could be articulated into large-scale farming as well. With data being the fundamental input to acquiring the machine learning algorithms to perform optimally, future data collection on a larger scale would be explored. This would also allow opportunities for many valuable parameters within the soil to be also explored that will assist in monitoring soil within the agricultural sector.

References

1. Bai ZG, Dent DL, Olsson L, Schaepman ME (2008) Proxy global assessment of land degradation. Soil Use Manag 24:223–234
2. Liu J, You L, Amini M et al (2010) A high-resolution assessment on global nitrogen flows in cropland. PNAS 107:8035–8040
3. Mafongoya PL, Kuntashula E, Sileshi G (2006) Managing soil fertility and nutrient cycles through fertilizer trees in southern Africa. In: Uphoff N, Ball AS, Fernandes E et al (eds) Biological approaches to sustainable soil systems. Taylor & Francis, Boca Raton, FL, pp 273–289
4. (2015) World Economic Forum; World Bank; African Development Bank; Organisation for Economic Co-operation and Development. The Africa Competitiveness Report 2015.World Economic Forum. https://openknowledge.worldbank.org/handle/10986/22014 License: CC BY-NC-ND 3.0 IGO.". Geneva
5. Kibblewhite MG, Miko L, Montanarella L (2012) Legal frameworks for soil protection: current development and technical information requirements. Curr Opin Environ Sustain 4(5):573–577
6. Ward KJ, Chabrillat S, Neumann C, Foerster S (2019) A remote sensing adapted approach for soil organic carbon prediction based on the spectrally clustered LUCAS soil database. German Research Centre for Geosciences. Telegrafenberg, Germany
7. Suchithra MS, Maya LP (2020) Improving the prediction accuracy of soil nutrient classification by optimizing extreme learning machine parameters. Information Processing in Agriculture 7, Amrita School of Arts and Science, India. pp 72–82
8. Eigenberg RA, Doran JW, Nienaber JA, Ferguson RB, Woodbury BL (2002) Electrical conductivity monitoring of soil condition and available N with animal manure and a cover crop. Agric, Ecosyst, Environ 88:183–193
9. De Neve S, Van De Steene J, Hartman R, Hofman G (2000) Using time domain reflectometry for monitoring mineralization of nitrogen from soil organic matter. Eu J Soil Sci 51:295–304
10. Aarthi R, Sivakumar D (2020) An enhanced agricultural data mining technique for dynamic soil texture prediction. Procedia Comput Sci 171:2770–2778

11. Vakhtang S, Gerrit H (2020) A new approach to clustering soil profile data using the modified distance matrix. Comput Electron Agric 176
12. Pulkit S (2019) The most comprehensive guide k-means clustering. Analytics Vidhya
13. Lebanon G, El-Geish M (2018) Learning Python and a few more things. In: Computing with data. Springer, Cham
14. Golagani PP, Beebi SK, Mahalakshmi TS (2020) Using K-means clustering algorithm with Python programming for predicting breast cancer. In: Fiaidhi J, Bhattacharyya D, Rao N (eds) Smart technologies in data science and communication. Lecture notes in networks and systems, vol 105. Springer, Singapore
15. Awange J, Paláncz B, Völgyesi L (2020) Clustering In: Hybrid imaging and visualization. Springer, Cham
16. Calogero C, Savastano S, Zezza A (2013) Fact or artifact: the impact of measurement errors on the farm size—productivity relationship. J Dev Econ 103
17. Barman U, Choudhury RD, Talukdar N, Deka P, Kalita I, Rahman N (2018) Prediction of soil pH using HSI colour image processing and regression over Guwahati, Assam, India. J Appl Natural Sci 10(2):805–809
18. Barman U, Choudhury RD (2020) Soil texture classification using multi-class support vector machine. Inf Process Agric 7(2): 318–332. ISSN 2214-3173
19. Peng J, Loew A (2017) Recent advances in soil moisture estimation from remote sensing. Water 9(7):530. https://doi.org/10.3390/w9070530
20. Serrano D, Ávila E, Barrios M, Darghan A, Lobo D (2020) Surface soil moisture monitoring with near-ground sensors: performance assessment of a matric potential-based method. Measurement 155. ISSN 0263-2241
21. Ma X, Asano M, Tamura K, Zhao R, Nakatsuka H, Wang T (2020) Physicochemical properties and micromorphology of degraded alpine meadow soils in the Eastern Qinghai-Tibet Plateau. Catena 194:104649. https://doi.org/10.1016/j.catena.104649
22. Nascimento CM, de Sousa Mendes W, Quiñonez Silvero NE, Roberto Poppiel R, Sayão VM, Dotto AC, Valadares dos Santos N, Accorsi Amorim MT, José AMD (2021) Soil degradation index developed by multitemporal remote sensing images, climate variables, terrain, and soil attributes. J Environ Manage 277:111316. ISSN 0301-4797
23. Baude M, Meyer BC, Schindewolf M (2019) Land use change in an agricultural landscape causing degradation of soil-based ecosystem services. Sci Total Environ 659:1526–1536. https://doi.org/10.1016/j.scitotenv.2018.12.455
24. Yu GH, Chen CM, He XH, Zhang XZ, Li LN (2020) Unexpected bulk density and microstructures response to long-term pig manure application in a Ferralic Cambisol Soil: implications for rebuilding healthy soil. Soil Tillage Res 203:104668. https://doi.org/10.1016/j.still.104668
25. Villamil BV, Amiotti NM, Peinemann N (2001) Soil degradation related to overgrazing in the semi-arid Southern Caldenal area of Argentina. Soil Sci 166:441–452. https://doi.org/10.1097/00010694-200107000-00002
26. Kumar S, Sharma B, Sharma VK, Poonia RC (2018) Automated soil prediction using bag-of-features and chaotic spider monkey optimization algorithm. Evol Intell 1–12. https://doi.org/10.1007/s12065-018-0186-9
27. Oshunsanya SO (2018) Introductory chapter: Relevance of soil pH to agriculture, soil pH for nutrient availability and crop performance, IntechOpen, https://doi.org/10.5772/intechopen.82551. https://www.intechopen.com/books/soil-ph-for-nutrient-availability-and-crop-performance/introductory-chapter-relevance-of-soil-ph-to-agriculture
28. László Phd M (2009) Main parameters of soil quality and its management under changing climate 1400

Chapter 33
Distributed Association Mining for Discovering Interesting Rules for Tours and Travel Company

Manoj Sethi and Rajni Jindal

1 Introduction

Association rules mining has many applications and it is now widely used in sales, tourism, medical, marketing. Globalization has opened new challenges and opportunities for the researchers in association rule mining due to the increased volume of data and changed characteristics of data. The data is now distributed amongst different sites or locations of the companies around the globe. The size and characteristics of data generated at different sites may differ and it may yield some interesting facts which can help in improving the business. Different industries like boutique, retail stores, online stores selling grocery, households, fashion, etc. are having multiple outlets at different locations and large amount of data is being generated daily at each location. It may not be feasible or required to have a centralized data in such cases, due to volume and need, therefore data is to be analysed in the distributed manner using the distributed data mining techniques (DDM). DDM helps management to know the interesting facts from the distributed data stored at different locations, to take decisions and maximize the profit by better planning.

Distributed Association Rule Mining

Distributed association rule mining is a popular distributed data mining technique which finds interesting associations between different items in the transactions. In distributed association rule mining (DARM), there is no centralized database rather data is stored at different locations. The number of transactions stored at different locations vary in number. The focus is to find the frequent itemsets having association

M. Sethi (✉) · R. Jindal
Department of CSE, Delhi Technological University, Bawana Road, Delhi 110042, India
e-mail: manojsethi@dce.ac.in

R. Jindal
e-mail: rajnijindal@dce.ac.in

G. Mathur et al. (eds.), *Proceedings of 2nd International Conference on Artificial Intelligence: Advances and Applications*, Algorithms for Intelligent Systems,
https://doi.org/10.1007/978-981-16-6332-1_33

in the entire database globally and drive association rules. As the data are distributed so initially local frequent itemsets are found and then these local frequent itemsets are aggregated to find the global frequent itemsets, which produces some association rules. These are useful to know customer buying behaviour, increase sales, give different offers to the customers, maintain inventory of items, optimize purchase pattern and much more. These techniques not only help management to take decisions to increase the sale but also customers to save time and money. This work is motivated by the increasing use of the distributed data, changing nature of data and more and more application of DDM in the commercial applications. The objective of the study is to develop a distributed data mining application model to find interesting relationships between different parameters in the data generated by the tour and travel company which is not much explored.

The rest of the paper is organized as follows: next section is review of the related work, Sect. 3 explains the methodology used in the research, Sect. 4 shows the implementation and result analysis and finally the paper is concluded in Sect. 5. In brief, in this paper, DARM is applied on the travel transactional dataset to generate association rules for the distributed data.

2 Related Work

The most popular algorithm for association rule discovery from market, Sales databases is Apriori algorithm. This algorithm performance is good with sparse datasets. Distributed computing is important in data Mining as it requires huge amounts of resources and work distributed is required for safety, scalability and resource sharing [1]. Association rules are used to examine the customer buying behaviour for different items purchased from the supermarket transaction data. It describes how often the items are purchased together [2]. Researcher used uninorms [3] to aggregate support and confidence in market basket analysis for UK grocery. Knowing the sales pattern by the customer can help to make potential strategies to increase the sales and also for recommendations [4].

The main purpose of the research conducted by [5] on Market Basket Analysis of Beauty Products was to find the interrelation of the products in a beauty shop and use the outcome for marketing and sales promotion of the products. The researcher used the apriori algorithm for this research. It is observed that there are two types of customers, casual who buy a few random items and planned buy more. The support and confidence value used in this research were 0.1 and 0.4, respectively [6]. A case study on analysing market basket was conducted by [7] on an electronic store data using FP-Growth and apriori algorithm and generated useful rules. [8] also performed research for customer basket and found items with highest correlations. Sales Analysis is performed by [9] same as market basket analysis to target prospective customers for the optimizing the profit. The work used Product Rating in Data Mining Techniques. [10] applied the association mining in finding association between different parameters causing the road accident. This study is useful for finding hidden association and improving the road safety.

Norulhidayah [11] proposed applied DDM on daily sales analysis on the data of Chan furniture and proposed the strategy for business promotion and marketing. A periodic mining due to the dynamic nature of data in the Market Basket Data is proposed [12] by automating the threshold values by working on the change modelling concept. Study is conducted by [13] and a technique is proposed to find associations amongst items and determine better showcase to appeal to customers and increase sales. Map-reduce-based consistent and inconsistent rule detection (MR-CIRD) algorithm is proposed [14] to detect the consistent and inconsistent rules from big data and provide useful and actionable knowledge to the domain experts. These pruned interesting rules also give useful knowledge for better marketing strategy as well. Whereas distributing data and mining in parallel is different than the real-world distributed data where data is generated at different locations and data size [15] at different sites varies. Merouane [16] presented a model for distributed association rules mining by combining it with the multi-agent system to run it in parallel and distributed manner from the centralized ERP database to provide perfect response time. There are many risks also in the supply chain mining due to the different technologies used. Liu [17] proposed DDM model "DRHPDM—Data source Relevance-based Hierarchical Parallel Distributed data mining Model" for e-business. Its focus was to "improve the openness, cross-platform ability, and intelligence of the DDM system". [18] proposes hybrid architecture by combining parallel and distributed mining which reduces redundant database and improves the performance. A framework is developed by [19] to analyse and manage the supply chain risk and gives better insight for better decisions.

3 Methodology

3.1 Terminology

In this section terminologies used in the case study are explained.

Itemset. The sets of items appear frequently in the transactions of the dataset. If I {I1, I2 ... In} be a set of distinct items in the dataset DS. Itemset is a set of items, A subset of I. An itemset A with k distinct items is referred to as k-itemset [20].

Support. Frequency of occurrence of an itemset in the transaction. It is the percentage of transactions in the dataset DS that contain both itemsets A and B. The support count A/B is defined as [21, 22],

$$\text{Support}\ (A \rightarrow B) = \text{Support}\ (A \cup B)$$

Confidence. The confidence is relationship between two items. It is the percentage of transactions in the dataset DS with itemset A that also contains the itemset B. The confidence is calculated using the conditional probability which is further expressed in terms of itemset support [21, 22].

$$\text{Confidence (A} \rightarrow \text{B)} = \text{Support (A} \cup \text{B)/Support (A)}$$

Here, Support (A ∪ B) is the number of transactions that contain both itemsets A and B, and Support (A) is the number of transactions that contain itemset A.

Local and Global Frequent Itemset. The frequent itemset generated from the local partition of database available on a site, having support count more than the local support threshold value is local frequent itemset. Whereas the frequent itemset aggregated from local itemsets representing the entire database and having support count greater than the global support threshold is global frequent itemset.

Association rule. Finding interesting relationship or correlation amongst the items in the dataset. In a dataset, with n number of transactions containing a set of items. An association rule is represented by A/B, where A and B are distinct itemsets [22].

3.2 Distributed Mining of Association Rules

Distributed association rule mining is mainly finding the globally large itemsets from the distributed data for finding association rules. In the distributed data mining, data DSi ($i = 1, 2, \ldots, n$) are created or gathered or generated at different sites Si ($i = 1, 2, \ldots, n$). Let's assume, a virtual dataset DS = ∪ DSi, |DSi| is the number of transactions in DSi. |DS| is the number of transactions in dataset, for any item A, A.sup is the support count of A in DS, and A.supi is the support count of A in DSi. Given an itemset A, if A.sup ≥ min sup × |DS|, A is said to be globally frequent itemset; if A.supi ≥ minsup × |DSi|, where A is locally frequent itemset on site S.

After finding the local frequent itemsets in DARM, global frequent itemsets are generated by merging them and then to find the confidence which generates the association rules.

Assumptions

- Data are not centralized rather partitioned and distributed amongst sites
- Data are created, gathered or captured at different sites or locations of the company
- Data size at various sites vary in size and number of transactions
- There is no centralized Database and it is not required or feasible to store data at one centralized location
- There is a coordinating unit having one or more processing units
- All sites participate in the mining process to improve the performance

Algorithm

1. At each site data are scanned to find each combination of parameters age group and destination called itemset
2. At each site, support count is calculated for each found combination in step 1
3. Pruning of the itemsets are done by removing the itemsets having support count less than the threshold support count
4. The remaining itemsets called candidate sets are communicated to the coordinating site for the assignment of polling site for finding the global frequent itemsets from all candidate itemsets
5. Coordinate site(s) assigns the polling site to each local frequent itemset and communicate to all sites
6. On receiving the polling site details, all sites communicate the local count for the itemset to the polling sites
7. Polling sites calculate the global support count by aggregating the local support counts received from all sites for the assigned itemsets.
8. At the Polling sites, the itemsets having support count less than the global support threshold are removed from the list of frequent itemsets to find the final global frequent itemsets, called global pruning.
9. Polling site then calculate the confidence for the itemsets, perform global pruning by removing the itemsets having confidence less than the minimum confidence threshold.
10. Association rules are created from the global frequent itemsets generated in the last step (Fig. 1).

Data Mining Architecture. Many DARM architectures are proposed in the literature. The most suitable architecture for the tour and travel company for distributed

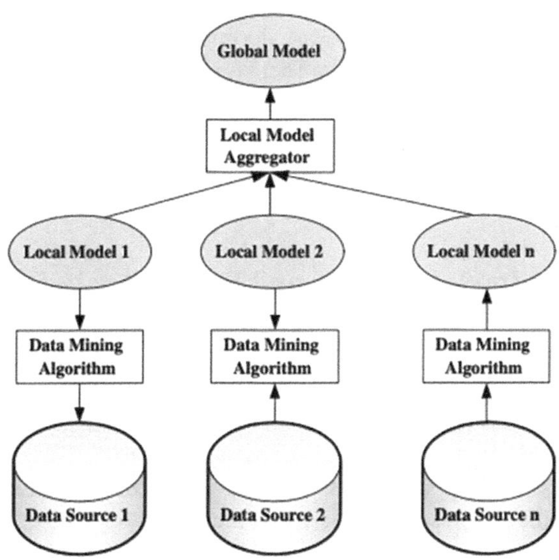

Fig. 1 A general purpose DARM architecture

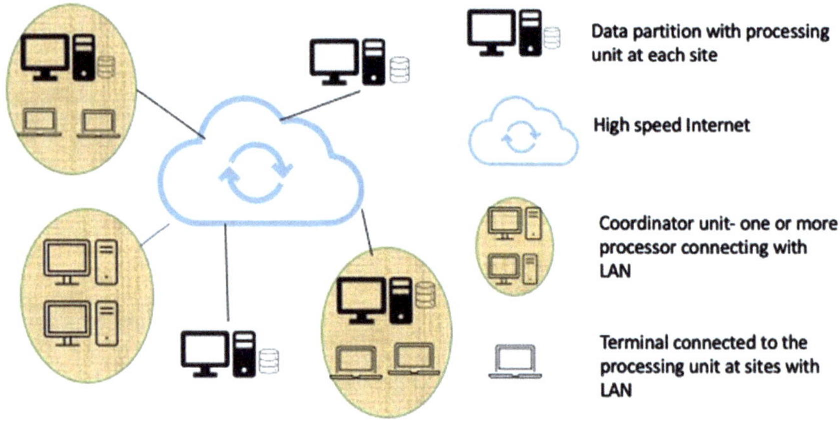

Fig. 2 Proposed model for DARM

data mining is given in figure, which was proposed by [23] as a general purpose DARM architecture. It consists of distributed data at various sites and local mining is performed and then these are aggregated to the global model. It is suitable for small as well as large scale distributed systems.

Proposed Model. Proposed model for the tour and travel industry with distributed data collection and maintenance at various sites without any centralized data repository and without any centralized data mining unit is given in the following Fig. 2. There is no centralized database and no centralized data mining unit. Data is partitioned and distributed amongst various sites, where data is collected and stored. Local mining is done at various sites and frequent itemsets are found. Coordinating unit assigns the polling site for the global processing. Number of processing units in coordinating unit may vary depending on the load based on the size of the overall data and number of sites. There is no centralized processing unit to process the global mining rather all sites participate in global mining. This is best suitable model for a large setup and large data size with a good amount of load distribution amongst the available resources.

4 Implementation and Results Analysis

This paper considers a Tour and Travel company "Voyagers Beat" for case study and implementation of the DARM techniques. The tour and travel companies can be classified as small, medium and large, based on different parameters given in Table 1.

This paper considers the medium-sized organization for the study with booking office at multiple location and multiple designations within the country having few

Table 1 Tour and travel companies' classification

Location	Destinations	Database type	Yearly transactions
Single	Single	Centralized	Few thousand
Single	Multiple	Centralized	Few thousand
Multiple	Single	Centralized	Few thousand
Multiple	Multiple	Centralized	Few thousand
Multiple	Multiple	Distributed	Few thousand to lakhs
Multiple	Inter-national	Distributed	Few lakhs
Inter-national	Inter-national	Distributed	Lakhs

lakhs of transactions. The Company has head office in Delhi and has booking offices in 13 cities for organizing package tours for various destinations in India. For the analysis purpose, 3 booking sites or distributed database locations are considered. A few destinations are chosen and are grouped into three broad destinations. For the experiments, two years 2017 and 2018 booking data are considered. Data tables given in Table 2 are taken for analysis.

Attributes Used. Above tables are combined and three attributes are taken for the analysis from the above tables. Final Attributes taken for analysis from three data sites are given in Table 3.

ID. Identification of the tourist, who is travelling. This attribute is used for reference only and not participating in the mining process.

Age. This attribute is the age of the tourist. This attribute is considered for the analysis by grouping age into three categories, i.e. 18 years to 30 years named as A1 group, 31 years to 45 years named as A2 and age greater than 45 is a third group A3.

Table 2 Data table—booking master and tourists-details

Booking master table	Tourists_details table
Booking_ID	Booking_ID
Date	Name
Booking_Name	Age
Address	
Phone_no	
Destination	
Date_from	
Date_to	
Tot_Amount	
Advance_received	
Mode_Payment	

Table 3 Data attributes taken for analysis

Attributes	Values	Descriptions
ID	Integer	Booking-ID
Age	A1(18–30), A2(31–45), A3(>45)	Tourist-age
Destination	D1(Himachal), D2(Rajasthan), D3(UP)	Place-visited

Table 4 **a** Sample data table **b** Sample data table after transformation

ID	Age	Destination
(a)		
1	24	Kasol
2	54	Agra
3	25	Pushkar
4	48	Kanpur
5	22	Manali
6	35	Agra
7	29	Shimla
8	38	Manali
9	47	Jaipur
ID	Age	Destination
(b)		
1	A1	D1
2	A3	D3
3	A1	D2
4	A3	D3
5	A1	D1
6	A2	D3
7	A1	D1
8	A2	D1
9	A3	D2

Destination. There are many destinations where the company operates. Three broad destinations are considered Himachal which includes Kasol, Manali, etc. Rajasthan includes Jaipur, Pushkar, Jodhpur, etc. and third destination UP by grouping Kanpur, Agra, etc.

Some of the destinations are preferred by youngsters and some by elderly people. The study is an attempt to find the association between these two attributes age and destination. Accordingly, the data is taken for the analysis and the sample data before and after transformation are shown in the following Table 4.

Table 5 Dataset available at various sites

	Site 1	Site 2	Site 3	Total
Total Transactions	20,872	14,273	18,370	53,515

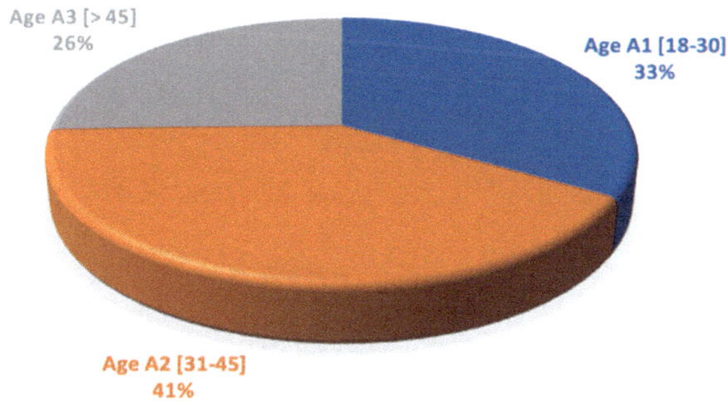

Fig. 3 Age distribution—support count

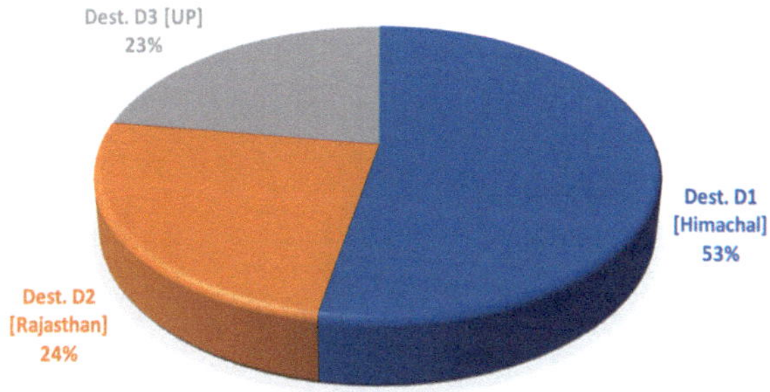

Fig. 4 Destinations frequency—support count

For the analysis, data from the three sites or locations are chosen. The number of transactions at each site and the total transactions in the database for this research are given in the following Table 5. Data mining technique is applied independently at each site and the local frequent 1-itemset are found.

The support of each age group and destination is calculated. Figure 3 shows age group A2 is 41%, A1 is 33% and A3 is 26% only. Whereas most preferred destination is D1 at 53%, then D2 at 24% and D3 at 23% shown in Fig. 4 below. D2 and D3 are almost equally preferred, whereas D1 is the most visited destination. Then, the 2-itemset with the combination of age groups and destination groups and their local support counts at all sites are found in Table 6.

To test the model and develop the association rules two experiments are conducted with minimum threshold support 0.200 and 0.100 given in Table 7.

Table 6 Local support count (%) at each site

	Site 1 support	Site 2 support	Site 3 support
Age A1 - Dest. D1	0.261	0.198	0.159
Age A1 - Dest. D2	0.068	0.072	0.106
Age A1 - Dest. D3	0.030	0.014	0.016
Age A2 - Dest. D1	0.281	0.221	0.150
Age A2 - Dest. D2	0.078	0.055	0.094
Age A2 - Dest. D3	0.042	0.029	0.081
Age A3 - Dest. D1	0.025	0.073	0.045
Age A3 - Dest. D2	0.044	0.063	0.078
Age A3 - Dest. D3	0.127	0.098	0.158

Table 7 Threshold values

	Support	Confidence
Experiment 1	0.200	0.500 & 0.300
Experiment 2	0.100	0.500 & 0.300

Experiment 1: Minimum support threshold = 0.200.

Local pruning. In this experiment, local frequent itemsets are calculated. The itemsets whose support count on at least one of the sites is more than 0.200 are the candidate sets and are communicated for finding the global frequent itemsets as under in Table 8.

Global Pruning. The itemsets having global support count less than 0.200 are eliminated. The global support and confidence are calculated for the global frequent itemsets, given in Table 9.

Association rules for confidence threshold 0.500

Table 8 Candidate sets for support count threshold = 0.200

	Site 1 support	Site 2 support	Site 3 support
Age A1—Dest. D1	0.261		
Age A2—Dest. D1	0.281	0.221	

Table 9 Global support and confidence

	Global support	Global confidence
Age A1 → Dest. D1	0.209	0.625
Age A2 → Dest. D1	0.220	0.536
Dest. D1 → Age A1	0.209	0.441
Dest. D1 → Age A2	0.220	0.464

These rules Table 10 show that the destination D1 (Himachal) is the preferred destination for age groups A1 (18–30) and A2(31–45).

Association rules for confidence threshold 0.300

The rules in Table 11 show that the destination D1 (Himachal) is the preferred destination for age groups A1 (18–30) and A2(31–45). Secondly, the destination D1 (Himachal) is mostly visited by the young tourists and not by elderly tourists.

Experiment 2: Minimum support threshold = 0.100

Local pruning. In this experiment local frequent itemsets, whose support count on at least one of the sites is more than 0.100 are communicated for finding the global frequent itemsets shown in Table12.

Global Pruning. The itemsets having global support count less than 0.100 are eliminated. The global support and confidence are calculated for the global frequent itemsets are given in Table 13.

*Association rules for confidence threshold 0.*500

These rules show that the destination D1 (Himachal) is a preferred destination for age groups A1 (18–30) and A2 (31–45). Also, age A3 prefers to go to destination D3 and D3 is mostly visited by A3 shown in Table 14.

Table 10 Association rules for support threshold 0.200 and confidence threshold 0.500

	Global Support	Global Confidence
Age A1 → Dest. D1	0.209	0.625
Age A2 → Dest. D1	0.220	0.536

Table 11 Association rules for support threshold 0.200 and confidence threshold 0.300

	Global support	Global confidence
Age A1 → Dest. D1	0.209	0.625
Age A2 → Dest. D1	0.220	0.536
Dest. D1 → Age A1	0.209	0.441
Dest. D1 → Age A2	0.220	0.464

Table 12 Candidate sets for support threshold = 0.100

	Site 1 support	Site 2 support	Site 3 support
Age A1—Dest. D1	0.261	0.198	0.159
Age A1—Dest. D2			0.106
Age A2—Dest. D1	0.281	0.221	0.150
Age A3—Dest. D3	0.127		0.158

Table 13 Global support and confidence

	Global support	Global confidence
Age A1 → Dest. D1	0.209	0.625
Age A2 → Dest. D1	0.220	0.536
Age A3 → Dest. D3	0.130	0.510
Dest. D1 → Age A1	0.209	0.441
Dest. D1 → Age A2	0.220	0.464
Dest. D3 → Age A3	0.130	0.641

Table 14 Association rules for support threshold 0.100 and confidence threshold 0.500

	Global support	Global confidence
Age A1 → Dest. D1	0.209	0.625
Age A2 → Dest. D1	0.220	0.536
Age A3 → Dest. D3	0.130	0.510
Dest. D3 → Age A3	0.130	0.641

Association rules for confidence threshold 0.300

These rules in Table 15 clearly indicate that there is a strong association between Age A1, A2 and Destination D1 where confidence is high implies tourists up to middle age like to visit destination Himachal also Himachal is mostly visited by youngsters

Table 15 Association rules for support threshold 0.200 and confidence threshold 0.300

	Global support	Global confidence
Age A1 → Dest. D1	0.209	0.625
Age A2 → Dest. D1	0.220	0.536
Age A3 → Dest. D3	0.130	0.510
Dest. D1 → Age A1	0.209	0.441
Dest. D1 → Age A2	0.220	0.464
Dest. D3 → Age A3	0.130	0.641

and middle-aged tourists. On the other side A3 age group preference is UP and vice versa as there is a good association between Age A3 and Destination D3.

The relationship between the local support, global support and confidence are also shown in Figs. 5, 6, and 7 and discussed below. It also shows some interesting

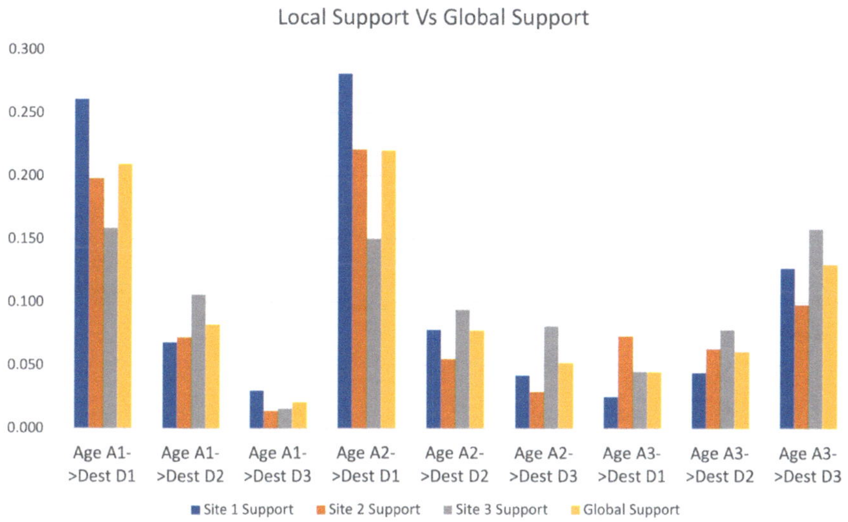

Fig. 5 Local support count at various sites and global support

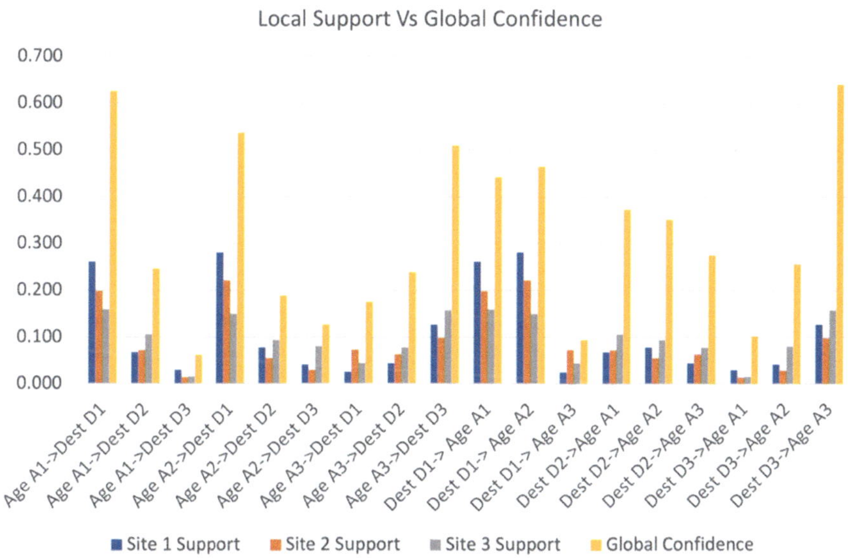

Fig. 6 Local support count at various sites and global confidence

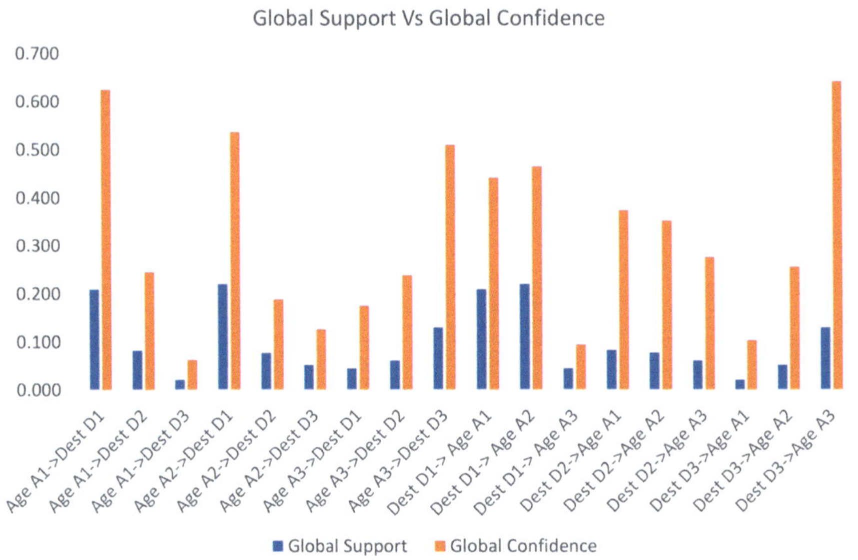

Fig. 7 Global support count and global confidence

relationships amongst them.

Figure 5 shows that global support is proportional to the average of the local supports at different sites, whereas global confidence can be high for low as compared to support count as the case of D3 → A3 in Fig. 7. It is observed in Fig. 6 that support count of A1 → D1 is high at site 1 and 2 and not on site 3 but the confidence is high. Whereas support D3–A3 is little high at site 3 only but confidence is quite high. Similarly, support D2 → A1is not high at all sites. This shows that the support at each site can't be sufficient enough to make the global rules. At the same time if the company needs the local association rules for each site that can be generated from the local frequent itemsets and this may be different from the global ones. Local rules are also important to make local strategies in different states or locations.

This case study developed the distributed data association rule mining model for the tour and travel company. Based on the study many association rules are generated. The results show that the tourists of age group A1 and A2 are more as compared to A3 hence gives the travelling habit based on the age. Similarly, the destination D1 is preferred over destination D2 and D3 by the tourists. Whereas destination D3 is visited more by age group A3.

5 Conclusion and Future Work

In this research, distributed association rule mining model is proposed for tour and travel industry to find the interesting facts and relations between the age and the destination visited by the tourists. This study helped the company to take decision about opening new booking office and also focused on the age of the tourists and adding new destinations. This study helped the company to expand its business. A medium-level company is chosen for the study having a few offices in the country. This research may help in expanding the business by adding different destinations, more booking offices in different locations and targeting the potential customers. This study shows that the mining is useful in the tourism industry and can give useful information. There are many more parameters which can affect the tourism industry like weather, gender, customer place of living, etc., which can also be considered for mining new facts. The mining can be applied to the bigger sized international company.

References

1. Ali AA, Varacha P, Krayem S, Zacek P, Urbanek A (2018) Distributed data mining systems: techniques, approaches and algorithms. In: MATEC web conf. 22nd international conference on circuits, systems, communications and computers (CSCC 2018), vol 210, p 04038
2. Ramesh Babu I, Pavani V (2019) Three level cloud storage scheme for providing privacy preserving using edge computing. Int J Adv Sci Technol 28(16):1929–1940
3. Moodley R, Chicilana F, Caraffini F, Carter J (2019) Application of uninorms to market basket analysis. Int J Intell Syst 34(1)
4. Kurnia Y, Isharianto Y, Giap YC, Hermawan A (2018) Study of application of data mining market basket analysis for knowing sales pattern (association of items) at the O! Fish restaurant using apriori algorithm. J Phys Conf Ser 1175:012047
5. Gancheva V (2013) Market basket analysis of beauty products. Financial Economics Erasmus University Rotterdam
6. Isa N, Kamaruzzaman NA, Ramlan MA, Mohamed N, Puteh M (2018) Market basket analysis of customer buying patterns at corm café. Int J Eng Technol 7(4.42)
7. Mehay A, Singh K, Sharma N (2013) Analyze market basket data using fp-growth and apriori algorithm. Int J Recent Innov Trends Comput Commun 1(9):693–696
8. Kouzis-loukas M (2014) Analysing customer baskets—a business to business case study. Financial Economics
9. Bhagwat S, Jethliya V, Pandey A, Islam L (2015) Sales analysis using product rating in data mining techniques. Int J Res Eng Technol 4(02):189–191
10. Ait-Mlouk A, Gharnati F, Agouti T (2017) An improved approach for association rule mining using a multi-criteria decision support system: a case study in road safety. Eur Transp Res Rev 9(40)
11. Isa N, Mohd Yusof NS, Atif Ramlan M (2019) The implementation of data mining techniques for sales analysis using daily sales data. Int J Adv Trends Comput Sci Eng 8(1.5):74–80
12. Kaur M, Kang S (2016) Market basket analysis: identify the changing trends of market data using association rule mining. In: International conference on computational modeling and security (CMS 2016). Procedia Computer Science 85:78–85
13. Ahn K-I (2012) Effective product assignment based on association rule mining in retail. Expert Syst Appl 39:12551–12556
14. Dinesh J. Prajapati, Sanjay Garg, Chauhan NC (2017) Interesting association rule mining with consistent and inconsistent rule detection from big sales data in distributed environment. Futur Comput Inform. J 2:19–30
15. Sethi M, Jindal R (2020) Distributed frequent itemset mining using size based assignment technique. Int J Emerg Trends Eng Res 8(10):5765–6773

16. Zoubeidi M, Okba Kazar, Saber Benharzallah, Nadjib Mesbahi, Abdelhak Merizig, Djamil Rezki (2019) A new approach agent-based for distributing association rules by business to improve decision process in ERP systems. Int. J. Inf. Decis. Sci 12(1)
17. Liu B, Cao SG, He W (2011) Distributed data mining for e-business. Inf Technol Manag 12:67–79
18. Vasoya A, Koli N (2016) Mining of association rules on large database using distributed and parallel computing. Procedia Computer Science 79:221–230
19. Kara ME, Fırat SÜO, Ghadge A (2020) A data mining-based framework for supply chain risk management. Comput Ind Eng 139:105570
20. Srikumar K, Bhasker B (2005) Metamorphosis: mining maximal frequent sets in dense domains. Int J Artif Intell Tools 14(3):491–506
21. Han J, Kamber M, Pie J (2011) Data mining concepts and techniques. Morgan Kaufmann Publishers, San Francisco
22. Agrawal R, Imielinski T, Swami A (1993) Mining association rules between sets of items in large databases. In: Proceedings of international conference of ACM-SIGMOD on management of data, pp 207–216
23. Cesario E, Talia D (2008) Distributed data mining models as services on the grid. In: IEEE international conference on data mining workshops Pisa, TBD, Italy, pp 486–495

Chapter 34
Estimation of Wave Overtopping Discharge at Quarter Circle Breakwater Using LSSVM

Haritha Sasikumar, Vishwanatha Mane, and Subba Rao

1 Introduction

Global warming and climate change have caused a rise in seawater levels. These changes have raised the necessity to accurately estimate the wave overtopping discharge for the construction of coastal structures. Wave overtopping occurs when water passes over the crest of the coastal structure, when the height of the structure is less than the wave height. Accurate prediction of wave overtopping discharge is required to prevent hazards. There are various empirical formulas and numerical modeling methods available for the estimation of wave overtopping. But they are often complex to solve. Though physical model studies give accurate results, they are expensive and time-consuming. Soft computing techniques can be used in the preliminary design stage of coastal structures.

The model considered here is a Quarter Circle Breakwater (QCB) (Fig. 1). It consists of a quarter circular surface facing the incoming wave, a rear vertical wall, and a base slab resting on a rubble mound foundation. They are of submerged and emerged type. The crest level of Emerged Quarter Circle Breakwater will be above still water level. The seaside surface of QCB may be unperforated or with perforations. For the perforated seaside surface, most of the wave energy will be absorbed. The paper discusses mean overtopping discharge which is function of wave height (H_i), water depth (d), time period (T), crest height with respect to water level (R_c), and percentage perforations (p).

Soft computing techniques have become widely popular in the field of coastal engineering. They are techniques based on artificial intelligence. They employ computers to solve complex problems using the knowledge from experimental data thereby utilizing the speed and robustness of machines. Support Vector Machines come under

H. Sasikumar (✉) · V. Mane · S. Rao
Department of Water Resources and Ocean Engineering, National Institute of Technology, Surathkal, Karnataka, India

G. Mathur et al. (eds.), *Proceedings of 2nd International Conference on Artificial Intelligence: Advances and Applications*, Algorithms for Intelligent Systems,
https://doi.org/10.1007/978-981-16-6332-1_34

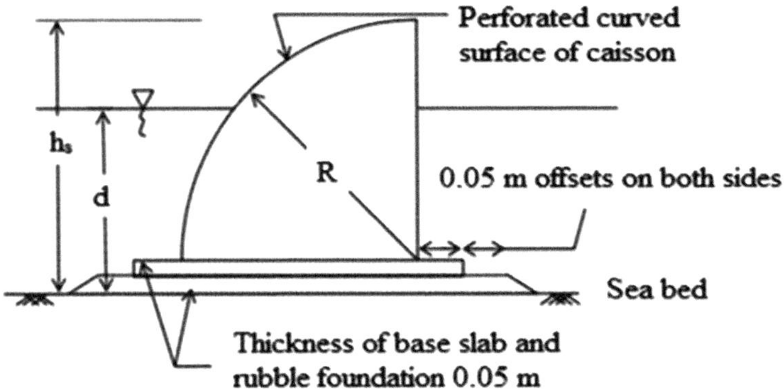

Fig. 1 Section of quarter circle breakwater

the category of supervised learning technique. Least Squares Support Vector Machine (LSSVM) proposed by Suykens and Vandewalle gives a least-squares version of SVM. LSSVM method is found to maintain the strengths of Artificial Neural Network (ANN), and it is able to overcome some of its deficiencies [10]. The LS-SVMlab toolbox in MATLAB provides simple code for solving classification and regression problems. Training and testing of LSSVM models with different kernel functions can be performed. Kernel functions such as Linear, Polynomial and Radial Basis Function (RBF) are considered here. Following are the steps of creating LSSVM model using LS-SVMlab in MATLAB.

1. The command initlssvm is carried out using object-oriented interface.
2. The parameters are tuned using tunelssvm.
3. The tuned parameters are used to train the model using trainlssvm.
4. The performance of model on test data can be simulated using simlssvm.

The parameters to be optimized in LSSVM include the regularization parameter (γ) and for RBF kernel is the squared kernel bandwidth (σ^2). A smaller γ would ensure better fitting of the model. As bandwidth increases the hyperplane would become flatter which would decrease the capability of the kernel to fit the function [8]. The parameters are optimized using grid search.

2 Methodology of LSSVM

Support Vector Machines is a supervised learning technique that has been successfully applied in solving classification and regression problems. SVM maps input vectors into a high-dimensional feature space through some non-linear mapping. An optimal hyperplane is created in this space [15]. SVM was developed mainly for classification problems but later it was successfully applied for regression problems as well. In Support Vector Regression Vapnik's ε-sensitive region is introduced. Here, an ε-tube is formed around the estimated function such that the absolute values of error less than ε are ignored [2].

LSSVM solves a set of linear equations, whereas in SVM quadratic programming is used. The performance of the model depends on the accurate optimization of the parameters. The different kernel functions are available to create different types of hyperplane. The kernel functions most commonly used are radial basis function, polynomial, and linear. The regularization parameter γ is common for all kernels. The linear kernel function is expressed as

$$K(x_i, x_j) = x_i^T x_j \tag{1}$$

The polynomial kernel function is given as

$$K\left(x_i, x_j\right) = \left(x_i^T, x_j + t\right)^d, t \geq 0 \tag{2}$$

where t is the intercept and d is the degree of polynomial.

RBF kernel function is expressed as below

$$K\left(x_i, x_j\right) = \mathrm{e}^{-\left(\frac{||x_i - x_j||^2}{2\sigma^2}\right)} \tag{3}$$

where σ^2 is the squared kernel bandwidth. It represents the variance of the Gaussian kernel. The various commands for developing LSSVM model in LS-SVMlab is given in Fig. 2.

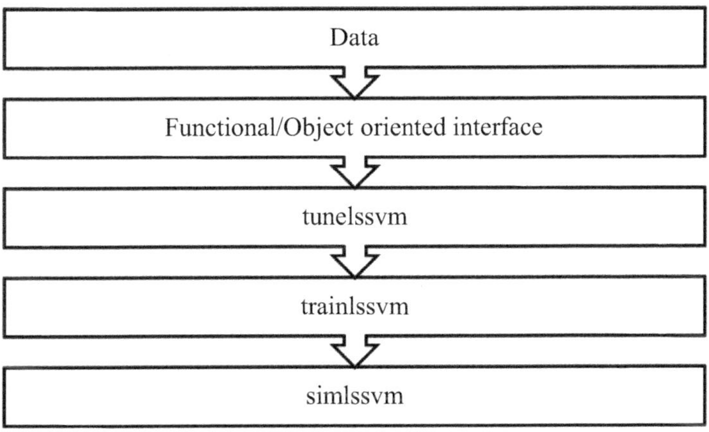

Fig. 2 Commands in LS-SVMlab [9]

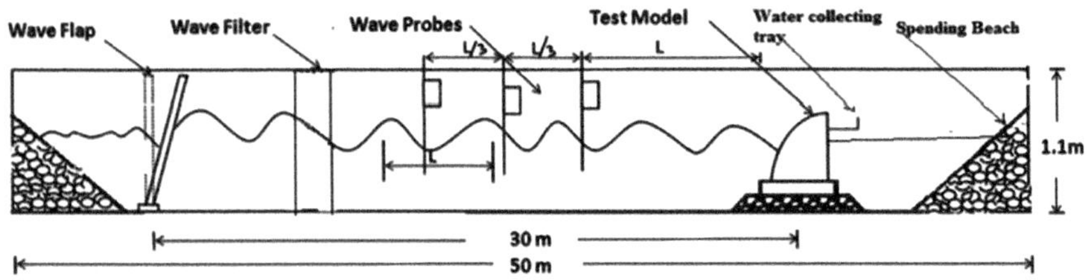

Fig. 3 Longitudinal section of the wave flume

3 Data Collection

The data for the present study is collected from the previous physical model studies carried out in the Marine Structures Laboratory of the Department of Water Resources and Ocean Engineering, NITK Surathkal, India.The longitudinal section of the wave flume is given in Fig. 3. The dimensions of wave flume are as follows: Length = 50 m, width = 0.71 m, and depth = 1.1 m. It has a deep chamber with length = 6.3 m, width = 1.5 m, and depth = 1.4 m at one end where the bottom-hinged flap (wave paddle) generates waves.

The physical model studies were conducted on QCB models of different radii with percentage of perforations 0, 1.25, 4, 5, and 20%. In the investigation, wave overtopping was measured as the amount of water overtopped over the structure, which is collected in a tray. It is attached to the lee side of the breakwater model. A burst of five waves is generated to obtain five overtops. Hence, overtop discharge per meter crest length of the breakwater is calculated by dividing the total volume of water by the time period, and crest length of breakwater.

The experimental data comprises 468 data points. The data is randomly segregated to include points of every range in both train and test data. The train and test consist of 355 (75%) and 113 (25%) data points, respectively. Dimensional analysis was carried out in the physical model study using Buckingham's π theorem. The parameters for training the network are given as non-dimensional parameters. Following are the parameters used for training the network: H_i/gT^2, d/gT^2, R_c/H_i, perforation percentage, and q/gTH_i (Table 1).

4 Results

LS SVM models with kernel function Linear, Polynomial, and Radial Basis Function are developed. The parameters γ, σ^2, and t are tuned using grid search in LS-SVMlab. The optimized parameters are given in Table 2. The statistical parameters obtained for different kernel functions are listed in Table 3. It is observed that the RBF kernel function produced better results compared to Linear and Polynomial kernel functions. The scatter plots for train and test data of RBF kernel are given in Figs. 4 and 5.

Table 1 Parameter description

Parameter	Description of the parameter
H_i	Wave height
d	Depth of water
T	Wave period
R_c	Crest height with respect to water level
p	Percentage of perforations
q	Wave overtopping discharge
H_i/gT^2	Non-dimensional wave steepness
d/gT^2	Non-dimensional depth
R_c/H_i	Non-dimensional crest height
q/gTH_i	Non-dimensional wave overtopping

Table 2 Optimized parameters

Parameters	RBF	Linear	Polynomial
γ	25.7615	11.6213	0.2915
σ^2	0.2606	–	–
t	–	–	113.84
d	–	–	3

Table 3 Statistical parameters for different kernel functions

Statistical parameters	Linear		Polynomial		RBF	
	Train	Test	Train	Test	Train	Test
RMSE	0.00035	0.00036	0.00027	0.00029	0.00019	0.00025
NSE	0.45261	0.43426	0.65826	0.63396	0.83338	0.72759
CC	0.67279	0.66318	0.81138	0.79812	0.91434	0.85509
SI	0.725	0.84393	0.5728	0.67883	0.39999	0.585604

5 Summary and Conclusions

5.1 Summary

LSSVM has been successfully applied for predicting mean overtopping discharge at quarter circle breakwater for varying radii and perforations. The data is collected from the previous experimental studies conducted at the NITK wave flume laboratory. Non-dimensional parameters H_i/gT^2, d/gT^2, R_c/H_i and p are given as input and q/gTH_i as output. Overtopping values equal to zero are not considered in this study. LSSVM models were developed using the LS-SVMlab toolbox in MATLAB.

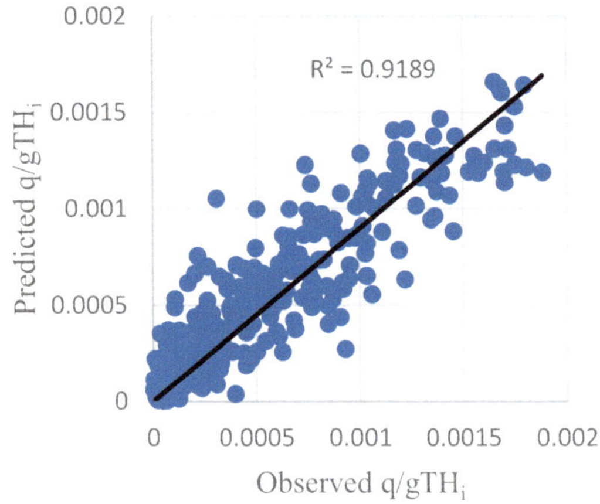

Fig. 4 Scatter plot of train data for RBF Kernel

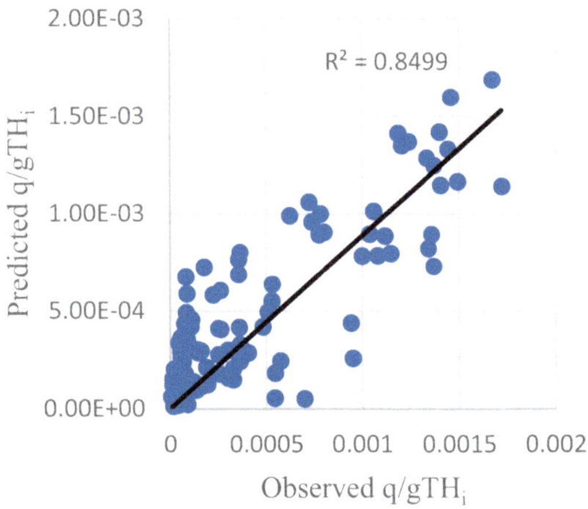

Fig. 5 Scatter plot of test data for RBF Kernel

Models with kernel function Linear, Polynomial, and RBF were developed. The hyperparameters are optimized using grid search in LS-SVMlab. The models are assessed and evaluated using statistical parameters RMSE, NSE, CC, and SI.

5.2 Conclusions

- The results of statistical parameters show that the LSSVM model with RBF kernel performed best compared to others.
- The optimized values of γ and σ^2 are 25.7615 and 0.2606, respectively.
- The LSSVM model with RBF kernel function shows good correlation of 0.91434 during training phase with RMSE = 0.00019, NSE = 0.83338, and SI = 0.39999.

- The LSSVM model with RBF kernel function shows good correlation of 0.85509 during testing phase with RMSE = 0.00025, NSE = 0.72759, and SI = 0.585604.
- The scatter plots for train and test give regression 0.9189 and 0.8499, respectively.

References

1. Hegde AV, Sharhabeel PS, Mohan S (2015) Stability of a perforated quarter circle breakwater. Int J Ocean Clim Syst 6(4):185–194
2. Awad M, Khanna R (2015) Support vector regression. In: Efficient learning machines. Apress, Berkeley, CA
3. Binumol S, Subba Rao, Arkal Vittal Hegde (2015) Runup and Rundown characteristics of an emerged seaside perforated quarter circle breakwater. In: International conference on water resources, coastal and ocean engineering. Aquat Procedia 4:234–239
4. Budime Raju, Hegde AV, Chandrashekar O (2015) Computational intelligence on hydrodynamic performance characteristics of emerged perforated quarter circle breakwater. In: Proceedings of 8th international conference on Asian and pacific coasts. Procedia Eng 116:118–124
5. Kuntoji G, Rao S, Manu, Mandal S (2017) Application of support vector machine technique for damage level prediction of tandem breakwater. Int J Earth Sci Eng 10(03):633–638
6. Kuntoji G, Rao S, Manu, Mandal S (2017) Performance evaluation of ANFIS and SVM model in prediction of wave transmission over submerged reef of tandem breakwater. Int J Ecol Dev 32(2)
7. Rubio G, Pomares H, Rojas I, Herrera LJ, Guillen A (2009) Efficient optimization of the parameters of LS-SVM for regression versus cross-validation error. Alippi C et al. (eds) ICANN 2009, Part II, LNCS 5769, pp 406–415
8. Zeng J, Tan Z-H, Matsunaga T, Shirai T (2019) Generalization of parameter selection of SVM and LS-SVM for regression. Mach Learn Knowl Extr 1:745–755
9. De Brabanter K, Karsmakers P, Ojeda F, Alzate C, De Brabanter J, Pelckmans K, De Moor B, Vandewalle J, Suykens JAK (2011) LS-SVMlab: a MATLAB/C toolbox for least squares support vector machines, ESAT-SCD-SISTA K.U. Leuven
10. Harish N, Mandal S, Rao S, Patil SG (2015) Particle swarm optimization based support vector machine for damage level prediction of non-reshaped berm breakwater. Appl Soft Comput 27:313–321
11. Gedik N (2018) Least squares support vector mechanics to predict the stability number of rubble-mound breakwaters. Water 10(10)
12. Samui P, Kothari DP (2011) Utilization of a least square support vector machine (LSSVM) for slope stability analysis. Sci Iran A 18(1):53–58
13. Huang T-M, Kecman V, Kopriva I (2006) Kernel based algorithms for mining huge data sets. Stud Comput Intell (SCI) 17:1–9
14. Van Gestel T, Suykens JAK, Baesens B, Viaene S, Vanthienen J, Dedene G, De Moor B, Vandewalle J (2004) Benchmarking least squares support vector machine classifiers. Mach Learn 54:5–32
15. Vapnik V (1995) The nature of statistical learning theory. Springer, New York
16. Yusof Y, Mustaffa Z (2016) A Review on optimization of least squares support vector machine for time series forecasting. Int J Artif Intell Appl (IJAIA) 7(2)

Chapter 35
Hybrid CNN-LSTM for Traffic Flow Forecasting

V. Rajalakshmi and S. Ganesh Vaidyanathan

1 Introduction

Traffic congestion is a critical issue faced on roads [1]. It is caused when vehicles travel very slow due to a decrease in road capacity or more number of vehicles on roads. It may happen due to unexpected events such as accidents and road closing or due to situations like festivals, seasonality, periodic changes, etc. The congestion caused due to unexpected events cannot be predicted and are unavoidable, whereas the situations that are time bound can be forecasted earlier and their impact can be reduced. Traffic flow forecasting provides a solution to handle these congestion issues. It is a great evolution in intelligent transportation systems [2] nowadays. It enables to monitor traffic conditions in order to reduce the travel delays on roads. The main objective of such a system is to predict accurate traffic flow information in order to avoid congestion on roads.

Many historical methods were used initially which considered only the past data for identical situations and also restricted its prediction to a specific location. Later various statistical methods such as Autoregressive Moving Average (ARMA), Integrated ARMA (ARIMA) [3] and Seasonal ARIMA (SARIMA) [4] were introduced to forecast the traffic density. These methods used the past data and applied some statistical computation to predict better than historical methods. Recently, machine learning and deep learning models are used in traffic flow forecasting.

V. Rajalakshmi (✉)
Department of Computer Science and Engineering, Sri Venkateswara College of Engineering, Chennai, India
e-mail: vraji@svce.ac.in

S. Ganesh Vaidyanathan
Department of Electronics and Communication Engineering, Sri Venkateswara College of Engineering, Chennai, India
e-mail: gvaidyas@svce.ac.in

G. Mathur et al. (eds.), *Proceedings of 2nd International Conference on Artificial Intelligence: Advances and Applications*, Algorithms for Intelligent Systems,
https://doi.org/10.1007/978-981-16-6332-1_35

The rest of the paper is outlined as follows. Section 2 discusses the literature review. Section 3 examines about the deep learning models. Section 4 describes the proposed hybrid CNN-LSTM framework. Section 5 discusses the results in comparison with various models. Finally, Sect. 6 concludes by discussing the efficacy of the system.

2 Literature Review

Lamyaa Sadouk proposed time series CNN approaches [3] for time series classification. The frequency-domain data was derived from time-domain data using the Stockwell transformation. Later adaptive convolutional filter was used in CNN for time series classification. Shangyu Sun et al. introduced a kind of multi-branch deep learning model [5] to interpret the short-term traffic flow in the city. This model uses external factors such as weather and accidents and spatiotemporal features as its input and predicts the future traffic flow.

Yashan Wang et al. discussed the approaches [6] to extract spatial features through Graph Convolutional Network (GCN) and temporal features through Long Short-Term Memory (LSTM). The GCN-LSTM model is proposed in this paper for traffic flow forecasting which produced better results compared to other deep learning models. Di Yang et al. proposed a multifeature-based CNN model (MF-CNN) that predicts the traffic flow using the external factors (holidays and weather) and the spatiotemporal features [7]. The five state-of-the-art models are implemented in order to compare the results with the proposed model and prove the obtained accuracy was quite good.

Hailong Zhu et al. proposed a new strategy based on RNN-GCN and BRB for traffic flow prediction [8]. To extract the new traffic data, data fusion is done using a Belief Rule Base (BRB). The time correlation from the traffic data is obtained by the RNN-GCN model. The RNN-GCN showed better performance compared with ARIMA, GCN and LSTM. Wenjie et al. proposed a CNN-LSTM model [9] for stock price prediction. The proposed model brings in the highest accuracy while comparing the other models such as CNN, MLP, LSTM, RNN and CNN-RNN. Hassan et al. discusses various deep learning models [10] specific to time series forecasting. In the paper, the author applied various models in different applications for a univariate dataset and 12 multivariate datasets.

Xianglong Luo et al. proposed KNN-LSTM [11]-based traffic flow prediction model. Here, KNN captures the spatial features and the LSTM captures the temporal features. The model showed better performance than the ARIMA [12], ANN, SVR, DBN-SVR and LSTM models. Aniekan et al. proposed Long Short-Term Memory (LSTM) Stacked Autoencoder (SAE) model [13] which predicts the bi-directional traffic flow information from the weather data, traffic data and tweet messages. The proposed model improves the prediction accuracy compared to the existing models. Wentain et al. proposed the Temporal Convolutional Network (TCN) model [14] for traffic prediction. The model showed good results compared to the deep learning models LSTM, GRU, SAE and CNN models.

A Convolutional LSTM NN model was developed by Zhang et al. [15] for multi-lane traffic prediction. The Unique features of the proposed model were the traffic conditions were characterized by applying multiple features and the rolling-prediction approach was used in predicting the multi-step forecast. The above study on the various deep learning models brings in the motivation to propose a hybrid CNN-LSTM model for time series forecasting.

3 Deep Learning Models

In this section, a detailed working of the deep learning models Convolutional Neural Networks (CNN) and Long Short-Term Memory (LSTM) is discussed.

3.1 Convolutional Neural Networks (CNN)

CNN is superior to other networks because it is capable of providing local connectivity. Each and every node is connected to the input local region alone. This is implemented by introducing [7] convolutions instead of weighted sums. In every layer, a feature map is created by convoluting the input with the weight matrix. The number of learning parameters is reduced considerably due to the local connectivity and the concept of shared weights in CNNs. Hence, CNN achieves efficient training and learning capability.

The Deep CNN architecture can be formed by stacking the convolution, ReLU and max-pooling layers on top of one another. The output from these layers acts as input for the fully connected layer and for the activation. On processing the L layers, the resultant output of the network is denoted by the matrix f^L. Finally, the model is trained recursively to reduce the error between the actual and the predicted output.

3.2 Long Short-Term Memory (LSTM)

In deep learning, Long Short-Term Memory (LSTM) is a variant of the artificial Recurrent Neural Network (RNN) architecture. The uniqueness of LSTM is that it has feedback connections and it is able to process ordered patterns of data points. It is suitable for applications that are related to time series data.

The traffic flow at time t denoted as x_t is passed as input to nodes of hidden layer h_t. The nodes output from one hidden layer is passed as input to nodes of the next hidden layer and so on. Finally, the output of the nodes from the last hidden layer is given as input to the nodes at the output layer y_t. The weights between the different layers are denoted as W_i and the bias at each layer is denoted by b. The activation function such as sigmoid, the hyperbolic tangent (tanh) and the Rectified Linear Unit (ReLU) is selected based on the necessity.

A LSTM architecture comprises of a cell, a forget gate, an input gate and an output gate [9]. The cell is capable of remembering information related to time series intervals. The three gates of LSTM are capable of regulating the flow of time bound information to and from the cell. LSTM consists of a cell memory unit whereas RNN consists of a recurrent unit. The advantage of LSTM is the cell vector which will forget the stored memory contents and will add the new contents.

4 Proposed Hybrid CNN-LSTM

4.1 Data Preprocessing and Network Initialization

The input data is collected and preprocessed before training the model. The preprocessing is done using the z-score normalization method. The procedure is implemented by the formula in Eq. 1:

$$p_i = \frac{d_i - \overline{d}}{s} \tag{1}$$

where the standardized value is denoted as p_i, the input data is denoted as d_i, the average of the input data is denoted as $\overline{d}$, and the input data's standard deviation is denoted as s.

The weights and the biases of various layers are initialized for the network.

4.2 Hybrid CNN-LSTM Model

The structure of the CNN-LSTM model is described in Fig. 1. The preprocessed data p_i is passed on to the convolution and max-pooling layers of CNN successively. The feature extraction is carried out and the output is passed on to LSTM. The LSTM processes the output of CNN to improve the feature extraction and produces the output. This is further moved to the fully connected and dense layers. The resultant output is compared to the actual output to calculate the error of the model. The error is backpropagated into the network to update the weights and biases at the various layers. These above processes are repeated until the loss from the model is negligible.

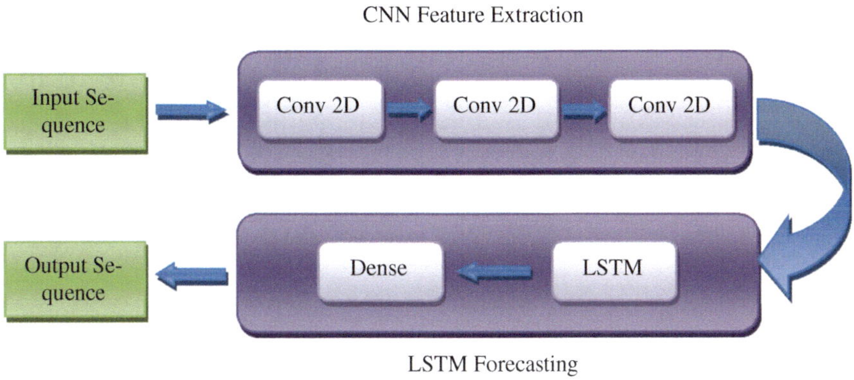

Fig. 1 Proposed CNN-LSTM model

4.3 Model Evaluation

The trained model is then used for forecasting the future traffic flow. The model is evaluated with the test data using the various metrics such as Mean Square Error (MSE) Rate, Mean Absolute Error (MAE) Rate, Mean Absolute Percentage Error (MAPE) Rate and Goodness of fit which can be measured by R-squared (R^2).

5 Results and Discussion

The time series data is extracted from the MIDAS Site. The data consists of traffic flow from January 2020 to November 2020. The peak hour data on weekdays from 8 to 9 AM is considered for prediction. The data retrieved in total is for 204 days. The extracted data needs to be preprocessed before applying it to time series forecasting models. Figure 2 shows the traffic flow pattern before preprocessing. The data is normalized using the Z-Score Normalization method. Figure 3 shows the traffic

Fig. 2 Before preprocessing

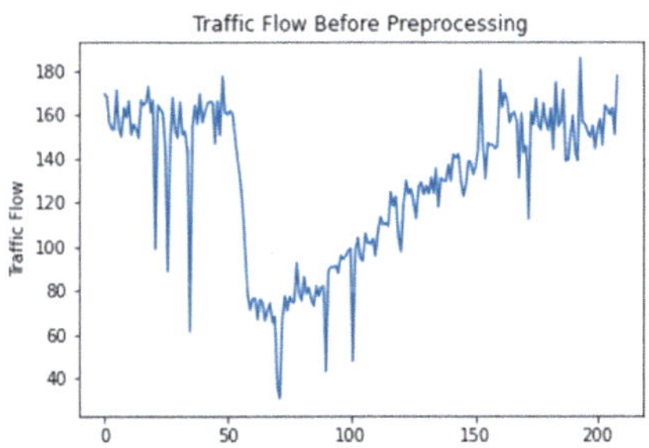

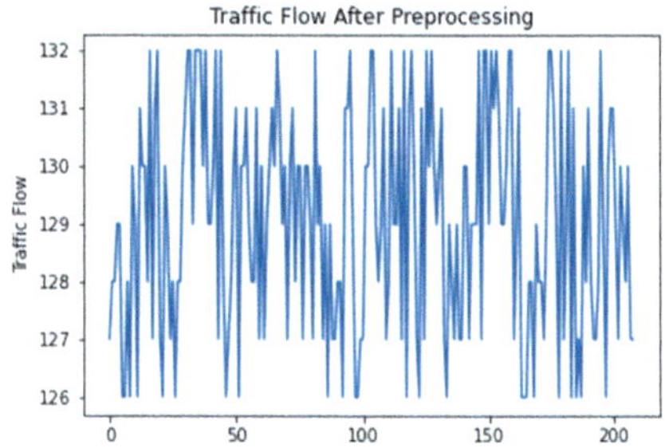

Fig. 3 After preprocessing

Table 1 Performance of CNN, LSTM and Hybrid CNN-LSTM models

	CNN	LSTM	Hybrid CNN-LSTM
MSE	0.3096	0.2123	**0.0568**
R^2	0.9266	0.9544	**0.9878**
MAE	0.4011	0.3379	**0.1434**
MAPE	28.35	26.17	**11.07**

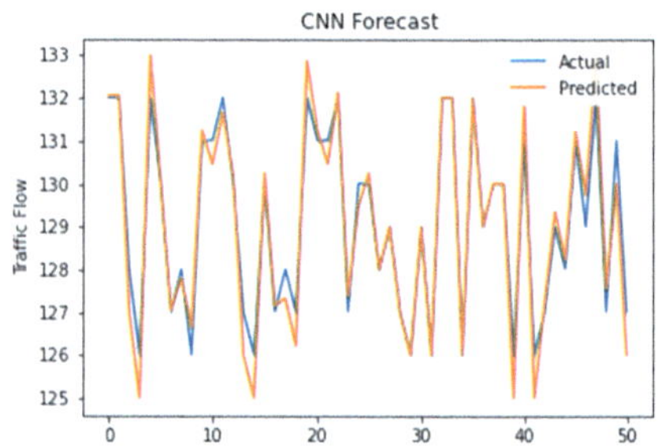

Fig. 4 CNN forecast

flow pattern after preprocessing. After preprocessing, the series has turned to be a stationary series.

The CNN model, LSTM model and CNN-LSTM model are trained with 153 days (75%) data and are tested with 51 days data (25%). Table 1 compares the performance of the models with the metrics MSE, MAPE, MAE and R^2. From the results, CNN-LSTM makes the forecasting better than the CNN and LSTM models. Figure 4 shows the CNN forecast, Fig. 5 shows the LSTM forecast and Fig. 6 shows the CNN-LSTM forecast.

6 Conclusion

Traffic flow forecasting is a research on accurate prediction of the number of vehicles that may flow through a particular lane at a specified period in the future. The accurate estimation of future traffic flows is quite challenging. A hybrid model CNN-LSTM

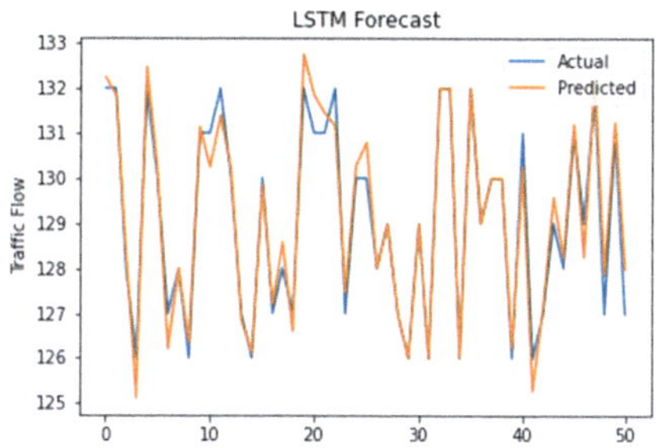

Fig. 5 LSTM forecast

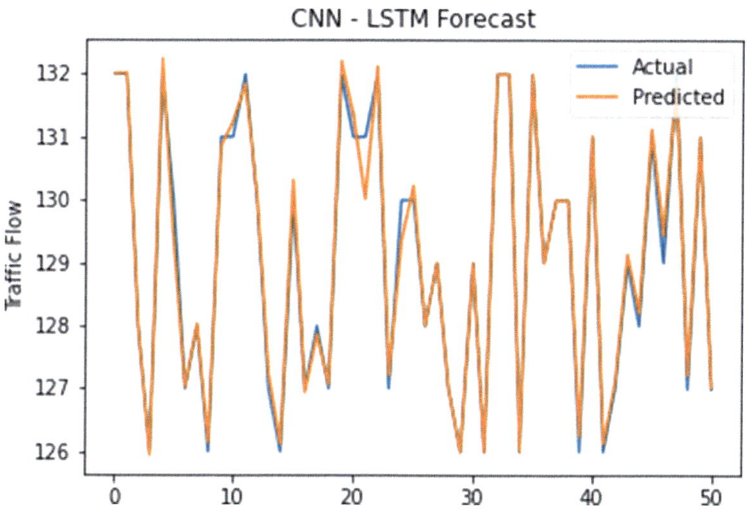

Fig. 6 CNN-LSTM forecast

proposed in this paper efficiently captured all the spatial and temporal features in the data. The results prove that the CNN-LSTM performs better prediction compared to the CNN and LSTM models. Especially, the R^2 score for the CNN-LSTM model is 0.98 which is close to 1 proves that the model is efficient.

Acknowledgements We would like to express our gratitude to our colleagues at our institution for offering deeper perspectives and knowledge that greatly aided our research.

References

1. Dataset. http://tris.highwaysengland.co.uk/detail/trafficflowdata#site-collapse-2011
2. Ye S (2012) Research on urban road traffic congestion charging based on sustainable development. Phys Procedia 24(Part B):1567–1572. https://doi.org/10.1016/j.phpro.2012.02.231
3. Sadouk L (2019) CNN approaches for time series classification. In: Ngan C-K (ed) Time series analysis—data, methods, and applications. IntechOpen
4. Kabbilawsh P, Sathish Kumar D, Chithra NR (2020) Trend analysis and SARIMA forecasting of mean maximum and mean minimum monthly temperature for the state of Kerala, India. Acta Geophys 68:1161–1174. https://doi.org/10.1007/s11600-020-00462-9
5. Sun S, Wu H, Xiang L (2020) City-wide traffic flow forecasting using a deep convolutional neural network. Sensors 20:421. https://doi.org/10.3390/s20020421
6. Wang Y, Guo Y, Wei Z, Huang Y, Liu X (2019) Traffic flow prediction based on deep neural networks. In: 2019 international conference on data mining workshops (ICDMW), Beijing, China, pp 210–215. https://doi.org/10.1109/ICDMW.2019.00040
7. Yang D, Li S, Peng Z, Wang P, Wang J, Yang H (2019) MF-CNN: traffic flow prediction using convolutional neural network and multi-features fusion. IEICE Trans Inf Syst E102.D:1526–1536. https://doi.org/10.1587/transinf.2018EDP7330
8. Zhu H, Xie Y, He W, Sun C, Zhu K, Zhou G, Ma N (2020) A novel traffic flow forecasting method based on RNN-GCN and BRB. J Adv Transp 2020(Article ID 7586154):11. https://doi.org/10.1155/2020/7586154
9. Lu W, Li J, Li Y, Sun A, Wang J (2020) A CNN-LSTM-based model to forecast stock prices. Complexity 2020(Article ID 6622927):10. https://doi.org/10.1155/2020/6622927
10. Fawaz HI, Forestier G, Weber J, Idoumghar L, Muller P-A (2019) Deep learning for time series classification: a review. Data Min Knowl Discov. arXiv:1809.04356 [cs.LG]
11. Luo X, Li D, Yang Y, Zhang S (2019) Spatiotemporal traffic flow prediction with KNN and LSTM. J Adv Transp 2019(Article ID 4145353):10. https://doi.org/10.1155/2019/4145353
12. Roy S, Bhunia GS, Shit PK (2020) Spatial prediction of COVID-19 epidemic using ARIMA techniques in India. Model Earth Syst Environ. https://doi.org/10.1007/s40808-020-00890-y
13. Essien A, Petrounias I, Sampaio P, Sampaio S (2020) A deep-learning model for urban traffic flow prediction with traffic events mined from twitter. World Wide Web. https://doi.org/10.1007/s11280-020-00800-3
14. Zhao W, Gao Y, Ji T, Wan X, Ye F, Bai G (2019) Deep temporal convolutional networks for short-term traffic flow forecasting. IEEE Access 7:114496–114507. https://doi.org/10.1109/ACCESS.2019.2935504
15. Ma Y, Zhang Z, Ihler A (2020) Multi-lane short-term traffic forecasting with convolutional LSTM network. IEEE Access 8:34629–34643. https://doi.org/10.1109/ACCESS.2020.2974575

Chapter 36
Classification of Driving Behaviour Using Machine Learning Methods at Signalized Intersections in Urban and Suburban Roads

Soni Lanka Karri, Liyanage C. De Silva, Daphne Teck Ching Lai, and Shiaw Yin Yong

1 Introduction

The World Health Organization (WHO) estimates that around 1.35 million people died in road accidents during 2018 worldwide [1]. Although traffic rules and signals have been executed to improve traffic safety, accidents regularly happen. Most road accidents occur at signalized junctions by disputing traffic movements in space and time, and intersections continue to be one of the major accident-prone areas [2, 3]. As the yellow signal lasts only 3 s [4], an improper decision will result in major accidents before it turns red signal. The right-angle crashes have received significant attention from these accidents because of the fatalities and severities affected [5]. Although red-light countermeasures have received optimistic effects, as research shows, such as reducing right-angle collisions; however, they have also caused an increase in back-end accidents [6]. These back-end accidents, also known as right-angle and rear crashes, are common outcomes of this problem zone, referred to as the dilemma zone [7]. There are many published articles, which suggested that the red-light running is influenced by features like (i) change of time interval [6], (ii) driver's driving experience [8], and (iii) different road conditions [9]. The contributions of these factors were also used to describe both intended and unintentional red-light crossing drivers' behaviour. The intentional red-light crossing is associated with

S. L. Karri (✉) · D. T. C. Lai · S. Y. Yong
Faculty of Science, Universiti Brunei Darussalam, Bandar Seri Begawan, Brunei Darussalam

L. C. De Silva
Faculty of Integrated Technologies, Universiti Brunei Darussalam, Bandar Seri Begawan, Brunei Darussalam

D. T. C. Lai
Institute of Applied Data Analytics, Universiti Brunei Darussalam, Bandar Seri Begawan, Brunei Darussalam

G. Mathur et al. (eds.), *Proceedings of 2nd International Conference on Artificial Intelligence: Advances and Applications*, Algorithms for Intelligent Systems,
https://doi.org/10.1007/978-981-16-6332-1_36

aggressive driving behavior, as it is linked to the driver's judgment and behaviour in the problem zone [10, 11].

Various researchers have shown their interest in this topic and focussed on studying drivers' behaviour during yellow intervals at intersections due to the uncertainty of stop/go decisions. At the onset of the yellow signal, a binary decision problem occurs—either slow down or speed ahead and cross. This area is referred to as the dilemma zone. Lu et al. [12] applied a regression model to illustrate the association between drivers' behaviour and traffic parameters during a yellow signal. They concluded that the number of drivers with 'go' decisions is correlated with traffic volume. Savolainen et al. [13] investigated the impact of driver's behaviour and suggested using a panel data random probability method on the onset of yellow signal, which displays warning flasher, yellow interval duration, and camera enforcement at the traffic signal.

The yellow light onset distance to the intersection was found to have the strongest effect on the probability of stopping for drivers [14]. Papaioannou [3] concluded that the probability of stopping depends on factors like vehicle approaching speed and distance to the intersection besides driver age and gender. Higher approaching speed and shorter distance to the stop line reduce the probability of stopping [15]. Based on many research studies, to mitigate the road accidents at the traffic signal, policymakers in various countries installed a countdown timer at the intersections along with a red-light crossing camera. Also, in some countries, flashing yellow lights are installed a few metres away from intersections, to alert the drivers that they are approaching a signal intersection. It was found that the driver's driving behaviour is different at intersections which have/don't have countdown timer and red-light camera. Even though these extra features have reduced the accidents at the intersections, however, few drivers ignore these and drive as usual, which is still leading to major accidents.

Different aspects and factors were analysed to understand the driver's decision at the intersection. Statistical techniques and stochastic modelling approaches to comprehend the driver's decision as they come close to the intersections were the most common implantation by researchers. However, with the advancement of high computing devices, very recently, researchers are applying different Machine Learning (ML) techniques. Researchers used ML techniques such as Random Forest (RF), Artificial Neural Network (ANN), and Convulsion Neural Network (CNN) to develop driver behaviour models. For a binary decision problem (stop/go), a classification approach is a good tool to classify the label. In this direction, approaches like a classification tree or Support Vector Machine (SVM) will be a good technique. Researchers chose to classify driving behaviour using these classification tools [16, 17]. The classification is based on the driver aggressiveness/road conditions/approach speed/distance to the intersection.

The main motivation of this researcher's study is to identify the problem zone concept and understand the factors leading the drivers in this zone. These outcomes will provide a basic understanding of driver behaviour in the dilemma zone approaching a signalized intersection. Then, categorizing the outcome of driving behaviour is based on unsafe stopping and safe stopping. The real-time data is difficult to obtain with all the driving factors and privacy concerns to keep this data in the public domain, and hence, a driving simulator is used. The simulator vehicle is a fixed interactive car with all the controls. The simulated test environment is designed to span rural, suburban, and urban roads to mimic the real-life scenarios and understand the driver's driving behaviour in different environments.

2 Explanation of a Problem or Dilemma Zone

The dilemma zone is close to signal intersection, where both halting and crossing actions are more troublesome. Figure 1 shows the dilemma zone, where a driver approaching the intersection faces the end of the green signal. Here d_i ($i = 0, 1, 2$) presents different distance zones before the stop line. If the driver travelling at a constant speed sees the yellow signal, he can safely stop before the stop line d_0. However, if the driver is going the past distance, d_2, he will have the option to cross the intersection safely. A dilemma zone exists between distances d_1 and d_2, which will enter into a situation where he/she can neither stop safely nor cross the intersection safely [18]. This 'problem zone' as shown in Fig. 1, somewhere in the range of d_1 and $d_{2,}$ is the zone inside which 'a driver may not have the option to stop safely after the beginning of yellow signal nor have the option to cross the intersection before the end of the yellow signal'. The zone (d_1–d_0) is defined as the 'the critical distance to stop safely (x_c)' and (d_2–d_0) is defined as 'the maximum distance to cross without acceleration (x_0)'. The problem zone arises when $x_0 < x_c$, where the short duration of 3 s of yellow light duration seems too short.

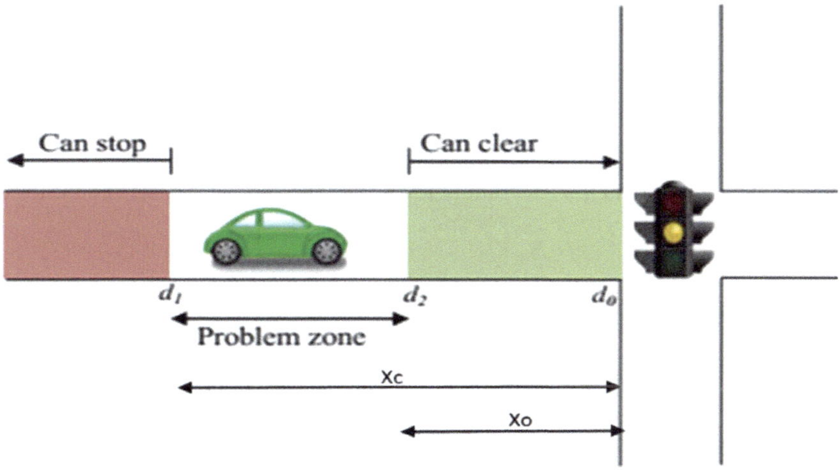

Fig. 1 Diagram representation of the problem zone with respect to distance to the stop line

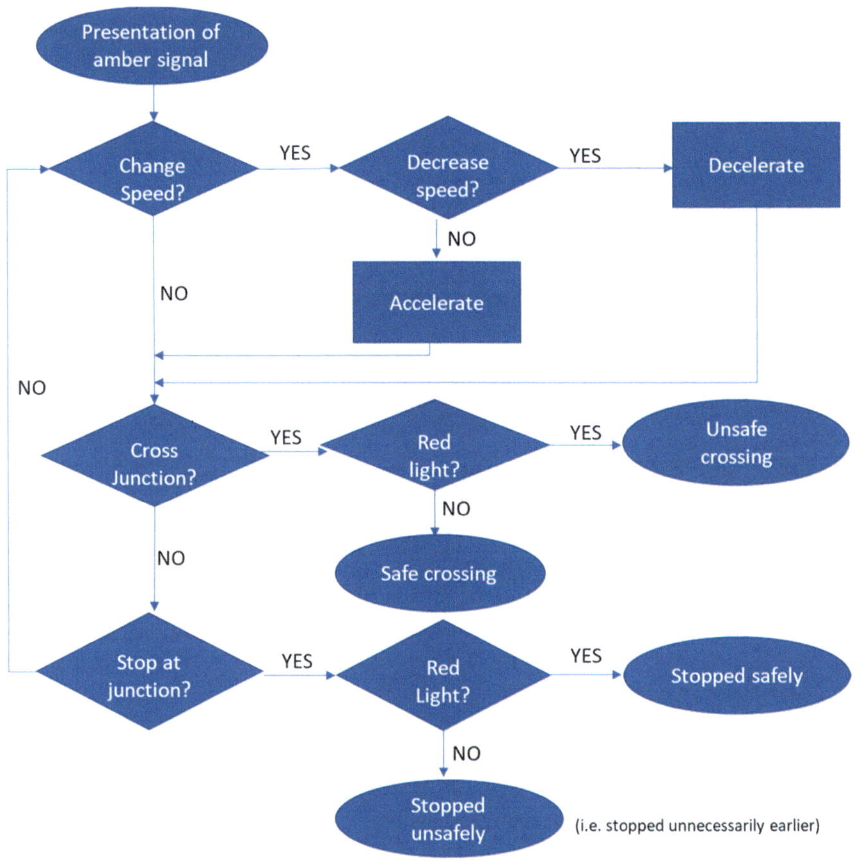

Fig. 2 Logistic methodology to develop a driver decision behaviour model

Logistic methodology to develop a driver decision behaviour model on the yellow signal's onset is given in Fig. 2. Acceleration and driver's speed are the important driving factors that signify the driving conditions. Hence, these factors are closely monitored to understand the uncertainty level and label safe and unsafe conditions.

3 Driving Simulator

An interactive fixed-base car driving simulator from Southampton University provided the data for this study. This shows a useful tool for studying driving behaviour on the straight road in this study [19]. Figure 3 shows the complete operational driver controls of a Jaguar XJ saloon (the simulator vehicle).

To imitator the real-time scenarios (see Fig. 4), the simulated test environment is

Fig. 3 The driving simulator with a virtual environment at Southampton University [19]

Fig. 4 The driving scenarios screenshots show for the **a** rural, **b** suburban, and **c** urban junctions [19]

designed to span rural, suburban, and urban roads to mimic the real-life scenarios with a total distance of 20 km. There are five junctions designed in the whole range so that three are located in suburban, one in urban, and one in rural. The speed limits in the rural, suburban, and urban regions are fixed as 50 mph, 40 mph, and 30 mph, respectively. To analyse distinctive driving practices, 50 drivers were voluntarily recruited to drive on the test system.

3.1 Drivers

Using a driving simulator, we record the car's parameters and analysed the driver's driving behaviour. Also, the real-time data is difficult to obtain with all the driving

Table 1 List of parameters monitored in the driving data which are collected as output from the driving simulator (units in FPS system—UK data)

Parameters	Units
Time in seconds	(s)
Driver total distance travelled	(ft)
Longitudinal acceleration	(ft/s^2)
Longitudinal velocity	(ft/s)
Signal status at the intersection	–

factors and privacy concerns to keep this data in the public domain. So, to mimic the real-time scenarios and understand the driver's driving behaviour in different environments, the above simulator vehicle car with all the controls is used. The 50 drivers (26 female and 24 male drivers) were voluntarily chosen among the students and staff community to obtain the driving behaviour. All drivers had been detained to have a valid driving license and driving experience for more than two years. Table 1 shows the parameters monitored in the driving data collected as output from the driving simulator.

At the yellow signal's onset, the driver enters into a problem zone where the vehicle could stop safely within the stop line or cross the junction within the signal 3 s yellow duration [19]. The vehicles at these dilemma zones usually end up with rear-end collisions or right-angle crashes. This zone attracts a high chance of accident/collision, keeping the driver's life at risk, so in this research study, our focus is to understand the driver's behaviour at the yellow signal. Therefore, from the driving data recorded at every time interval of 1 ms and lasted for approximately 20 min (average), crossing through 26 signalized junctions, the data is extracted. The traffic status is recorded yellow.

4 Results and Discussion

To categorize the driver driving behaviour as safe or unsafe stopping, different machine learning techniques like SVM and KNN are implemented in this study [20, 21]. As stated earlier, the driver's data for two intersections (urban and suburban) is extracted and investigated in this study. There are 60 total instances for urban and suburban intersections recorded for 50 driver's driving data. The different scenarios of model validation are used to check the ability of the SVM and KNN approaches as follows.

4.1 The Classification Performance Using 70:30 Rule

Initially, the data is segregated into two datasets of 70% (training) and 30% (testing) [22]. 42 instances of driver's data make up the training dataset, and the test dataset

Table 2 Comparison of SVM and KNN for classification for different scenarios

Data segregation	Methods	Training	Testing
70–30%	SVM	88.1%	72.2%
	KNN	83.3%	61.1%
Random 5 CV and 30 iterations	SVM	89.6% (avg)	91.6% (avg)
	KNN	85.4% (avg)	66.6% (avg)

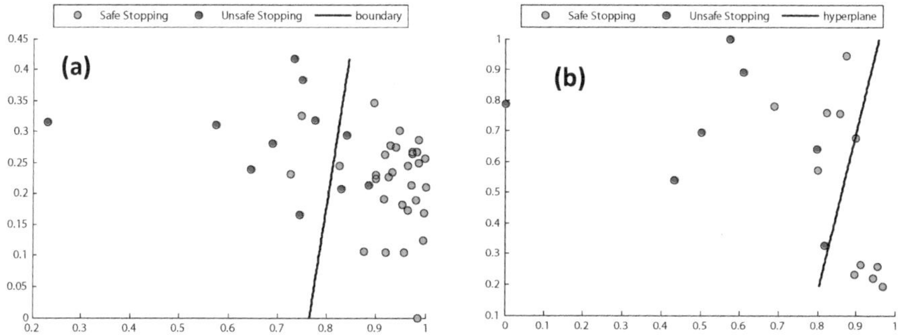

Fig. 5 SVM classification result for 70–30% rule **a** training, **b** testing

is made up of 18 instances of driver's data. Table 2 presents the results of the classification model, which has been adjusted based on the 70% training dataset. The classification performance is measured in terms of accuracy, shown in Table 2 for SVM and KNN. Forty two instances of whole data had been used for training, and the classfication accuracy of driving behaviour were around 88.1%, as shown in Fig. 5. The remaining 18 instances of data driving behaviour's testing accuracy resulted in 72.2% accuracy shown in Fig. 5. Among 18 instances of driver's data used for testing, five instances that represent safe stopping are inaccurately classified as unsafe stopping. However, the remaining data were accurately classified under safe and unsafe stopping. However, the KNN approach has 83.3 and 61.1% accuracy for training and testing. Therefore, these outcomes induce that the SVM classifier figures out how to choose the best classification boundary to separate safe and hazardous halting drivers' information.

4.2 The Performance of Classification with Multiple Repetitions Using Random Segregation

To verify the effect of randomization in choosing training and testing datasets, random segregation is evaluated. The classification process is performed on 60 instances of data using five cross-validations (CV), allowing complete randomization of input data [23]. The simulation runs are repeated 30 times (iterations) further to comprehend

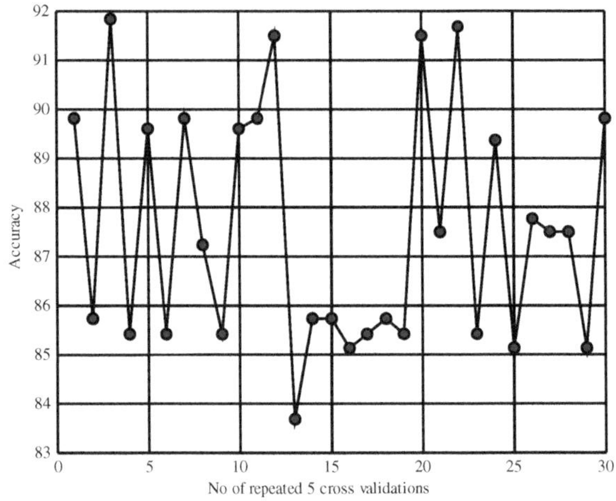

Fig. 6 SVM classification result using 5 CV and 30 iterations on training data

the impact of randomization on the grouping precision. Figure 6 shows the accuracy vs the number of repeated 5 CVs using the SVM approach. During each simulation run, a completely different dataset is used for training the model. Hence, the accuracy is also different for each iteration. The significance of the model will represent each iteration and shows the variation of accuracy. The average accuracy for the training of the 30 repetition runs shows 89.6% accuracy, in Table 2. Using the best model (one of the highest accuracy models), the model is validated on the remaining data, which resulted in 91.6% accuracy. Figure 7 shows the results of SVM classification using the model obtained at 30th iterations.

The KNN is also tested for the same 5 Cross-Validation and repeated for 30 repetitions process. The accuracy for training is 85.4% and testing is 66.6%. Again, comparing both the methods, the SVM method was able to categorize the safe and hazardous halting driver's data. These results prove that the SVM approach is useful for determining the precise driving behaviour consequence at intersections [16, 17].

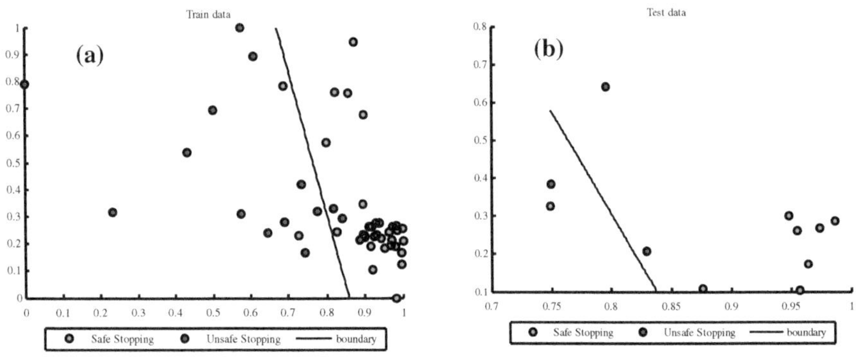

Fig. 7 SVM classification model obtained at 30th iterations data using on **a** training and **b** testing

5 Development of Pre-warning System Based on the Distance Approaching the Intersection

Among the total 60 driving instances data recorded in the urban and suburban regions, there are 7 instances in Suburban and 13 instances in the urban region which are marked as unsafe stopping, and the remaining 40 instances are labelled as safe stopping. To understand the driver's driving pattern in the data recorded as unsafe stopping, these data are further analysed to determine the suitable distance from the intersection. Based on the driver's velocity and longitudinal acceleration, a warning alert can be given. In view of the driver's speed and longitudinal acceleration increase, a warning alarm can be given. In such a manner, a provisional caution sign can be given to drivers crossing a specific distance as indicated by their driving velocity at an intersection, as demonstrated in Figs. 8 and 9. In these figures, the green zone represents safe driving, and the red zone represents hazardous driving scenarios. The blue triangle (Δ) shows a warning light to the driver when the driving scenario changes from safe to risky situation as the driver moves towards the intersection at a specific speed. It can be seen that the warning alert (Δ) for both Urban and Suburban roads are showing up at a different distance from the crossing point. For most cases, the warning alert is around 20–40 m. Consequently, these outcomes can undoubtedly separate the driving scenario, leading from safe to hazardous halting as they were moving towards the signal. This analysis gives the outline to distinguish the essential

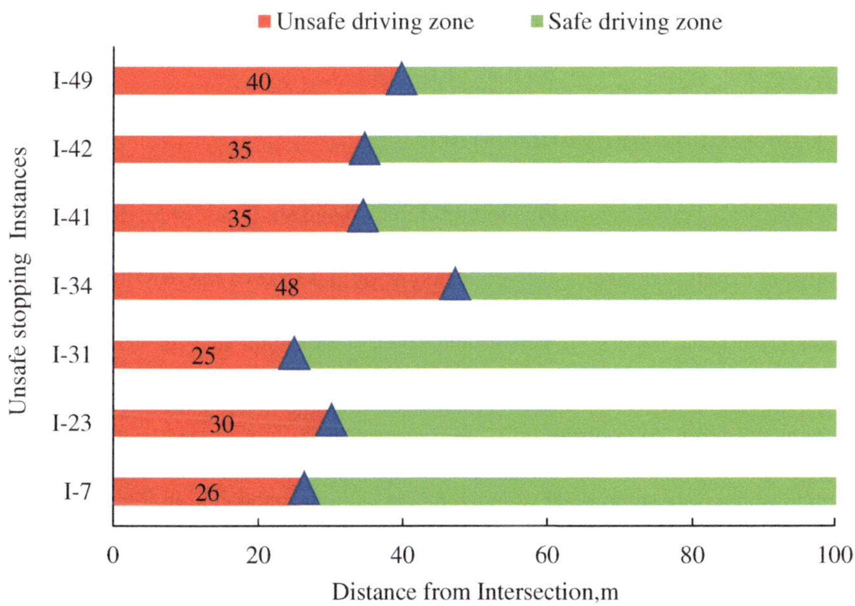

Fig. 8 Pre-alarming system for suburban data

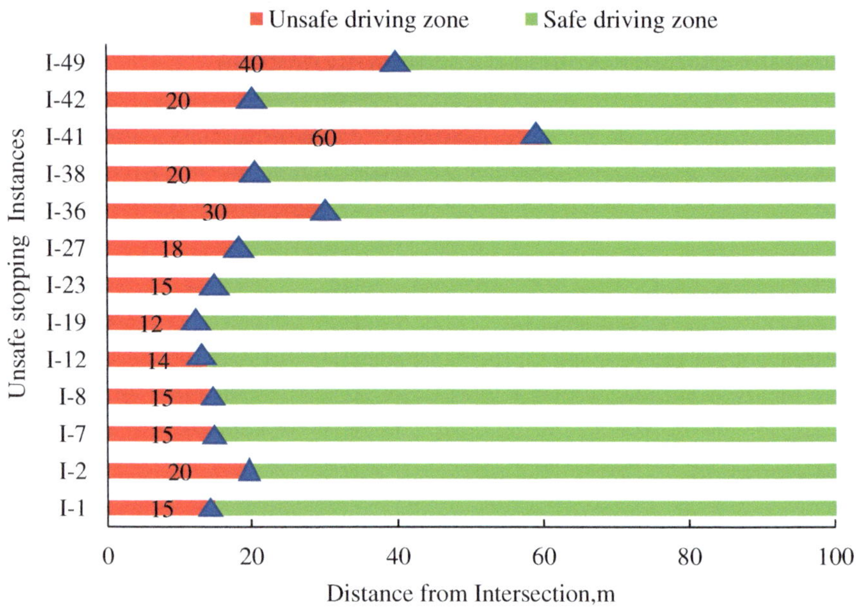

Fig. 9 Pre-alarming system for urban data

distance, where the caution light ought to be kept set up if the vehicle speed is higher than the recommended speed cutoff of that street.

6 Conclusions and Future Work

In this study, driver's driving behaviour is collected and analysed for 50 driver's data, especially at the traffic signals covering the rural, suburban, and urban regions. Seven significant driving parameters are collected as output data of the driving simulator at every 0.1 s. In addition, the driver's data is extracted for two intersections (urban and suburban), and the driving performance data of individual drivers were analysed. These parameters are captured using a driving simulator as there was no data publicly available that had recorded the car's parameters and the driver's driving behaviour. Thus, in this study, two different machine learning techniques are useful to study the drivers driving behaviour's classification performance regarding two labels safe stopping and unsafe stopping at signalized intersections at the yellow signal's arrival.

To assess the SVM and KNN approaches' performance, two different scenarios were designed to validate the data's randomization. In the first scenario, the data is segregated into two datasets of 70–30%. From the obtained results, it was found that the SVM approach succeeded in selecting the best classification boundary to separate the safe stopping and unsafe stopping driver's data and resulted in better classification. To verify the effect of randomization in choosing training and testing datasets, random segregation is evaluated. The 60 instances of data had been randomized and performed the classification process using 10 cross-validations (CV), allowing complete randomization of input data. The significant results show that the SVM method can conclude the driver's driving behaviour data with a high-performance level. The identified model can be beneficial for determining the truthful outcome of driving behaviour at the intersection.

As a future study, we would like to compare the classification results for all the intersections across rural, suburban, and urban areas. We are also willing to implement and compare with other techniques like Tree and Linear discriminant and different kernels of SVM.

Funding 'Soni Lanka Karri's Ph.D. is funded by Universiti Brunei Darussalam under the University Bursary Scholarship. This research is also partially funded by Universiti Brunei Darussalam's research grants (UBD/PNC2/2/RG/1(311) and UBD/RSCH/1.11/FICBF/2018/002)'.

Conflict of Interest The authors declare that there is no conflict of interest.

References

1. World Health Organization Violence and Injury Prevention (2018) Global status report on road safety 2018. World Health Organization
2. Montella A, Aria M, D'Ambrosio A, Galante F, Mauriello F, Pernetti M (2011) Simulator evaluation of drivers' speed, deceleration and lateral position at rural intersections in relation to different perceptual cues. Accid Anal Prev. https://doi.org/10.1016/j.aap.2011.05.030
3. Papaioannou P (2007) Driver behaviour, dilemma zone and safety effects at urban signalised intersections in Greece. Accid Anal Prev 39(1):147–158. https://doi.org/10.1016/j.aap.2006.06.014
4. Dong S, Zhou J (2020) A comparative study on drivers' stop/go behavior at signalized intersections based on decision tree classification model. J Adv Transp 2020. https://doi.org/10.1155/2020/1250827
5. Sacchi E, Sayed T, Deleur P (2013) A comparison of collision-based and conflict-based safety evaluations: the case of right-turn smart channels. Accid Anal Prev. https://doi.org/10.1016/j.aap.2013.06.002
6. Retting RA, Ferguson SA, Farmer CM (2008) Reducing red light running through longer yellow signal timing and red light camera enforcement: results of a field investigation. Accid Anal Prev. https://doi.org/10.1016/j.aap.2007.06.011
7. Hurwitz DS, Wang H, Knodler MA, Ni D, Moore D (2012) Fuzzy sets to describe driver behavior in the dilemma zone of high-speed signalized intersections. Transp Res Part F Traffic Psychol Behav. https://doi.org/10.1016/j.trf.2011.11.003
8. Elmitiny N, Yan X, Radwan E, Russo C, Nashar D (2010) Classification analysis of driver's stop/go decision and red-light running violation. Accid Anal Prev. https://doi.org/10.1016/j.aap.2009.07.007
9. Jensupakarn A, Kanitpong K (2018) Influences of motorcycle rider and driver characteristics and road environment on red light running behavior at signalized intersections. Accid Anal Prev. https://doi.org/10.1016/j.aap.2018.02.007

10. Pathivada BK, Perumal V (2017) Modeling driver behavior in dilemma zone under mixed traffic conditions. Transp Res Procedia 27:961–968. https://doi.org/10.1016/j.trpro.2017.12.120
11. Sahithi G, Prabhanjan N, Yadav GS, Poongudi K, Kiran MS, Lanka S (2019) Comparision of transition curves in terms of lateral acceleration and lateral jerk
12. Lu G, Wang Y, Wu X, Liu HX (2015) Analysis of yellow-light running at signalized intersections using high-resolution traffic data. Transp Res Part A Policy Pract 73:39–52. https://doi.org/10.1016/j.tra.2015.01.001
13. Savolainen PT, Sharma A, Gates TJ (2016) Driver decision-making in the dilemma zone—examining the influences of clearance intervals, enforcement cameras and the provision of advance warning through a panel data random parameters probit model. Accid Anal Prev 96:351–360. https://doi.org/10.1016/j.aap.2015.08.020
14. Gates T, Noyce D, Laracuente L, Nordheim E (2007) Analysis of driver behavior in dilemma zones at signalized intersections. Transp Res Rec J Transp Res Board. https://doi.org/10.3141/2030-05
15. Elhenawy M, Rakha H, El-Shawarby I (2014) Enhancing driver stop/run modeling at the onset of a yellow indication using historical behavior and machine learning techniques
16. Karri SL, De Silva LC, Lai DTC, Yong SY (2021) Identification and classification of driving behaviour at signalized intersections using support vector machine. Int J Autom Comput 1–12
17. Karri SL, De Silva LC, Lai DTC, Yong SY (2021) Classification and prediction of driving behaviour at a traffic intersection using SVM and KNN. SN Comput Sci 2(3):1–11
18. Elhenawy M, Jahangiri A, Rakha HA, El-Shawarby I (2015) Classification of driver stop/run behavior at the onset of a yellow indication for different vehicles and roadway surface conditions using historical behavior. Procedia Manuf 3:858–865. https://doi.org/10.1016/j.promfg.2015.07.342
19. Yong SY (2013) Investigating driver's behaviour on approaching a junction at the end of green time. University of Southampton
20. Yu B, Wang YT, Yao JB, Wang JY (2016) A comparison of the performance of ANN and SVM for the prediction of traffic accident duration. Neural Netw World. https://doi.org/10.14311/NNW.2016.26.015
21. Iranitalab A, Khattak A (2017) Comparison of four statistical and machine learning methods for crash severity prediction. Accid Anal Prev. https://doi.org/10.1016/j.aap.2017.08.008
22. Thanh Noi P, Kappas M (2017) Comparison of random forest, k-nearest neighbor, and support vector machine classifiers for land cover classification using Sentinel-2 imagery. Sensors (Basel). https://doi.org/10.3390/s18010018
23. Wong TT (2015) Performance evaluation of classification algorithms by k-fold and leave-one-out cross validation. Pattern Recognit. https://doi.org/10.1016/j.patcog.2015.03.009

Chapter 37
A Granular Intuitionistic Fuzzy Meta Clustering Algorithm and Its Application

B. K. Tripathy and Urmi Bhambhani

1 Introduction

The most popular approach to data mining is applying an algorithm using the best features from the collected data. The focus of researchers has always been a two-pronged effort to improve (a) the machine learning algorithm and/or (b) finding the reduced set of most useful features from the collected dataset. While there have been ample improvements in the efficacy of these approaches, we can think of newer ways to solving older problems. One such way is to integrate granular computing concepts in the data mining technique. Informally, granular computing refers to looking at a given dataset from different points of view. Formally, it means that we can look at any dataset from different level of abstraction or level of detail. This level is called a granule in information system parlance. Hence, granular computing provides us with ways to look at raw information at different granular levels or abstractions. It also refers to looking at the same information with different objects as frame of reference. This means that, for example, if we have an entire university's data, we can look at the data from students' point of view. In this way, a data point is important only if it pertains to a single student or group of students. Similarly, we can look at the same data from the point of view of professors. Now, data point such as personal information of students is not relevant from the professors' point of view. Hence, it would not be considered when we look at the data from the granularity of professor. Thus, granular computing provides us different ways to look at the same data.

B. K. Tripathy (✉)
School of Information Technology and Engineering, VIT, Tamil Nadu Vellore-632014, India
e-mail: tripathybk@vit.ac.in

U. Bhambhani
School of Computer Science and Engineering, VIT, Tamil Nadu Vellore-632014, India

G. Mathur et al. (eds.), *Proceedings of 2nd International Conference on Artificial Intelligence: Advances and Applications*, Algorithms for Intelligent Systems,
https://doi.org/10.1007/978-981-16-6332-1_37

When we combine the granular computing concept with traditional data mining techniques, new techniques emerge. We can possibly improve the accuracy of mining algorithms by making use of this context [2].

In this paper, we introduce a new algorithm called intuitionistic fuzzy meta clustering. We apply it to a real-world dataset of Yelp.com and see improvements in the clustering performance.

2 Literature Review

2.1 Hard k-means

Hard k-means (also called Forgy-Lloyd algorithm) is one of the most effective clustering algorithm used for clustering [11]. Here, we need to divide these 'n' objects into 'k' groups or clusters. We start by as-signing 'k' randomly chosen objects into 'k' clusters. Then, we calculate the centroids of each of the 'k' groups. All the subsequent objects are assigned to one of the 'k' groups after calculating the distances of the object and centroids of all the groups. The cluster centroid with the lowest distance from a given object is chosen, and the object is assigned to the corresponding cluster. The centroid is then updated.

There are many extensions of k-means algorithm, like modified k-means [35], k-means+ + [13], k-means based on divide and conquer paradigm [24] and hierarchical clustering [28].

2.2 Fuzzy c-means

After the introduction of Fuzzy set as a generalization of the crisp set by Zadeh [42], k-means was generalized to Fuzzy c-means. In Fuzzy c-means instead of taking binary values of 1 or 0, the membership u_{ij} can have a range of 0 and 1 with real values included. This means that a data point can now belong to more than one cluster to varying degrees. The higher the degree of membership, the closer is the value to 1. Of course, the sum total of all degree of membership values for a data point must be 1.

Dunn introduced Fuzzy c-means [9]. But Bezdek [3] gave the form of the algorithm that is in use today. Some extensions of FCM are proposed and applied to study sequential data [33], in the study of recommender systems [7], used for medical data analysis [31], and depth computation of leukemia images [37].

2.3 Intuitionistic Fuzzy c-means

Intuitionistic Fuzzy set [1] generalizes Fuzzy set to a setting where the non-membership value is not exactly the one's complement of the membership value, but their sum lies in the interval [0, 1] providing a space for the hesitation value, which provides greater uncertainty and more reality. Thus Fuzzy c-means was further generalized to Intuitionistic Fuzzy c-means (IFCM) [6]. Some extensions and applications of IFCM are in big data clustering [32, 34].

2.4 Rough c-means

Rough set introduced by Pawlak [21] is yet another model of uncertainty. It uses the boundary of uncertainty concept of Frege and is having the notion of granularity inherent in it. The rough c-means was proposed by Lingras [16]. Rough clustering has been applied to knowledge discovery [38].

2.5 Hybrid Clustering Methods

It has been observed that hybrid models are more efficient than their individual components if the hybridization is made suitably by taking the strong points of the individual models. Rough Fuzzy [8] and Rough Intuitionistic Fuzzy [23] models are used to develop clustering algorithms [17, 19], and it has been established that Rough Intuitionistic Fuzzy c-means algorithm [30] is the best in the series. Clustering approaches based upon rough fuzzy [29, 36] and rough intuitionistic fuzzy methods have been studied [34]. A good compilation of hybrid means in clustering and classification for big data is in [25].

2.6 Granular Computing

Granular computing refers to different levels of abstraction or granularity in processing raw data and problem solving in general. A set of 5 tuples (U, D, L, R, and J) denotes s granular computing formulation.

Here,

U = universe or context of problem

D = data set

L = operator in one direction

H = operator in opposite direction

J = restriction imposed on L and H.

When it was conceived granular computing work was focused on fuzzy and rough sets as well as cluster analysis and interval analysis. Zadeh [44] was the pioneer of information granulation in 1979. Although initially thought to be alternative to fuzzy set theory, the theory of rough set introduced by Pawlak also dealt with partitioning the data using equivalence relations. This is another form of granular computing. Bargiela and Pedrycz [2] wrote the first book, which dealt with the subject of granular computing. A paper by Zadeh [43] that dealt with information granulation in de-tail preceded this book.

Yao [39, 41] is credited with studying granular computing from Artificial Intelligence perspective comprehensively. Various aspects of AI ranging from computer vision and image processing to top-down problem solving can take advantage of granular computing concepts [22]. There is a body of work [22] that provides detailed review of granular computing research.

Yao [40] came up with theoretical framework and structured foundations that helped form the basis for granular computing research. The triarchic theory was thus formed. This strong theoretical foundation led to developments in algorithms based on granular computing concepts and real world applications that make use of these. The 3 foundations viz. philosophy, computation, and algorithms of granular computing can be used in intelligent system applications to solve real-world problems.

Gacek and Pedrycz [12] extended these concepts to manipulate time-series data leading to granular interval concept.

Granular computing has been applied for rule generation [27] and soft granular computing is used for classification of data [20].

2.7 *Meta Clustering*

Research has shown that clustering in parallel can prove advantageous in many situations. A 2-phase clustering method was first proposed [26] for document classification problem. This was extended to the general form of the algorithm for recursive clustering [10]. Fuzzy clustering was used in meta clustering to give even more generalized view of meta clustering technique [4]. It has been shown [4] that meta clustering can give more relatable and meaningful understanding of the clustering results.

Bi-clustering [5, 14, 18] was introduced as a generalization of meta clustering. This was extended to n-clustering eventually. Bi clustering refers to clustering the rows and columns in parallel. Tri clustering [15], by extension, is used in case of 3-dimensional data. This method has been shown to be applicable and effective for various real-world applications.

2.8 Proposed Algorithm

In our meta-clustering algorithm, we first take the raw features of the in the dataset and perform clustering over that data. On every iteration that follows, we calculate the dynamic features of the data. The dynamic features are calculated based on the clustering done at the previous iteration. After that, we concatenate the raw or static features of the data with the dynamic features thus obtained and use this combined feature set to again perform clustering. This forms 1 iteration of meta clustering. This keeps repeating cyclically. At each cycle, we check to see if the clustering results obtained from current iteration have converged with the previous results. When convergence is obtained, the algorithm halts. This is illustrated in Fig. 1.

In this algorithm, there are different ways of determining how the dynamic variables are calculated. In our algorithm, we use the concept of granular computing. In that, we take 2 or more different granules of information from the data and perform clustering based on those granules in parallel. To calculate the dynamic features for meta clustering of one particular granule, we use the clustering result information obtained from clustering of other granules. This process is repeated for every granule that we are clustering. And this happens simultaneously for each granular clustering. So, if granule x is at its ith clustering iteration, all other granules y, $y \neq x$ are also at their ith clustering iteration.

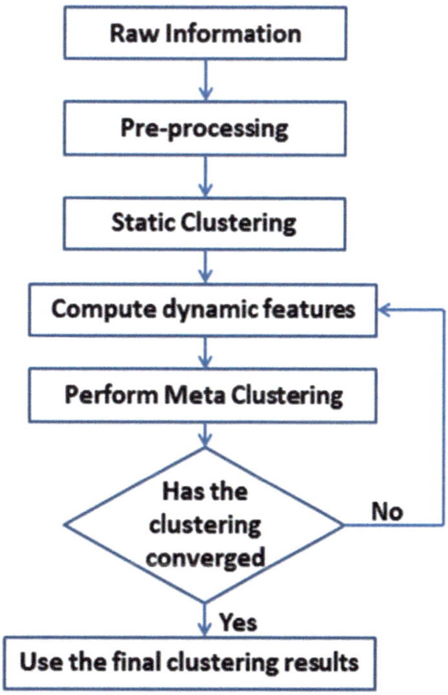

Fig. 1 Meta clustering algorithm

For each iteration of clustering, we have used intuitionistic fuzzy clustering. Intuitionistic fuzzy clustering is a superset of fuzzy clustering. In that, along with assigning a membership value for each data point to a cluster, we also assign a hesitation degree such that the sum of membership value and the hesitation degree is equal to non-membership value subtracted from 1. That is,

$$\mu + \pi = 1 - \upsilon \tag{1}$$

where μ is the membership value,

π is the hesitation degree,

V is the non-membership value.

In fuzzy set, we only consider the membership value, while in Atanassov's intuitionistic fuzzy set, we also generate a hesitation value.

An intuitionistic fuzzy set A can be expressed as

$$A = \{(x, \mu_A(x), \upsilon_A(x))|x \in X\}. \tag{2}$$

Here, $\mu_{\mathrm{A}}(x)$ is the membership degree of x in the intuitionistic set A and $\nu_{\mathrm{A}}(x)$ is the non-membership degree of x in the intuitionistic set A.

Also,

$$\begin{aligned} &0 \leq \mu_{\mathrm{A}}(x) \leq 1, \\ &0 \leq \upsilon_{\mathrm{A}}(x) \leq 1, \\ &0 \leq \mu_{\mathrm{A}}(x) + \upsilon_{\mathrm{A}}(x) \leq 1.. \end{aligned} \tag{3}$$

For each member in Intuitionistic fuzzy set, there is a hesitation degree associated with the inability to determine the membership value of that member. This is given by,

$$\pi_{\mathrm{A}}(x) = 1 - \mu_{\mathrm{A}}(x) - \upsilon_{\mathrm{A}}(x) \tag{4}$$

of course, if the hesitation degree becomes 0 for all members of the intuitionistic set, then,

$$\upsilon_{\mathrm{A}}(x) = 1 - \mu_{\mathrm{A}}(x) \tag{5}$$

And the set becomes a general fuzzy set where non-membership value only depends on the actual membership value.

For generation of intuitionistic fuzzy set, we have to make use of an intuitionistic fuzzy generator. It is defined as follows:

A continuous decreasing and increasing function Γ(x) defined over the interval [0,1] can be treated as a fuzzy intuitionistic generator if

$$\Gamma(x) \leq (1-x) \text{ for all } x \in [0,\ 1] \text{ and } \Gamma(0) \leq 1 \text{ and } \Gamma(1) \leq 0. \tag{6}$$

In this paper, we have used Sugeno intuitionistic fuzzy generator, which is defined as follows:

$$c_\lambda(a) = \frac{1-a}{1+\lambda a} \lambda \epsilon(-1, \infty). \tag{7}$$

Here, $c_\lambda(a)$ is the non-membership value for a given membership value a.
So, the intuitionistic fuzzy set now becomes

$$A_\lambda = \{(x,\ \mu_A(\mathrm{x}), (1-\mu_A(\mathrm{x}))/(1+\lambda\mu_A(\mathrm{x})))|\, x \epsilon X\}. \tag{8}$$

For fuzzy clustering, following is our objective function:

$$J_m(U, v, X) = \sum_{i=1}^{c} \sum_{k=1}^{n} (u_{ik})^m d^2(x_k, v_i), \tag{9}$$

where

$d(x_k, v_i)$ = distance (typically Euclidean) between x_k i.e. the kth data point and v_i that is the cluster center of the ith cluster,

u_{ik}= membership value of kth data point in the ith cluster.

The membership value gets updated using the equation,

$$u_{ik} = \frac{1}{\sum_{i=1}^{c} [\frac{d_{ik}^2}{d_{jk}^2}]^{\frac{1}{m-1}}}, \tag{10}$$

and to update the cluster centers, following equation is used:

$$v_i = \frac{\sum_{k=1}^{n} u_{ik}{}^m x_k}{\sum_{k=1}^{n} u_{ik}{}^m} \forall i, k, i = 1, 2, \ldots c, k = 1, 2, \ldots n. \tag{11}$$

Now, to add the intuitionistic aspect to this method, we use a modified membership value according to the following equation:

$$u_{ik}{}^* = u_{ik} + \pi_{ik}. \tag{12}$$

Here, for the calculation of hesitation value, we use this equation,

$$\pi_{ik} = 1 - u_{ik} - (1-u_{ik})/(1+\lambda u_{ik}) \tag{13}$$

where we have taken $\lambda = 2$.

Now, using the new membership value u_{ik}^*, the calculation of cluster centers will now be done using the following equation:

$$v_i^* = \frac{\sum_{k=1}^{n} u_{ik}^* x_k}{\sum_{k=1}^{n} u_{ik}^*} \forall i, k, i = 1, 2, \ldots c, k = 1, 2, \ldots n. \quad (14)$$

The cluster centers are calculated at every iteration as the membership values change.

The algorithm halts when there are no changes in the membership values.

3 Results and Discussion

To illustrate this algorithm, we have used dataset released by Yelp in 2013 as part of RecSys 2013 challenge organized by ACM. Yelp is a website that lists different businesses in localities. Users can check in to a particular business and rate it. The rating for businesses ranges from 1 to 5 stars. Our data set contains data about 11,537 businesses, 43,873 users, 2, 29,907 reviews, and 8,282 check-ins. and write a review about their experience with the business. Other users can see those reviews and also vote on how useful, funny, or cool the reviews are.

For multi-granularity, we shall use "businesses" and "users" as 2 granularities based upon which we shall perform clustering.

We have used the following indices to measure the performance of clustering.

(i) Partition Co-efficient. It is given by:

$$\mathrm{PC} = \sum_{i=1}^{c} \sum_{k=1}^{n} (\mu_{ik})^2 \quad (15)$$

(ii) Partition Entropy as defined by Bezdek. That is given by:

$$\mathrm{PE} = \sum_{i=1}^{c} \sum_{k=1}^{n} \mu_{ik} log_2(\mu_{ik}) \quad (16)$$

(iii) Cluster validity index by Fukuyama and Sugeno calculated according:

$$\mathrm{FS} = \sum_{i=1}^{c} \sum_{k=1}^{n} u_{ik}{}^m ||x_k - v_i||^2 - \sum_{i=1}^{c} \sum_{k=1}^{n} u_{ik}{}^m ||v_i - \overline{v}||^2 \quad (17)$$

where $\overline{v} = \sum_{i=1}^{c} v_i/c$ ..

(iv) Cluster validity index by Xie and Beni calculated according:

Table 1 Optimal values for cluster validity indices

Serial no.	Index	Optimal value
1	Partition coefficient	High
2	Partition entropy	Low
3	Fukuyama and Sugeno Index	Low
4	Xie and Beni Index	Low

$$\mathrm{XB} = \frac{\sum_{i=1}^{c}\sum_{k=1}^{n} u_{ik}{}^{m}||x_k - v_i||^2}{n min_{i,k}||x_k - v_i||^2} \tag{18}$$

Table 1 shows the optimal values for various indices listed above. Optimal value of high means higher the value of the index, better is the clustering. Similarly, the opposite is true for low optimal value. Lower the value better is the clustering (Figs. 2, 3, 4, 5, and 6).

The resultant trends of various indices are shown below for both user and business clustering. This is shown for the case when intuitionistic fuzzy meta clustering is used as well as when simply fuzzy meta clustering is done (Figs. 7, 8, 9, 10, 11, 12, 13, 14, 15, 16, and 17).

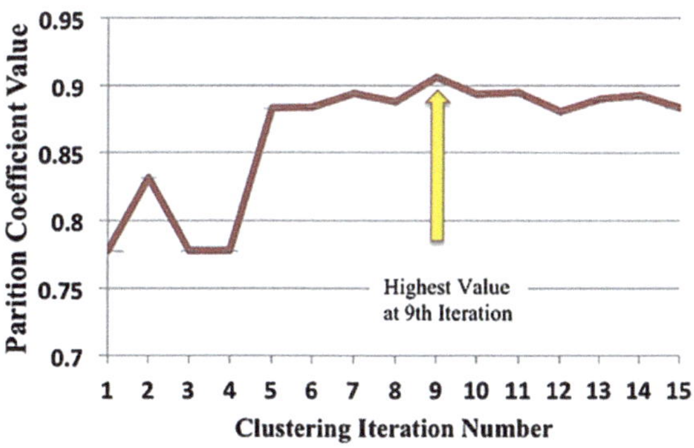

Fig. 2 Partition coefficient for business clusters with IFCM

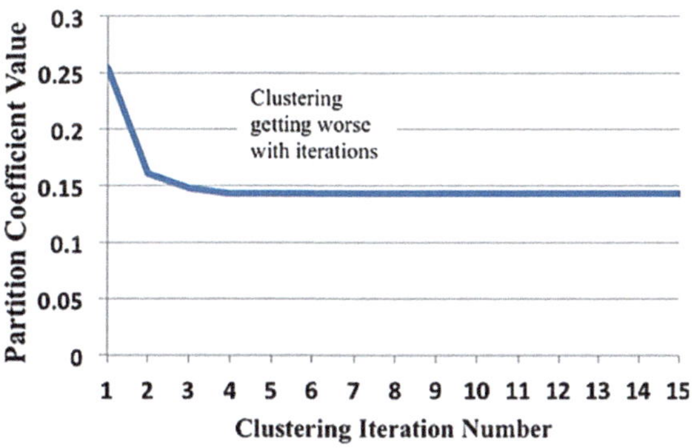

Fig. 3 Partition coefficient for business clusters with FCM

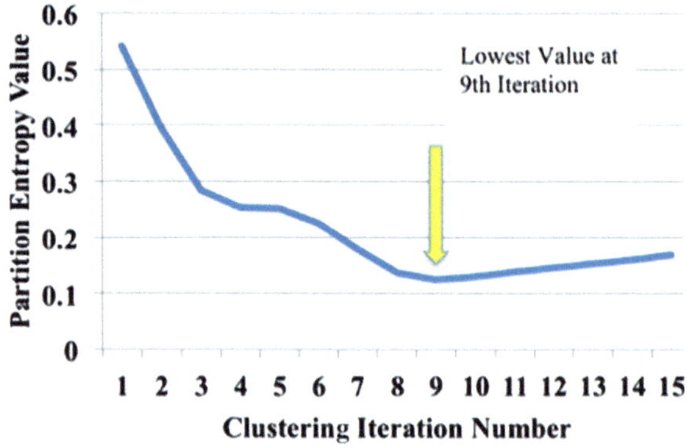

Fig. 4 Partition entropy for business clusters with IFCM

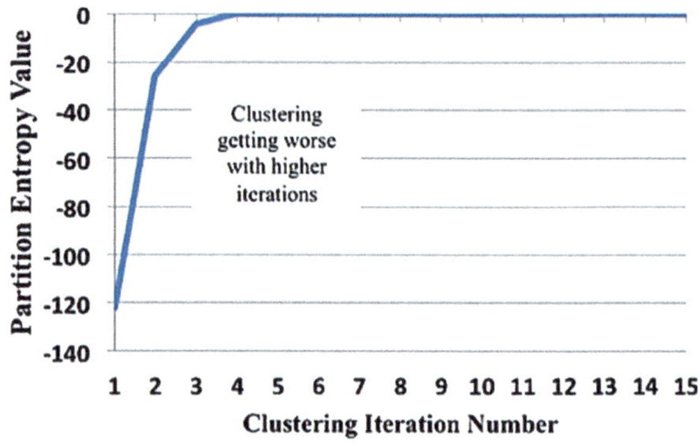

Fig. 5 Partition entropy for business clusters with FCM

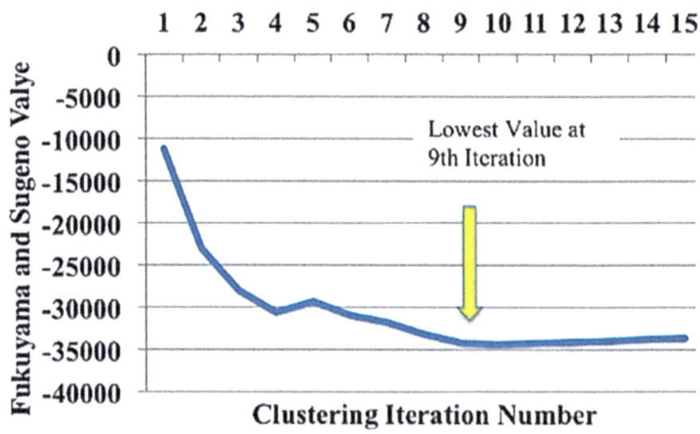

Fig. 6 Fukuyama and Sugeno Index for business clusters with IFCM

As we can see from the trends, the clustering for users reaches an optimal point at 9th iteration of intuitionistic meta clustering. The clustering for businesses reaches an optimum at 15th iteration of intuitionistic meta clustering.

We also see that for the case when only fuzzy meta clustering is used, the clustering for businesses does not improve at all. While for users, it does become better with

Fig. 7 Fukuyama and Sugeno Index for business clusters with FCM

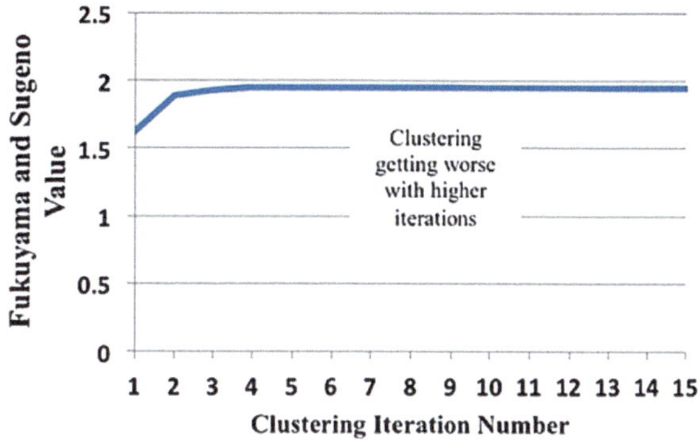

Fig. 8 Xie and Beni Index for business clusters with IFCM

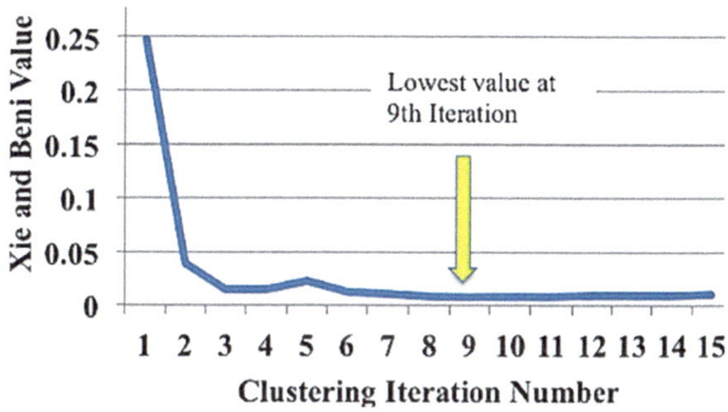

Fig. 9 Xie and Beni Index for business clusters with FCM

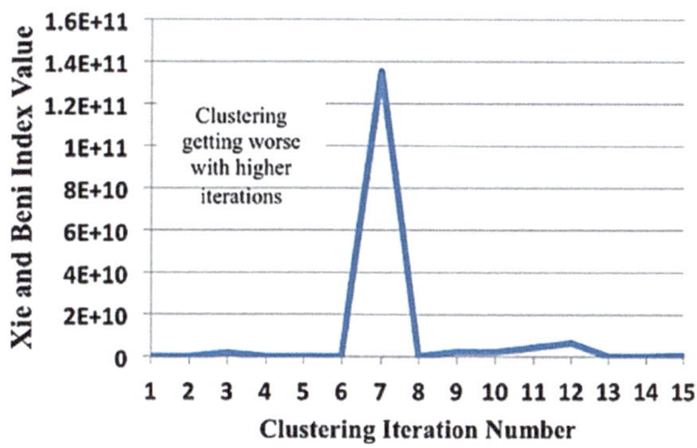

higher iteration. It reaches an optimum at 9th iteration. If we were to compare index values for fuzzy meta clustering and intuitionistic fuzzy meta clustering, it is clear that intuitionistic fuzzy meta clustering works much better.

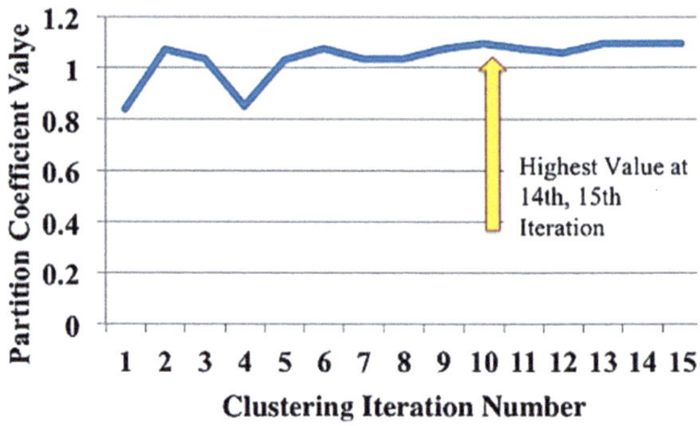

Fig. 10 Partition coefficient for user clusters with IFCM

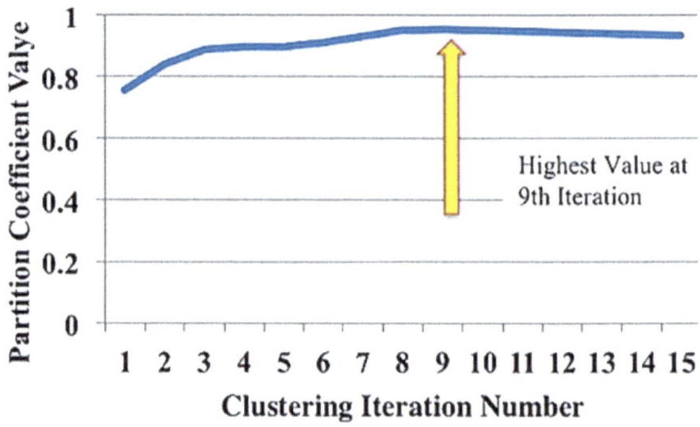

Fig. 11 Partition coefficient for user clusters with FCM

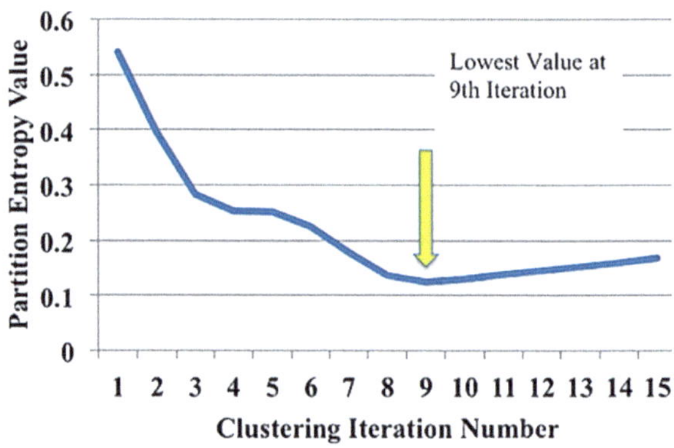

Fig. 12 Partition entropy for user clusters with IFCM

4 Conclusions

In this paper, we introduced a new algorithm called intuitionistic fuzzy meta clustering which combines granular computing concept with intuitionistic fuzzy meta

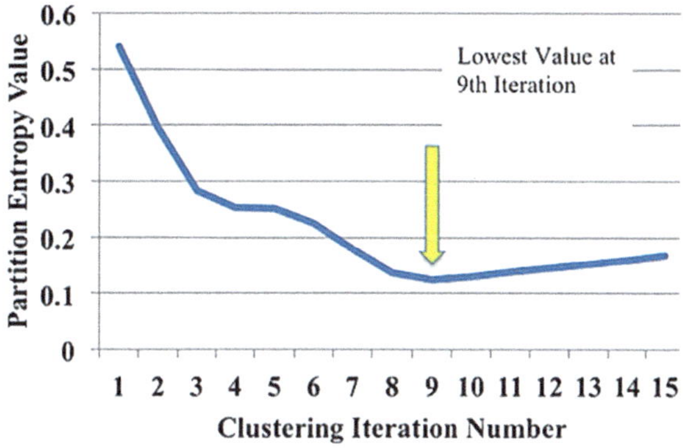

Fig. 13 Partition entropy value for user clusters with FCM

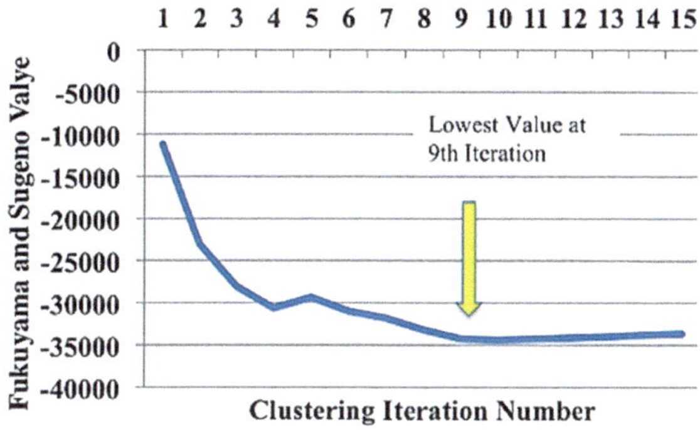

Fig. 14 Fukuyama and Sugeno Index for user clusters with IFCM

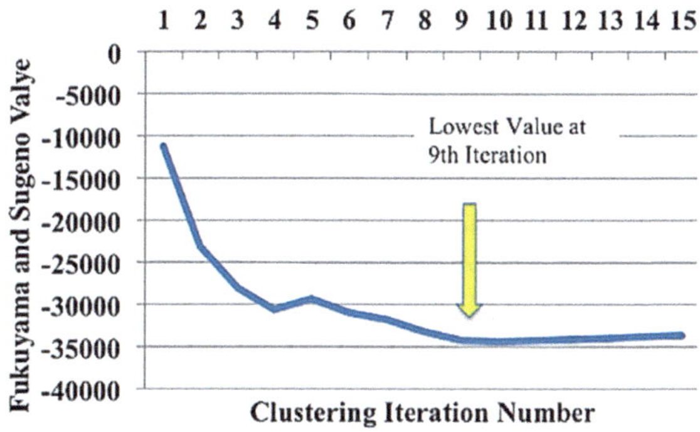

Fig. 15 Fukuyama and Sugeno Index for user clusters with FCM

clustering algorithm for improved performance. We applied the algorithm to a real-world dataset of yelp.com and showed how this new algorithm performs better than fuzzy c-means algorithm using various cluster validity indices.

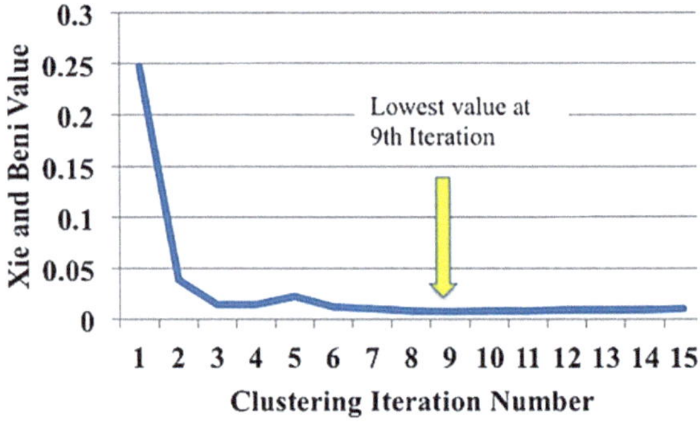

Fig. 16 Xie and Beni Index for user clusters with IFCM

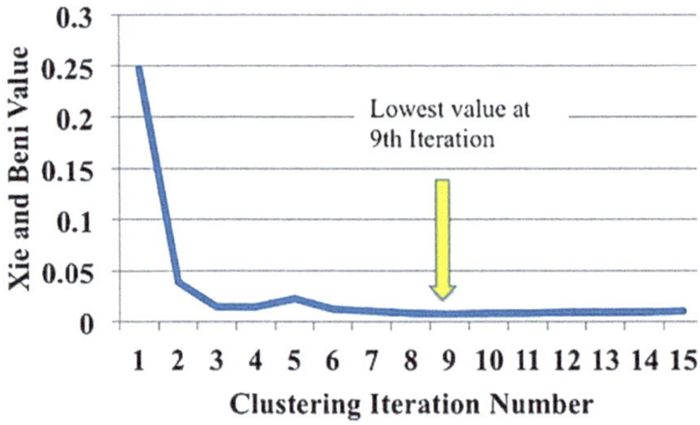

Fig. 17 Xie and Beni Index for user clusters with FCM

References

1. Atanassov KT (1986) Intuitionistic fuzzy sets. Fuzzy Sets Syst 20(1):87–96
2. Bargiela A, Pedrycz W (2002) Granular computing: an introduction. Kluwer Academic Publishers, Boston
3. Bezdek JC (1981) Pattern recognition with fuzzy objective function algorithms. Kluwer Academic Publishers, Norwell
4. Caruana, R., Elhaway, M., Nguyen, N., Smith, C.: Meta clustering. In: 6th International Conference on Data Mining, 2006. ICDM'06, pp. 107–118. IEEE, Hong Kong, China (2006).
5. Castellano G, Fanelli AM, Mencar C (2002) Generation of interpretable fuzzy granules by a double-clustering technique. Arch Control Sci 12(4):397–410
6. Chaira T (2011) A novel intuitionistic fuzzy c means clustering algorithm and its application to medical images. Appl Soft Comput 11:1711–1717
7. Debgupta R, Saha A, Tripathy BK (2020) A Faster Fuzzy Clustering Approach for Recommender Systems. In: Bhateja V, Satapathy S, Zhang YD, Aradhya V (eds) Intelligent Computing and Communication, ICICC 2019, Advances in Intelligent Systems and Computing, vol 1034. Springer, Singapore, pp 315–324
8. Dubois D, Prade H (1990) Rough fuzzy sets and fuzzy rough sets. Int J Gen Syst 17(2):191–200
9. Dunn. J.C.: A fuzzy relative of the ISODATA process and its use in detecting compact well-separated clusters. Cybernetics 3(3), 32–57 (1973).
10. doi:https://doi.org/10.1080/01969727308546046
11. El-Yaniv. R., Souroujon, O.: Iterative double clustering for unsupervised and semi-supervised learning. Machine Learning: ECML 2001. Springer, Freiburg, Germany, pp. 121–132 (2001).

12. Forgy EW (1965) Cluster analysis of multivariate data: efficiency versus interpretability of classifications". Biometrics 21(3):768–769
13. Gacek A, Pedrycz W (2015) Clustering granular data and their characterization with information granules of higher type. IEEE Trans Fuzzy Syst 23(1):850–860. https://doi.org/10.1109/TFUZZ.2014.2329707
14. Gupta, A., Tripathy, B. K.: Implementing GloVe for context based K-means++ clustering, Proceedings of the ICISS 2017, 7–8 Dec. 2017, 1041–1046, (2018).
15. Ignatov, D.I., Kuznetsov, S.O., Poelmans, J.: Concept-based biclustering for internet advertisement. In: 2012 IEEE 12th International Conference on, Data Mining Workshops (ICDMW), IEEE, Brussels, Belgium, pp. 123–130 (2012).
16. Ignatov DI, Kuznetsov SO, Poelmans J, Zhukov LE (2013) Can triconcepts become triclusters? Int J Gen Syst 42(6):572–593
17. Lingras P, West C (2004) Interval Set Clustering of Web Users with Rough K-means. Journal of Intelligent Information Systems 23(1):5–16
18. Maji P, Pal SK (2007) RFCM: A Hybrid Clustering Algorithm Using Rough and Fuzzy Sets. Fund Inform 80(4):475–496
19. Mirkin B (1996) Mathematical classification and clustering. Kluwer Academic Publishers, Boston
20. Mitra S, Banka H, Pedrycz W (2006) Rough Fuzzy Collaborative Clustering. IEEE Transactions on Systems Man and Cybernetics Part B-Cybernetics 36:795–805
21. Panda M, Abraham A, Tripathy BK (2016) Soft granular computing based classification using hybrid fuzzy-KNN-SVM. Intelligent Decision Technologies 10(2):115–128
22. Pawlak Z (1982) Rough Sets. Int J Comput Inform Sci 11(5):341–356
23. Pedrycz W, Skowron A, Kreinovich V (2008) Handbook of Granular Computing. Wiley, New York
24. Saleha, R, Haider J.N., Danish N.: Rough Intuitionistic Fuzzy Sets, Conference: Proceedings of the 6th Joint Conference on Information Science, March 8–13, (2002).
25. Sandhu, S.S., Jadhav, A. R., Tripathy, B. K.: Comparison of centroid-based clustering algorithms in the context of divide and conquer paradigm based FMST framework, ICRCICN 2017, 3–5 Nov. (2017).
26. Seetha, H., Tripathy, B. K., Murthy, M. K.: Modern Technologies for Big Data Classification and Clustering, IGI Edited volume, (2017)
27. Slonim, N., Tishby, N.: Document clustering using word clusters via the information bottleneck method. In: 23rd annual international ACM SIGIR conference on Research and development in information retrieval, Athens, Greece, pp. 208–215 (2000).
28. Tripathy, B.K., Acharjya, D.P.: Association Rule Granulation using Rough Sets on Intuitionistic Fuzzy Approximation Spaces and Granular Computing, Anale Seria Informatica, Vol.IX, fasc.1, 125 -144 (2011).
29. Tripathy BK, Anuradha J (2014) Hierarchical Clustering Algorithm based on Attribute Dependency for Attention Deficit Hyperactive Disorder. International Journal of Intelligent Systems and Applications (IJISA) 6(6):37–45
30. Tripathy, B.K.,Anuradha, J.: An optimal rough fuzzy clustering algorithm using PSO, Int. Jour. of Data Mining, Modeling and Management, 7(4), 257 (2015).
31. Tripathy, B. K., Bhargava, R, Verma, E., Swarnalatha, P., Tripathy, A., Dhull, R.: Rough intuitionistic fuzzy C-means algorithm and a comparative analysis, Pro-ceedings of the 6th ACM India Computing Convention on - Compute '13, (2013).
32. Tripathy, B.K. and Deepthi, P.H.: Uncertainty based Clustering Algorithms for Medical Image Analysis, Edited Volume: Hybrid Intelligence for Image Analysis and Understanding, Wiley Publications, (Editors: S. Bhattacharya, I. Pan, A. Mukherjee, P. Dutta), Wiley Publications, (2017), Chapter-12, pp.283–295.
33. Tripathy BK, Deepthi PH, Mittal D (2016) Hadoop with Intuitionistic Fuzzy C-means for clustering in Big Data. Adv Intell Syst Comput 438:599–610
34. Tripathy BK (2018) Rahul: Fuzzy Clustering of Sequential Data. International Journal of Intelligent Systems and applications, MECS publishers 11(1):43–54

35. Tripathy, B. K., Seetha, H., Murthy, M. K.: Uncertainty Based Clustering Algorithms for Large Data Sets, In: Modern Technologies for Big Data Classification and Clustering, IGI Edited volume, Chapter-1, pp.1–33, (2017).
36. Tripathy BK, Sharma A, Natarajan P, Swarnalatha P, Lakkshmanan A, Krishnan N (2013) Image Transformation using Modified K-means clustering algorithm for Parallel saliency map. IJET 3(5):347–353
37. Tripathy, B.K., Sobti, S., Shah, V.: A refined rough fuzzy clustering algorithm, proceedings of the IEEE ICCIC2014, 776–779 (2014).
38. Tripathy BK, Swarnalatha P (2015) A Comparative analysis of Depth computation of Leukemia Images using a refined Bit Plane and Uncertainty based clustering Techniques. Cybernetics and Information Technologies 15(1):126–146
39. Tripathy, B.K., Tripathy, H. K.: A Rough set approach for Clustering the Data Using Knowledge Discovery in World Wide Web for E-Business", Proc. of IEEE Int. Conf. on e-Business Engg., 24–26 Oct., Hong Kong, CHINA, pp.717–722 (2007).
40. Yao, Y.Y.: Granular Computing: Past, Present, and Future. In: Rough Set and Knowledge Technology (RSKT 2008). Springer- Verlag, Berlin, Germany, vol. 5009 in LNAI, pp. 27–28 (2008b).
41. Yao, J.T.: Human-inspired Granular Computing. In: Yao J (ed) Novel developments in granular computing: applications for advanced human reasoning and soft computation. Information Science Reference, Herskey, pp. 1–15 (2010a).
42. Yao, Y.Y.: Artificial Intelligence Perspectives on Granular Computing. In: W P, S-M C (eds) Computing Granular, Systems Intelligent. Design with Information Granules of Higher Order and Higher Type. Springer, Berlin, pp 17–34 (2011).
43. Zadeh LA (1965) Fuzzy Sets. Inf Control 8:338–353
44. Zadeh LA (1997) Towards a theory of fuzzy information granulation and its centrality in human reasoning and fuzzy logic. Fuzzy Sets Syst 90(1):111–127
45. Zadeh LA (1979) Fuzzy Sets and Information Granularity. In: Gupta NRR, Yager R (eds) Advances in Fuzzy Set Theory and Applications. North-Holland, Amsterdam, pp 3–18

Chapter 38
A Dynamic Web Data Extraction from SRLDC (Southern Regional Load Dispatch Centre) and Feature Engineering Using ETL Tool

J. Dhanalakshmi and N. Ayyanathan

1 Introduction

Web extraction, otherwise called web information extraction, is the way toward recovering or "extraction" information from a website. The way toward eliminating data from operative structures and to establish the data appropriation focus is called ETL (Extraction, Transformation and Loading) [1].

The large amount of data is collected from the websites, and the data have been preprocessed and stored in the data source. From the data source, the retrieved data are adopted, and the patterns and trends are used to analyze the data for efficient storage in big data analysis [2].

The enhancement is emerging in various platforms. The automatic web data extraction is mainly used in unstructured and semistructured data. These data have grown exponentially with the new tools, and it can easily transform the extracted data in a new technology development by using ETL data extraction [3].

1.1 Problem Definition

The undertaken research work is the first phase in the design and development of big data framework for the power grid operations and time-series data clustering of Indian southern region power distribution system. The dynamic web data extraction from SRLDC is performed using Python ETL data extraction tool. The proposed system provides feature engineering on the select explanatory variables, like thermal, hydro, gas/diesel/naptha, of SRLDC with specific data ingestion strategy.

J. Dhanalakshmi (✉) · N. Ayyanathan
B.S. Abdur Rahman Crescent Institute of Science and Technology, Chennai, Tamil Nadu, India

G. Mathur et al. (eds.), *Proceedings of 2nd International Conference on Artificial Intelligence: Advances and Applications*, Algorithms for Intelligent Systems,
https://doi.org/10.1007/978-981-16-6332-1_38

2 Literature Review

The big data strategy is a huge volume of data such as structured, unstructured and semistructured of various applications. It cannot be preprocessed or analyzed using tools and methodologies to increase the challenges in the way of automatic web extraction [3]. The PDF toolkit originating from the pyPdf project is PyPDF2. Presently, it is maintained by Phaseit, Inc. In order to construct a new file, PyPDF2 can extract data from PDF files or manipulate existing PDFs [4].

The study examines that the exploration will attempt to locate a formal model for extracting the ETL measures that map the approaching information from various data sources to be in a reasonable design [5, 6]. Beginning from the prerequisite of adaptability, we planned and executed our framework as an ETL (Extract, Transform, and Load) measure [7, 8]. Another is to extract information from PDFs. This is exceptionally troublesome, as PDF is a language for printers and doesn't hold a lot of data on the design of the information that is shown inside a record [9].

A Python library for pulling data from HTML and XML files is Beautiful Soup. It usually saves programmers hours or days of work. The Beautiful Soup object represents the parsed text and the methods mentioned in Tree Navigation and Tree Search [10]. Beautiful Soup also describes classes called Stylesheet, Script and TemplateString for embedded CSS stylesheets, embedded Javascript and HTML templates [11].

2.1 Quantitative Review Summary

The systematic literature review was done by accessing and extracting relevant research areas focusing on the following key concepts: focus on extracting the relevant data sources, transforming the various perspectives of PDF data, and it has been identified and understanding the loading concept of ETL framework and combining the various perspectives of dynamic web data extraction. The information data sources of relevant research articles from 2016 to 2020 were searched. The various following databases: IEEE Explore [5%], Science Direct [5%], Springer [3%] and some of the international journals [10%] were accessed, and period was also included based on the research study. The screening of the papers is based on the title, abstract and keywords and the inclusion and exclusion principle. The study selection of research literature methodology was carried out in a predefined manner according to the study's need and relevance. The searches are retrieved keywords such as Dynamic Web Data Extraction, Big Data, ETL and Beautiful Soup used from the scientific databases.

2.2 Research Questions Formulation

RQ1: What are the different data sources identified?
RQ2: What are the different ETL tools?
RQ3: What is the main research based on data extraction?
RQ4: How is information on the subject retrieved from different file formats?

2.3 Search Query Keywords

Q1: The key sources are identified from Indian Southern Regional Load Dispatch Centre (SRLDC) grid operations.
Q2: Dynamic web extraction using beautiful soup.
Q3: The main research is based on PDF document, extraction and transformation.
Q4: The subject retrieved from demand data of SRLDC day-wise and month-wise data.

2.4 Research Gap

The outcome of review summary reveals that the following research gaps are identified.

(a) Web crawling or web scraping is professionally accepted as data collection method.
(b) The implementation of web scraping is done in the cloud platform GoogleColab.

3 Proposed Model for Dynamic Web Data Extraction

3.1 Design Process of Extraction

The main focus of the extraction procedure is to acquire, with the least resources, all the necessary data from the source system as far as possible. The extract method should be constructed as it does not have a negative effect on the source system in terms of presentation of two methods of extraction exist like Logical Design Extraction and Physical Design Extraction [12].

During the extraction procedure, the data needed are examined and extracted from POSOCO database systems in the form of an application. Depending entirely on the source system and market conditions, the scope of the derived data will often be changed [13–18]. The main focus of the extraction procedure is to acquire, with the least resources, all the necessary data from the source system as far as possible [12, 19–22].

3.2 Transformation Design

The transformation of data is carried out in this process using SRLDC into the required format that can be easily stored in a dynamic database system. SRLDC site data have been retrieved from the transformation process and then it is in the PDF format. Using Beautiful Soup, this PDF format information is converted and formatted [23–26]. The process of data transformation also includes correction of data, cleaning, data deletion of inaccurate and redundant data, the incomplete creation of data, fixing and correcting data bugs, data integrity and incompatible data formatting until loading it into a web transformation framework design.

3.3 Process Loading

The last ETL strategy is to stack data into the objective's multidimensional design. The dimensionals are constructed by end-clients and application frameworks. Referential uprightness ought to be kept up through the ETL cycle to guarantee congruity [27].

4 Proposed Model for Dynamic Web Data Extraction

The input–output cycle of the proposed automation system is as depicted in Fig. 1. It elaborates the ETL methodologies for extracting, transformation and loading the various data sources. The first phase is used to extract the PDF file from the HTML web page. Then the PDF file has been transformed into excel csv format and finally the last phase is to load the excel csv data for various energy data analyses.

5 Implementation Metrics

The functionalities used to identify the dynamic web data extraction is done by implementing tabula-py in Google cloud platform. Tabula-py is a Python library package used for reading table in PDF. Using this tabula-py, the pdf file can be read and converted into a panda's data frames. It is also used to convert pdf file into csv/tsv/json file. The following steps are initiated using GoogleColab.

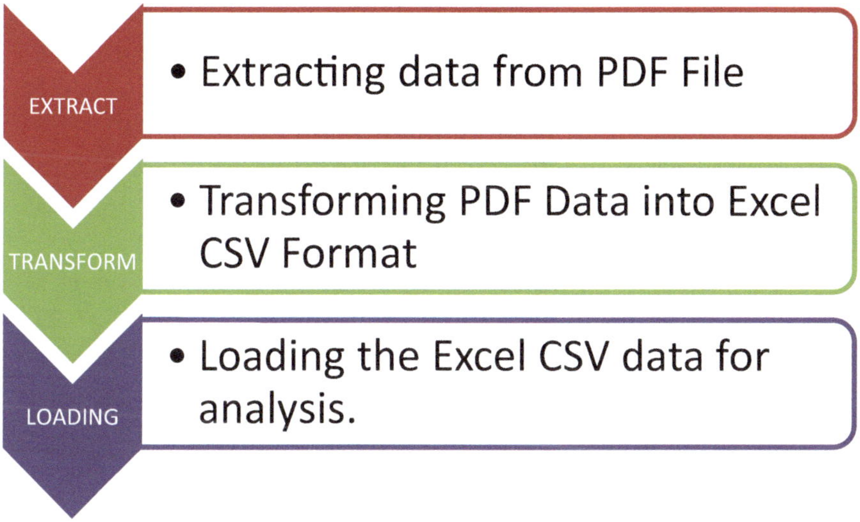

Fig. 1 Data extraction automation process

Step 1: Log into a Google account for accessing the cloud platform GoogleColab.
Step 2: Import Beautiful Soup library package and run the code for verifying the library files.
Step 3: Once the library package is installed import tabula-py for reading the PDF file of HTML data.
Step 4: The PDF file has been fetched and it can convert into csv/tsv/json file format.
Step 5: Finally, the new data have been generated using the proposed automation procedure model.

6 Generated Dataset Using Proposed Automation Procedure

The dynamic web data extraction is done from SRLDC data using beautiful soup implementation. This dataset variable contains the date, state, thermal, hydro, gas/diesel/naphtha, wind and others. First web extraction is done using all these variables and generated the web extraction results of various energy data. The various states have been demonstrated and measured based on the daily scheduling of load details. By using this web extraction methodology, the data have been extracted based on table structure on web page and give the measurement of each and individual values shown in Fig. 2.

Date	State	Thermal	Hydro	Gas/Diesel/Naptha	Wind	Others
01-01-2019	AndhraPradesh	81.31	7.41	9.24	2	10.07
01-01-2019	Telangana	43.67	0.51	0	0	10.13
01-01-2019	Karnataka	84.3	21.31	3.12	9.07	17.93
01-01-2019	Kerala	0	12.91	0.33	0.09	0.22
01-01-2019	Tamilnadu	77.9	14.64	6.78	2.56	39.35
01-01-2019	Pondycherry	0	0	0	0	0
02-01-2019	AndhraPradesh	82.07	8.87	7.32	1	11.77
02-01-2019	Telangana	42.13	1.27	0	0	9.99
02-01-2019	Karnataka	84.41	30.96	3.63	4.06	19.1
02-01-2019	Kerala	0	11.94	0.33	0.01	0.2
02-01-2019	Tamilnadu	79.3	16.89	6.59	0.95	39.46
02-01-2019	Pondycherry	0	0	0	0	0

Fig. 2 Generated dataset

7 Conclusion and Future Enhancement

Thus the automation procedure using the required data extraction from SRLDC website has been successfully implemented using Python beautiful soup. The first phase of design and development of big data framework of power grid operations and time-series data clustering of Indian Southern Region Power Distribution System have been successfully implemented and deployed. The remaining data ingestion module is the scope of future enhancement.

References

1. Vyas S (2017) A comparative study of various ETL process and their testing techniques in data warehouse. J Stat Manage Syst Taylor and Francis 20(6)
2. Dhanalakshmi J, Ezil Sam Leni A (2017) Instance vehicle monitoring and tracking with internet of things using Arduino. Int J Smart Sens Intell Syst
3. Hassanien HED (2019) Web scraping scientific repositories for augmented relevant literature search using CRISP-DM. Appl Syst Innov MDPI
4. Dhanalakshmi J, Ayyanathan N (2019) An implementation of sustainable energy model using multilayer perceptron and EM algorithm. In: National conference on recent trends in computer science and mathematics' Apr. 2019
5. Galici R, Ordile L, Marchesi M, Pinna A, Tonelli R (2017) Applying the ETL process to blockchain data prospect and findings. Int J MDPI Energies 4(6)
6. El-Sappagh SHA, Hendawi AMA, El Bastawissy AH (2016) A proposed model for data warehouse ETL processes. J King Saud Univ –Comput Inf Sci 4(6)
7. Dhanalakshmi J, Ayyanathan N, Pandian NS (2019) Energy analytics and comparative performance analysis of machine learning classifiers on power boiler dataset. In: ICOAC'19. IEEE

8. Naquishbandi TA, Ayyanathan N (2019) Clinical big data predictive analytics transforming healthcare—an integrated framework for promise towards value based healthcare. In: Int. Conf. Emerg. Trends Eng
9. Ashiwal P, Tandan SR, Tripathi P, Miri R (2016) Web Information retrieval using Python and BeautifulSoup. Int J Res Appl Sci Eng Technol 4(6)
10. Dhawani R, Shukla M, Puvar P, Prajapati B (2015) A novel approach to web extraction technology. Int J Adv Res Comput Sci Softw Eng 5(5)
11. Dhanalakshmi J, Ayyanathan N (2019) An implementation of energy demand forecast using J48 and simple K means. In: ICONSTEM'19, IEEE
12. Mishra RN, Chaturvedi DK (2019) Current scenario of electricity sector in India and restructuring. IEEE Sponsored
13. Uchil P (2015) Use of social media API and data extraction for financial service on Twitter. Int J Comput Sci Mob Comput 4(10)
14. Shreesha M, Srikara SB, Manjesh R (2018) A novel approach for news extraction using web scrapping. Int J Eng Res Technol
15. Gupta M, Gupta S, Thakur T (2018) Implementation of new electricity regulatory norms for deviation settlement mechanism: a case study of India. Taylor and Francis
16. Patel H, Prajapati P (2018) Study and analysis of decision tree based classification algorithms. Int J Comput Sci Eng 6(10)
17. Latha PG, Anand SR, Imthias ATP, Sreejith PS (2013) Commercial impact and optimum capacity determination of pumped storage hydro plant for a practical power system. Int J Emerg Electr Power Syst
18. Sabria IAA, Mana M, Bakarb WAWA, Roseb NM (2019) Web data extraction approach for deep web using WEIDJ. Procedia Comput Sci
19. Muthu Kumar T, Anbunesan G, Kumar SP (2018) Low voltage issues and its mitigation in southern regional grid in India. POSOCO
20. Safiullah H, Hug G, Tongia R (2017) Design of load balancing mechanism for Indian electricity markets. Energy Syst
21. Weinberg AI, Last M (2019) Selecting a representative decision tree from an ensemble of decision-tree models for fast big data classification. J Big Data, Elsevier. Ltd
22. Ayyanathan N, Bavani Rekha A, Shilpa A, Pethru Raj (2016) Comparative performance analytics of machine learning classifiers in the countrywise foreign tourists' arrival pattern and demand modeling of Indian tourism. In: ICOAC'16. IEEE
23. Brunello A, Marzano E, Montanari A, Sciavicco G (2018) J48SS: a novel decision tree approach for the handling of sequential and time series data. Int J MDPI
24. Shahbazi F (2019) Using decision tree classification algorithm to design and construct the credit rating model for banking customers. IOSR J Bus Manag
25. Patel N, Singh D (2019) An algorithm to construct decision tree for machine learning based on similarity factor. Int J Comput Appl 111(10)
26. Panda D, Singh SN, Kumar V (2019) Frequency based risk assessment of a power producer in Indian electricity market. Int J Emerg Electr Power Syst
27. Prabhu Chakkaravarthy A, Pugalendhi R, Ramya Jayaraman, Dhanalakshmi J (2020) An automatic Classification of Covid with J48 and Simple K Means using Weka. Int J Futur Gener Commun Netw 13(3)

Chapter 39
A Systematic Study of Intelligent Face Scanning in Rare Disease Detection

Suksham Sharma and Deepti Malhotra

1 Introduction

There are around 6000–8000 rare diseases worldwide, new rare diseases being reported regularly. Ironically, rare diseases are of low prevalence and uncommon but altogether they affect a significant portion of population of a country [1]. The cause of many rare diseases remains unknown. Though, a considerable proportion of rare diseases are found to be caused due to gene mutations. These rare diseases, caused by genetic mutation, are called rare genetic diseases. About 8% of the world population is affected by such rare genetic diseases [2]. *Angelman, Apert, Cornelia de Lange, Down, Fragile X, Noonan, Turner, Treacher Collins, and Williams-Beuren* are some rare genetic diseases. The rarity of such diseases implies that not much clinical information is available to diagnose a person suffering from rare genetic disease. Not many people suffering from a rare genetic disease get a proper diagnosis. Diagnosis, as early as possible, hence becomes crucial. An early diagnosis will not only help clinicians but also preclude possible health problems, such as critical respiratory problems, cardiovascular dysfunction, developmental delays, and mental retardation [12].

Rare diseases are found to have distinct observable physical characteristics called phenotypes. Facial phenotypes can be an important aspect in the identification of a genetic disorder. About 30–40% affected of the 7,000 known genetic disorders develop a physical change manifesting abnormalities in the face and skull [4]. Many patients suffering from rare disease undergo various trials and tests until they are correctly diagnosed using traditional methods. Researchers have found that artificial intelligence can be used to make a comparatively quick and reliable diagnosis with

S. Sharma (✉) · D. Malhotra
Department of Computer Science and IT, Central University of Jammu, Jammu & Kashmir, India

D. Malhotra
e-mail: deepti.csit@cujammu.ac.in

G. Mathur et al. (eds.), *Proceedings of 2nd International Conference on Artificial Intelligence: Advances and Applications*, Algorithms for Intelligent Systems,
https://doi.org/10.1007/978-981-16-6332-1_39

facial analysis. The Human Phenotype Ontology (HPO), a standardized vocabulary of phenotypic abnormalities, lists 1007 distinct terms for defining different facial abnormalities [5]. A dysmorphologist, who studies congenital structural malformations or anomalies, commonly called birth defects, will often identify these abnormalities individually and create a clinical diagnosis. But because of the phenotypic similarity of genetically disparate syndromes, this clinical diagnosis is often incorrect. Here, AI can help out mapping the image-derived phenotypes and their probable genetic source.

Facial recognition systems are proficient at identifying even the minute details that human eyes might miss. Facial recognition-based diagnostic tools work in three phases as shown in Fig. 1.

Such facial frameworks have been reported to be able to detect genetic diseases like *Down's Syndrome* with a fair level of accuracy.

Presently, integrating artificial intelligence and facial analysis to generate a workflow for efficient detection of rare genetic diseases is being studied. The use of AI in rare genetic diseases will help in early diagnosis as well as improved screening accuracy. This research work presents the methods used for the detection of rare genetic diseases using facial analysis.

Rare Disease Detection with Machine Learning. Machine learning technologies are proving to be a boon in healthcare and can also help improve the diagnosis of rare genetic diseases.

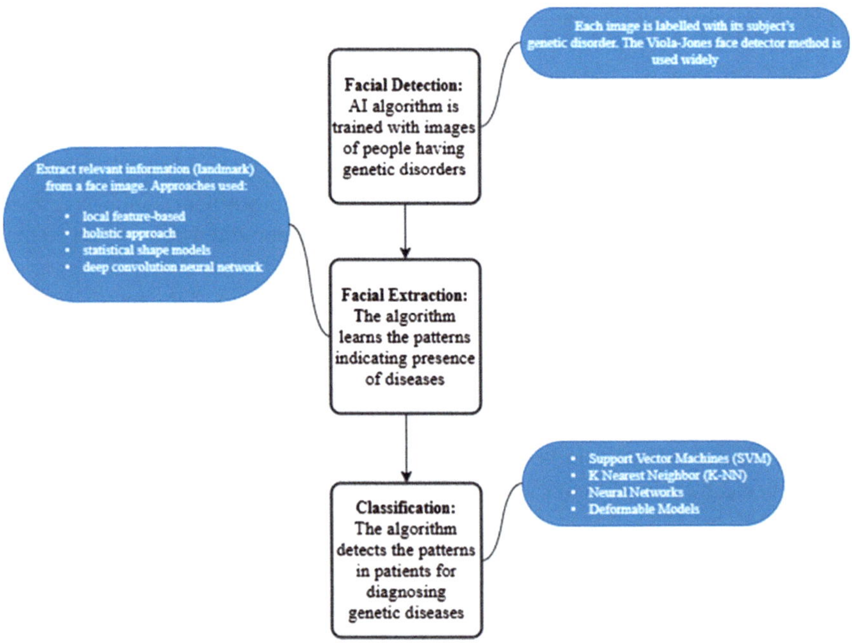

Fig. 1 Phases of face recognition-based diagnostic tools

Rare Disease Detection with Deep Learning. Mainly, deep learning methods focus on supervised algorithms; however, in case of rare disease, image data are often scarce. So, unsupervised or semi-supervised algorithms are preferred.

The organization of the paper is as follows: Section 1 gives a brief introduction of rare genetic diseases. Section 2 presents the work done by various researchers to detect rare genetic diseases. Section 3 gives the comparative analysis of recent works in AI-based rare disease detection. Section 4 covers the open gaps and challenges found. Finally, Section 5 sums up the paper.

2 Literature Review

Facial recognition software can help identify rare genetic diseases using two-dimensional and three-dimensional photographs of affected patients. Using AI algorithms, a tool's accuracy can be improved over time.

2.1 Detection Using Machine Learning

Machine learning algorithms can screen person suffering from rare genetic diseases on the basis of the facial data (2D and 3D) used to train them.

Using two-dimensional images. The existing face image algorithms are primarily for face recognition and Content-Based Image Retrieval (CBIR) [6]. In [6], the authors have used the concepts of symmetry and geometry of the human face for image-based screening for dysmorphic children from a frontal image of a face with Linear Discriminant Classifier (LDC) giving maximum classification accuracy of 89.19%. On comparison with other classifiers (RBF-SVM and QDC), in terms of classification accuracy and time, LDC proved to be better. Figure 2 shows some anatomical landmarks.

In [7], PCA and LCA algorithms are used to study the dimensions of different facial features and compare them with healthy child's facial dimensions to screen Progeria, also known as Hutchinson-Gilford syndrome. Here, PCA works on facial detection in newborns, and LCA is used for pattern recognition. Another work [8] uses fusion of deep features with the geometric and Oriented FAST and rotated BRIEF (ORB) representation, which enhanced the discriminative strength of deep features.

In [9], an approach to detect genetic diseases using transfer learning has been given. This work uses a pre-trained VGGFace model with Restnet50 architecture by fine-tuning the last layers and a SVM classifier. For fine-tuning the VGGFace model, three fully connected layers were added, with batch normalization and a 0.5 dropout rate to prevent overfitting, and an SVM classifier on the top; giving an accuracy of 97.66%. Alternatively, VGGFace model with four fully connected layers gave 97.45% accuracy.

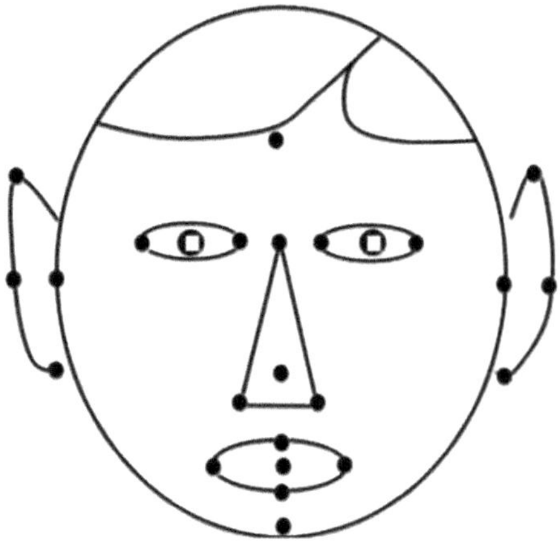

Fig. 2 Anatomical landmarks on face [6]

Using three-dimensional images. Another work [10] uses 68 different sets of 35,709 facial points along with VGGFace neural network model for feature extraction. It reconstructs the 3D face using 2D frontal image points. It achieved a classification accuracy of 92.5% using combined features (deep features and all 3D reconstruction points). It concludes that deep features provide a better classification accuracy than geometric features.

2.2 *Detection Using Deep Learning*

Reference [11] performed a study on 49 persons incubating causative variants in known Cornelia de Lange syndrome (CdLS) genes to gauge Face2Gene effectiveness, understanding and sensitivity in the clinical diagnosis of CdLS. It concluded that deep learning-based computer-aided image analysis could facilitate the clinical diagnosis of CdLS with their study giving 97.9% top five prediction accuracy.

In [3, 12], DeepGestalt, a facial image analysis framework, outperformed clinicians' ability to identify disease in three separate experiments. It uses computer vision and deep learning algorithms, which measures closeness to the multitude of syndromes from 2D images [3]. This framework was able to discern between different subtypes of the genetic disorder Noonan syndrome in 64% of the cases appropriately when compared with 20% accuracy of clinicians. Another work [13] uses Deep Convolutional Neural Network (DCNN) for feature extraction with fivefold cross-validation for the detection of developmental disorders. It resulted in 98.80% recognition accuracy.

3 Comparative Analysis

The comparative analysis of rare disease detection with face recognition using machine learning is presented in Table 1.

In Fig. 3, the comparative analysis of different machine learning-based classifiers used to screen rare genetic diseases is presented. It can be said that the use of classifiers along with PCA and LDA dimension reduction together can yield good results in terms of efficacy as well as accuracy.

Table 2 presents the comparative analysis of detection of rare diseases with facial recognition using deep learning.

In Fig. 4, the comparative analysis of machine learning-based rare genetic disease detection approaches is presented, which shows that the use of DCNN along with SVM classifier can give a high accuracy rate.

Figure 5 gives the percentage usage of different AI techniques presented in the study. Table 3 presents a relative comparison of machine learning and deep learning approaches on certain parameters involved in the detection of rare genetic diseases with facial recognition. From the comparison, it was found that 3D dataset-based deep learning approaches are yet to be explored. If considered, this can provide a more comprehensive facial dataset, which, in turn, reduces the chances of erring. Deep learning approaches provide the ease of automatic facial landmarking on contrary to manual landmarking associated with machine learning approaches. Moreover, the deep feature extractions by deep learning algorithms are better. The prediction accuracy provided by DCNN along with SVM outperforms the existing machine learning approaches and involves wider range of syndromes.

4 Open Issues and Challenges

The challenges involved in the detection of rare genetic diseases using facial analysis are:

- Age-specific facial dataset affects the results obtained using machine and deep learning approaches. It is found that detection using image-based approaches is more efficient in newborn than aged individuals.
- Data availability, especially images, in case of rare diseases, is an issue. Deep learning algorithms, therefore, need to shift from supervised deep learning field to unsupervised or semi-supervised. Efficiency of this shift needs to be considered and further explored.
- Unavailability of data can cause a disease to be underrepresented in dataset, causing class imbalance is another major issue.
- Lack of experts to validate the results is another problem.
- Work on analysis of 2D facial images has been done more in comparison to 3D facial images due to the dearth of supporting tools to acquire 3D facial data.

Table 1 Rare genetic disease detection using machine learning approaches

Author	Year	Proposed model	Description	Dataset	Number of syndromes considered	Performance evaluation
V. Kumov and A. Samorodov [10]	2020	Use of combined features (Geometric and Deepfeatures) to detect genetic diseases	3D face reconstruction obtained from a single 2D frontal image used for feature localization along with PCA and LDA dimension reduction tested on different classifiers	Facial images of 12 syndromes	8	*3D points, LDA classifier*: 90.1% recognition accuracy *3D points, LDA dimension reduction, Deep features, Logistic Regression:* 92.4% recognition accuracy *3D points + deep features, PCA + LDA dimension reduction, logistic regression:* 92.5% recognition accuracy
A. P. Ekbote and V. R. Ratnaparkhe [6]	2019	Image-based screening using basic concepts of symmetry and geometry of human face	Facial landmarks selected were used to extract linear and angular features extraction. Linear Discriminant Classifier (LDC) achieved maximum accuracy	Self-collected using digital camera	8	*Linear Features, LDC:* 89.19% accuracy *Angular features, LDC:* 83.78% accuracy *Combined features, LDC:* 86.49% accuracy
U. S. Singh, A. Kumar Gupta, T. Choudhury, and T. P. Singh [7]	2019	Work on facial coordinate system to develop a specific approach for detection of Progeria over existing generalized methods	Uses Principal Component Analysis (PCA) for facial detection and Linear Discriminant Analysis (LDA) for detecting patterns		1	Enables early stage detection of Progeria

(continued)

Table 1 (continued)

Author	Year	Proposed model	Description	Dataset	Number of syndromes considered	Performance evaluation
J. Y. R. Cornejo and H. Pedrini [8]	2019	Fusion of deep features with the geometric and Oriented FAST and rotated BRIEF (ORB) representation	Fused geometric and deep features with PCA + LDA feature reduction gave a reduced feature space, which was used by different classifiers	Diagnostically Relevant Facial Gestalt Information from Ordinary Photos Database	8	*Deep features + geometric features + ORB, PCA + LDA feature reduction, GNB classifier:* 90.29% accuracy
A. Singh and D. R. Kisk [9]	2018	Fine-tuning pre-trained model VGGFace	Transfer learning is employed to avoid overfitting and underfitting with tiny dataset. VGGFace model is tuned with ResNet50 architecture	eLife database	12	*VGGFace- 4 FC layers on top*: 97.45% accuracy *VGGFace- 3 FC layers and SVM classifier:* 97.66% accuracy

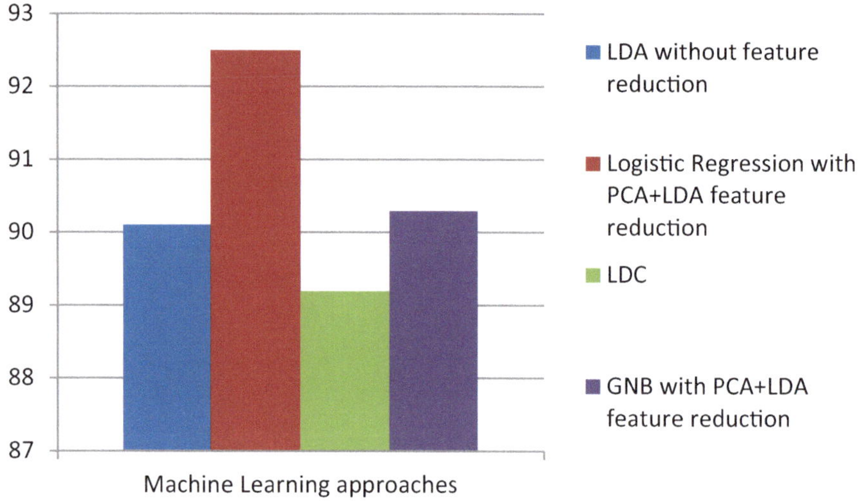

Fig. 3 Comparison of best possible accuracy rate given by machine learning-based classifiers

5 Conclusion

This research mainly focuses on analyzing the recent advances in intelligent powered facial recognition technology to detect genetic syndromes, which have observable distinct phenotypes. The study shows that with the integration of AI and facial technology, the detection of rare genetic diseases can be made efficient, faster and with a good accuracy rate. By analyzing different works given by the researchers, the research work concludes that deep learning approaches are a possible solution in comparison to machine learning approaches for rare disease detection using facial frameworks. The deep learning approaches not only surpass machine learning approaches in terms of accuracy but have also proved to outperform clinicians, in some scenarios, in terms of efficacy, accuracy with speed. Deep learning-based frameworks need to be explored further to improve the diagnostic rate of rare diseases, which will be vital aid to researchers as well as health professionals. Though, one of the prominent challenges that still remain is the availability of datasets with adequate amount of data for different genetic syndromes.

Table 2 Rare genetic disease detection using deep learning approaches

Author	Year	Proposed model	Description	Dataset	Number of syndromes considered	Performance evaluation
A. Latorre-Pellicer et al. [11]	2020	A phenotype study to evaluate the efficiency of Face2Gene in the identification of CdLs	Frontal images were analyzed using Face2Gene technology. The top-1 and top-5 sensitivity were measured	49 individuals already clinically and molecularly confirmed as CdLS	1	*Top-5 prediction accuracy:* 97.9% *Top-1 prediction accuracy:* 83.7%
Y. Gurovich et al. [3][3]	2019, 2018	Computer vision and deep-learning algorithm-based facial image analysis framework, *DeepGestalt*	Uses cascaded deep convolutional neural networks (DCNNs) for face detection and specialized DCNNs work on prediction of syndromes	CASIA Web-Face dataset aggregated from *Face2Gene*	215	*Top-10 accuracy*: 91%
P. Shukla, T. Gupta, A. Saini, P. Singh, and R. Balasubramanian [13]	2017	Deep Convolution Neural Network (DCNN) for feature extraction along with fivefold cross-validation	Division of facial image into four parts for extraction of facial information using DCNN and SVM for classification	Collected from Organisations in Uttarakhand, India and Internet	6	98.80% accuracy

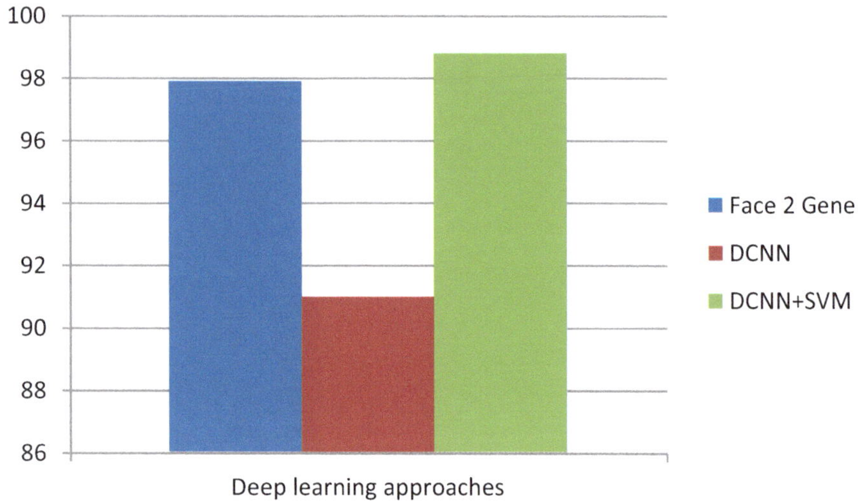

Fig. 4 Comparison of best possible accuracy rate given by deep learning-based classifiers

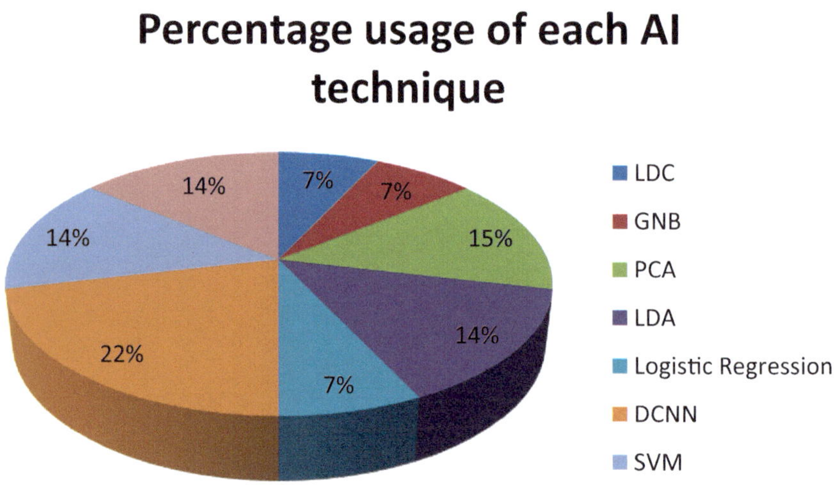

Fig. 5 Percentage usage of each AI technique based on the study

Table 3 Relative Comparison of Machine Learning and Deep Learning-based approaches

Parameter	Machine learning		Deep learning	
	Merits	Demerits	Merits	Demerits
Dataset	Can work on small datasets	Overfitting and underfitting can be caused due to tiny dataset	No overfitting and underfitting	Requires larger training dataset
Sample data	Work has been done with both 2D and 3D facial data 3D images are resilient to these variations due to expressions, illumination, makeup, pose and occlusions	2D data are sensitive to these variations	Variation due to 2D data does not impact the performance considerably	3D data have not been worked upon with
Landmarks	Manual landmarking	Time-consuming	Automatic landmarking	–
Feature extraction	Deep features are better than geometric features in classification accuracy generally	Geometric features are important in classification but lag behind deep features	Works on combined features (geometric and deep). Features extracted using DCNN are found to be better	–
Classifier	Classifiers like LDC, logistic regression, etc. provide good classification accuracy	Conventional classifiers are likely to miss valuable information from a full cranium profile perspective	DCNN gives high specificity and better accuracy against machine learning counterparts	Lack of comparison to other methods or human experts

References

1. National policy for treatment of rare diseases, Ministry of Health and Family Welfare, Government of India. https://main.mohfw.gov.in/sites/default/files/RareDiseases Policy FINAL.pdf.
2. Baird PA, Anderson TW, Newcombe HB, Lowry RB (1988) Genetic disorders in children and young adults: a population study. Am J Hum Genet
3. Gurovich Y et al (2019) Identifying facial phenotypes of genetic disorders using deep learning. Nat Med. https://doi.org/10.1038/s41591-018-0279-0
4. Hart TC, Hart PS (2009) Genetic studies of craniofacial anomalies: clinical implications and applications. Orthod Craniofac Res. https://doi.org/10.1111/j.1601-6343.2009.01455.x
5. Köhler S et al (2019) Expansion of the Human Phenotype Ontology (HPO) knowledge base and resources. Nucleic Acids Res. https://doi.org/10.1093/nar/gky1105
6. Ekbote AP, Ratnaparkhe VR (2019) Genetic syndrome identification: an image processing approach. IETE J Res. https://doi.org/10.1080/03772063.2019.1619488

7. Singh US, Kumar Gupta A, Choudhury T, Singh TP (2019) Developing a new framework using facial recognition system for the detection of Progeria. https://doi.org/10.1109/ICCIKE47802.2019.9004365
8. Cornejo JYR, Pedrini H (2019) Recognition of genetic disorders based on deep features and geometric representation. In: Lecture notes in computer science (including subseries Lecture notes in artificial intelligence and lecture notes in bioinformatics), vol 11401, LNCS. https://doi.org/10.1007/978-3-030-13469-3_77
9. Singh A, Kisku DR (2018) Detection of rare genetic diseases using facial 2D images with transfer learning. https://doi.org/10.1109/ISED.2018.8703997
10. Kumov V, Samorodov A (2020) Recognition of genetic diseases based on combined feature extraction from 2D face images. https://doi.org/10.23919/FRUCT48808.2020.9087456
11. Latorre-Pellicer A et al (2020) Evaluating face2gene as a tool to identify cornelia de lange syndrome by facial phenotypes. Int J Mol Sci. https://doi.org/10.3390/ijms21031042
12. Gurovich Y et al (2018) DeepGestalt—identifying rare genetic syndromes using deep learning. http://arxiv.org/abs/1801.07637
13. Shukla P, Gupta T, Saini A, Singh P, Balasubramanian R (2017) A deep learning frame-work for recognizing developmental disorders. https://doi.org/10.1109/WACV.2017.84

Chapter 40
Unlocking the Potential of Natural Language Processing and Healthchatbots in Healthcare Management

R. Sivarethinamohan and S. Sujatha

1 Introduction

The COVID-19 crisis has seriously affected the patient care paradigm and altered patient behavior and expectations [1]. Virtual care as opposed to a physical checkup is here to stay and can be a central part of the patients' experience with health care providers in addition to being a part of clinical trials to assess the usefulness and safety of medications or medical devices [2]. The patient survey of Accenture COVID-19 was conducted diagonally in China, France, Germany, Japan, the UK, and the USA. The survey revealed that patients willingly accept virtual visits that take place between patients and clinicians via communications technology virtual care and communications; however, traditional methods are followed at very high rates based on the quality of diagnosis and care. Healthcare providers can keep up or even improve the overall experience of patients with emerging technologies in healthcare [18]. They feel a sense of pressure to explore more intelligent and efficient ways to share information and uncover patterns. In fact, Artificial Intelligence (AI) powered NLP, Healthchatbots, and Machine Learning (ML) can digitalize entire patient medical records [9]. This advancement is now groundbreaking, unlike before. For instance, chronic diseases like type-2 diabetes, asthma, respiratory or pulmonary illness, or weak immunity among patients make them highly vulnerable to the virus, and hence this technology can be made available to them to save lives [23]. Availability of historical data and patient analysis can provide an individual's medical history of hereditary and genetic causes, thereby saving time to take down a detailed history [14]. Furthermore, reinvented EHR's and digital prescriptions are more efficient

R. Sivarethinamohan (✉)
Department of Professional Studies, CHRIST (Deemed to be University), Bengaluru, India

S. Sujatha
Department of Civil Engineering, K Ramakrishnan College of Technology, Trichy, Tamilnadu, India

G. Mathur et al. (eds.), *Proceedings of 2nd International Conference on Artificial Intelligence: Advances and Applications*, Algorithms for Intelligent Systems,
https://doi.org/10.1007/978-981-16-6332-1_40

than understanding handwritten ones and are more analytical and insightful [19]. In a nutshell, these technologies are used to empower people to obtain better health care and immunity during this pandemic. In the USA, businesses prioritize privacy in their AI strategies by enforcement of GDPR and state-level privacy laws [7].

2 NLP: The Real Opportunities for Health Care

Communication in medicine is a fundamental clinical skill to create an affiliation with the patient, paving a way to effective diagnosis and treatment [8] Effective physician–patient communication increases patient satisfaction, leading to greater patient understanding of health problems and treatments available. Patients expect faster responses, shorter wait times, and near-instant communication to obtain the information they need. A better understanding of patients' touch point preferences captivates and retains them. Strengthening emerging technology, such as ChatBots can provide the information to the patients' needs, within seconds [5]. A ChatBot can with no trouble access medical-related information such as hours of operations, driving directions, doctor or office contact information, etc. Patients can visit the website and quickly get the information without the support of a customer service representative. An automated messaging system sends regular appointment or medication refill reminders to connect with the patients and keep them healthy between visits. FAQ-styled pages on websites including the top asked questions can also easily navigate and manage patient inquiries. In automated or online appointment system, patients can make their own appointments, on their own time. ChatBots and online messengers can offer healthcare information [4]. Even patients resolve more complex health-related questions or concerns. AI-powered medical assistants can reserve appointments, observe patient health status, and carry out other time-intensive responsibilities such as inventory, billing, and claims management. Most hospitals offer translation services and treatment-related dialogue with patients. It implicates the better understanding of oral conversation or written communication through a personal computer [16].

Ai chatbots in health care.

Figure 1 explains the architecture of Ai chatbots in health care.

3 Healthchatbots: Reconnoitering Language Understanding in EHR Scenarios Using Chatbots

The first chatbot ELIZA, well-known as the mother of all chatbots, was created by Joseph Weizenbaum in the mid-1966 at MIT. It is surpassing into the territory of humans: empathy and feelings. Simple decision tree questions can be answered by ELIZA. It imitates a Rogerian therapist, largely by rephrasing the patient's statements

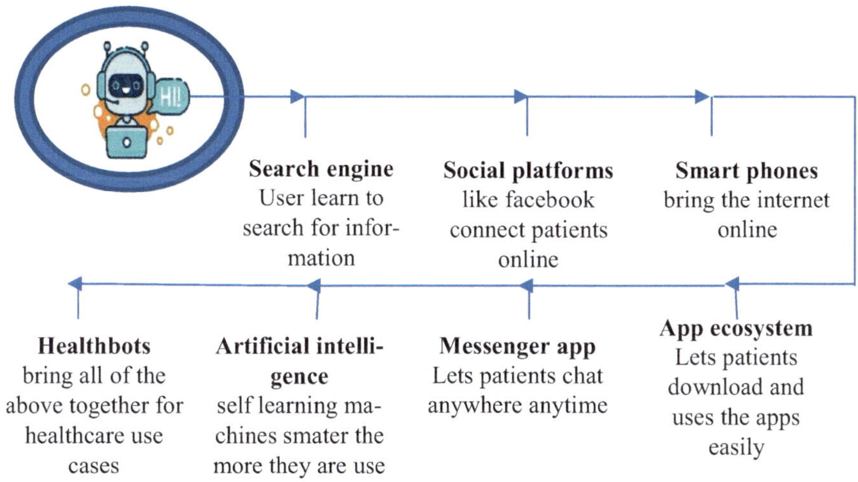

Fig. 1 Architecture of Ai chatbots in health care

as questions and posing them to the patient. A chatbot is a computer program, which is designed to simulate human conversation [10]. Chatbots are typically text-driven, with images and widgets. Patients communicate their issues with chatbots using a chat user interface centered on AI or via voice or interactive cards. Chatbots, as a personal nurse, interpret the words given to them by a patient and deliver a pre-set answer. It usually asks patients to know their symptoms and then communicate the information to doctors in primary care, who diagnose and prescribe medicine in real time [22]. Healthchatbots, computer programs, or smart algorithms conducting conversation via auditory or textual methods offer prompt conversational responses and help care providers to surpass patient expectations and improve patient outcomes [11]. Patients may perhaps senseless like using a personal computer and more like conversing with a person. Let's discuss a few Chatbots in use for EHR: Information Chatbot, Commerce chatbot, Enterprise Productivity Chatbot, and IoT devices (Fig. 2).

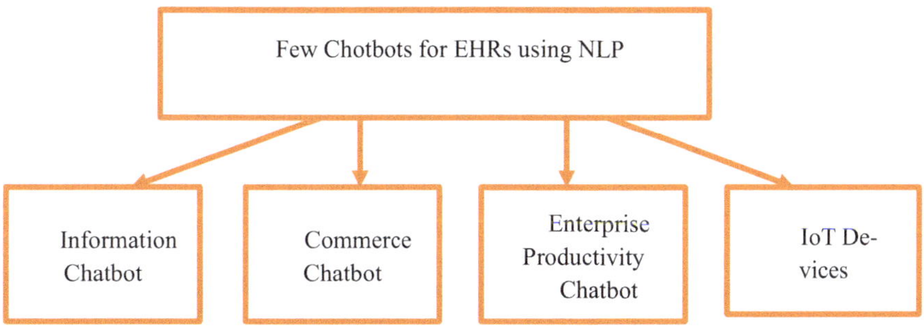

Fig. 2 Few Chatbots for EHRs using NLP

3.1 InformationChatbot

This Informational Bot responds and replies to statement well-defined as a question in a knowledge set or frequently asked questions by means of Cognitive Services Q&A Maker. It answers most questions that cannot be answered with a simple 'yes' or 'no', using Cognitive Search from a doctor who specializes in clinical work to persons who are receiving medical treatment [13]. Simple bot interaction provides quick answers to questions of patients and intelligently provides access to services offered by the health care provider. They are incredibly flexible and able to handle a few simple commands, or complex, artificial intelligence-powered, and natural language processing tasks [6].

Cincinnati Children's Hospital, recognized as a leader in pediatric healthcare, developed a family-friendly mobile app, Caren with chatbot in Microsoft Azure. It is a digital concierge that patients and families can interact with through a mobile app. It helps patients to sail across the large medical center as well as the care of their child, get answers to their questions, and delivers patient updates, facilitating healthier patient care and improved experiences for patients' families. It helps the hospital go much faster in the cloud infrastructure with Azure. By leveraging Microsoft Bot code and LUIS, the hospital could build out a language model [21]. The Microsoft chatbot SDK made it very easy to get off the ground. With Caren, parents can know what's going on with their child and help them be better mother to their child.

How does an Information Chatbot work?

A bot is a programmed software application to run according to its instructions without a human user needing to manually start them up every time. When developers build a bot on the Microsoft Bot Framework, they customarily select a .NET or JavaScript SDK. Microsoft ensures toolkits with abundant templates, tools, and tutorials. The Bot Builder home page on GitHub and the MS Virtual Assistant solution accelerator are the two right places to start for developers. Here, the information bot answers queries stated in a knowledge set or frequently asked questions by means of cognitive services QnA Maker and replies to answers in open text format via cognitive search.

The systematic procedure is exhibited in Fig. 3. First, a patient boots the application bot. Then Azure active directory authenticates the patient's uniqueness. The patient competent then requests the chatbot on the type of query reinforced. When a FAQ built with Q&A Maker is returned by cognitive services, the patient expresses an important question. The bot submits the questions to cognitive search, which, in turn, sends back information what they needed. Finally, application insights collect runtime telemetry to check the progress of Information Chatbot.

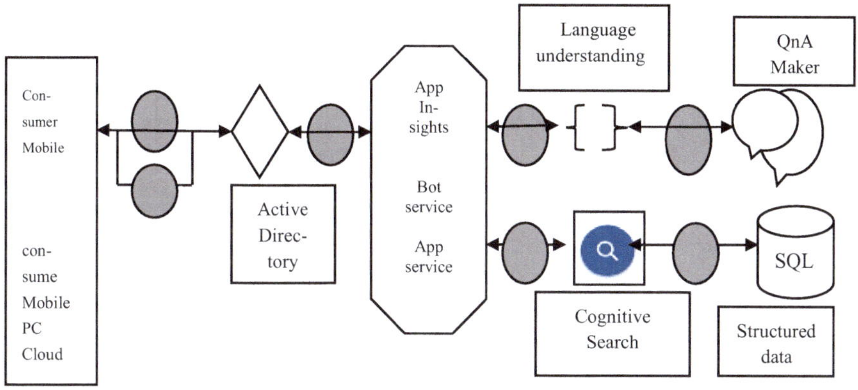

Fig. 3 Information Chatbot architecture

3.2 Commerce Chatbot

This type of chatbot uses Azure Bot Service by means of language understanding. It facilitates EHR developers to generate conversational interfaces for health care management. For example, a hospital's doorman or reception or ICUs can apply a chatbot to advance upon outdated e-mail and phone call interactions. It authenticates a patient over Azure active directory. Patient requests are processed by cognitive services via text and voice [3]. The inbuilt speech service can identify and upkeep voice commands.

How a commerce chatbot works in hospitals or health care units?

First a patient uses a mobile app and validates it using Azure Active Directory B2C. Then the patient makes a request for information using the custom application bot. Simultaneously, cognitive services assist and support the processing of the natural language request. The patient's response is reviewed by the authorized EHR developer to upgrade the question by means of natural conversation. When the patient gets satisfaction with the outcomes, the application bot keeps informed of the patient's booking or registration, and application insights collect runtime telemetry to monitor chatbot performance and practice.

The systematic workflow of a commerce chatbot is shown in Fig. 4

3.3 Enterprise Productivity Chatbots

Bot service simply uses Language Understanding to develop prevailing enterprise productivity bots, permitting hospitals to modernize routine work by Office 365 calendar, customer cases stored in Dynamics CRM, and others. It can use Azure Bot Service [17].

How Enterprise Productivity Chatbot works in hospitals and health care units?

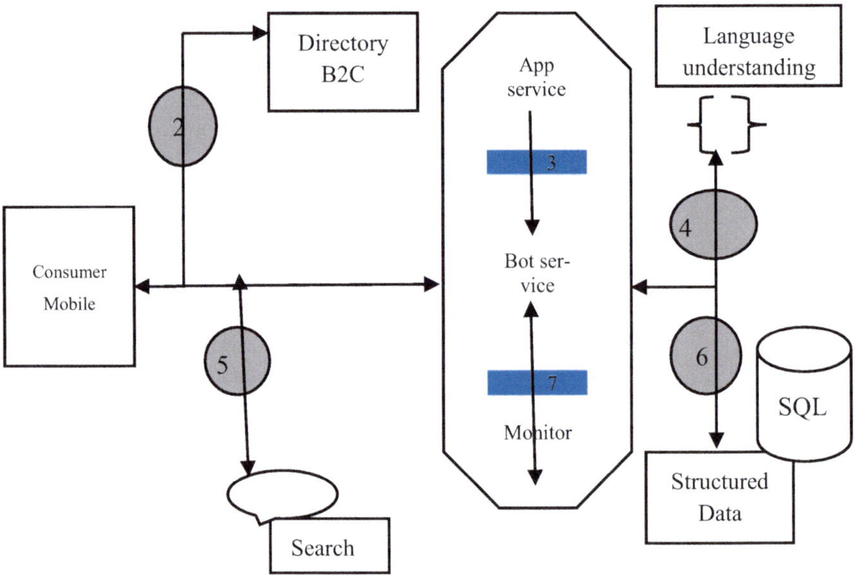

Fig. 4 Workflow chart of a commerce chatbot

The orderly flow of a commerce chatbot is presented in Fig. 5. Patients can start to access enterprise productivity bot as soon as Azure Active directory authenticates the patient's identity. The chatbot then requests the patient's Office 365 calendar via the Graph. Here, the bot retrieves case information in Dynamics CRM using data collected from the calendar. Further information is sent back to the patient who can filter the data without leaving the bot. Finally, application insights accumulate runtime telemetry to assist in developing the bot's progress and practice. In reality, entity-based privacy filtering and transformation are applied directly on the app (client) side.

3.4 IoT devices

IoT takes in extending internet connectivity beyond standard devices. Remote temperature is monitored for vaccines. Medical data transferring tools, air quality sensors, drug effectiveness tracking, vital signs data capturing, sleep monitor, medication refill reminder technology, remote care biometrics scanners are a few examples of IoT Medical Devices.

How IoT devices work in hospitals/health care units?

A well-ordered flow of IoT devices is displayed in Fig. 6. First, a patient signs in Skype and finds information on the IoT bot and then requests it by means of voice to turn on the lights via the IoT device. When the request is transmitted to

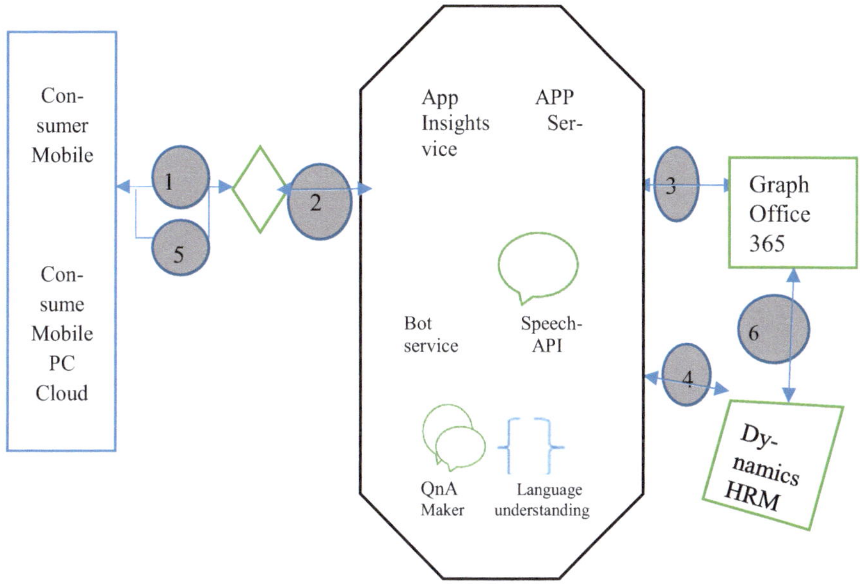

Fig. 5 Workflow chart of an Enterprise Productivity Chatbot

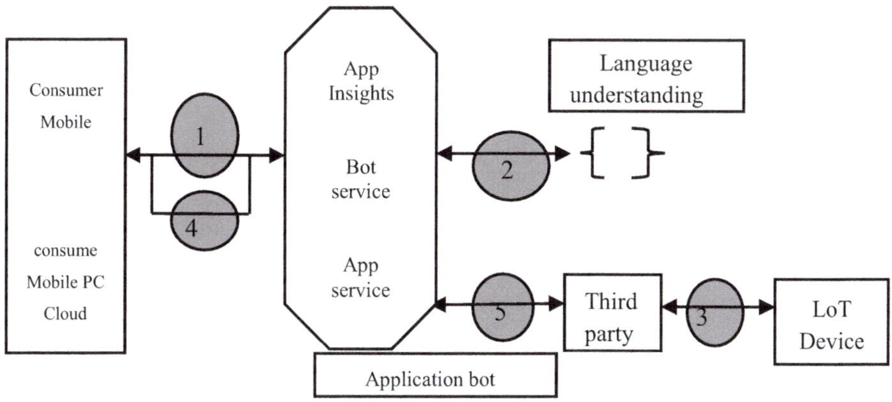

Fig. 6 Flowchart of IoT devices

a third-party service that concedes to the IoT device network, the outcomes of the command are sent back to the patient. Finally, application insights bring together runtime telemetry to support expansion with chatbot routine and procedure. In fact, Chatbots are as secure as you make them. However, there is the possibility of using a backdoor by hackers. Searchable Encryption can protect user chat privacy, however, not familiar with the chatbot design.

Let us discuss the boom of AI-Powered Health Chatbots in 2020 and are presented in Table 1.

Table 1 The Boom of AI-Powered Health Chatbots in 2020

Health chatbots	Company	Features	Intrinsic worth
OneRemission	Introduced by a New York-based company	It delivers complete details of diets, exercises, and post-cancer practices curated by Integrative Medicine experts for cancer patients and cancer survivors Patients need not depend on a doctor Patients may explore the dangers of cancer and understand the benefits of a healthy product that curing cancer-related diseases	The objective is to support cancer patients to have a comfortable life with adequate information to understand their emotions and empower them to care for their emotional health
Youper's AI	Developed by Dr. Jose Hamilton and a group of doctors, scientists, and engineers	It scans and progresses patients' emotional health with speedy custom-made conversations using psychological methods It offers personalized meditation and can track mood and observe emotional health If patients communicate with a chatbot, it will study them and make small adjustments to achieve the best or the desired performance suited to their needs	The objective is to know patients' emotions and assist patients anytime, anywhere

(continued)

Table 1 (continued)

Health chatbots	Company	Features	Intrinsic worth
Safedrugbot	David Hawig, Germany 2016	It is a chat service that offers real-time text transmission It provides suitable data to physicians about the use of drugs during breastfeeding This chatbot is a personal health assistant It recaps and remembers patients to eat their capsule or birth control medicines and helps them keep track of their body weight, mood, and other health issues Patients are able to chat with Florence inside a messaging platform, like Skype, Kik, or Facebook messenger	The objective is to provide informative guides for physician who care for pregnant and breastfeeding women
Babylon Health	Babylon's engineers, doctors, and scientists	First patients inform the signs of their infection or disease to the app. It examines the diseases against a database of diseases with the help of speech recognition and provides a suitable course of action. Doctors then listen and analyze the longsuffering and write medicaments/drugs or refer to an expert as necessary	The objective is to book same-day consultation and is perfect for appointments that don't require physical presence

(continued)

Table 1 (continued)

Health chatbots	Company	Features	Intrinsic worth
Florence	André Fialho, MIT Portugal Bioengineering PhD Alumni June 2016	It is a "personal nurse" named Florence in blue. It can operate on Facebook Messenger, Skype, or Kik "She" can help patients to recall and take their medicines. When elderly patients write the medicine name, the frequencies and time in which they need to take medicines, Florence sends a message in chat at the time a medication is to be taken	The objective is to keep track of patient health and work as a medication reminder
Your.MD	Founded in Oslo, Norway in 2013	It is a symptom checker obtainable on iOS, Android, Facebook Messenger, Slack, KIK, Telegram, as well as a browser version	The objective is to help and solve structural healthcare inequality
Sensely	Adam Odessky, CEO and co-founder of Sensely	It is a virtual medical assistant named Molly that uses text/r speech to transmit information It understands the user's symptoms and suggests an opinion when data are entered into its smart algorithm	The objective is to provide intensive care and assess a case's urgency
Ada Health	Daniel Nathrath, CEO of Ada Health	It is a usual diagnostic tool for surgeons. It carries a list of queries, which are personalized by its algorithm according to the reactions received from each end user. It then advises possible health issues and proposes next steps, for example making an appointment or going to an emergency room	The objective is to get expertise from actual doctors to help patients understand and manage their health [20]

(continued)

Table 1 (continued)

Health chatbots	Company	Features	Intrinsic worth
GYANT	Aeria Games in Europe	It requests a person who is receiving medical treatment to know their symptoms and then shows data to physicians, who diagnose and recommend drugs at present	The objective is to control not requiring much labor and pleasant experiences for the deceased person. that trade-off cost and time savings, and enhanced better patient conversion rates for health systems

4 Conclusion

This study revealed that NLP, AI, and ML are invariably effective technologies to solve tasks associated with the success of EHRs and can be further used to unlock ways to win the race against time especially during the current COVID-19 pandemic. A huge amount of clinical data is now accessible via electronic health records (EHRs), which can track and identify target areas for quality improvement in health care [15]. The pandemic has provided a diverse, splintered, and heterogeneous consumer base with a much broader and varied a set of demands for natural language technology. The future digital information security in healthcare Act (DISHA) protects digital health data of citizens and strictly prohibits the use, access, or disclosure of digital health data for commercial purposes [12]. As NLP (know-hows matures), it is the right time to consider the process of envisioning and revolutionizing its potentials and profitable in trade and industry in the present pandemic stricken world.

References

1. Gowda BNS, Lakshmikantha V (2020) User behavior prediction using a novel sentence N-gram model. In: 2020 2nd international conference on innovative mechanisms for industry applications (ICIMIA), Bangalore, India, 2020, pp 391–397. https://doi.org/10.1109/ICIMIA48430.2020.9074898
2. Baldry M, Cheal C, Fisher B, Gillett M, Huet V (1986) Giving patients their own records in general practice: experience of patients and staff. Br Med J (Clin Res Ed) 292(6520):596–598
3. Detmer D, Bloomrosen M, Raymond B, Tang P (2008) Integrated personal health records: transformative tools for consumer-centric care. BMC Med Inform Decis Mak 8:45
4. Fox BI, Hollingsworth JC, Gray MD et al (2013) Developing an expert panel process to refine health outcome definitions in observational data. J Biomed Inform 46:795–804
5. Gurwitz D, Pirmohamed M (2010) Pharmacogenomics: the importance of accurate phenotypes. Pharmacogenomics 11:469–470
6. Hashimoto DA, Rosman G, Rus D Meireles OR (2018) Artificial intelligence in surgery: promises and perils. Ann Surg 268:70–76
7. He J, Baxter SL, Xu J et al (2019) The practical implementation of artificial intelligence technologies in medicine. Nat Med 25:30–36. https://doi.org/10.1038/s41591-018-0307-0
8. Jiang F, Jiang Y, Zhi H et al (2017) Artificial intelligence in healthcare: past, present and future. Stroke Vasc Neurol 2. https://doi.org/10.1136/svn-2017-000101
9. Johnson KW et al (2018) Artificial intelligence in cardiology. J Am Coll Cardiol 71:2668–2679
10. Jones RB, McGhee SM, McGhee D (1992) Patient on-line access to medical records in general practice. Health Bull (Edinb) 50(2):143–150
11. Kermany DS et al (2018) Identifying medical diagnoses and treatable diseases by image-based deep learning. Cell 172:1122-1131.e9
12. Keselman A, Slaughter L, Smith CA, Kim H, Divita G, Browne A et al (2007) Towards consumer-friendly PHRs: patients' experience with reviewing their health records. AMIA annual symposium proceedings, pp 399–403
13. Keselman A, Smith CA (2012) A classification of errors in lay comprehension of medical documents. J Biomed Inform 45(6):1151–1163
14. Kho AN, Pacheco JA, Peissig PL et al (2011) Electronic medical records for genetic research: results of the eMERGE Consortium. Sci Transl Med 3:3–79

15. Kruse CS, Goswamy R, Raval Y, Marawi S (2016) Challenges and opportunities of big data in health care: a systematic review. JMIR Med Inform 4:e38
16. Kudyakov R, Bowen J, Ewen E et al (2012) Electronic health record use to classify patients with newly diagnosed versus preexisting type 2 diabetes: infrastructure for comparative effectiveness research and population health management. Popul Health Manag 15:3–11
17. Linder JA, Ma J, Bates DW et al (2007) Electronic health record use and the quality of ambulatory care in the United States. Arch Intern Med 167:1400–1405
18. Murdoch TB, Detsky AS (2013) The inevitable application of big data to health care. JAMA 309:1351–1352
19. Nazi KM, Hogan TP, McInnes DK, Woods SS, Graham G (2013) Evaluating patient access to Electronic Health Records: results from a survey of veterans. Med Care 51(3 Suppl 1):S52–S56
20. Newton KM, Peissig PL, Kho AN et al (2013) Validation of electronic medical record-based phenotyping algorithms: results and lessons learned from the eMERGE network. J Am Med Inform Assoc 20:e147–e154
21. Ong CJ, Orfanoudaki A, Zhang R, Caprasse FPM, Hutch M, Ma L et al (2020) Machine learning and natural language processing methods to identify ischemic stroke, acuity and location from radiology reports. PLoS ONE 15(6):e0234908. https://doi.org/10.1371/journal.pone.0234908
22. Peissig PL, Santos Costa V et al (2014) Relational machine learning for electronic health record driven phenotyping. J Biomed Inf. https://doi.org/10.1016/j.jbi.2014.07.007
23. Pyper C, Amery J, Watson M, Crook C (2004) Patients' experiences when accessing their on-line electronic patient records in primary care. Br J Gen Pract 54(498):38–43
24. Sitapati A et al (2017) Integrated precision medicine: the role of electronic health records in delivering personalized treatment. Wiley Interdiscip Rev Syst Biol Med 9:1–12

Chapter 41
Multi-objectives for TCSC Placement Using Self-adaptive Firefly Algorithm

Selvarasu Ranganathan, Ellappan Venugopal, and C. Kannan

1 Introduction

Nowadays, the load on the existing power system network is substantially increasing due to the growth of industries and modernization of human lifestyle. As a result, the existing power system is forced to operate in stressed conditions and leads to voltage instability. Also, it's a challenge to widen the existing power system to meet the ever-increasing power demand by installing new generation and transmission networks due to economic and environmental constraints. An alternate way to overcome this challenge is installing Flexible Alternating Transmission Systems (FACTs) controllers. The series, shunt controlled and combination of both are the classification FACT devices. TCSC is one of the classification of series-connected FACT device, which is capable of increasing power transfer capability, loss minimization, voltage profile (VP) improvement and enhancing voltage stability (VS) of power system network [1–3].

The power system researchers are carried out their research work for placement of FACT devices on power system network and reported in their publications. They used both conventional and nonconventional methods of an optimization algorithm for FACTs placement. Since optimization algorithms have many applications in the field of solving optimization problem, it is generally accepted by most of the power system researchers, and they have been considered like Differential Evolution (DE), GA, Honey Bees Algorithm (HBA), Ant Colony Optimization (ACO), PSO, Bacterial Foraging Algorithm (BFA) and Gravitational Search Algorithm (GSA) [4–13].

S. Ranganathan · E. Venugopal
School of Electrical Engineering and Computing, Adama Science and Technology University, Adama, Ethiopia

C. Kannan (✉)
Arunai Engineering College, Tiruvannamalai, Tami Nadu, India

G. Mathur et al. (eds.), *Proceedings of 2nd International Conference on Artificial Intelligence: Advances and Applications*, Algorithms for Intelligent Systems,
https://doi.org/10.1007/978-981-16-6332-1_41

Available transfer capability has been improved by the FACTs placement using HBA [4]. Evolutionary algorithms like DE and GA have been proposed to choose the feasible locations and parameters of TCSC for increasing the power flow, reduction of losses and enhancement of stability of the system and performances are tested in IEEE 14 bus system [5]. A Fuzzy lag-lead controller has been proposed to control TCSC and SVC for oscillation damping, stability enhancement and ACO applied for setting of parameters of TCSC and SVC [6]. Particle Swarm Optimization Algorithm (PSO) has been employed for improving power transfer capability and economic power system operation through proper SVC and TCSC placement [7]. A complete review has been reported the application of PSO for FACT placement in power system network [8].

Dr. Passino developed BFA to solve optimization problem than later it has been employed for minimizing loss and enhancing voltage profile through UPFC placement [9, 10]. Security enhancement has been achieved by the placement of SVC and TCSC through optimization. In addition, comparative solutions are presented with BBO, WISPO and PSO [11]. A cumulative Gravitational Search Algorithm (GSA) has been employed for loss minimization and VS enhancement through placement of FACT devices [12]. Enhancement of voltage stability of the system has been obtained by placement of FACTS through evolutionary algorithm [13]. Dr. Xin-She Yang proposed a Firefly Algorithm (FA) in 2007, for optimization problems solving [14]. FA has been applied to solve power system optimization problems like economic dispatch and unit commitment etc. [15, 16]. Still, due to improper selection of FA parameters affects the convergence and leads to sub-optimal solutions. Hence, SAFA has been employed for SVC and TCSC placement while considering loss minimization is an objective [17, 18]. Later, multi-objectives have been considered for placement of FACTs [19].

In these works, single- and multi-objectives are taken into consideration for optimal placement of TCSC in IEEE 30 bus system using SAFA.

2 Self-adaptive Firefly Algorithm

Firefly algorithm is a nature-inspired meta-heuristic algorithm for solving optimization problem [14]

It employs based on the light intensity among two fireflies.

The rth firefly light intensity is represented as

$$LI_r = Fitness\ (x_r) \tag{1}$$

$\beta_{r,s}$ is the absorption coefficient between two fireflies and is represented as

$$\beta_{r,s} = (\beta_{\max,r,s} - \beta_{\min,r,s}) \exp(-\gamma_r r_{r,s^2}) \tag{2}$$

$r_{r,s}$ = Cartesian distance of fireflies, which is represented by

$$r_{r,s} = \|x_r - x_s\| = \sqrt{\sum_{v=1}^{nd} \left(x_r^k - x_s^k\right)^2} \tag{3}$$

Firefly movement is based on light intensity. If *rth* firefly light intensity is less than the *sth* firefly light intensity, *r*th firefly will move toward sth firefly as per the following equations

$$x_r(k) = x_r(k-1) + \beta_{r,s}(x_s(k-1) - x_r(k-1)) + \alpha(rand - 0.5) \tag{4}$$

Firefly parameters such as random movement factor (α), absorption coefficient ($\beta_{\min}$) and attractiveness parameter (γ) are added with decision variables each firefly to avoid repetitive trial and error procedure in each iteration as per the following equation.

$$x_r = \left[x_r^1, x_r^2 \cdots, \ x_r^{nd}, \ \alpha_r, \beta_{\min,r}, \ \gamma_r\right] \tag{5}$$

Each firefly with its parameters undergoes whole search process. Eq. (2) is modified as follows,

$$\beta_{r,s} = (\beta_{\max,r,s} - \beta_{\min,r,s}) \exp(-\gamma_r r_{r,s}^2) + \beta_{\min,r,s} \tag{6}$$

The advantage of SAFA includes less computational effort, avoiding the sub-optimal solution and convergence enhancement.

3 Mathematical Modeling of TCSC

The power flows between *i* and *j* buses are represented by the following equations [18]

$$\text{Real power,} \ P_{ij} = \frac{V_i V_j}{x_{ij}} \sin\delta_{ij} \tag{7}$$

$$\text{Reactive Power,} \ Q_{ij} = \frac{1}{x_{ij}}(V_i^2 - V_i V_j \cos\delta_{ij}) \tag{8}$$

The TCSC is one of the classification of series-connected FACT device. The static model of TCSC is considered and it can be denoted as static reactance $-jX_{c,}$ added in series with line connecting buses *i* and *j* as presented in Fig. 1. The transfer reactance between buses *i* and *j* decreases resulting that power flow increases, as a result, there is reduction in reactive power loss and bus voltage improvement. The effective transmission line reactance amid buses *i* and *j* with series compensation is expressed as follows,

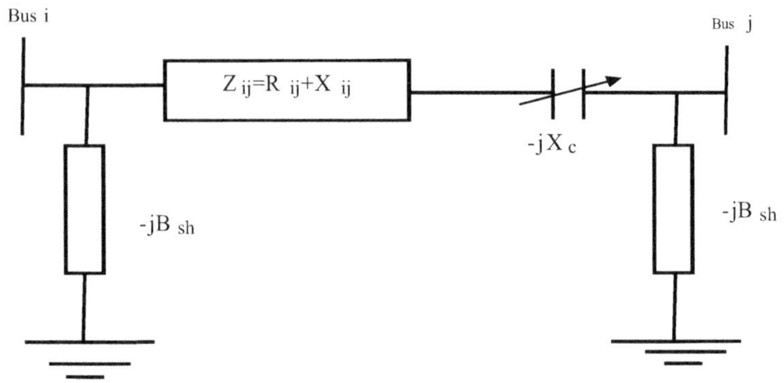

Fig. 1 Static model of TCSC with transmission line

$$x_{eff} = x_{ij} - x_c \tag{9}$$

$$x_{ij} = (1 - k)x_{ij} \tag{10}$$

where, $k = \frac{x_c}{x_{ij}}$ = transmission line compensation factor, whose value is chosen between -0.8 and 0.2 to evade over compensation and x_{ij} is transmission line reactance.

3.1 Problem Formulation

TCSC placement is chosen as an optimization problem and an objective is given by the following expression,

$$\text{Minimize } \Phi(x, y) \tag{11}$$

$$\text{Subjected to } g(x, y) = 0 \tag{12}$$

$$h(x, y) \leq 0 \tag{13}$$

Equality constraints and inequality constraints are expressed as $g(x, y)$ and $h(x, y)$, respectively. x is the state variable comprising real and reactive power generation at slack bus, VAR generation at PV buses and $P_{loss.}$ Control variable (y) includes location and parameters of TCSC.

The equality constraints, $g(x, y)$ are given by the expression as follows

$$P(V, \delta) - P^{sp} = 0 \text{ for PV and PQ bus} \tag{14}$$

$$Q(V, \delta) - Q^{sp} = 0 \text{ for load bus} \tag{15}$$

Equality constraints, $h(x, y)$, are defined as follows

$$-0.8 \leq k \leq 0.2 \text{ for TCSC} \tag{16}$$

$$Q_{Gi}^{\min} \leq Q_{Gi} \leq Q_{Gi}^{\max} \text{ at PV buses} \tag{17}$$

$$V_i^{\min} \leq V_i \leq V_i^{\max} \text{ t PQ buses} \tag{18}$$

The objective functions are formed with four different cases of three single objectives and one multi-objective. Single objectives considered are $P_{loss,}$ minimization, VP improvement, VS enhancement. Multi-objectives include $P_{loss,}$ minimization, simultaneously improve the VP besides enhancing the VS.

Case 1: Real power loss (P_{loss})

P_{loss} reduction is represented as follows [18]

$$\text{Minmize } J_1(x, y) = P_{loss} = \sum_{k \in \Im} g_{ij} \left(|V_i|^2 + |V_j|^2 - 2\,|V_i|\,|V_j| \cos \delta_{ij} \right) \tag{19}$$

Case 2: Bus voltage profile improvement

Voltage profile improvement is achieved by minimizing net voltage deviation (NVD) as follows [19]

$$\text{Minimize } J_2(x, y) = \sum_{j \in \Phi} |V_j - 1| \tag{20}$$

Case 3: Voltage stability enhancement

Voltage stability enhancement is obtained by minimizing L-index as follows [19].

$$\text{Minimize } J_3(x, y) = \sum_{j \in \Phi} L_j \tag{21}$$

$$\text{Where, } L_j = \left| 1 - \sum_{i = \Omega} F_{ij} \frac{V_i}{V_j} \right| \tag{22}$$

Case 4: P_{loss}, VP and VS

In this case, multi-objectives are considered for minimizing P_{loss} at the same time improving the VP also enhancing VS and the objective function as

Table 1 Firefly illustration for TCSC placement

L_1	L_2	L_3	…	L_{nF}
k_1	k_2	k_3	…	k_{nF}
α	$\beta_{\min}$	γ		

$$\text{Minimize,}\ J(x, y) = \sum_{i=1}^{nobj} w_i\ J_i \tag{23}$$

3.2 SAFA Variables Illustration

Firefly is illustrated in Table 1. The first row specifies the location for the TCSC, in which transmission line is to be placed (L_k). The second row denotes the TCSC compensation, and the third row shows the firefly parameters.

3.3 Fitness Function

Light intensity function (*LI*) is a fitness function, which is maximized for getting optimal solutions.

$$\text{Maximize}\ LI = \frac{1}{1 + \Psi} \tag{24}$$

$$\Psi = w_1\ J(x, u) + w_v \sum_{i=1}^{nload} V_{di} + w_Q \sum_{j=1}^{ngen} \left| Q_{Gi} - Q_{Gi}^{\text{limit}} \right|^2 \tag{25}$$

4 Simulation Solution and Discussion

The simulation is performed on Matlab software to analyze the feasibility of proposed SAFA for single and multi-objectives optimization through TCSC placement in IEEE 30 test system. Power flow solution is attained using N-R method during the optimization process [20]. The power loss reduction of case 1 is obtained by considering different number of TCSCs and is presented in Table 2. It can be seen from Table 2 that P_{loss} reduction is 17.4338 MW from 17.5028 MW when six TCSCs are placed

Table 2 P_{loss} reduction after TCSC placement

	Number of TCSC	Ploss (MW)
IEEE 30	0	17.5028
	4	17.4521
	5	17.4485
	6	17.4338
	7	17.4360

in IEEE 30 bus system. If we choose another number, the power loss reduction is insignificant. Therefore, six TCSCs are considered for remaining cases 2–4.

The performances obtained through simulation like P_{loss}, NVD, Maximum Voltage Stability Index (MVSI), and voltage magnitudes limits are presented in Table 3 for all four cases of before and multiple TCSC placements. In case 1, the objective considered is to minimize P_{loss}. It can be seen from the table that P_{loss} minimization is 17.4338 MW by SAFA, while HBA and BFA decrease the P_{loss} to 17.4437 and 17.4519 MW, respectively. Figure 2 indicates the P_{loss} savings after TCSC placement. The percentage of P_{loss} saving of HBA and BFA is 0.34% and 0.29%, correspondingly, though SAFA directs the P_{loss} savings to 0.39%. SAFA provides higher savings in losses compared with HBA and BFA.

The objective in Case 2 is VP improvement through minimizing NVD. It can be noted from Table 3 of case 2, NVD was reduced from 0.4562 to 0.4205 by SAFA, whereas HBA and BFA reduce NVD to 0.4297 and 0.4310 by TCSC placements. Figure 3 represents % voltage profile improvement. It is possible to see from the figure that SAFA offers 7.82% of VP improvement, while HBA and BFA provide

Table 3 Performance solutions for Cases 1–4

	Method	P_{loss} (MW)	NVD	MVSI	V^{low}/V^{high} (pu)
Base case		17.5028	0.4562	0.1420	0.989/1.082
Case 1	SAFA	17.4338	0.4748	0.1406	1.002/1.050
	HBA	17.4437	0.5501	0.1350	0.991/1.054
	BFA	17.4519	0.4570	0.1414	0.989/1.052
Case 2	SAFA	17.5547	0.4205	0.1422	0.989/1.050
	HBA	17.5093	0.4297	0.1434	0.988/1.052
	BFA	17.5278	0.4310	0.1457	0.991/1.050
Case 3	SAFA	17.6062	0.6083	0.0880	1.008/1.053
	HBA	17.6042	0.5848	0.0886	1.009/1.054
	BFA	17.6870	0.5176	0.0904	1.006/1.054
Case 4	SAFA	17.5012	0.4415	0.1322	0.989/1.051
	HBA	17.5019	0.4487	0.1402	0.987/1.051
	BFA	17.5036	0.4465	0.1358	0.989/1.049

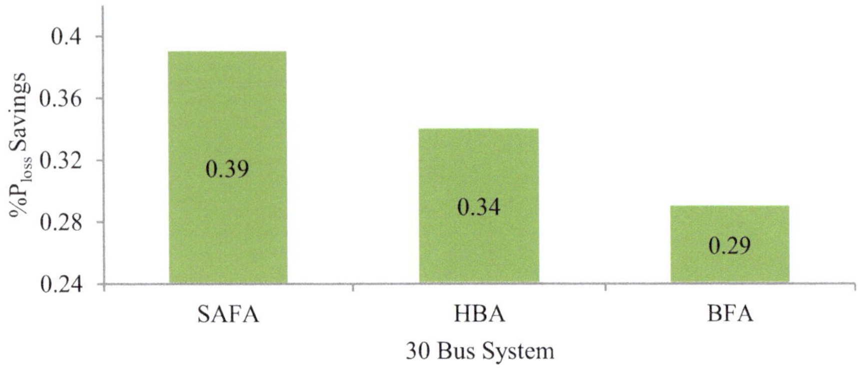

Fig. 2 Percentage P_{loss} savings after TCSC placement

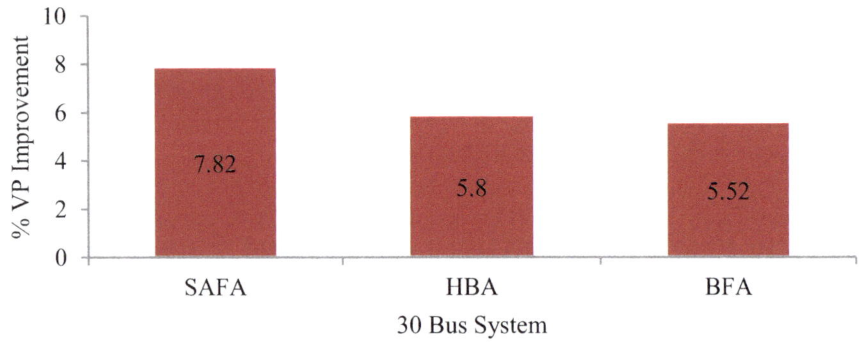

Fig. 3 Percentage VP improvement after TCSC placement

5.8% and 5.52%, respectively. It is obvious that SAFA offers the optimal placements for TCSC and their VAR support, while minimizing the TVD to improve the VP.

Voltage stability enhancement is achieved by minimizing MVSI in case 3. It is possible to see from Table 3, MVSI is reduced from 0.1420 to 0.0880, 0.0886 and 0.0904 by SAFA, HBA and BFA, respectively. Figure 4 indicates the voltage stability enhancement percentage. SAFA provides 38.02% VS enhancement, whereas BFA and HBA offer VS enhancement as 36.34% and 37.6%, respectively. It is obvious from the results that SAFA identifies optimal locations with suitable reactive power support for TCSCs, which minimize the MVSI to the minimum values, then those of BFA and HBA.

The multi-objectives are considered in case 4, and they are minimizing P_{loss}, simultaneously improving the VP also enhancing VS. It is observed from Table 3 (case 4) that self-adaptive firefly algorithm reduces P_{loss}, TVD and MVSI are 17.5012 MW, 0.4415 and 0.1322, respectively. The comparison of results obtained through SAFA, HBA and BFA for P_{loss} minimization, VP improvement and VS enhancement are given in Figs. 5, 6 and 7, respectively. The presented results obviously show that the

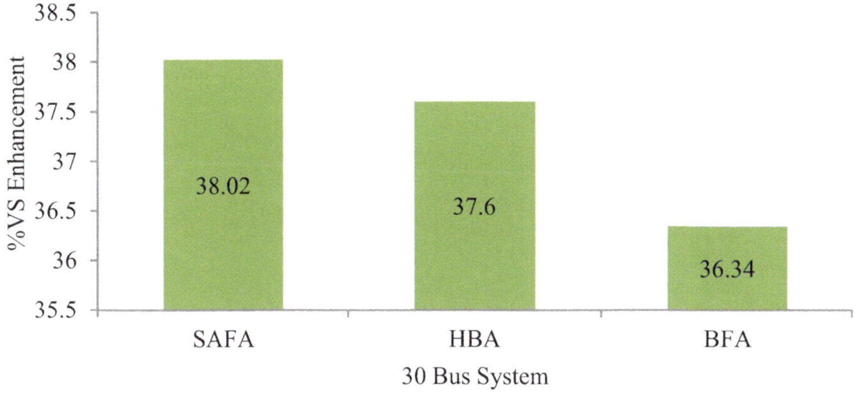

Fig. 4 Comparison of % VS enhancement after TCSC placement

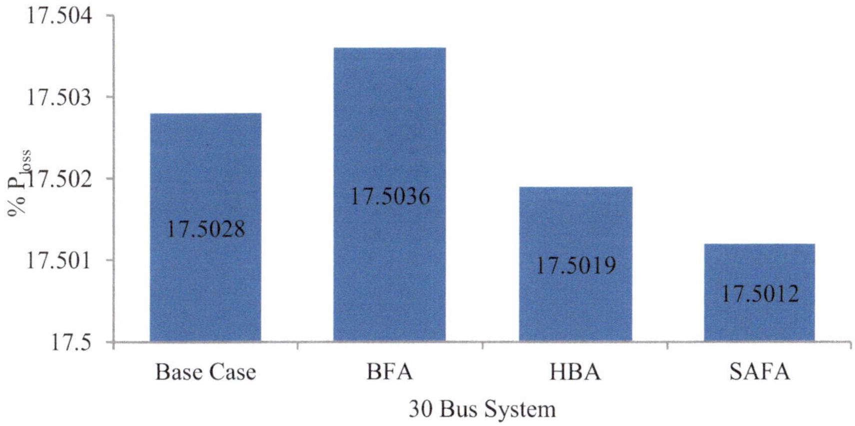

Fig. 5 P_{loss} Comparison obtained through case 4 after TCSC placement

SAFA presents the optimal positions with reactive power values for multiple TCSC placements, which minimize P_{loss}, improve the VP besides enhancing the VS.

Load bus voltages of IEEE 30 bus system attained through SAFA for considering all the four cases are presented in Fig. 8. Interestingly, the values of voltage magnitudes are found to lie between 0.95 and 1.1 p.u. It is possible to see from the results that SAFA offers optimal positions for TCSC placements in order to minimize P_{loss}, improve VP and enhance voltage stability. TCSC locations and their parameters attained by SAFA, HBA and BFA are given in Tables 4, 5 and 6, respectively.

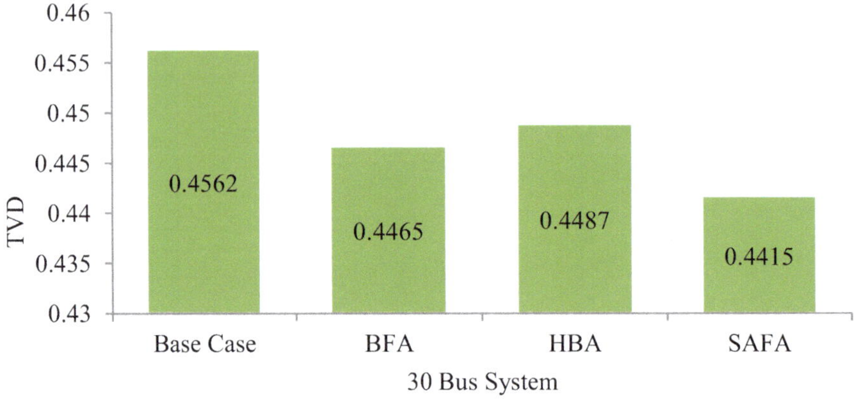

Fig. 6 Comparison of TVD obtained through case 4 after TCSC placement

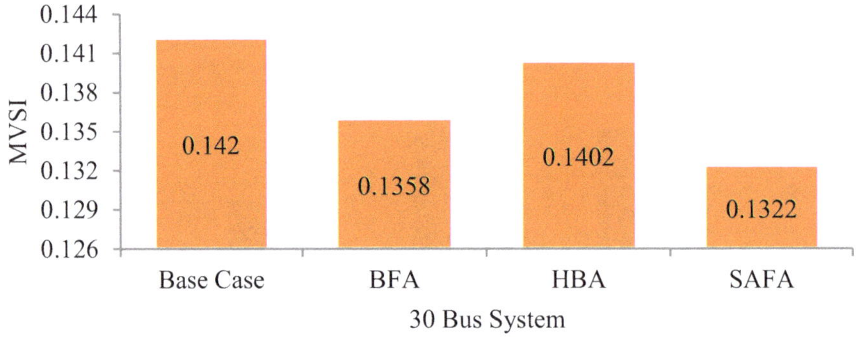

Fig. 7 MVSI Comparison obtained through case 4 after TCSC placement

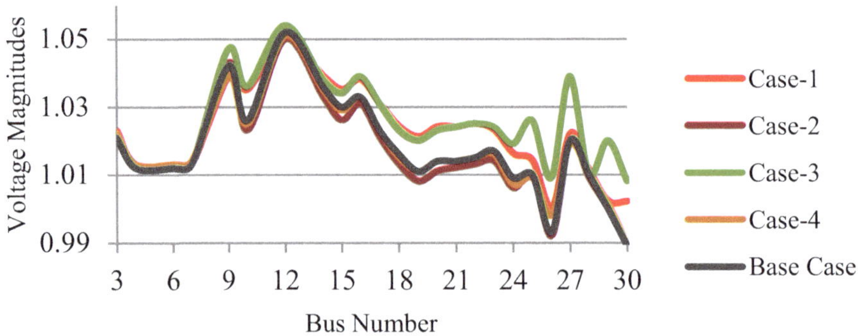

Fig. 8 IEEE 30 bus systems VP after TCSC placement

Table 4 TCSC location obtained by SAFA

TCSC no.		1	2	3	4	5	6
Case 1	Line location	4	9	19	29	31	18
	Compensation factor, (*k*)	−0.469	0.200	−0.479	0.200	−1.172	−0.240
Case 2	Line location	14	18	11	15	4	41
	Compensation factor, (*k*)	0.200	0.200	0.200	0.200	0.057	−0.800
Case 3	Line location	36	29	25	12	27	11
	Compensation factor, (*k*)	−0.800	−0.468	−0.575	−0.776	0.037	−0.067
Case 4	Line location	41	11	37	34	9	29
	Compensation factor, (*k*)	−0.408	−0.797	−0.722	−0.790	0.200	−0.721

Table 5 TCSC location obtained by HBA

TCSC no.		1	2	3	4	5	6
Case 1	Line location	28	18	25	4	14	7
	Compensation factor, (*k*)	−0.527	−0.175	−0.800	−0.680	−0.785	0.200
Case 2	Line location	14	25	34	19	18	30
	Compensation factor, (*k*)	0.141	−0.003	−0.621	0.163	0.200	0.005
Case 3	Line location	14	30	29	36	37	26
	Compensation factor, (*k*)	−0.561	−0.592	−0.416	−0.787	−0.600	−0.406
Case 4	Line location	11	34	23	4	35	30
	Compensation factor, (*k*)	−0.800	−0.525	−0.444	−0.163	0.057	−0.401

Table 6 TCSC location obtained by BFA

TCSC no.		1	2	3	4	5	6
Case 1	Line location	9	41	26	21	4	33
	Compensation factor, (*k*)	0.147	−0.361	−0.754	−0.005	−0.628	−0.314
Case 2	Line location	35	37	15	14	30	32
	Compensation factor, (*k*)	0.175	−0.569	0.195	0.200	0.136	−0.163
Case 3	Line location	20	41	26	25	36	4
	Compensation factor, (*k*)	−0.606	−0.687	−0.103	−0.559	−0.773	−0.466
Case 4	Line location	15	22	36	26	25	11
	Compensation factor, (*k*)	0.199	0.193	−0.098	0.200	−0.788	−0.770

5 Conclusions

Single- and multi-objectives of four cases are considered to seek the optimal positions and parameters of TCSC in IEEE 30 bus system. The performances in terms of P_{loss}, TVD, MVSI and VMs limits are presented and discussed in terms P_{loss} minimization,

VP improvement and VS enhancement. The SAFA offers optimal positions and parameters of TCSCs for existing power system performance improvement. It is very clear from the above discussion that the SAFA provides sensible solutions than HBA and BFA.

Conflicts of Interest No conflict of interest was declared by the authors.

References

1. Taylor CW (1994) Power system voltage stability. McGraw-Hill, New Delhi
2. Hingorani NG, Gyugyi L (1999) Understanding FACTS: concepts and technology of flexible AC transmission systems. Wiley, IEEE Press
3. Mathur RM, Varma RK (2002) Thyristor-based FACTS controllers for electrical transmission systems. IEEE Press, Piscataway
4. Mohamid Idris R, Khairuddin A, Mustafa W (2009) Optimal allocation of FACTS devices for ATC enhancement using Bees algorithm. World Acad Sci Eng Technol 30:313–320
5. Rashed GI, Sun Y, Shaheen HI (2012) Optimal location and parameter setting of TCSC for loss minimization based on differential evolution and genetic algorithm. Phys Procedia 33:1864–1878
6. Lu C-F, Hsu C-H, Juang C-F (2012) Coordinated control of flexible AC transmission system devices using an evolutionary Fuzzy lead-lag controller with advanced continuous Ant colony optimization. IEEE Trans Power Syst 28(1):385–392
7. Bhattacharyya B, Gupta VK (2012) Application of DE & PSO algorithm for the placement of FACTs devices for economic operation of a power system. WSEAS Trans Power Syst 7(4):209–2016
8. Rezaee Jordehi A (2015) Particle swarm optimization for allocation of FACTS devices in electric transmission systems: a review. Renew Sustain Energy Rev 52(C):1260–1267
9. Passino KM (2002) Biomimicry of bacterial foraging for distributed optimization and control. IEEE Control Syst Mag 22(3):52–67
10. Senthil Kumar M, Renuga P (2012) Application of UPFC for voltage stability improvement and loss minimization using Fast Voltage Stability Index (FVSI). Arch Electr Eng 61(2):239–250
11. Kavitha K, Neela R (2016) Comparison of BBO, WIPSO & PSO techniques for the optimal placement of FACTS devices to enhance system security. Procedia Technol 25:824–837
12. Packiasudha M, Suja S, Jerome J (2017) A new cumulative gravitational search algorithm for optimal placement of FACT device to minimize system loss in the deregulated electrical power environment. Int J Electr Power Energy Syst 84:34–46
13. do Nascimento S, Gouvea Jr MM (2017) Voltage stability enhancement in power systems with automatic facts devices allocation. Energy Procedia 107:60–67
14. Yang XS (2010) Nature-inspired meta-heuristic algorithms, 2nd edn. Luniver Press, Beckington
15. Yang XS, Hosseini SS, Gandomi AH (2012) Firefly Algorithm for solving non-convex economic dispatch problems with valve loading effect. Appl Soft Comput 12(3):180–186
16. Chandrasekaran K, Sishaj Simon P (2012) Network and reliability constrained unit commitment problem using binary real coded firefly algorithm. Int J Electr Power Energy Syst 43(1):921–932
17. Selvarasu R, Surya Kalavathi M, Christober Asir Rajan C (2013) SVC placement for voltage constrained loss minimization using self-adaptive Firefly algorithm. Achiev Electr Eng 62(4):649–661

18. Selvarasu R, Surya Kalavathi M (2015) TCSC Placement for loss minimization using self-adaptive firefly algorithm. J Eng Sci Technol 10(3):291–306
19. Ranganathan S, Kalavathi MS, Christober Asir Rajan C (2016) Self-adaptive firefly algorithm based multi-objectives for multi-type FACTS placement. IET Gener Trans Distrib 10(11):2576–2584
20. Saadat H (2004) Power system analysis, 2nd edn. McGraw-Hill, New Delhi

Chapter 42
Evaluation of Machine Learning Models for Sign Language Digit Recognition

K. Divya Lakshmi and S. R. Balasundaram

1 Introduction

The key challenge for a sign language recognition (SLR) system is the lack of common standards across regions. Like spoken languages, sign languages are local to a region and contain a set of symbols, vocabulary of words and grammar, which are represented through hand gestures and head movements. There is no exact estimate on the number of sign languages that exist across the world. Also, for regions with the same spoken language, the sign languages are different. Examples are the British Sign Language (BSL), American Sign Language (ASL), French Sign Language (FSL) that are observed in English-speaking countries. Though the alphabet is common for the spoken language, the sign language gestures vary for each letter in the above-mentioned sign languages. The Indian Sign Language (ISL) is similar to BSL with two hands used for the representation of various elements. A systematic review of the state-of-the-art sign language recognition systems can be found in [1]. A general framework of the SLR systems that can be trained and deployed for the different sign languages is shown in Fig. 1. The trainable SLR system has an offline phase during which a model is learnt using features extracted from labeled data and an online phase where the learnt model is used for the recognition. The accuracy of the system thus depends on the dataset, which is used for learning the model in addition to the choice of model, training parameters and features extracted.

K. Divya Lakshmi (✉)
PDF Scholar, Department of Computer Applications, National Institute of Technology, Tiruchirappalli 620012, India

S. R. Balasundaram
Professor, Department of Computer Applications, National Institute of Technology, Tiruchirappalli 620012, India
e-mail: blsundar@nitt.edu

G. Mathur et al. (eds.), *Proceedings of 2nd International Conference on Artificial Intelligence: Advances and Applications*, Algorithms for Intelligent Systems,
https://doi.org/10.1007/978-981-16-6332-1_42

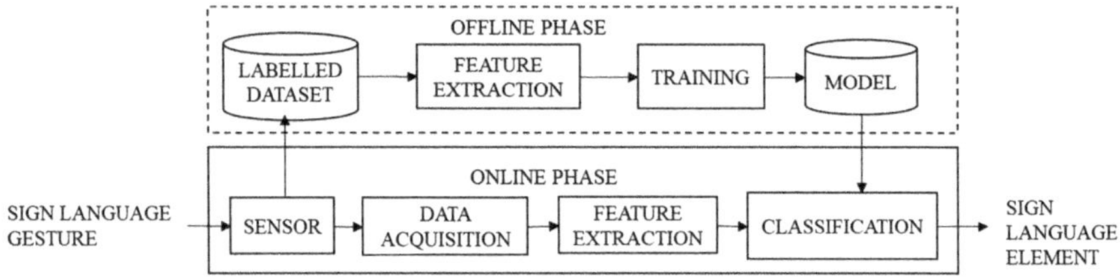

Fig. 1 SLR system that can be trained over various sign languages

While datasets are available for ASL, which is standardized, datasets are not yet available for ISL, which lacks a standard set of gestures and movements for words, letters and digits. Thus, an SLR module that detects and identifies the elements of sign language should be trainable with the dataset available from a region. This paper, hence, focuses on the study of various machine learning models for the sign language digit recognition.

2 Related Works

The signs that represent a letter or word can be decomposed into the following elements—Handshape, Orientation, Location, Movement and Facial Expressions [2]. Thus face, hand, and body postures, and movements must be analyzed simultaneously for the recognition of sign. The state-of-the-art recognition systems can be classified based on the sensor it uses to capture the posture and movements. The sensor for capture includes accelerometers, visual camera, Kinect camera, which has other sensors in addition to the optical lens. Literature contains algorithms that focus on static recognition from images and dynamic recognition from video [3].

A number recognition system using HMM (Hidden Markov Model) over data collected using color-coded hand gloves is proposed in [4] for Filipino Sign Language (FiSL). An average accuracy of 82.5% has been obtained for FiSL number recognition. A 10-layered convolutional neural network (CNN) with residual learning is proposed in [5] for detecting sign with the hand, rotated at any angle between 0 and 360 degrees. The network when trained and tested over MU_HandImages_ASL dataset with rotation augmentations achieved an accuracy of 97%. It was found to misclassify digits for hand gestures with similar positions of the thumb. An improved recognition network is proposed in [6], which incorporates the power of Support Vector Machines (SVM) to the CNN. The performance of the system is tested for various zooming and rotation over the MU_HandImages_ASL [7] and the standard SLD dataset [8]. Misclassification was observed for similar ASL signs like that of digits 7 and 3.

Recognition using multi-class SVM over histogram of gradient (HOG) features extracted from Bangla digit sign language (BdSL) images in binary format can be found in [9]. An accuracy of 95% has been reported for the recognition from the binary images. A deep learning model with similar accuracy is proposed in [10] for the recognition of BdSL images in grayscale format [11]. A CNN-based recognition system is tested over Bhutanese sign language (BhSL) dataset of 20,000 images, collected from 21 Thai students in [12]. The signs of the digits in BhSL are same as that of ASL, and test accuracy of 97% was achieved by CNN over the dataset.

3 Methods and Materials

The datasets used for the evaluation and the details of their sign images are presented in Table 1.

A preview of SLD dataset and BdSL digit dataset is shown in Fig. 2. While the signs of the first two datasets in Table 1 are the same as in Fig. 2a, the signs are different for BdSL, which is shown in Fig. 2b.

Table 1 Details of hand gesture sign images used for evaluation

Dataset	SL	Image type	Pixels	Images count
SLD [7]	ASL/ISL/BSL	RGB images	100 × 100	200 sign images per digit
MU HandImages ASL[a] [8]	ASL/ISL/BSL	Segmented RGB images	200 × 300	85 sign images for digit 0 and 95 images for the rest of the digits
BdSL digits dataset [9]	BdSL	Binary images	128 × 128	Around 100 sign images for each digit

[a]Digit images were extracted from the dataset for the experiments

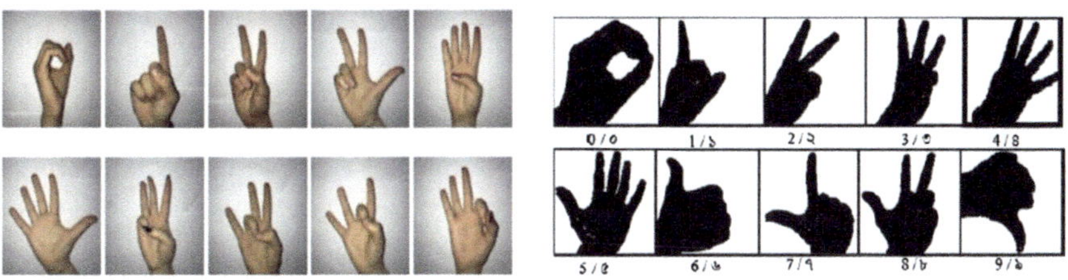

(a) (b)

Fig. 2 Images from sign language digit dataset in **a** SLD **b** BdSL (left to right 0–4 (top) 5–9 (bottom))

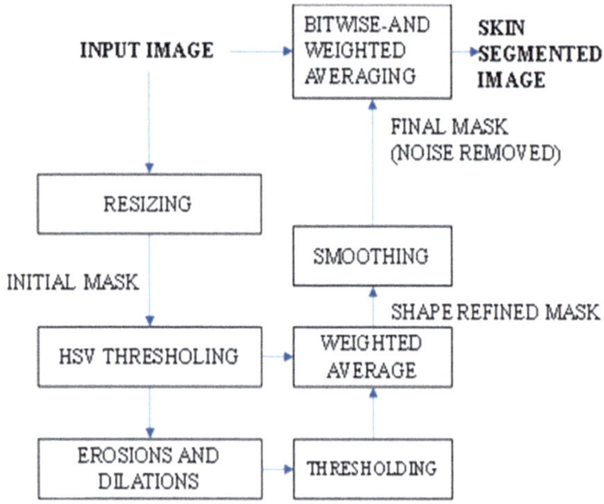

Fig. 3 Skin segmentation procedure for an input sign image

3.1 *Skin Segmentation*

Figure 3 shows the procedure to separate the hand region from the background of the sign images of the SLD dataset. The images are resized to a lower resolution for improved time complexity at the rest of the phases of segmentation, and for faster training and testing of the machine learning models.

The input image is in RGB color space, which comprises the intensity values of the red, green and blue color components in the images. The color space is converted into HSV color space, which includes the hue, saturation and value components. The positions in the input image where HSV values lie within the range of (0, 40, 30) and (43, 255, 254) are initially identified as the skin mask. The mask is further refined by morphological operations of erosion and dilation using elliptical filters of size 3. The initial mask and shape-refined masks are combined using weighted average with equal weights for both the masks. Sharp discontinuities in the mask are removed by smoothing with a median filter of size 5. The final noise-removed mask and the input image are combined using a series of bitwise AND operations and weighted averaging, resulting in a final image containing only the hand region with the background region replaced by black pixels [13].

3.2 *Learning Models*

The models are classified as machine learning and deep learning models based on the methodology of training and the number of parameters. While machine learning models are fitted over training dataset, deep learning models have a higher number of parameters and are fitted using training dataset and tuned using validation datasets.

Also, deep learning modes require a higher amount of training data and very high computation resources for learning.

Machine learning models. The classifiers based on decision trees, regression, and probabilistic approaches were considered for evaluation.

- *Tree-based classifiers.* The decision tree learns the conditions for a class by splitting the records based on values of the attributes and by measuring the quality of split. The parameters of the classifier include the number of levels in the tree starting from the root node, the quality measure used for computing the split of the data created by the values of an attribute. Entropy-based information gain and Gini index are the widely used measures for the construction of the decision trees. While information gain is chosen for attributes with a small set of possible values, Gini index based on sum of square of probabilities facilitates attributes with wider distributions such as image intensities. Random forest classifier is an ensemble of several decision trees where decision from trees with different parameters is combined for the determination of the class of an image.
- *Regression-based classifiers.* While the above-mentioned classifiers are based on measures over attributes such as entropy and Gini index, regression-based classifiers fit functions that estimate the probability of the samples belonging to various classes. Logistic regression fits sigmoid functions to estimate the probability that a model belongs to various classes. Gradient boosting is an ensemble of several regression-based classifiers, which learns several optimizable loss functions. While logistic regression estimates the function that returns the probability of a data that belong to a class; Support Vector Machines find a hyperplane that is at a maximum distance from the data of the different classes [14].

Deep learning-based classifier. The convolutional neural network tested for SLD dataset in [15] is considered for learning from the ASL and BdSL digit datasets.

The layers of the model, which accepts a 64×64 sign image, along with the details of type, dimensions of the output and number of parameters of each layer are shown in Table 2.

A CNN model starts with convolutional layers, which convert images to feature maps. This is followed by non-linearity layer, which generates the activation maps from the feature maps. The layers are further followed by rectification units, which eliminate the cancellation effects at the subsequent layers of the network. The other layers of the model include drop out layers, which avoid overfitting of the model to the training data, pooling layer that reduces the spatial size of the activation maps, and fully connected layers or dense layers that accept the response maps and return the class in the form of one hot encoded vector. Batch normalization layers of the CNN enable a layer to learn independently of the other layers of the network by standardizing the output of a layer through normalization of the values, which also helps in avoiding overfitting of the model to the training dataset.

Table 2 Layers of the CNN model used for evaluation

S. no.	Layer	Type	Output shape	Param #
1	conv2d_1	Conv2D	(None,64,64,64)	1664
2	batch_normalization_1	Batch Normalization	(None,64,64,64)	256
3	conv2d_2	Conv2D	(None,64,64,64)	102,464
4	batch_normalization_2	Batch	(None,64,64,64)	256
5	max_pooling2d_1	MaxPooling2D	(None,32,32,64)	0
6	dropout_1	Dropout	(None,32,32,64)	0
7	conv2d_3	Conv2D	(None,32,32,32)	32,800
8	batch_normalization_3	Batch Normalization	(None,32,32,32)	128
9	conv2d_4	Conv2D	(None,32,32,32)	16,416
10	batch_normalization_4	Batch Normalization	(None,32,32,32)	128
11	max_pooling2d_2	MaxPooling2D	(None,16,16,32)	0
12	dropout_2	Dropout	(None,16,16,32)	0
13	conv2d_5	Conv2D	(None,16,16,16)	4624
14	batch_normalization_5	Batch Normalization	(None,16,16,16)	64
15	max_pooling2d_3	MaxPooling2D	(None,8,8,16)	0
16	dropout_3	Dropout	(None,8,8,16)	0
17	flatten_1	Flatten	(None,1024)	0
18	dense_1	Dense	(None,512)	524,800
19	dropout_4	Dropout	(None,512)	0
20	dense_2	Dense	(None,1024)	525,312
21	dropout_5	Dropout	(None,1024)	0
22	dense_3	Dense	(None,10)	10,250

3.3 *Training and Testing*

75% of the sign images in each category are used for training a machine learning model, and the rest of the sign images are used for validating the model learnt. The hue, saturation and value components of the images are converted to a single grayscale value. The images are further downsampled, and the resulting two-dimensional values are flattened to a linear structure resulting in a feature vector. The label value of a feature vector is obtained from the digit to which it belongs. The training and test vectors for the machine learning models are thus formed from the entire dataset [16]. For training the CNN, images are resized to 64×64 resolution. The training images were augmented with rotated, scaled and translated images. 80% of the resulting extended dataset was used for training the network weights for a set of hyper-parameter values, and 20% of the training images were used as validation dataset for tuning the hyper-parameters of the model. The model, thus, built after training and validation was finally tested over the initial training and test images obtained from the dataset.

4 Evaluation

The image processing steps were implemented using OpenCV (open-source package for computer vision). Sklearn (machine learning library) was used for training and testing the machine learning models.

Table 3 shows the test accuracy values obtained by the models over the datasets listed in Table 1. It can be observed that the segmentation of SLD images has improved the accuracy of the tree-based and regression-based ensemble classifiers. Support Vector Machines have returned the maximum accuracy over the SLD with an accuracy of 81.59%. It can also be observed that the accuracy of the model has been improved when trained over skin segmented sign images. Random Forest classifiers trained over skin segmented sign images have the maximum accuracy among the machine learning models. The accuracy of training and testing is maximum for the deep learning-based CNN model, which has a higher number of parameters and elaborates the training process compared with the machine learning models.

As discussed, the machine learning models have returned higher test accuracy over MU_HandIMages_ASL compared with their performance over SLD segmented sign images. The dataset images are acquired with controlled illumination conditions, and the region below the wrist is covered with color same as that of the background, which has resulted in segmented images of higher quality [7]. Thus, it can also be inferred that the learning can be improved in machine learning models by improving the segmentation quality. Among the machine learning models, SVM has returned the highest accuracy over the BdSL digits dataset. It can be inferred from the table that the binary format images have returned models of lesser accuracy compared with the RGB hand-segmented images, which are converted to grayscale images with highly distributed intensity levels.

The recognition rate of individual digits for the random forest classifier over SLD dataset is shown in Fig. 4. It can be observed that the skin segmentation has improved the recognition rate of almost all the classes of the sign images.

The accuracy of the model learnt at the end of each epoch is shown in Fig. 5. An epoch is a step where the model is built using the entire training and validation dataset. The final model was achieved after 30 epochs. It can be observed that the

Table 3 Test accuracy (%) of machine learning models over the datasets

Model/Dataset	SLD	SLD + seg	ASL	BdSL
Random forest	77.33	85.08	98.72	79.18
SVM	81.59	83.14	98.72	88.10
CNN	99.03	98.84	99.15	94.80

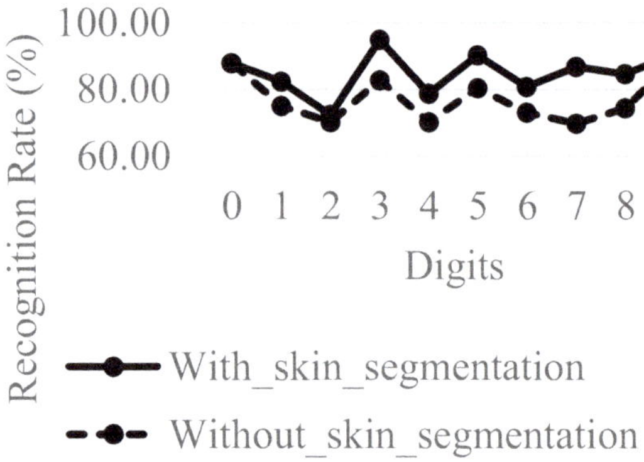

Fig. 4 Recognition rate of digits for random forest classifier in SLD dataset

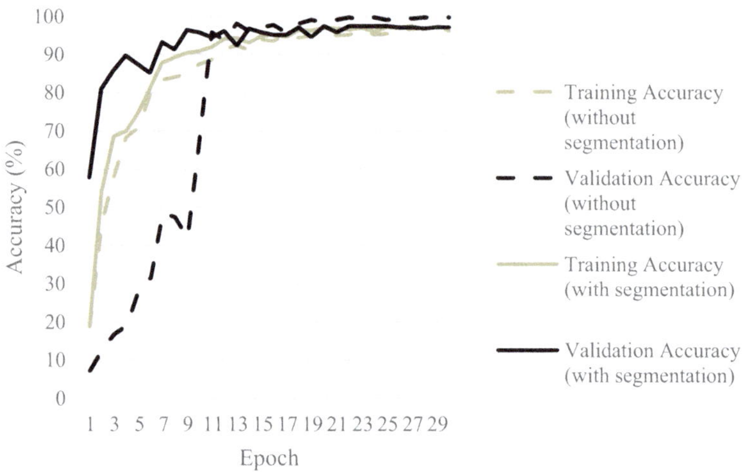

Fig. 5 Training and validation accuracy at successive epochs during CNN model fitting

accuracy is high for the learning on skin-segmented sign images compared with the raw sign images at earlier epochs. Thus, skin-segmented sign images can be used where a model must be learnt in a lesser amount of time.

5 Conclusions and Future Works

Sign languages vary among different regions and hence machine learning algorithms that can be trained using limited available dataset were chosen for the study. The decision tree-based, regression-based and deep learning-based classifiers were trained and tested over the publicly available datasets of ASL and BdSL digit sign images. It was found that among machine learning models trained using skin-segmented SLD dataset images, decision tree-based ensemble classifier, Random Forest exhibited the highest accuracy. SVM-based classifier exhibited superior performance over unsegmented SLD images and the binary images of BdSL digit dataset. Deep learning-based CNN with a higher number of parameters and hyper-parameter tuning has achieved the maximum accuracy over the machine learning models. The segmentation of hand region was also observed to impact the progress of learning in CNN. Improving the recognition rate through other image features like salient image key points, sign language recognition from streaming video, extension of recognition to alphabets, words are potential future works discussed in this paper.

References

1. Wadhawan A, Kumar P (2019) Sign language recognition systems: a decade systematic literature review. Arch Computat Methods Eng
2. Stokoe WC, Casterline DC, Croneberg CG (1976) A dictionary of American sign language on linguistic principles, 2nd edn. Linstok Press, Gallaudet, Washington
3. Cheok MJ, Omar Z, Jaward MH (2019) A review of hand gesture and sign language recognition techniques. Int J Mach Learn Cybern 10:131–153
4. Sandjaja IN, Marcos N (2009) Sign language number recognition. In: NCM 2009—5th international joint conference on INC, IMS, and IDC. IEEE, pp 1503–1508
5. Kalam MA, Mondal MNI, Ahmed B (2019) Rotation independent digit recognition in sign language. In: 2nd international conference on electrical, computer and communication engineering, ECCE 2019. IEEE, pp 1–5
6. Alom MS, Hasan MJ, Wahid MF (2019) Digit recognition in sign language based on convolutional neural network and support vector machine. In: 2019 international conference on sustainable technologies for industry 4.0, STI 2019. IEEE, pp 1–5
7. Sign Language Digits Dataset. https://github.com/ardamavi/Sign-Language-Digits-Dataset. Last accessed Jan 24 2021
8. Barczak ALC, Reyes NH, Abastillas M, Piccio A, Susnjak T (2011) A new 2D static hand gesture colour image dataset for ASL gestures. Res Lett Inf Math Sci 15:12–20
9. Hasan MM, Ahsan SMM (2019) Bangla sign digits recognition using HOG feature based multi-class support vector machine. In: 2019 4th international conference on electrical information and communication technology (EICT). IEEE, pp 1–5
10. Islam MS, Mousumi SSS, Azad Rabby AKMS, Hossain SA, Abujar S (2018) A potent model to recognize bangla sign language digits using convolutional neural network. Procedia Comput Sci 143:611–618
11. Alom MS, Hasan MJ, Wahid MF (2019) Digit recognition in sign language based on convolutional neural network and support vector machine. In: 2019 international conference on sustainable technologies for industry 4.0 (STI). IEEE, pp 1–5
12. Wangchuk K, Riyamongkol P, Waranusast R (2020) Real-time Bhutanese sign language digits recognition system using convolutional neural network. ICT Express (in press)
13. Indian Sign Language Recognition. https://github.com/imRishabhGupta/Indian-Sign-Language-Recognition. Last accessed Jan 07 2021
14. Mitchell T (1997) Machine learning, 1st edn. McGraw-Hill Inc., United States
15. Sign Language digit recognizer. https://www.kaggle.com/cemutku/sign-language-digit-recognizer-cnn. Jan 24 2021
16. Sign Language Digits classification. https://medium.com/analytics-vidhya/sign-language-classification-64fe8ad0fc2c. Last accessed Jan 07 2021

Chapter 43
Parking Lot Occupancy Detection Using Hybrid Deep Learning CNN-LSTM Approach

Bui Thanh Hung and Prasun Chakrabarti

1 Introduction

Smarter parking management based on IoT is an indispensable requirement of smart city, in which car parking occupancy detection is a critical task. Determining available parking spaces in real time and communicating directly with commuters will help greatly in reducing congestions both on the roadways and in the parking slots, allowing drivers to identify the vacant parking spots in a more efficient and time-effective manner.

Many researchers have proposed methods to resolve this problem that can be divided into three groups based on counter, sensor, and image. In the first group, based on counting the cars entering and exiting the parking slots, the car management systems guide the driver toward a vacant space. However, it cannot guide the driver to the exact location; instead, it can only notify the total number of unoccupied spaces. The second group uses sensor to manage vacant parking space. However, building the system is too expensive because of using a huge number of sensors. Finally, the image-based method can overcome the shortcomings of previous methods. Because deep learning has many advantages, that's why many researchers solve the issue using deep learning approach. In this research, we focus on the image-based method. The proposed solution for detecting vacant parking spaces based on image method follows the trend of applying advanced and modern techniques such as Deep Learning, specifically CNN.

B. T. Hung (✉)
DAAI Laboratory, Engineering-Technology School, Thu Dau Mot University, Thu Dau Mot, Vietnam
e-mail: hungbt.cntt@tdmu.edu.vn

P. Chakrabarti
Techno India NJR Institute of Technology, Udaipur, Rajasthan 313003, India

G. Mathur et al. (eds.), *Proceedings of 2nd International Conference on Artificial Intelligence: Advances and Applications*, Algorithms for Intelligent Systems,
https://doi.org/10.1007/978-981-16-6332-1_43

In this paper, we present a method to detect parking lot occupancy using hybrid deep learning approach by combining the superior feature CNN-LSTM methods. The organization of this paper consists of five sessions. Section 2 discusses related works. How to detect parking lot occupancy using hybrid deep learning approach is described in Sect. 3, and the experiments are shown in Sect. 4. Finally, Sect. 5 summarizes our work and gives future directions.

2 Related Works

Detecting vacant parking spaces based on image method have been conducted by many researchers. During the beginning time, many researchers focused on using machine learning techniques and hand craft features: support vector machine (SVM) and color vector features [1], color histogram across three spaces and SVM, Bayes and three features: edges, corners, and wavelet [2], multiple textural features and SVM [3]. However these methods got the low detection accuracy as well as executed with limited dataset.

There are two main streams of automated parking detection methods: straight-based and angular-based flow depending on the original visual extracted features. Jung el al. proposed clustering of maximum pairs in the Hough space. The separation of final marker segments is identified by matching a T-shaped pattern [4]. Huang used HoG characteristic surface to detect vacant parking spaces. A surface-based hierarchy combines information of 3-D blocks with each side of 3-D blocks to infer an empty parking space [5].

Deep learning, with its advantages, has been flowered and become the effective method to solve this problem. Using visual features and neural network is proposed in [6] to determine parking demand and occupancy status. Chen et al. used a multi-CNN to detect vehicles. Amato et al. proposed Deep CNN for parking lot monitoring. They used Alexnet and mLenet to detect vacant parking spaces with the smart camera. They did experiments on two their datasets: Pklot and CNRPark-Ext [7]. However their system requires to be trained in significant timeframe by a strong computer.

CNN can extract spatial features well and LSTM works well in long-term memory. However, almost researches only focus on using deep learning approach independently for parking lot occupancy detection problem. We introduce a new method, Hybrid-PLO for parking lot occupancy detection in our research. We will discuss more detail the proposed model in the next session.

3 Methodology

3.1 Hybrid-PLOD Model

We named our proposed model as Hybrid-Parking Lot Occupancy Detection (Hybrid-PLOD). Hybrid Deep Learning Method (CNN-LSTM) and Parking lot occupancy detection are two main parts of the model. This model is described in Fig. 1.

3.2 Working

Convolutional Neural Network (CNN)

CNN [8] is one of the deep learning models. A CNN has an input layer, a number of convolution and pooling layers and an output layer.

A CNN on our own is created with eight layers: three convolutional layers, three max-pooling layers and two fully connected layers. Our CNN structure is shown in Fig. 2.

Long Short-Term Memory

Long Short-Term Memory (LSTM) is an advanced modification of the Recurrent Neural Network [9]. In LSTM, each cell unit representing as the memory has six main elements: an input gate I, a forget gate F, a hidden state H, a candidate layer $\overline{C}$, a memory state C and an output gate O. Figure 3 shows a structure of a LSTM cell.

Hybrid-PLOD Working

Donahue et al. proposed the original hybrid deep learning approach [10]. We build Hybrid-PLOD model based on that [10–15]. Spatial features are extracted by CNN, and the temporal features will be learnt by LSTM. Combining superior features of each model CNN, LSTM makes an effective hybrid deep learning model. Figure 4

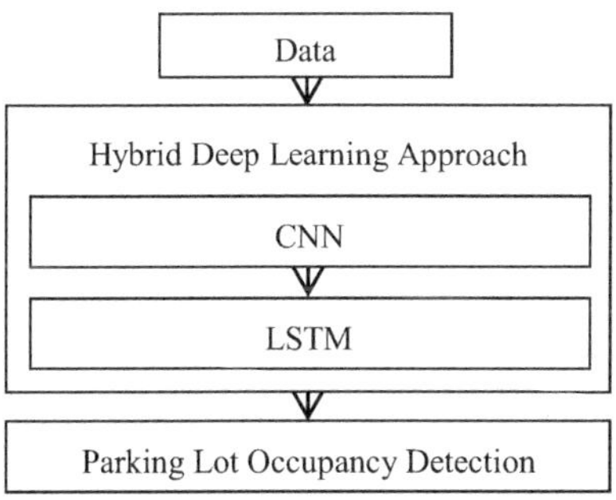

Fig. 1 The hybrid-PLOD model

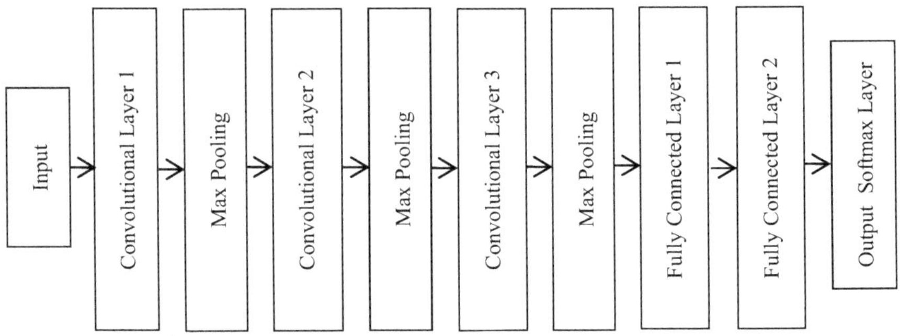

Fig. 2 Our CNN structure

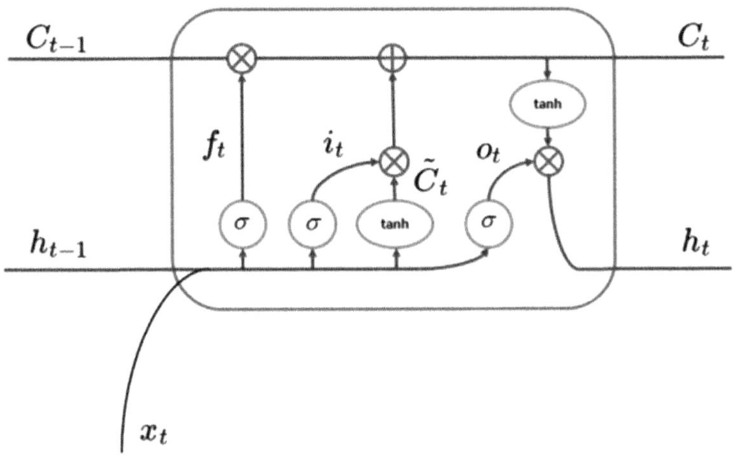

Fig. 3 The long short-term memory cell

shows the architecture of this model (Fig. 5).

4 Experiments

4.1 Dataset

Two datasets: CNRPark and CNRPark-EXT [7] are used in our experiments. CNRPark has 12,000 labeled images of occupied and vacant parking spaces from two cameras. CNRPark-EXT consists of 150,000 labeled images from nine cameras in different weather conditions: Sunny, Rainy and Overcast. Overview of pictures from the CNRPark and CNRPark-EXT datasets is shown in Figs. 6 and 7. The CNRPark

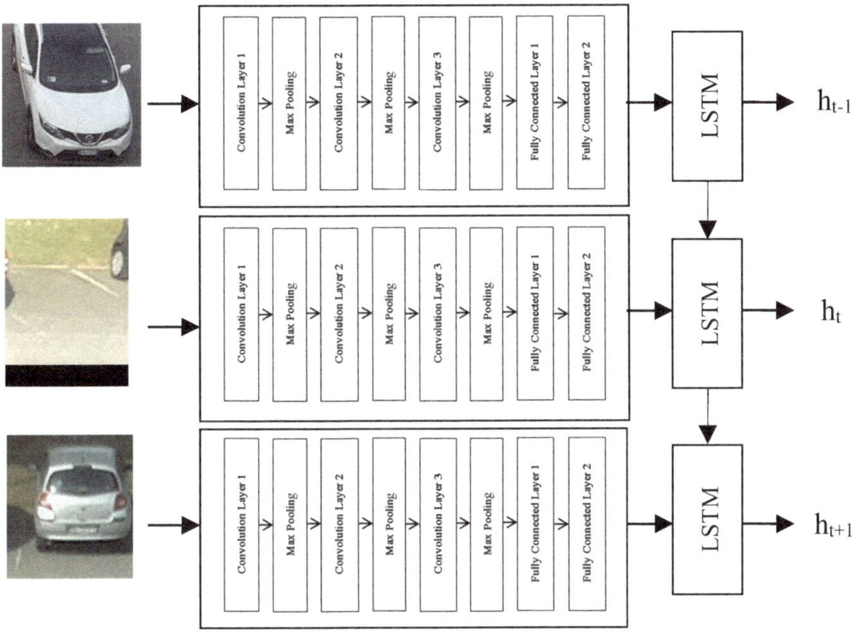

Fig. 4 The hybrid deep learning model

Fig. 5 Example photos from the CNRPark dataset in cameras A and B

Fig. 6 Examples of photos taken under sunny, overcast, rainy weather in the CNRPark-EXT dataset

Fig.7 Some pictures in dataset in vacant and occupied spaces

dataset is described in detail as shown in Table 1.

We did three experiments in the CNRPark dataset. Table 2 presents more detail about the training, testing in this dataset.

For the CNRPark-EXT dataset, we did the fourth experiment with the training, testing dataset as shown in Table 3.

Some pictures in dataset in vacant and occupied spaces are shown in Fig. 7. The first row is in occupied space, and the second row is in occupied space.

Table 1 The CNRPark dataset

Camera	Vacant space	Occupied space
A	2550	3621
B	1632	4781

Table 2 CNRPark dataset in our three experiments

Experiments	Training	Test
ExCNRPark 1	A	B
ExCNRPark 2	B	A
ExCNRPark 3	1/2 (A+B)	1/2 (A+B)

Table 3 CNRPark-EXT dataset in our fourth experiment

Data type	Vacant space	Occupied space
Test	13,546	18,272
Training	46,783	47,602

Table 4 The result of the hybrid-PLOD and CNN models

Experiments	CNN	Hybrid-PLOD
ExCNRPark 1	89.6	92.5
ExCNRPark 2	84.3	88.6
ExCNRPark 3	90.1	93.2
ExCNRPark-EXT	91.4	95.8

4.2 Experiment Result

We used Keras deep learning library [16], TensorFlow [17] to install the model. To operate image, we used OpenCV [18]. The results are evaluated by the Acc metric as follows:

$$Acc = \frac{TN + TP}{TN + FN + TP + FP}$$

where TN is the true occupied parking spaces, FN is the false occupied parking spaces, TP is the true vacant parking spaces and FP is the false vacant parking spaces.

The spatial information is extracted by CNN, and LSTM is used for classification. We used parameters as follows: 64 neurons of fully connected layer, activation function: ReLu, 128 LSTM units/1 LSTM layer. We used binary cross-entropy and Adam optimization in our hybrid deep learning model. We compared the result of the proposed model with the CNN model. Table 4 and Fig. 8 show the result.

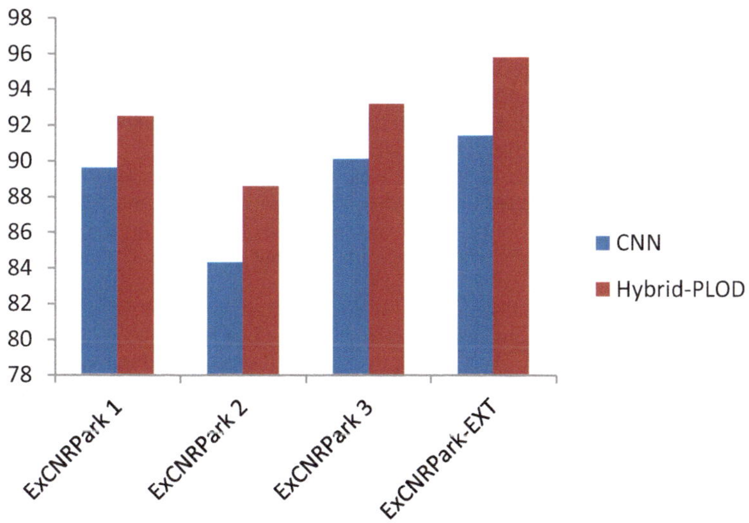

Fig. 8 The results of our experiments with Hybrid-PLOD and CNN models

Table 4 and Fig. 8 show that the Hybrid-PLOD got the best performance. The hybrid CNN+LSTM model gave more accuracy than the CNN. Because of the advantages of LSTM in predicting the output of the spatial extracted features, the proposed model gave better accuracy. CNN is effective in extracting spatial features, and LSTM can memorize previous input because of its internal memory. So that LSTM takes training less time and also has high accuracy.

5 Conclusion

We proposed the Hybrid-PLO model to detect parking lot occupancy. Our proposed method has been evaluated on the public CNRPark and CNRPark-Ext datasets and compared the result with the others in four experiments: ExCNRPark 1, ExCNRPark 2, ExCNRPark 3 and ExCNRPark-EXT. The experiment shows that our proposed method has achieved the best results. In the future, based on our proposed model, we will develop an application that detects parking lot occupancy in real time from camera. We would like to train our proposed model in large data and apply to other deep learning models to find the best model for detecting parking lot occupancy problems.

References

1. Dan N (2002) Parking management system and method. US Patent App 10/066, 215
2. Tsai L-W, Hsieh J-W, Fan K-C (2007) Vehicle detection using normalized color and edge map. Image Process IEEE Trans 16:850–864
3. De Almeida PR, Oliveira LS, Britto AS, Silva EJ, Koerich AL (2015) Pklot-a robust dataset for parking lot classification. Expert Syst Appl 42:4937–4949
4. Jung HG, Kim DS, Yoon PJ, Kim J (2006) Parking slot markings recognition for automatic parking assist system. Proc IEEE Intell Veh Symp: 106–113
5. Huang C, Dai Y, Wang S (2012) A surface-based vacant space detection for an intelligent parking lot. In: 12th International Conference on ITS Telecommunications (ITST), pp 284–288
6. Jermsurawong J, Ahsan U, Haidar A, Haiwei D, Mavridis N (2014) One-day long statistical analysis of parking demand by using single-camera vacancy detection. J Transp Syst Eng Inf Technol 14:33–44
7. Chen X, Xiang S, Liu C-L, Pan C-H (2014) Vehicle detection in satellite images by hybrid deep convolutional neural networks. Geosci Remote Sens Lett IEEE 11:1797–1801
8. LeCun Y, Kavukcuoglu K, Farabet C (2010) Convolutional networks and applications in vision. In: Proceedings of 2010 IEEE international symposium on circuits and systems, IEEE, pp 253–256
9. Hochreiter S, Schmidhuber J (1997) Long short term memory. Neural Comput 9(8):1735–1780
10. Donahue J, Hendricks LA, Rohrbach M, Venugopalan S, Guadarrama S, Saenko K, Darrell T (2017) Long-term recurrent convolutional networks for visual recognition and description. IEEE Trans Pattern Anal Mach Intell 39:677–691
11. Hung BT (2018) Vietnamese keyword extraction using hybrid deep learning methods. In: Proceedings of the 5th NAFOSTED conference on information and computer science

12. Hung BT (2019) Document classification by using hybrid deep learning approach. In: Lecture notes of the institute for computer sciences, social informatics and telecommunications engineering—LNICST, vol 298. Springer, pp 167–177
13. Hung BT, Tien LM (2020) Facial expression recognition with CNN-LSTM. In: Research in intelligent and computing in engineering. Springer series in advances in intelligent systems and computing, pp 549–560
14. Hung BT, Semwal VB, Gaud N, Bijalwan V (2021) Violent video detection by pre-trained model and CNN-LSTM approach. In: Proceedings of integrated intelligence enable networks and computing. Springer series in algorithms for intelligent systems, pp 979–989
15. Tekouaboua SCK, Alaouibc EAA, Cherifd W, Silkana H (2020) Improving parking availability prediction in smart cities with IoT and ensemble-based model. J King Saud Univ Comput Inf Sci
16. Open source computer vision library, OpenCV: https://www.opencv.org (2015)
17. Keras https://keras.io/
18. Abadi M, Barham P, Chen J, Chen Z, Davis A, Dean J, Devin M, Ghemawat S, Irving G, Isard M, Kudlur M, Levenberg J, Monga R, Moore S, Murray DG, Steiner B, Tucker P, Vasudevan V, Warden P, Wicke M, Yu Y, Zheng X (2016) Tensorflow: a system for large-scale machine learning. Tech Rep Google Brain. arXiv preprint

Chapter 44
Adaptive Exon Prediction Using Maximum Error Normalized Algorithms

Md. Zıa Ur Rahman, Baligodugula Vishnu Vardhan, Lakkakula Jenith, Veeramreddy Rakesh Reddy, Sala Surekha, and Putluri Srinivasareddy

1 Introduction

Genomics is a massive subject wherein regions that code for proteins is known the utilization of clever AEP-based totally gadget provided here. DNA regions have a position to play with inside the evaluation of sicknesses and drug design. Intergenic and sequence sections are protected in polymer collection [1]. The shape of the range one super molecule segments is studied to help each DNA sections tertiary additionally secondary shape, once figuring out this for usual DNA segments, any abnormalities are probably to be identified, and heal sicknesses [2]. The concept of three base periodicity and its impact on DSP-based exon prediction methods is discussed in [3]. Whole stay beings retain to be alienated as prokaryotes further to eukaryotes. Protein secret writing segments are nevertheless recognized as exons in a part of eukaryotes, on the identical time as introns are mentioned as non-protein secret writing sections. Simply 3% of the eukaryotic human factor collection has secret writing regions and relaxation retains to be non-coding regions. Consequently, it is far associated crucial mission to stumble on coded sections during a polymer collection [4, 5]. Therefore, several methodologies to DNA popularity in the literature rely upon diverse sign process techniques [6–10].

AEP strategies are totally based on clever conversation machine in a few iterations and also on prolonged sequences. Our gift day painting affords a brand new reconciling deoxyribonucleic acid predictor (AEP) with adaptive MENLMLS algorithms. MENLMLS with its signed variations is taken into consideration for mending AEP efficiency over LMS. LMS drawbacks are resolved through manner of approach of ENLMLS algorithm, thereby will increase velocity and capacity of deoxyribonucleic

Md. Zıa Ur Rahman · B. V. Vardhan · L. Jenith · V. Rakesh Reddy · S. Surekha (✉) · P. Srinivasareddy
Department of ECE, Koneru Lakshmaiah Education Foundation, Green Fields, Vaddeswaram, Guntur 522002, India

G. Mathur et al. (eds.), *Proceedings of 2nd International Conference on Artificial Intelligence: Advances and Applications*, Algorithms for Intelligent Systems,
https://doi.org/10.1007/978-981-16-6332-1_44

acid tracking. Excess mean rectangular error (EMSE) is likewise decreased for the amount of deoxyribonucleic acid identification [11]. Sign-based totally algorithms additionally lessen the signal characteristic through manner of approach of amount of multiplication calculations [12]. Several mistakes additionally do now no longer meet the tracking desire fact of the data-unbiased table bound step-length algorithms [13]. Lower EMSE and bigger step length are essential for the great convergence rate. Disadvantages of LMS are triumph over through manner of approach of the use of MENLMLS-based totally strategies.

The bundle nonheritable in iteration technique indicates times of different step area adaptational techniques in [14–18] wherever step length is banned. They are many ways to achieve higher than contrary numbers of Least Mean Squares (LMS) technique. We have a tendency to merge ENLMLS set of rules with signal rules to lower method complexity. Hybrid variations of projected AEPs embody maximum blunders normalized least suggest index squares (MENLMLS), maximum blunders normalized signal regressor MENSRLMLS (ENSRLMLS), most blunders normalized signal ENLMLS (MENSLMLS), in addition as maximum blunders normalized signal LMLS(MENSSLMLS) techniques. Projected AEPs are evaluated in phrases of average performance victimization real normal genomic know-how units acquired from the gene information of NCBI [19]. The presentation sizing to evaluate the effectiveness of various AEPs is the individually of the method, the characteristics of, exactitude (Pr), specificity (Sp), additionally sensitivity (Sn). Various strategies for coding DNA identity are mentioned inside the creative writing [20–22]. In the ensuing sections, the concept of adaptive techniques [23–27] and findings of AEPs are additionally mentioned, which additionally talk average performance of awesome.

2 Adaptive Algorithms for Exon Prediction

The first step is to understand the genetic sequences taken from the NCBI sequencing database based on the nucleotide dimer density, converted into the digital notation in the proposed AEP, so that only digital or signaling can be used. Discrete signal for signal processing, it is still a major task in genome processing, where the sequence is converted to binary information describing the four binary streams using the binary mapping. Digital modification is an important task in gene sequence processing, as signal processing techniques can only be used for these signals.

Therefore, 1 denotes the present of nucleotides and 0 denotes the absent of nucleotides. They result chain remains suitable for the adaptive process a i/p. CAs are generated using adaptive signal processing technique (Fig. 1).

Let **i,** M (n) as a **mapped** numerical sequence, x (n) is a DNA series of, d (n) is a **TBP-compliant** gene series of, y (n) represents the results that are formed by using Adaptive technique as e (n) represents **a** signal of the response weighting coefficients forming in the response loop. Mostly range in LMS technique is still review by "T".

Depending on present value of 'P', the following weight **factors** can be measured **by** this algorithm **with** the current value h (n) and **are** included in the binary map

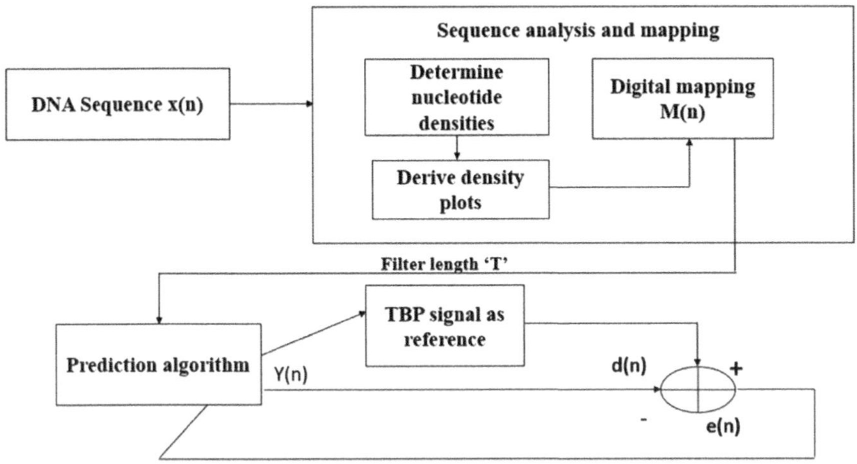

Fig. 1 Schematic drawing of AEP

sequence shown as M (n) current. Always displayed, [12) **Statistical analysis**. Diagram 1 conveys a block representing AEP.

The weight declaration in LMS adaptive methodology is written as

$$h(n+1) = h(n) + P\,x(n)\,e(n)(1) \tag{1}$$

Adaptive algorithms should have a small process problem in applications related to the detection of deoxyribonucleic acid drawing in nano applications based largely on bioinformatics. This is often stated in [18] the strategies in this cause. Three signed types are included in those techniques.

The signum notations is expresses and given below.

$$C\{x(n)\} = \left\{ \begin{array}{c} 1 : x(n) > 0 \\ 0 : x(n) = 0 \\ -1 : x(n) < 0 \end{array} \right\} \tag{2}$$

These variations prefer to deduct the problems of the LMS calculation. LMS has higher calculate problems and differentiate on those forms. [x (n)] is still switch for x (n), while the sign notation C will be utilized for x (n) on a component-by-component basis.

The mass renovate equation for DCLMS parameter as follows

$$h(n+1) = h(n) + PC\{x(n)\}e(n) \tag{3}$$

The weight connection for ECLMS algorithm constant form through modifying e(n) by signed way as

$$h(n + 1) = h(n) + Px(n)C\{e(n)\} \tag{4}$$

As well as, weight connection for DECLMS is concluded by thru exchange of x(n), e(n) by using of signed bureaucracy as

$$h(n + 1) = h(n) + PC\{x(n)\}C\{e(n)\} \tag{5}$$

Due to its robustness as well as simplicity, the standard adaptive LMS technique is suitable for exon forecast. In order to choose parameter of step size for the convergence as well as stability, understanding of preceding input power level rate is required for LMS filter. As one of the statistical unknown levels is generally the input power level, it will normally be assessed through information prior start of adaptation process. The vector of the input information is proportionate to weight update process. Other one being its step size is fixed. Both these remain two setbacks of LMS.

An algorithm must be designed so that weak as well as strong signals can be handled in real time. Therefore, the tap coefficients must be adapted accordingly on the basis of filter changes in input as well as output. So, for large input data vectors, the LMS technique is subjected to the gradient noise amplification limit (Fig. 2).

The normalized LMLS algorithm is considered as a unique LMS algorithm application that takes into account signal level changes at the filter output and selects a stable adaptation algorithm as well as a log-normalized cost function leading to faster convergence. Here, to increase AEP efficiency, we use MENLMLS and SRA-based adaptive algorithms to overcome the LMS shortcomings, increase the ability to identify exons, and provide quick merging if error at maximum. It also detects surplus EMSE in the exon identification process. This MENLMLS adaptive algorithm is used to develop AEP to cope with the compute difficulty of AEP in actual implementation, where mathematical modeling of the MENLMLS algorithm is clearly illustrated in Fig. 4.

It also uses a sign function to reduce the complexity and uses the sign function for MENLMLS to derive three simplified signed versions of the MENSRLMLS, MENSLMLS and MENSSLMLS algorithms.

The weight relationships are MENSRLMLS, MENSLMLS and MENSSLMLS techniques become

$$h(n + 1) = h(n) + \frac{P'}{\varepsilon + max(||e(n)||)^2} C[x(n)]e(n)[\frac{\alpha(e(n))^2}{1 + \alpha(e(n))^2}] \tag{6}$$

$$h(n + 1) = h(n) + \frac{P'}{\varepsilon + \max(||\boldsymbol{e}(n)||)^2} x(n)C\left[e(n)\left[\frac{\alpha(e(n))^2}{1 + \alpha(e(n))^2}\right]\right] \tag{7}$$

$$h(n + 1) = h(n) + \frac{P'}{\varepsilon + max(||\boldsymbol{x}(n)||)^2} C[x(n)]C\left[e(n)\left[\frac{\alpha(e(n))^2}{1 + \alpha(e(n))^2}\right]\right] \tag{8}$$

Mathematical Modeling of ENLMLS Algorithm

Parameters: T = number of taps (i.e. filter length), μ = step size parameter

Let the tap input be x(n) and filter length T is moderate to large

Initialization: Set w(0) = 0 as initial condition

Data: Given x(n) = T-by-1 tap input vector to filter n2 at time n = $[x(n), x(n-1) \ldots\ldots x(n-T+1)]^T$, w(n) is the tap weight vector of adaptive filter, d(n) is desired response at time n, e(n) is the error signal, ω_o is an unknown vector, μ′ is adaptation constant, $(.)^T$ is the transpose of (.)

To be computed: w(n+1) = estimate of tap-weight vector at time n+1

Computation:

The FIR filter output is given by $y(n) = x^T(n)\, w(n) = w^T(n)\, x(n)$

The conventional cost function of the error signal e(n) is $F[e(n)] = E\left[(e(n))^2\right] = E[|e(n)|]$

The LMS recursion expression from (4.1) is $w(n+1) = w(n) + 2\beta(n)\, x(n)\, e(n)$ (6)

where $\beta(n)$ is the time varying step size parameter, that is selected so that a posteriori error e*(n)

$e^*(n) = (1 - 2\beta(n)\, x^T(n)\, x(n)\, e(n))$ (7) is minimized in magnitude

From $\beta(n)$ in (6), resultant expression is obtained as

$$w(n+1) = w(n) + \frac{1}{2\, e^T(n)\, e(n)} e(n) x(n) \quad (8)$$

The bound on step size for mean square convergence of LMS is given as $0 < \beta' < \frac{2}{x^T(n)x(n)}$ (9)

Setting $\beta(n)$ in case of ENLMLS algorithm, it becomes $\beta(n) = \frac{\beta'}{e^T(n)\, e(n)} = \frac{\beta'}{||e(n)||^2}$ (10)

A small positive constant ε is added to prevent denominator is too low, $\beta(n) = \frac{\beta'}{\varepsilon + ||e(n)||^2}$ (11)

By replacing μ in the LMS weight vector update equation with $\beta(n)$ from (11), the weight update relation for ENLMLS is written as $w(n+1) = w(n) + \frac{\beta'}{\varepsilon + ||e(n)||^2} x(n) \left[\frac{\alpha(e(n))^2}{1+\alpha(e(n))^2}\right]$ (12)

Thus, the weight update relation for maximum ENLMLS for $e_{Li} \neq 0$ and $\varepsilon \neq 0$ becomes

$$w(n+1) = w(n) + \frac{\beta'}{\varepsilon + \max(||e(n)||)^2} x(n) \left[\frac{\alpha(e(n))^2}{1+\alpha(e(n))^2}\right] \quad (13)$$

Fig. 2 Mathematical modeling of MENLMLS Algorithm

We further considered advanced four AEPs with those methods will decrease the effects with AEP the use of LMS. P is actual evaluation, it is used to represent Sensitivity, Specificity and Precision displaying that MENSRLMLS is simply beneath its signal-loose regression variant. Thus, MENSRLMLS is Maximum, and it is signed type, touching on to metrics for usual production is likewise a computational difficulty, among numerous strategies selected for the application.

3 Computational Complexities and Convergence Issues

Quantity of multiplication required is specifically set on by metrics for estimating and comparing algorithmic complexity, and it focuses on accuracy of unique MENLMLS-form adaptation technique, rather than an accurate analysis of the complexity to perform the calculations. Sign functions do not have the necessary multiplication

Table 1 Computations required for LMS and various ENLMLS-based AEPs

S. no.	Algorithm	Multiplications	Additions
1	LMS	T+1	T+1
2	MENLMLS	2T+7	2T+2
3	MENSRLMLS	2T	2T+2
4	MENSLMLS	2T+5	2T+2
5	MENSSLMLS	T+2	T+2

calculations for exon finding and implementation. For example, LMS requires a T+1 multiplication calculation with ones adding while calculating the mass improved expression, only 2T+ while calculating 'Se(n)', the MENSRLMLS-based adaptive transformation requires 1 multiplication.

For two other signed MENLMLS algorithms, 2T+1 multiplication computations are still required. With MENSRLMLS, less multipliers with less computational difficulty are required to prepare for other ENLMLS-based techniques. Table 1 illustrates the calculate problem of LMS and MENLMLS-form alternatives. The proposed AEPs that depend on MENLMLS provide less complexity for performing positioning calculations of the desired genome position in the genome input sequence and can be used in nanodevices [5].

Where convergence function and signed variant of the proposed ENLMLS are shown in Fig. 3, and all proposed ENLMLS-based tunable algorithms converge clearly faster than LMS-based AEP.

Therefore, the ENSRLMLS adaptive algorithm is considered better, based on computing difficulty as well as convergence efficiency in contrast to LMS and its other signed algorithms, among the algorithms considered for AEP implementation. It was obvious that ENSRLMLS converges quicker compared with ENSLMLS and ENSSLMLS-based AEPs from convergence features.

4 Results and Discussion

We discuss to show the comparisons in different AEPs. The overview of AEP is shown in Fig. 1. Various AEPs for ENMLS-based technique using sign alternative is considered. Invent an LMS-form relative study of AEP. It uses 10 genomic datasets from the NCBI Databank. For evaluation purposes, the output performance is determined by considering metrics such as correctness (Pr), sensitivity (Sn) and specificity (Sp) of [13, 19], and their equations and theories are presented. The resulting output is a measurement based on changes in the threshold level of the output spectrum used for comparison as part of the exon segment determination using the result DSP method of the various algorithms discussed in Table 3.

There are few values. The amount of nucleotides is located like an exon intron. The localization phase remains described as True Negative (TN), but the correctly recognized exon region is described as True Positive (TP), for example. In addition,

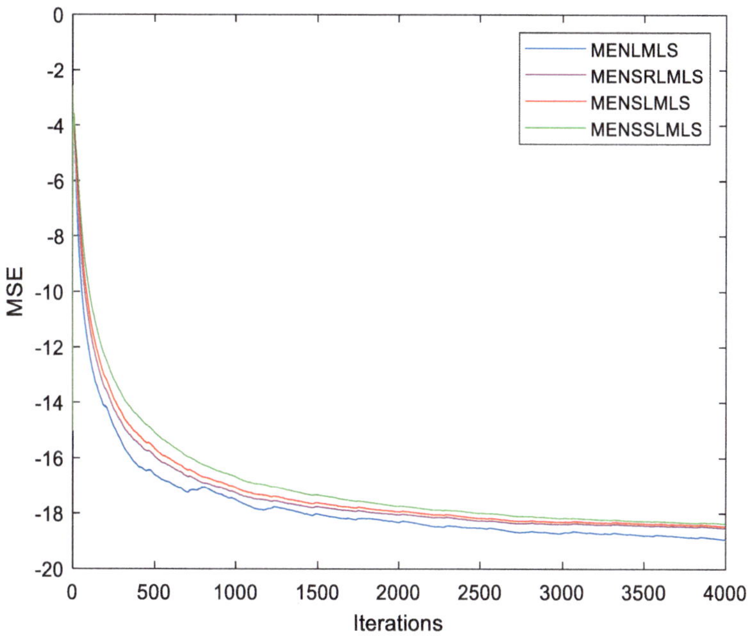

Fig. 3 Convergence curves of MENLMLS method and moreover it signed form

the amount of the exon region, positioned as the intron region, has been shown to be false negative (FN) compared with the actually expected amount of intron, such as the exon region, which is a false positive (FP). I will. NCBI's 10 genetic datasets continue to be considered to evaluate and to give accurate of various techniques. The ascending for these sequences follows X59065.1, E15270.1, U01317.1, X77471.1, AF009962, X92412.1, AB035346.2, AJ223321.1, AJ225085. 0.1, and X51502. 1 each as shown in Table 2.

Expressions for performance metrics are

$$\mathrm{Pr} = (\mathrm{TP} + \mathrm{TN})/(\mathrm{TP} + \mathrm{FP} + \mathrm{TN} + \mathrm{FN})$$

$$\mathrm{Sp} = \mathrm{TP}/(\mathrm{TP} + \mathrm{FP})$$

$$\mathrm{Sn} = \mathrm{TP}/(\mathrm{TP} + \mathrm{FN})$$

Most number of exons mostly identified in some of the exon sections remains specific (Sp), but the amount of exons as properly predicted is planed the sensitivity (Sn). For the exon finding result and gene series using the MNLMLS-dependent method, the value of the figure threshold is selected from 0.4 to 0.9 at intervals of 0.05. The efficiency of the metrics Pr, Sn and Sp is evaluated using these values. Exxon predictions are accurate when the threshold is 0.

Table 2 Gene datasets from NCBI gene databank

Seq. no.	Accession no.	Sequence definition
1	E15270.1	Human gene for osteoclastogenesis inhibitory factor (OCIF) gene
2	X77471.1	Homo sapiens human tyrosine aminotransferase (TAT)
3	AB035346.2	Homo sapiens T-cell leukemia/lymphoma 6 (TCL6) gene
4	AJ225085.1	Homo sapiens Fanconi anemia group A (FAA) gene
5	AF009962	Homo sapiens CC-chemokine receptor (CCR-5) gene
6	X59065.1	Homo sapiens human acidic fibroblast growth factor (FGF) gene
7	AJ223321.1	Homo sapiens transcriptional repressor (RP58) gene
8	X92412.1	Homo sapiens titin (TIN) gene
9	U01317.1	Human beta globin sequence on chromosome 11
10	X51502.1	Homo sapiens gene for prolactin-inducible protein (GPIPI)

Consequently, the measures for performance at the 0.8 threshold value are outlined in Table 3. Operations with a large percentage of A + T nucleotides in DNA sequencing usually show the components of intergenic series, while min A + T nucleotides and larger G + C show potential genes. Primarily, max CG dinucleotide found is often found in advance for a gene.

Functions of statistics for a gene sequence remain beneficial for determining whether the input gene sequence has protein-coding segments. For DNA sequence having accession AF009962, Fig. 4 shows a standard nucleotide density plot.

Its dimer distribution is displayed as a bar chart using MATLAB software. Figure 4 shows that TT base pair dimers, in contrast to all dimers, are abundant in this gene sequence 5. It is composed of 680 TT base pair dimers. The gene sequence also contains 527 ATs and 70 GC dimers. The G + C content has been shown to be less than the A + T dimer content. This indicates that the number of genes is low.

AEP: The steps are instructed below:

(a) The survey has the genomic dataset of the NCBI i/p using the density graph in Fig. 5, and genes assess the present in the accessory gene sites, Top of Form. It demands on ester density base pairs of G + C and A + T dimers. By modifying this sequence, it is converted to digital notation once analysis by digital mapping techniques, but the AEP i/p same the encoded data.
(b) Next sequence. This is given as i/p to suggest AEP after the review, given as the recommendation signal for the proposed ENLMLS-dependent AEPs.
(c) For the filter factor update, the e (n) feedback signal derived from Fig. 1 was used. (d) When this signal becomes minimal, the genes from the DNA sequence are precisely positioned with the graph of PSD. (e) The graph of the desired exon regions is shown in PSD. Output performance is calculated by considering performance metrics Sp, Pr and Sn. Index of 5 series of succession AF009962 for marked ground.

Simulations are done using MATLAB. From Table III, it is clear that metrics measures the performance of ENLMLS and it depends on AEP are there inferior of

Table 3 Performance measures of various ENLMLS and their INCRIBE-based AEPS with respect to Sn, Sp Pr calculations

Algorithm	Metric	Serial number of gene sequence									
		1	2	3	4	5	6	7	8	9	10
LMS	Sn	0.6281	0.6385	0.6458	0.6272	0.6480	0.6162	0.6197	0.6248	0.6269	0.6203
	Sp	0.6434	0.6629	0.6586	0.6406	0.6518	0.6323	0.6526	0.6285	0.6455	0.5966
	Pr	0.5923	0.5895	0.5935	0.5857	0.5905	0.5784	0.5895	0.5852	0.5819	0.5762
MENLMLS	Sn	0.7914	0.7675	0.7912	0.7925	0.7907	0.7903	0.7882	0.7982	0.7873	0.7978
	Sp	0.7903	0.7789	0.7901	0.7877	0.7905	0.7857	0.7971	0.7875	0.7909	0.7856
	Pr	0.7921	0.7901	0.7888	0.7907	0.7917	0.7919	0.7874	0.7912	0.7890	0.7900
MENSRLMLS	Sn	0.7789	0.7451	0.7777	0.7781	0.7678	0.7749	0.7650	0.7788	0.7677	0.7753
	Sp	0.7672	0.7566	0.7684	0.7879	0.7785	0.7665	0.7791	0.7676	0.7783	0.7675
	Pr	0.7674	0.7703	0.7763	0.7686	0.7712	0.7787	0.7677	0.7743	0.7671	0.7719
MENSLMLS	Sn	0.7668	0.7343	0.7571	0.7547	0.7559	0.7545	0.7451	0.7560	0.7467	0.7521
	Sp	0.7571	0.7330	0.7559	0.7561	0.7540	0.7450	0.7591	0.7458	0.7471	0.7479
	Pr	0.7462	0.7593	0.7546	0.7536	0.7561	0.7534	0.7470	0.7588	0.7502	0.7551
MENSSLMLS	Sn	0.7456	0.7015	0.7335	0.7301	0.7322	0.7366	0.7236	0.7349	0.7230	0.7335
	Sp	0.7237	0.7119	0.7791	0.7230	0.7354	0.7349	0.7363	0.7212	0.7277	0.7249
	Pr	0.7361	0.7202	0.7379	0.7359	0.7348	0.7361	0.7250	0.7366	0.7264	0.7351

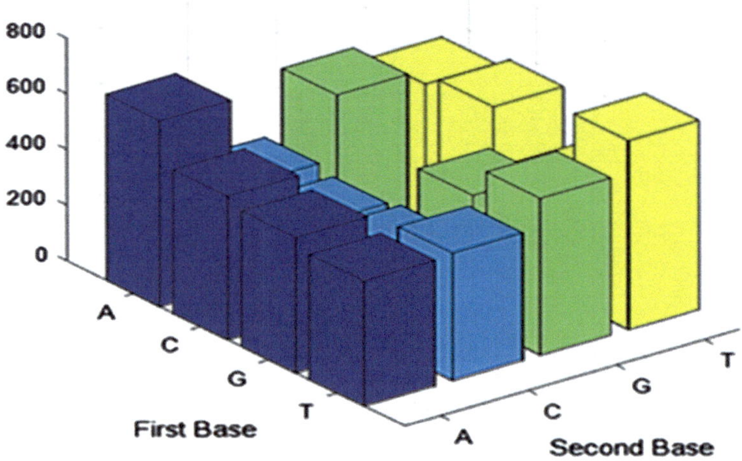

Fig. 4 Plot for nucleotide dimer density for gene series 5 with Accession number AF009962

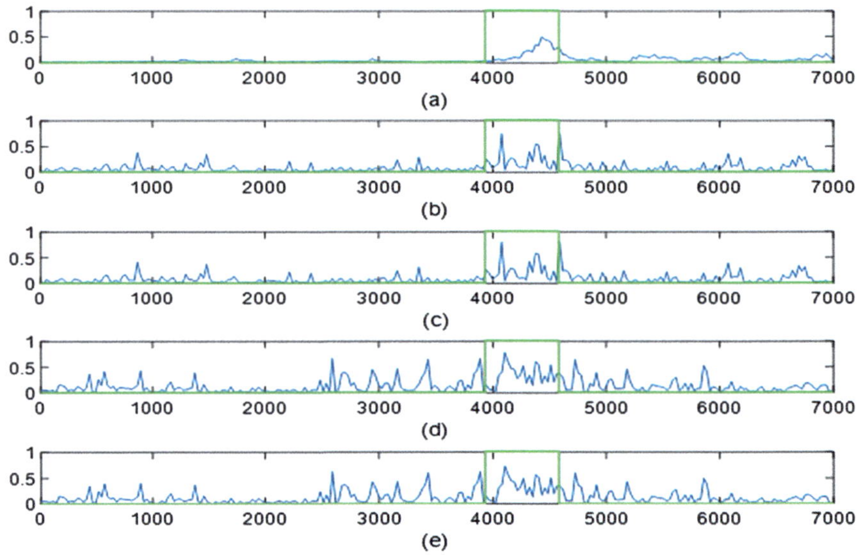

Fig. 5 PSD plots with position of exon (3934–4581) located for genomic sereies having ascent AF009962 through numerous AEPs, **a** AEP of LMS, **b** AEP of MENLMLS, **c** AEP of MENSRLMLS, **d** AEP of MENSLMLS, **e** AEP of MENSSLMLS. *(The relative base position is taken on the x-axis and the power spectrum on the y-axis)*

ENSRLMLS-built AEP limited times more repetition because of least problems and ability to locate exon by exact positioning in 3934-4581. High intensity and sharp peaks in PSD diagram are viewed.

In such cases, the ENSRLMLS algorithm becomes efficient due to low complexity in performing the computations and in terms of exon locating ability. The gesture function present in all signed versions of ENLMLS reduces the computational complexity, and thus all signed versions predict the exon locations more accurately.

Of all these algorithms, ENSRLMLS-based AEP is effective in terms of accurate exon prediction when compared with LMS, ENLMLS and its other signed variants with Specificity Sp, 0.7890 (78.90%), Sensitivity Sn 0.7789(77.89%) also Precision, Pr 0.7806 (78.06%), respectively. At 0.8 threshold value, the exon prediction appears to be better for ENSRLMLS-based AEP. These PSD constructs for ENLMLS and the accented variants are given in data 5 (b), (c) and (d), idiomatic expressions. Finally, all proposed ENLMLS-based AEPs are more effective to discover exon areas in genomic sequences compared with the prevailing LMS technique.

5 Conclusion

If matter-supported observation of the deoxyribonucleic acid place an exceeding order is visible during Wrapper. A unique process mistreatment sensible transmission primarily based system is bestowed here for reconciling deoxyribonucleic acid identification. ENLMLS-based techniques are accustomed method for multiple desoxyribonucleic acid sequences to handle the higher than problem. The local properties of deoxyribonucleic acid are evident in Table 3, and PSD plots are given in Fig. 5. Properly administered AEPs shows deoxyribonucleic acid location at 3934-4581 with towering dynamic structure in the PSD. ENSRLMLS gives higher showing supported complexness to perform computations, executing metrics earned mistreatment sequence five having accession AF009962 at threshold of zero. Eight stays below ENLMLS prices based on AEP, however, find exons. ENSRLMLS {based | based mainly | primarily based on} AEP will be used as such in SOC and LOC and bioinformatics and cloud-based exon forecasting systems. Gene sequence analysis enables us to generate more efficient drugs, better food products also to attain valuable insight into our own body functioning. Other areas of future interest may include extending our research on DNA sequences to produce disease-free crops, healthier farm animals, etc. Future scope of work related to bioinformatics can also be aimed at the discovery of disease mechanisms, improved disease diagnosis and finding prevention strategies to improve human health.

References

1. Ning LW, Lin Ding H, Huang J, Rao N, Guo FB (2014) Predicting bacterial essential genes

using only sequence composition information. Genet Mol Res 13:4564–4572
2. Min L, Qi L, Gamage Upeksha G, Jian Xin W, Fang Xiang W, Pan Y (2014) Prioritization of orphan disease-causing genes using topological feature and go similarity between proteins in interaction networks. Sci China Life Sci 57:1064–1071
3. Singh AK, Kumar Srivastava V (2020) The three base periodicity of protein coding sequences and its application in exon prediction. In: 2020 7th international conference on Signal Processing and Integrated Networks (SPIN), Noida, India, 2020, pp 1089–1094. https://doi.org/10.1109/SPIN48934.2020.9071068
4. Maji S, Garg D (2013) Progress in gene prediction: principles and challenges. Curr Bioinform 8:226–243
5. SrinivasareddyP, Zia Ur Rahman Md (2016) New adaptive exon predictors for identifying protein coding regions in DNA sequence. ARPN J Theor Appl Sci 11:13540–13549
6. Ahmad M, Jung LT, Bhuiyan A (2017) From DNA to protein: why genetic code context of nucleotides for DNA signal processing? a review. Biomed Signal Process Control 34:44–63
7. Wazim Ismail M, Yuzhen Y, Haixu T (2014) Gene finding in metatranscriptomic sequences. BMC Bioinform 15:01–08
8. Ghorbani M, Hamed K (2015) Progress in gene prediction: principles and challenges. Bioinf Approaches Gene Find 4:12–15
9. Scalzitti N, Jeannin-Girardon A, Collet P et al (2020) A benchmark study of ab initio gene prediction methods in diverse eukaryotic organisms. BMC Genomics 21(293). https://doi.org/10.1186/s12864-020-6707-9
10. Azuma Y, Onami S (2014) Automatic cell identification in the unique system of invariant embryogenesis in caenorhabditis elegans. Biomed Eng Lett 4:328–337
11. Guangchen L, Yihui L (2014) Identification of protein coding regions in the eukaryotic DNA sequences based on marple algorithm and wavelet packets transform. Abstr Appl Anal 2014:1–14
12. Simon Haykin O (2014) Adaptive filter theory, 5th edn. Pearson Education Ltd, pp 320–380
13. Saberkari H, Shamsi M, Hamed H, Sedaaghi MH (2013) A fast algorithm for exonic regions prediction in DNA sequences. J Med Signals Sens 3:139–149
14. Nagesh M, Prasad SVAV, Rahman MZ (2016) Efficient cardiac signal enhancement techniques based on variable step size and data normalized hybrid signed adaptive algorithms. Int Rev Comp Soft 11:1–13
15. Sayin MO, Vanli ND, Kozat SS (2014) A novel family of adaptive filtering algorithms based on the logarithmic cost. IEEE Trans Signal Process 62(17):4411–4424
16. Gogineni VC, Mula S (2017) A family of constrained adaptive filtering algorithms based on logarithmic cost. IEEE Trans Signal Process: 1–14
17. Mula S, Gogineni VC, Dhar AS (2017) Algorithm and architecture design of adaptive filters with error non-linearities. IEEE Trans VLSI Syst 25(9):2588–2601
18. Paula Diniz SR (2013) Adaptive filtering, algorithms and practical implementation, 4th edn. Springer Publishers
19. National Center for Biotechnology Information. www.ncbi.nlm.nih.gov/. Accessed Jan 25 2019
20. Srinivasareddy P, Zia Ur Rahman Md, Chandra Sekhar A, Nagireddy P (2019) New exon prediction techniques using adaptive signal processing algorithms for genomic analysis. IEEE Access 7:80800–80812
21. Putluri SR, Zia Ur Rahman Md (2018) Identification of protein coding region in DNA sequence using novel adaptive exon predictor. J Sci Ind Res 77:1–5
22. Srinivasareddy P, Zia Ur Rahman Md, Fathima SY (2018) Cloud based adaptive exon prediction for DNA analysis. IET Healthc Technol 5(1):1–6
23. Sulthana A, Ur Rahman MZ (2018) Efficient adaptive noise cancellation techniques in an IOT enabled telecardiology system. Int J Eng Technol (UAE) 7(2):74–78
24. Sulthana A, Rahman MZU, Mirza SS (2018) An efficient kalman noise canceller for cardiac signal analysis in modern telecardiology systems. IEEE Access 6:34616–34630
25. Salman MN, Trinatha RP, Ur Rahman MZ (2017) Adaptive noise cancellers for cardiac signal enhancement for IOT based health care systems. J Theor Appl Inf Technol 95(10):2206–2213

26. Gopisettryi GKD, VaddiKasulu K, Kamal KR, Pranith G, Rahman MZU (2019) Significant node tracking effective reception networks using influential checkpoints. Int J Innov Technol Explor Eng 8(7):57–60
27. Gayathri NB, Thumbur G, Rajesh Kumar P, Rahman MZU, Reddy PV, Lay-Ekuakille A (2019) Efficient and secure pairing-free certificateless aggregate signature scheme for healthcare wireless medical sensor networks. IEEE Internet Things J 6(5):9064–9075

Chapter 45
A Novel Approach for Wavelength Optimization in GPON Quad Play

Md. Hayder Ali and Mohammad Hanif Ali

1 Introduction

Nowadays, it is going to be a fundamental right to access the Internet. Last-mile users are not bothered about waiting time, network delay, page load delay, they demand it faster. For novel facilities and bandwidth-hungry applications, internet speed has improved pointedly in the past decade to retain pace with the demand of users. Video on demand, IPTV, high-definition TV, multimedia content-based e-commerce, social media, online gaming, etc. are mainly included as demands from internet users. Researches and telco operators are trying to introduce better technology to satisfy the users' demand, gigabit Ethernet, sensor network, mobile internet, etc. are most used technologies. Considering easier, faster and comparatively low cost, Gigabit passive optical networks (GPON) is one of the high-speed access technologies which offers triple-play service (i.e. data, voice and video). Ethernet, ATM and TDM (PSTN, ISDN, E1 and E3) traffic is transferable by GPON. Many operators are concerned about bandwidth exaltation, facility bearer as well as passive network maintenance and GPON is commonly recognized by the ISP (Internet Service provider). High bandwidth size, high excellence and multi-play facilities through a single station grant a robust professional chance for telecom operatives as it affords sustenance that embraces voice, Ethernet, ATM, leased lines and others. Since provisioning abilities, conservation, direction and simple process, GPON offers end-to-end service organization services also. It not only transports meaningfully advanced efficiency as a carriage network, but also transports effortlessness and exceptional scalability for forthcoming development in backup supplementary facilities.

GPON is a point-to-multipoint last-mile solution. Its key representation is the use of passive splitters in the fiber distribution network, empowering one single feeding fiber from the operator's CO (central office) to attend multiple homes and minor

Md. Hayder Ali (✉) · M. H. Ali
Department of Computer Science and Engineering, Jahangirnagar University, Dhaka, Bangladesh

G. Mathur et al. (eds.), *Proceedings of 2nd International Conference on Artificial Intelligence: Advances and Applications*, Algorithms for Intelligent Systems,
https://doi.org/10.1007/978-981-16-6332-1_45

industries. The GPON normal diverges from other PON ideals in that it realizes higher bandwidth and higher efficiency using larger, variable-length packets. GPON offers effectual packing of user traffic, with frame separation permitting higher quality of service (QoS) for delay-sensitive voice and video transports traffic.

SDH (E1) traffic with existing triple play explanation is the mandate of today's high-tech explanation. It is much more significant to emphasize on some strictures like productivity signal variety, broadcast wavelength, power inexpensive and compassion edge for merging explanation. In this paper, the last end receiving sensitivity is calculated and compared with both triple play and proposed quad play.

The rest of this paper is organized as follows. In Sect. 2, literature review is presented. Existing GPON triple play and the proposed quad play are stated in Sect.3. Simulation scenario is in Sect. 4 and performance analysis between GPON triple play and the proposed quad play is stated in Sect. 5. Finally, Sect. 6 draws a conclusion to this paper.

2 Literature Review

GPON technology is a recent technology but it is blessed with lots of investigative work. Over the years, a lot of research works have been carried out to study the dynamic bandwidth allocation (DBA) algorithm of the GPON network structure. Some of them are discussed below.

Assi et al. (1) have proposed to practice the multipoint regulator procedure distinct by the IEEE 802.3ah task force to decide the communication of dissimilar operators, and they described dissimilar dynamic bandwidth allocation (DBA) algorithms to allot bandwidths effectually and equally between end operators. [1]. Shami et al. (2005) have stated a novel dynamic bandwidth allocation (DBA) algorithm that effectually assigns bandwidths between end users [2]. They presented a DBA algorithm for EPON Network which can sustenance distinguished facilities in a network with mixed traffic. Elias et al. (2006) have introduced a novel provision prototypical that affords numerical per-flow bandwidth pledges [3], where operators contribute for a definite frequency; furthermore, the network sporadically individuates vacant bandwidth and propositions short-term indentures where extra-bandwidth was owed and failsafe completely to operators who can adventure it to diffuse at a proportion higher than their contributed rate [4]. I-Shyan Hwang et al. (2011) have introduced an adaptive precedence forecast combined with bi-partition dynamic bandwidth allocation (APS-BDBA) apparatus to increase the discriminated amenities over GPON and in the inter-ONU arrangement, the B-DBA instrument to divide the sequence time by sectionalization of the ONUs into two clusters with some timing intersection to implement interwove bandwidth apportionment, and corrects the bandwidth vigorously between the original cluster for high consequences T-CONT 1–2 and the second group for low significances T-CONT 3–4 to eradicate the idle retro tricky and reduce bandwidth leftover [5]. Tania Panayiotou et al. (2019) have proposed a data-driven bandwidth apportionment (BA) context for sometimes and vigorously reconfiguring an adaptable optical network bestowing to analytical bandwidth apportionment (PBA) representations [10].

3 Existing GPON Triple Play and Proposed Quad Play

Triple play provision permits providing voice, video and data in a solitary entree contribution. The most common requests are telephony, TV and high-speed internet access. The projected bandwidths which have been used for the diverse requests [6]:

- VoIP: For voice over IP, it is needed 15Kbps with G.723 voice codec.
- TV: For television, used 8Mbps for high-definition channel and 1,5Mbps for typical explanation channel with MPEG4 solidity.
- Internet Access: Used from 6Mbps up to 10Mbps for downstream and 1Mbps for ambitious.

In case of the proposed triple play, SDH (synchronous digital hierarchy) could be received from ONT (optical network terminal). That is proposed quad paly could be used for VoIP, TV, Internet Access and SDH [7].

4 Simulation Design

The simulation was designed in two ways: (i) GPON triple play and (ii) GPON quad play. This simulation was designed by the packet simulation software OptSIM. OptSIM recommends ordered assembly of decoration for generating new imitation situations.

General GPON triple play: In this scenario, the video signal (VS) and ISP (Internet Service Provider) connection directly enter into OLT (optical line terminal). One output line from PON (passive optical network) port enters into a 1:32 optical splitter [8]. In a practical scenario, splitter is situated in LDP (local distribution point). The optical fiber is 12 km long. From the optical splitter, one output fiber was entered into ONT–1 (optical network terminal). Similarly, to the other ONTs (ONT-2, ONT-3,…, ONT–32) [9]. In this simulation, up to ONT for all connections, optical fiber (single mode) is used. From ONT's Ethernet port-1, video signal and similarly, data and voice signal are received from Ethernet port-2 and 3. In this simulation, 1310 nm to 1610 nm wavelengths was used and all output data was captured and all output signals were analyzed (Figs. 1 and 2).

Proposed GPON Quad play: In this scenario, video signal (VS) and ISP (Internet Service Provider) connection directly enter into a MUX (multiplexer) and the signal is modulated into it. This modulated signal was entered into OLT [13]. All other designs remained the same as general triple play. From ONT's Ethernet port-1, the

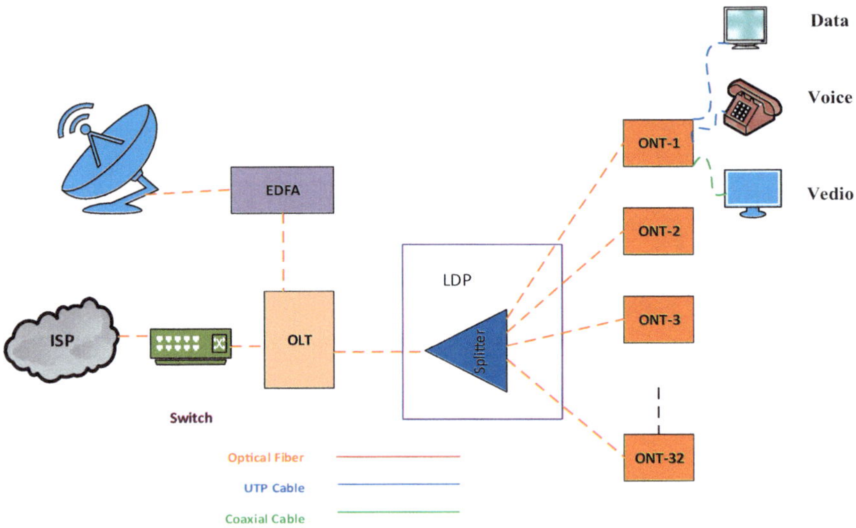

Fig. 1 Simulation design of triple play architecture [11].

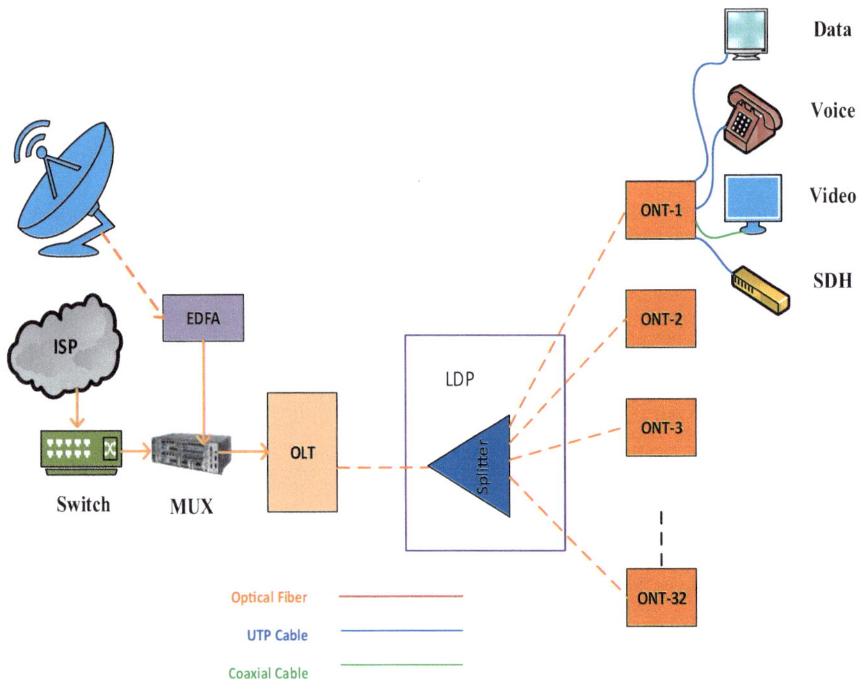

Fig. 2 Proposed GPON quad-play simulation design [12].

SDH signal and video, data and voice signal are received from Ethernet port-2, 3 and ,4 respectively. Eye diagram was analysed for SDH traffic [14].

5 Result Analysis

The input and output signal of general triple-play architecture and the proposed quad-play architecture are given below.

Data and Voice signal for wavelength-1310 nm.

At first data, voice and video signals combinedly enter into OLT and here, all signals are in the same wavelength (1310 nm). In the general triple play, the signal will be modulated at OLT but in the proposed quad-play architecture, the combined signal will be modulated at MUX and the modulated signal will propagate through OLT.

General Triple Play (TP):

Proposed Quad Play (QP):

In the signal graph (Figs. 3, 4, 5 and 6), it is shown as power losses in dBm versus wavelength. From the figure, we see that the input signal propagates at 0dBm loss and the output signal gets at approximately −30dBm loss. But in the proposed quad-play architecture, the output signal gets at approximately −25dBm. Here, the output signal's receiving sensitivity is approximately −30dBm at general triple play whereas approximately -25dBm at proposed quad play (Figs. 7, 8, 9 and 10).

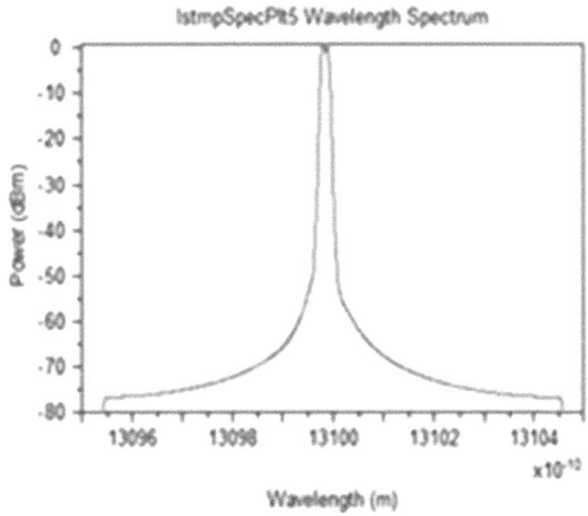

Fig. 3 Input signal (1310 nm) (Triple Play)

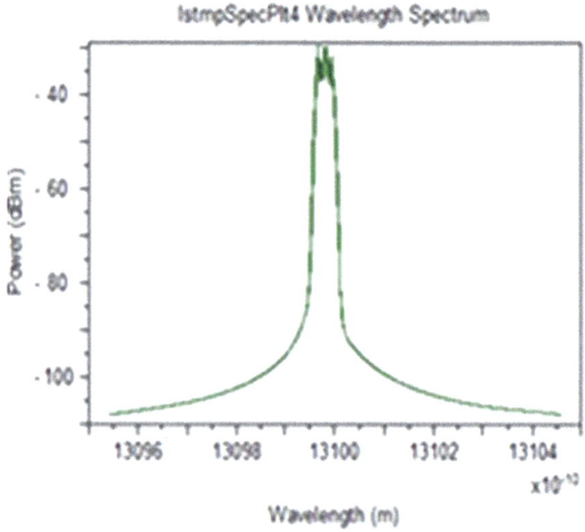

Fig. 4 Output signal (1310 nm) (Triple Play)

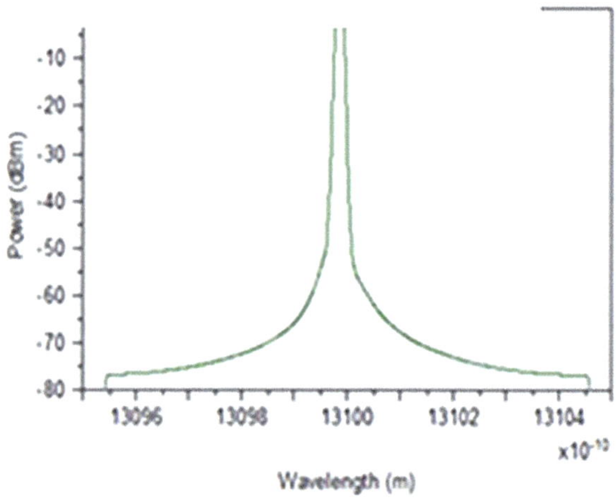

Fig. 5 Input signal (1310 nm) (Quad Play)

Video signal for Wavelength 1310 nm.

General Triple play (TP):

Proposed Quad Play (QP):

The video input signal propagates at −10dBm through the OLT and its output signal is at approximately −30dBm at general triple play while approximately −26dBm is at proposed quad play. Here, we are getting better output signal.

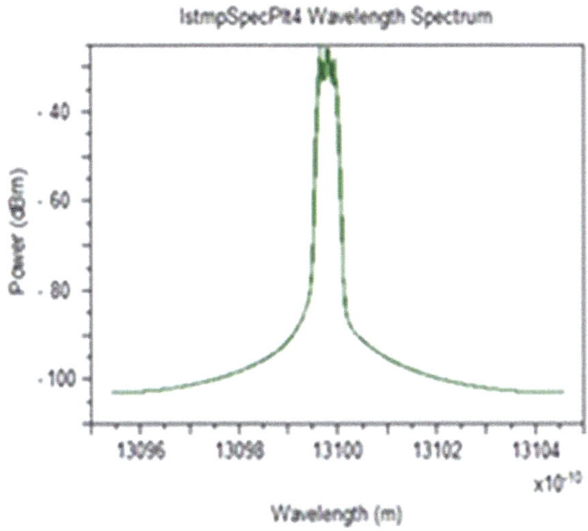

Fig. 6 Output signal (1310 nm) (Quad Play)

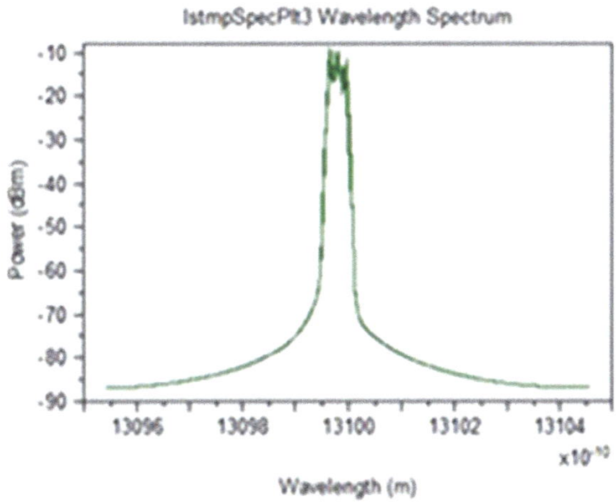

Fig. 7 Video input signal (1310 nm) (TP)

SDH Traffic:

In Fig. 11, an eye diagram shows the SDH traffic. In the general triple-play architecture, the SDH traffic is absent but in the proposed quad-play architecture, the SDH traffic could be measured by the ONT's Ethernet port.

Data and Voice signal for Wavelength 1490 nm.

At first data, voice and video signals combinedly enter into OLT and here all signals are in the same wavelength (1490 nm). In the general triple play, the signal will be modulated at OLT, but in the proposed quad-play architecture, the combined signal will be modulated at MUX and the modulated signal will propagate through OLT.

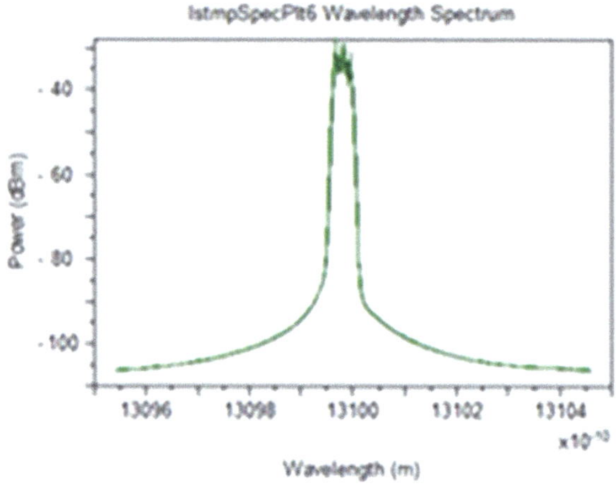

Fig. 8 Video output signal (1310 nm) (TP)

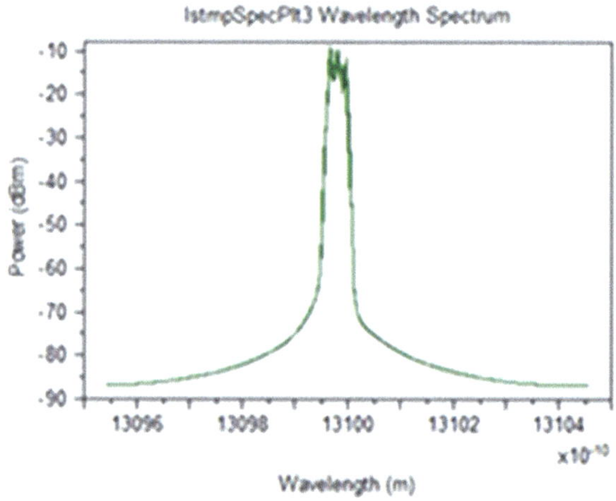

Fig. 9 Video input signal (1310 nm) (QP)

General Triple Play (TP):

Proposed Quad Play (QP):

The signal graph is shown in Figs. 12, 13, 14 and 15 as power losses in dBm versus wavelength. The figure shows that the input signal propagates at 0dBm loss and the out signal gets at approximately −30dBm loss. But, in the proposed quad-play architecture, the output signal gets at approximately −25dBm. Here, the output signal's receiving sensitivity is approximately −30dBm at general triple play whereas −30 dBm at proposed quad play (Figs. 16, 17, 18 and 19).

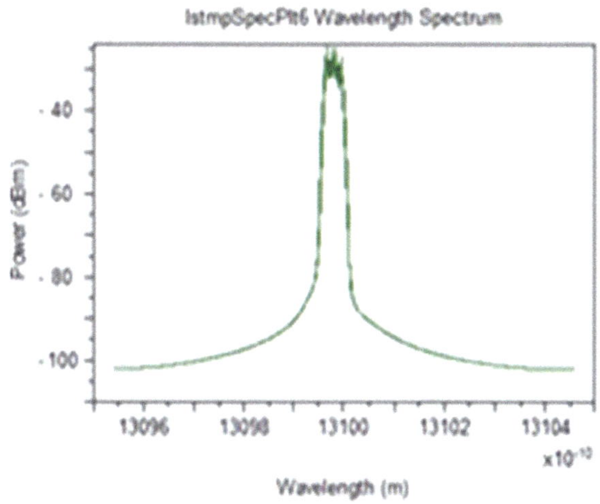

Fig. 10 Video output signal (1310 nm) (QP)

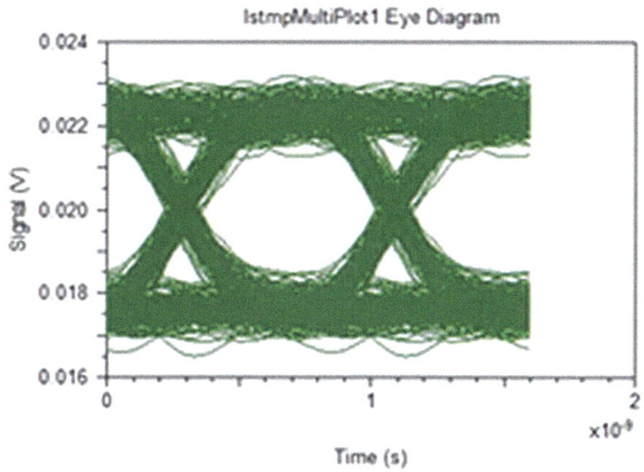

Fig. 11 Eye diagram of SDH traffic (QP)

Video signal for Wavelength 1490 nm.

General Triple Play (TP):

Proposed Quad Play (QP):

The video input signal propagates at −10dBm through the OLT and its output signal is at approximately -32 dBm at general triple play while approximately -30dBm is at the proposed quad play. Here, we are getting a better output signal.

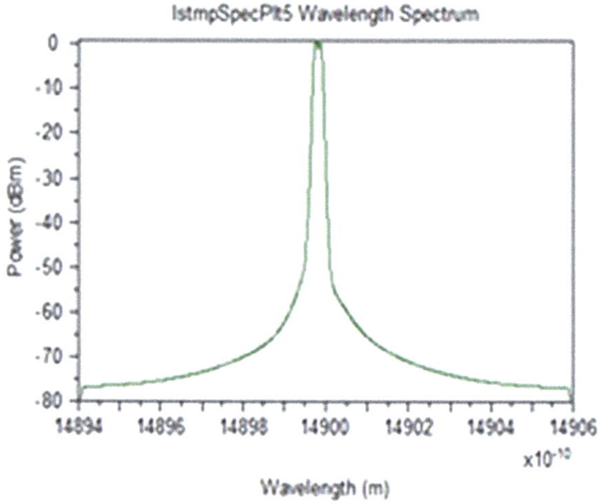

Fig. 12 Input signal (1490 nm) (TP)

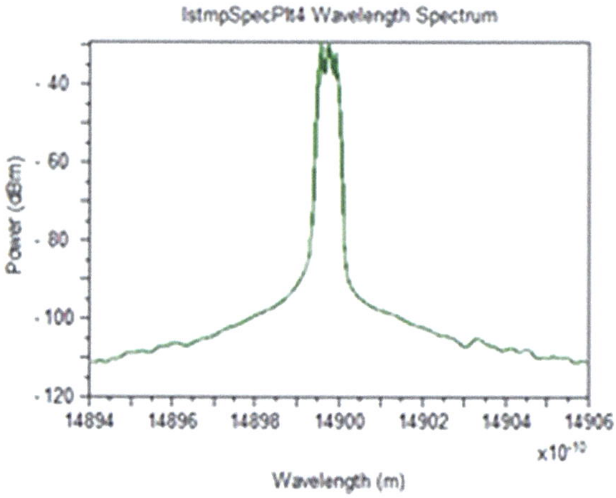

Fig. 13 Output signal (1490 nm) (TP)

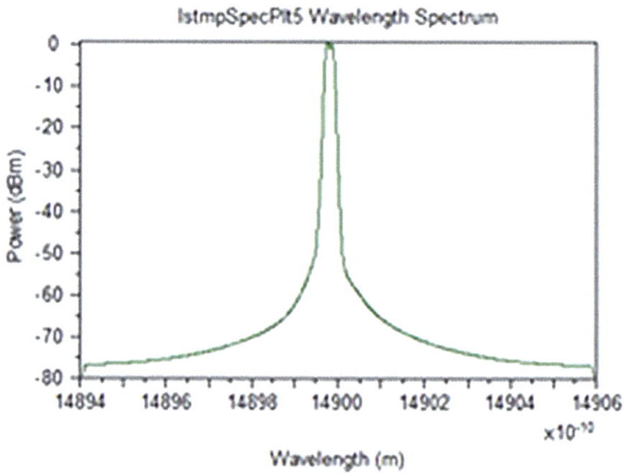

Fig. 14 Input signal (1490 nm) (QP)

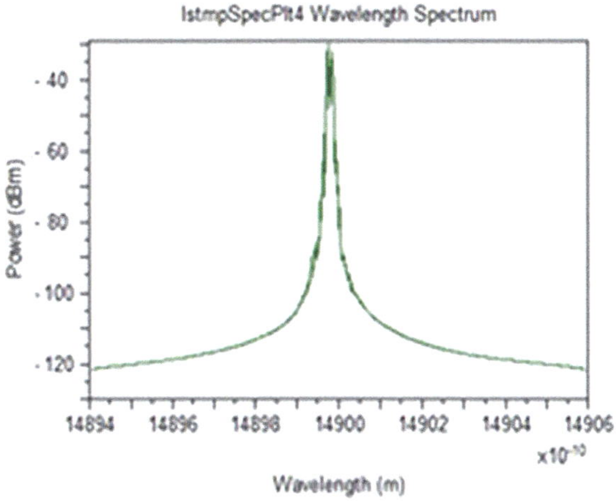

Fig. 15 Output signal (1490 nm) (QP)

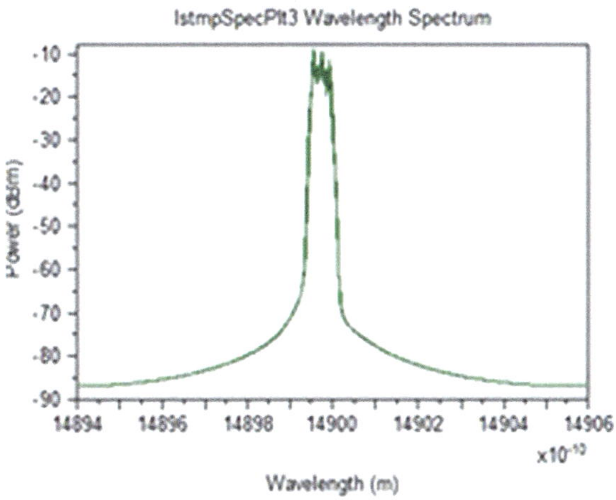

Fig. 16 Video input signal

SDH Traffic:

In Fig. 20, an eye diagram shows the SDH traffic. In the general triple-play architecture, the SDH traffic is absent but in the proposed quad-play architecture, the SDH traffic could be measured by the ONT's Ethernet port.

Data and Voice signals for Wavelength 1550 nm.

At first data, voice and video signals combinedly enter into OLT and here, all signals are in the same wavelength (1550 nm). In the existing triple play, the signal will be modulated at OLT, but in the proposed quad-play architecture, the combined signal will be modulated at MUX and the modulated signal will propagate through OLT.

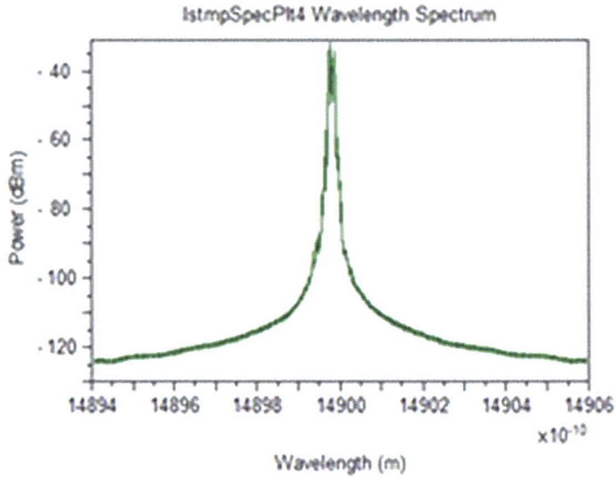

Fig. 17 Video output signal (1490 nm) (TP)

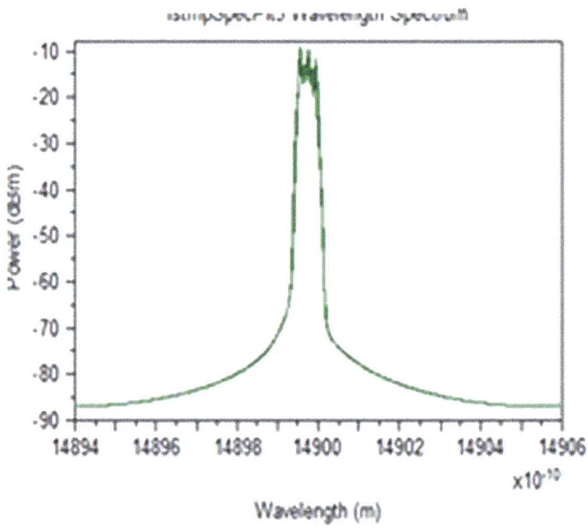

Fig. 18 Video input signal (1490 nm) (QP)

General Triple Play:

Proposed Quad Play (QP):

The signal graph is shown in Figs. 21, 22, 23 and 24 as power losses in dBm versus wavelength. Figures show that the input signal propagates at 0dBm loss and the output signal gets at -30dBm loss. But in the proposed quad-play architecture the output signal gets at approximately −25dBm. Here, the output signal's receiving sensitivity is approximately −30dBm at general triple play whereas approximately −25dBm at proposed quad play (Figs. 25, 26, 27 and 28).

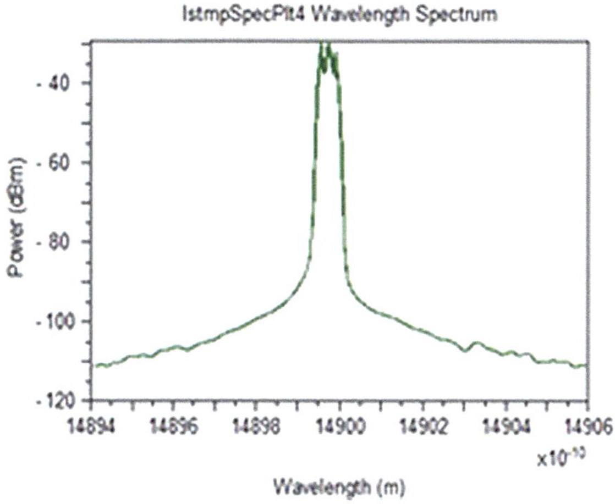

Fig. 19 Video output signal (1490 nm) (QP)

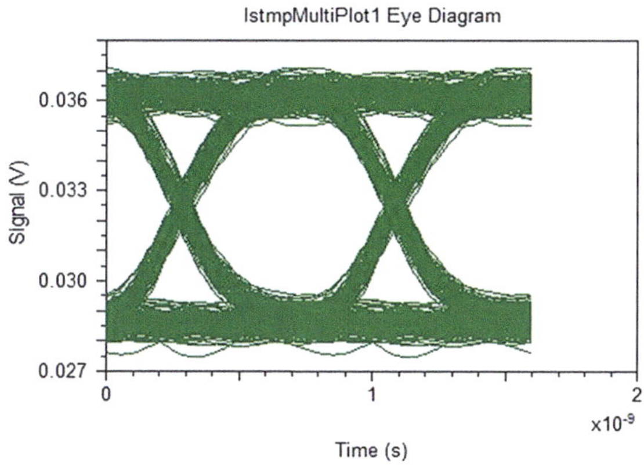

Fig. 20 Eye diagram of SDH traffic (QP)

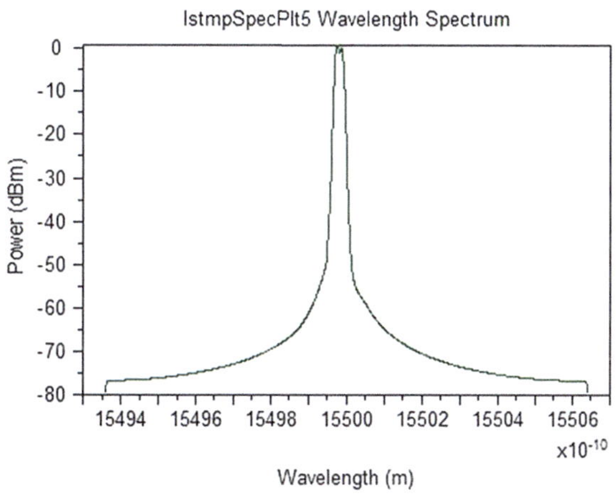

Fig. 21 Input signal (1550 nm) (TP)

Fig. 22 Output signal (1550 nm) (TP)

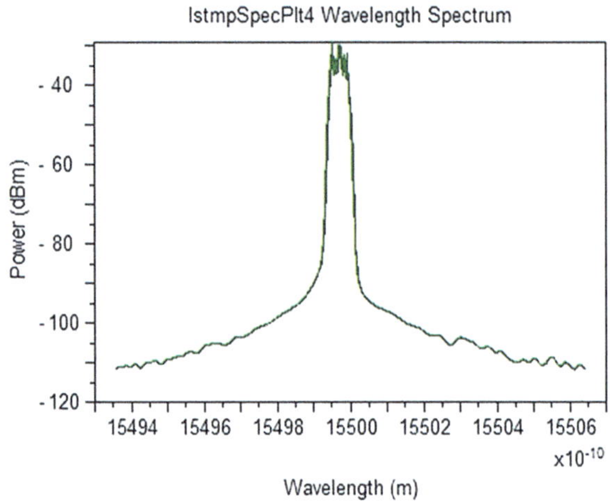

Fig. 23 Input signal (1550 nm) (QP)

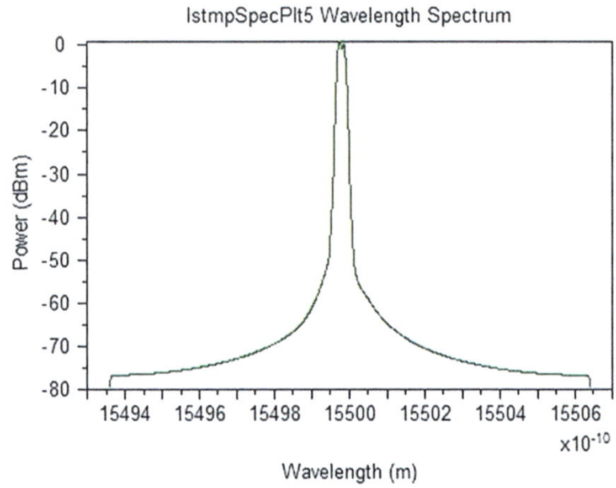

Video signal for Wavelength 1550 nm.

General Triple Play:

Proposed Quad Play (QP):

The video input signal propagates at −10dBm through the OLT and its output signal is at approximately −32dBm at general triple play while approximately −29dBm at the proposed quad play. Here, we are getting a better output signal.

SDH Traffic:

In Fig. 29, the SDH traffic is shown by an eye diagram. In the general triple-play architecture, the SDH traffic is absent but in the proposed quad-play architecture, the

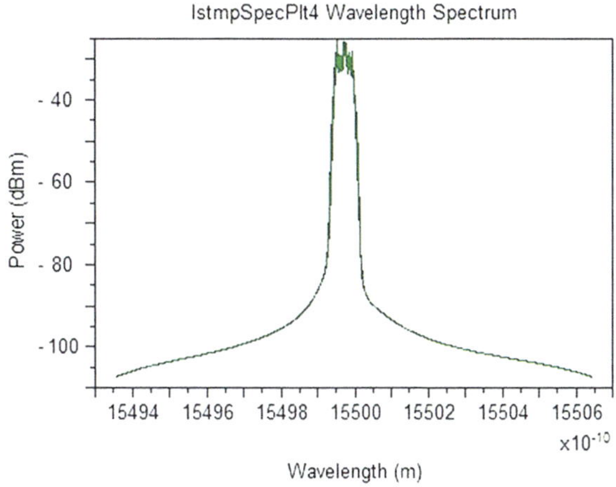

Fig. 24 Output signal (1550 nm) (QP)

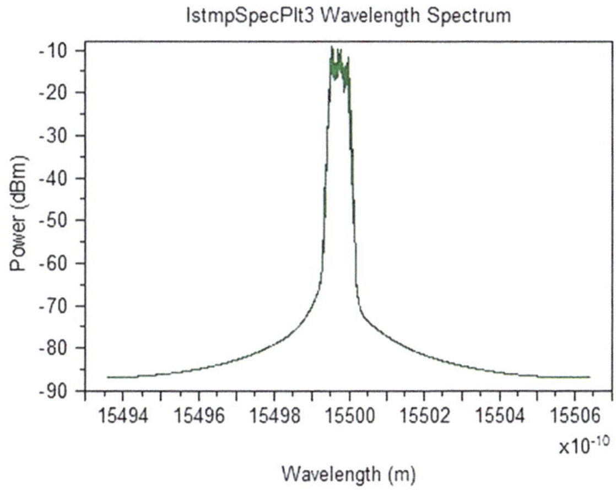

Fig. 25 Video input signal (1550 nm) (TP)

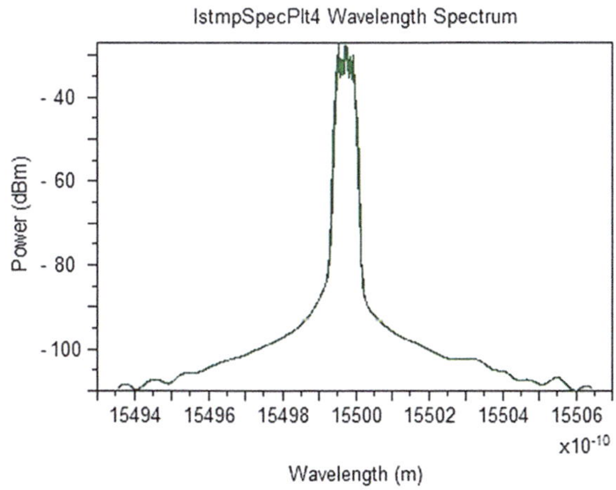

Fig. 26 Video output signal (1550 nm) (TP)

Fig. 27 Video input signal (1550 nm) (QP)

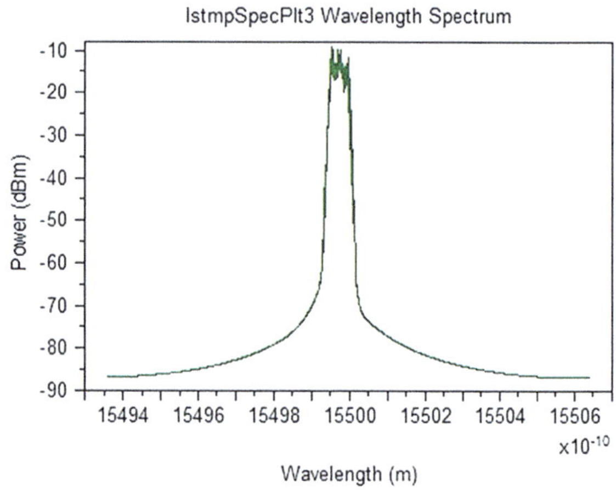

Fig. 28 Video output signal (1550 nm) (QP)

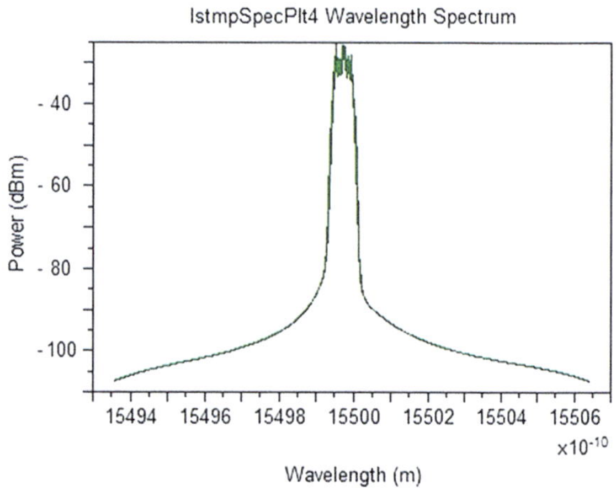

Fig. 29 Eye diagram of SDH traffic (QP)

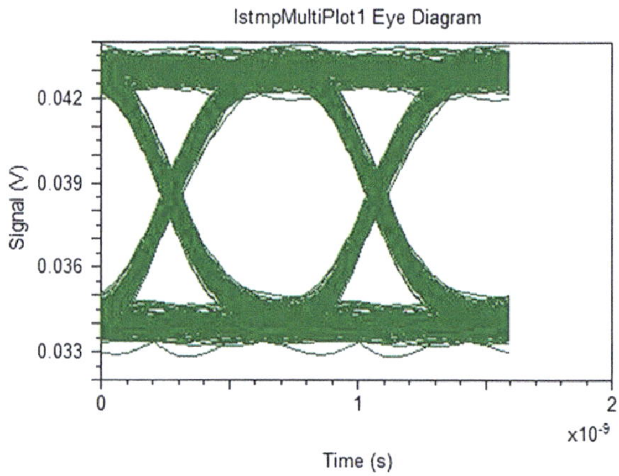

Fig. 30 Input signal (1490 nm) (TP)

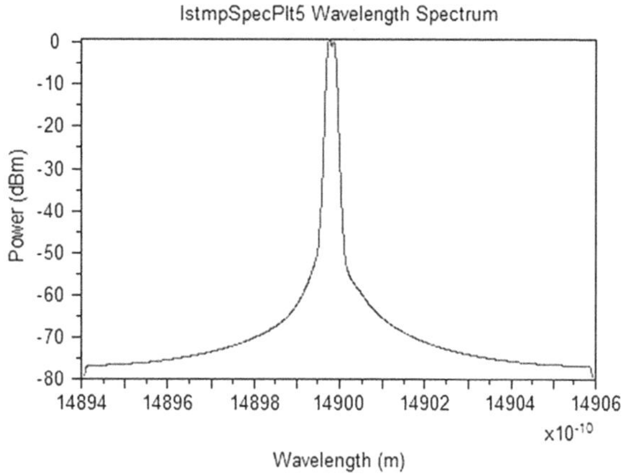

SDH traffic could be measured by the ONT's Ethernet port. If the noise increased, then the eye will become smaller.

Data and Voice signal for Wavelength 1490 nm and video signal for wavelength 1310 nm (Combinedly).

In this scenario, the voice and data signals are at 1490 nm and the video signal at 1310 nm. The signal will be modulated at OLT in the general triple-play system, but in proposed quad-play system, it is modulated inside the MUX. From the proposed quad-play system, there is an SDH output taken from the ONT's Ethernet port.

Data and Voice (1490 nm):

General Triple Play (TP):

Proposed Quad play (QP):

The signal graphs are shown in Figs. 30, 31, 32 and 33 as power losses in dBm versus wavelength. Figure 31 shows that the input signal propagates at 0dBm loss and the output signal gets at approximately −34dBm loss. But in the proposed quad-play architecture, the output signal gets at approximately -32dBm. Here, the output signal's receiving sensitivity is approximately -34dBm at general triple play whereas approximately -32dBm at the proposed quad play (Figs. 34, 35, 36, 37, 38, 39, 40, 41, 42, 43, 44, 45 and 46).

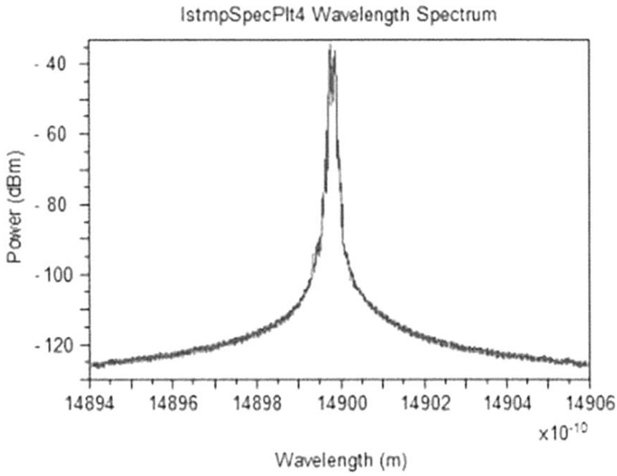

Fig. 31 Output signal (1490 nm) (TP)

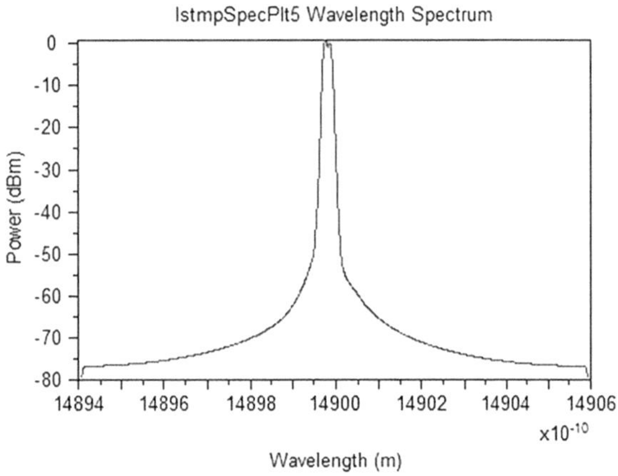

Fig. 32 Input signal (1490 nm) (QP)

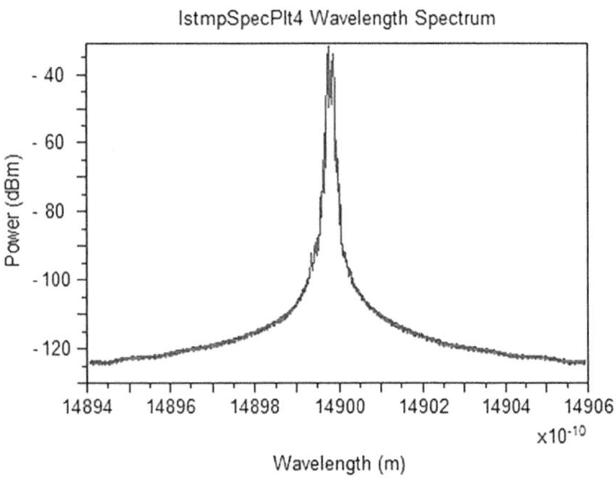

Fig. 33 Output signal (1490 nm) (QP)

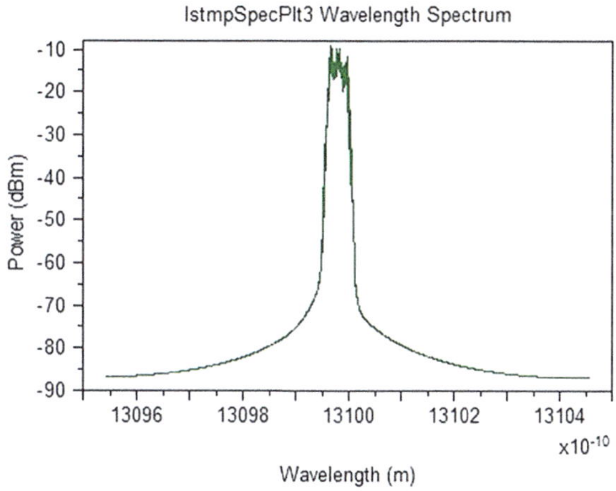

Fig. 34 Video input signal (1310 nm) (TP)

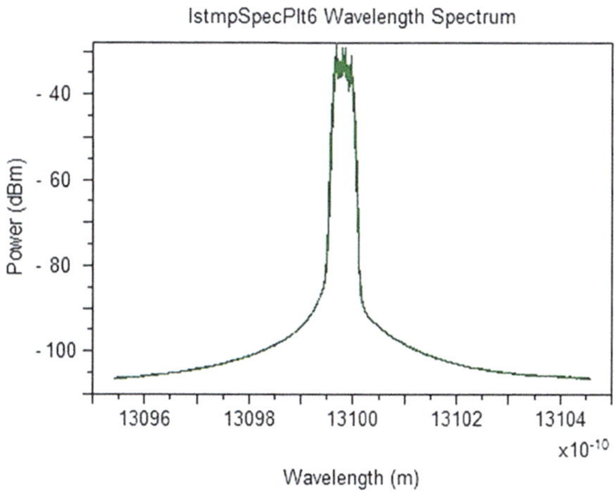

Fig. 35 Video output signal (1310 nm) (TP)

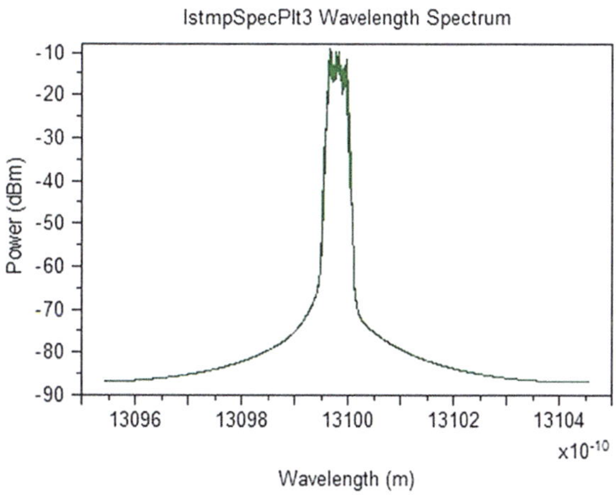

Fig. 36 Video input signal (1310 nm) (QP)

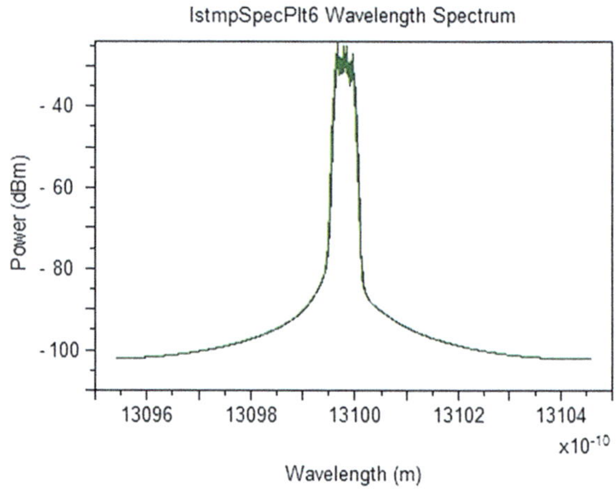

Fig. 37 Video output signal (1310 nm) (QP)

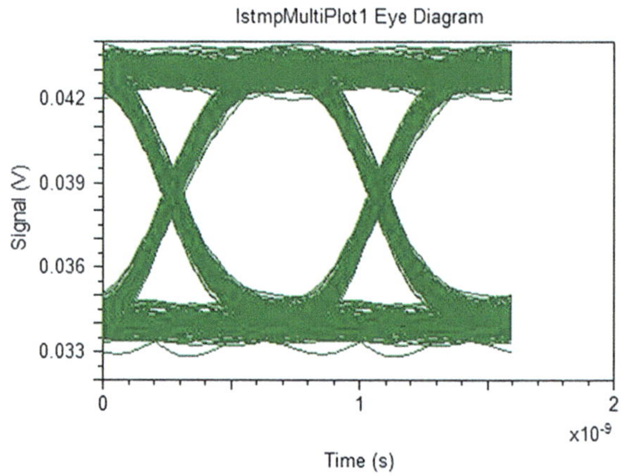

Fig. 38 Eye diagram of SDH traffic (QP)

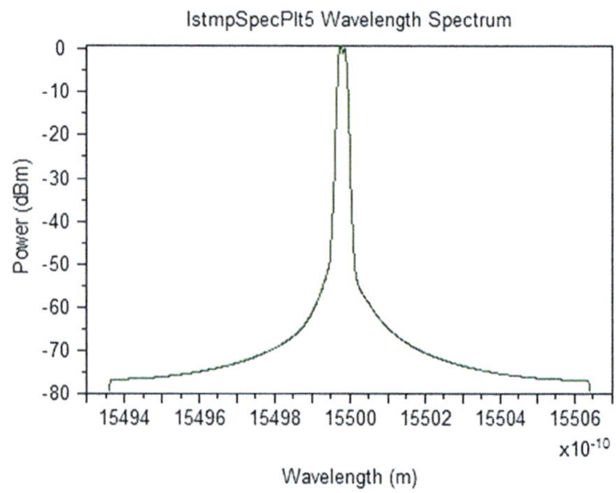

Fig. 39 Input signal (1550 nm) (TP)

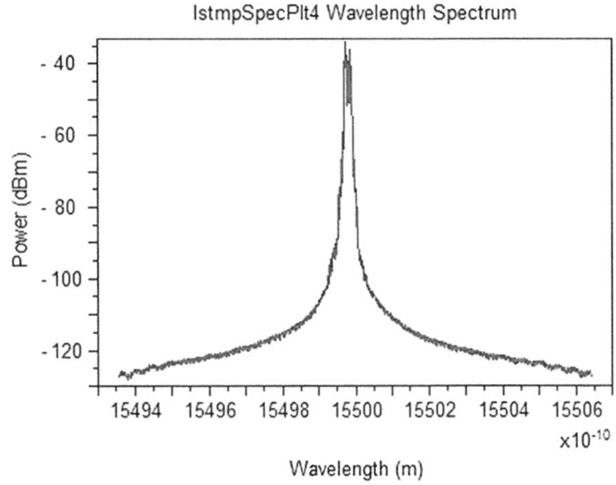

Fig. 40 Output signal (1550 nm) (TP)

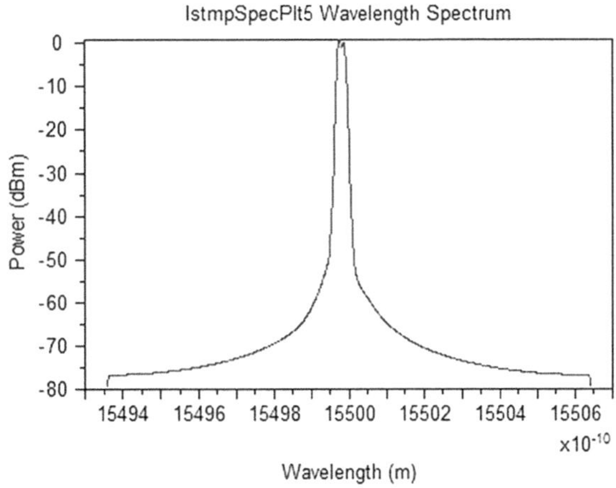

Fig. 41 Input signal (1550 nm) (QP)

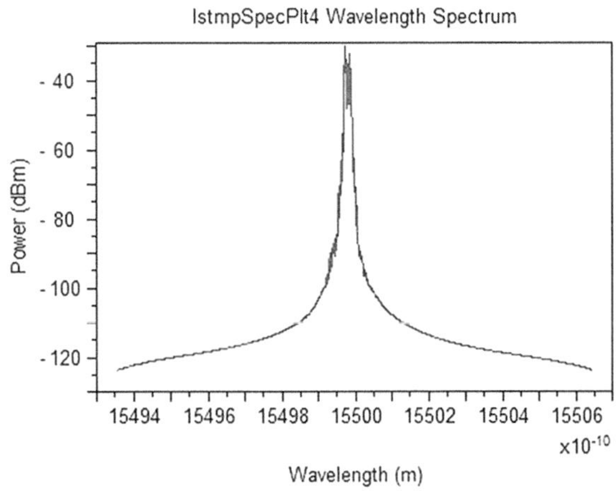

Fig. 42 Output signal (1550 nm) (QP)

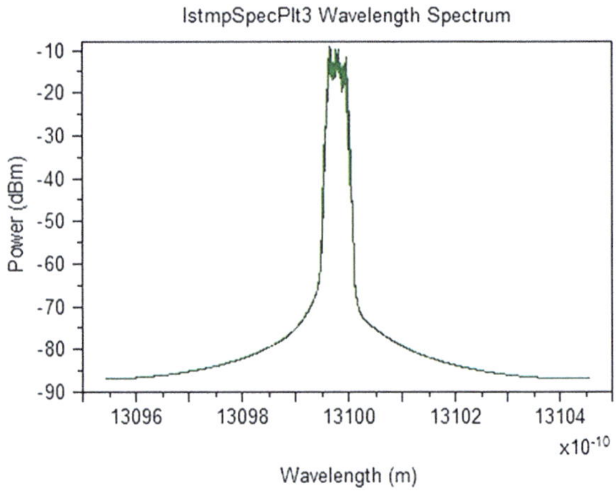

Fig. 43 Video input signal (1310 nm) (TP)

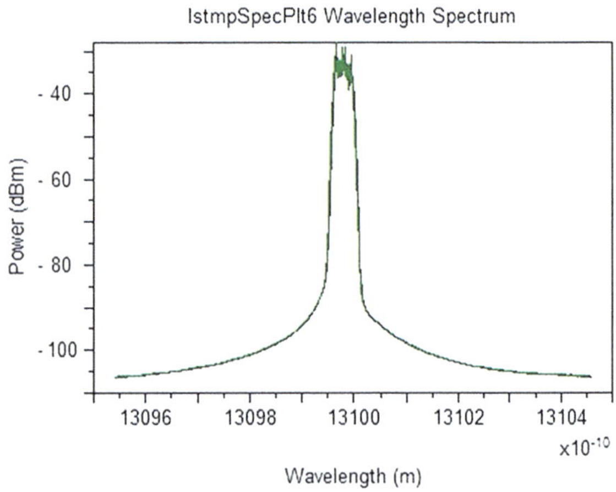

Fig. 44 Video output signal (1310 nm) (TP)

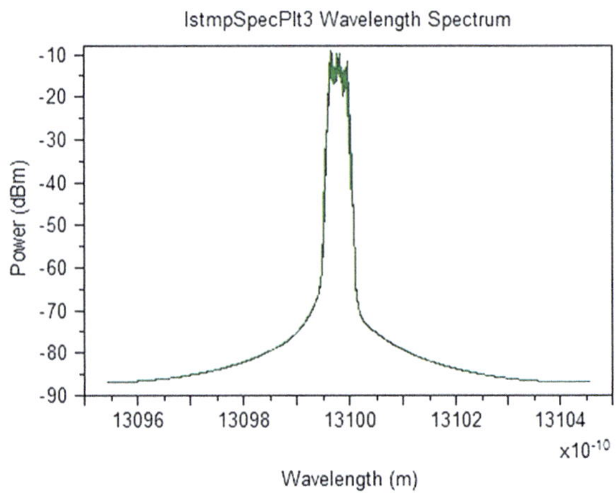

Fig. 45 Video input signal (1310 nm) (QP)

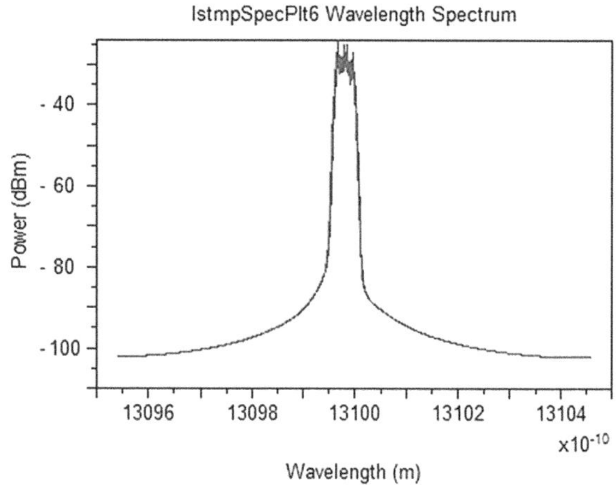

Fig. 46 Video output signal (1310 nm) (QP)

Video (1310 nm):

General Triple Play (TP):

Proposed Quad play (QP):

The video input signal propagates at approximately −10dBm through the OLT and its output signal is at approximately −29dBm at general triple play while approximately −35 dBm at proposed quad play. Here, we are getting a better output signal.

SDH traffic:

If the eye becomes shorter, the noise increased. From the above Fig. 38, it could be said that the SDH traffic carries a few noises. The eye is sharper.

Data and Voice signal for Wavelength 1550 nm and video signal for wavelength 1310 nm (Combinedly).

In this scenario, the voice and data signals are at 1550 nm and the video signal is at 1310 nm. The signal will be modulated at OLT in the general triple play system, but in MUX, at the proposed quad-play system. From the proposed quad-play system, there is an SDH output taken from the ONT's Ethernet port.

Data and Voice (1550 nm):

General Triple Play (TP):

Proposed Quad play (QP):

The signal graph is shown as power losses in dBm versus wavelength. The figure shows that the input signal propagates at 0dBm loss and the out signal gets at − 35dBm loss. But in the proposed quad-play architecture, the output signal gets at approximately −30dBm. Here, the output signal's receiving sensitivity is approximately −35dBm at general triple play whereas approximately −30dBm at proposed quad play.

Video (1310 nm):

General Triple Play (TP):

The video input signal propagates at approximately −10dBm through the OLT and its output signal is at approximately −29dBm at general triple play while approximately −27dBm is at proposed quad play. Here, we are getting a better output signal.

SDH traffic:

If the eye becomes shorter, the noise increases. From the Fig. 47, it be could said that the SDH traffic carries a few noises. The eye is sharper. It is an almost noise-free eye diagram (Figs. 48, 49, 50, 51, 52, 53, 54 and 55).

Data and Voice Signal for Wavelength 1490 nm and Video Signal for Wavelength 1550 nm (Combinedly).

In this scenario, the voice and data signals are at 1490 nm and the video signal at 1550 nm. The signal will be modulated at OLT in the general triple-play system, but in MUX, at the proposed quad-play system. From the proposed quad-play system, there is an SDH output taken from the ONT's ethernet port.

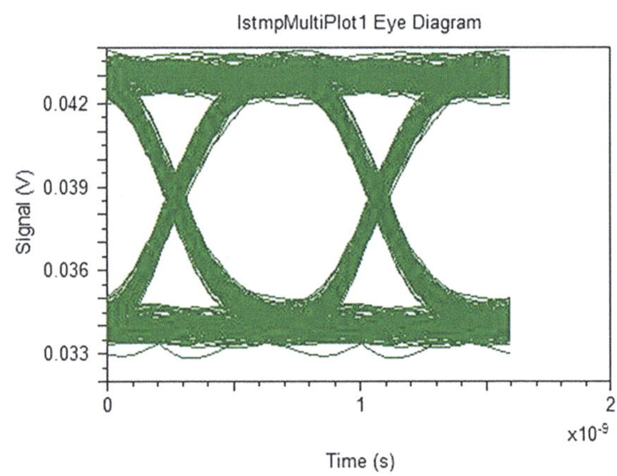

Fig. 47 Eye diagram of SDH traffic (QP)

Fig. 48 Input signal (1490 nm) (TP)

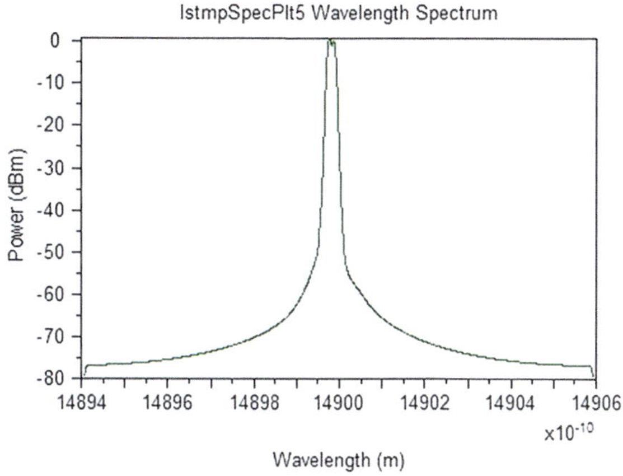

Fig. 49 Output signal (1490 nm) (TP)

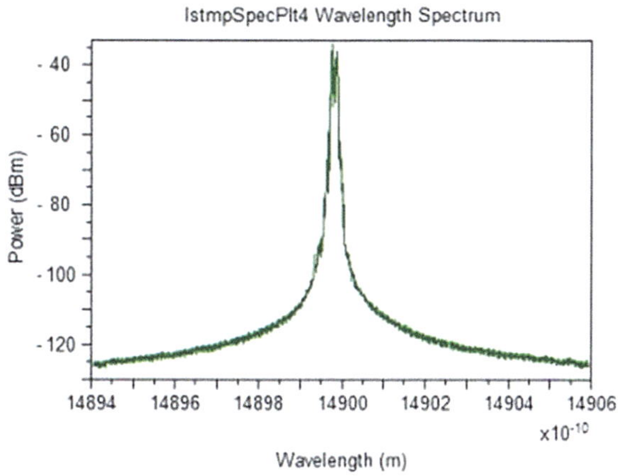

Fig. 50 Input signal (1490 nm) (QP)

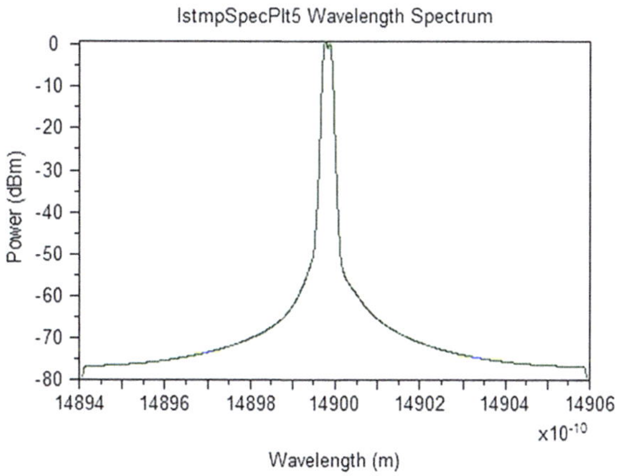

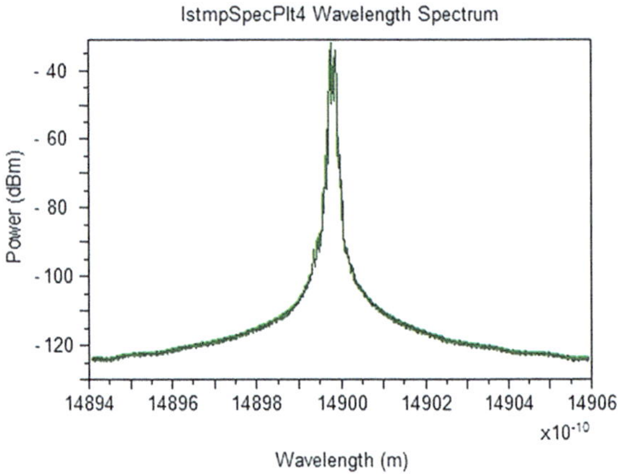

Fig. 51 Output signal (1490 nm) (QP)

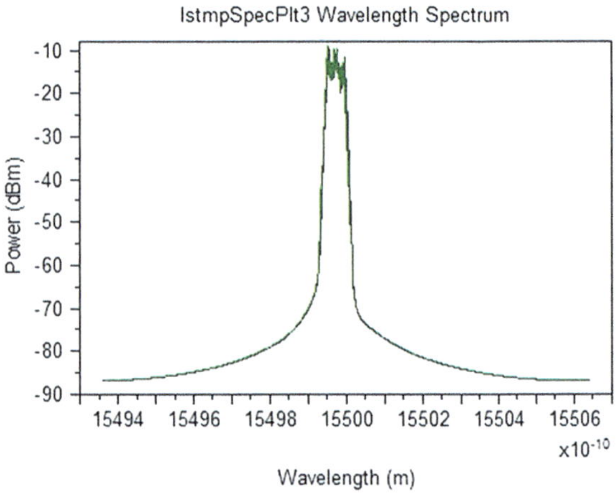

Fig. 52 Video input signal (1550 nm) (TP)

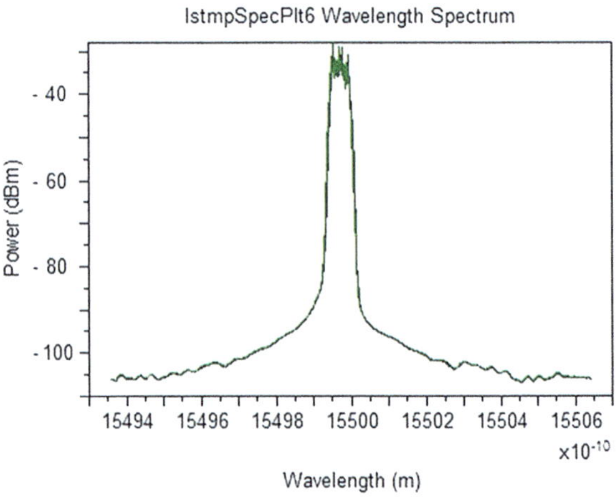

Fig. 53 Video output signal (1550 nm) (TP)

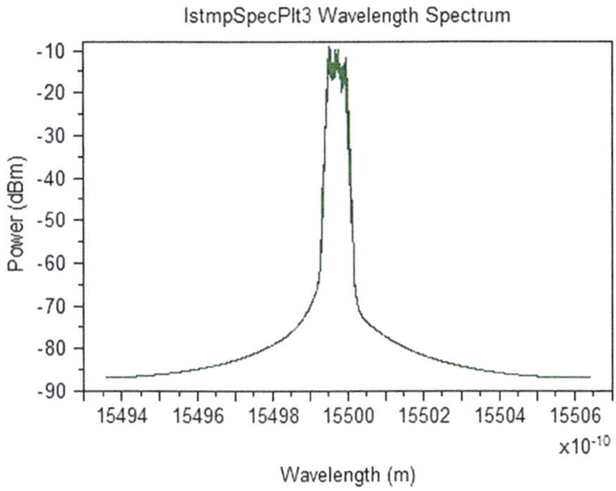

Fig. 54 Video input signal (1550 nm) (QP)

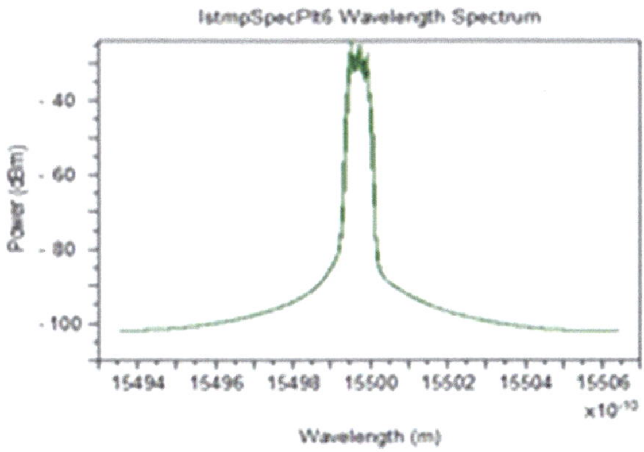

Fig. 55 Video output signal (1550 nm) (QP)

Data and Voice (1490 nm):

General Triple Play:

Proposed Quad play (QP):

The signal graph is shown as power losses in dBm versus wavelength. The figure shows that the input signal propagates at 0dBm loss and the output signal gets at approximately −35dBm loss. But in the proposed quad-play architecture, the output signal gets at approximately -38dBm. Here, the output signal's receiving sensitivity is approximately -35dBm at general triple play whereas approximately −38dBm at the proposed quad play.

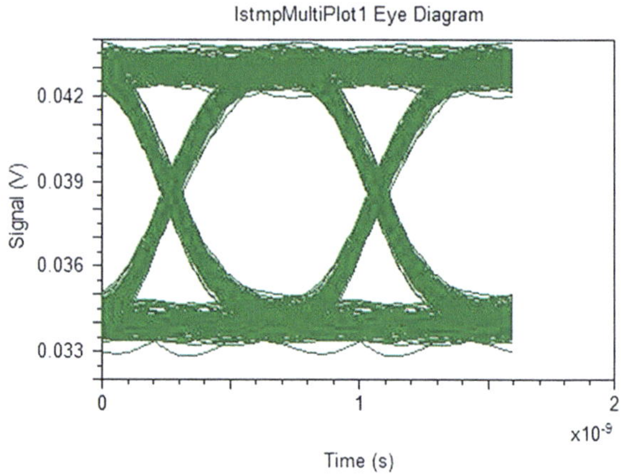

Fig. 56 Eye diagram of SDH traffic (QP)

Video (1550 nm):

General Triple Play (TP):

Proposed Quad play (QP):

The video input signal propagates at approximately −10 dBm through the OLT and its output signal is at approximately −34dBm at general triple play while approximately −27dBm is at proposed quad play. Here, we are getting a better output signal compared to general triple play.

SDH traffic:

In general triple-play facilities, it is difficult to measure the eye diagram for SDH traffic, but in projected quad play, from ONU/ONT's Ethernet port, MUX could connect and easily eye diagram could capture. Almost noise-free traffic has been captured in quad-play architecture (Fig. 56).

6 Conclusion

The primary consideration was to acquire the lowest receiving sensitivity for both triple play and quad play. In recent days, the typical triple-play capability explanation is functioning by a wide variety of wavelengths. Here, 1310 to 1550 nm wavelength is used for triple-play simulation connection. Occasionally, it makes some interruptions within users. It was intense to discover a single wavelength as an alternate of frequent wavelengths. Analysing all probable blends, it is originated that for quad-play construction, 1550 nm wavelength provides a better eye diagram for SDH traffic and better presentation in data, voice, and video package.

References

1. Assi CM, Yinghua Y, Sudhir D, Ali MA (2003) Dynamic bandwidth allocation for quality-of-service over Ethernet PONs. IEEE J Sel Areas Commun 21(9):1467–1477
2. Shami A, Bai X, Ghani N, Assi CM, Mouftah HT (2005) QoS control schemes for two-stage Ethernet passive optical access networks. IEEE J Sel Areas Commun 23(8):1467–1478
3. Santos-Boada G, Domingo-Pascual J (2006) Quality of service in multioperator GPON access networks with triple play services. Trans Eng, Comput Technol 69–73
4. Elias J, Martignon F, Capone A (2006) An efficient dynamic bandwidth allocation algorithm for quality-of-service networks. IFIP International Federation for Information Processing, AN 2006, LNCS 4195, 132–145
5. I-Shyan H, Jhong-Yue L (2011) Adaptive priority scheduling integrated with B-DBA for revenue optimization with QoS and CoS guarantees in GPON. In: Proceedings of the 13th Asia-Pacific Network Operations and Management Symposium, pp 21–23
6. Sultan DMS, Md Taslim A (2011) GPON, the Ultimate pertinent of next generation triple-play bandwidth resolution. J Telecommun Inf Technol 2:53–60
7. Irfan M, Qureshi S, Zafar S (2015) Evaluation of advanced modulation formats using triple-play services in GPON based FTTH. In: Proceedings of the International Conference on Cloud Computing (ICCC), pp 1–6
8. Frnda J, Voznak M, Fazio P, Rozhon J (2015) Network performance QoS estimation. In: Proceedings of the 38th International Conference on Telecommunications and Signal Processing (TSP), pp 1–5
9. Li L, Sh X, De-Gong D (2016) Research of DBA schemes and QoS in PON system. In: Proceedings of the 2nd IEEE International Conference on Computer and Communications (ICCC), pp 2148–2153
10. Tania P, Konstantinos M, Sotirios Ch, Georgios E (2019) A data-driven bandwidth allocation framework with QoS considerations for EONs. J Lightwave Technol 37(09):1853–1864
11. Ali MH, Ali MH (2017) H-DBA for Gigabit passive optical network with performance analysis. In: Proceedings of the 1st International Conference on Computer, Electrical & Communication Engineering (ICCECE-2017), pp 1–5
12. Ali MH, Ali MH (2021) A dynamic bandwidth allocation algorithm for gigabit passive optical network for reducing packet delay and bit error rate, Algorithm for Intelligent System (AIS), Springer of Nature, pp 197–207
13. Ali MH, Ali MH (2021) Wavelength optimization in gigabit passive optical network by proposed quad play architecture, Algorithm for Intelligent System (AIS), Springer of Nature, pp 53–64
14. Ali MH, Ali MH (2021) A priority based deficit weighted round robin queuing for dynamic bandwidth allocation algorithm in gigabit passive optical network, Algorithm for Intelligent System (AIS), Springer of Nature, pp 65–71

Chapter 46
An Enhanced DBA for Supporting Maximum User with Minimum Delay

Md. Hayder Ali and Mohammad Hanif Ali

1 Introduction

High bandwidth and easy access of network could be provided by PON (passive optical network) system. PON system is connected by point to multipoint through the optical network. It is basically connected by an OLT (optical line terminal), splitters, and ONTs (optical network terminals). Splitters are connected through the PON port and splitting ratios are 1: 2^n (Here, n = 1,2,3,4,5,6,7) [1].

Customer requirement for new service is increasing and for that bandwidth requirement is also increasing. Transaction delay minimization with multiple services becoming a basic requirement for smart users [2]. Service operators are bound to find about the novel approach that will allocate the bandwidth among the consumers and afford high network consistency. GPON could be the solution for all that requirements. Formation of the networks in which each associated consumer acquires the best QoS, despite the variation of payment is a too expensive solution. Dynamic bandwidth allocation (DBA) profile is an entity for increasing QoS. An enhanced DBA could minimize network delay and could maintain QoS and SLAs [3].

The rest of this paper is ordered as follows. The literature review is presented in Sect. 2 while proposed DBA is stated at Sect. 3. In Sect. 4, simulation design is briefed and result analysis is itemized in Sect. 5. Finally, Sect. 6 stated a conclusion to this paper.

Md. H. Ali (✉) · M. H. Ali
Department of Computer Science and Engineering, Jahangirnagar University, Dhaka, Bangladesh

G. Mathur et al. (eds.), *Proceedings of 2nd International Conference on Artificial Intelligence: Advances and Applications*, Algorithms for Intelligent Systems,
https://doi.org/10.1007/978-981-16-6332-1_46

2 Literature Review

GPON is a current knowhow but it is blessed with lots of research work. Over the years, many researchers have been working out with the dynamic bandwidth allocation (DBA) algorithm of GPON network construction. Some of them are described below.

C.M. Assi et al. (2003) have proposed to practice the multipoint regulator procedure distinct by the IEEE 802.3ah task force to decide the communication of dissimilar operators, and they described dissimilar dynamic bandwidth allocation (DBA) algorithms to allot bandwidths effectually and equally between end operators. [4]. The proposed DBA algorithms are also enlarged to provision discriminated facilities, a vital prerequisite for a congregated broadband access network with mixed traffic. Chengwei Xiao et al. (2004) have addressed a procedure grounded on circulation estimate is projected, in bid to forecast immediate traffic in ONU, chaperon higher-priority bandwidth apportionment, and avoid "T + 2" queuing delay [5]. Their proposed algorithm was guaranteeing the efficiency and impartiality of the evidence, was quite able to improve delay and jitter in higher-priority facilities as well but this proposal was only for EPONs.

Onn Haran et al. (2008) have described bandwidth distribution in the PON system and highlighted the recompenses of dynamic bandwidth allocation (DBA) [6]. A well-defined DBA algorithm can meaningly progress network presentation, afford resources of compliantly adapting network receptiveness and qualify a facility breadwinner to create more income from their FTTH networks without increasing fresh bandwidth by cumulative the ratio of adequate oversubscription [7]. C-DBA is an algorithm grounded on T-CONT, which allots a static bandwidth to TCONT1 [8]. It guarantees the requirement of high-priority, but is relaxed to spare bandwidth in low load, foremost to a deprived presentation of low-priority, so it cannot encounter the equality prerequisite.

3 Proposed DBA Algorithm

OLT allocates priority traffic and it is the main advantage between existing all DBA and proposed DBA (Fig. 1). It also can calculate the required traffic that needs for the grand frame [9]. To calculate the priority traffic, ONU/ONT sends the grand frame to OLT. OLT then analyzes traffic pattern accept the grand frame for transmitting to ONU/ONT [10].

As per the explanation of different class of T-CONT, the proposed DBA policy mutually encounters the GPON approvals G.984.X and guarantee the precedence of the QoS necessities, which can successfully progress the system performance [10].

P denotes the type of traffic:

P $\in$ { T - $CONT_1$,T - $CONT_2$,T - $CONT_3$, T - $CONT_4$} ;

For class-P traffic,$ONT_i/ONU_i ONT_i/ONU_i$ in the $n^{th} n^{th}$ cycle-

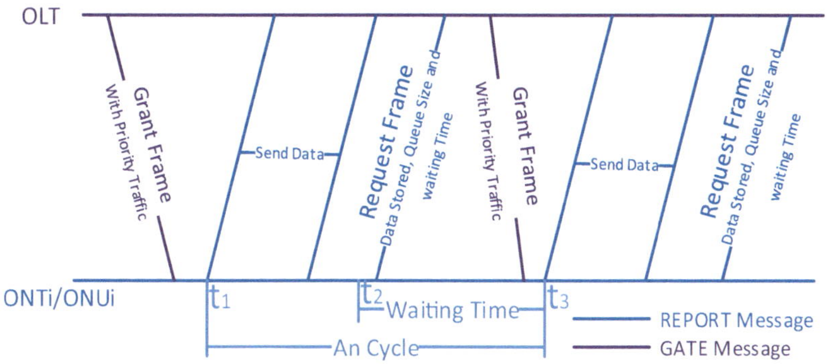

Fig. 1 Proposed DBA pulling mechanism

$Sb^{P}_{i,n}$ → the amount of stored data in the buffer while appeal frame is sent.
$Rw^{P}_{i,n}$ → the amount of data received in waiting time.
$Aw^{P}_{i,n}$ → the amount of prediction arriving in waiting time.
$\mathrm{Re}q^{P}_{i,n}$ → the amount of traffic request.
$Au^{P}_{i,n}$ → the amount of traffic authorization.
M → the arithmetic mean.
Min →T −CONT$_3$'s minimum bandwidth;
Max →Max → T - CONT$_3$'s maximum bandwidth;
Bc_{remain} → current cycle's remaining bandwidth;

T−CONT$_1$*:* It is assigned for bandwidth arrangement and should be finished of T−CONT$_1$. Therefore, the authorization of T−CONT$_1$ circulation is:

$$Au^{T-CONT_1}_{i,n} = Sb^{T-CONT_1}_{i,n} + Aw^{T-CONT_1}_{i,n}$$

T−CONT$_2$*:* Is assigned for flexible bit rate traffic and submissions with both packet delay and required throughput necessities. T−CONT$_1$ and T−CONT$_2$ are the same as for the traffic prediction method. So, the approval traffic of T−CONT$_2$ is:

$$Au^{T-CONT_2}_{i,n} = Sb^{T-CONT_2}_{i,n} + Aw^{T-CONT_2}_{i,n}$$

T−CONT$_3$*:* T−CONT$_3$ is offering better exertion for service delivery with a guaranty of minimum bandwidth (MIN); any extra bandwidth is allocated only on demand and obtainability. the sanction of T−CONT$_3$ traffic is:

$$Au^{T-CONT_3}_{i,n} = Min + \min\Bigg\{\mathrm{Re}q^{T-CONT_3}_{i,n-1} - Min,$$
$$\frac{(\mathrm{Re}q^{T-CONT_3}_{i,n-1} - M) * Bc_{remain}}{\sum_i \mathrm{Re}q^{T-CONT_3}_{i,n-1} - M}, Max - Min\}$$

T−CONT$_4$: T−CONT$_4$ is projected for virtuously finest sweat facilities, consequently, the provision for T−CONT$_4$ is leading when individuals for extra advanced significances is completed and accessible. So, the sanction of T−CONT$_4$ circulation is

$$Au_{i,n}^{T-CONT_4} = \min\left\{\mathrm{Req}_{i,n-1}^{T-CONT_4}, \frac{\mathrm{Req}_{i,n-1}^{T-CONT_4} - M * Bc_{remain}}{\sum_i \mathrm{Req}_{i,n-1}^{T-CONT_4}}\right\}$$

4 Simulation Design

The simulation was designed in three ways, only changing the splitting ratio. (i) Using 1:32 Splitter (ii) 1:64 Splitter and (iii) 1:128 Splitter. This simulation was designed by the OptSIM simulator (Fig. 2).

Video signal (VS) enters into EDFA combiner and ISP (Internet Service Provider) connection directly enters into Layer-2 switch. From the EDFA and Switch, the connections enter into OLT (Optical Line Terminal) [11]. One output line from PON (Passive Optical Network) port enters into a 1:32 optical splitter for first simulation, and then 1:64 optical splitter and next 1:128 optical splitter for the second and third simulation, respectively. In a practical scenario, splitter is situated in LDP (local

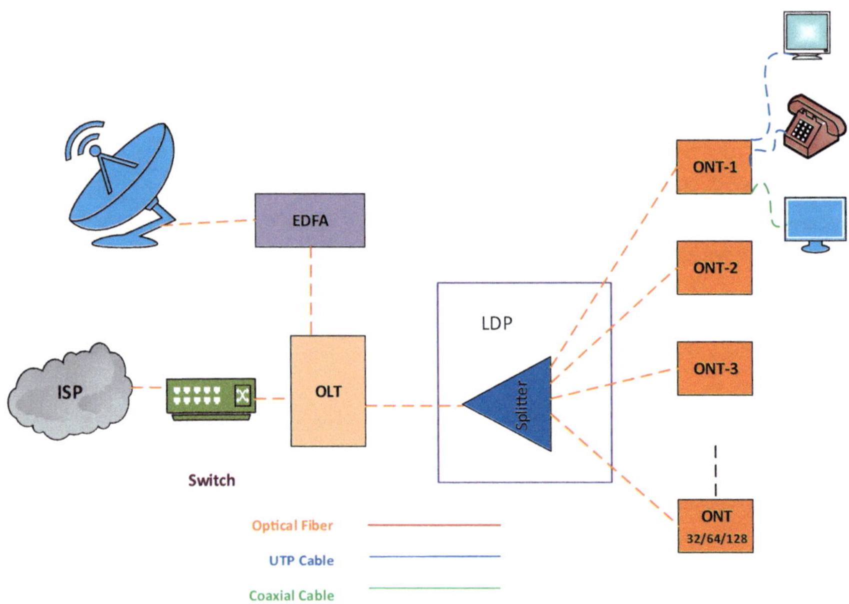

Fig. 2 Simulation Design in OptSIM simulation Software

distribution point). The optical fiber is 12 km long. From the optical splitter, one output fiber was entered into ONT-1 (optical network terminal) [12]. Similarly, to the other ONTs (ONT-2, ONT-3, …., ONT-32/ ONT-64/ ONT-128). In this simulation, up to ONT for all connections, optical fiber (single mode) is used. From ONT's Ethernet port-1, video signal, and similarly, data and voice signal are received from Ethernet port-2 and 3. In this simulation, C-DBA, P-DBA, and the proposed DBA are used for simulation and captured data is analyzed by MATLAB coding (Fig. 3).

Components of OLT: Sine wave generator (2), summer (1), spectrum plot (4), BER tester (1), electrical generator (1), signal plot (4), DML (2), and pre-amplifier (2). One optical MUX is used for modulating the signals.

Components of ONT: Optical filter (2), electrical filter (2), receiver (2), BER tester (1), spectrum plot (3), and multiplot (3).

One optical splitter is used between OLT and ONT [13].

5 Result Analysis

The simulation is designed in three steps and all output data are captured and analyzed by MATLAB coding.

5.1 1:32 Splitting Ration (Up to 32 End User)

Figure 4 shows that while using a 1:32 splitting ratio (maximum 32 users), the existing C-DBA has a maximum delay of 6.8 ms, P-DBA occupied 5.2 ms. But the proposed DBA has less delay and it is 3.5 ms.

5.2 1:64 Splitting Ration (Up to 64 End User)

Figure 5 shows that while using an 1:64 splitting ratio (maximum 64 users), the existing C-DBA has a maximum delay of 10.4 ms, P-DBA occupied 8.0 ms. But the proposed DBA has less delay and it is 5.2 ms.

5.3 1:128 Splitting Ration (Up to 128 End User)

Figure 6 stated that while using an 1:128 splitting ratio (maximum 128 user), the existing C-DBA has a maximum delay of 17.1 ms, P-DBA occupied 12.8 ms. But the proposed DBA has less delay and it is 8.8 ms. Table 1 summuraised the delay comparison between existing DBAs and propossed DBA.

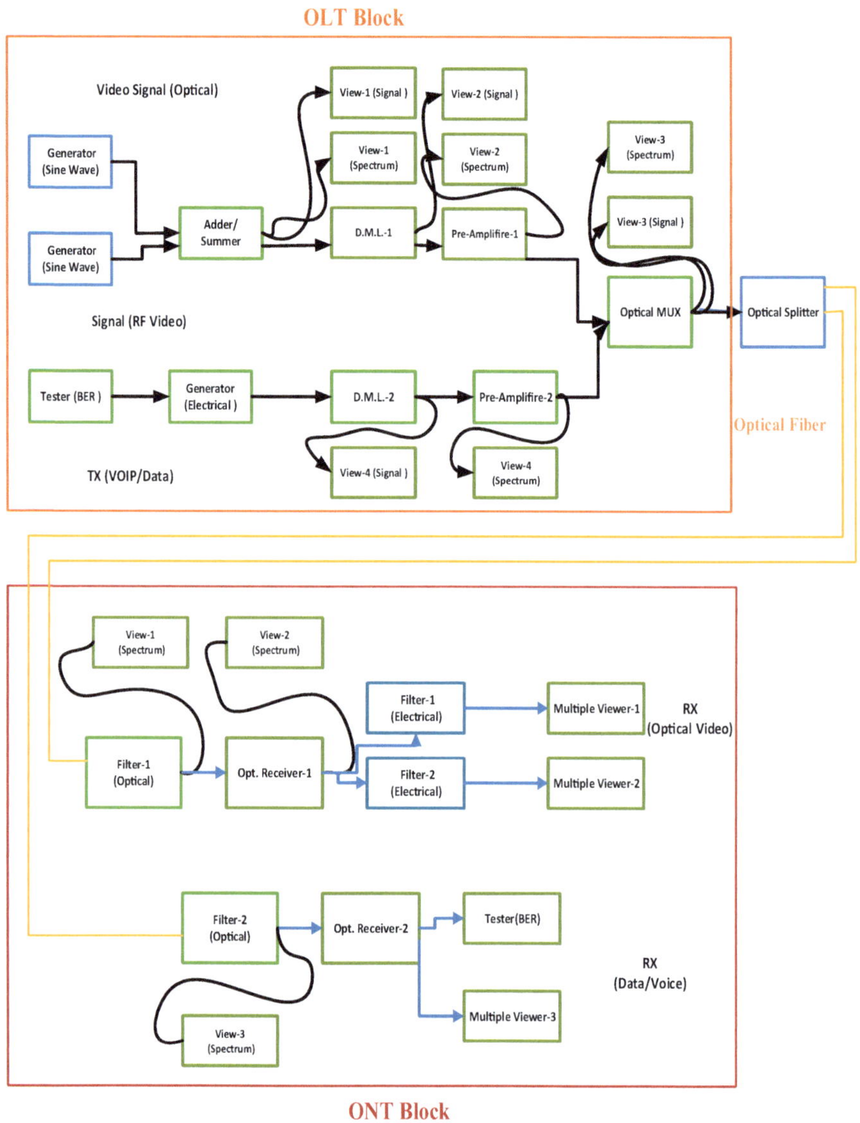

Fig. 3 Block Diagram and Component of OLT and ONT

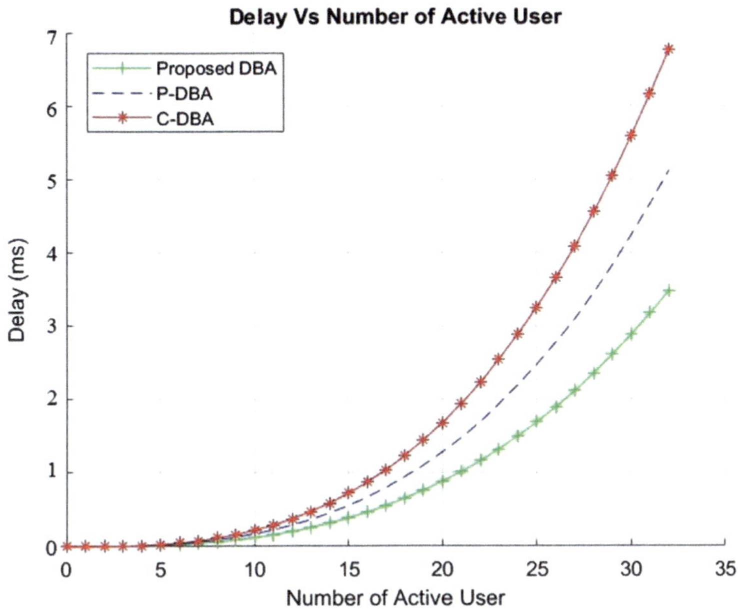

Fig. 4 Delay versus Number of Active users using 1:32 Splitter

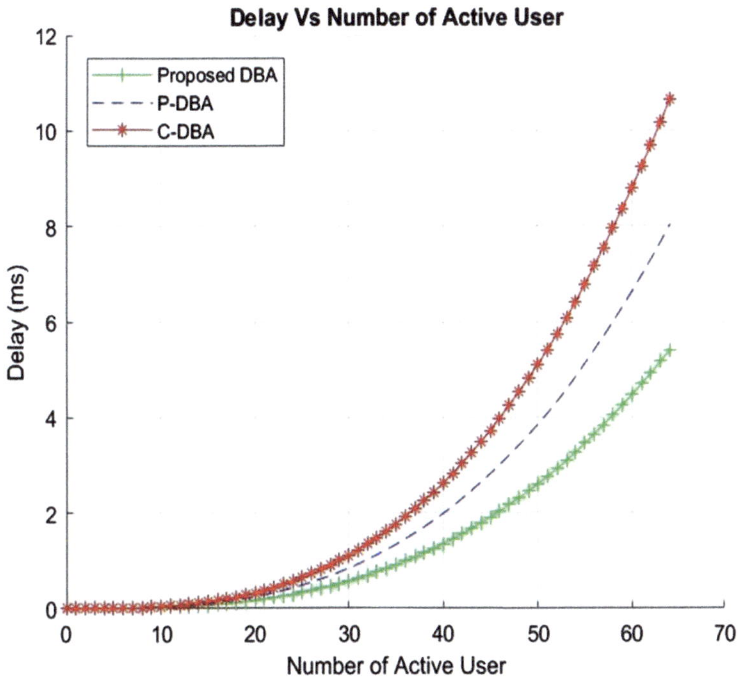

Fig. 5 Delay versus Number of Active users using 1:64 Splitter

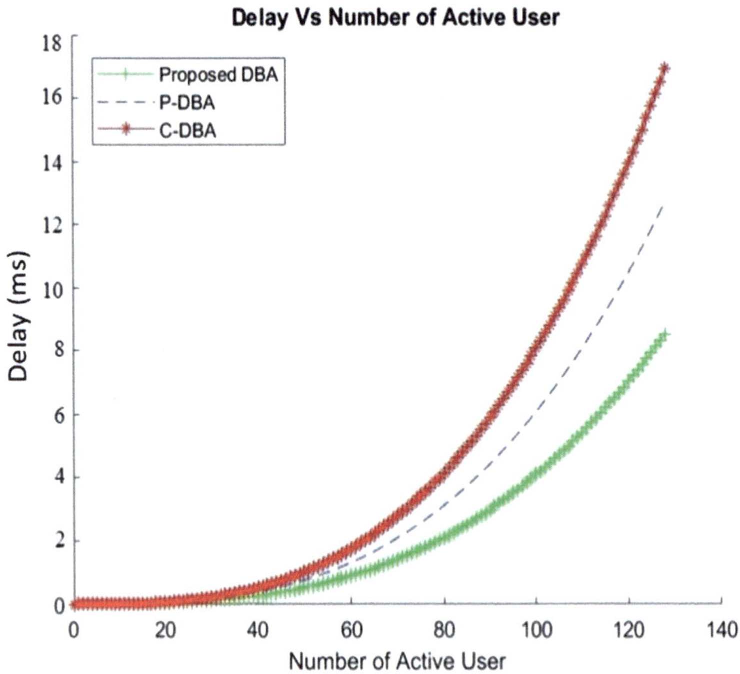

Fig. 6 Delay versus Number of Active users using an 1:128 Splitter

Table 1 Comparison of different splitting ratios and corresponding delay

DBA Profile	1:32 Splitting Ratio (Up to 32 Users) Delay (ms/user)	1:64 Splitting Ratio (Up to 64 Users) Delay (ms/user)	1:128 Splitting Ratio (Up to 128 Users) Delay (ms/user)
C-DBA	6.8	10.4	17.1
P-DBA	5.2	8.0	12.8
Proposed DBA	3.5	5.2	8.8

6 Conclusion

In this research, the main focus was to develop an enhanced DBA which could minimize delay during service to user. Compared to the existing C-DBA and P-DBA, the proposed DBA could support the maximum users with minimum delay. For the 1:32 splitting ratio (for 32 users), C-DBA has 6.8 ms, P-DBA has 5.2 ms, and the proposed DBA has a 3.5 ms delay. For the 1:64 splitting ratio (for 64 user), C-DBA has 10.4 ms, P-DBA has 8.0 ms, and the proposed DBA has a 5.2 ms delay. For 1:128 splitting ratio (for 128 users), C-DBA has 17.1 ms, P-DBA has 12.8 ms, and the proposed DBA has an 8.8 ms delay. Maintaining GPON propriety QoS and SLAs, the proposed DBA has better performance than existing C-DBA and P-DBA.

References

1. Matthew A, Simon CB, Francis D, Predrag RJ, Krishnan K, Ramakrishnan KG, Philip AW (1998) Dynamic bandwidth allocation algorithms for high-speed data wireless networks. Bell Labs Tech J 31–49
2. Rudy H, Kristiaan JV, Saso S (1999) Dynamic bandwidth allocation on SuperPON. In: Proceedings of the SPIE Conference on All-Optical Networking: Architecture, Control and Management Issues, pp 1–8
3. Joanna O, Sarah R, Lars D, Wessing H, Smolorz S (2001) Dynamic bandwidth allocation in GPON networks. J Recent Adv Circuits, Syst, Signal Telecommun 182–187
4. Assi CM, Ye Y, Sudhir D, Ali MA (2003) Dynamic bandwidth allocation for quality-of-service over Ethernet PONs. IEEE J Sel Areas Commun 21(9):1467–1477
5. Chengwei X, Hong F, Dongjing L (2004) Dynamic bandwidth allocation algorithm for ethernet passive optical networks based on traffic prediction. Adv Comput Sci Its Appl 279:649–654
6. Elias J, Martignon F, Capone A (2006) An efficient dynamic bandwidth allocation algorithm for quality of service networks, IFIP International Federation for Information Processing, AN 2006. LNCS 4195:132–145
7. Onn H, Amir S (2008) The importance of dynamic bandwidth allocation in GPON networks. White Paper, PMC-Sierra Inc 1:01–22
8. Xie H, Xiaodong T, Li Z (2005) An algorithm to implement dba of GPON, Access network (06), pp 58–59
9. Qi-yu Z, Bin L, Run-ze W (2012) A dynamic bandwidth allocation scheme for GPON based on traffic prediction. In: Proceedings of the 9th International Conference on Fuzzy Systems and Knowledge Discovery (FSKD 2012), pp 2043–2046
10. Ali MH, Ali MH (2017) H-DBA for Gigabit passive optical network with performance analysis. In: Proceedings of the 1st International Conference on Computer, Electrical & Communication Engineering (ICCECE-2017), pp 1–5
11. Ali MH, Ali MH (2021) A dynamic bandwidth allocation algorithm for gigabit passive optical network for reducing packet delay and bit error rate, Algorithm for Intelligent System (AIS), Springer of Nature, pp 197–207
12. Ali MH, Ali MH (2021) Wavelength optimization in gigabit passive optical network by proposed quad play architecture, Algorithm for Intelligent System (AIS), Springer of Nature, pp 53–64
13. Ali MH, Ali MH (2021) A priority based deficit weighted round robin queuing for dynamic bandwidth allocation algorithm in gigabit passive optical network, Algorithm for Intelligent System (AIS), Springer of Nature, pp 65–71

Chapter 47
Malarial Parasite Detection Based on Smartphone Microscopic Imaging Using Deep Learning Approach

R. Breesha and S. Alagu

1 Introduction

Malaria is a contagious and life-threatening disease normally caused by parasites called plasmodium which are spread by affected female anopheles mosquitoes to humans. Malaria is caused by plasmodium species, namely Plasmodium falciparum, Plasmodium vivax, Plasmodium ovale, and Plasmodium malariae. Malaria disease mostly affects young children and pregnant women easily and sometimes leads to death. Malaria causes great danger to the future generation. Malaria infection initiates when an infected female anopheles mosquito bites a human, it actually injects parasites called plasmodium in sporozoites form. The sporozoites quickly pass into the human being's liver and it multiplies in the cells of the liver for about seven to ten days. Next, the parasites form merozoites and it will pass through the blood, then to erythrocytes and there it will multiply more and more and at last, the cell will burst. The above process undergoes a complex cycle of the mosquito's carrier to another healthy human.

The symptoms of malaria start after the bite of an infected mosquito for about two weeks. Symptoms include high fever, shivering, headache, abdominal pain, profuse sweating, nausea, anemia, vomiting, organ failure, anemia, coma, diarrhea, convulsions, low blood sugar, bloody stools, and muscle pain. According to the world report for malaria released by the World Health Organization (WHO) in December 2019, there were around 228 million cases affected by malaria and around four lakh death over the world in 2018. In that, India accounts for 4% of malaria cases [7].

With the usual manual examination of microscopic blood smears, it is a rigorous physical process requiring proper proficiency in classifying the parasitized and healthy cells. Typically, this examination may not work effectively and might cause problems if the right expertise is not present in particular places throughout the world.

R. Breesha (✉) · S. Alagu
Anna University, Chennai, India

G. Mathur et al. (eds.), *Proceedings of 2nd International Conference on Artificial Intelligence: Advances and Applications*, Algorithms for Intelligent Systems,
https://doi.org/10.1007/978-981-16-6332-1_47

Some improvements have been done in already developed models to take out hand-engineered attributes and build classification models based on machine learning. Anyhow, the models are not flexible with a large number of data being accessible for training and designated the actuality that hand-engineered attributes consume more time.

Among a large variety of computer vision tasks, the model of deep learning, called convolutional neural networks (CNNs) has shown to be more effective. Shortly, the CNN model's important layers are convolution and pooling layers. CNNs support us with automated attribute engineering. Also, at the end of the model, the fully connected layers enable to perform tasks like detection and classification. Automated detection of malaria using the models of deep learning includes CNNs that could be very cheap, scalable, and effectual especially with the emergence of pre-trained models and transfer learning which works remarkably well still with limitations like less number of data [6].

Increasing the depth of the model leads to an increase in the network's accuracy, as much as over-fitting is regarded. The complication is, as the number of layers in the model increases, at some point new weight value will match the old weight value during backpropagation and this is called vanishing gradient. Residual neural networks (ResNet) allow the training of such networks which are deep, by constructing the network through residual models. For effective prediction and to obtain high accuracy, ResNet is used in the proposed method. Some of the recent research works are discussed below.

Pattanaik et al. [1] proposed a multi-magnification deep residual neural network that automatically classified the thin blood smear images acquired by a smartphone into parasitized and uninfected erythrocytes. It provided better results even in low-quality images, resolved the vanishing gradient problem, and achieved 98.08% accuracy.

Feng Yang et al. [2] proposed two methods such as an iterative global minimum screening (IGMS) method, which was used for quick screening of the candidates affected by parasites, and a customized convolutional neural network used to categorize the blood smear images into malarial infected and uninfected RBC cells.

Chowdhury et al. [3] proposed a convolutional neural network and performed a complete blood cell count on thick blood smear images. The goal was to implement on an embedded computer, so they have used modified yolo architecture. Finally, the whole software was executed on a Raspberry Pi 3 single-board microcomputer and also detected the malarial pathogens.

Pattanaik et al. [4] proposed a computer-aided diagnosis and detected the presence of the malarial parasite in microscopic blood smear images. The variables are pre-trained by functional link artificial neural network (FLANN) and followed by another technique called sparse autoencoder (SSAE).

Umer et al. [5] proposed a novel stacked convolutional neural network architecture to automatically detect the presence of malaria. It provided better feature extraction and achieved 99.96% of accuracy. The achieved results proved that convolutional layers can take out different abstract level attributes for classification, with varying sizes of filter and depth.

Shekar et al. [6] proposed a new model of machine learning based on a CNN which automatically classified and predicted the affected RBCs in thin blood smear images. Three types of CNN models were compared based on their accuracy and selected the precise accuracy—VGG-19 Frozen CNN, Basic CNN, and VGG-19 Fine Tuned CNN. Finally, concluded that fine-tuned CNN had provided the higher accuracy.

Nayak et al. [7] compared various deep learning models which are used to identify the presence of malaria. These models have been observed experimentally with respect to the better accuracy and faster resolution and finally concluded that ResNet50 model had provided the highest accuracy of 97%.

Karthik et al. [8] proposed an system that classified the erythrocytes affected by Plasmodium falciparum, Plasmodium malariae, Plasmodium ovale, and Plasmodium vivax in microscopic thin blood smear images. The experimentation had been implemented in MATLAB in which pre-processing including segmentation and classification had been performed.

Dong et al. [9] estimated three types of popular CNNs, including the GoogLeNet, AlexNet, and LeNet. Simulation results exhibited that all these deep CNNs achieved accuracies for classification of over 95%, higher than the accuracy of about 92% achievable by using the support vector machine (SVM) method.

Delas Peñas et al. [10] proposed such an intelligent system to detect malaria parasites through thin blood smear images and used a CNN which achieved an accuracy of 92.4% and sensitivity of 95.2% for detection of the malaria parasite and an accuracy of 87.9% for identifying the two species called Plasmodium vivax and Plasmodium falciparum.

The rest of the work is assembled as follows: Sect. 1 briefs the introductory part of malaria and some of the literary work related to malaria detection, Sect. 2 explains the methodology, Sect. 3 describes the results and discussion and Sect. 4 states the concluding remarks.

2 Methodology

The methodology section describes the malaria detection model from the smartphone microscopic blood smear samples. Figure 1 shows the block diagram of malaria detection. The microscopic blood smear images acquired by smartphones are reshaped into the required size. The reshaped image is passed to convolutional layers and it consists of residual units. Each residual unit is defined by ReLu and batch normalization and passed to a fully connected layer. Finally, the given microscopic images are predicted either as malarial infected or uninfected images.

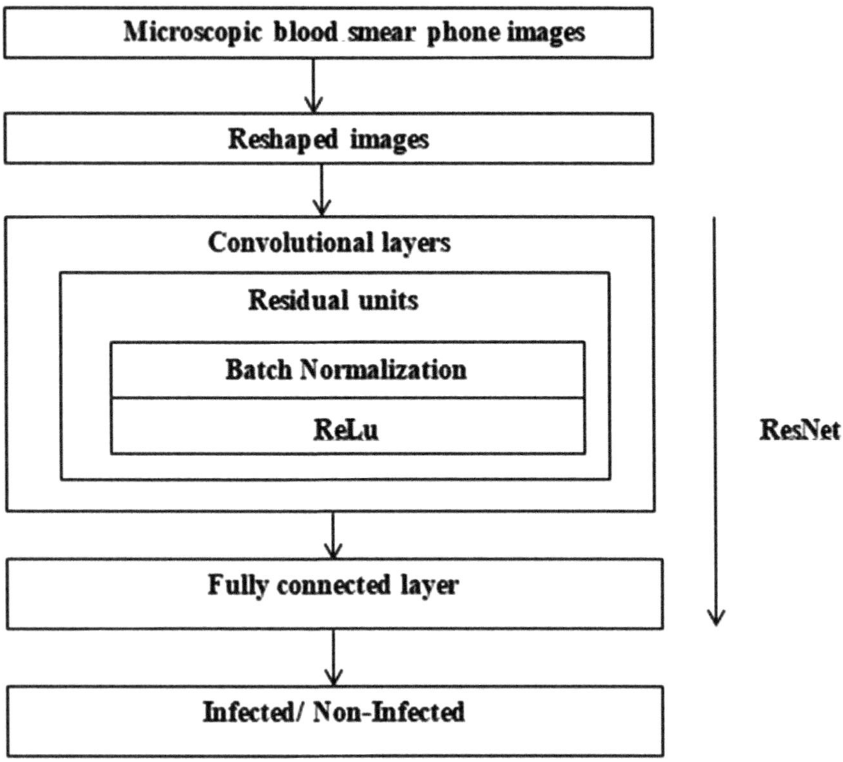

Fig. 1 Block diagram of Malaria Detection using ResNet

2.1 Image Acquisition

Smartphones can produce quantifiable microscope images. The input image datasets are acquired by the smartphone's camera which is affixed to the ocular lens of the microscope. The microscope is adjusted to identify the target area in the blood smear and images are taken. The malarial infected images are collected from a public database and collected real-time uninfected images from Government Mohan Kumaramangalam medical college hospital, Salem.

2.2 Pre-processing

The images in the database are subjected to the pre-processing stage. As the microscopic blood smear images are acquired by smartphones, they are unequal in size. The images are reshaped into equal size.

2.3 Convolutional Layer

The convolutional layer plays an indispensable role in how Convolutional Neural Networks operate. The convolutional layer consists of residual units. Each residual unit consists of batch normalization and Rectified Linear unit (ReLu).

2.4 Residual Neural Network

ResNet allows to train the exceedingly deep neural networks with more than 150 layers successfully. Deep neural networks are difficult to train due to the problem called vanishing gradient, as the weight is propagated backward to previous layers, replicated multiplication makes the weight very small. As the deepness of the network becomes more, the network's performance gets saturated or begins to degrade fastly. ResNet has a major advantage called skip connection. Skip connection means adding the actual input to the convolution block's output. But it can be added only if they possess identical size. Hence whether the convolution along with batch normalization operations are performed in a way that size of the output is the same as the input size, it can be simply added. Skip connection is used before rectified linear unit (ReLu) activation for best results. The mathematical formulation of ResNet is expressed as

$$F_{m+1} = g_c(F_{l\to m}, K_{l\to m}) + F_l \quad (1)$$

$$F_{m+1} = g_a(F_{m+1}) \quad (2)$$

Where

In Eq. (1), $k_{l\to m}$ indicates the kth processing unit which is the filter, whereas l-> m indicates the residual blocks with one or more concealed layers. Actual input F_l is summed up to transformed signal $g_c(F_{l\to m},K_{l\to m})$ through shortcut pathway, and thus provides an aggregated output F_{m+1}, which is passed to the upcoming layer after using the activation function $g_a(.)$. Whereas, $(F_{m+1} - F_l)$ gives a information of residual, which is used to carry out weights optimization. ResNet recommended that functions of residual are simple to optimize and accuracy can be gained for increased depth. In ResNet, skip connections or shortcut pathway connections are introduced within the layers to allow cross-layer connectivity. In highway networks, the layers represent non-residual functions when a shortcut pathway is closed. Shortcut connections accelerate the confluence of deep networks, thus offering ResNet the capability to avoid the problem called vanishing gradient.

The major advantage of ResNet is ResNet resolves the vanishing gradient problem by using skip connection. It adopts batch normalization after every convolution which helps to accelerate the training process and it makes the computation faster [12].

2.5 Batch Normalization

During training, one of the weights in the network may be drastically larger than another weight. This large weight causes the output from its corresponding neuron to be extremely large and this imbalance causes instability in the neural network. Normalization occurs per batch basis, so it is called batch normalization. It makes the network more stable during training and speeds up the learning process.

2.6 Rectified Linear Unit

The activation function used in the proposed method is ReLu. In backpropagation, when the output is greater than zero, then the derivative value of ReLU is one. If the output is less than zero, then the derivative value of ReLU is zero. The weights will get converge easily and it will not have a vanishing gradient problem.

$$\text{Relu (y)} = \max(0, \text{y}) \tag{3}$$

2.7 Fully connected layer

The fully connected layer consists of neurons, which are linked straightly to the neurons in the two adjoining layers, without being linked to any other layers within them [11]. It gives the predicted output either as a malarial infected image or an uninfected image.

3 Results and Discussions

The malaria parasite detection is done for the microscopic blood smear images which are acquired by smartphone using a convolutional neural network. The input images are trained and tested using the ResNet model. The images are resized for compatibility with the proposed model. The optimization technique is used for the purpose of fine-tuning the model. The optimization technique which is used here is the SGD optimizer. In the total image dataset, 28% have been used for validation and 24% used for testing, and the majority of the dataset have been used training. The images are then classified based on the features. The results of classified images are discussed.

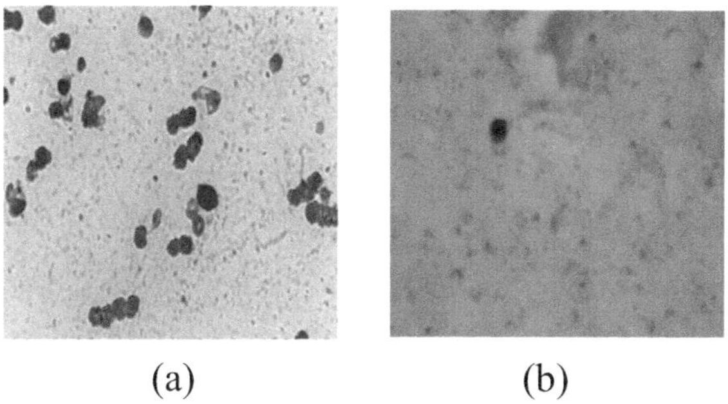

Fig. 2 Blood smear images (**a**) Healthy Blood Smear image (**b**) Malarial infected Blood Smear image

3.1 Inputs

The malarial infected images are collected from an online database and collected real-time uninfected images from the hospital. The malaria-infected thick blood smear images (450 images) are taken from (http://air.ug/downloads/plasmodiumimages.zip). The uninfected thick blood smear images (175 images) are taken from Government Mohan Kumaramangalam medical college hospital, Salem.

In Fig. 2, the blood smear images of healthy and malaria red blood affected cells are shown. In an uninfected thick blood smear image, healthy red blood cells can be noticed. In an infected thick blood smear image, the red blood cells are ruptured. The accole ring form of trophozoites can be noticed. Those structures are called plasmodium falciparum.

3.2 Pre-processing

Pre-processing is the image resizing process. The images are acquired by smartphone, each image is different in size. In order to uniform the sizes of all input images, changing all the microscopic blood smear image sizes into equal size of 1024.

3.3 Prediction

Accuracy and Loss

Figure 3 shows the accuracy graph for the training and validation dataset and Fig. 4 shows the accuracy graph for the training and validation dataset. For training and validating the data, all the images in each set are processed 30 times individually of forward and backward to the network which means 30 epochs. After the 20th epoch, training and validation accuracy started to increase. From Figs. 3 and 4, it is

Fig. 3 Accuracy plot

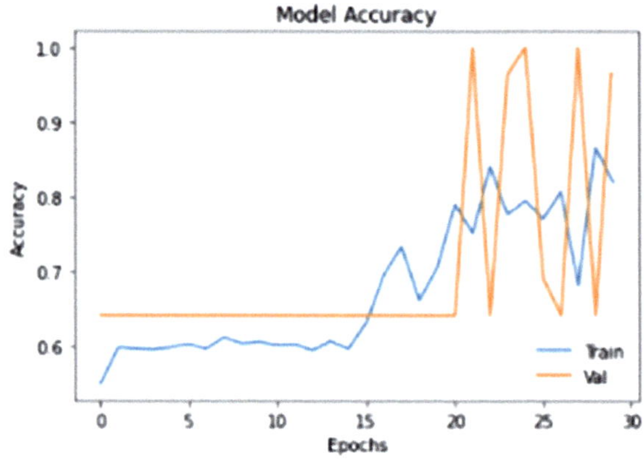

Fig. 4 Loss plot

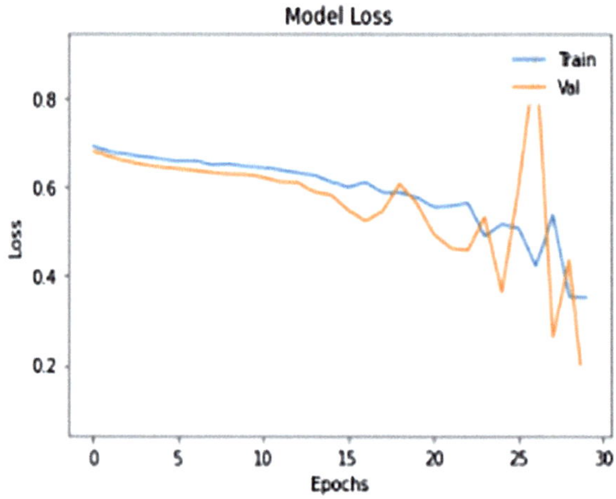

inferred that as the training and validation accuracy increases its corresponding loss decreases or vice-versa.

Prediction output

Figure 5 shows the predicted output of microscopic blood smear phone images such as malaria-infected and non-infected. This method of classification using a convolutional neural network provides more accuracy and is easily detectable.

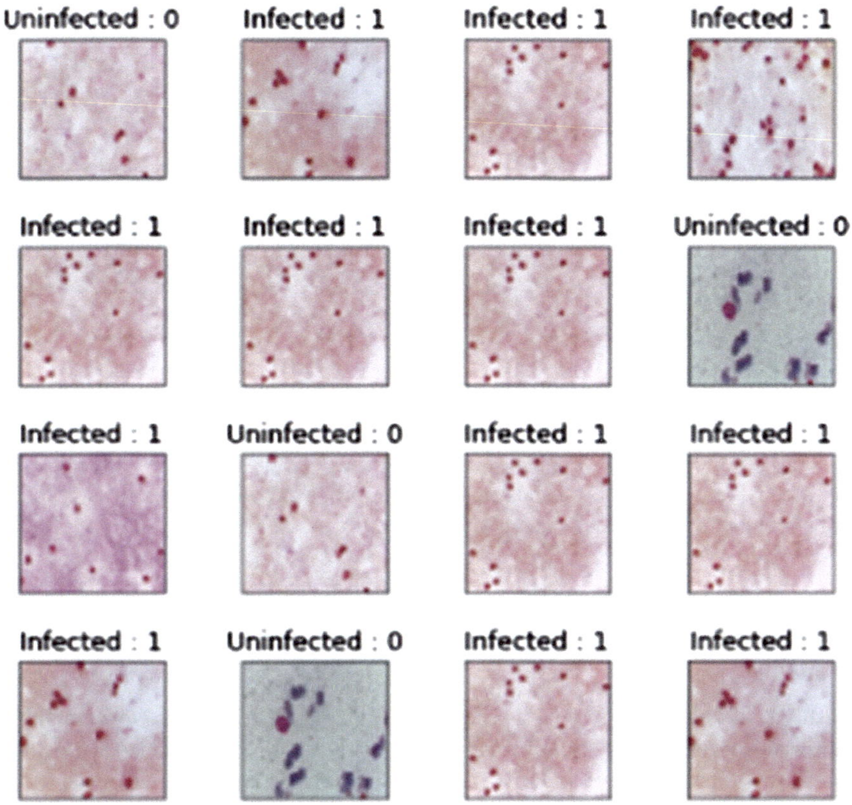

Fig. 5 Prediction output

3.4 Performance Metrics

Confusion Matrix

Confusion Matrix provides an output matrix that represents the total model's performance. Figure 6 infers that 38 samples are predicted correctly as negative, 6 samples are predicted falsely as positive, 16 samples are predicted falsely as negative, and 90 samples are correctly predicted as positive.

Precision

Precision is the proportion of the amount of correct malaria positive results to the amount of malaria positive results predicted by the classifier. In the proposed work, 86.41% of data are predicted as positive.

$$\text{Precision} = \frac{True\,positives}{Truepositives + False\,positives} \tag{4}$$

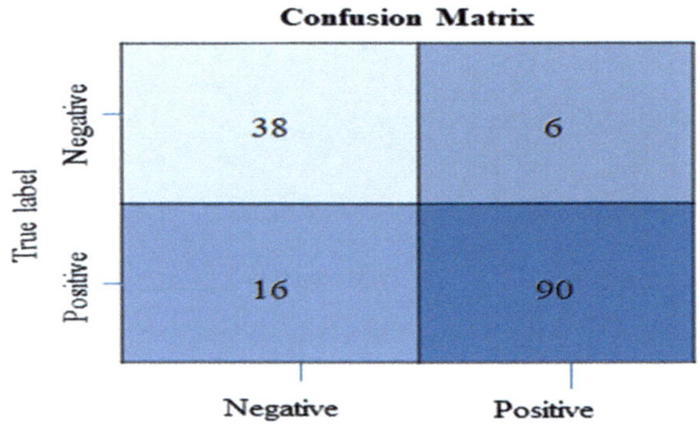

Fig. 6 Confusion matrix of testing

F score

F1 Score is used to measure a model's test accuracy. It shows how precise the used classifier is, as well as how robust it is. In the proposed work, F-score is 85.65% which means the test model's accuracy is high.

$$\text{Fscore} = 2 * ((\text{precision} * \text{sensitivity})/(\text{precision} + \text{sensitivity})) \quad (5)$$

Sensitivity

Sensitivity is otherwise known as the true positive rate. It is the ratio of malaria positive results that are accurately regarded as positive, with correspond to all positive results. In the proposed work, 84.90% of infected images are predicted correctly as positive.

$$\text{Sensitivity} = \frac{True\ positives}{True\ positives + False\ negatives} \quad (6)$$

Specificity

Specificity is otherwise known as the true negative rate. It is the ratio of malaria negative results that are accurately regarded as negative, with correspond to all negative results. In the proposed work, 86.31% of non-infected images are predicted correctly as negative.

$$\text{Specificity} = \frac{True\ Negative}{True\ Negative + False\ Positive} \quad (7)$$

Accuracy

Accuracy measures the proportion of correct malaria predictions over the total number of samples estimated. In the proposed work, 86.14% of data has been predicted correctly.

Table 1 Comparison of performance metrics with different learning rates

Learning Rate	Accuracy (%)	Precision (%)	F Score (%)	Sensitivity (%)	Specificity (%)
0.001	66.04	66.40	65.46	64.52	66.75
0.1	74.28	74.82	73.83	72.85	74.26

$$\text{Accuracy} = \frac{True\,positive + True\,negative}{True\,positive + True\,negative + False\,positive + False\,negative} \quad (8)$$

The above performance metrics have been evaluated for a learning rate of 0.01. By changing the learning rate to 0.001, 0.1, the performance metrics have been evaluated for the proposed model which is shown in Table 1 and Fig. 7. From the comparison, it is observed that at a learning rate of 0.01, the proposed model's performance is better than the other learning rates.

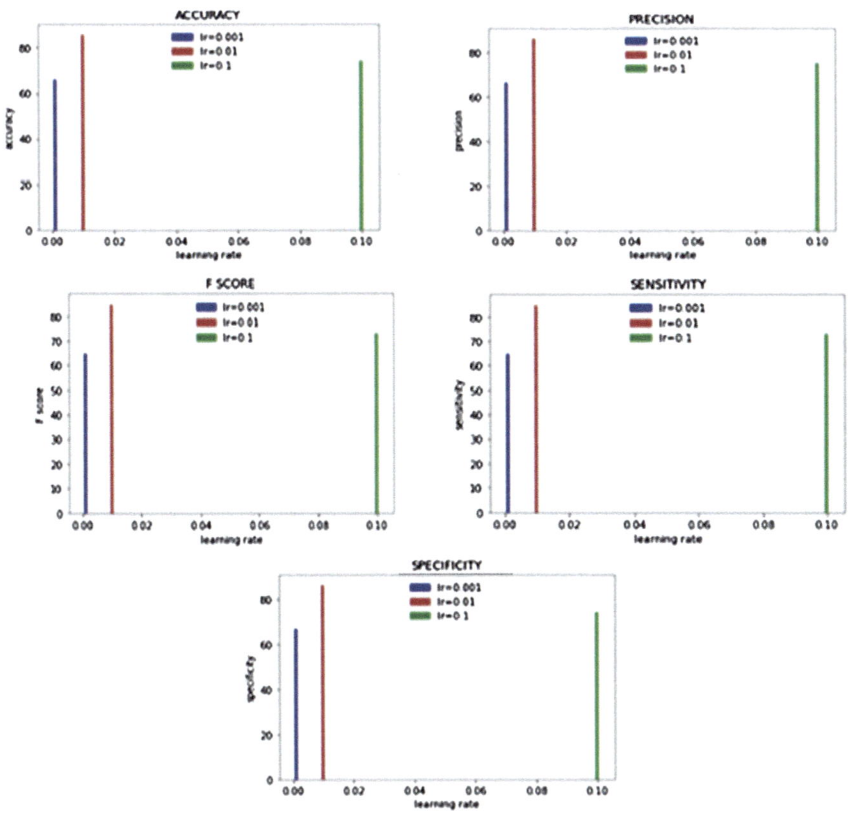

Fig. 7 Graph for comparison of performance metrics with different learning rates

Table 2 Comparison of different baseline methods

Methodology	Accuracy (%)	Precision (%)	F Score (%)	Sensitivity (%)	Specificity (%)
AlexNet	83.78	83.92	82.63	81.42	83.68
VGG-16	81.13	81.87	80.14	79.94	81.72
DenseNet	84.42	84.14	83.48	82.36	84.78

The proposed work has been compared with other baseline methods such as AlexNet, VGG-16, DenseNet. From the performance metrics of ResNet and Table 2, it is inferred that the ResNet model has provided better accuracy than other baseline methods.

4 Conclusion

To detect malarial parasitized images, the ResNet model has been proposed. In the proposed work, the images are passed through a convolutional layer consisting of residual units which are defined by ReLu and Batch normalization. Finally, proceeded by fully connected layer to give the predicted output either malarial infected or uninfected images. On training and testing the data, loss and accuracy graphs have been plotted. The images are predicted as malarial infected and non-infected with moderate accuracy. The performance metrics of the model have been evaluated. Further, this model can be used to detect various diseases.

References

1. Pattanaik PA, Mittal M, Khan MZ, Panda SN (2020) Malaria detection using deep residual networks with mobile microscopy. J King Saud Univ–Comput Inf Sci ISSN 1319–1578. https://doi.org/10.1016/j.jksuci.2020.07.003
2. Yang F et al (2020) Deep learning for smartphone-based malaria parasite detection in thick blood smears. IEEE J Biomed Health Inform 24(5):1427–1438. https://doi.org/10.1109/JBHI.2019.2939121
3. Chowdhury AB, Roberson J, Hukkoo A, Bodapati S, Cappelleri DJ (2020) Automated complete blood cell count and malaria pathogen detection using convolution neural network. IEEE Robot Autom Lett 5(2):1047–1054. https://doi.org/10.1109/LRA.2020.2967290
4. Pattanaik PA, Mittal M, Khan MZ (2020) Unsupervised deep learning CAD scheme for the detection of malaria in blood smear microscopic images. IEEE Access 8:94936–94946. https://doi.org/10.1109/ACCESS.2020.2996022
5. Umer M, Sadiq S, Ahmad M, Ullah S, Choi GS, Mehmood A (2020) A novel stacked CNN for malarial parasite detection in thin blood smear images. IEEE Access 8:93782–93792. https://doi.org/10.1109/ACCESS.2020.2994810
6. Nayak S, Kumar S, Jangid M (2019) Malaria detection using multiple deep learning approaches. In: Proceedings of 2019 2nd International Conference on Intelligent Communication and Computational Techniques (ICCT), Jaipur, India. pp 292–297. https://doi.org/10.1109/ICCT46177.2019.8969046

7. Karthik G, Muttan S, Saravanan MP, Seetharaman R, Vignesh V (2019) Automated malaria diagnosis using microscopic images. In: Proceedings of 2019 Third International Conference on Inventive Systems and Control (ICISC), Coimbatore, India. pp 514–517. https://doi.org/10.1109/ICISC44355.2019.9036458
8. Shekar G, Revathy S, Goud EK (2020) Malaria detection using deep learning. In: Proceedings of 2020 4th International Conference on Trends in Electronics and Informatics (ICOEI) (48184), Tirunelveli, India. pp 746–7. https://doi.org/10.1109/ICOEI48184.2020.9143023
9. Dong et al Y (2017) Evaluations of deep convolutional neural networks for automatic identification of malaria infected cells. In: Proceedings of 2017 IEEE EMBS International Conference on Biomedical & Health Informatics (BHI), Orlando, FL. pp 101–104. https://doi.org/10.1109/BHI.2017.7897215
10. Delas Peñas KE, Rivera PT, Naval PC (2017) Malaria parasite detection and species identification on thin blood smears using a convolutional neural network. In: Proceedings of 2017 IEEE/ACM International Conference on Connected Health: Applications, Systems and Engineering Technologies (CHASE), Philadelphia, PA. pp 1–6. https://doi.org/10.1109/CHASE.2017.51
11. O'Shea K, Nash R (2015) An Introduction to Convolutional Neural Networks, Department of Computer Science, Aberystwyth University, Ceredigion. arXiv:1511.08458 [cs.NE]
12. Khan A, Sohail A, Zahoora U, Qureshi AS (2015) A survey of the recent architectures of deep convolutional neural networks. Published in Artificial Intelligence Review, arXiv:1901.06032 [cs.CV]
13. Joshi AM, Das AK, Dhal S (2020) Deep learning based approach for malaria detection in blood cell images. In: Proceedings of 2020 IEEE region 10 conference (TENCON), Osaka, Japan, pp 241–246. https://doi.org/10.1109/TENCON50793.2020.9293753
14. Gouda N, Amudha J (2020) Skin Cancer Classification using ResNet. In: Proceedings of 2020 IEEE 5th International Conference on Computing Communication and Automation (ICCCA), Greater Noida, India. pp 536–541. https://doi.org/10.1109/ICCCA49541.2020.9250855
15. Mohammed HA, Abdelrahman IAM (2017) Detection and classification of Malaria in thin blood slide images. In: Proceedings of 2017 International Conference on Communication, Control, Computing and Electronics Engineering (ICCCCEE), Khartoum. pp 1–5. https://doi.org/10.1109/ICCCCEE.2017.7866700

Chapter 48
LoRa-Based Sensing Network Setup and IoT Integration for Smart Agricultural Management

Aruna Singh and Roopam Gupta

1 Introduction

IoT technology provides the interconnection of sensors with an object built with computing, communication, and sensing capabilities like in the case of multiple-intelligent sensing node networking that we have proposed in the paper. The development of the IoT is highly facilitated as its execution in different industries such as smart agriculture, smart city, smart factory, smart health care, etc. LoRa is designed as one of the important communication technologies in the field of IoT applications, specifically for long range, low- power communications, high bandwidth, etc. It provides long-range connectivity by using the LoRa mesh network. agriculture has advanced with the extension of IoT in the direction of automation, intellectualization and smart communication. Smart agriculture system with advance communication technology can allow utilization of water efficiently, soil moisture condition, sensor data coordination in the precision place at the appropriate time when needed and in the right amount. It also helps to optimize power consumption and labor costs. Whereas, the smart agriculture system provided with GPRS has the major problems of high battery power consumption and high maintenance and deployment cost. Contrarily, smart systems which use WiFi and Zigbee have the problem of network coverage. As a whole, LoRa Mesh Network has the capacity to cover a range up to 5–10 km in an open area which ensures the high coverage of the irrigation system. As LoRa device consumes low power, it can be easily and efficiently operated on a battery. With help of LoRa smart and durable system can be develop which brings great benefits such as water-saving, smart communication and minimum costs of maintenance and system deployment. Therefore, a smart agriculture system based on LoRa technology with the integration of IoT is proposed here. It will be a great solution to all those

A. Singh (✉) · R. Gupta
Department of Computer Science & Engineering, UIT, Rajiv Gandhi Proudhyogiki Vishwavidyalaya, Bhopal, M.P, India

G. Mathur et al. (eds.), *Proceedings of 2nd International Conference on Artificial Intelligence: Advances and Applications*, Algorithms for Intelligent Systems,
https://doi.org/10.1007/978-981-16-6332-1_48

problems which are mentioned above. The proposed Lora and IoT-based system will be capable of communication between irrigation devices like sensors, controllers, etc., and mobile applications through LoRa module and its mesh network. The main purpose behind the development of the smart system is to allow applications that can control the agriculture system through cloud Agriculture nodes, which will send its status information to the gateway, and all the status information of each node in the system will be forwarded to cloud for further processing and storage. Applications can send various commands to control the agriculture system with the help of cloud-based Application Programming Interfaces (APIs). Furthermore, relays and solenoids in agriculture nodes can be operated from send commands. In the proposed work, the implementation and design part of the LoRa-based smart agriculture system is explained in more detail. Reports of the International Benchmarking Network for Water and Sanitation Services (IBNET) and other surveys show that about 35% of usable water is lost due to inappropriate channelization and conventional distribution methods. The Utilization of water can be classified broadly into three categories: Residential, Industrial, and Agricultural, out of which agriculture consumes a maximum percentage of available water. Water utilization for the agricultural sector is highly unorganized and also maintenance, monitoring, and controlling in real time become tedious as the consumers are located on high ground areas that are very far from supply stations. Technology applications and computational methodologies are needed to be indulged for managing, optimizing, monitoring, and controlling the network. In general, multiple automated and control systems are already available through which several applications can be designed including water management. In different time zones, with gradual improvement in technology, updated systems have been introduced, but in the past few decades, technology has been changed abruptly and water crises have arisen at an unnatural pace. Currently, more efficient system demand has taken place for water supply management in the agricultural sector. It has been observed that almost all commercially available systems are based on wired technology which is comparatively bulky with the majority of components installed on the field, therefore, it requires close monitoring and a lot of maintenance. Programmable logic controller (PLC) is widely accepted worldwide for automation systems. On the one side, it carries all networking applications and other processing capabilities, but on the second side, it is the most expensive technology and mostly works in the local area with wired configuration. Internet of thing (IoT) is the most suitable tool that is capable of connecting clients, users, and administrators with precise wireless sensing information and access methodologies. Transfer of information between electronic devices is possible through IoT only and is helpful for control applications. Around 2.1 billion people in the world suffer from safe drinking water and today agricultural demands are much higher than this as per report data shown by WHO in 2017. So from IoT, it is expected that developers and researchers ensure the quality and wastage of water in agricultural land and distribution network. A lot of traditional ways of measuring the water distribution parameters by collecting water datasets are available, but in case of real time, data collection from water distribution network system is hard to manage and monitor, and still required more research due to several factors such as: the data is heterogeneous, time investment is needed for the collection of data, requires processing energy and connectivity, and coverage of all the nodes in the proposed network. IoT has the capability to combine many other readily available technologies such as WSN, AI, Machine learning, etc. Instant information and alert make users capable to take action on quick notice. A smart water distribution network for agricultural needs should be planned in initial from its architecture to its functional execution by studying, analyzing and comparing multiple technologies, hardware equipment, associated cost and methods to build a smart water system. The paper defines the framework of smart water management for the agriculture sector with integrated monitoring and controlling. All of this is possible because of an innovative concept of IoT and their application that is cost-efficient, reliable, fastest dual side data communication possibilities, and also capable enough to involve a network of sensors to collect data automatically at node points defined strategically. These nodes are facilitated with communication devices for long-range data propagation independent of any interference and dependency on wiring systems. Programming adaptability makes the system more flexible and allows it to adjust with field requirements in a more appropriate way. Microcontrollers are being suggested for this purpose,which are responsible for establishing connectivity between sensing nodes and end-use functional devices, that too without any human intervention. The best part is that the activity can be monitored by authorities at any time and the possibility to change the functionality according to current situations is always remaining open. Water distribution network for agricultural activities is comparatively unorganized and highly inefficient in India. We are utilizing 108 m ha out of 139.89 m ha total potential of irrigation. An average of 1829 cubic meters is the per capita availability of water annually at present and it is continuously declining. Irrigation carries an 80 percent share of the total water supply in the country. We have a storage capacity of 207 CUM per capita, whereas in China, it is 1,111 CUM. So it is more crucial to put efforts toward a better, efficient water distribution system concentrated on agricultural needs. Agricultural fields are mostly located on the outer area of any residential zone. So the environmental conditions are not very favorable for human-designed systems to fulfill various tasks. Most wired systems have been developed and fully functional at the commercial level till now. These systems have very high cost as compared to the economic status of the majority of farmers in the country. Also, it requires heavy maintenance. So it is worth researching on wireless system for the purpose and bringing commercially viable products. The problems that have been sensed while designing the wireless systems are security, reliable sensory feedback integration, communication, and easy to access infrastructure.

2 Literature Survey:

[1] "An experimental setup of multi-intelligent control system (MICS) of water management using the Internet of Things (IoT)".

In this paper, a new system named as multi-intelligent control system also known as MICS has been introduced and used for water management and distribution in the agricultural sector. The design of MICS has been discussed and its testing done successfully. This system consists of three main control systems which cover the control system of the water reservoir, the control system of electro-pump, and the control system of the water pump station. All these three control systems were based on IoT technology and used to control and monitor the water consumption and the level of reservoir water by advanced electronic and intelligent components. The paper structure described briefly about water management background with multiple control systems. Also, described about internet layout of MICS for proposed infrastructure and mathematical structure in order to find out the electro-pump capacity, multiple electronic used components and their respective control systems. Here five-layer structure has been defined which are as follows: smart internet layouts, smart communication, smart security, smart monitoring, smart water level indicator, and smart alerts in the form of alarm.

[2] "Monitoring of Large-Area IoT Sensors Using a LoRa Wireless Mesh Network System: Design and Evaluation"IEEE TRANSACTIONS ON INSTRUMENTATION AND MEASUREMENT 2019

In this paper, it is shown that the PDR ratio increased in a mesh network without additionally installing more than one gateway node. Also, analyzed the study of PDR and RSSI and PDR and SNR. Results show that the suggested threshold for SNR and RSSI is very important for maintaining the communication quality. But in case of LoRa network, such a threshold is impractical. Additional gateway nodes increase the deployment cost of the system, so here trying to reduce its cost by using a single gateway node to achieve the ideal RSSI and SNR. The practical evaluation of mesh network for long-range area is tested here and also explored the deployment of IoT sensors on basis of factors like potential and data collection and data monitoring required Long-Range transmissions. Here also focused on parameters like spreading factor and bandwidth to increased PDR radio during transmission of data. Also, worked to minimized delay in long-range data transmission.

[3] "IoT-based low cost and intelligent module for smart irrigation system" © 2019 Elsevier

Proposed work uses IoT to make devices to connect and talk with each other. In this paper, they worked on one-time setup for smart agriculture system, user interface, decision-making based on neural and data monitoring remotely. A crop test bed has been chosen as a sample for neural-based decision in a smart irrigation system. Neural network plays a role by providing intelligence to current sensor devices used in irrigation for better efficiency of irrigation system. To inform about the crop situation, HTTP and MQTT protocol is used for user interface when crop field is far away. Following parameters are informed to the user: soil moisture value, perform evaporate transpiration and decision based on neural network, alarm alert for activation of zone wise watering, and finally the last sensor data send to cloud through MQTT for remotely accessing the data.

[4] "Optimizing Operation-mode Selection for IoT Edge Devices" ACM Transactions on Internet Technology, Vol. 19, No. 2, Article 22. Publication date: March 2019

In this paper, efficient utilization of battery power is focused to manage the system and increase the system's performance by sharing resources of all the nodes. In the proposed work, the framework is discussed to manage IoT devices and all the limited shared resource shown operation mode of IoT devices at runtime with constraints of system devices and the gateway. The proposed framework also supports a wide range of optimization goals and helps in significant reduction in runtime overhead of operation mode selection.

[5] "Water Management in Agricultural: A Survey on current challenges and technological solutions" IEEE Access February 2020

In this paper, four challenges have been discussed and these four challenges are like reuse of water and water monitoring, wireless sensor network, IoT, and cloud-related technologies that are basically used for pollution detection and making wastewater management system safer and efficient by connecting physical objects to Internet. So objects can be monitored and used at very large scale with more effectiveness. Also, ensure that water distribution should be homogeneous to obtain a complete crop, and this required water pipeline complete monitoring. Leakage and damage in water distribution pipeline can be caused by some factors like overpressure in pipe, age of pipe, installation of pipe improper, natural disaster, malfunctioning of actuator, etc. Leakage in the water pipeline is the major problem that cause insufficient water supply for the growth of crop and reduce the yield production. Some control mechanism based on real time monitoring is introduced here, which helps to overcome all the above issues. Another major challenge is to maintain water moisture level in soil and to maintain salinity concentration level in the roots of plants, salt level needs to be controlled to the right amount in the soil. IoT-based smart system consists of different devices which include gateway, sensors, battery, and also consist of the database for stored data and further processing at the server end. IoT integration with cloud storage is cost-effective in case of data storage and processing. GSM is used to send data collected from sensors to remote servers and which can be retrieved at the user end on mobile phones from where the user can send control commands for the actuators to operate a particular work.

[6] "Towards an IoT Based water management system for a campus" IEEE Access 2015

This paper describes the design of both the network and the sensor. They have used sub- GHz radios to build a wireless network to transfer the sensory information to a gateway. Then the gateway uploads the data to a cloud. They have used an ultrasound-based system that allows us to mount the sensors outside the water and also use a pair of ultrasonic transducers that are mounted under the lid of the tank. The ultrasound pulses reflect off the surface of the water and are picked up by the receiving transducer. The traveling time between transmitted and received ultrasound is measured to obtain the distance between the tank top and the water surface. Data was collected every minute from the tank and encapsulated into packets then transmitted to the relay. The relay then forwards it to the gateway and the gateway unpacks the received packet and uploads the level data online to a Google App Engine application.

[7] "Leak detection in water distribution networks: an introductory overview" 2019 Springer

The history of leakage detection is enriched with a lot of technical development. The paper has described a few of them with details like wireless micro-electro-mechanical systems (MEMS): It's an electromechanical system associated with systems and works with multiple analysis devices like accelerometers, acoustic, and thermal applications. A viewer software is being mentioned to visualize the output related to leakages. Few other, very effective, technologies like Infra-red thermography have been featured in the paper which detects the changes in soil and its thermal variation due to leakage effects. Similarly, ground penetrating radar has been used on a vast level to detect changes in leak location via imaging techniques. All these have been developed with time and researched to acknowledge the nearest location. Further, the static leak detection system has been presented that apart from dynamic leak detection used to detect a leak on early stage without much human interference.

[8] "A smart farming concept based on smart embedded electronics, internet of things and wireless sensor network" 2020 Elsevier Internet of Things

This paper explains about agriculturally-based smart systems integrating embedded technology, IoT, WSN, and Cloud storage for agricultural farms. Also, the paper explained about different protocols, electronic circuits, hardware, control monitoring system for user interface, etc. Here explained about different available technology for smart system in agriculture or farming sector with their future scope.

[9] "A Self-Stabilizing Publish/Subscribe Middle ware for IoT Applications" ACM Transactions June 2018

In this paper, IoT-based protocol is discussed and its working is explained. IoT-based smart system consists of the following functions like transferring sensor data to the gateway, analyzing the collected data, and sending commands accordingly. Working of protocol between client and server end (publish/subscribe) is optimized for communication links when the subscription is continuously changed by system nodes. The role of size and maintenance cost of the routing table is considered as per the length of the routing path. Also explained the routing table scalability and overhead of forwarding messages. Real development is usable against the theoretical concept of simulation.

[10] "MAAT: Mobile Apps As Things in the IoT" ACM Transaction December 2019

In this paper, the author proposes dynamic programming and working keywords. The main focus of the paper was on enabling the dynamic interaction of things or objects. The paper presents the implementation part of the mobile app which is based on android platform programming along with plug in of android studio. In this paper, the author conducted a study to evaluate the implementation of the idea and its response. Also studied about plug in features with end user and understand the exact need which has to be improved. To reduce the overhead of developer its is suggested that need to provide decisions for database format for broadcast and keywords which describe it for mobile app. Programming and the concept of actionable keywords were introduced to build the link between mobile app and interaction logic. Data source is represented by keywords set within the mobile app, which is presented in this paper.

[11] "Study of Wireless Communication Technologies on Internet of Things for Precision Agriculture" Springer 2019

In this paper, the drawback of communication technologies and their study is presented. Limited bandwidth of communication technologies like ZigBee is focused and compared with other technologies. LPWA, NB-IoT, LoRa are considered as the best choice in case if the distance is not less than 5 km and up to 15 km. Also work on Power consumption of communication technologies. It is clear in the paper that all these above technologies consume less power with the transmission of data up to a large distance. NB-IoT and LoRa technology drawback are disclosed is the maintenance cost which is monthly paid to subscriber.

[12] "A Survey: Internet of Things (IoT) Technologies, Applications and Challenges" IEEE International Conference on Smart Energy Grid Engineering 2016

The main focus of this paper is technology-related issues, their transmission range, and protocols as IoT is the integration of multiple devices and technologies. RFID, smart sensors, and internet protocols are empowered in IoT development. Smart and advanced sensors are capable of dealing with the application without the participation of human beings or any external factors. Different phases of IoT is explained with respect to recent developments like smart mobile phone, machine to machine application, and internet. Also worked on technical and architectural part of IoT with their protocols, application, and issues. Working of different protocol, application framework and connection between other technologies & IoT. In this paper, relationship between technologies like big data analytic, cloud computing, and IoT is explained.

[13] Varsha Radhakrishnan and Wenyan Wu "Iot technology for smart water system" IEEE International Conference on Smart City 2018

In this paper, the requirement of IoT in smart water management setup, IoT basic architecture, and its applications are explained. It shows that to develop a smart and intelligent water system, selection of basic architecture is done after comparing different technology, cost of equipment, and different methods for development. In this survey paper, it is mentioned that the parameter selection for IoT-based system isvery important for the implementation of smart system.

3 Noteworthy Contribution

MICS for smart agricultural system with IoT- based application [1] LoRa: Use of Long Range and low power wireless platform [2] Interfacing of SX1278 LORA Module with Arduino to make to transmitter and receiver circuit. Without consuming large amount of battery power, LoRa module can transmit bidirectional information upto range of 15–20 km. This LoRa module can plug in easily into existing infrastructure and operate the IoT-based application with low cost and low consumption of battery. Basically, two LoRa modules and two Arduino boards will be used. On the transmitter side, Arduino Nano will be used to send data from one board, and at the receiver side, Arduino Uno will be used to receive it on the other board. Use of advanced sensors in IoT like Ultrasonic sensor HCSR04 for water level, FC-28 sensor for soil moisture, LDR light sensor module, PIR motion detector sensor HC-SR501 module and smoke detector sensor module MQ2 for better results and performance [8]. Make use of Long-Range mesh network wirelessly which is based on LoRa PHY for data collection from IoT Sensor by using single LoRa Gateway and acting as a data router. In a wireless mesh network, the system node is connected to each other as much as possible. To prevent communication problem, it is required to route the information efficiently and directly. Measure of the LoRa technology scalability on parameters like spreading factor, the size of the packet payload and the multiple nodes [2]. In a wireless mesh network, by increasing hop-count, Packet Delivery Ratio can be enhanced in order to deliver sensor data to the Gateway node. Without using an additional Gateway node, an increase in PDR is possible through LoRa mesh networking which helps in avoiding additional costs of the infrastructure associated with the data collection through IoT sensors [5]. A simple and straightforward method can be used to disable relaying of data through multiple nodes and send and receive data wirelessly at the trans receiver end while utilizing battery power efficiently as it is limited in the system. However, this required modifying the protocol. Another approach is to keep the Spreading Factor at a suitable value [7]. A voltage converter is required for converting 5 V into 3.4 V because Arduino operates on 5 V, whereas the LoRa SX1278 does not work on 5 V, therefore it cannot supply 5 V directly to LoRa Module, otherwise the LoRa board will get damaged [2]. MQTT stands for Message Queue Telemetry Transport and it is a messaging protocol for data sending which flows from multiple sensors to applications and is best suitable where bandwidth is limited, latency is high, and when the network is unreliable. It is suitable for IoT connectivity and is extremely simple to use. This protocol is suitable where memory is low, power is low, cheap and small devices. Complexity at the receiver and transmitter end should be less.

4 Research Objective

1. Understand and implement MICS setup with the integration of LoRa Technology to support IoT applications.
2. Improve the system performance by:

    - Use of IoT system with low GSM signal reachability
    - Use of advanced sensors
    - Prevent access water loss in agricultural sector
    - Minimize the power consumption
    - Detection of Water leakage
    - Reducing deployment cost

Following parameters are considered in the proposed work:

- Maximize the PDR ratio
- Minimize Battery Power Consumption
- Long-Range communication.

5 Proposed Methodology

We are developing a system which is comprised of hardware unit and software system. The hardware consists of four sensors and processing boards which are regulated through programming inputs. For communication through IoT tactics, we use GSM800 Modules that are responsible for internet availability. We have to provide SIM on the board for GSM network connectivity and GPRS activation. The sensors are made for different purposes like water level detection. Live status of water level and leakage of the reservoir can be analyzed with water level ultrasonic sensor. For safety with unauthorized persons, we put on a proximity sensor that can be capable of distinguishing between humans and all other movements like animals, objects, etc. On the software end, we are using MQTT-based android application. To achieve Long-Range communication, Lora Module works with MQTT protocol.

6 Proposed System Architecture

A. **Hardware Device**

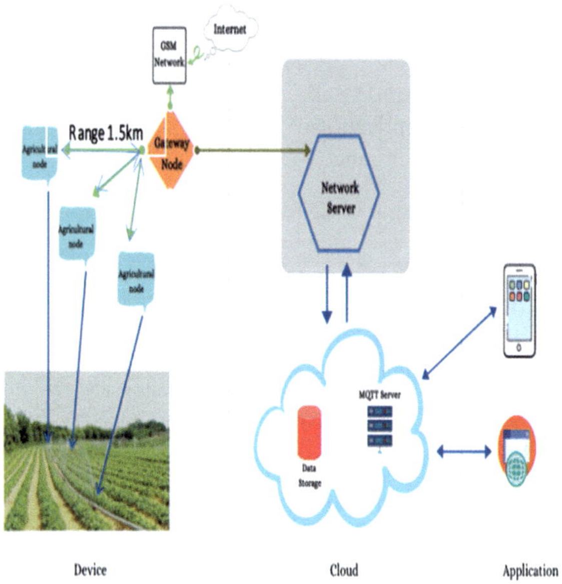

- Couples of devices are defined in the described system based on its functionality. Node available at agricultural field to report status of conditions and control over relay, valves etc. Another node is named as Gateway node that is used as a centralized node.

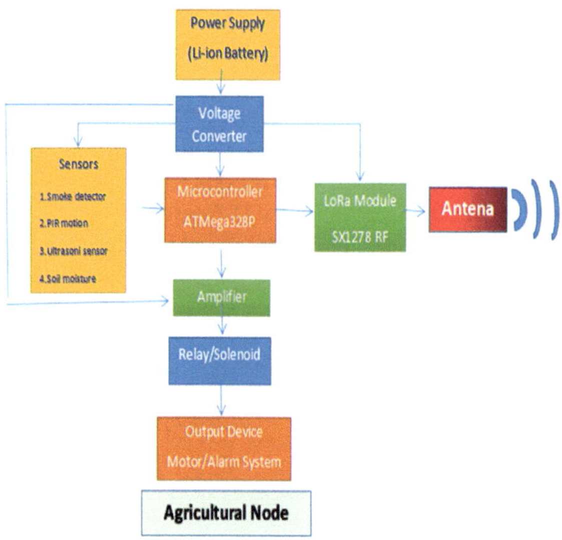

- 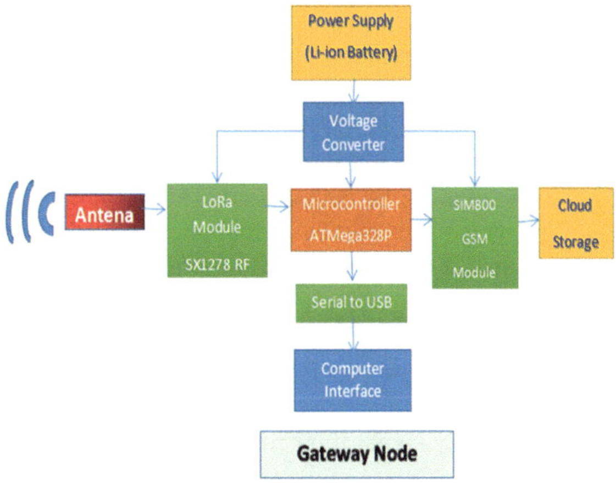

  Agricultural node consists of four main modules: Sensors, Microcontroller Module, LoRa module, and power module.
- In the proposed system, the architecture describing the system design consists of two devices, i.e., Agricultural and Gateway node. The Agricultural node shown in this figure is responsible for reporting and controlling the information of solenoid valve/Relay Module Proposed System Architecture Block Diagram status. Solenoid valve/relay module is the core component of the system.

Sensor:

- HCSR0 Ultrasonic Sensor: Higher sound frequency than the limitation of human hearing capability is emitted by sensors known as ultrasonic sensors. It works on the time gap difference between the emitted and reflected waves that come from the object responsible for the calculation of distance.
- MQ2 Sensor: High Sensitivity is observed in the presence of smoke. MQ2 sensor changes its resistance according to the availability of gases. It has the potentiometer function inside to set the sensitivity level of the sensor to respond to variable environmental conditions. As an outcome, we get the analog voltage range between 0 V and the Vcc we provide to the sensor circuit. For a stable output generation, it takes around 20 s.
- HC-SR501 Sensor: Motion detection is done using the sensor. PIR Sensor is the secondary name given to it. Different kind of sensitivity is used to detect humans or animals. We provide it Vcc and ground on two of its pins and receive output on the third pin that is compatible with microcontrollers.
- SEN0193 Sensor: The sensor is dedicatedly used to measure the moisture of the soil. The capacitive electrode plays a very important role for sensing purposes. Both analog and digital type of sensor output is available with the sensor. Also, it is used to interface with a microcontroller for various applications.
- Microcontroller Module: Arduino ATmega328P is used for our purpose. It is an 8- Bit Microcontroller that can execute its function on 1.8 V, and on the other side, it can bear a maximum voltage of 5.5 V. 1 MHz, frequency is optimum on which

microcontroller consumes minimum power. two 8-bit Timer/Counters, 6-channel 10-bit ADC and 2 K Bytes Ram.

- LoRa Module: RF communication is widely used for a long time. Range enhancement is always a matter of research in this field so that data can be communicated as long as possible. Along with this, minimization of power consumption is the next task. LoRa is found to be best suitable for this purpose, where the data is communicated at a larger range. Also, with the growing field of IoT applications, LoRa became very useful where sensor outputs can be carried to the available nodes which are connected to the Internet or IoT. This LoRa (Long-range Radio) module help in IoT. Security and wide bandwidth help to use module in number of applications. SX1278 IC is used to frame the LoRa module by keeping 433 MHz of frequency. Frequency hopping will cover a range of 420–450 MHz and will give a balance of quality signal transmission. The technology behind this IC is completely perfect where range and strength is required. LoRa module SX1278 RF 430 MHz transceiver Frequency Range is considering Sensitivity: -136dBm, Output Power: + 20dBm, Data Rate: < 300kbps and Interface is available like SPI.
- Power module: Li-ion battery, which can be charged by the 5 V.
- UART Protocol is converted in USB with the use of CP2102. It is majorly used in gateways and processes data to the application interfaces.
- Gateway node consists of the main module as follows: ATmega328P 8-Bit Microcontroller which can be programmed with Arduino software. For the internet or GPRS application, we have SIM800. 850/900/1800/1900 MHz. It can communicate information through both messages and the internet. It can be connected with a Li-ion battery to regulate the circuit and interface with LoRa SX1278 RF 430 MHz transceiver.

B. Cloud

The major parts of LoRa cloud server can be categorized as follows: Network Server (NS): It majorly deals with the mess network data. Once the data is collected by the gateway node, it further processes it to the application server. After this, Message Queuing Telemetry Transport (MQTT) protocol comes into the picture to publish the data. More than that same network server plays the responsibility of delivering commands sent by the user to control the devices on the field. This is known as subscriber functionality.

Application Server (AS): Application server is very crucial in the system as it manipulates the data by encrypting and decrypting it. Also, it verifies the processing data and extracts information from it. The device registered on the network and its presence or active status is being continuously checked by the application server. So the configuration of each and every node is handled by the application server by communicating through MAC command. Header packets are already circulated to join the network before actual data transfer. Both gateway and sensor nodes initiate in the same way.

Network Controller: The transmission (TX) and reception (RX) data setting is done by the network controller. It can also be assumed as the part of the Network server. The routing of data is managed by the Network controller including the dynamic configuration of nodes on which the MAC commands are communicated and executed. The device and application management is important to maintain that all depend on APIs and data storage. All this is managed by cloud servers that may integrate multiple technologies for its functioning.

III. Application

Android Application is developed to handle and control the proposed system. It is the easiest way to make a user's task easy by the development of a graphical user interface (GUI). On the one side, it deals with the cloud server, and on the other side, operations are handled by the user manually. Device status of working and controls are the factor managed by the android application.

The application has Message Queuing Telemetry Transport (MQTT) protocol on the back to transceiver the data. For security purposes, development of password-based login has been done to prevent unauthorized access to the system. Time response is not that much important in our case but reliability should be assured. The data is sent through publish and topic subscription lead to data fetching of MQTT. API's has been developed to make the task easy for the user to communicate between application and server.

Framework.

Among commercially available methods, IoT capabilities are most suitable for our purpose. Cost-effective and modular system can be framed with IoT algorithms that allows simultaneous and dual side communicative data flow between server, sensors, and actuators.

Sensor Network.

In IoT, wireless sensing devices are embedded in a network that is specified to sense the environmental conditions and help to record the field data with human intervention. Further, these data get connected with the internet or cloud and reach the back-end processing units for analysis and are used in the desired application. Sensors can be powered through batteries, solar or other power supply methods. Mostly they are implanted to work continuously for a longer period. Safety precautions have to be taken accordingly at the time of installation.

Communication.

Execution is always considered as the most important unit of any remote data analysis and access system. IoT infrastructure follows the same so that accurate communication on the dual side is more critical since it carries major applications in industries like Oil refineries, Power grid, Automobile, etc., and also in civil applications. This can be understood with more clarity in the following ways:

- Large number of sensors is integrated into a common network and communicate millions of different analytical data simultaneously.

- Encryption and other protocol flow is maintained throughout the process of execution
- Control commands with specified addressing and execution status are communicated properly in the system

Communication should be simple enough and faster in execution.

7 Expected Outcome

Development of Multi-Intelligent control system (MICS) and its application for irrigation sector. Our proposed work defines efficient technical aspects for agricultural management, water distribution, and security for the needs that are planned initially. The work presented is from the architecture to its functional execution after analyzing and comparing different wireless technologies and adopting Lora Mesh Networking by modifying it according to agricultural uses. Optimize Equipment's, lowering the cost and reducing power consumption are the prime work shown to build a multi network system. The proposed framework of proposed MICS & Smart water management with integrated monitoring and controlling. We use the innovative concept of IoT, Lora wireless networking, MQTT protocol, and their application to achieve cost efficiency, reliability, fastest dual side data communication possibilities, and also develop enough capability to involve a network of sensors to collect data automatically at node points defined strategically.

Reservoir plays a vital role while designing water supply system for agricultural farms, leakage, water level maintenance, and more physical involvement of farmers for system functioning are some of the major challenges that are faced commonly in most of the setup, so we have worked to resolve this issue. Following are the expected outcomes of the proposed system: Successfully implementation of MICS setup with new parameters, Increase PDR Ratio, successful long-range transmission and good quality communication, minimize the power consumption, minimize access water loss and minimize system cost.

8 Conclusion

The paper describes the establishment of MICS with its IoT integration and innovative tactics for wireless system design to handle irrigation requirements, water distribution, and wireless data network management against conventionally proven systems defined in the reference papers. We have defined the process to overcome its disadvantages. The scenario is taken of agricultural sector as it consumes highest percentage of available water resource than any other sector. Managing water of bigger farmland is also more typical since it usually lies apart from residence area where human interference is minimal. Also, the internet and mobile network is always a major challenge for these areas. Advance LoRa networking devices and IoT tool is found to be useful for purposes which rely on large sensor network, processing hardware, and end user devices for execution with feedback collection strategies. IoT helped us to ensure the efficient utilization of network shared resources and their runtime constraints, low powered devices, optimized framework, and fast communication. The experimental setup is likely to be built in the future to bring the defined system in the operational platform and relevant data analysis in order to reach the commercially viable, robust, secure, and user-friendly system.

References

1. Hadipour M, Derakhshandeh JF, Shiran MA (2019) An experimental setup of multi-intelligent control system (MICS) of water management using the Internet of Things (IoT) 0019–0578/© ISA. Published by Elsevier, Accepted 25 June 2019 https://doi.org/10.1016/j.isatra.2019.06.026
2. Lee HC, Ke KH (2018) Monitoring of large-area IoT sensors using a LoRa wireless mesh network system: design and evaluation 0018–9456 ©. IEEE Trans Instrum Meas 2018, Digital Object Identififier https://doi.org/10.1109/TIM.2018.2814082, http://www.ieee.org/publications_standards/publications/rights/index.html
3. Nawandar NK, Satpute VR (2019) IoT based low cost and intelligent module for smart irrigation system July. J Comput Electron Agric 162:979–990. Available online 28 May 2019, 0168–1699/ © 2019 Elsevier
4. Samie F, Tsoutsouras V, Bauer L, Xydis S, Soudris D, HENKEL J (2019) Optimizing operation-mode selection for IoT edge devices. ACM Trans Internet Technol 19(2), Article 22. Publication date: March
5. Saad A, Benyamina AEH, Gamatie A (2020) Water management in agricultural: a survey on current challenges and technological solutions. IEEE Access, 8:38083, publication February 19, 2020, date of current version March 3, 2020 Digital Object Identifier. https://doi.org/10.1109/ACCESS.2020.2974977
6. Verma P, Kumar A, Rathod N, Jain P (2015) Towards an IoT based water management system for a campus. IEEE Conference
7. El-Zahab S, Zayed T (2019) Leak detection in water distribution networks: an introductory overview. Springer, Accepted: 21 May 2019, Published online 11 June 2019, https://doi.org/10.1186/s40713-019-0017-x
8. Mahbub M (2020) A smart farming concept based on smart embedded electronics, internet of things and wireless sensor network. Elsevier Internet of Things (2020) https://doi.org/10.1016/j.iot.2020.100161, Internet Things 9:100161, Available online 14 January 2020
9. Siegemund G, Turau V (2018) A self-stabilizing publish/subscribe middleware for IoT applications. AACM Trans Cyber-Phys Syst 2(2), Article 12. Publication date: June, © 2018 ACM 2378–962X/2018/06-ART12 https://doi.org/10.1145/3185509
10. Helal S, Khaled A, Kotonya G, Lee J, Lindquist W (2019) Lancaster University, United Kingdom "MAAT: Mobile Apps As Things in the IoT. Proc ACM Interact Mob Wearable Ubiquitous Technol 3(4), Article 143
11. Feng X, Yan F, Liu X (2019) Study of wireless communication technologies on Internet of Things for precision agriculture Springer 2019. Wirel Pers Commun. https://doi.org/10.1007/s11277-019-06496-7
12. Shah SH, Yaqoob I (2018) A survey: Internet of Things (IOT) technologies, applications and challenges. In: Proceedings of IEEE International Conference on Smart Energy Grid Engineering

13. Wu W, Radhakrishnan V (2018) Iot technology for smart water system. In: Proceedings of IEEE International Conference on Smart City
14. Nolan KE, Guibene W, Kelly M (2016) An evaluation of low power wide area network technologies for the Internet of Things. In: Proceedings of International Wireless Communications and Mobile Computing Conference
15. Datta SK, Bonnet C, Nikaein N (2014) An IoT gateway-centric architecture to provide novel M2M services. In: Proceedings of the World Forum on Internet of Things (WF-IoT'14), pp 514–519
16. Dehghan M, Massoulie L, Towsley D, Menasche D, Tay YC (2016) A utility optimization approach to network cache design. In: Proceedings of the International Conference on Computer Communications (INFOCOM'16). pp 1–9
17. Hu P, Dhelim S, Ning H, Qiu T (2017) Survey on fog computing: architecture, key technologies, applications, and open issues. J Netw Comput Appl 98(2017)
18. Liu Z, Zhou Q, Hou L, Xu R, Zheng K (2019) Design and implementation on a LoRa system with edge computing. In: IEEE Coference
19. Li W, Delicato FC, Pires PF, Lee YC, Zomaya AY, Miceli C, Pirmez L (2014) Efficient allocation of resources in multiple heterogeneous wireless sensor networks (2014). J Parallel Distrib Comput 74(1):1775–1788
20. Kwak J, Kim Y, Lee J, Chong S (2015) DREAM: Dynamic resource and task allocation for energy minimization in mobile cloud systems. IEEE J Select Areas Comm 33:12

Chapter 49
Cataract Detection Using Textural Features and Machine Learning Algorithms

Kaushal Chande, Piyush Jha, Kamaljit Kaur Aulakh, and Swapnil Shinde

1 Introduction

A cataract is a white cloud formation that gradually hardens and forms a yellow sheet on the normally transparent lens of the eye. Cataracts occur when the normally clear lens of the eye becomes cloudy, restricting the amount of light that reaches the retina, leading to a decrease in vision.

The World Health Organization conducted a survey on world sight day 2019 under which 36 million people were identified as lined and 217 million had moderate or severe distance vision impairment (MSVI), out of these cases of blindness and MSVI about 65 million people were affected by cataract [1]. Thus, cataract is responsible for or 48% of blindness cases. Even though cataract develops very slowly it can cause long-term problems as it moves from covering a very small portion to spreading over the eye lens causing a vision loss since light can no longer reach the retina due to the cloud that is formed on the lens (Fig. 1).

Among the various signs and symptoms of cataract few of which are prominent include sensitivity to light, double vision, need for brighter light to read and difficult to see at night [3]. The various causes include diabetes, hypertension, smoking previous eye surgeries, significant alcohol consumption, a family history of cataract, and also so age as one of important factors. In this project, we have taken into consideration the three major types of cataracts; one of which is nuclear cataract which at the initial stages of development causes nearsightedness, however, as time follows the lens becomes densely yellow and also brown in some cases. Moving to the next type which is cortical cataract this begins with the white wedge-shaped opacities on the outer edge of the lens which further moves to the central portion of the lens. Lastly, posterior subcapsular cataract that starts with a small; however, black-colored portion

K. Chande (✉) · P. Jha · K. K. Aulakh · S. Shinde
Department of Information Technology, Ramrao Adik Institute of Technology Navi, Mumbai, India

G. Mathur et al. (eds.), *Proceedings of 2nd International Conference on Artificial Intelligence: Advances and Applications*, Algorithms for Intelligent Systems,
https://doi.org/10.1007/978-981-16-6332-1_49

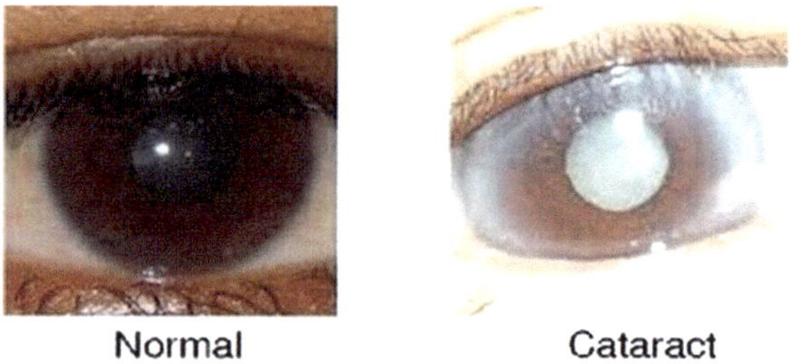

Fig. 1 Normal eye versus cataract eye [2]

occupying the center of the lens blocking the path of the light to the retina this type of cataract generally grows faster in comparison to the others [3].

The current system of conventional practices followed to detect cataracts or other similar diseases includes a visual acuity test under which the eyesight of the patient is checked with the help of a chart or a viewing device that has letters in decreasing order of its size [3]. This detection technique helps to check the vision accuracy of the patient another method that is majorly used by doctors and ophthalmologist is a slit-lamp examination to see the structure of the front of the eye with the bright illumination of light on the cornea, iris, and lens all under high magnification. Another method is the retinal exam in which using a slit lamp or an ophthalmoscope, the doctor examines the lens for signs of cataract.

2 Related Work

In one methodology studied, they proposed a smart computer system to detect and further categorize the cataract using slit-lamp images as their source input. The proposed system performs the following steps in the given order [4].

A. Lens Localization: In this method, images are taken through a digital camera that is placed on the eyepiece of a slit lamp. This helped to create a data set of 400 images with classes as early, intermediate, and mature types of cataracts, which were later cropped for further process. This cropping is done such that the image now contains only the part of the iris. The resizing is done to about 65px and lastly Hough circle detection is one of the famous algorithms for detecting the region of interest as the circle is used.

B. Detection of nuclear cataract using SVM. The images contain intensity values similar to that of a random distribution; hence, it is expressed by using statistical values of mean, energy, homogeneity, smoothness, and correlation. These features of about 20 images are used as an input to the SVM classifier. This gives the results with an accuracy of 90.25%. The results also state that the score of sensitivity could be enhanced by using some other feature extraction method along with the circle detection algorithm.

Another paper studied works upon using the algorithms of machine learning and image processing together [5]. The system's methodology has been divided into four steps shown below:

(a) Gathering of images,
(b) Pre-processing of data,
(c) Classification using KNN, and
(d) Validation of developed application.

(a) Gathering of Images: The first step is to gather images which were done with the help of google database. The Google API along with AJAX is being used in order to download the pictures of cataracts and also the pictures of normal eyes.

(b) Pre-processing of Data: The pictures are then cropped in order to obtain only the interesting area of the image. The cropped image is now shown to the user to verify that the image has been cropped appropriately. The image is now fed to the Orange Tool for the process of feature extraction. About 2048 features of the image are extracted with help of this tool. After that, the dataset is categorized into 2 categories namely Cataract and Normal. After that, both of them are then merged into a one.csv file.

(c) Classification using KNN: After the feature extraction process, there is no explicit training phase. The KNN model works on Similarity features which are calculated by using distance formula, for example, Euclidean. The KNN is used in order to decrease the computational work. The testing image is classified on the basis of similarity as Cataract or Normal.

(d) Validation of Developed Application: This is the last stage where the proper working of the mobile application is validated by the users and the ophthalmologist. The model is implemented using the Orange tool. Accuracy results of the KNN, SVM, and Naïve Bayes were stated as 83.07, 72.5, and 76.67%, respectively. The model for cataract detection, classification, and grading is mainly divided into 3 phases which are described as follows:

(a) Pre-processing phase: In this step, the main goal is to remove all the unnecessary information from the image. This is done by resizing the image, cleaning the images. This includes removing glare from the images and also converting them into grayscale form.

(b) Feature Extraction phase: The main feature is extracting the image of the eye lens. Therefore, an invisible circle is made and only the part of the image containing the pupil is extracted in this process. After the extraction, it is then given to the wavelet transform with Log Gabor filters to obtain the feature vectors.

(c) Classification phase: This is the last phase of the proposed method wherein the feature vectors which are obtained after performing the wavelet transform are sent to both the support machines SVM and ANN. Both of these machines classify the images into three classes: normal, early, and advanced stage. Here the first class refers to the normal eyes having no sign of cataract, the second class refers to the eyes which are having cataracts but it is in the early stages and the last class refers to the eyes in which the cataract has reached the advanced stage. Classification results show the accuracy of the SVM classifier as 96.8% and the accuracy of the ANN classifier as 92.3% on a dataset of 120 images [6].

Also, in one of the proposed methods, digital image processing is used on eye images in this research paper to assess the existence of a cataract along with its severity. A dataset consisting of images of eyes having differing degrees of cataract is applied to two separate image processing algorithms.

The first method uses an automatic detection algorithm based on Feature Extraction. This acknowledges the Image pre-processes it, extracts data on the strength of the mean value, and makes decisions on classification into the healthy eye, mild, and severe cataracts.

In an unhealthy eye, the second method is based on measuring the region of the pupil and the cataract. The percentage of cataracts is measured as the relative measure of cataracts with respect to the pupil area which is the affected part. Further, the image is resized and then converted to a grayscale version of the original image, also a low pass filter is added, and finally by applying an opening of morphological type, the image is preprocessed. The transformation of Hough is then used in the picture to define and isolate circles. The transformation of Hough is then used in the picture to define and isolate circles. For detection of the affected region inside or even near the pupil, contours are used. If the infected region lies within the pupil, a cataract is confirmed by the algorithm. Also, the percentage of cataracts is measured with the formula: (cataract area / (pupil + cataract area))*100. [7].

In this research work, an algorithm is proposed for cataract screening based on texture features: uniformity, intensity, and standard deviation. In preprocessing, images are converted to grayscale images, applied Gaussian filter with image resizing of 256 X 256. In feature extraction, they calculated the uniformity of the image based on histogram which will be further used to compute the mean intensity and standard deviation of the image.

Results [8]:

1. Healthy eye—uniformity value $> =$ 1.59 and mean intensity is less than 125.31 and more than 12.93.
2. Mild cataract—uniformity value in between 0.09 and 0.11 and mean intensity value is between 125.3 and 134.
3. Severe cataract—uniformity value.

3 Automatic Cataract Diagnoses Model

The following steps are involved in designing and training the model (Fig. 2).

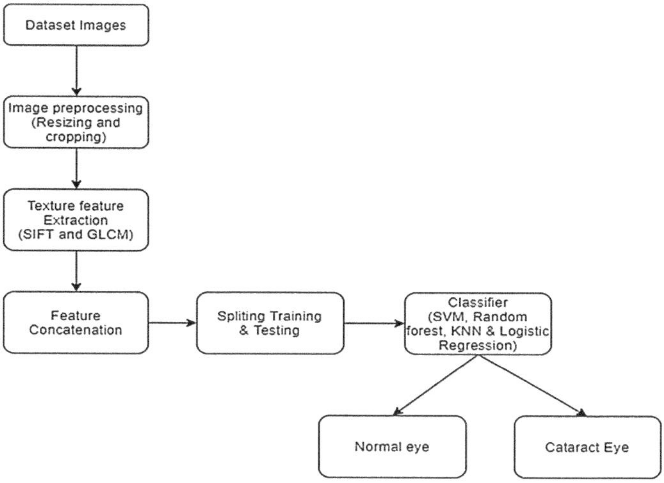

Fig. 2 Flowchart of the proposed system

3.1 Preprocessing

The dataset consists of 97 images that are cropped manually in the initial stage and then they are resized to 128 pixels followed by grayscale conversion of the images which are then used for texture feature extraction.

3.2 Feature Extraction Using GLCM

GLCM which is an abbreviation of Gray-Level Co-occurrence Matrix is a method of determining texture by calculating the statistical relationship in terms of space between the pixels. Characterization of the texture of the image is done by GLCM by calculation of how frequently pixels with specific values and a specific relationship occur together in a pair and thus creating a matrix and finally extracting statistics from it [9].

Several statistics can be computed from the gray co-matrix. The computed statistical values provide us with essential information regarding the texture of the image. Some of the statistics are listed below (Table 1).

Table 1 Statistics obtained from GLCM

Statistic	Description
Contrast	It is used to measure the local changes, i.e., the intensity contrast between a pixel and its neighbor over the entire image
Correlation	It determines the gray-level linear dependency of the specified pairs of pixels
Energy	It determines angular second moment, i.e., it provides the sum of square in GLCM
Homogeneity	Evaluates how close the elements are distributed in the GLCM with respect to the GLCM diagonal

3.3 *Feature Extraction Using SIFT*

Scale-Invariant Feature Transform (SIFT) is mainly used to draw out those features from an image that are independent of the scale of an image or any kind of rotation. The SIFT algorithm is divided into 4 steps [9].

3.3.1 Construction of the Scale-Space

This step focuses on identifying important key points which can be viewed equally from a different orientation. For this, a Gaussian function is used. It generates various parts of the same image by successively blurring and scaling down the image. Different techniques can be used to find the locations of stable key points in the scale space. One of those techniques is the Difference of Gaussian, the scale-space extrema can be found, $D(x, y, \sigma)$ by calculating the difference between two images in the octave and consecutively blurring using gaussian blur to reduce the noise between the pair images.

3.3.2 Keypoint Localization

In this step, we detect the maxima and minima in the DoG images we have created in the first step and compare the neighboring pixels in the current scale, the scale above and below it. Then, to remove additional key points by detecting whether points have enough contrast or they lie around the edge by Taylor expansion of the image around keypoint finding those having less contrast or are poorly present around an edge. The location of extremum, z, is expressed as

$$\mathrm{Z} = \partial 2D\ \partial x2 - \partial \mathrm{D} \partial \mathrm{x} \quad (1)$$

To detect the edges and to retrieve corners, a Hessian matrix is used. In this, two eigenvalues are calculated and their ratio is considered.

3.3.3 Orientation Assignment

The main aim of this step is to focus on assigning an orientation to the keypoints based on the image's gradient and orientation properties. Then histogram of magnitude and orientation is created. Using histogram, we identify the most important orientation. If there is only one peak, then it is the most prominent orientation and if there are multiple peaks above the 80% mark, they are all converted into new keypoints.

3.3.4 Keypoint Descriptor

Keypoint descriptors usually use a set of 16 histograms, which are aligned in a 4×4 grid and each grid consists of 8 orientation bins, one is used for the main compass directions and the other is used for each of the midpoints of these directions. This helps in generating a feature vector consisting of 128 elements.

3.4 Making a New Feature Vector

The features obtained from the above two procedures are then combined to form the resultant vector, which is used in the further steps to train the model. This was done to improve the accuracy of the result by including more features in our vector. The dataset is then divided into training and testing as 75% and 25%, respectively.

3.5 Detection and Classification

The classifiers used to obtain the results and for the comparative study are as follows.

3.5.1 Support Vector Machine

SVM is a type of supervised machine learning method which is preferred in classification and sometimes used for regression problems in which support vectors are the coordinates of the individual observation. SVM mainly focuses on dividing the datasets into classes in order to find the most optimistic hyperplane and it can be done in the two steps as follows [10]:

- The algorithm generates hyperplanes repetitively that classifies the classes.
- Then, it will select one of the hyperplanes that separate the classes with minimum outliers.

SVM uses a kernel that takes an input data vector and transforms it into a higher dimensional vector.

3.5.2 Linear Kernel

Linear Kernel is used when data is linearly separable, i.e., it uses straight lines to classify the classes. The formula is stated as follows:

$$\mathrm{K}\,(\mathrm{x},\ \mathrm{y}) = \mathrm{sum}\,(\mathrm{x} * \mathrm{y}) \tag{2}$$

From the above formula, we can see that the product between two vectors x and y is the addition of the dot product of the input number.

3.5.3 Radial Basis Function (RBF) Kernel

It maps the input space in higher dimensional space like quadratic, cubic, etc. The following formula is used to express the relationship

$$K\,(\mathrm{x,xi}) = \exp(-\mathrm{gamma} * \mathrm{sum}(\mathrm{x}-\mathrm{xi}^2)) \tag{3}$$

Here, Gamma values lie between 0 and 1. We can set Gamma values in the SVM algorithm manually [11].

3.5.4 Logistic Regression

Logistic regression is a classification algorithm that is mostly used in binary classification problems. As a function of x, the logistic regression model predicts P(y = 1). The Logistic regression equation can be derived from the Linear Regression equation. The steps to get Logistic Regression equations are given below: Equation of the straight line is written as

$$\mathrm{y} = \mathrm{b0} + \mathrm{b1x1} + \mathrm{b2x2} + \mathrm{b3x3} + \ldots + \mathrm{bnxn} \tag{4}$$

In Logistic Regression y can be between 0 and 1 only, so dividing the above equation by (1−y):

$$\mathrm{y}/\,(\mathrm{y}-1) : 0 \text{ for y } = 0 \text{ and infinity for y } = 1 \tag{5}$$

But we need range between −[infinity] and + [infinity], hence taking the logarithm of the equation it will become

$$\mathrm{Log}\left[\mathrm{y}\ /1--\mathrm{y}\right] = \mathrm{b0} + \mathrm{b1x1} + \mathrm{b2x2} + \mathrm{b3x3} + \ldots.. + \mathrm{bnxn} \tag{6}$$

The above equation is the final equation for Logistic Regression [12].

3.5.5 Random Forest

Random forest is the supervised learning method that can be used for both classifications as well as regression. This algorithm produces multiple decision trees and then gets the predictions from all of these trees and eventually chooses the final answer by the majority voting technique. It used multiple decision trees for its decision-making and that is why it is called an ensemble algorithm, which results in over-fitting reduction [12].

Step 1—Initially it starts with a selection of random fragments from the dataset.

Step 2—It constructs a decision tree for each fragment and gets the forecast from each decision tree.

Step 3—Performs voting for the whole forecast result.

Step 4—Lastly, select the most chosen forecast as the final prediction decision.

3.5.6 K Nearest Neighbor

KNN majorly uses "similarity between the features" to predict the values of the latest points. In simple words, it means that the new data point is given a new value based on how accurately it matches the points in the training set [13].

Step 1 —ividing into training and testing dataset.

Step 2—Next, we need to choose number K, i.e., number of neighbors.

Step 3 —For each of the data point i, the algorithm performs the following:

3.1 Calculating distance between test data and each row of training data. Euclidean distance is a commonly used method for calculating distance.

3.2 Sort them in ascending order with respect to their distance.

3.3 Assigning a category to the test point based on which the number of neighbors is maximum.

4 Performance of the Proposed Model

Accuracy is one of the important metrics that help us to determine the performance of our classification or prediction model. It is possible to describe accuracy formally as the number of correct predictions divided by the total number of predictions.

In this research paper, we have made an attempt to evaluate the accuracy when different feature extraction algorithms are applied as classification parameters to the machine learning models trained with different algorithms such as SVM-Linear, SVM-Radial basis function, Random forest Logistic regression, and KNN. The comparative study of the accuracies can be studied through the tables given below.

Table 2 Accuracy rates of various classifiers with SIFT features

Classifier	SIFT (%)
1. SVM-LINEAR	92
2. SVM-RBF	92
3. Random Forest	84
4. Logistic Regression	92
5. KNN	68

Table 3 Accuracy rates of various classifiers with GLCM features

Classifier	GLCM (%)
1. SVM-LINEAR	76
2. SVM-RBF	56
3. Random Forest	64
4. Logistic Regression	80
5. KNN	60

4.1 Phase 1: SIFT with Various Classifier Algorithms

The table below describes the accuracy of various classification algorithms with the use of SIFT features with the SVM algorithm having the highest classification accuracy of 92% along with Logistic Regression. The random forest algorithm achieved an accuracy of 84% (Table 2). Among all the algorithms used, KNN achieved the lowest accuracy, i.e., 68% for cataract classification.

4.2 Phase 2: GLCM with Various Classifier Algorithms

The table below describes the accuracy of various classification algorithms with the use of various GLCM features. Logistic Regression algorithm has achieved the highest classification accuracy of 80% (Table 3). The SVM-Linear kernel algorithm achieved an accuracy of 76% while SVM-RBF achieved the lowest accuracy, i.e., 56%. The random forest algorithm achieved an accuracy of 64%. The KNN algorithm has achieved an accuracy of 60% for cataract classification.

4.3 Phase 3: Combination of GLCM and SIFT Feature Vectors and Applied on Various Classifier Algorithms

From the above two feature sets used for classification, it can be observed that the algorithms have achieved better accuracies when used with the SIFT features. Also,

Table 4 Accuracy rates of various classifiers with SIFT and GLCM features

Classifier	SIFT + GLCM (%)
1. SVM-LINEAR	96
2. SVM-RBF	96
3. Random Forest	96
4. Logistic Regression	96
5. KNN	56

the GLCM features have also provided considerable accuracy, which incepts the idea of using a new feature set by concatenating the SIFT and GLCM features.

Further, in the attempt to improve the accuracies, we trained our final model by combining the features from SIFT and G LCM which were 2303 and 5, respectively, giving a total of 2308 features in our final vector set (Table 4).

The above table shows accuracies obtained by various classification algorithms by using the combined feature set of GLCM and SIFT features. The highest accuracy achieved is 96% by SVM, Random Forest and Logistic Regression and KNN algorithm having lowest accuracy of 56%.

5 Discussion and Conclusion

In this project, various machine learning algorithms have been proposed for cataract detection. For this, we have used texture features. We have used a new set of features by concatenating two sets of features: SIFT (Scale--Invariant Feature Transform) and GLCM (Gray-Level Co-occurrence Matrix). Thus, a new set of features have been obtained and used for Cataract prediction. The various algorithms proposed in our project are SVM(Linear), SVM(RBF), KNN, Logistic Regression, Random Forest. Out of which the highest accuracy achieved is 96%. In conclusion, the combined set of features have obtained better results than the individual features.

References

1. Blindness and vision impairment (2017) World Sight Day https://www.who.int/blindness/world_sight_day/2017/en
2. Supriyanti MR, Habe MH, Kidode MM (2012) Normal Eye vs Cataract Eye. https://www.intechopen.com/books/ocular-diseases/utilization-of-portable-digital-camera-for-detecting-cataract
3. Cataracts Mayo Clinic. https://www.mayoclinic.org/diseases-conditions/cataracts/symptoms-causes/syc-20353790. Accessed 23 Jun 2018
4. Jagadale A, Sonavane S, Jadav DV (2019) Computer aided system for early detection of nuclear cataract using circle hough transform. In: Proceedings of 2019 3rd International Conference on Trends in Electronics and Informatics (ICOEI), pp 1009–1012

5. Agarwal V, Gupta V, Vashisht VM, Sharma K, Sharma N (2019) Mobile application based cataract detection system. In: Proceedings of 2019 3rd International Conference on Trends in Electronics and Informatics (ICOEI), pp 780–787
6. Tawfik HR, Birry RA, Saad AA (2018) Early recognition and grading of cataract using a combined log gabor/discrete wavelet transform with ANN and SVM)
7. Patwari A, Chowdhury M, Nurul M, Arefin A, Imam MI (2011) Detection, categorization, and assessment of eye cataracts using digital image processing
8. Jindal I, Gupta P, Goyal A (2019) Cataract detection using digital image processing. In: Proceedings of 2019 Global Conference for Advancement in Technology (GCAT), pp 1–4
9. Texture Analysis Using the Gray-Level Co-Occurrence Matrix (GLCM) MATLAB & Simulink MathWorks India. https://in.mathworks.com/help/images/texture-analysis-using-the-gray-level-co-occurrence-matrix-glcm.html
10. SIFT Image Features http://homepages.inf.ed.ac.uk/rbf/CVonline/LOCAL_COPIES/AV0405/MURRAY/SIFT.html
11. ML Support Vector Machine (SVM) https://www.tutorialspoint.com/machine_learning_with_python/machine_learning_with_python_classification_algorithms_support_vector_machine.htm
12. Machine Learning Logistic Regression https://www.tutorialspoint.com/machine_learning_with_python/machine_learning_with_python_classification_algorithms_logistic_regression.htm
13. Classification Algorithms–Random Forest https://www.tutorialspoint.com/machine_learning_with_python/machine_learning_with_python_classification_algorithms_random_forest.htm
14. KNN Algorithm-FindingNearest Neighbors https://www.tutorialspoint.com/machine_learning_with_python/machine_learning_with_python_knn_algorithm_finding_nearest_neighbors.htm

Chapter 50
Discrete Wavelet-Based Multi-Classifier Approach for Recognition of Offline Handwritten Hindi Numerals

Danveer Rajpal and Akhil Ranjan Garg

1 Introduction

In the era of Artificial Intelligence (AI), Machine Learning (ML), and Deep Learning (DL), it needs to accelerate the automation of natural language processing (NLP) so that machines can read and understand our languages efficiently. One of the aspects of NLP is the machine-based identification of handwritten characters and numerals. It is a three-step process, which includes pre-processing, feature extraction, and classification. The major challenges, one has to face in designing such systems are writing styles, type of pen, ink, ink-color, and paper, writer's physical and mental conditions, etc. Most of the recent works have addressed these challenges by employing Deep Convolutional Neural Networks (DCNN). Although DCNNs are excellent in solving such problems, they need bulky network architectures, large memory spaces, and heavy computations. The objective of the proposed research work is to achieve comparable results without DCNN and support NLP automation of the Hindi language.

The numeral set of Hindi script is given in Fig. 1.

The upcoming subsections will include a brief theoretical base of techniques and classifiers used in the proposed scheme.

1.1 Discrete Wavelet Transform

The DWT can be used to separate low-frequency and high-frequency contents of an image with the help of high-scaled and low-scaled wavelet functions, respectively.

D. Rajpal (✉) · A. R. Garg
Department of Electrical Engineering, Faculty of Engineering, J.N.V University, Jodhpur 3420001, India

G. Mathur et al. (eds.), *Proceedings of 2nd International Conference on Artificial Intelligence: Advances and Applications*, Algorithms for Intelligent Systems,
https://doi.org/10.1007/978-981-16-6332-1_50

Fig. 1 Numeral set of Hindi script

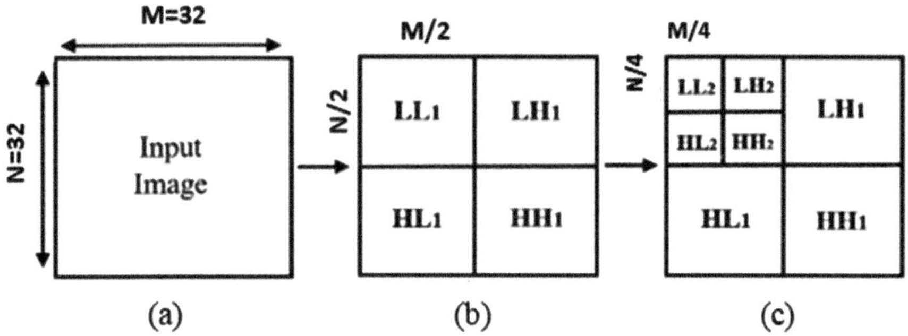

Fig. 2 Level-2 DWT decomposition structure

Multi-resolution filter banks with special wavelet filters are used for the decomposition of an image with multiple resolutions [1]. DWT can transform an image into Approximation and Detail coefficients. The Detail coefficients carry horizontal, vertical, and diagonal details of the given image. Bi-orthogonal wavelet gives a degree of flexibility over orthogonal one; as in the former case, the related transform will be invertible without the need of being orthogonal. Figure 2 shows the pattern of level-2 decomposition of an image using DWT.

The upcoming subsections will include a brief theoretical base of techniques and classifiers used in the proposed scheme.

In Fig. 2, the terms LL, LH, HL, and HH represent Approximation, Horizontal, Vertical, and Diagonal details sub-band images.

1.2 Classifiers

This section includes a brief theoretical base of classifiers used in the proposed scheme; namely: MLP, SVM, and KNN. All three classifiers are based on supervised-learning algorithms and able to solve the problems of classification as well as regression [2].

MLP Classifier.

MLP network has three types of layers; namely: Input, Hidden, and output layer. Depending upon the applications; the classifier may have one or more hidden layers. Nodes of hidden layers are governed by a non-linear activation function. MLP

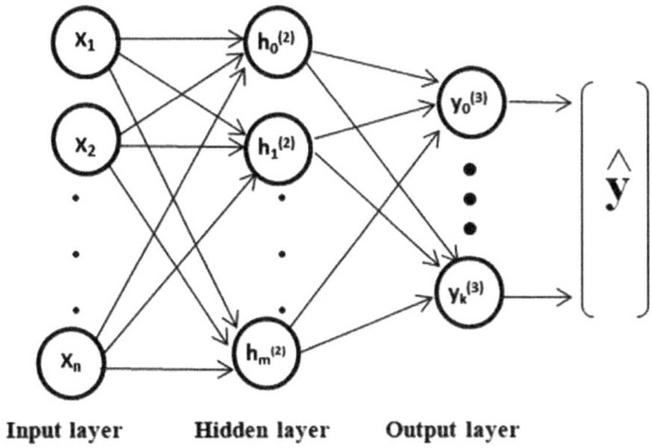

Fig. 3 Construction of MLP classifier

network has the tremendous ability to classify non-linearly separable data. The construction of the MLP classifier is displayed in Fig. 3.

SVM Classifier.

SVM can be used to find out the optimum decision boundary in N-dimensional space to separate distinct classes. Here, N represents the number of features. Figure 4a gives some possible decision boundaries to separate two different classes and Fig. 4b represents support vectors and optimum decision boundary that holds maximum possible distance from data points of distinct classes.

KNN Classifier.

The algorithm determines the distance between the query sample and the current sample from the data and arranges them in ascending order of distances. Depending upon the value of integer K; the labels of the top K number of data points are observed and the most frequent one is assigned to query data point. In Fig. 5, the black dot represents the query data point, and as per the KNN algorithm, label of class-2 will be assigned to it.

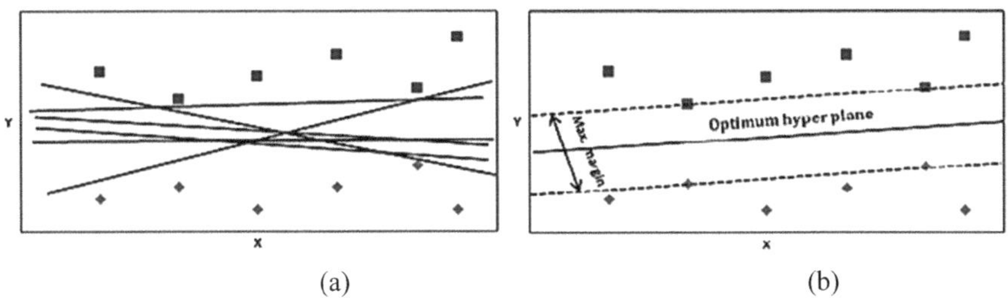

Fig. 4 **a** Some possible decision boundaries to separate out distinct classes **b** Optimum decision boundary determined by SVM

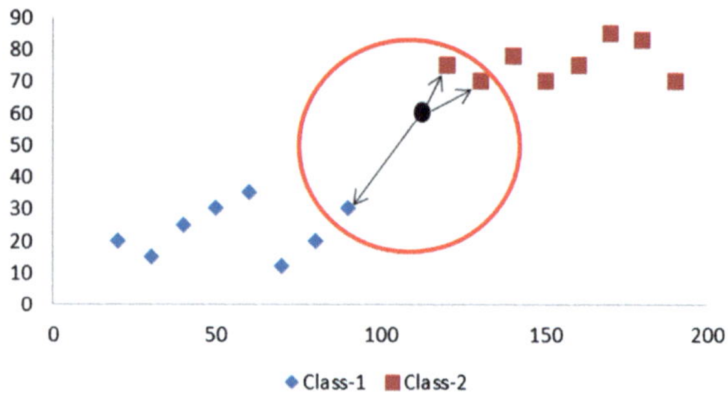

Fig. 5 Representation of KNN algorithm for K = 3

2 Related Work

Several research papers have been published to address the problem of handwritten alphabet recognition. In recent review articles [3–5], quality research work has been arranged systematically based on results and methods used for feature extraction and classification of various Indic and non-Indic language scripts like Devnagari, Bengali, Gurumukhi, Kannada, Telugu, Gujarati, Oriya, Tamil, and Malayalam etc. Feature extraction based on Deep Convolutional Neural Networks and their variants [6–11], geometric methods [12], wavelet transform [13], and PCA method [14] have been implemented extensively with different classifiers, namely, MLP, SVM, and KNN for recognition of handwritten numerals.

It has been found that most of the recent research works were based on CNN [6–11]. Such models need huge datasets, multiple layered network architectures, large memory space, and complex computations. The proposed scheme has tried to address these problems using Discrete Wavelet Transform with multiple classifiers for feature extraction and classification of handwritten digits.

3 Methodology

This section will describe all major stages of the proposed model. The complete design is depicted in Fig. 6.

3.1 Dataset Preparation

The dataset was prepared by collecting handwritten documents of Devnagari script from mass people belonging to different age groups, localities, and professions. The manuscripts were scanned and individual digits were cropped manually. Acharya and Pant have developed the whole dataset [11] and made it accessible in the public domain. For the proposed scheme, 5000 samples were selected.

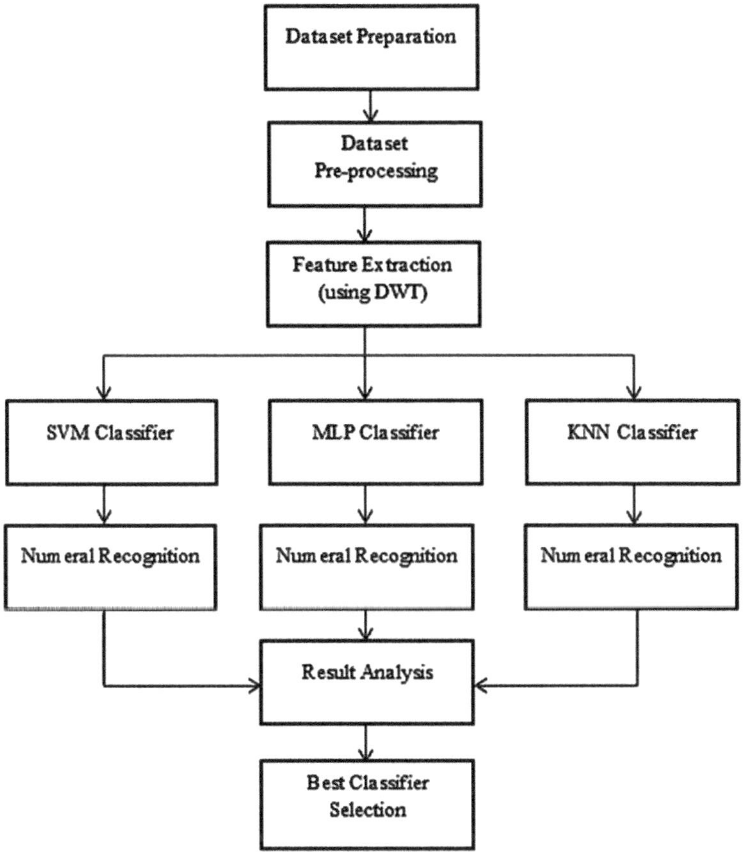

Fig. 6 Design of the proposed model

3.2 Dataset Pre-Processing

All the digit images of the dataset were labeled properly and checked for a uniform size of 32 by 32. The images were converted into a grey level format to achieve smoothness of brightness levels; as it was needed by discrete wavelet transform to produce fine results.

3.3 Feature Extraction

The discrete wavelet transform technique was used to extract salient features from pre-processed images. The Bi-orthogonal wavelet-1.3 (Bior-1.3) was implemented for the task. Various responses of Bior-1.3 are given in Fig. 7.

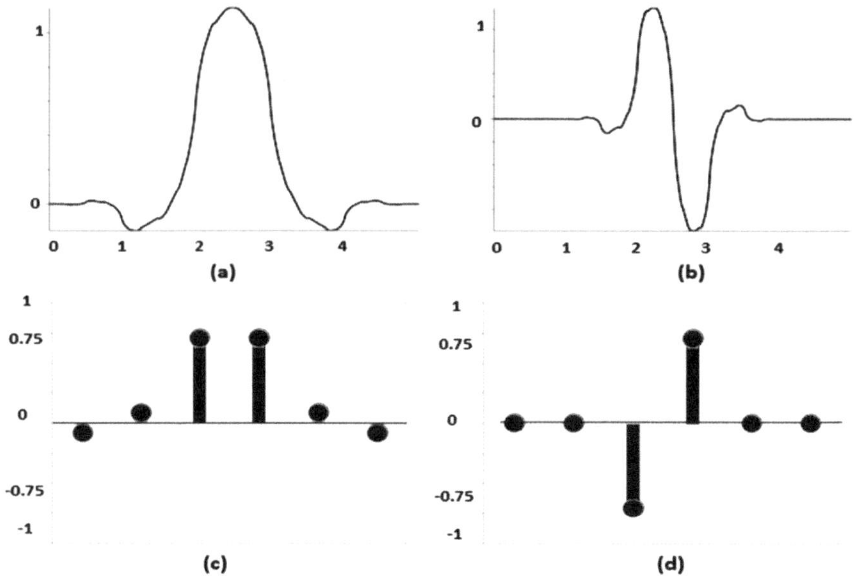

Fig. 7 Bior-1.3 decomposition responses (**a**) scaling, (**b**) wavelet, (**c**) low pass filter, and (**d**) high pass filter

The images of the dataset were decomposed into four sub-band images, namely, Approximation (A), Horizontal (H), Vertical (V), and Diagonal (D) up to level-2, as shown in Fig. 8.

The two-dimensional coefficients of sub-band images were transformed into a one-dimensional feature map for further use. The feature map format is given in Fig. 9.

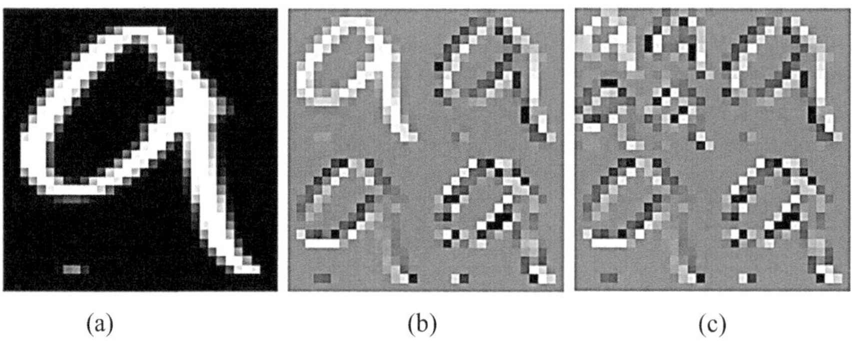

Fig. 8 Level-2 wavelet decomposition of digit image by proposed scheme; (**a**) Image, (**b**) level-1 decomposition, and (**c**) level-2 decomposition

Approximation sub-band coefficients (size:121)	Horizontal detail sub-band coefficients (size:121)	Vertical detail sub-band coefficients (size:121)	Diagonal detail sub-band coefficients (size:121)

Fig. 9 Format of the one-dimensional feature map

3.4 Classification

The whole dataset was split into Train and Test-set with a ratio of 75:25. The proposed scheme has employed three classifiers, namely, MLP, SVM, and KNN for the classification task. The associated parameters are summarized in Tables 1, 2, and 3.

Table 1 Major parameters of MLP classifier

Parameter	Value
Number of hidden layers	1
Units in hidden layer	100
Activation function in hidden layer unit	Rectified linear unit (Relu)
Solver	Adaptive moment(Adam)
Learning rate	0.001
Number of iterations	3000

Table 2 Major parameters of SVM classifier

Parameter	Value
Kernel	Linear
Cache size	200 MB
Max iterations	-1 (no limit)
Decision function shape	One versus rest (ovr)
Probability	False

Table 3 Major parameters of KNN classifier

Parameter	Value
Nearest neighbors	3
Weights	Uniform
Algorithm	Auto (will attempt to decide the most appropriate algorithm out of ball_tree, kd_tree and brute algorithms)
Distance metric	Minkowski
Number of parallel jobs to run neighbor search	1

The proposed scheme has implemented a cross-validation strategy to avoid the over-fitting problem of classifiers.

3.5 Recognition

The proposed scheme was devised to generate classification reports with associated confusion matrixes.

The confusion matrix is a technique to evaluate predictive model performance in terms of Accuracy, Precision, Recall, and F1-score. It gives a true interpretation of the result by mentioning a total number of correct and incorrect predictions for the given class. Mentioned metrics are determined in terms of True Positive (TP), True Negative (TN), False Positive (FP), and False Negative (FN).

Accuracy represents the ratio of the number of correct predictions to the total number of predictions made. It does not give information about how many events were falsely predicted as positive and negative. Mathematically, it is represented by Eq. (1).

$$\text{Accuracy} = \frac{\text{TP} + \text{TN}}{\text{TP} + \text{TN} + \text{FP} + \text{FN}} \tag{1}$$

Precision represents the ratio of correctly predicted positive events to total predicted positive events. High precision ensures a low false-positive rate. It is given by Eq. (2).

$$\text{Precision} = \frac{\text{TP}}{\text{TP} + \text{FP}} \tag{2}$$

The recall is the ratio of correctly predicted positive events to all events of the given class. High recall ensures a low false-negative rate. It is given in Eq. (3).

$$\text{Recall} = \frac{\text{TP}}{\text{TP} + \text{FN}} \tag{3}$$

F1-Score takes precision and recall both into account because individually they do not provide complete information about predictions. For example, high precision can be obtained with very low recall or vice versa; the F-measure includes both into a single score. It is given in Eq. (4).

$$F1 - \text{score} = 2.\frac{\text{Precision.Recall}}{\text{Precision} + \text{Recall}} \tag{4}$$

4 Results And Discussions

4.1 Results

Test results of the proposed model are compiled in Tables 4, 5, and 6. The classification report contains precision, recall, f1-score, and support values for each digit class.

The comparative analysis of three classifiers in terms of Precision, Recall, and F1-score are plotted in Fig. 10.

Table 4 Classification report of MLP classifier

Digit class	Precision	Recall	F1-score	Support Value
0	0.99	0.96	0.97	113
1	0.98	0.99	0.98	136
2	0.89	0.95	0.92	135
3	0.94	0.91	0.92	130
4	0.98	0.98	0.98	116
5	0.96	0.98	0.97	127
6	0.96	0.96	0.96	128
7	0.97	0.97	0.97	121
8	0.98	0.94	0.96	123
9	0.98	0.99	0.98	121
Overall recognition accuracy (%)		96.64		

Table 5 Classification report of SVM classifier

Digit class	Precision	Recall	F1-score	Support Value
0	1	1	1	131
1	1	1	1	134
2	0.99	1	0.99	123
3	1	0.99	0.99	118
4	1	1	1	129
5	1	1	1	126
6	1	1	1	122
7	1	1	1	122
8	1	1	1	134
9	1	1	1	111
Overall recognition accuracy (%)		99.84		

Table 6 Classification report of KNN classifier

Digit class	Precision	Recall	F1-score	Support Value
0	0.90	0.97	0.94	131
1	1	0.97	0.98	134
2	0.95	0.95	0.95	123
3	0.95	0.96	0.95	118
4	1	0.99	0.99	129
5	0.96	0.96	0.96	126
6	1	0.93	0.96	122
7	1	0.95	0.97	122
8	0.94	0.97	0.96	134
9	1	1	1	111
Overall recognition accuracy (%)		97.04		

The proposed classifiers; MLP, SVM, and KNN have been managed to score percentage recognition accuracies of 96.64, 99.84, and 97.04. The graphical representation of recognition accuracy is given in Fig. 11.

The relevant confusion matrixes are displayed in Fig. 12. For a given digit class, the terms TP, TN, FP, and FN can be easily determined from it.

To have a better visualization of decision boundaries drawn by various classifiers; the proposed scheme has determined two principal components from each feature map. Thereafter the Classifiers were made to train over these components for generating decision boundaries in two-dimensional space as shown in Fig. 13.

4.2 Discussions

Referring to Tables 4, 5, and 6; all the classifiers have produced excellent results for digits 1, 4, and 9, while comparatively low results for digits 2 and 3. This might be due to the uniqueness of shapes in the former case and the presence of a degree of similarity in the latter case.

Referring to Fig. 10; the SVM classifier has achieved the highest scores for all digit classes. It has attained 100% of the F1-score for all the digit classes except 2 and 3. The scores of MLP and KNN classifiers were comparable to each other.

Referring to Fig. 11, the SVM classifier has outperformed the rest of the classifiers in overall recognition accuracy.

Referring to Fig. 13, decision region boundaries developed by the SVM classifier has more crisp than the rest of the two.

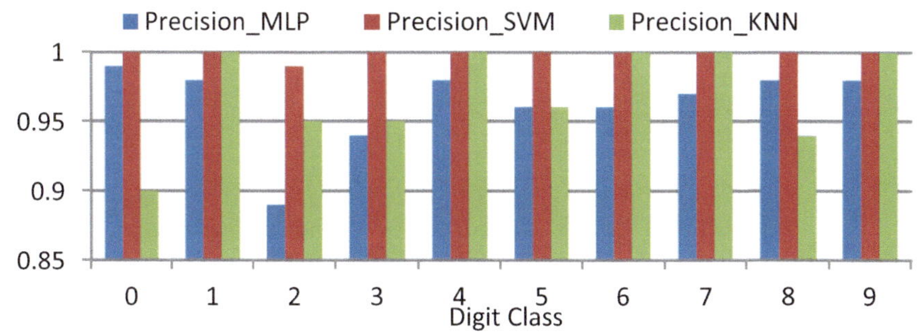

(a)

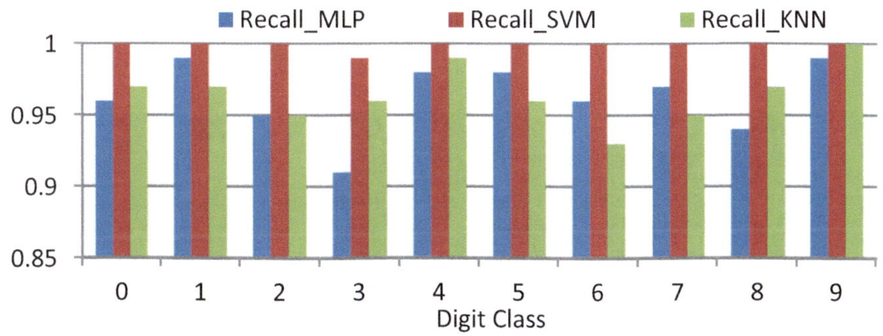

(b)

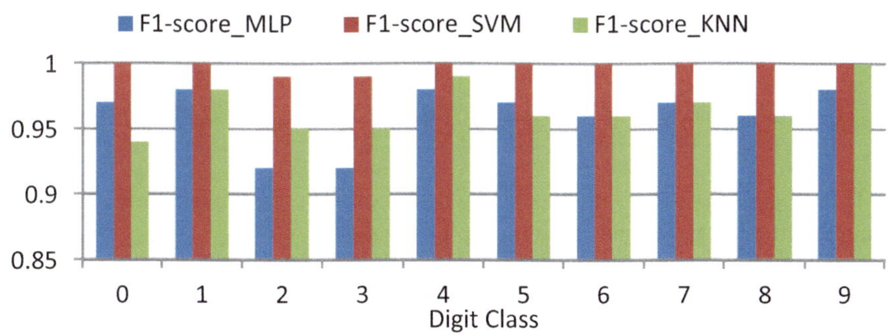

(c)

Fig. 10 Comparative analysis of proposed classifiers; (**a**) Precision, (**b**) Recall, and (**c**) F1-score

5 Conclusion

The proposed scheme has achieved a maximum of 99.84% recognition accuracy without DCNN. Results are comparable to some state-of-the-art existing research work; as summarized in Table 7.

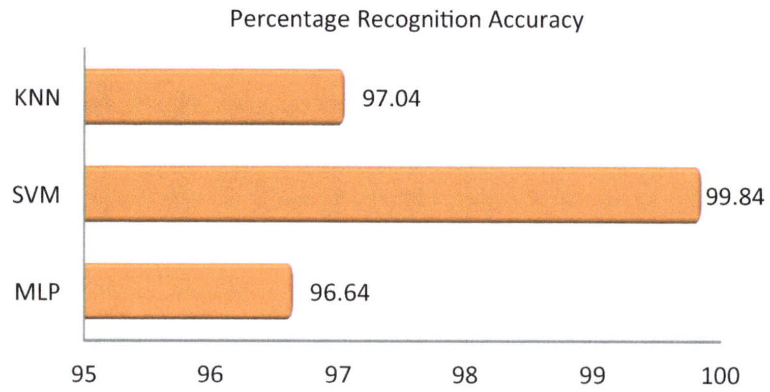

Fig. 11 Recognition Accuracy of proposed classifiers

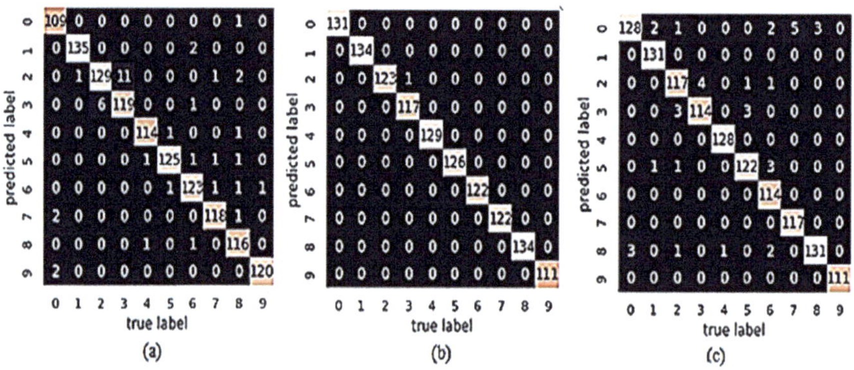

Fig. 12 Confusion matrix of (**a**) MLP classifier, (**b**) SVM classifier, and (**c**) KNN classifier

The proposed scheme may be extended in the future by implementing multiple feature extraction techniques with multiple classifiers.

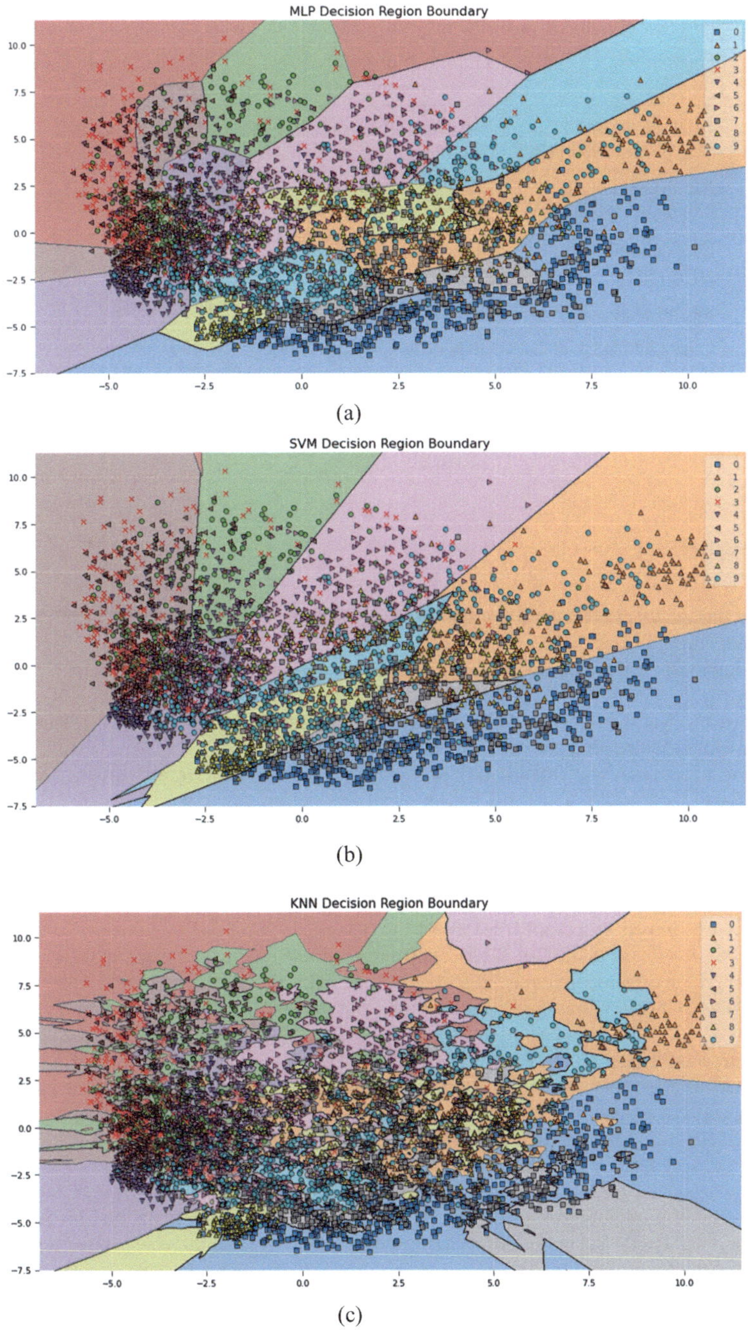

Fig. 13 Decision region boundaries; (**a**) MLP classifier, (**b**) SVM classifier, and (**c**) KNN classifier

Table 7 Summary of some state-of-the-art existing techniques with the proposed one

Sl. No	Particular	Technique used for Recognition	Max. Recognition Accuracy achieved (%)
1	Bhattacharya and chaudhuri, 2009 [13]	Wavelet transform with MLP classifier	65.20
2	Acharya, Pant and Gyawali, 2015 [11]	DCNN	98.47
3	Dhakad and Soni, 2017 [14]	PCA with KNN and SVM classifier	93.21
4	Sufian et.al., 2020 [8]	Densly connected CNN	99.78
5	Gupta and Bag, 2021 [6]	CNN with SVM classifier	99.50
6	Proposed scheme	Bi-orthogonal wavelet with MLP,SVM, and KNN classifiers	99.84

References

1. Lawgali A (2015) Handwritten digit recognition based on DWT and DCT. Int J Database Theory Appl 8(5):215–222
2. Amancio DR et al (2014) A systematic comparison of supervised classifiers. PLoS ONE 9(4):1–13
3. Sharma R, Kaushik B (2020) Offline recognition of handwritten Indic scripts: a state-of-the-art survey and future perspectives. Comput Sci Rev 38:100302
4. Yadav M, Purwar RK, Mittal M (2018) Handwritten hindi character recognition: a review. IET Image Process 12(11):1919–1933
5. Memon J, Sami M, Khan RA, Uddin M (2020) Handwritten Optical Character Recognition (OCR): A Comprehensive Systematic Literature Review (SLR). IEEE Access 8:142642–142668
6. Gupta D, Bag S (2021) CNN-based multilingual handwritten numeral recognition: a fusion-free approach. Expert Syst Appl 165:113784
7. Aly S, Almotairi S (2020) Deep convolutional self-organizing map network for robust handwritten digit recognition. IEEE Access 8:107035–107045
8. Sufian A, Ghosh A, Naskar A, Sultana F, Sil J, Rahman MMH (2020) BDNet: Bengali handwritten numeral digit recognition based on densely connected convolutional neural networks. J King Saud Univ - Comput Inf Sci, Article in press
9. Chaurasia S, Agarwal S (2018) Recognition of handwritten numerals of various indian regional languages using deep learning. In: Proceedings of 2018 5th IEEE Uttar Pradesh Sect Int Conf Electr Electron Comput Eng. UPCON 2018. Uttar Pradesh, India, pp 1–6
10. Kumar S, Aggarwal RK (2018) Augmented handwritten devanagari digit recognition using convolutional autoencoder. In: Proceedings of the International Conference on Inventive Research in Computing Applications, ICIRCA 2018. Koimbatore, India, pp 574–580
11. Acharya S, Pant AK, Gyawali PK (2015) Deep learning based large scale handwritten Devanagari character recognition. In: Proceedings of SKIMA 2015 - 9th International Conference on software, knowledge, information management and applications, 2015. Kathmandu, Nepal, pp 1–6
12. Ahlawat S, Rishi R (2017) Off-line handwritten numeral recognition using hybrid feature set - a comparative analysis. Procedia Comput Sci 122:1092–1099
13. Bhattacharya U, Chaudhuri BB (2009) Handwritten numeral databases of Indian scripts and multistage recognition of mixed numerals. IEEE Trans Pattern Anal Mach Intell 31(3):444–457

14. Dhakad R, Soni D (2017) Devanagari digit recognition by using artificial neural network. Int Conf Energy Comm, Data Analytics and Soft Computing, ICECDS 2017, Chennai, India, pp 2754–2757

Chapter 51
Performance Evaluation and Comparison of Optical Amplifiers in Nonlinear Effects for WDM Long-Haul Transmission System

Tsegaye Menber Belay and Pushparaghavan Annamalai

1 Introduction

The most important transmission impairments associated with long-distance high-Bit Rates optical fiber communication systems include fiber dispersion, fiber Kerr nonlinearities, the terms linear and nonlinear, in optics, refer to intensity independent and dependent phenomena, respectively. When the incident optical power of a fiber exceeds a certain value, the optical power density in the fiber core will become excessively high because the effective cross-sectional area of a fiber ranges only from 50 to 80 μm^2.

When a high-intensity pulse enters into the fiber, the region of the fiber that interacts with the peak of the pulse is having high refractive index while the region that interacts with the trough of the pulse is having low refractive index. Because of this continuous change of refractive index each, of the components of the pulse propagates at a different velocity which induces a nonlinear phase shift in the pulse which results broadening of a pulse. The change in the refractive index of the medium with optical intensity and inelastic scattering gives rise to nonlinear effects in optical fibers. The intensity dependence of the refractive index is responsible for the Kerr effect. Depending upon the type of input signal, the Kerr nonlinearity has three different effects such as SPM, XPM, and FWM [1, 2]. As refractive index of fiber

T. M. Belay (✉)
Lecturer, Communication Engineering, Faculty of Electrical and Computer Engineering, Gondar University, Gondar, Ethiopia
e-mail: tsegu.menbe@gmail.com

P. Annamalai
Professor, Communication Engineering, Faculty of Electrical and Computer Engineering, Bahir Dar Institute of Technology, Bahir Dar, Ethiopia
e-mail: aprshamu@gmail.com

G. Mathur et al. (eds.), *Proceedings of 2nd International Conference on Artificial Intelligence: Advances and Applications*, Algorithms for Intelligent Systems,
https://doi.org/10.1007/978-981-16-6332-1_51

core is now power dependent, it affects chromatic dispersion in the fiber, which changes the pulse broadening rate throughout the fiber.

2 Fiber Kerr Nonlinearity Effects

A. Self-Phase Modulation (SPM)
SPM arises due to phase alteration of a pulse by its own intensity. Optical fiber is an intensity-dependent medium. The refractive index varies when the intensity of the pulse varies. This causes an induced nonlinear phase shift that is proportional to the intensity of the pulse. Thus, different parts of the pulse undergo different phase shifts, which give rise to chirping of the pulses. The SPM-induced chirp affects the pulse broadening effects of chromatic dispersion and thus, it is important to consider for high-Bit-Rate systems that already have significant chromatic dispersion limitations [4–8].

B. Cross-Phase Modulation (CPM)
When two optical pulses pass through the same fiber, the intensity variation of one of the high-power intensities alters the second pulse, which causes phase modulation. This is also referred to as the walk over effect since two pulses at varying Bit Rates or group velocities walk over each other. So, power fluctuation in one of the optical pulses that causes the phase fluctuation in the co-propagating pulse [5, 8].

C. Four-Wave Mixing (FWM)
It is an intermodulation phenomenon in nonlinear optics, whereby interactions between two wavelengths produce four or more new wavelengths. Unlike SPM and XPM, which result in nonlinear phase shift in the optical field, the FWM process causes an energy transfer between co-propagating channels. This leads to power depletion, which degrades the performance.

In addition, FWM yields inter-channel cross talk if the generated signal falls into other co-propagating channels. This results in significant system performance degradation due to cross talk among channels. FWM depends on the fiber dispersion and channel spacing [8]. Thus, increasing the fiber dispersion limits the interactions between signals and reduces the power transfer to the newly generated signals. Increasing the channel spacing decreases the FWM effect as well.

3 System Model

Figure 1 shows the system block diagram and model for performance of long-haul transmission with optical amplifiers using four-channel WDM system. Its performance has been measured with Q-factor and Bit Error Rate (BER), by varying the Bit Rate, input power, and length of the Single-Mode Fiber (SMF).

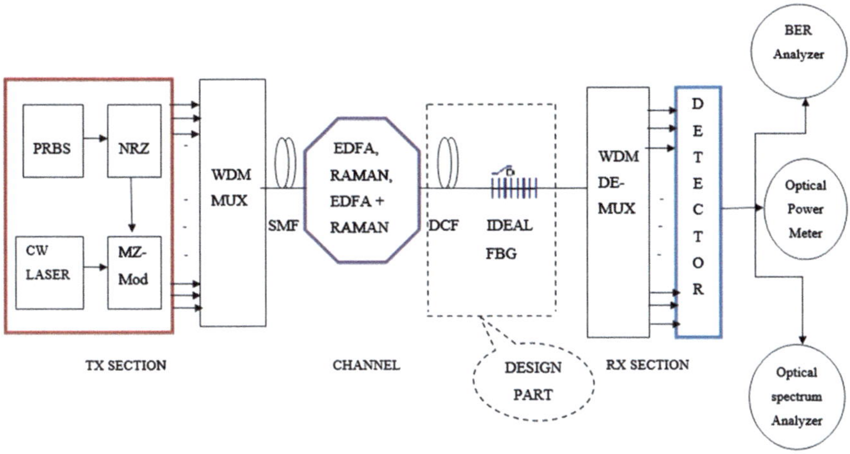

Fig. 1 Block Diagram for WDM with Optical Amplifiers in Long-Haul transmission

3.1 Optical Transmission Link Design for Long-Haul Transmission System

The following three Transmission Link design helps to measure the optical signal power at each consecutive node for Linear Optical Amplifier (EDFA), non-Linear optical amplifier (RAMAN), and hybrid (EDFA + RAMAN) optical amplifier to determine the total received power at the detector.

(a) Optical Transmission Link Design for Linear Optical Amplifier(EDFA)
The following Fig. 2 shows Linear Optical Amplifier (LOA) link design for 4 × 10 Gb/s optical network, with four channels and the received power measured at each stage. The Bit Rate of each channel is 10 Gb/s with 100 GHz channel spacing, the total transmission span 120 km (115 km SMF and 5 km DCF) is considered for the transmission distance.

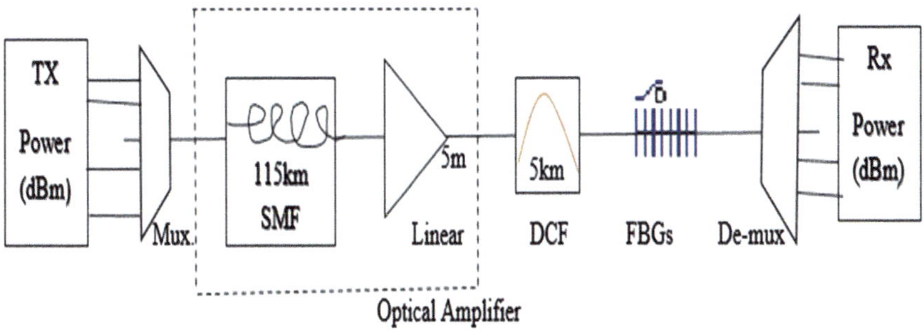

Fig. 2 Transmission Link Design Using Linear Optical Amplifier (LOA)

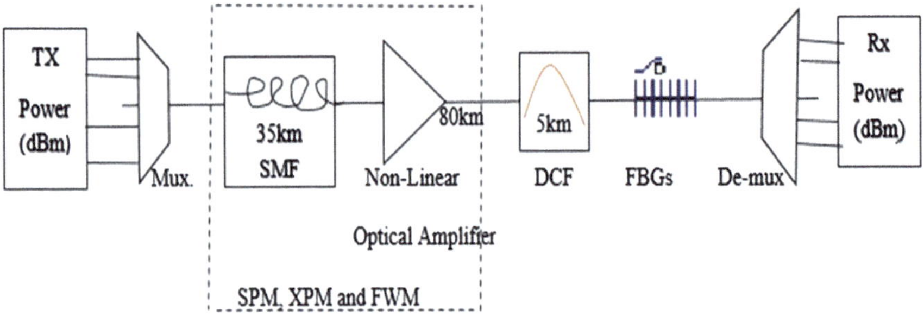

Fig. 3 Transmission Link Design Using Non-Linear Optical Amplifier (NLOA)

The following Eq. (1) helps to calculate the total received power Linear Optical Amplifier (EDFA) link design and calculated the received power, P_R:

$$P_{R,total} = P_T - MZM_{loss} - (\alpha_{SMF} + (SPM + XPM)_{loss}) - \alpha_{Mux} - \alpha_{DeMux} - DCF_{insertionloss} - FBG_{loss} \quad (1)$$

$$P_R = 15.96\,dB.$$

(b) Optical transmission link Design for Non-Linear Optical Amplifier (NLOA) Fig. 3 shows Non-Linear Optical Amplifier(NLOA) link design for 4×10 Gb/s optical network.

The following Eq. (2) helps to calculate the total received power for non-Linear Optical Amplifier (RAMAN) link design and calculated the received power P_R:

$$P_{R\,total} = P_T - MZM_{loss} - (\alpha_{SMF} + (SPM + XPM)_{loss}) - RAMAN_{dis} - \alpha_{Mux} - \alpha_{DeMux} - DCF_{insertionloss} - FBG_{loss} \quad (2)$$

$$P_R = -103.64\,dB.$$

(c) Optical Transmission Link Design for Hybrid Optical Amplifier (LOA + NLOA)
Fig. 4 shows Hybrid configuration of Linear Optical Amplifier (LOA) and Non-Linear Optical Amplifier(NLOA) link design for 4×10 Gb/s optical network.

Below Eq. (3) helps to calculate the total received power Linear and non-Linear Optical Amplifier (EDFA + RAMAN) link design and found out the received power, $P_{R:}$

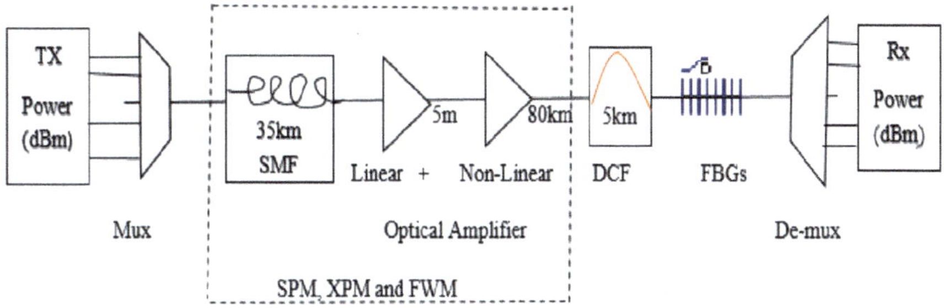

Fig. 4 Transmission Link Design Using Linear Optical Amplifier and Nonlinear Optical Amplifiers (LOA + NLOA)

Table 1 SMF and DCF parameters

SMF parameter	Value
Attenuation (a)	0.2 dB/km
Dispersion parameter (D)	16.75 ps/nm km
Dispersion slope (S)	0.075 ps/nm^2 km
Effective area (A_{eff})	80μm^2
DCF parameter	Value
Attenuation (a)	0.5 dB/km
Effective area (A_{eff})	22μm^2

$$P_{\text{Rtotal}} = P_{\text{T}} - \text{MZM}_{\text{loss}} - (\alpha_{\text{SMF}} + (\text{SPM} + \text{XPM})_{\text{loss}}) - \text{RAMAN}_{\text{loss}} - \alpha_{\text{Mux}} - \alpha_{\text{DeMux}} - \text{DCF}_{\text{insertionloss}} - \text{FBG}_{\text{loss}} \quad (3)$$

$$P_{\text{R}} = -10.635\,\text{dB}.$$

The following Tables 1 and 2 show dispersion and dispersion slope parameters for Single-Mode Fiber (SMF) land DCF [11].

4 Results and Discussion

i. *Four-Channel Performance Comparison Between Optical Amplifiers Performance (EDFA, RAMAN, EDFA + RAMAN) for various Bit Rates in Gbps*

   We can see in Figs. 5 and 6 that performance comparison of EDFA, RAMAN, EDFA + RAMAN in the presence of SPM and XPM, from the result we can conclude that the use of EDFA adds more noise signals, which causes distortion of signals and affects signal flatness. Therefore, to achieve signal flatness,

Table 2 Calculated D_{DCF} and S_{DCF} values for respective SMF length

Calculated D_{DCF} and S_{DCF} values for respective SMF length DCF = 5 km		Optical amplifiers		
	EDFA, RAMAN, EDFA + RAMAN	EDFA, RAMAN, EDFA + RAMAN	EDFA, RAMAN, EDFA + RAMAN	EDFA, RAMAN, EDFA + RAMAN
SMF length (km)	55	115	175	235
Dispersion of DCF (D_{DCF}) parameter	$L_{SMF} \times D_{SMF} = -D_{DCF} \times L_{DCF}$ [10–12]			
	−184.25 ps/nm km	−385.25 ps/nm km	−586.25 ps/nm km	−787.25 ps/nm km
Dispersion slope for DCF (S_{DCF})	$S_{DCF} = S_{SMF} \times \left(\frac{-D_{DCF}}{D_{SMF}}\right)$ [12]			
	−0.825 Ps/nm^2 km	−1.725 Ps/nm^2 km	−2.625 Ps/nm^2 km	−3.525 Ps/nm^2 km

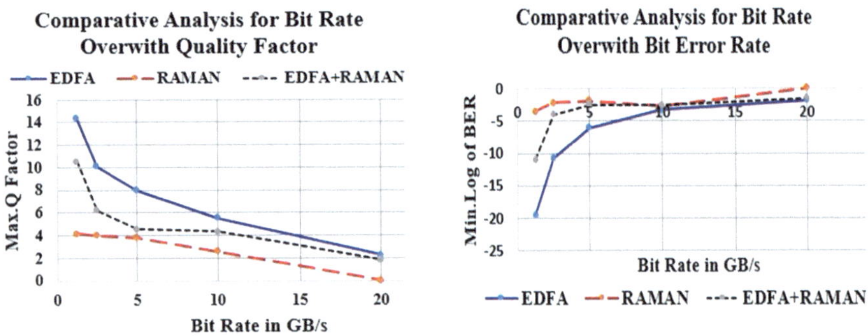

(a). Q-factor vs. Bit Rate (b). BER vs. Bit Rate

Fig. 5 Plot for comparison of EDFA, RAMAN, and EDFA + RAMAN. **a** Q-factor vs. Bit Rate and **b** Bit Error Rate (BER) vs. Bit Rate in the presence of SPM

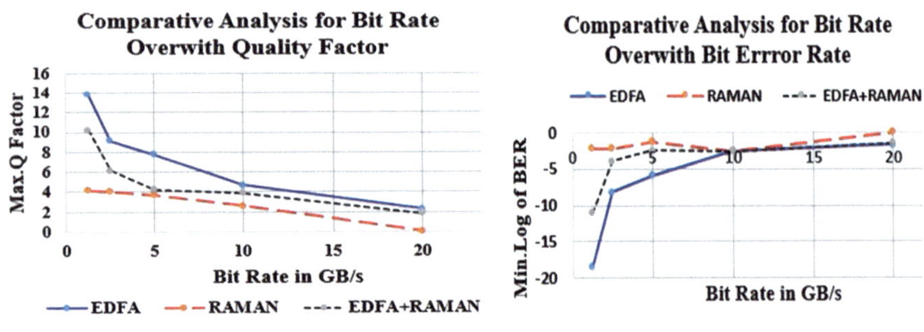

(a). Q-factor vs. Bit Rate (b). BER vs. Bit Rate

Fig. 6 Plot for comparison of EDFA, RAMAN, and EDFA + RAMAN. **a** Q-Factor vs. Bit Rate and **b** BER vs. Bit Rate in the presence of XPM

RAMAN amplifier can adequately produce flat gain output with appropriate pumping wavelength which depends on the pumping power (Figs. 7 and 8).

Eventually, the Bit Rate increases. EDFA is seriously affected by SPM and XPM more than RAMAN, as a result the Q-factor degrades and BER increases. But RAMAN has almost a linear change because of less noise, hence combining the two can reduce the change.

ii. *Four-Channel Performance Comparison Between Optical Amplifiers Performance (EDFA, RAMAN, EDFA + RAMAN) for various Input Power in dBm*

For low input power, EDFA always has high Q-Factor and low BER, as the input power increases the Q-Factor degrades and BER increases, on the other hand, RAMAN amplifier has low Q-Factor and highest BER at low input power (0dBm) and high Q-Factor and low BER at 15dBm.

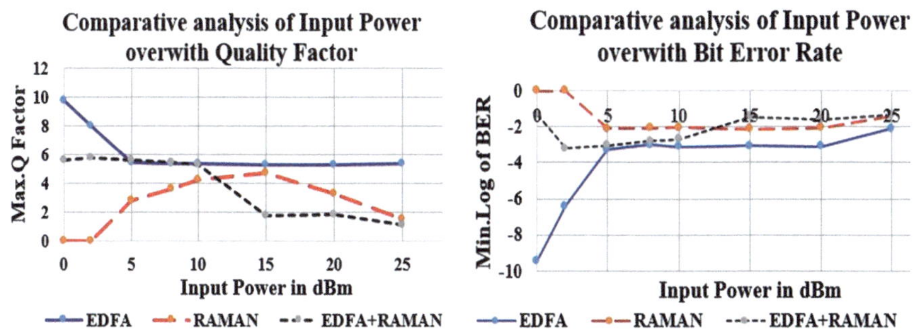

(a) Q-factor vs. input power (b). BER vs. input power

Fig. 7 Plot for comparison of EDFA, RAMAN, and EDFA + RAMAN. **a** Q-Factor vs. input power and **b** BER vs. input power in the presence of SPM

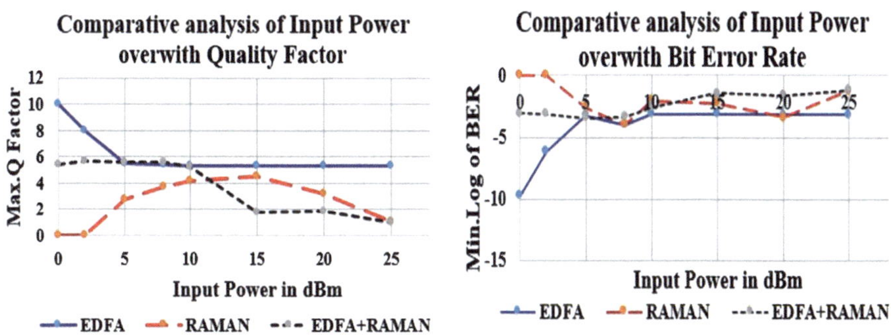

(a) Q-factor vs. input power (b). BER vs. input power

Fig. 8 Plot for comparison of EDFA, RAMAN, and EDFA + RAMAN. **a** Q-Factor vs. input power and **b** BER vs. input power in the presence of XPM

iii. *Four-Channel Performance Comparison Between Optical Amplifiers Performance (EDFA, RAMAN, EDFA + RAMAN) for various fiber lengths in km*

From Figs. 9 and 10, we see that for short distance EDFA always has high Q-Factor and low BER, as the optical fiber length increases the Q-Factor degrades and BER increases. When we compare EDFA with RAMAN amplifier, EDFA has high Q-Factor and low BER than RAMAN Amplifier for long-haul transmission. Using both EDFA + RAMAN amplifier, it is possible to build the high gain low noise and attenuation WDM transmission system because RAMAN amplifier has a low noise that can reduce the high noise of EDFA, because RAMAN amplifier can support EDFA to increase the transmission span because of its low noise and distributed

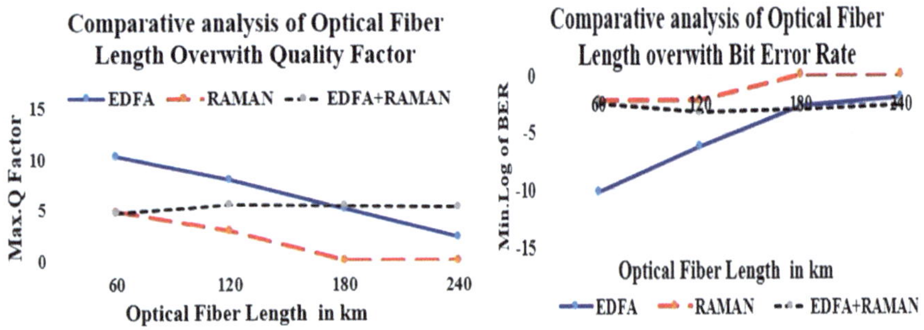

(a) Q-Factor vs. Distance (b).BER vs. Distance

Fig. 9 Plot for comparison of EDFA, RAMAN, and EDFA + RAMAN. **a** Q-Factor vs. distance and **b** BER vs. distance in the presence of SPM

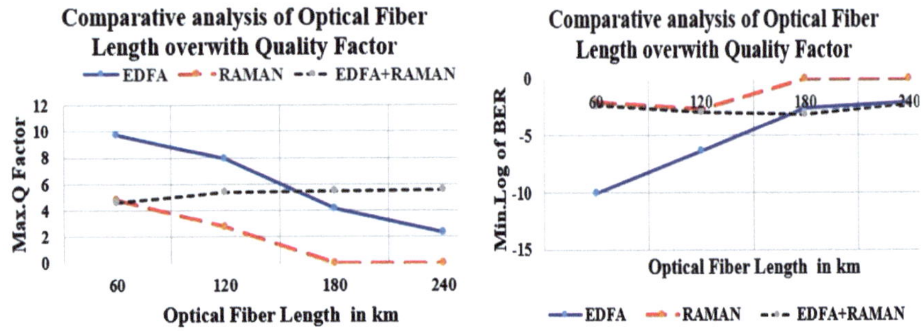

(a) Q-Factor vs. Distance (b). BER vs. Distance

Fig. 10 Plot for comparison of EDFA, RAMAN, and EDFA + RAMAN. **a** Q-Factor vs. distance and **b**. BER vs. distance in the presence of XPM

gain. As a result, EDFA + RAMAN can enhance the transmission span compared to EDFA and RAMAN using alone.

5 Conclusion and Further Study

The performance of both LOA and NLOA for WDM long-haul transmission with the approaches of FCS and DCS has been investigated. We analyzed various cases of Four-Channel EDFA, RAMAN, and EDFA + RAMAN, without SPM, with SPM, and with XPM with the support of DCF and FBGs as a Dispersion Compensation. In overall performance analysis, it has been evaluated and compared for different Iterative parameters such as: Bit Rate (1.25, 5, 10, 20) Gb/s, Input Power (0, 2, 5, 8, 10, 15, 20, 25) dBm, Optical Fiber Length (60, 120, 180, 240) km, and DCS (12.5, 25, 50, 100, 200) GHz. Hence, comparison is performed by varying one parameter and keeping constant others simultaneously (Bit Rate, Input Power, Optical Fiber Length, and the channel spacing).

Mainly, Four-Channel WDM/DWDM system for DCS has been investigated in two cases; high Bit Rate and low Bit Rate (1.25 and 10 Gb/s). As optical amplifiers comparison of Output Optical Signal Power (OOSP), Output Optical Noise Power (OONP), and Output Optical Signal-to-Noise Ratio (OOSNP) at high Bit Rate (10 Gb/s) show that the OOSP in EDFA is higher than both RAMAN and hybrid optical amplifier, to compare the values (EDFA > EDFA + RAMAN > RAMAN) which is 17.36 dBm, -24.51dBm, and 2.94 dBm, respectively, by taking the highest values of each.

For OONP, EDFA has higher noise power than both RAMAN and EDFA + RAMAN.

Amplifiers and their output noise powers are: (−79.03dBm, −100dBm, and −100dBm), respectively, taking the minimum noise powers for each. However, OOSNR is the difference between OOSP and OONP.

From the three optical amplifiers, EDFA + RAMAN has the highest OOSNR value than EDFA and RAMAN, to compare the values (EDAF + RAMAN > EDFA > RAMAN) which is, 102.94 dB, 95.11 dB, 75.49 dB, respectively. Therefore, EDFA + RAMAN gives better OOSNR, than EDFA and RAMAN amplifiers because of noise compensation using RAMAN amplifiers.

In WDM long-haul transmission system, optical amplifiers boost the weak Signal, DCF enhances the skew/distorted signal whereas FBG act as Optical Filter to block the undesired signal wavelengths. In future work, we recommend that FBGs along with DCF can enhance the transmission efficiency more and reduce signal degradation and nonlinear impairments, by cascading multiple gratings with different wavelengths and different exposures.

References

1. Bhusari Shraddha N, Deshmuku Vikas U (2016) Analysis of SPM, XPM, and FWM in fiber optic communication using OptiSystem
2. Study of fiber nonlinear effects on fiber optic communication systems a Ph. D. dissertation. Presented to the Faculty of School of Engineering and Applied Sciences University of Virgini (August 2015)
3. Sidhik S (2014) Performance analysis of SPM, XPM and FWM in an optical fiber using Optisystem
4. Sihval NK, Garg AK (2015) Simulation and analysis of self phase modulation fiber non linearity
5. de Sousal FB, de Sousa FM (2017) Simulations of SPM, XPM and FWM in single-mode fiber optic networks
6. Shraddha NB, Deshmuku Vikas U (2016) Analysis of SPM, XPM, and FWM in fiber optic communication using OptiSystem
7. S. P. Singh and N. Singh, Nonlinear Effects in Optical Fibers: Origin, Management and Applications.
8. Abdelkerim Amari, Octavia A. A survey on fiber non-linearity compensation for 400 Gbps and beyond optical communication systems
9. Kiaee MS, Seraji FE (2016) Design of a 32 × 5 GB/s DWDM optical network over a distance of 1000 km
10. Asiwal A, Dhawa D (2017) Comparative analysis of dispersion compensating fiber (DCF) and multiple optical phase conjugation (OPC) modules used for dispersion compensation and nonlinearity mitigation
11. Kumar S (2015) Comparative analysis of different configurations of optical amplifiers (EDFA, RAMAN and EDFA + RAMAN) for Intensity Modulated WDM System
12. Sadat Kiaee M, Seraji FE (2016) Design of a 32 × 5 GB/s DWDM optical network over a distance of 1000 km

Chapter 52
Sentiment Analysis Through Machine Learning: A Review

Meenu Bhagat and Brijesh Bakariya

1 Introduction

The sentimental analysis aims to derive information based on the opinions of the speaker or writer towards a particular issue, e.g., a review about a movie or product. The word sentiment analysis was first used in papers as a keyword around a decade ago, but the discipline has emerged in the middle of the nineteenth century. It can be classified as a supervised learning problem with three different kinds of classes, i.e., neutral, negative, and positive. It differs from topic-based text classification which classifies documents into predefined subject categories like sports, law, politics, etc. Opinion words (also known as sentiment words) that signify negative and positive opinions, such as nice, sad, excellent, fantastic, brilliant, horrible, awful, worst, etc., are used instead of topic-related words in Sentiment classification. Naïve Bayesian classification and Support vector machines (SVM) are examples of supervised learning methods that can be applied for the classification of sentiments. Sentiment Analysis can be further categorized into three classification levels, i.e., Document level, Sentence level and Aspect level. Document Level is used for expressing whether a document is expressing a positive or negative opinion. The entire document is treated as a single unit. In Sentence Level, we express the sentiments at the sentence level. They can be negative or positive. Aspect Level classifies the different sentiments based on the various aspects of an entity, e.g., "The sound quality of this video is good but picture could be better", so the first step is to find entities and their different aspects. This paper is organized as follows: Sect. 1 provides an overview about sentimental analysis. In Sect. 2, a thorough review of the literature is carried out. Section 3 elaborates various data sources that are being used in sentimental analysis and provides details about the processing steps to be followed in

M. Bhagat (✉) · B. Bakariya
Department of Computer Science & Engineering, I.K. Gujral Punjab Technical University, Jalandhar, Punjab, India

G. Mathur et al. (eds.), *Proceedings of 2nd International Conference on Artificial Intelligence: Advances and Applications*, Algorithms for Intelligent Systems,
https://doi.org/10.1007/978-981-16-6332-1_52

sentiment analysis. Section 4 discusses sentiment analysis methods such as lexicon-based, machine-leaning, and hybrid approaches. The comparative analysis is defined in Sect. 5, and the paper's conclusions are presented in Sect. 6.

2 Literature Review

Electronic word-of-mouth (e-WOM) has been emerged as a modern mode of communication, replacing the traditional word of mouth (WOM), used earlier as the oral form of communication. Malo et al. [1] focused on developing advanced methods for analyzing emotions in the financial domain, but domains aren't always distinct, and a set of terms that works in one domain does not work in another. Li and Zong [2] introduced a multi-domain sentiment classification approach for overcoming the previous shortcomings handled in the case of domain-specific data. Using Twitter, Akcora et al. [3] attempted to recognise emotional and word patterns in order to influence public opinion. Sun et al. [4] used Facebook fan groups to try to explain diffusion trees. As three separate indicators of impact, Kwak et al. [5] compared page-ranks, retweets, and number of followers. They discovered that ranking users by page rank and number of followers produces similar results, but results differed when the ranking criteria was changed to "number of tweets". Cha et al. [6] compared the number of retweets, followers, and mentions, and came to the conclusion that it is not appropriate to rank on all other metrics by the most followed users.

Rui Yao et al. [7] suggested a model for getting good accuracy of revenue of box office sales using online review of movie data. Weng et al. [8] compared the number of page-ranks and followers for a given topic with a change in the page-rank, and came to the conclusion that the ranking was influenced by prominent Twitter users. Collier et al. [9] proposed methods for classifying text data into negative and positive polarity using SVM, and they worked on extracting value phrases, giving sentiment orientation values through point-wise mutual information.

Cheong et al. [10] developed a sentiment analysis model, to estimate the terrorism scenarios by giving graphical visualizations based on sentiments collected from Twitter. Yates and Paquette [11] had also worked on social media technologies and their effects on decision making and sharing of knowledge during the Haitian Earthquake. Lee et al. [12] worked on the Twitter usage behaviors of journalists. In a case study, Crannel et al. [13] explored the behaviors of cancer patients while using Twitter for expressing their feelings, and it was observed that cancer patients expressed their feelings openly on Twitter. Carley et al. [14], examined the reactions of Twitter users during Tsunami warnings in Padang Indonesia.

Nagamma P et al. [15] worked on different types of data mining methods to predict the box office collection of the movies. By classifying movie reviews with the clustering method was used to improve accuracy. Ahmed et al. [16] analyzed the political campaigns during general elections in India and worked on how first-time voters are affected by the Internet.

3 Data Collection and Process [17]

User opinions collected through various methods like review sites, blogs, datasets, and microblogs are the major sources criterion for enhancing the quality of services rendered.

3.1 Review Sites

Opinions expressed by different users can be a deciding factor before making any purchase. A large volume of data is present on the Internet containing reviews of different users regarding different products and services. The database used in the sentimental analysis for reviewing is generally collected from e-commerce sites like www.Amazon.com, www.eopinions.com (Popesscu & Etzioni 2005 [18]; Hu, B. Liu, 2006 [19]), etc.

3.2 Blogs

Blogs are used as a major source of collecting opinions in sentimental analysis. They are used to record emotions and opinions about a product, event, or any issue. They are used in the sentimental analysis as a source of opinion.

3.3 Dataset

Mostly, we use movie reviews data in the field for classification. Datasets containing review data are available on http://www.cs.cornell.edu/People/pabo/movie-review data) [20] and multi-domain datasets, e.g., (http://www.cs.jhu.edu/mdredze/datasets/sentiment) [21]. Multi-domain datasets contain reviews on different products like kitchen appliances, electronics, DVDs and books, etc., with positive as well as negative reviews.

3.4 Microblogging

Twitter is one of the quite frequently used microblogging services, where users send status messages, called "tweets" for expressions of their opinions. These tweets are used to express opinions about different issues, e.g., sports, beauty, politics, health, etc. In most of the research studies, tweets have been used as a major data source for the classification of various sentiments about any product or any event.

General steps to be followed in the sentimental analysis process are shown in Fig. 1.

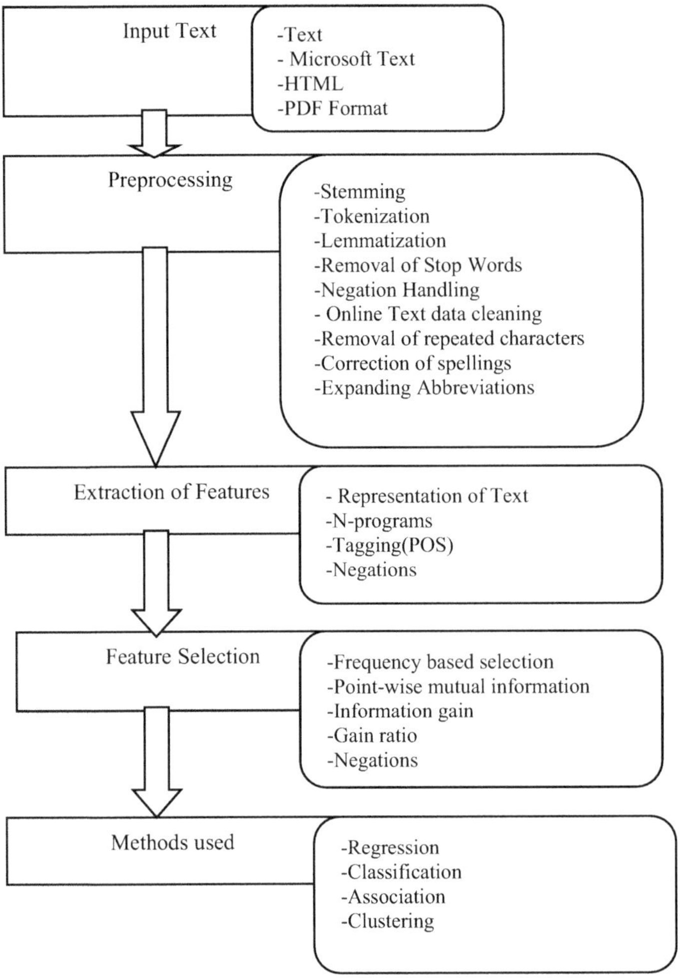

Fig. 1 Sentimental analysis process

4 Sentimental Analysis Techniques

4.1 Lexicon-Based Approach

In this approach, we build a sentiment lexicon of terms for calculating the sentiments for a given text and further counting the number of positive and negative words in the text. It can be divided into two types of approaches: Dictionary based and Corpus based. We first look for the opinion words from the review text in dictionary-based approach. In corpus based, we find the opinion words in a context-specific orientation approach and they require a big labeled dataset.

4.2 Machine Learning

It is the most prominent technique used by researchers due to its accuracy. It consists of the following stages:

1. Data Collection: Data is gathered from a variety of sources like social networking sites, blogs, etc.
2. Data Preprocessing: In this stage, cleaning of data is done and keywords and symbols are extracted, e.g., removing white spaces, correcting upper and lower case to one case, etc.
3. Training: In this stage, hand-tagged data is prepared and that acts as fuel for classifier.
4. Classification: In this stage, SVM or Naïve Bayes methods are used for classification.
5. Results: In this stage, results are plotted on charts/graphs or any other representation method used.

Classification of Machine Learning Techniques

There are three major approaches to machine learning [Fig. 2]: Supervised Learning, Un-supervised Learning, and Reinforcement Learning.

Supervised Learning: It is the most widely used method of performing machine learning operations. It is called supervised, as it seems as if a teacher is guiding the entire learning process. In this case, dataset is labeled it contains input as well as desired outputs. The algorithm recognizes all the features explicitly and prediction and classification of data are done accordingly. The main purpose of supervised learning is to find the correct label for any given new set of inputs based on its training experience with the training dataset. Decision tree, Linear regression, Naive Bayes, Random Forest, and SVM are common supervised machine learning algorithms [Fig. 3].

Unsupervised Learning: The dataset is not specifically classified into different groups in the case of an unsupervised learning algorithm, i.e., output variables

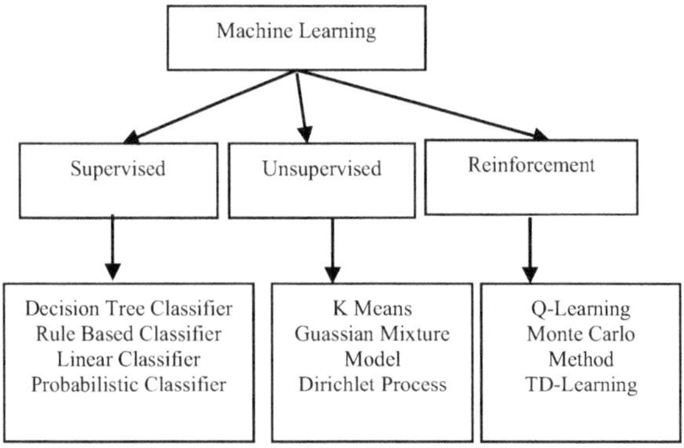

Fig. 2 Classification of machine learning techniques

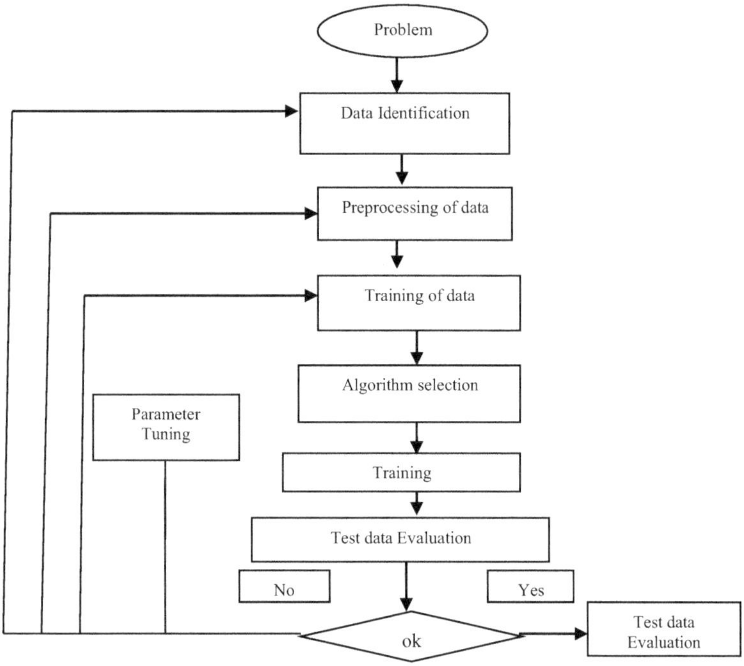

Fig. 3 Processing steps in supervised machine learning

are not predefined for the given input variables. The learning process is based on finding implicit patterns or clusters. The data is identified using unsupervised learning algorithms based on their structures, densities, related segments, and other similar features. K-means, Hierarchical model, Mixture model are a few examples of Unsupervised machine learning algorithms.

Reinforcement Learning: In this type of learning, an agent learns by interacting with its environment. An agent is rewarded by performing correctly and gets penalties for incorrect performance. The agent learns without any supervision of a human by minimizing its penalty and maximizing its reward. It is used in robotics, text mining, health care, games, trade execution, etc. Q-Learning, Temporal Difference method, and Monte Carlo method are a few examples of reinforcement machine learning algorithms.

In addition to the above machine learning techniques, another type of Machine Learning, called "Semi-Supervised Learning" is also used in which a learning classifier is developed on the basis of labeled as well as on unlabeled data.

4.3 Hybrid Approach

This approach is the combination of lexicon based and machine learning approach. We can achieve improved accuracy and recall by combining both approaches. M.Govindarajan [22] has done sentiment analysis of movies reviews using a hybrid approach. In his research, a combination of Naïve Bayes and genetic algorithms was carried out to analyze performance on the basis of "accuracy" and it was concluded that the output of hybrid NB-GA is efficient and among NB and GA, GA is more efficient than NB.

4.4 Related Work

Sentiment Analysis on Twitter Done by Nazan Öztürk and Serkan Ayvaz [23] **on the Syrian Refugee Crisis**

In this study, a series of Turkish and English tweets about the Syrian war and refugee crisis were collected and analyzed. A total of 2,381,297 tweets (English and Turkish) were collected for analysis. Out of the total number of tweets, 1,027,930 were in Turkish and 1,353,367 were in English. It was found that people of different languages shared their opinions about refugee children. There were many duplicate tweets found, which were removed. In this study "emoji" characters were removed during data cleaning. It was observed that English-speaking people shared more neutral and negative opinions in comparison to positive tweets. On the other hand, in Turkish language, the number of positive tweets was more than the neutral and negative.

Sentimental Analysis of Hollywood movies on Twitter done by Umesh Rao Hodeghatta [24]

In this research, an analysis of Hollywood movie tweets was carried out for understanding the emotions of people on Twitter. Tweets were collected across different parts of the world. Twitter messages were classified as positive, negative and cognitive. There were two types of Machine learning methods used in this research, i.e., Naive Bayes and Max Entropy. The classification tests were carried out both for Unigram and Bigrams. The accuracy for the Naïve Bayes for Unigram and Bigrams was found to be 79% and 64%, respectively. In the case of Unigram data, the Max Entropy classifier had shown an accuracy of 84%.

Comparative Study of Machine Learning Algorithms and Open-Source Technologies done for Sentiment Analysis on Movie Reviews [25]

They worked on the comparison between Apache Hadoop and Naïve Bayes Classifier for doing sentimental analysis for calculating the performance of algorithms and concluded that Apache Hadoop gives better results than Naïve Bayes classifier. They used an exclusive tool called Natural Language Tool Kit (NLTK).

Sentiment Analysis Using Machine Learning Classifiers: Evaluation of Performance [26]

They worked on the sentiments expressed through tweets on Twitter and evaluate the polarity of the tweets, whether they are negative or positive. They selected the Naive Bayes Classifier (NBC) for testing and training of the features words and evaluated the sentiment polarity of tweets. Naïve Bayes is a family of algorithms based on the Bayes theorem. Accuracy, precision, and time were considered as the major parameters for evaluation and comparison between three machine learning classifiers, i.e., Random Forest, Naive Bayes, and Support Vector Machine (SVM). Support vector machine is a supervised machine learning method for classification based on two groups of data. It works on drawing the data points on a hyperplane using a decision boundary where in one class of data points lies on one side of the plane and the other class of data points lies on the other side of the plane. It was observed that execution time of Random Forest was much higher than NB and SVM and accuracy of Random Forest, Naive Bayes, and Support Vector Machine (SVM) was found to be 60–70% for different sets of data but a slight increase in accuracy was observed when numbers of tweets were increased.

5 Discussion

Statistical analysis of 23 research papers has been done considering the type of dataset, the technique used, accuracy, precision, and recall as major factors. Analysis has also been done on accuracy achieved while considering movie reviews as the major dataset. Comparative analysis of accuracy has also been done considering, type of machine learning technique used by a particular author. Figure 4 shows the comparative analysis while taking the movie review dataset for sentimental analysis.

Figure 5 depicts the accuracy of sentiment analysis while using different machine learning methods (Naïve Bayes, SVM, Maximum Entropy, ANN) consideration. The performance was measured by the metrics like Accuracy, Precision, and Recall [27].

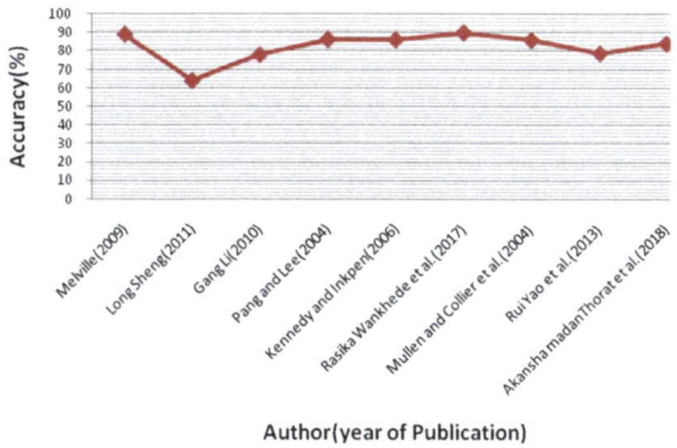

Fig. 4 Movie review dataset sentiment analysis

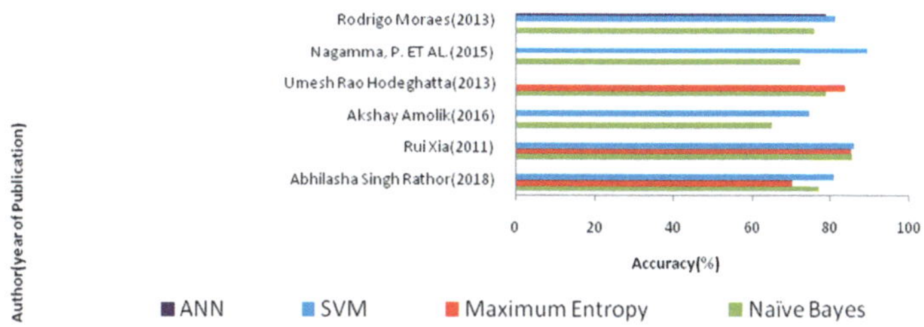

Fig. 5 Accuracy of sentiment analysis considering the machine learning techniques

(a) Precision—This is the ratio of true positive classifications (True Positives) from the cases that are predicted positive
(b) Recall—This is the ratio of true positive classifications from the cases that are actually positive.
(c) Accuracy—It is the ratio of true classifications (True Positive and Negative) from the total number of cases.

Table 1 shows the comparative analysis of various machine learning techniques (SVM, Naïve Bayes, Maximum Entropy, ANN, etc.) implemented by various researchers in the field of sentimental analysis. It has been found that different features along with different classification algorithms can be combined for overcoming the drawbacks of individual classifiers and enhancing the overall performance.

6 Conclusions

Sentimental analysis has a variety of applications in summarizing and classifying user reviews. Although there are many classification algorithms used in giving good results and many of the algorithms are combined together to overcome their individual

Table 1 Research work done in the area of sentiment analysis

S. no.	Author	Technique	Data source	Accuracy	Precision	Recall
1	Melville [28]	ID3, SVM, Hybrid	Movie review	89%	–	–
2	Long [29]	BPN	Movie review	64%	60%	98%
3	Rui Xia [30]	Naïve Bayes, Maximum Entropy, SVM	Movie review, multi-domain dataset, Amazon	NB-85.8% ME-85.4% SVM-86.4%	–	–
4	Abhilasha Singh Rathor, Amit Agarwal, Preeti Dimri [31]	Naïve Bayes, Maximum Entropy, SVM	Amazon reviews	NB-77.42% ME-70.35% SVM-81.20%	–	–
5	Rodrigo Moraes, Joao Francisco Valiati, Wilson P. Gaviao Neto [32]	SVM and ANN, Naïve Bayes	Amazon reviews, movie review dataset	NB-76.1% ANN-79.2% SVM-81.4% For Books dataset And 5000 number of terms	NB-0.75:0.77 ANN-0.79:0.8 SVM-0.78:0.85 For Books dataset And 5000 number of terms	NB-0.78:0.74 ANN-0.81:0.78 SVM-0.87:0.76 For Books dataset And 5000 number of terms
6	Gang Li [33]	K Means Clustering	Movie review	78%	–	–
7	Kaiquan Xu [34]	Multiclass SVM	Amazon reviews	61%	61.9%	93.4%
8	Pang and Lee [35]	Naïve Bayes and SVM	Movie review	86.4%	–	–
9	Gamgarn somprasti [36]	Maximum Entropy	Amazon reviews	–	72.6%	78.7%
10	Gagandeep Kaur, Kamaldeep Kaur [37]	SVM	Punjabi news articles	90%	88.5%	89.3%
11	Akshay Amolik, Niketan Jivane, Mahavir Bhandari, Dr. M.Venkatesan [38]	SVM Naïve Bayes	Twitter posts of movie reviews	SVM-75% Naïve Bayes-65%	–	–
12	Kennedy and Inkpen [39]	SVM terms counting	Movie review	86.2%	–	–

(continued)

Table 1 (continued)

S. no.	Author	Technique	Data source	Accuracy	Precision	Recall
13	Umesh Rao Hodeghatta [24]	Naïve Bayes Maximum Entropy	Hollywood movie tweets	Naïve Bayes-79% Maximum Entropy-84%	–	–
14	Rasika Wankhede and Prof. A.N. Thakare [40]	Random Forest	Movie review	90%	–	–
15	Mullen and Collier et al. [9]	SVM	Movie review	86%	–	–
16	Pang et al. [41]	NB, SVM, ME Unigrams and Bigrams	Movie reviews	77–82.9%	–	–
17	Prabowo et al. [42]	Hybrid SVM	Movie review, product review, my space comments	87.3%	–	–
18	Gamon [43]	SVM	Customer feedback	77.5%	–	–
19	Dave et al. [44]	Scoring, Smoothing, NB, ME, SVM(Unigrams, Bigrams and Trigrams)	Product reviews	88.9%	–	–
20	Konig and Brill [45]	Pattern based, SVM, Hybrid	Movie reviews customer feedback	Movie reviews >91% Customer feedback <72%	–	–
21	Rui Yao et al. [7]	Naïve Bayes	Movie review	78.75%	–	–

(continued)

Table 1 (continued)

S. no.	Author	Technique	Data source	Accuracy	Precision	Recall
22	Nagamma P, Pruthvi HR, Nisha KK, Shwetha NH [46]	SVM Naïve Bayes	Movie review	SVM-89.65% Naïve Baye-72.41%	–	–
23	Akanksha Madan Thorat, R. Vishnu Priya [47]	Naïve Bayes	Movie reviews	84.13%	Positive reviews-68.34% Negative reviews-98.47%	Positive reviews-97.6% Negative reviews-77.4%

drawbacks, but still there are a lot of problems in this field that are yet to be solved. The main challenges are the use of sentimental analysis in other languages, complexity of sentences, and summarization of user opinions based on the features of different products.

References

1. Malo P, Sinha A, Takala P, Ahlgren O, Lappalainen I (2013) Learning the Roles of Directional Expressions and Domain Concepts in Financial News Analysis. In Proceedings of IEEE International Conference on Data Mining Workshops (SENTIRE-2013), IEEE Press
2. Li S, Zong C (2008) Multi-domain sentiment classification. In Proceedings of the 46th Annual Meeting of the Association for Computational Linguistics on Human Language Technologies: Short Papers, Association for Computational Linguistics, pp 257–260
3. Akcora CG, Bayir MA, Demirbas M, Ferhatosmanoglu H (2010) Identifying breakpoints in public opinion.In ACM Proceedings of the First Workshop on Social Media Analytics, July, pp 62–66
4. Sun E, Rosenn I, Marlow C, Lento T (2009) Gesundheit! modeling contagion through Facebook news feed. In Proc. Of International AAAI Conference on Weblogs and Social Media, May, pp 22
5. Kwak H, Lee C, Park H, Moon S (2010) What is twitter, a social network or a news media?. In ACM Proceedings of the 19th International Conference on World Wide Web, April, pp 591–600
6. Cha M, Haddadi H, Benevenuto F, Gummadi KP (2010) Measuring user influence in Twitter: the million follower fallacy. In 4th international AAAI Conference on Weblogs and Social Media (ICWSM), May, vol 14(1)
7. Yao R, Chen J (2013) Predicting movie sales revenue using online reviews. In Proceedings - 2013 IEEE International Conference on Granular Computing, GrC, pp 396–401. https://doi.org/10.1109/GrC.2013.6740443
8. Weng J, Lim EP, Jiang J, He Q (2010) Twitter rank: finding topic sensitive influential twitterers. In Proceedings of the Third ACM International Conference on Web Search and Data Mining, February, pp 261–270
9. Mullen, Tony, Nigel Collier (2004) Sentiment Analysis using SVM with Diverse Information Sources. EMNLP vol 4
10. Cheong M, Lee VCS (2011) A micro blogging-based approach to terrorism informatics: Exploration and chronicling civilian sentiment and response to terrorism events via Twitter. Inf Syst Front 13(1):45–59
11. Yates D, Paquette S (2011) Emergency knowledge management and social media technologies: a case study of the 2010 Haitian earthquake. Int J Inf Manage 31(1):6–13
12. Lee NY, Kim Y, Sang Y (2017) How do journalists leverage Twitter? Expressive and consumptive use of Twitter. Social Sci. J. 54(2):139–147
13. Crannell WC, Clark E, Jones C, James TA, Moore J (2016) A pattern-matched twitter analysis of US cancer-patient sentiments. J Surg Res 206(2):536–542
14. Carley KM, Malik M, Landwehr PM, Pfeffer J, Kowalchuck M (2016) Crowd sourcing disaster management: the complex nature of Twitter usage in Padang Indonesia. Saf Sci 90:48–61
15. Nagamma P, Pruthvi HR, Nisha KK, Carlos Soares (2015) An Improved Sentiment Analysis of Online Movie Reviews. IEEE, International conference on Computer and Information Technology.
16. Ahmed S, Jaidka K, Cho J (2016) The 2014 Indian elections on Twitter: a comparison of campaign strategies of political parties. Telematics Inform 33(4):1071–1087
17. Vinodhini G, Chandrasekaran RM (2012) Sentiment Analysis and Opinion Mining: A Survey vol 2, Issue 6, June 2012, IEEE paper

18. Popescu AM, Etzioni O (2005) Extracting Product Features and Opinions from Reviews. In Proc. Conf. Human Language Technology and Empirical Methods in Natural Language Processing, Vancouver, British Columbia, pp 339–346
19. Hu, Liu (2006) Opinion extraction and summarization on the web. AAAI, pp 1621–1624
20. http://www.cs.cornell.edu/People/pabo/movie-reviewdata.
21. http://www.cs.jhu.edu/mdredze/datasets/sentiment
22. Govindarajan M (2013) Sentiment Analysis of Movie Reviews using Hybrid Method of Naive Bayes and Genetic Algorithm. Int. J. Adv. Comput. Res 3(4) Issue-13
23. Ozturk N, Ayvaz S (2018) Sentiment analysis on Twitter: A text mining approach to the Syrian refugee crisis. Telematics Inform 35(1):136–147
24. Hodeghatta, Umesh Rao (2013) Sentiment Analysis of Hollywood Movies on Twitter. In Proceedings of the 2013 IEEE/ACM International Conference on Advances in Social Networks Analysis and Mining, August 2013, pp 1401–1404
25. Narendra B, Uday Sai K, Rajesh G, Hemanth K, Chaitanya Teja MV, Deva Kumar K (2016) Sentiment Analysis on Movie Reviews: A Comparative Study of Machine Learning Algorithms and Open Source Technologies. Int. J. Intell. Syst. Appl. (IJISA) 8(8): 66–70. https://doi.org/10.5815/ijisa.2016.08.08.
26. Shamantha RB, Shetty SM, Rai P (2019) Sentiment Analysis Using Machine Learning Classifiers: Evaluation of Performance. IEEE 4th International Conference on Computer and Communication Systems (ICCCS), https://doi:https://doi.org/10.1109/ccoms.2019.8821650.
27. Vakili M, Ghamsari M, Rezaei M (2020) Performance Analysis and Comparison of Machine and Deep Learning Algorithms for IoT Data Classification. arXiv preprint arXiv:2001.09636.
28. Melville, Wojciech G, (2009) Sentiment analysis of blogs by combining lexical knowledge with text classification KDD 09, Paris, France. 978-1-60558-495-9/09/06.
29. Chen LS, Liu CH, Chiu HJ (2011) A neural network based approach for sentiment classification in the blogosphere. J Informetrics 5: 313–322
30. Xia R, Zong C, Li S (2011) Ensemble of feature sets and classification algorithms for sentiment classification. Inf Sci 181:1138–1152
31. Rathor AS, Agarwal A, Dimri P (2018) Comparative study of machine learning approaches for amazon reviews. Procedia Comput Sci 132:1552–1561. https://doi.org/10.1016/j.procs.2018.05.119
32. Moraes R, Valiati JF, Neto WP (2013) Document-level sentiment classification: an empirical comparison between SVM and ANN. Expert Syst Appli 40: 621–633. https://doi.org/10.1016/j.eswa.2012.07.059
33. Li G, Liu F (2010) A clustering-based approach on sentiment analysis, 978-1-4244-6793-8/10 ©2010 IEEE
34. Xu K, Liao SS, Li J, Song Y (2011) Mining comparative opinions from customer reviews for competitive intelligence. Decis Support Syst 50:743–754
35. Pang B, Lee L (2004) A sentimental education: sentiment analysis using subjectivity summarization based on minimum cuts. ACL
36. Somprasertsri G, Lalitrojwong P (2010) Mining feature-opinion in online customer reviews for opinion summarization. J Univers Comput Sci, 16(6):938–955
37. Kaur G, Kaur K (2017) Sentiment detection from Punjabi text using support vector machine. Int J Sci Res Comput Sci Eng 5:39-46. https://doi.org/10.26438/ijsrcse/v5i6.3946
38. Amolik A, Jivane N, Bhandari M, Dr Venkatesan (2016) Twitter sentiment analysis of movie. Int J Eng Technol (IJET) 7(6)
39. Kennedy A, Inkpen D (2006) Sentiment classification of movie reviews using contextual valence shifters. Comput Intell 22:110–125
40. Wankhede R, Thakare AN (2017) Design approach for accuracy in movies reviews using sentiment analysis. In: International conference of Electronics Communication and Aerospace Technology (ICECA), 6-11. https://doi.org/10.1109/ICECA.2017.8203652
41. Pang B, Lee L, Vaithyanathan S (2002) Thumbs up? Sentiment classification using machine learning techniques. EMNLP. 10.10.3115/1118693.1118704

42. Prabowo R, Thelwall M (2009) Sentiment analysis: a combined approach. J Informetrics 3:143–157
43. Gamon M (2004) Sentiment classification on customer feedback data: noisy data, large feature vectors, and the role of linguistic analysis. In Proceedings of the 20th international conference on Computational Linguistics (COLING '04). Association for Computational Linguistics, USA, 841–es
44. Dave K, Lawrence S, Pennock DM (2003) Mining the peanut gallery: opinion extraction and semantic classification of product reviews. Proceedings of WWW03, 12th International Conference on the World Wide Web, pp 519–528. Budapest, HU: ACM Press
45. König AC, Brill ED (2006) Reducing the human overhead in text categorization. Proceedings of the 12th ACM SIGKDD international conference on Knowledge discovery and data mining (KDD '06). Association for Computing Machinery, New York, NY, USA, 598–603
46. Nagamma P, Pruthvi HR, Nisha KK, Shwetha NH (2015) An improved sentiment analysis of online movie reviews based on clustering for box-office prediction. International Conference on Computing, Communication and Automation, ICCCA 2015. 933–937
47. Thorat AM, Priya RV (2018) Sentiment analysis of movie review using text mining. Int J Pure Appl Math, 119(16)

Chapter 53
Optimization of Fractional-Order PID Controller (FOPID) Using Cuckoo Search

Tarun Varshney, Vikas Singh Bhadoria, Pravin Sonwane, and Nitin Singh

1 Introduction

In the field of control system, Integer-Order PID (IOPID) controller plays a vital role in achieving the desired performance but in many situations, it fails to give optimized results in terms of transient/frequency characteristics because it opts integer value for controlling parameter and sometimes integer value don't converge the solution to the optimal value. So, the FOPID was introduced to improve the steady state and dynamic performance of the system. The FOPID was generally described by fractional order of the differential equation. It is mostly used in industrial, residential, different engineering applications [1–4]. There were limited numbers of methods for solving integer order of PID controller due to this deficiency. The fractional PID controller has an enormous number of methods for solving the differential equation based on fractional value. In principle, compared with classical PID controllers, which have three tuning parameters, the FOPID controllers give a greater reliability in the design process, being based on five tuning parameters, but their tuning is more cumbersome. FOPID controller has a flexible range of freedom for selecting the parameters [3, 5]. A lot of research related to frequency domain approach, inter-model control,

T. Varshney (✉)
Department of Electrical Electronics and Communication Engineering, School of Engineering & Technology, Sharda University, Greater Noida, UP 201310, India

V. S. Bhadoria
Department of Electrical and Electronics Engineering, ABES Engineering College, Ghaziabad, UP, India

P. Sonwane
Department of Electrical Engineering, Poornima College of Engineering, Jaipur, Rajasthan, India

N. Singh
EED, MNNIT, Allahabad, UP, India
e-mail: nitins@mnnit.ac.in

G. Mathur et al. (eds.), *Proceedings of 2nd International Conference on Artificial Intelligence: Advances and Applications*, Algorithms for Intelligent Systems,
https://doi.org/10.1007/978-981-16-6332-1_53

state-space design approach, were used in literature for the development of FOPID controllers. Moreover, various toolboxes such as CRONE, FOMCON, etc. were also developed for the design and implementation of FOPID controller for various systems [6, 7]. The performance of the FOPID controller for robotics, nonlinear, minimum phase, time-delay systems, etc. has been analyzed for a given desired performance [8–13]. In literature, it is shown that fractional PID controllers performed better in comparison to integer PID controllers [14–16]. In the recent era, many contributions have been made to develop nature-inspired optimization techniques. These are quite advantageous over the conventional optimization techniques. These algorithms are very diverse including Clonal Selection Algorithm, Differential Evolution (DE), Particle Swarm Optimization (PSO), Simulated Annealing (SA), Genetic Algorithms (GA), and so on [17, 18]. These recent optimization techniques mainly use the hit and trail method to tune the parameters of the controller and subsequently find the solution to complex problems. The wide range usage of FOPID controllers and their advantages along with optimization techniques opened a research area in which new nature-inspired optimization algorithms could be used to tune the parameters of FOPID controllers. In 2009, Yang and Deb have proposed a new intelligent algorithm namely Cuckoo Search (CS) based on the cuckoo's behavior of searching nests to lay eggs [19]. It is simpler, easy to understand, and implement, which has a fast convergence rate and strong robustness [20, 21]. Moreover, CS does not require making strong assumptions on the continuity and differentiability of the objective or constrained functions. In the proposed work, the CS algorithm is used to determine the FOPID controller's parameter.

The complete procedure and novelty of the work are described in the following points:

(a) Due to the extra two degrees of freedom of FOPID over IOPID, FOPID controllers are chosen.
(b) An estimated system (Practical fractional-order plant) is considered for simulation studies. The system is not being linearized so that the exact characteristics of the practical system can be utilized for study.
(c) Current idea reveals that CS is quite much better than GA and PSO owing to the fact that DE, PSO, and SA could consider as the special cases of CS. It can be said that CS is a solid and mixed reliable combination of these algorithms into one algorithm. In addition, the numbers of parameters in CS are reduced up to a certain extent than other parameters used in any met heuristic techniques. The involvement of fewer parameters in CS shows the faster converging nature of FOPID parameters [4].
(d) Finally, The performance of both IOPDS and FOPIDs controllers are compared for most popular conventional Nelder–Mead's, Integer point algorithms and nature-inspired CS optimization, in terms of MSE. The value of MSE for CS-based FOPID controller is found to be minimum.

The structure of the rest paper is organized as follows: Sect. 2 shows the broad idea for fractional calculus. In Sect. 3, FOPID has been quite explained. Section 4 is dedicated to the CS algorithm and Sect. 5 deals with the simulation results, and at the end, conclusions are presented.

2 Fractional Calculus: A Brief Introduction

The concept of a FOPID controller originates from the concept of calculus. Fractional-order differential and integral are defined by two authors as [2].

The Grunwald–Letnikov (GL) definition:

$$\alpha D_t^r f(t) = \lim_{h \to 0} h^{-r} \sum_{j=0}^{\left[\frac{t-\alpha}{h}\right]} (-1)^j \binom{r}{j} f(t-h) \tag{1}$$

where $[\cdot]$ means the integer term.

The Riemann–Liouville (RL) definition:

$$\alpha D_t^r f(t) = \frac{1}{\Gamma(n-r)} \frac{\partial^n}{\partial t^n} \int_\alpha^t \frac{f(\tau)}{(t-\tau)^{r-n+1}} d\tau \tag{2}$$

for $(n - 1 < r < n)$ and $\Gamma(.)$ means Gamma function.

Many approximation techniques are quite available in the literature to convert the fractional value to integer value for the FOPID controller. The most frequently used technique for such a purpose is Oustaloup's approximation algorithm. It is the most important technique to convert the order of fractional value to an integer by approximation within a particular band of frequency range [22].

3 Fractional-Order PID Controller

Classical PID controller is a weighted addition of derivative and integral operators. On other hand, a FOPID controller is also a weighted addition of derivative and integral operators but of fractional magnitude in nature, and its transfer function is given in Eq. (3).

$$C(s) = K_p + K_I s^{-\lambda} + K_D s^{\mu}, \ (\lambda, \ \mu > 0) \tag{3}$$

where λ and μ are the order of fractional integral and differential control. All conventional PID controllers can be realized in the special case of the FOPID controller as depicted in Fig. 1.

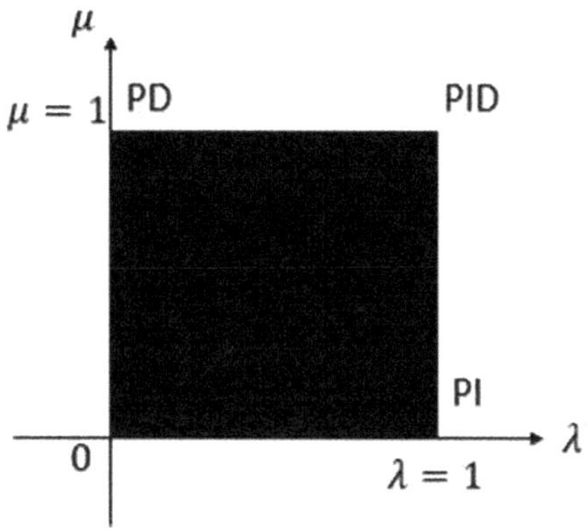

Fig. 1 Realization of PID controllers with FOPID controller

The main objective is to find the unknown undefined parameters of the controller to achieve desired performance. To perform such an operation, an optimization framework based on an error signal is developed and is analyzed using the CS algorithm. The performance index is taken to be Integral Time Absolute Error (ITAE).

4 CS Algorithm

CS is one of the nature-inspired metaheuristic algorithms which rely on the self-parasitism of different cuckoo species [23]. Some species such as the ani and Guiro cuckoos produce eggs in communal nests, although they might remove other eggs to enhance the existence of their self-eggs. More number of engaged species to obligate brood parasitism by producing their eggs in the nests of other birds. Other birds can be of different species. For illustrating the standard Cuckoo Search, the main three assumptions are considered as given below.

Individual cuckoo produces single egg at a time, and keep at chosen nest randomly;

The best nests with highly profiled-good eggs will be transferred to the next immediate generations;

The available number of host nests is kept fixed, and the egg produced by a cuckoo is identified by the host bird with a probability pa $\geq$ (0, 1). In such as case, the host bird could get free either of egg or simply remove the nest and construct a new nest completely.

For consideration of the implementation of CS, the following simple representations can be used such that every egg in the nest shows a solution, and each and individual cuckoo produce only 1 egg (which representing the 1 solution), the main target is to utilize newly produced good potential solutions (cuckoos) to remove and put a bad solution in its nests. Obviously, CS can utilize for solving the problem of a complex system, where every nest produces many eggs which shows solutions of different sets. The simplest way is used where every nest produces only one egg. In such a situation, there are no differences among nest, egg, or cuckoo, as an individual nest corresponds to 1 egg which also represents the each of cuckoo. In CS algorithm represents a valid solution for the designed problem according to its fitness value. The flowchart of the CS algorithm is represented in Fig. 2.

The procedure of the CS algorithm is as follows:

1. Initiate Cuckoo habitats with a selection of randomly selected points on the optimized function.
2. Assigned a few eggs to every cuckoo.
3. Describe the radius of produced eggs for every cuckoo.

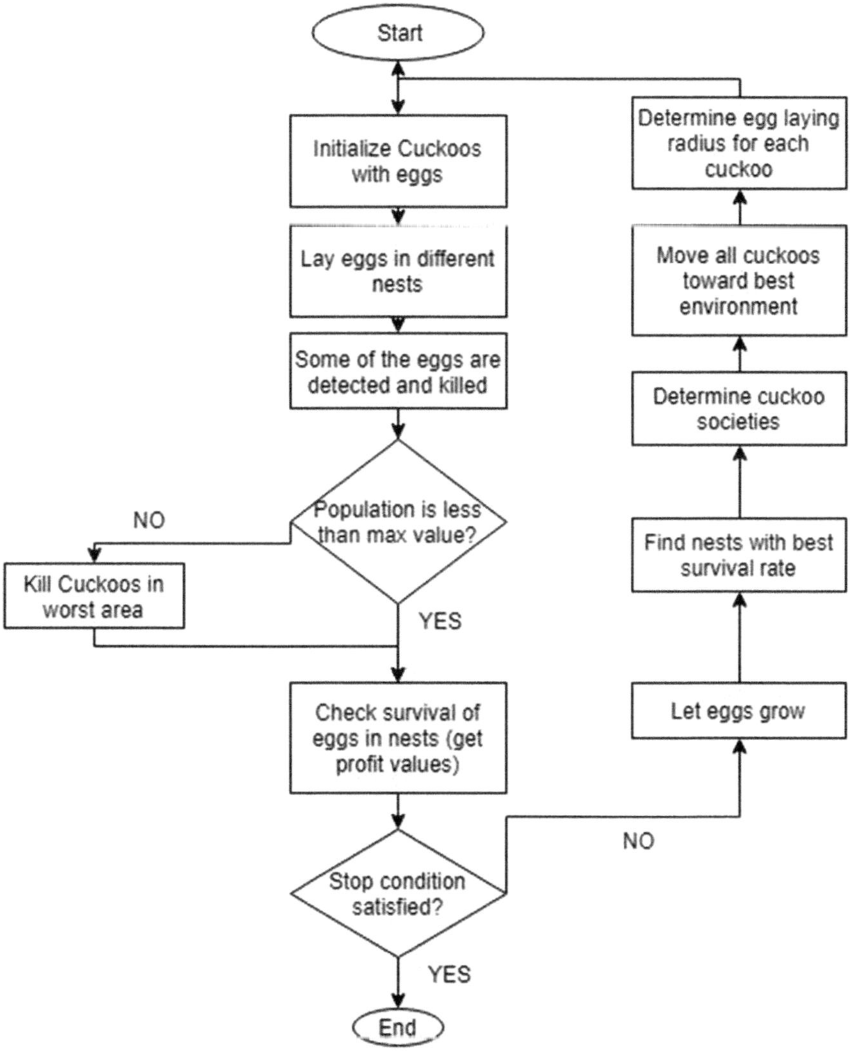

Fig. 2 Flowchart for the CS algorithm

4. Let Cuckoo laid their eggs in the prescribed radius of generated eggs
5. Kill the all shown eggs that are being identified by hosting birds.
6. Let the eggs planned up to grow up the chicks.
7. Calculate the abode of every newly grown Cuckoo.
8. Try to set Cuckoos' in more in no. as per environment condition and remove all existing hot abode in the worst place
9. Collect Cuckoos and select the best quality group and identify the habit of goal.
10. Let the newly identified cuckoo search population be heading towards target.
11. If it is possible to get stop condition then stop, if not go to step 2.

5 Simulation Results

All simulation studies have been carried out with the MATLAB platform by using Intel.

(R) Core (TM) i3-6006U CPU@2.00 Ghz machine. To test the effectiveness of the FOPID controller and CS algorithm, studies have been carried out for a thermal object [4]. The second-order integer transfer function of the thermal object is given by

$$G_{Ip}(s) = \frac{1}{73043s^2 + 4893s + 1.93} \tag{4}$$

and fractional-order model is given by

$$G_{Fp}(s) = \frac{1}{14994s^{1.31} + 6009.5s^{0.97} + 1.69} \tag{5}$$

The negative feedback control scheme is shown in Fig. 3. Initially, the parameters of IOPID controllers were optimized, then keeping these values of gains the same, exponents of FOPID controller are optimized and finally the optimization

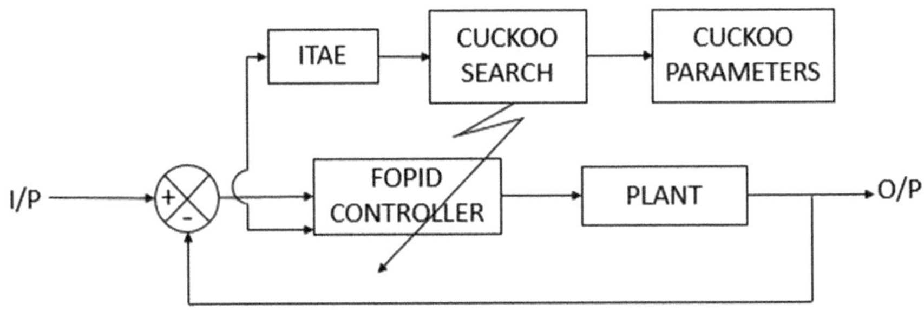

Fig. 3 Configuration of the control scheme

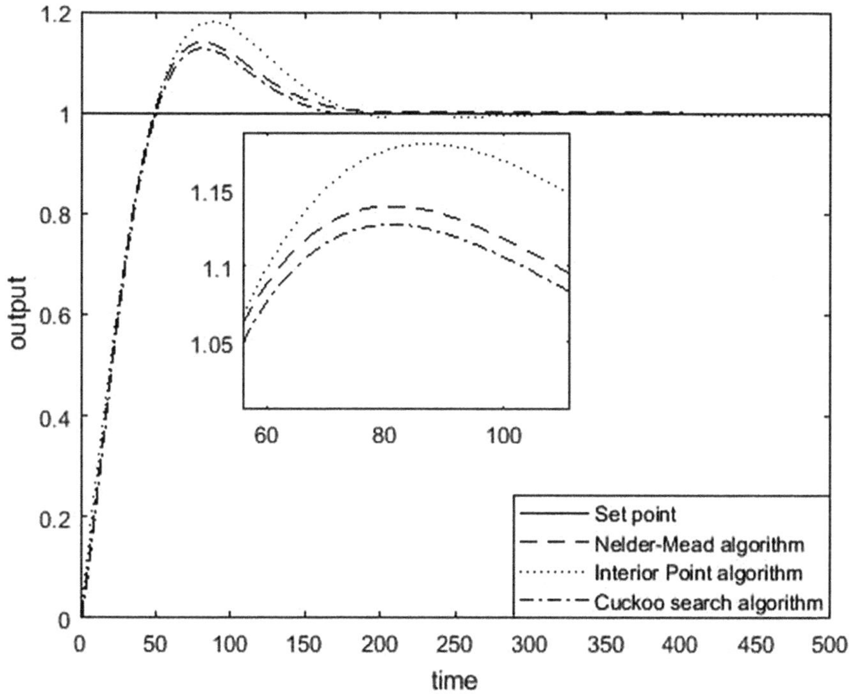

Fig. 4 Step response for Category III

of all five parameters of FOPID controller is performed. Three algorithms, namely Nelder–Mead's, Interior Point, and CS algorithms are used for the optimization of the controller's parameters. ITAE has been considered as an optimization function.

Simulation studies for all categorized controller have been carried out but keeping the objective in mind step response for category III has been shown in Fig. 4.

In addition, the performance of closed-loop mechanism is compared and shown in Table 1. It is found that output with CS-Based FOPID controller is best as the value of MSE is minimum in this case. It is also evident from the response that the value of MSE is significantly reduced with the CS algorithm as compared to other considered algorithms in all categories. But the minimum value of MSE is corresponding to the category III controller which indicates the superiority of the controller.

6 Conclusions

In this work, a study on estimated fractional-order thermal objects is carried out. IOPID and FOPID controller's performance have been fairly compared for conventional optimization algorithms, namely Nelder's–Mead, Integer point algorithms, and

Table 1 Comparison of controller's performance

Category	Controller	Nelder–Mead	Interior-Point	CS	
I	IOPID with variable gains and fixed λ and μ	$K_{\rm P}$	244.87	77.075	180.55
		$K_{\rm I}$	1.8969	0.013677	1.1593
		$K_{\rm D}$	500	246.9	316.7
		λ	1	1	1
		μ	1	1	1
		MSE	0.0164	0.0323	0.0115
II	FOPID with fixed gains and variable λ and μ	$K_{\rm P}$	244.87	77.075	180.55
		$K_{\rm I}$	1.8969	0.013677	1.1593
		$K_{\rm D}$	500	246.9	316.7
		λ	0.99428	0.19708	0.73578
		μ	0.0164	0.68288	0.3674
		MSE	0.0151	0.0263	0.0127
III	FOPID with variables gains and variable λ and μ	$K_{\rm P}$	112.66	0.034006	101.94
		$K_{\rm I}$	0.34261	229.45	0.478
		$K_{\rm D}$	500	490.76	475.60
		λ	1.2572	0.011612	1.3057
		μ	0.55258	0.08578	0.4789
		MSE	0.0135	0.0156	0.0090

nature-inspired CS algorithm. Firstly, the gains of IOPID controllers have been optimized then secondly keeping the optimized values of these gains the same, exponents of FOPID controller are optimized and finally, all of the five parameters of FOPID are being optimized. Due to the extra two degrees of freedom, it is observed that the CS algorithm offers lesser MSE as compared with considered conventional algorithms and hence validates the proposed CS algorithm. It is concluded that the best results are obtained with the FOPID controller while the controller parameters are being optimized using nature-inspired CS algorithm.

References

1. Li H, Luo Y, Chen Y (2010) A fractional order proportional and derivative (FOPD) motion controller: tuning rule and experiments. IEEE Trans Control Syst Technol 18:516–520
2. Podlubny I (1999) Fractional-order systems and PI/sup/spl lambda//D/sup/splmu//-controllers. IEEE Trans Autom Control 44:208–214
3. Zhong J, Li L (2014) Fractional-order system identification and proportional-derivative control of a solid-core magnetic bearing. ISA Trans 53:1232–1242
4. Chen Y, Petras I, Xue D (2009) Fractional order control—a tutorial. In: 2009 American control conference, pp 1397–1411

5. Valério D, Costa JSD (2010) A review of tuning methods for fractional PIDs. In: Proc. 4th IFAC workshop Fract. Differ. Appl, pp 1–5
6. Oustaloup A, Mathieu B, Lanusse P (1995) The CRONE control of resonant plants: application to a flexible transmission. Eur J Control 1:113–121
7. Tapljokov A, Petlenkov E, Belikov J (2011) FOMCON: a MATLAB toolbox for fractional-order system identification and control. Int J Microelectron Comput Sci 2(2):51–62
8. Xu W, Torrance J, Chen B, Potgieter J, Bronlund J, Pap J (2008) Kinematics and experiments of a life-sized masticatory robot for characterizing food texture. IEEE Trans Industr Electron 55:2121–2132
9. Su YX, Duan BY, Zheng CH (2004) Nonlinear PID control of a six DOF parallel manipulator. Proc. Inst. Elect. Eng.-Control Theory Appl 151(1): 95–102
10. Merrikh-Bayat F, Karimi-Ghartemani M (2010) Method for designing PIλDμ stabilisers for minimum-phase fractional-order systems. IET Control Theory Appl 4:61–70
11. Hamamci S (2007) An algorithm for stabilization of fractional-order time delay systems using fractional-order PID controllers. IEEE Trans Autom Control 52:1964–1969
12. Ataşlar-Ayyıldız B, Karahan O (2018) Tuning of fractional order PID controller using CS algorithm for trajectory tracking control. In: 6th international conference on control engineering & information technology (CEIT), Istanbul, Turkey
13. Essa MEM (2019) Identification and temperature control for thermal model of a house based on model predictive control tuned by cuckoo search algorithm. In: 15th international computer engineering conference (ICENCO), Cairo, Egypt, pp 144–149
14. Yeroglu C, Tan N (2011) Note on fractional-order proportional–integral–differential controller design. IET Control Theory Appl 5:1978–1989
15. Das S, Das S, Gupta A (2011) Fractional order modeling of a phwr under step-back condition and control of its global power with a robust ${\rm PI}^{\lambda} {\rm D} ^{\mu}$ controller. IEEE Trans Nucl Sci 58:2431–2441
16. Erkol H (2018) Optimal PI$^{{\lambda}}$ D$^{{\mu}}$ controller design for two wheeled inverted pendulum. IEEE Access 6:75709–75717
17. Bhadoria V, Pal N, Shrivastava V (2019) Artificial immune system based approach for size and location optimization of distributed generation in distribution system. Int J Syst Assur Eng Manag 10:339–349
18. Yang XS (2011) Nature-inspired metaheuristic algorithms. Luniver Press, United Kingdom
19. Yang XS, Deb S (2009) Cuckoo search via Levy flights. In: Proceedings of world congress on nature & biologically inspired computing (NaBIC India), pp 210–214
20. Sharma R, Gaur P, Mittal AP (2015) Performance evaluation of cuckoo search algorithm based FOPID controllers applied to a robotic manipulator with actuator. In: International conference on advances in computer engineering and applications (ICACEA), Ghaziabad, India, pp 356–363
21. Hussain M, Apama V, Banu US, Lakshmanaprabu SK (2017) Automation of two interacting conical tank process with multiobjective optimization of fractional PID controller with a meta-heuristic optimizer, the cuckoo search algorithm. In: Trends in industrial measurement and automation (TIMA), Chennai, pp 1–6
22. Monje C (2010) Fractional-order systems and control fundamentals and applications. Springer, London
23. Yang, X (2006) Cuckoo search and firefly algorithm. Theory Appl. Springer

Chapter 54
Deep Model for Robust Tomato Disease Detection on Low-Resolution Leaf Images

Siddhant Baldota, Rubal Sharma, Nimisha Khaitan, and E. Poovammal

1 Introduction

Despite the advancements in agricultural technology, food, and crop security have remained a major cause of concern for nations across the globe. Crops can be protected from diseases, provided the manifestation of plant diseases is detected at an inchoate stage. Early detection of diseases allows the farmers to adopt curative and preventive measures in order to avoid the further damages. The symptoms indicating the onset of crop disease are typically observed on the leaves of the plant. Analysis of every leaf by humans can be a long-drawn process. Automation of disease detection in plants has proven to be beneficial in this aspect. To expedite the process of disease detection through plant leaves several techniques like image processing, computer vision, and deep learning have been used over the past decade. We propose a fine-tuned architecture using residual deep convolutional neural network, ResNet101 as our baseline model. The model works quite well with low-resolution tomato leaf images, achieving near human-level performance.

2 Related Works

Aravind et al. [1] propose to use two deep learning-based architectures, namely, AlexNet and VGG16 net on images of tomato leaves having 6 categories of diseases and one healthy category. The original input images of dimension 256 × 256 were augmented to 227 × 227 for AlexNet and 224 × 224 for VGG16 net. Transfer learning was implemented for the classification of images. 0.0001 was used as the

S. Baldota (✉) · R. Sharma · N. Khaitan · E. Poovammal
Department of Computer Science and Engineering, SRM Institute of Science and Technology, Kattankulathur 603203, India
e-mail: poovamme@srmist.edu.in

G. Mathur et al. (eds.), *Proceedings of 2nd International Conference on Artificial Intelligence: Advances and Applications*, Algorithms for Intelligent Systems,
https://doi.org/10.1007/978-981-16-6332-1_54

global learning rate. They achieved an accuracy of 97.49% and 97.23% for AlexNet and VGG16net, respectively. For evaluation, mini batches of the same number of images were introduced. With regard to computation, best accuracy was obtained when a mini batch containing 373 images was used. In comparison to VGG16 net, AlexNet was doing better with less execution time. Santosh et al. [2] presented the use of image processing for tomato plant diseases. Dataset was extracted from the internet and captured using a camera. It contains 520 images in 3 categories of plant diseases which were annotated manually. Python and the OPENCV library are used to manipulate raw input images. The designed network is based on YOLO and AlexNet. 0.1 was set as the learning rate. At 10–3 learning rate for 10 epochs, the images were downsized to 448*448 from 580*580. Linear regression was used for detection and finally the ones with higher confidence scores were used. An overall accuracy of the system is found to be 89% based on plant village dataset. The limitation of this system was that it can only detect 3 diseases and more training is required for the detection of other classes.

We further reviewed and widened our learning area to plant diseases and its detection in general. Sharada et al. [3] propose deep learning models and algorithms for plant disease detection and classification. They analyzed large data which is labeled. The dataset contains 54306 images which have 38 class labels and resized them to 256×256. They experimented first with color, then in grayscale and finally where leaves were segmented. They implemented AlexNet and GoogleNet and used scratch and transfer learning. Learning rate decreased by a factor of 10 every 30/3 epochs and weight decay was 0.0005. The model achieved a best accuracy of 98.21% when train test split ratio was 1:4. Using another model Haijing et al. [4] propose VGG-16 and VGG-19 models for classifications on insects on leaves. A 3-label classifier was used and fine-tuning was performed on VGG-19 and VGG-16 models. VGG-16, VGG-19, InceptionV3, and ResNet 50 were trained on the vegetable images which were infected with pests and to let the models adapt to it. The experiment showed that both models of VGGNet achieved accuracy of more than 99%. Augmentation methods such as horizontal flipping, vertical flipping, random horizontal flipping, and color change were used.

Krishna et al. [5] explored deep learning for classification of images. They discussed the architecture of AlexNet and the basics of artificial neural networks. From the database, Sea Anemone, Barometer, Stethoscope, and radio interferometer were taken for training, testing, and classification. The first layer has 96 11×11 filters. The output volume is $55 \times 55 \times 96$. They observed that even partial images were classified correctly showing the efficiency of deep learning models. To further our study on image classification we reviewed deep residual networks [6]. Kaiming et al. tested various deep neural networks for image classification. A random crop of 224×224 and augmentations like horizontal flip were performed on the benchmark ImageNet [7] dataset. They proposed a novel architecture for deep residual learning using skip connections (ResNet). Roslidar et al. [8] discussed the detection of Breast Cancer in thermal images using CNN models like, DenseNet, ShuffleNetV2, MobileNetV2, and ResNet101. Their objective was to classify the thermal images as healthy or cancerous. The data is split into 80% for training and 20% for validation.

The use of ResNet101 reduced the training error by 3.5%. ResNet101 and DenseNet provided 100% validation accuracy on all 10, 20, and 30 epochs.

Muhammad et al. [9] proposed VGG-19, PCA, and SVD to classify diabetic retinopathy from fundus images. The dataset images were converted into grayscale and their ROI was extracted using vessel segmentation techniques. PCA and SVD were used to reduce the dimensions. For better efficiency the images were scaled down to 224 × 224 pixels. More than 98% accuracy was achieved for the two classes. PCA, along with SVD was the primary reason for the exaltation of accuracy. Heechung et al. [10] used ResNets for vehicle classification and localization using real time traffic recordings consisting of images of shape, 224 × 224 × 3. For better results photometric distortions and color augmentation were applied. Using fully joint fine-tuning better classification rates could be obtained. They applied various versions of ResNets for convolutional feature extraction and region proposal networks for ROI-based extraction. ResNet-18 yielded an accuracy of 97.93%, whereas their joint fine-tuning and DropCNN model achieved 97.75%.

Srikanth [11] used the VGG-16 architecture for transfer learning using image data generators to aid the augmentation process. A validation accuracy of 72.40% was obtained using a CNN trained from scratch. They fine-tuned this model and achieved better accuracy. Lastly, they leveraged the power of VGG-16 by fine-tuning it, achieving an accuracy of 95.40%. Christian et al. [12] trained their network using TensorFlow with 20 replicas. As a result, the model could be trained without partitioning the replicas. They introduced Inception-ResNet-v2, a hybrid version of Inception and ResNets, which is costlier in comparison to others but has increased recognition performance and Inception-v4 which doesn't include any residual connections. Francois [13] replaced Inception with depthwise separable convolutions in neural architecture. They tested their model on a very large dataset obtained from Google which contains millions of images. They concluded that depthwise separable convolution offers similar characteristics like Inception but is easier to use. This new neural network was named Xception, having depthwise separable convolutional layers stacked linearly along with residual connections.

Traditionally, in order to balance classes pertaining to binary classification, upscaling of smaller class or downscaling of the larger class was used. Zhu et al. [14] proposed a better method for Multi-class cost sensitive learning. It was observed that if rescaling was applied directly to the multi-class problem using consistent cost, then the results were much better. However, when costs were inconsistent, the multi-class problem was reduced to binary class sub-problems. Subsequently binary class data was rescaled and fed to a classifier. A combination of results led to the best approach for multi-class cost sensitive learning.

Mohammed et al. [15] proposed a learning algorithm for classification of tomato diseases, for which they claimed exceptional accuracy. They used the dataset which contained 14828 images distributed among 9 categories of diseases. Transfer learning using state-of-the-art neural networks like GoogleNet and AlexNet, pretrained on Imagenet was implemented. They compared their performance with ML algorithms like SVM. The pretrained deep learning models gave the best results, with GoogleNet giving an accuracy of 99.18%. However, the validation errors and F1 scores were not accounted for. The high accuracy is not the only indicator for performance. Metrics such as recall, precision and F1 score are also needed to be considered. It was observed that the sizes of pretrained AlexNet and GoogleNet were well above 200MB. Moreover, healthy tomato leaves were misclassified into some disease as training was done only on classes with disease. We intend to perform the detection of the healthy class as well as the nine diseases using lesser storage space without compromising with performance. Furthermore, we intend to use the upgraded plant village dataset consisting of higher number of images per class and account for imbalanced class using class balancing, thus reducing the bias of the deep learning model toward the larger class.

3 Proposed Methodology

3.1 Dataset

The dataset used in this paper is acquired from Kaggle, which contains over 50,000 expertly curated images from the PlantVillage Dataset. For our purpose, we use a total of 16065 images comprising various categories of diseases in tomato plants. There are 10 categories in total including nine different diseases and 1 healthy category.

3.2 Preprocessing

Our dataset consists of 16065 images of 10 categories of tomato diseases (nine) and healthy (one) leaves, which includes 14253 training images and 1812 cross-validation images. We use an Image Data Bunch to augment the images with a rescaling of each pixel by a factor of 1/255. Transformation methods such as shearing, warping, and horizontal flipping were also applied. Another important factor which is to be taken into consideration is the class imbalance. The imbalance in the dataset leads to model bias toward the class with the greater frequency. Class weights penalize each class, assigning them a priority. The class with the lowest frequency gets the highest weights and vice versa. After class balancing, class bias is eliminated. Note that class balancing is only applied to the training dataset (Figs. 1 and 2).

3.3 Models Used

In this paper, we have used state-of-the-art DCNN architectures performing transfer learning on models pretrained on Imagenet weights. As DCNN architectures become deeper, the problem of vanishing gradients arises. This means that the gradi-

Fig. 1 Classes of tomato disease (left to right)—bacterial spot, early blight, late blight, leaf mold, septoria leaf spot, two-spotted spider-mite, target spots, tomato mosaic virus and yellow leaf curl virus

Fig. 2 Healthy tomato leaves

ent becomes very small in order to prevent weights from changing their value. This leads to non-convergence of the cost function, resulting in poor training. The ResNet block shown in Fig. 3, introduced by Kaiming He et al. changed the course of deep learning. ResNet blocks are basically a combination of identity mapping and residual mappings instead of the normal direct mapping

If the output mapping is represented formally by $H(x)$, the non-linear layers are stacked to fit a residual mapping denoted by $F(x) = H(x) - x$, and the original mapping is reformed into $F(x) + x$. The expression $F(x) + x$ can be considered as feedforward neural networks having "shortcut connections". The ResNet block is a part of some of the state-of-the-art deep learning architectures like Xception which

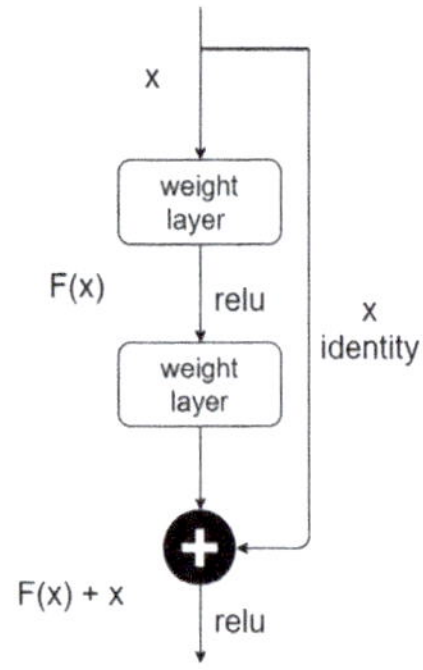

Fig. 3 A ResNet block

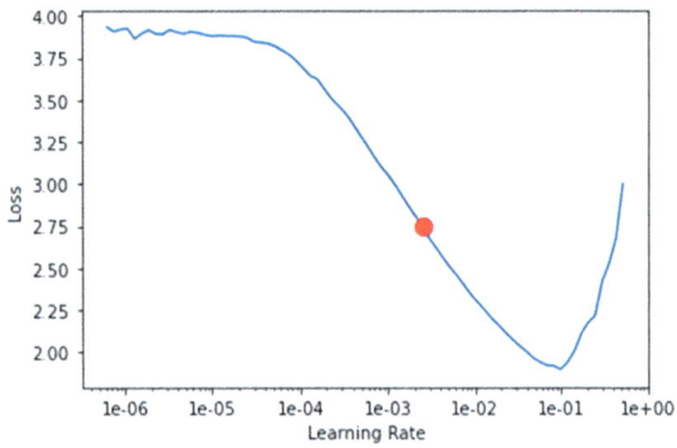

Fig. 4 Graph for finding the optimum learning rate for training ResNet101

consists of Depthwise Separable Convolutions. To obtain a greater boost in metrics, we intend to shift toward purer ResNet architectures which are based on residual networks with skip connections. Relatively shallow architectures like ResNet50 may cause instability in loss. Nevertheless, we anticipate that ResNet50 will yield higher accuracy. To obtain greater stability, we can implement the ResNet101 architecture. We hypothesize that increasing the number of hidden layers would increase the stability of the model.

4 Training

For the training procedure, we used a batch size of 50 images. We used Adam optimizer instead of Stochastic Gradient Descent in order to update the network weights iteratively. The minimum numerical gradient was chosen as the optimum learning rate for the model. We found the optimal learning rate for all the five models. Figure 4 shows the optimal learning rate for ResNet101 with the red dot indicates the optimal learning rate.

The loss function was set as categorical cross-entropy and the evaluation metric was set as accuracy. Training was carried out on Google Colaboratory with a hardware

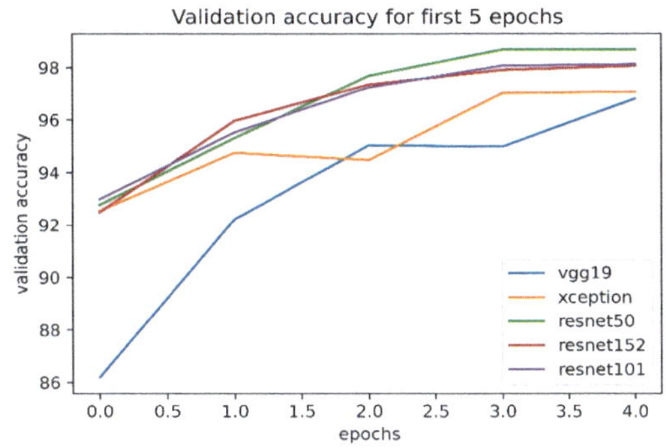

Fig. 5 Validation accuracy over 5 epochs for all five models

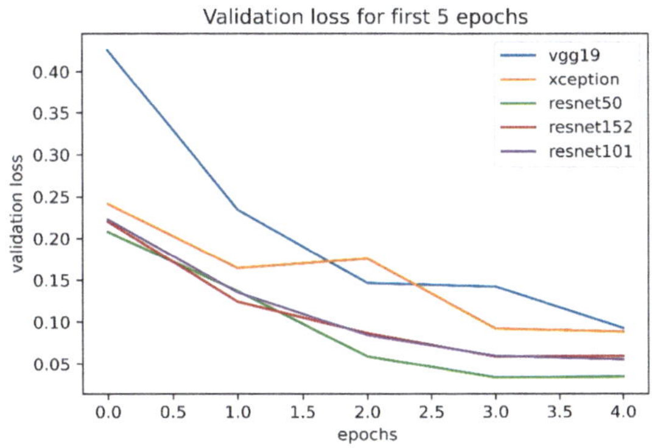

Fig. 6 Validation loss over 5 epochs for all five models

accelerator as Nvidia K80 with 12 GB GPU memory having a performance of 4.1 TFLOPS. We trained each model for five epochs and noticed that the first epoch took comparatively more time.

5 Results

For all five models, the validation accuracy versus epoch graph was plotted over the total of five epochs. The changes in accuracy and loss over epoch were observed as shown in Figs. 5 and 6. It can be noticed that the pure ResNet models (ResNet50, ResNet101, and ResNet152) yielded the highest accuracies as shown in Table 1. Out of these, ResNet50 gave the maximum accuracy.

In order to differentiate between the pure ResNet models, we trained them for 10 more epochs. Over a total of 15 epochs, we found that ResNet50 gave maximum accuracy, whilst ResNet101 gave the second highest accuracy as shown in Fig. 7. Although ResNet50 was more accurate than ResNet101, ResNet101 remained more stable as shown in Fig. 8. Moreover, accuracy and training time are not the only

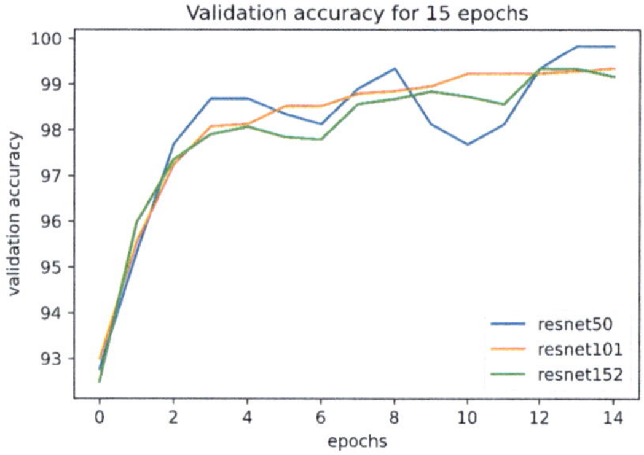

Fig. 7 Validation accuracy of ResNet models over 15 epochs

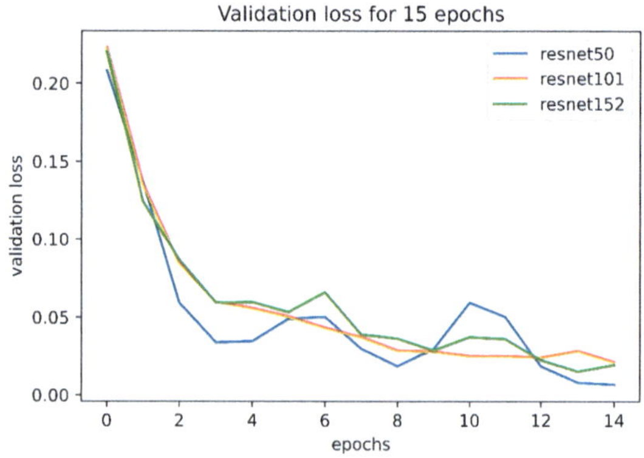

Fig. 8 Validation loss of ResNet models over 15 epochs

Table 1 Validation accuracy after 5 epochs in %

Model used	Validation accuracy in %
VGG-19	96.7991
Xception	97.0751
ResNet50	98.6755
ResNet101	98.1236
ResNet152	98.0684

parameters. we need to check. A few other metrics are Recall, Precision, and F1 Score.

From Table 2, we notice that ResNet50 yields the maximum accuracy out of all the three models. Though it outperforms ResNet101 in precision, recall, and F1 score too, the training time taken was much higher. Moreover, ResNet101 had a much smoother decrease in loss while ResNet50 faced the problem of exploding gradients. Thus, it purely comes down to preference of stability over a minimal increase in accuracy. Hence, ResNet101 is taken as the baseline architecture to perform transfer learning on.

Table 2 Accuracy, precision, recall, and F1-scores after 15 epochs for ResNet models

Model used	Accuracy in %	Precision	Recall	F1 score
ResNet50	99.8344	0.9988	0.9989	0.9988
ResNet101	99.34	0.99337	0.99337	0.9933
ResNet152	99.1722	0.9866	0.9930	0.9898

6 Conclusion

It can be seen that our model detects both healthy and diseased tomato leaf images, improving on the existing work where only diseases were detected. Thereby, we have reduced the misclassification of a healthy tomato leaf image into disease categories. From the results, we conclude that ResNet101 was the ideal model for our application. Since the accuracy of ResNet101 obtained is 99.34%, it has achieved near human-level performance. Despite the infinitesimal higher accuracy of ResNet50 (99.83%) in comparison with ResNet101, we choose ResNet101 because the absence of exploding gradients which gave ResNet101 higher stability. In the future, we look forward to working on more categories of plants and detecting diseases. Furthermore, we plan to add noise to the data to check whether the model built for low-resolution images could perform well for noisy images. We intend to work specifically on images obtained from developing nations which tend to have low-resolution images.

References

1. Rangarajan AK, Purushothaman R, Ramesh A (2018) Tomato crop disease classification using pre-trained deep learning algorithm. Procedia Comput Sci 133:1040–1047. https://doi.org/10.1016/j.procs.2018.07.070
2. Adhikari S, Saban Kumar KC, Balkumari L, Shrestha B, Baiju B (2018) Tomato plant diseases detection system using image processing. In: 1st KEC conference on engineering and technology, Lalitpur, vol 1, pp 81–86
3. Mohanty SP, Hughes DP, Salathé M (2016) Using deep learning for image-based plant disease detection. Front Plant Sci 7:1419. https://doi.org/10.3389/fpls.2016.01419
4. Ye H, Han H, Zhu L, Duan Q (2019) Vegetable pest image recognition method based on improved VGG convolution neural network. J Phys: Conf Ser 1237(3):032018 (IOP Publishing). https://doi.org/10.1088/1742-6596/1237/3/032018
5. Krishna MM, Neelima M, Harshali M, Rao MVG (2018) Image classification using deep learning. Int J Eng Technol 7(2.7):614–617. https://doi.org/10.14419/ijet.v7i2.7.10892
6. He K, Zhang X, Ren S, Sun J (2015) Deep residual learning for image recognition. In: Proceedings of the IEEE conference on computer vision and pattern recognition, pp 770–778. https://doi.org/10.1109/CVPR.2016.90
7. Russakovsky O, Deng J, Su H, Krause J, Satheesh S, Ma S, Huang Z, Karpathy A, Khosla A, Bernstein M, Berg AC (2015) Imagenet large scale visual recognition challenge. Int J Comput Vis 115(3):211–252. https://doi.org/10.1007/s11263-015-0816-y
8. Roslidar R, Saddami K, Arnia F, Syukri M, Munadi K (2019) A study of fine-tuning CNN models based on thermal imaging for breast cancer classification. In: IEEE international con-

ference on cybernetics and computational intelligence (CyberneticsCom). IEEE, pp 77–81. https://doi.org/10.1109/CYBERNETICSCOM.2019.8875661

9. Mateen M, Wen J, Song S, Huang Z (2018) Fundus image classification using VGG-19 architecture with PCA and SVD. Symmetry 11(1):1. https://doi.org/10.3390/sym11010001
10. Jung H, Choi MK, Jung J, Lee JH, Kwon S, Young Jung W (2016) ResNet-based vehicle classification and localization in traffic surveillance systems. In: Proceedings of the IEEE conference on computer vision and pattern recognition workshops, pp 61–67
11. Tammina S (2019) Transfer learning using VGG-16 with deep convolutional neural network for classifying images. Int J Sci Res Publ 9(10):143–150. https://doi.org/10.29322/IJSRP.9.10.2019.p9420
12. Szegedy C, Ioffe S, Vanhoucke V, Alemi A (2017) Inception-v4, inception-resnet and the impact of residual connections on learning. In: Proceedings of the AAAI conference on artificial intelligence 31(1). arxiv.org/abs/1602.07261
13. Chollet F (2017) Xception: deep learning with depthwise separable convolutions. In: IEEE conference on computer vision and pattern recognition (CVPR). https://doi.org/10.1109/CVPR.2017.195
14. Zhou ZH, Liu XY (2010) On multi-class cost-sensitive learning. Comput Intell 26(3):232–257. https://doi.org/10.1111/j.1467-8640.2010.00358.x
15. Mohammed B, Kamel B, Abdelouahab M (2017) Deep learning for tomato diseases. Classification and symptoms visualization. Appl Artif Intell 2–17. https://doi.org/10.1080/08839514.2017.1315516

Chapter 55
Privacy-Preserving Dynamic Task Scheduling for Autonomous Vehicles

S. Muthurajkumar, C. Ajay Karthikeyan, K. Pradeep, and A. Hariharan

1 Introduction

Many new technological advancements have been brought to the automobile industries recently, implementing automation and computerization to car driving. A Self-Driving Vehicle (SDV) is a vehicle which can drive itself without human efforts. This kind of vehicle has become an embodied reality and shall make the way for upcoming systems where automation's take over the art of driving. Connected SDV are all around the corner. A huge number of deep-pocketed and innovative companies around the globe are competing to bring the SDV to the market and that too for an economical reason: the social and economic gains that they are about to generate will be huge. But there are always new challenges and risks that are being created in any transforming technology in addition to benefits. And this is no exclusion. Vehicular cybersecurity will become the largest threat that the society will face as the transportation is transforming itself in the upcoming years. For those who are there at the revolutionary automobile industry, vehicular cybersecurity is still an unsolved topic; vehicular connectivity is the newest evolving astonishment and the technology hold on to improve more rapidly than ever. Conscientiously, no serious malevolent cyber-attack has ever occurred as of now on a vehicle. But recently, when two computer security specialist hackers secluded took control of a Jeep of the indigenous people of United States of America and dissect its transmission on an active highway, which was part of a research initiative, illustrated the conjectural danger dramatically. This led Chrysler to recall its 1.4 million vehicles after the well-publicized incident.

Autonomous vehicles should reduce accidents dramatically, assist and manage traffic congestion, reduce duration and travel energy consumption, and provide critical mobility to the disabled and elderly. Self-driving cars (also known as the driver-

S. Muthurajkumar (✉) · C. A. Karthikeyan · K. Pradeep · A. Hariharan
Department of Computer Technology, Anna University, MIT Campus, Chrompet, Chennai 600044, India

G. Mathur et al. (eds.), *Proceedings of 2nd International Conference on Artificial Intelligence: Advances and Applications*, Algorithms for Intelligent Systems,
https://doi.org/10.1007/978-981-16-6332-1_55

less cars, or automated cars, or robot-driven cars) are no longer committed to the tasks of science knowledge. There are already many vehicles on the road currently, which complies with the Automated Driving Assistance Module (ADAM) that maintains speed, brake, and other maneuver with literally limited or no driver enhancement in driving.

Building on these technological advancements, the completely self-driving cars are on the border, along with development and extensive testing initiatives taking place around the world. It will surely make the vehicle sharing more convenient, popular, and important to fundamentally change the fate of the transportation modules. The automated software and hardware that are required for AVs include the inclusive addition and adaption of new sensing technology and communication for guidance, and application for each automotives. These equipment and sensors are relatively expensive and significantly impact the cost rate of these cars, and which eventually hamper a capacious manufacturing and the accessibility of the SDVs. As an example, the LIDAR or light detection and ranging is a technology which uses 64 laser spins at about 950 rotation per minute (rpm) to give a 360 degree view to the driver. All this is to develop an elaborated three-dimensional mapping of the neighborhood area and to view all obstacles in real time in the environment. It is mainly used in most of the recent day's self-driven vehicles which cost 40000–90000. So, the distribution of SDVs is a promising and interesting answer to hike the efficiency of vehicular usage and remove the rate issue by bringing the SDVs under profit for the true owner of the vehicle and bisecting the rate price over different passengers or users.

Timing, availability, and the human effort for driving are necessary for manually driven vehicles for the transportation from one place to another place. This enables the family member for the many number of vehicles that are not effectively utilized all the times by the users and are usually parked for long duration in parking lots and garages. A recent study indicates that vehicles are almost parked 85% of the entire time. These types of situations will be changed by self-driving transforming to become more popular and practical among its users. In a time-sharing case, after a self-driving vehicle finishes one passenger's task, the AV drives the vehicle away to perform other requests without any need for extra help or other human effort. The drop-off and pickup places are discretion as the requester task. And correspondingly, a vehicle owner's work is explained as the place and time of accessibility of the autonomous vehicle and the place and time the SDV is hence needed back forth completing the task of the requester. And with this operation, the high cost can be distributed among a certain section of SDV users. But, in order to easily coordinate the distributed services, the users have to disclose their destinations and origin explicitly and the number of times of their travel. Certainly, many sensitive locations such as future location and information will be acquired by the coordination of AV usage, which is surely sensitive data. The locations such as home residences, their working location, places of entertainment, and other usual paths of travel.

One of the major impacts in sharing the self-driven vehicles pave the way to the big safety problem. The leakage of locality data has clear and direct negative consequence for passengers, e.g., travel information like the time users get picked-up and dropped-off to their respective locations of residence may be helpful for the criminals and impact the users negatively and harm them. Moreover, information about frequency of visits to hospitals or pharmacy stores can be used to effectively find one's health state which may reveal the user's choices of preferences while visiting any place of amusement. The common methods for privacy preservation, such as false hiding of location and optimization of location, are really not efficient enough in these situation of distributing the drop-off and pickup places for path routing, since even a minute difference in the coordinates may tend to make some huge difference in paths to be selected as the travel route because of the congestion in traffic and factor across the neighborhood. And due to methods like these, there are situations, where the privacy preservation that is achieved is not enough and secure because by intersecting the randomly chosen locations or generalized places with the highest possibility of user presence, the user locations are not identifiable. In a time-sharing service, we need to effectively find and calculate the time duration and displacement from the SDV location to that of the requester's location, then to the requester's destination and finally again back to the owner's live location. We develop a scheme based on the location privacy preservation scenarios and its usability and have drawn a conventional privacy-preserving work matching assignment and work matching method over the encrypted information. The work matching algorithmic scheme matches the requester and AV owners without having their live paths and the times of their travel. Therefore, the live locations of the user cannot be intercepted by the server. We propose an ID or intermediate-destination-assited work matching technique which uses the travel times and distances from these live locations to a set of predefined highly busy locations for matching, instead of using the users' real location information

In this paper, we propose a transportation scheme with the following contributions:

- An efficient matching scheme algorithm that matches autonomous vehicles with the feasible requester.
- A dynamic task scheduling scheme to assign autonomous vehicle to the requester without user's location and route, which would eventually improve the transportation system.

2 Related Work

B. Ying, D. Markakis, and Z. Hou proposed in [1] that the privacy of location is an important matter of concern in vehicular networks as the information about a vehicle location can lead to the leaks in some sensitive information. A vehicle should change its mix zone which consists of their designation in deliberate regions to make sure the location privacy of these vehicles is protected. However, self-driven vehicles may not change their mix zone because of less availability of resources (such as bandwidth and deliberates). This may expose and venture the location privacy of the vehicles that are in need of changing their mix zone. To overwhelm vehicles to communicate and cooperate in changing their mix zones, a method called Motivation for Protecting Selfish Vehicle's Location Privacy (MPSVLP) had been proposed. When the expiration is close vehicles could form a mix zone in optimized manner when their deliberates and can also earn reputation of "credit" by implementing a deliberate change in MPSVLP. The simulations have shown that MPSVLP motivated more self-centered vehicles to communicate and cooperate among each other while at the same time ensuring the privacy of the location at a high level.

To provide some contextualized service according to the locations in the group and not just individuals, the usage of location-based services by using locations from a group of users was explained by the complementing the Location-Sharing-Based Services (LSBS) [2]. However, day by day the misuse of location data and the other information by the third parties are increasing, which enables the need for more privacy control mechanism in the location-based services. The main problem of privacy in the LSBSs by providing practical and effective solutions to solve this issue in one such service, that is, the fair rendezvous point (FRVP) determination service had been proposed. The privacy-preserving FRVP (PPFRVP) problem thoroughly interprets the privacy requirements and computations in LSBSs. The two privacy-preserving algorithms for the FRVP problem analytically evaluate their privacy in both passive and active adversarial scenarios which has been proposed in this paper. The attempt to understand the popularity, the privacy, and acceptance of the proposed solutions are done by the means of a targeted user study.

C. K. Brownell proposed an explanation in [3] about the provided Anonymous Taxi Network (ATN). The major points that have been taken are: (1) a solution to solve the congestion of traffic issue, (2) improvements in safety over established human-driven cars, (3) a low carbon emission and reduction in pollution over the environment, (4) be reliable economically, and (5) convenience and comfort to rival the automated vehicles. A potential solution to the vehicles's growing issue has presented itself: an Automated Cab Network (ACN), given the recent advancements in the field of automobile infrastructure. The traditional Personal Rapid Transit system as well as Mark Gorton's idea of Smart Para-transit, two potential designs for an ATN are discussed and compared among each other, and along with the usability of the ATN concept as a whole is explored using state-wide transportation demand from the state of New Jersey has been drawn.

H. Kido, Y. Yanagisawa, and T. Satoh proposed in [4] that one of the most biggest issues is the protection of location privacy in location sharing service. To overcome this anonymous communication technique has been used to protect the location privacy of the users of location-based services. The user provides the several false information (dummy information) to send to service providers with the true position data of users. By that the service-provide does not distinguish the true position of the data, i.e., user location privacy is protected. Many experiments are conducted on the practical positioning data. As a result of the experiment, this technique protects the location privacy of the people and sufficiently reduces the cost of the communication so that this technique has been used in practical location sharing services.

As the popularity of the Mobile Social Networks (MSNs) grew, it increased the danger of breaches in privacy due to user location exposures. This was explained by H. Li, H. Zhu, S. Du, X. Liang, and X. S. Shen in [5]. The location privacy leakage due to the Mobile Social Networks (MSNs) had been quantified by conducting the 3-week experiments by both indirect location sharing and direct location sharing revealing user's real point of interests (POIs) and observing user's exposed location profiles. The experiment reveals that attacker can effectively predict demographic attributes about user's shared location. To resolve such attack, Smart Mask, a context-based system-level privacy protection solution is proposed which automatically learns user's preferences in privacy under many different contexts and provides a transparent location for MSNs users.There are many works in the literature that discuss about vehicular social network and retrieval methods [11–15].

For the utilization in various applications such as Vehicular Social Networks (VSNs), i.e., the chatting between drivers or sharing the common information between them, the wireless communication and the capabilities of Vehicular Ad hoc Networks (VANETs) have been explained in [6]. The one advantage of VSNs is sharing of common interested information such as location, fuel capacity, and so on among nearby car drivers. Drivers may benefit from identifying neighbors that have degree of interests (DOIs) that is common with them. But, a few issues related to privacy should be addressed. Revealing the nature and degree of interests (DOIs) of drivers can be violation of their privacy. The author proposed an efficient chatting scheme among the drivers nearby but it should comprise some privacy policies or privacy scheme. For anonymous common interest verification and homomorphic encryption technique for anonymous DOI verification, they used the attribute-based encryption (ABE) technique.

L. Liao, D. J. Patterson, D. Fox, and H. Kautz proposed [7] proposed an introduction to a hierarchical Markov model in which user's daily movements over the social community had been learned and inferred. On account to develop the gap between raw Global Positioning System sensor calculations and multi-level information, the model used high levels of abstraction such as a user's type of transportation or her destination. The decisive locations such as destinations or locations where the user frequently changes type of transportation are learned from GPS database without incurring any human effort. They had shown that to detect unwanted behaviors (e.g., getting into a congested road) by frequently tracking his day-by-day actions with a well-trained and an earlier model. Many experiments had been conducted to evaluate the learning and inferring transportation activities of some user's daily activities by Global Positioning System. Experiments had shown that the prediction of the destination of a person and the recognition of situations in which the user performs unknown activities were done by the model accurately.

An unconventional distributed anonymizing protocol based on peer-to-peer architecture has been proposed in [8]. Each mobile junction was responsible for anonymizing a specific zone. The mobile junction collaborates without the required need to get access to any data in anonymizing the queries. In this the request would be sent with arbitrary chosen ticket. And the encrypted response that was created by the server was sent to the particular mobile junction (intermediate junction). The user queries the response from the intermediate junction. Except for the fields mandatorily needed for the anonymizer and the intermediate junction, all parts of the message are encrypted by some encryption algorithms. This would secure the package exchange over the peer-to-peer network. Many experiments are executed and tested successfully in this paper, and the results of the experiments stated that it could be deployed efficiently to get the car user privacy in location-based services.

The sharing of the self-driven vehicles has the prospective to be the complete explanation for overcoming the cost defect of these type of vehicles to underlying changed transportation schema which had been introduced in [9]. The self-driven owners shared their AVs at those times when they do not need them by the time-sharing services where AVs enabled. This sharing could significantly reduce the cost by allowing the owner to share the cost of the car with other users. But, these services produce an uncompromised privacy concern due to shared location and path information of the users that are surely confidential and sensitive. The privacy-preserving time-sharing scheme allows requester and owner to perform a privacy-preserving matching according to their needs over server without disclosing their paths and location until the matching is free of issues and efficient. To do so, set of point of interests (POIs) are used as intermediate destinations in the travel ways.

The ride-sharing services (RSSs) are discussed in [10], which assists the drivers of the vehicle to locate conventional riders for unoccupied seats on the path, providing evolutionary vehicle occupancy and the demanding benefits of shared ride cost, which have changed the business of transportation, as witnessed by the success of Ola cabs, Zoom cars, and FastTrack Carpool. Selecting proper ride-share partners for drivers based on rider's trip data is essential for RSSs, but at the same time, there is some exposure of riders' and drivers' next locations and paths. To preserve the privacy during partner selection individually, a privacy-preserving ride-matching system for selecting attainable ride-share partners in RSSs had been proposed by the author. A spatial region-based selection algorithm was designed at first, in which the ride-sharing server (RS server), which previously choose riders in the matched regions with drivers without revealing their pointed current origins and destinations, was allowed. Second, having the encrypted parameters of drivers and riders, the RS server further selects prospective ride-share partners according to the travel time saving (TTS) and the feasibility of time schedules.

3 Proposed Work

3.1 System Overview

To improve the system of transportation, sharing of AVs among different requesters can possibly be the answer to this emerging problem. We have situations involving real-time cases where the vehicles are idle at the owner's place and that can be utilized to share the autonomous vehicle with others at the required place of origin. And when such a service is initiated due to sharing, the requester's origin and other important information can be compromised and may cause privacy issues. So, here we propose a novel scheme for services of autonomous vehicles and sharing them in real time using privacy preservation and task scheduling.

Initially, our first step is to develop a scheme for matching the available autonomous vehicle with the required requester. Secondly, using various deciding parameters, we assign the the feasible requester to the autonomous vehicle using a scheduling scheme. This includes the various approaches to implement the scheme. Without affecting the efficiency of the processes involving both the schemes, the sensitive information involving user's details is not compromised, eventually improving the profit for the autonomous vehicle owner and decreasing the cost of operation.

The scheme utilizes a matching server in order to identify the intermediate destination (ID) present across the travel route based on the given requester and autonomous vehicle. Based on the matching result provided by the server, the one with the highest efficiency is chosen as the appropriate route and only then the least required information are shared with the two parties involved to complete the process.

The service feasibility is checked along with the required verification on the server. This also involves the calculations for the upcoming processes. With traffic being considered for the simulation, along with the patterns found in the area, the intermediate destinations selected randomly are outperformed by the one selected by our scheme.

3.2 System Architecture

The architecture diagram (Fig. 1) clearly states the architecture of the proposed work. The diagram involves many components of the system, which includes the key candidates—the requester and the owner, who are responsible for initiating the process. The matching server essentially includes the other two components called the temporal information manager and the access control manager inside it. The privacy service provider maintains the important cryptographic operations among others. And finally, the ID manager is held responsible for the location operation along with the database to store the requester information on the user's interface once the request has been successfully processed and commissioned.

3.3 ID Selection and Matching Scheme

The concept of the intermediate destinations and the extensive advanced encryption and decryption methodologies for the implementation of the privacy preservation scheme is of major importance.

The main concept of the intermediate destination lies in the fact that we need to establish a network of closely related points on the closest distances around the required spot, which is essential to determine the path between the requester and the owner vehicle (also called the autonomous vehicle).

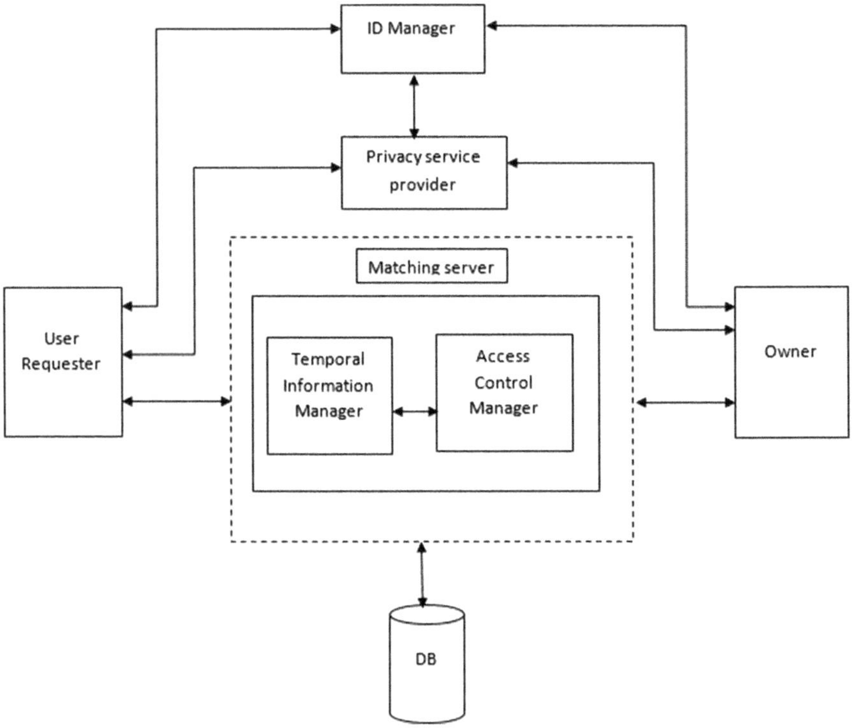

Fig. 1 Architecture diagram for privacy preservation and task scheduling

In order to obtain the required set of intermediate destinations, we need to establish or mark-up a certain range of destinations in a region and try to predict all the best possible cases while choosing the desired result. This range of destinations is constructed using various sets of polygon shapes, that is, like a rectangle, a square, a pentagon, a hexagon, or any other shape which is deemed essential in that respective area of acquisition. The expected results may vary according to the number of points being chosen in these shapes and it also depends on other minor factors. The next set of problems involves the high level of security arrangements made for ensuring the privacy preservation scheme is achieved is the extensive calculations involving the encryption and decryption of a set of task values provided from the previous functions of the intermediate destination. The tasks involve the usage of some of the sensitive contents, which may have the chance of being compromised in case of any third-party attacks, for example, the preference of a person's type of leisure activity place, etc. These choices of the user must not be made public or being made available to others as these details may risk the compromise of the user's details also. And in the case of autonomous vehicles and other carpooling companies, until the requester or the user gives the consent of his/her information regarding the sensitive details of one's, the company does not have the right to disclose these private information to others, and in some cases may even attract legal actions being taken against the respective institution or organization.

So, based on the above extensively described problems and other cases, it is essential to make sure, we use the best possible or available cryptographic algorithm to curb the attacks by the third parties and also combat the compromise of the user/requester's data. And therefore, we have chosen Paillier's cryptosystem, with some modification to make sure the encryption and decryption procedure takes place carefully. The usage of large primes and other required algorithm makes it an effective scheme to communicate among the servers and the requester/vehicles or vice versa.

3.4 Dynamic Task Scheduling

The immediate work to be done after all the required parameters and other factors are being determined is the matching of the requester and the autonomous vehicle with the best possible case scenario to make sure the ride is carried smoothly. And then comes the dynamic scheduling of the the path to be taken by the autonomous vehicle and disclose the essential travel information about the requester and location to the autonomous vehicle once the matching is done. Since the earlier section provides us with information about the set of intermediate destinations which are eligible for the matching scheme, we need to consider some other parameter before finalizing which intermediate destination to be taken. So, among the different parameters to be considered, the most important of them all is the traffic. The traffic plays a crucial role in determining which intermediate destination is to be taken. In order to get that ability to determine the intermediate destination, we need extensive knowledge about the real-time data about the traffic, which is bound to change constantly with time and the number of vehicles in a particular region. And after the matching scheme is complete, we are ready to share the information of the requester, which are deemed to be sensitive to the requester, such as the location and time to the allotted autonomous vehicle. Among the most important thing to be noted down here is the fact that the information about the requester is never shared with the autonomous vehicle owner at any cost. The autonomous vehicle owner till the end remains a spectator to the scheme as far as the requester is concerned, but the owner participates in the process of proclaiming the autonomous vehicle or the retrieval procedure of the respective vehicle.

As far as the role of the autonomous vehicle's owner is concerned, the person just provides the necessary details about the vehicle initially, such as the location and time at which the vehicle is free, and finally at which point the vehicle is bound to meet the owner or get parked automatically around the location provided in the respective region. Hence, once the trip between the requester and autonomous vehicle is completed successfully, the vehicle, without any compromise, is bound to get back to the location where the owner has requested his/her vehicle until any further notice regarding its next trip is made. In this way, the integrity of the system is maintained with maximum security and the desired result is obtained.

4 Proposed Scheme

4.1 Matching Scheme

Initially, we develop a privacy-preserving task matching scheme across the travel data that is encrypted, which matches the SDV owner and requester. The system evaluates both the owner and requester sets and produces a value for each according to a cost service, all without having any of the user's most confidential places data. Further, here we take two varying situations of random and traffic-based obtained IDs.

On order to select the intermediate destinations, there is a need to select a predefined region. This region of selection will be based on the requester location. The most widely used and popular locations around the region may be included in the selected region. As part of the algorithm, sorting of these selected regions occurs based on a number of parameters, such as time and distance. The boundary of each selection is also taken from a predefined set of locations around it.

Finally based on the these calculations, we determine the IDs to be added into the list of possible and best intermediate destinations. This list is then sent to the required function for its usage (Fig. 2).

Algorithm 1 Selection of Intermediate Destinations (ID)

Input: Locations and distances from the requester

```
procedure SELECTLOCATIONS()
    regions = selectRegion()
    procedure PRESORT()
    procedure BOUNDARY()
    while regions in boundary do
        IDlist = regions.add()
    procedure SENTREGIONS()
```

4.2 Cryptographic Processing

The encryption of the location and the other parameters of the user is done by the homomorphic encryption. Homomorphic encryption is an encryption technique with some additional evaluation. It computes over the encrypted data without the secret key accession. The result of the computation of homomorphic encryption remains encrypted. Homomorphic encryption of two types, namely, symmetric encryption and asymmetric encryption. Homomorphic is defined as the homomorphism in algebra: the encryption and decryption operations can be as homeomorphisms between plaintext and ciphertext spaces. Paillier cryptosystem algorithm which satisfies homomorphic properties consists of three algorithms. The asymmetric algorithm

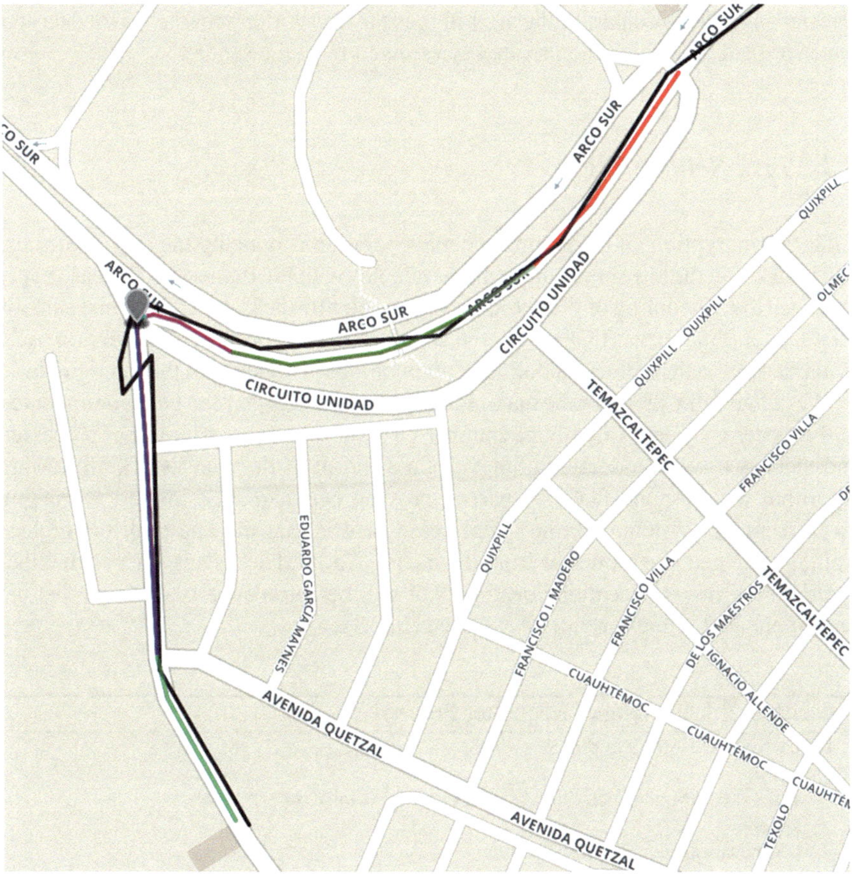

Fig. 2 Matching of routes

has two keys, namely, public key and private key. Private key is used for obtaining plaintext from ciphertext by decryption. Public key is used for the encryption of the plaintext. The first part of the Paillier cryptosystem algorithm is key generation which generates public and private keys by randomly choosing the two prime numbers. The sender who sends the confidential data will have the public key for the encryption and the receiver who receives the confidential data will have the private key for decryption of the data. The second part of the algorithm is encryption scheme which encrypts the data by using the public key. The reason behind the usage of the two keys, namely, public and private keys are to maintain data privacy stronger. As the usage of the single key value for both encryption and decryption is not as much good because while sending the encrypted data to the sender we want to send the key along with the encrypted data which leads the attacker easily to the data while transferring. The third part of the algorithm is the decryption scheme which decrypts the encrypted data using the private key received from the sender.

4.3 Task Scheduling

After the encryption and matching scheme operation is done by the server, the next one is task scheduling of the autonomous vehicles with the requester. The base paper proposed the scheduling of the autonomous vehicles by the static algorithms, namely, First Come First Serve (FCFS) and the greedy-based algorithms. But we propose a dynamic approach for the scheduling of autonomous vehicles with the requester task.

We define a threshold for the maximum number of tasks that can be accommodated in the system. It means that huge number of requesters cannot be left unprocessed and there is a need to decrease the wait time for all of those tasks. The following algorithm will respond faster as part of proving better quality, instead of making profit from the system. So, the simple case here is that the one with the highest waiting time gets the requester from the queue. And if there is any mismatch either based on the time or location constraints, it will be automatically declared that the immediate next requester will be assigned (Fig. 3).

Algorithm 2 Scheduling of Requester Priority

```
for each present AutoVeh do
    iter=0;
    while var=var-queue.peak(iter) match AutoVeh) is false do
        iter++;
        var=var-queue.peak(iter)
    Assign var to AutoVeh;
    var-queue.remove(iter);
```

The major disadvantage of the static algorithm is that in first come first serve scheduling the server does not schedule the autonomous vehicles optimally. It only looks which autonomous vehicle comes first and allocates it to the requester. It does not look for another autonomous vehicle which has the minimum cost between the requester and itself.

As similar to what we declared for the requester priority, the definition of owner priority is same. Just some minor changes are required for defining it. Any extensive idleness noticed in the owner's side will immediately be followed by assigning a requester according to their time of waiting throughout the system. Again, we set a threshold for the number of owners that can be accommodated in the system. And if there is any mismatch either based on the time or location constraints, it will be automatically declared that the immediate next owner will be assigned.

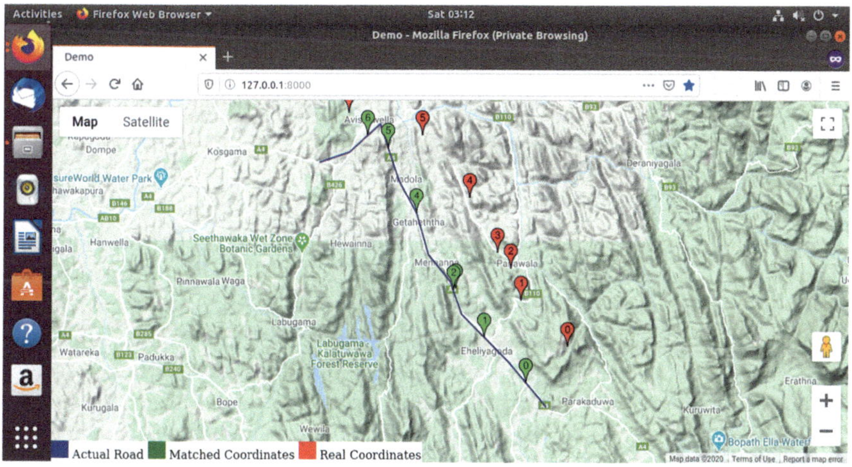

Fig. 3 Scheduled AV and requester

Algorithm 3 Scheduling of Owner Priority

```
for each present var do
    iter=0;
    while AV=AutoVeh-queue.peak(iter) match AV) is false do
        iter++;
        AV=var-queue.peak(iter)
    Assign var to AV;
    AutoVeh-queue.remove(iter);
```

---

5 Result and Performance Analysis

5.1 *Overall Result Analysis*

The scheme was developed, so that, no confidential data of either the autonomous vehicle or the requester is present. And only with the help of the server, the requester and AV owners can reveal their request and autonomous vehicle tasks to each other. And using the local traffic servers and the ID locations, the removal of sensitive location information was done, along with the encryption of all the parameter, making it better in terms of security. Most importantly, location information of the autonomous vehicle is never disclosed to the requester, and only when the entire scheduling process is done and the index of the scheduled requester is revealed to the autonomous vehicle, both the parties will then be in contact immediately, and requester will share the correct locations with the owner.

5.2 Overheads and Evaluation

Considering the computations involved in the scheme, we have—If m = number of owners, n = number of requesters, d = number of IDs, Then, basically, we have $(m + n) * (4 * d + 1)$ number of calculations along with the same number of calculations for encryption in the user's device interface. The matching server also performs extensive calculations for encryption and other factors. So, based on the above factors, an additional $(2 * m * n)$ number of minimum operations are needed on d number IDs, in order to get the result. Finally, the cost and requester index value is $(m * n)$.

5.3 Analysis

Based on the traffic-based scheme, comparison between added travel duration and distance is made, when using several locations as intermediate destination and the normal direct origin to the destination. The selection of ID based on the origin-destination method on average among 20 pairs increase by a large percentage compared to the direct route and eventually decrease when the number of IDs is increased.

Based on Fig. 4, it can referred that as we try to increase the number of miles, the errors too increase, followed by giving us more intermediate destinations. There is a sharp decrease in the percentage of IDs being reduced, leading to poor accuracy. This ultimately affects the performance of the design. But the algorithm is far better considering the overheads it handles.

But, the traffic-based performs much better in the worst case, and that eventually shows that the accuracy increases as the number of the IDs also increases, because the possibility of having a suitable ID will increase. So, it's noted that the efficiency is hit hard based on the calculations and other contributing factors. So, in order to get desired privacy preservation of places and time, we introduce intermediate destinations based on the real-time traffic patterns and allow the server to execute the required matching algorithm used for the autonomous vehicle and the requester. Finally, based on the modules and the algorithms developed, the scheduling scheme matches the requester task to the AVs.

```
Out[614]: <matplotlib.text.Text at 0x2b4ebefd0>
```

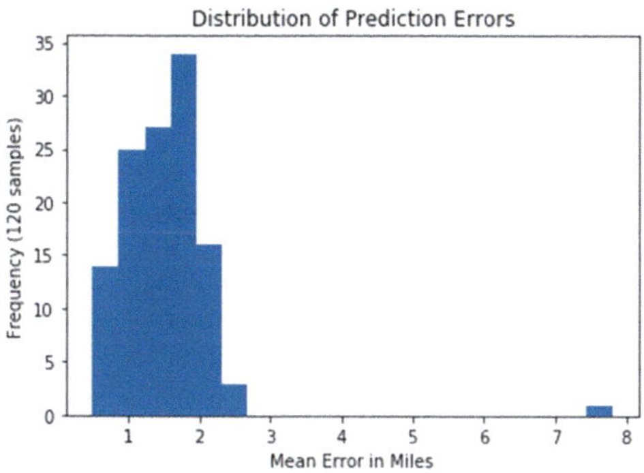


```
In [615]: print 'p-value: %s' %(scs.ttest_1samp(a=total_errors, popmean=2)[1])
          p-value: 1.12606893086e-09
```

Fig. 4 Mean errors in routes

We also include the number of false negatives to the scheme.The time required for any trip based on the traffic reaches false negative ratios about 0.032 for an average of 3500 observations. The addition of IDs significantly improves the performance, but decreases the independence on time consumption.

And this non-interference of user's travel information is which includes the time and location by the server is the main motive behind this scheme. Eventually, this is an improvised scheme to outperform the existing scheme for better efficiency and accuracy.

6 Conclusion

We introduce a novel privacy-preserving dynamic scheduling scheme for the time-sharing service of autonomous vehicles. In order to reach the goal of the privacy preservation of location and time, we choose Intermediate Destination (ID) based on traffic information and use these IDs to allow the server to execute a privacy-preserving matching task for autonomous vehicle owners and autonomous vehicle requester. Along with this, we propose a set of different scheduling schemes to assign the matched requester tasks to the autonomous vehicles. We concluded that the semi-trusted scheduling server is not able to intercept the user travel times and places. We conducted various experiments on real-time data sets to demonstrate improved scheme with traffic information consideration that experimentally outperforms the existing scheme.

References

1. Yuing B, Markarkis D, Huoou Z (2015) Motivation for protecting selfish vehicles' location privacy in vehicular networks. IEEE Trans Veh Technol 64(12):5631–5641
2. Bilogrevic I, Jadliwalai M, Kalkani K, Hubaux J-P, Aad I (2011) Privacy in mobile computing for location-sharing-based services. In Proc. Int. Symp. Privacy Enhancing Technol., pp 77–96
3. Brownell CK, Bellamnd J (2013) Shared autonomous taxi networks: an analysis of transportation demand in NJ and a 21st century solution for congestion. Canbera University, Apr. 2013
4. Kidoyon H, Yanagisawar Y, Satoh T (2005) An anonymous communication technique using dummies for location-based services. In: Proc. Int. Conf. Pervasive Serv., 2005, pp 88–97
5. Li H, Zhu H, Du S, Liang X, Shen XS (2018) Privacy leakage of location sharing in mobile social networks: attacks and defense. IEEE Trans Dependable Secur Comput 15(4):646–660 Aug
6. Rabieh K, Mahmoud M, Siraj A, Misic J (2015) Efficient privacy preserving chatting scheme with degree of interest verification for vehicular social networks. In: Proc. Global Commun. Conf., 2015, pp 1–6
7. Liao L, Patterson DJ, Fox D, Kautz H (2007) Learning and inferring transportation routines. Artif Intell 171(5/6):311–331
8. Ghaffari M, Ghadiri N, Manshaei MH, Lahijani MS (2017) p4qs: a peer-to-peer privacy preserving query service for location-based mobile applications. IEEE Trans Veh Techno 66(10):9458–9469 Oct
9. Hadian M, Altuwaiyan T, Liang X (2017) Privacy-preserving time-sharing services for autonomous vehicles. In: Proc. IEEE 86th Veh. Technol. Conf., 2017, pp 1–5
10. He Y, Ni J, Wang X, Niu B, Li F, Shen XS (2018) Privacy-preserving partner selection for ride-sharing services. IEEE Trans Veh Technol 67(7):5994–6005 Jul
11. Hadian M, Altuwaiyan T, Liang X, Zhu H (2019) Privacy-preserving task scheduling for time-sharing services of autonomous vehicles. IEEE Trans Veh Technol 68(6):5260–5270
12. Keane JR, Forrest AL, Johannsson H, Battle D (2020) Autonomous underwater vehicle homing with a single range-only beacon. IEEE J Ocean Eng 45(2):395–403
13. Zhang Y, Liu L, Lu Z, Wang L, Wen X (2020) Robust autonomous intersection control approach for connected autonomous vehicles. IEEE Access 8:124486–124502
14. Yao Q, Tian Y, Wang Q, Wang S (2020) Control strategies on path tracking for autonomous vehicle: state of the art and future challenges. IEEE Access 8:161211–161222
15. Chae H, Yi K (2020) Virtual target-based overtaking decision, motion planning, and control of autonomous vehicles. IEEE Access 8:51363–51376

Chapter 56
Introduction of PMI-SO Integrated with Predictive and Lexicon Based Features to Detect Cyberbullying in Bangla Text Using Machine Learning

Md. Tofael Ahmed, Maqsudur Rahman, Shafayet Nur, AZM Touhidul Islam, and Dipankar Das

1 Introduction

Cyberbullying is the intentional and persistent harm caused by the use of electronic media [15]. The increased use of various social media sites has greatly worsened this situation. Survey shows that nearly 3.6 billion individuals were using social media [11]. As of January 2020, this usage is also rising at a rate of 49% [11]. This escalation of online activity of people is causing cyberbullying to occur more frequently than before. As cyberbullying causes huge psychological and physical distress and even suicide [16, 24], it should be eliminated as early as possible. Different detection and prevention approaches have been proposed by researchers which detects and

Md. T. Ahmed (✉)
Department of Information and Communication Technology, Comilla University, Kotbari, Bangladesh
e-mail: tofael@cou.ac.bd

S. Nur
Department of Computer Science & Engineering, Port City International University, Chattogram, Bangladesh

A. T. Islam
Department of Electrical & Electronics Engineering, University of Rajshahi, Rajshahi, Bangladesh
e-mail: touhid.eee@ru.ac.bd

D. Das
Department of Information and Communication Engineering, University of Rajshahi, Rajshahi, Bangladesh
e-mail: dipankar@ru.ac.bd

M. Rahman
Department of Information and Communication Technology, Comilla University, Comilla, Bangladesh

G. Mathur et al. (eds.), *Proceedings of 2nd International Conference on Artificial Intelligence: Advances and Applications*, Algorithms for Intelligent Systems,
https://doi.org/10.1007/978-981-16-6332-1_56

prevents cyberbullying including machine learning based automated cyberbullying detection.

Machine Learning has become a huge field of research in recent days. COVID-19 pandemic outbreak prediction in Bangladesh [26] and identification and monitoring of vehicles from real-time video streams [25] are two examples of recent work in this field. Machine learning is successfully deployed for COVID-19 analysis [21], prediction [29], and forecasting [6]. Machine Learning algorithms have also been effective in predicting breast cancer [5] and classifying diabetic and nondiabetic patients [19]. Text classification is another area where machine learning is very useful. The authors of [17] used Machine Learning techniques to conduct text classification, and in [7], authors used Machine Learning techniques to evaluate sentiment in English, Dutch, and French texts. Moreover, Semberecki et al. [27] presented method of classification of text documents using deep neural network. Also, Kowsari et al. [20] used hierarchical Deep Learning to classify texts. Performance of these techniques can be improved by transforming data [28] and by using nature inspired algorithms [10].

These techniques can be used for classifying cyberbullying texts. In [13], Haider et al. proposed a cyberbullying detection approach that was 71.25% accurate in both English and Arabic. Similarly, in [22], cyberbullying was detected from Spanish texts using Machine Learning classifiers with 93% accuracy and in [4], cyberbullying was detected with 80% accuracy.

There are many more approaches available for detecting cyberbullying discussed in Sect. 2 which uses normal text classification models. But still there is a huge lacking of available works in feature-based cyberbullying detection, especially for Bangla text. For that purpose, we conduct this research to detect cyberbullying through feature-based Machine Learning classifiers by introducing a new feature called pointwise mutual information-semantic orientation (PMI-SO). We also build an algorithm to generate this feature. Along with this feature, we used some more features such as likes and replies count of each text. We also used lexicon based features such as bullying and non-bullying words count of each text.

The remainder of the paper is divided into the sections following. The related works are discussed in Sect. 2 and our methodology is presented in Sect. 3. In Sect. 4, we discuss the classifiers performance in terms of precision, recall, f1 score, accuracy, and roc area and also showed some snaps of our web page. Finally, Sect. 5 brings the paper to a conclusion.

2 Related Works

To stop the devastating effects of cyberbullying, the detection of cyberbullying has become very important. Many researchers are finding good methods of cyberbullying detection in several languages. In this section, we discuss the existing works some of which can be improved more by applying feature-based techniques.

By scrapping texts from Facebook and Twitter, Haidar et al. proposed a solution for cyberbullying detection that was only for English and Arabic [13]. For the datasets, they used Machine Learning classifiers with Nave Bayes achieving 90.1% overall Precision and SVM 93.4%. Dalvi et al. [12] have proposed a Machine Learning model that uses SVM and Nave Bayes classifiers to identify and avoid cyberbullying on Twitter. They used twitter data and the highest accuracy was 71.25%, but it could be better.

León-Paredes et al. retrieved Spanish texts from Twitter and used Machine Learning algorithms to achieve a 93% accuracy rate [22]. Ali et al. [4] used three datasets and Machine Learning techniques to achieve an average of 80% accuracy using SVM. In addition, Haider et al. [14] proposed deep learning methods for detecting cyberbullying in Arabic. The highest accuracy they could get was 91.17%. This study does not use feature-based cyberbullying detection, which we addressed in our study. Deep learning algorithms were also used by Iwendi et al. [18] to detect cyberbullying. Their best obtained accuracy was 82.18%. This accuracy score is quite satisfactory, but in our research, we obtained 10.82% more accuracy than them.

For Bangla texts, there are some notable works in this domain. Akhter et al. [3] used Machine Learning algorithms in two phases to analyze 2400 Bangla public posts from social media. They used only texts in the first phase of the training and SVM had the highest accuracy of 95.40%. They combined some user details with the texts in the second phase and trained the models again, with SVM having the highest accuracy of 97.27%. This study needs additional data, which we addressed in our research.

Chakraborty et al. used MNB, SVM, CNN, and LSTM to analyze Bangla texts [8]. The texts contained emoticons and Unicode Bangla characters. In their research, SVM performed best with 78% accuracy, which is 15% less than the best accuracy obtained in our study. Similarly, Ahammed et al. collected Bangla texts from Facebook and used Machine Learning to achieve a maximum accuracy of 72% [1], which is 21% lower than ours.

Furthermore, Ahmed et al. used Machine and Deep learning algorithms and detected cyberbullying from Bangla and Romanized Bangla text [2]. A comparative analysis was also presented by the authors. In their research, they found best of 76% accuracy for Bangla text using conventional machine learning techniques.

All these research are very good on their own. But there is still a lacking of feature-based approach to detect cyberbullying, especially for Bangla text. For that purpose, we conduct this research to detect cyberbullying in Bangla text. In this research, we introduced a new feature called pointwise mutual information-semantic orientation (PMI-SO). We used this feature along with some other features which we elaborated in Sect. 3. We believe the introduction of pointwise mutual information-semantic orientation (PMI-SO) as an input feature to detect cyberbullying in Bangla text is first of this domain. The detailed methodology is discussed in Sect. 3.

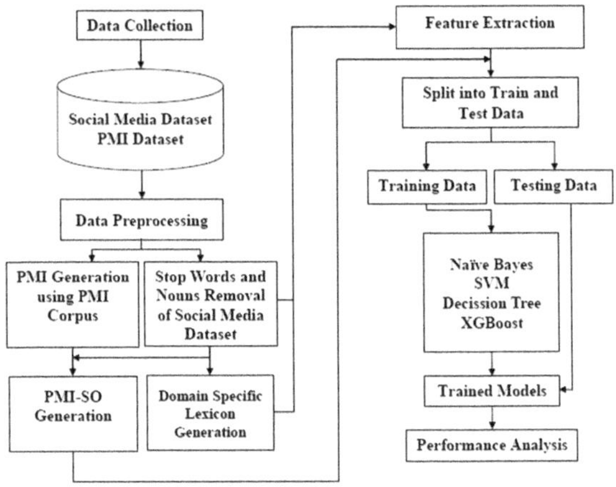

Fig. 1 Proposed methodology

3 Methodology

This research aims to detect cyberbullying using PMI-SO. Figure 1 shows the methodology.

3.1 Dataset

In this research, we used two datasets. They are denoted as Social Media Dataset and PMI Dataset. Both the datasets were created using YouTube API version 3.0. The Social Media Dataset contained 5000 Bangla texts with features such as likes count and replies count. We manually labelled the Social Media Dataset into two classes, bullying and not bullying. The PMI Dataset contained 10277 Bangla texts. We used these texts to generate PMI-SO for the Social Media Dataset.

3.2 Preprocessig & Phrase Extraction

It was necessary to clean the data since both datasets had a large number of noisy texts. We first removed all the duplicate data from both datasets. Punctuation marks, links, urls, digits, emoticons, user tags, elongated words and mentions were then omitted from the dataset.

After preprocessing, the phrases that contained the most significance were extracted from the Social Media Dataset. For that purpose, we removed the stop words and nouns from all the texts of the Social Media Dataset. The remaining phrases were used to generate PMI-SO for each text.

3.3 PMI

Pointwise Mutual Information (PMI) is used to calculate the probability of co-occurrence of two words. While calculating PMI between two words, it is taken into account that their co-occurrence might be caused by the frequency of the single words. Equation 1 shows the PMI between two words, a and b.

$$PMI(a, b) = log_2[\frac{p(a\&b)}{p(a)p(b)}] \quad (1)$$

Here, p(a & b) is the probability that a and b co-occur. If the words occur independently, then their probability is given by the product of p(a) and p(b). The log of the ratio between p(a & b) and product of p(a), p(b) is the Pointwise Mutual Information of the words a and b.

In our study, we calculated PMI using the PMI dataset which contained 10277 texts. The reason for using a second dataset containing more texts than the other dataset is that it helps to get more accurate PMI score.

3.4 PMI-SO Generation

The SO or Semantic Orientation of a phrase is calculated using Eq. 2.

$$SO(Phrase) = PMI(Phrase, word1) - PMI(Phrase, word2) \quad (2)$$

Here, word1 and word2 are reference words form the negative and positive classes. In our study, PMI-SO is calculated by subtracting the PMI of the target word with a non cyberbullying class from the PMI of the target word with a cyberbullying class. Based on this, we calculated the average PMI-SO of each text in the Social Media Dataset. This technique was successfully used by Turney in [31] and is also known as Turney's technique. We used the Algorithm 1 to calculate PMI-SO for our Social Media Dataset.

3.5 Generating Domain Specific Lexicon

Domain specific lexicon were created to extract more features. In our research, we developed three dictionaries. Each dictionary contained words that were present only in the cyberbullying class, non cyberbullying class, and also in both the classes, respectively.

Algorithm 1: Calculation of PMI-SO

initialization:

1. for each sentence ϵ corpus
2. set TotalPmiSo, AveragePmiSo, WordCount = 0
3. Tokenize each sentence into words
4. for each word ϵ tokens
5. PmiScore= PMI(word, 'reference word of positive class') - PMI(word, 'reference word of negative class')
6. TotalPmiSo = TotalPmiSo + PmiScore
7. WordCount = WordCount + 1
8. AveragePmiSo = TotalPmiSo/WordCount

3.6 Feature Extraction

The Domain specific lexicon were used to extract features like cyberbullying word count, non cyberbullying word count, and count of words that are present in both the classes. These features along with the PMI-SO and number of likes and replies in a text were used as input for the Machine Learning classifiers.

3.7 Machine Learning Classifiers

To predict categorical data, machine learning classifiers are widely used. There are numerous Machine Learning classifiers, but we used four in our research, Nave Bayes, SVM, Decision Tree, and XGBoost.

3.7.1 Naïve Bayes

The Bayes theorem is used to construct Nave Bayes algorithms. It uses Bayes theorem [33] to calculate the probability of each category, and the category with the highest probability is considered the output. The Bayes theorem calculates the probability P(c|x), where c is the class of possible outcomes and x is the given instance that must be categorized based on unique characteristics.

$$P(c \mid x) = P(x \mid c) * P(c)/P(x) \tag{3}$$

3.7.2 Support Vector Machine (SVM)

SVM is the most commonly utilized Machine Learning algorithm. It categorizes data by locating a hyperplane in an N-dimensional space [32]. Here, N is the number of features. SVM selects the hyperplane with the largest margin out of a large number of possibilities.

3.7.3 Decission Tree

The Decision Tree is a supervised learning method. When it comes to classification issues, the Decision Tree is often used. It's a tree-structured classifier in which internal nodes describe dataset characteristics, branches represent decision rules, and each leaf node represents the outcome.

3.7.4 XGBoost

XGBoost is a Gradient Boosting algorithm. It is based on decision trees that implements a Gradient Boosting algorithm [9]. XGBoost is also an ensemble of decision trees, which are weak models on their own but become more efficient when combined. XGBoost iteratively constructs simple decision trees, known as slow learners because of their high bias. The iteration continues by creating a new tree. Each tree is a correction of the previous tree in each iteration. When a stopping condition is reached, iteration comes to an end.

3.8 Performance Analysis

We evaluated the classifiers using the testing set and analyzed the parameters after they were trained. We evaluated the performance in terms of confusion matrix, precision, recall, f1 score, accuracy, and roc area [23, 30].

The Confusion Matrix is a mixture of four different expected and actual values such as TP, FP, FN, and TN that is used to evaluate the efficiency of machine learning classification models. The accuracy, precision, recall, and f1 score are calculated using Eqs. 4, 5, 6 and 7.

$$Accuracy = \frac{TP + TN}{TP + TN + FP + FN} \tag{4}$$

$$Precision = \frac{TP}{TP + FP} \tag{5}$$

$$Recall = \frac{TP}{TP + FN} \tag{6}$$

$$F1 - score = 2 * \frac{Precision * Recall}{Precision + Recall} \tag{7}$$

4 Results and Discussion

The accuracy results are discussed in depth in this section. The confusion matrices of all classifiers on the Social Media Dataset are shown in Table 1. Table 1 shows that XGBoost was the most accurate classifier. XGBoost classified 457 among 507 Cyberbullying texts and 469 among 493 non Cyberbullying texts correctly. SVM performed beneath all the classifiers. SVM classified 362 among 507 Cyberbullying texts and 487 among 493 non Cyberbullying texts correctly. To learn more about the classifier which correctly classified the most instances and which classifier has the fewest instances incorrectly classified, Table 2 shows the number of instances of all classifiers that were correctly and incorrectly categorized.

From Table 2, it is evident that XGBoost has the highest number of correctly classified instances (92.6%) than other classifiers. Also XGBoost has the lowest numbers (7.4%) in terms of incorrectly classified instances. On the other hand, the lowest performance among all algorithms was given by SVM. SVM has the lowest numbers in terms of correctly classified instances and highest (15.1%) in terms of incorrectly classified instances. In addition, Table 3 shows the detailed accuracy of all the classifiers by class. Table 3 again demonstrates the efficiency of XGBoost. From Table 3, it is evident that XGBoost got precision score of 0.93, 0.92, respectively, for both positive and negative classes. Also, XGBoost got recall score of 0.93, 0.93 and f1 score of 0.93, 0.93, respectively, for both positive and negative classes. Again the lowest performance was given by SVM. SVM obtained precision score of 0.79 and

Table 1 Confusion Matrix of all the classifiers

Applied algorithm	TP	FP	FN	TN
Naïve Bayes	352	141	1	506
SVM	362	131	20	487
Decission Tree	444	49	36	471
XGBoost	457	36	38	469

Table 2 Number of correctly and incorrectly classified instances of all the classifiers

Applied Algorithm	Naïve Bayes (%)	SVM (%)	Decission Tree (%)	XGBoost (%)
Correctly Classified Instances	858 (85.8%)	849 (84.9%)	915 (91.5%)	926 (92.6%)
Incorrectly Classified Instances	142 (14.2%)	151 (15.1%)	85 (8.5%)	74 (7.4%)
Total Number of Instances	1000	1000	1000	1000

Table 3 Detailed accuracy by class of all the classifiers

Class	Precision				Recall				F1 score			
	Naïve Bayes	SVM	Decission Tree	XGBoost	Naïve Bayes	SVM	Decission Tree	XGBoost	Naïve Bayes	SVM	Decission Tree	XGBoost
0	1.00	0.95	0.93	0.92	0.71	0.73	0.90	0.93	0.83	0.83	0.91	0.93
1	0.78	0.79	0.91	0.93	1.00	0.96	0.93	0.93	0.88	0.87	0.92	0.93

Table 4 Error matrix of all the classifiers

Classifiers	Naïve Bayes	SVM	Decission Tree	XGBoost
Kappa Statistics	0.715	0.697	0.834	0.852
MAE	0.142	0.151	0.083	0.074
RMSE	0.377	0.388	0.288	0.272
MSE	0.142	0.151	0.083	0.074

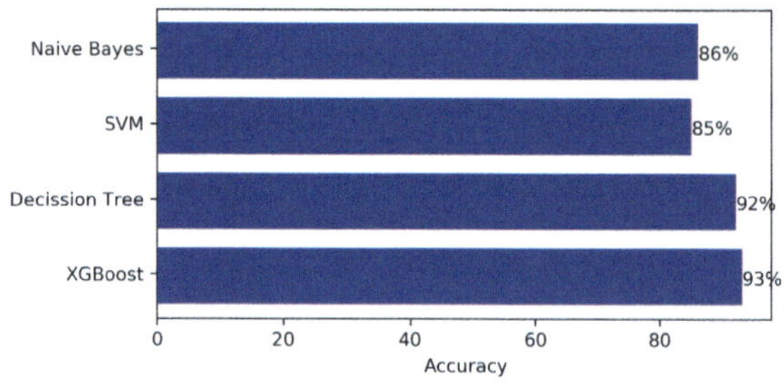

Fig. 2 Average accuracy of all the classifiers

0.95 along with recall score of 0.96, 0.73 and f1 score of 0.87, 0.83 for both positive and negative classes, respectively. Now to get more valuable insights on our model, we showed some error metrics in Table 4.

The kappa statistics reveals the balance percentage of classes in a dataset. Higher Kappa Statistics means that the classifier is doing better. From Table 4, we can see that XGBoost has the highest Kappa statistics of 0.852 among other algorithms. It indicates that XGBoost classifier is actually giving better predictions. Also, XGBoost has the lowest MAE, RMSE, and MSE error values and SVM has the highest which again indicates the better performance of XGBoost. Furthermore in Fig. 2, we showed the average accuracy of all classifiers.

Figure 2 shows that XGBoost obtained best accuracy of 93%. The performance of XGBoost was followed by Decission Tree (92%), Naïve Bayes (86%) and SVM (85%). Figure 3 depicts the ROC area of the best performing algorithms for all three datasets. As higher ROC area means better classification model, Fig. 3 clearly shows Naïve Bayes as the best performing algorithm and SVM as the lowest performing algorithm.

In this research, we also developed a web page using our model which predicts occurrence of cyberbullying in Bangla text. The web page takes as input the Bangla text, the likes and reply counts of that text and predicts if that text contains cyberbullying or not. The additional features are created in the background to feed into the model for prediction. This web page can be used by anyone, especially who is not familiar with Bangla language. Figure 4 shows snap of our web page.

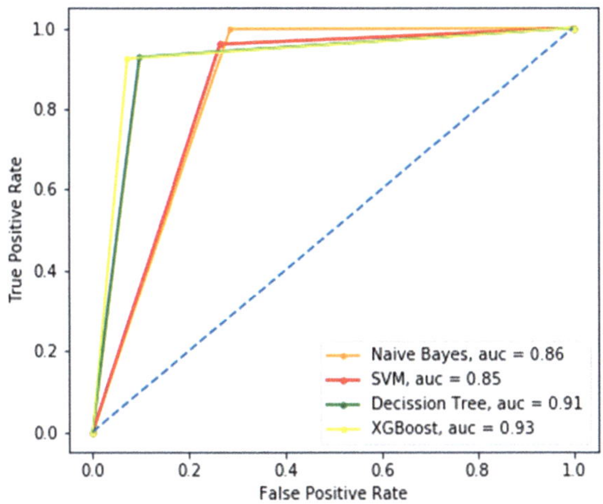

Fig. 3 Roc curves of all the classifiers

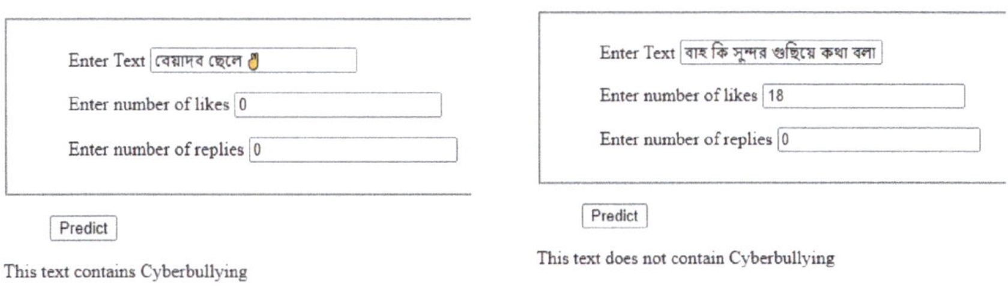

(a) Prediction of Cyberbullying text

(b) Prediction of Non Cyberbullying text

Fig. 4 Prediction of texts from web page

5 Conclusion and Future Work

This study used PMI-SO as a new input feature to create a feature-based model that detects cyberbullying in Bangla text. Along with PMI-SO, some other features such as likes and reply count of a particular text and also lexicon based features were used. We developed two datasets, one of which we used for classification and the other for generating PMI-SO. The extracted features were fed into the Machine Learning classifiers before their performance was measured. Our research revealed that XGBoost classifier classified texts with 93% accuracy which is the highest among all the other classifiers that we used. Also, we found out that SVM performed lowest among all the other classifiers by obtaining accuracy of 85%. We hope that our findings will be useful to the research community in conducting future studies in this area. We want to be able to detect the severity level of cyberbullying in each Bangla text in the future.

References

1. Ahammed S, Rahman M, Niloy MH, Chowdhury SMH (2019) Implementation of machine learning to detect hate speech in bangla language. In: 2019 8th international conference system modeling and advancement in research trends (SMART). IEEE, pp 317–320
2. Ahmed, M.T., Rahman, M., Nur, S., Islam, A., Das, D.: Deployment of machine learning and deep learning algorithms in detecting cyberbullying in bangla and romanized bangla text: A comparative study. In: 2021 International Conference on Advances in Electrical, Computing, Communication and Sustainable Technologies (ICAECT), pp. 1–10. IEEE (2021)
3. Akhter, S., et al.: Social media bullying detection using machine learning on bangla text. In: 2018 10th International Conference on Electrical and Computer Engineering (ICECE), pp. 385–388. IEEE (2018)
4. Ali A, Syed AM (2020) Cyberbullying detection using machine learning. Pak J Eng Technol 3(2):45–50
5. Asri H, Mousannif H, Al Moatassime H, Noel T (2016) Using machine learning algorithms for breast cancer risk prediction and diagnosis. Procedia Comput Sci 83:1064–1069
6. Bhatnagar V, Poonia RC, Nagar P, Kumar S, Singh V, Raja L, Dass P (2020) Descriptive analysis of covid-19 patients in the context of India. J Interdiscip Math:1–16
7. Boiy E, Moens MF (2009) A machine learning approach to sentiment analysis in multilingual web texts. Inf Retr 12(5):526–558
8. Chakraborty P, Seddiqui MH (2019) Threat and abusive language detection on social media in Bengali language. In: 2019 1st international conference on advances in science, engineering and robotics technology (ICASERT). IEEE, pp 1–6
9. Chen T, Guestrin C (2016) Xgboost: a scalable tree boosting system. In: Proceedings of the 22nd ACM SIGKDD international conference on knowledge discovery and data mining, pp 785–794
10. Chugh A, Sharma VK, Kumar S, Nayyar A, Qureshi B, Bhatia MK, Jain C (2021) Spider monkey crow optimization algorithm with deep learning for sentiment classification and information retrieval. IEEE Access 9:24249–24262
11. Clement J (2020) Number of social media users worldwide 2010–2021. [https://rb.gy/hdjcko
12. Dalvi RR, Chavan SB, Halbe A (2020) Detecting a twitter cyberbullying using machine learning. In: 2020 4th international conference on intelligent computing and control systems (ICICCS). IEEE, pp 297–301
13. Haidar B, Chamoun M, Serhrouchni A (2017) A multilingual system for cyberbullying detection: Arabic content detection using machine learning. Adv Sci, Technol Eng Syst J 2(6):275–284
14. Haidar B, Chamoun M, Serhrouchni A (2018) Arabic cyberbullying detection: using deep learning. In: 2018 7th international conference on computer and communication engineering (ICCCE). IEEE, pp 284–289
15. Hinduja S, Patchin JW (2008) Cyberbullying: An exploratory analysis of factors related to offending and victimization. Deviant Behav 29(2):129–156
16. Hoff DL, Mitchell, SN (2009) Cyberbullying: causes, effects, and remedies. J Educ Adm (2009)
17. Ikonomakis M, Kotsiantis S, Tampakas V (2005) Text classification using machine learning techniques. WSEAS Trans Comput 4(8):966–974
18. Iwendi C, Srivastava G, Khan S, Maddikunta PKR (2020) Cyberbullying detection solutions based on deep learning architectures. Multimed Syst:1–14
19. Kaur H, Kumari V (2020) Predictive modelling and analytics for diabetes using a machine learning approach. Appl Comput Inform (2020)
20. Kowsari K, Brown DE, Heidarysafa M, Meimandi KJ, Gerber MS, Barnes LE (2017) Hdltex: Hierarchical deep learning for text classification. In: 2017 16th IEEE international conference on machine learning and applications (ICMLA). IEEE, pp 364–371
21. Kumari R, Kumar S, Poonia RC, Singh V, Raja L, Bhatnagar V, Agarwal P (2021) Analysis and predictions of spread, recovery, and death caused by covid-19 in india. Big Data Min Anal 4(2):65–75

22. León-Paredes GA, Palomeque-León WF, Gallegos-Segovia PL, Vintimilla-Tapia PE, Bravo-Torres JF, Barbosa-Santillán LI (2019) Presumptive detection of cyberbullying on twitter through natural language processing and machine learning in the spanish language. In: Paredes-Pinos MM (ed) 2019 IEEE CHILEAN conference on electrical, electronics engineering, information and communication technologies (CHILECON). IEEE, pp 1–7
23. Narkhede S (2018) Understanding auc—roc curve. https://towardsdatascience.com/understanding-auc-roc-curve-68b2303cc9c5
24. Patchin JW, Hinduja S (2010) Cyberbullying and self-esteem. J Sch Health 80(12):614–621
25. Rahman M, Ahmed MT, Islam AZMT (2020) On detection and tracking of the vehicles from the real time video stream using background subtraction process with blob tracker algorithm, https://doi.org/10.5281/zenodo.4042914.
26. Rahman M, Ahmed MT, Nur S, Islam A (2021) The prediction of covid-19 pandemic outbreak on bangladesh perspective using multiple machine learning algorithms: a comparative study. Accepted
27. Semberecki P, Maciejewski H (2017) Deep learning methods for subject text classification of articles. In: 2017 federated conference on computer science and information systems (FedCSIS). IEEE, pp 357–360
28. Shekhawat SS, Sharma H, Kumar S, Nayyar A, Qureshi B (2021) BSSA: Binary salp swarm algorithm with hybrid data transformation for feature selection. IEEE Access 9:14867–14882
29. Singh V, Poonia RC, Kumar S, Dass P, Agarwal P, Bhatnagar V, Raja L (2020) Prediction of covid-19 corona virus pandemic based on time series data using support vector machine. J Discret Math Sci Cryptogr:1–15
30. Solutions E, Name (2016) Accuracy, precision, recall & f1 score: Interpretation of performance measures (2016). https://blog.exsilio.com/all/accuracy-precision-recall-f1-score-interpretation-of-performance-measures/
31. Turney PD (2002) Thumbs up or thumbs down? semantic orientation applied to unsupervised classification of reviews. arXiv preprint cs/0212032
32. Wu Q, Zhou DX (2006) Analysis of support vector machine classification. J Comput Anal Appl 8(2)
33. Xu S, Li Y, Wang Z (2017) Bayesian multinomial Naïve Bayes classifier to text classification. In: Advanced multimedia and ubiquitous engineering. Springer, pp 347–352

Chapter 57
Graph-based data analysis in Big Data Computing Environment: An investigation of Flight Network Datasets

Naishadh Mehta, Anand Ruparelia, Jai Prakash Verma, and Manoj Kumar Khinchi

1 Introduction

The airline industry is constantly expanding its horizons with the use of technologies that are enhancing business management as well as decision-making throughout the enterprise. The industry is dynamic and requires continuous upgrades in terms of operational planning and execution. Many factors govern the administration of airlines and all of them are taken into consideration for carrying out operations on a day-to-day basis. These operations are to be done to maintain the air traffic in an effective manner which should take into consideration the safety and regulatory norms of the International Air Transport Association (IATA). The airline industry is meant to be the fastest mode of transportation for people as well as for goods and so the primary goal should be to improvise the services by maintaining the air space well through effective route planning [4] as it also adds more value to the economy of the world by contributing to globalization. There are a variety of costs an airline industry has to take up for its operation, and even after taking up those costs, the airline industry misses to achieve the targeted profit. Many departments [21] are a part of the aviation paradigm, the technical department ranges from crew management, flight maintenance till the ticketing systems, and the non-technical department ranges

N. Mehta (✉) · A. Ruparelia · J. P. Verma
Institute of Technology, Nirma University, Ahmadabad, India
e-mail: 19mced07@nirmauni.ac.in

A. Ruparelia
e-mail: 19mced12@nirmauni.ac.in

J. P. Verma
e-mail: jaiprakash.verma@nirmauni.ac.in

M. K. Khinchi
Mody University, Lachhmangarh, India
e-mail: scholar.mkhinchi@gamil.com

G. Mathur et al. (eds.), *Proceedings of 2nd International Conference on Artificial Intelligence: Advances and Applications*, Algorithms for Intelligent Systems,
https://doi.org/10.1007/978-981-16-6332-1_57

from identifying potential customers to providing an enriching experience to them which increases the customer base of the airline firm.

The deeper we dig into the business model of the airline industry, the more we would be able to analyze the core factors behind the industry's weakness. In our study, we could find many of those factors and the prime among them is fuel cost. According to IATA (International Air Transport Association) 2012 report, airlines spend up to 33% of their operating costs on the fuel itself [21]. This factor being non-deterministic due to the variation in global prices of oil cannot be controlled like other deterministic factors such as labor costs, sales and marketing, aircraft maintenance, depreciation, and many more [6]. The cost-cutting in fuel expenses could only be achieved by the appropriate planning of the flight route and that can be achieved through well-formed analysis of the flight network data.

Proper utilization and analytics of data related to the airline industry have many benefits. The benefits may span across improved planning, strategic decision-making, and maintaining the critical performance metrics. With the advent of technologies in online ticket-booking systems, a lot of data is generated. Meaningful analytics of ticket-booking data can help to target passengers with personalized offers. Often are the times where unexpected airplane maintenance leads to delay and cancelation of flights, predictive analytics can help in planning the maintenance beforehand.

Risk management is also a potential area where data analytics can help in identifying and mitigating risks before any mishap. Detecting the abnormal flights in form of an anomaly detection from the airline data parameters has also shown encouraging results. Hence, it is evident that data analytics employed in the right form and place prove to be a boon [9].

1.1 Background

Planning a flight requires lots of safety-critical aspects such as calculation of required fuel, ensuring the safety of aircraft from source to the destination as well as several compliances with the Air Traffic Control norms which would minimize the risk of mid-air collision during the journey. The flight planners try to normally minimize the cost of flight through the best choice of the route along with height, speed, and accurate weather reports, and minimum necessary fuel is also to be taken into account [6], and as a result, thousands of calculations are to be done with the data for successfully executing a single flight plan. So, cost optimization seems challenging here due to various elements involved but there is an opportunity for data availability. To compensate for the issues that the airline industry is facing, the major and dominating components should be concentrated on, and one among them is the effective utilization of flight network data. The amount of data generated is surely huge in size and due to the "Volume" of data, "Big data" [1, 15] has to be incorporated.

1.2 Motivation

Considering the data statistics for India, one of the airline operators operates around 1500 flights per day [11]. This essentially translates to the fact that the airline operators in every country would be having data that contains the raw information on the network of flights being operated between different airports, and that needs to be utilized for better decision-making. The ultimate aim for any industry would be to generate more and more revenue and gain profit in the business that they are into; however, due to some of the flaws in the current system, the industry is unable to achieve the same. Considering the issues of the airline industry, several options can be considered to curb the bad situation and those options align perfectly with the current trends of big data analytics. Big data analytics gives justice by analyzing and systematically extracting information from that data where the traditional data processing systems would take the defeat. The term "Big data" in itself represents a huge quantity of data as well as the data that grows exponentially to time [6, 12]. This data brings value to the business if it's employed in correct analytics and the flight network data goes appropriate with the graph-based analytics and algorithms.

1.3 Incorporating Graph Data Analytics

Generally, the flight network data is composed of various attributes that are suitable for performing data analytics using a graph. These graph-based analytics are based on the graph theory that usually works with a set of vertices and edges. The inherent nature of the flight network data suits the requirement of the former, where the airports would act as vertices and the connectivity between them acts as edges. Graph-based analytics proves to be a versatile way of expressing and finding out answers to various queries that are arising from a complex system of the dataset [3]. Graph-based analytics [16] uses various graph theory-based algorithms and tries to define a relationship between the objects. Graphs are very powerful structures that can help us to represent the complex dependencies of our data in a simple form. It also tries to form a structural characteristic of the graph by considering it as a whole. The applications of graph-based data analysis are speech/text parsing, debugging complex software cycles, webpages, communication networks, path routing, etc. but the most common example in which graph-based analytics is employed is in social network analysis where each person in the network is considered as a vertice and the connectivity between the person and his/her connections is determined by edges [14].

1.4 Paper Organization

The research paper is divided into multiple sections based on the research work carried out. The second section shows the literature review of various research works done on flight network data. The third section is about the proposed research model and system process flow. The fourth section majorly focuses on the concepts of various graph-based algorithms. The fifth section is all about the execution and implementation of the algorithms on the selected datasets. The sixth section shows the results of the proposed model. The seventh section provides the conclusion to the research and future work discussion.

2 Related Work

The flight network data helps to achieve different business objectives for the airline industry. Well, the data needs to be processed and analyzed with different algorithms. Results from these algorithms help to make effective business decisions from the underlying insights. Boominathan et. al. [2] proposed the Fuzzy Routing algorithm for the flight network data, to find the shortest path and lowest price from one place to another by using graph theory as an application of route planning. The research study done in [20] utilized the big data environment and applied the Improved Backtracking Algorithm and Node Compression Algorithm for route optimization. Kasturi et.al [6] have also tried to optimize the route by applying various heuristic techniques such as the Firefly algorithm, Bat algorithm, and Cuckoo algorithm. An advanced approach to the former namely the Hybrid BCF algorithm is also proposed by them. A comparative analysis of research work done by different researchers is shown in Table 1.

3 Proposed Research Work

The proposed approach applies algorithms such as PageRank and Label Propagation algorithm (LPA) on flight network datasets. PageRank algorithm finds the most influential airports, and LPA extracts communities among the airports from the given flight network. The results of the proposed model can apply to develop a Recommendation System. The recommendation system can be much helpful for airline operators. It can suggest new flight routes from various sources and destinations. Also, recommend the busiest airport and optimized routes.

Table 1 Comparative analysis of various Graph-based Data Analytics with pros and cons

Approach	Objective	Methodology	Pros	Cons
Bat, Cuckoo, and Firefly algorithms [6]	Minimize the total expected expense or maximize profit per airliner per route	In Bat algorithm, optimum route is derived, and the route allocation and passenger allocation are made between multiple airlines	The Bat algorithm gives better optimized results in comparison with the other algorithms	Performance analysis such as CPU utilization, Network utilization, and Cluster load are not mentioned
Hybrid BCF algorithm [7]	Optimize the airline route and seat allotment in the airline	Three different algorithms used for generating the particles to deal with the airliners	Optimal flight path based on constraints such as departure time and availability	The dataset used is limited to a small number of records. The performance on big data was not predicted
Cloud to integrate and mine the airline data [8]	Measure airline performance review, competitive benchmarking, airport operations, and schedule	Methodology that includes Hybrid cloud architecture is included to integrate data from various sources	Resolved the challenges of size, scalability of data, data formats, and transparency issues	Optimization issues are not handled
Node Compression and Improved backtracking algorithm [20]	Find the shortest path from one source to destination	Graph-based techniques are employed for the purpose of route optimization	Time complexity is reduced compared to conventional approaches	The efficiency of IBA and NCA decreases drastically as the number of nodes increases
Fuzzy Routing algorithm [2]	Reduce path-request blocking and increase overall utilization	Modified Djikstra's Single source shortest path algorithm	Determining the addition of two edges to compare distance between two different paths	The implementation is done on a very small scale of graph, i.e., with less number of nodes

3.1 System Architecture and Process Flow

Figure 1 shows the proposed model process flow and system architecture. The initial step would be to choose the dataset that contain the attributes as one three-letter IATA code of the airports, origin city of the airport, destination city of the airport, and distance between the airports. The next step would be to mold the dataset into the required form as per the Apache Spark Graphframe. A dataset is basically divided into two files such as vertices and edges. Vertices contain attribute names such as

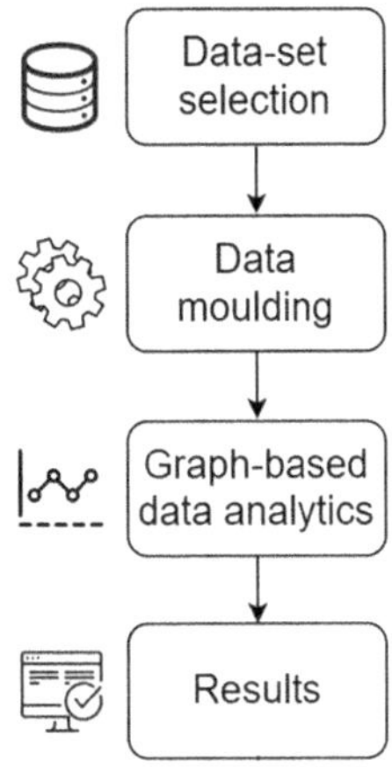

Fig. 1 System architecture and process flow

id and Name. The Edges contain source, destination, and relationship (the names of attributes are given according to the naming conventions of Apache Spark Graph-frame). After the dataset is transformed into vertices and edges format, graph-based algorithms are applied, and in the last phase, results are obtained that can be useful for effective decision-making.

4 Graph-Based Algorithms for Flight Network Systems

Various graph-based algorithms can be used for analyzing the domain of flight network systems but we are mainly focusing on the PageRank algorithm and Label Propagation algorithm (LPA)

4.1 PageRank Algorithm

Looking into the history of this algorithm, it was introduced by Sergey Brin and Larry Page in 1999 [13]. Although it was developed for ranking the websites. Later after several years, it was utilized in various fields such as social network analytics, community detection, and also in ranking the genes [19]. As the name of the algorithm suggests, it is highly employed in finding the ranks of the webpages (here, rank means importance). It does by counting the number of pages linked to it, which are termed as backlinks. If the backlink is coming from an important page, then that backlink is assigned higher weightage than other ones [5]. The core study from the nature of the algorithm brings us to apply the same algorithm in finding out the most influential airports from the selected dataset. It makes us understand the complex structure of our network analysis more easily.

A simple way to find out the most influential or important airport is by counting the number of flights handled by the airport. But just counting the number of flights in and out of the airport does not give us the importance of that airport in the flight network system because there are several important junctions in the route for the flights and they have to be included in the list of influential airports even though the number of flights handled by them is less [10]. So, the best way to get the most influential airports is through the application of the PageRank algorithm. It evaluates the flights coming from different airports, and if the flight is coming from a major airport, it lays importance to that airport and likewise, we can get the required outcome. The mathematical representation of the algorithm is as shown in equation-1 [5]

$$PageRank(Airport1) = (1 - d) + d(PageRank(Ti)/C(Ti) + ... + PageRank(Tn/C(Tn)). \tag{1}$$

In above Eq. (1), We are calculating the PageRank of Airport 1 denoted by PageRank(Airport 1)

- PageRank(Ti....Tn) represents the PageRank of airports, which have a connection (direct flights) to Airport 1.
- d is the damping factor. It is used to stop other airports from having too much influence. And so, the total vote is "damped down" by multiplying it to d (here, d = 0.85).
- C(Ti...Tn) is the total number of outgoing flights from the Airport(Ti...Tn).

The PageRank gives us the probability distribution over the airports so the final sum of the PageRank of all the airports would be summing up to 1.

4.2 Label Propagation Algorithm

This algorithm was introduced in the early 2000s [22] and was mostly used to detect the communities from the network. The main advantage of this algorithm over other algorithms that were used for community detection is its linear time complexity [18]. The basic workflow of the algorithm is by propagating the labels across the network and forming communities that are based on the process of Label Propagation. Within the complex structures, this algorithm finds communities, and the other advantage apart from its linear time complexity is the minimal requirements, i.e., the amount of information required beforehand the usage of this algorithm is very minimal. The main intuition behind the algorithm works is a single label would quickly be dominant in a dense graph but that would not be the similar case in a sparsely connected graph because as the labels propagate inside a densely connected graph, the nodes inside the graph would get the same label after the algorithm finishes its execution and then those nodes would be the part of the similar community at the end [17].
The steps that this algorithm follows for our research study are represented as follows:

(Here, Airports are considered as vertices, and the direct flight connection between those airports is considered as edges. Hence, a graph-like network is to be visualized):

- Step 1- Every airport is initialized with a unique label (an identifier).
- Step 2- These identifiers propagate through all the airports in our flight network graph.
- Step 3- At the end of each iteration, of the propagation, the airport in our graph updates its identifiers to the one that the maximum number of neighbors (neighboring airports) in the graph belongs to.

The convergence would be reached when each airport in the graph has a majority label of its neighbors. It can also converge if the number of iterations mentioned explicitly is reached.

5 Execution and Implementation

5.1 Dataset Description

In this paper, two datasets are used for experimental analysis. Also, the performance of the process flow of the proposed approach is compared with both of the datasets. The first dataset used for the research experiment here is having around 3.5 million instances (Size: 509.02 MB) and the dataset is of all the domestic flights of the USA from 1990 to 2009 (Publicly available on Kaggle). It has been taken from the OpenFlights website which has a huge database of different traveling mediums across the globe. There are 15 attributes in the dataset but only a few of those are used for our research and those are mentioned as follows: Origin-airport: Origin-airport code in IATA format. Origin-city: Name of the origin city of the airport. Destination-airport: Destination-airport code in IATA format. Destination-city: Name of the destination city of the airport. Distance: Distance between two airports. The second dataset used for the research experiment is having around 1.8 million instances (Size: 393.04 MB) and it is of the delayed domestic flights of the USA for the month of January, August, November, and December of the year 2016 (Publicly available on Kaggle). There are 43 attributes in the dataset but we have only used a few of them which are mentioned as follows: Origin: Name of the origin city of the airport in three letters. Origin-city-name: Name of the origin city of the airport. Dest: Destination-airport code in IATA format. Dest-city-name: Name of the destination city of the airport. Distance: Distance between two airports.

5.2 Execution

As discussed in the previous section, the research dataset was transformed (molded) into vertices and edges format, and once we had the data in the required form, we applied algorithms such as PageRank and LPA on a cluster formed with the help of Databricks Cloud Platform for execution. The cluster configuration is as follows:

Memory: 15.3 GB,

Runtime environment: 6.3 ML with 2.4.4 Apache Spark and 2.11 Scala, 2 cores. 1 DBU (Data Bricks Unit).

The parameters shown in Table 2 represent the ID of the airport, the name of the airport, and the calculated PageRank of the airport. As we can see from the results below, the higher the PageRank of airports, the more influential it is in the flight network of airports.

The PageRank algorithm is applied on both the datasets and the results are shown in Table 2. The most influential airport in the first dataset is YIP airport, Detroit, and in the second dataset, it is ATL airport, Atlanta. We can get many insights from the results such as effective planning of the flight route, deciding the connecting airports in the route, and suggesting a new flight route.

The parameters shown in Table 3 represent the ID of the airport, the name of the airport, and the calculated Label of the airport. As we can see from the results above, there are clusters of airports formed from the datasets, and the airports which share the same label number are part of a similar cluster.

Table 2 PageRank algorithm results for Dataset 1 and Dataset 2 (displaying top 10 rows)

Dataset 1			Dataset 2		
ID	Name	Pagerank	ID	Name	Pagerank
YIP	Detroit, MI	8.92670223	ATL	Atlanta, GA	12.5074428
MCI	Kansas City, MO	5.77178857	ORD	Chicago, IL	9.69949799
SHV	Shreveport, LA	5.44243344	DFW	Dallas/Fort Worth...	9.44619488
MSP	Minneapolis, MN	5.20930848	MSP	Minneapolis, MN	8.40849695
DFW	Dallas, TX	5.1634221	DEN	Denver, CO	8.33119868
AUS	Austin, TX	4.99847623	PHX	Phoenix, AZ	7.3506779
MEM	Memphis, TN	4.9574065	DTW	Detroit, MI	7.18301432
STL	St. Louis, MO	4.76604645	IAH	Houston, TX	6.84498195
LRD	Laredo, TX	4.67304068	SLC	Salt Lake City, UT	5.90858336
OKC	Oklahoma City, OK	4.46604843	SFO	San Francisco, CA	5.41648325

Table 3 LPA results for dataset 1 and dataset 2 (displaying top 10 rows)

Dataset 1			Dataset 2		
ID	Name	Label	ID	Name	Label
JSE	Juneau, AK	652835028992	CDV	Cordova, AK	1640677507072
BGM	Binghamton, NY	489626271744	WYS	West Yellowstone, MT	1606317768709
GTR	Columbus, MS	489626271744	OME	Nome, AK	1606317768706
WBR	Big Rapids, MI	489626271744	BGR	Bangor, ME	1597727834112
LCI	Laconia, NH	489626271744	APN	Alpena, MI	1486058684416
TCM	Tacoma, WA	489626271744	INL	International Fal...	1425929142272
JZU	Bennington, VT	489626271744	SMX	Santa Maria, CA	1391569403905
MSI	Grenada, MS	489626271744	MVY	Martha's Vineyard...	1322849927169
IKK	Kankakee, IL	489626271744	AKN	King Salmon, AK	1245540515840
BTM	Butte, MT	489626271744	EFD	Houston, TX	1082331758592

The **Label Propagation algorithm** is applied on both the datasets and the results are shown in Table 3. For instance, in dataset 1, the airports, BGM, GTR, WBR, LCI, TCM, JZU, MSI, IKK, and BTM are parts of one cluster. We can make communities of different airports using this algorithm and that can help us to include different airports under one cluster and that cluster formation can help us to manage flight traffic efficiently. We can also make use of these communities of the airports when there is a pandemic or epidemic spread across the world which would help us in converging our route plan of flights to lessen the spread of the disease. The same communities can also be used while there is any natural disaster in a few parts of the world and at those parts, we need to optimize our operations.

6 Conclusion

As proposed in the research study, deriving the crisp analysis of flight network data is through graph-based analytics by incorporating algorithms such as PageRank and Label Propagation. It shows that the results can be used to develop recommendation systems for the airline operators by finding out the most influential airports and communities of airports in the flight network through which business objectives such as effective flight route plan, fuel-cost optimization, and suggestion of new routes can be accomplished and that can be an effective solution to resolve the pitfalls of the airline industry. The future work of the study may focus on dealing with other major issues of the airline industry and solving them through the use of graph-based analytics under big data environments.

References

1. Victor Emmanuell Badea, Alin Zamfiroiu, Radu Boncea (2018) Big data in the aerospace industry. *Inf Eco* 22(1):17–24
2. Boominathan P, Arora K (2014) Routing planning as an application of graph theory. Int J Sci Technol Res 3:61–66
3. Farrahi Amir H, Goldberg Alan T, Leonard Bagasol, Jaewoo Jung (2017) Applying graph theory to problems in air traffic management. In *17th AIAA Aviation Technology, Integration, and Operations Conference*, p 3775
4. Hausladen Iris, Schosser Maximilian (2020) Towards a maturity model for big data analytics in airline network planning. J Air Transp Manag 82:101721
5. Jain Rekha, Purohit Dr GN (2011) Page ranking algorithms for web mining. Int J Comput Appl 13(5):22–25
6. Kasturi E, Prasanna Devi S, Vinu Kiran S, Manivannan S (2016) Airline route profitability analysis and optimization using big data analyticson aviation data sets under heuristic techniques. *Proc Comput Sci* 87:86–92
7. Kousik NV, Tapas Kumar, Rameshkumar C, Vetrivendhan L. Flight plan route optimization and increase the profit in airline industry by using hybrid bcf algorithm
8. Tulinda Larsen (2013) Cross-platform aviation analytics using big-data methods. In *2013 Integrated Communications, Navigation and Surveillance Conference (ICNS)*, pp 1–9. IEEE
9. Lishuai Li, Santanu Das, Hansman R John, Rafael Palacios, Srivastava Ashok N (2015) Analysis of flight data using clustering techniques for detecting abnormal operations. *J Aerosp Inf Syst* 12(9):587–598
10. Michael (2019, Dec) Applying the pagerank algorithm to airline flight data
11. Mihir Mishra (2019, Dec) Indigo operates 1500 flights daily; up 22 % annually
12. Neumann T (2016) Routing planning as an application of graph theory with fuzzy logic. *TransNav, Int J Marine Navigation and Safety od Sea Transp* 10(4)
13. Page Lawrence, Brin Sergey, Motwani Rajeev, Winograd Terry (1999) The pagerank citation ranking: bringing order to the web. Technical report, Stanford InfoLab
14. Sun Jiawen, Vandierendonck Hans, Nikolopoulos Dimitrios S (2020) Fast load balance parallel graph analytics with an automatic graph data structure selection algorithm. Future Gen Comput Syst 112:612–623
15. Verma Jai Prakash, Mankad Sapan H, Sanjay Garg (2018) Big data analytics: performance evaluation for high availability and fault tolerance using mapreduce framework with hdfs. In *2018 Fifth International Conference on Parallel, Distributed and Grid Computing (PDGC)*, pp 770–775. IEEE
16. Verma Jai Prakash, Mankad Sapan H, Sanjay Garg (2018) A graph based analysis of user mobility for a smart city project. In *International Conference on Next Generation Computing Technologies*, pp 140–151. Springer
17. Wang Tao, Chen Shanshan, Wang Xiaoxia, Wang Jinfang (2020) Label propagation algorithm based on node importance. Phys A Statist Mechan Appl 551:124137
18. Yan Xing, Fanrong Meng, Yong Zhou, Mu Zhu, Mengyu Shi, Guibin Sun (2014) A node influence based label propagation algorithm for community detection in networks. *Sci World J*, 2014
19. Kecheng Xu, Teng Long, Shaojie Qiao, Yating Zheng, Nan Han (2017) A global flight networks analysis approach using markov clustering and pagerank. In *2017 IEEE International Conference on Big Knowledge (ICBK)*, p 96–102. IEEE

20. Lifeng Yang, Liangming Chen, Ningwei Wang, Zhifang Liao (2017) Routing optimization algorithms based on node compression in big data environment. *Scientific Programming*, 2017
21. Anastasiia Zamiatina (2019, May) 9 incredible ways data analytics is transforming airlines
22. Xiaojin ZhuΓ́, Zoubin GhahramaniΓ́H (2002) Learning from labeled and unlabeled data with label propagation

Chapter 58
Key Exchange Using Tree Parity Machines: A Survey

Ishak Meraouche and Kouichi Sakurai

1 Introduction

Cryptography is an important building block of network communications. Without it, many online services will not be available anymore. The reason behind that is that networks were not built with security in mind. Therefore, any data exchanged between clients and servers is publicly visible and anyone connected to that network physically or wirelessly can see all the data exchanged between the parties communicating. As the data exchanged might contain sensitive information such as passwords and bank details, malicious users eavesdropping the communication might steal it and use it in a non-authentic way.

As a result, there are many protocols that add a security layer to the communication process and allow to hide the data from eavesdroppers and make it readable to the authorized entities only. Among the cryptography protocols, we are going to focus on key exchange, Which is a crucial part in order to use encryption methods like the Rijndael Algorithm (AES) [7].

Key exchange is a challenging task especially when the two parties are far from each other and cannot exchange a key physically. There are many math-based techniques that allow to agree on a key through a public network. The most used one is the Diffie–Hellman protocol [4].

While most techniques are based on mathematical problems, there are other techniques that are based on Machine Learning (neural networks). Research in neural networks-based cryptography has not seen a lot of a success in the past, and therefore the research flow in that area is facing a lack of interest. Our focus is going to be

I. Meraouche (✉) · K. Sakurai
Kyushu University, Fukuoka, Japan
e-mail: meraouche.ishak.768@s.kyushu-u.ac.jp

K. Sakurai
e-mail: sakurai@inf.kyushu-u.ac.jp

G. Mathur et al. (eds.), *Proceedings of 2nd International Conference on Artificial Intelligence: Advances and Applications*, Algorithms for Intelligent Systems,
https://doi.org/10.1007/978-981-16-6332-1_58

on key exchange using neural networks. A model [8] has been introduced in 2001 by kanter et al. which demonstrates a fast way to exchange a secret key between two parties. However, the model has been broken in another research paper [10] using three different techniques. After that, the Tree Parity Machine has seen many improvements and security analysis to try to build a stronger model.

2 The Tree Parity Machine Algorithm

The tree parity machine [8] was introduced in 2002. Kanter et al were among the first researchers to employ machine learning in the aim of achieving cryptography. The research paper they published [8] aims to calculate a secret key mutually just like the Diffie–Hellman key exchange protocol [4]. However, the approach is based on neural networks and not on mathematics.

The model consists of two neural networks with the same structure that can mutually calculate a secret key over a public channel. It is assumed that eavesdroppers can freely read the communication without being able to modify the messages or stopping them from being sent.

The structure of the two neural networks consists of three layers: an input layer with $K \cdot N$ input neurons, one hidden layer with K neurons, and a single neuron output layer as shown in Fig. 1.

The inputs to the neural network can take three values: $x_{i,j} \in \{-1, 0, 1\}$. The W_{KN} values can take a value between $-L$ and $+L$ and if an updated value exceeds those boundaries it will be replaced by the nearest value (either $-L$ or $+L$).

The protocol consists into generating two random inputs of size N, pass them through the neural networks, and get the output O. We compare the outputs of the two neural networks and update them according to a Hebbian learning rule. If the outputs are equal, this means that the two neural networks are synchronized (have the same weights) [8] and can use their weight vectors as a secret key for a symmetric encryption algorithm like AES [7] or as a seed for a pseudorandom number generator.

We implemented and experimented the synchronization time of the two tree parity machines. We used the following parameters: $K = 150, N = 10, L = 8$, and the two

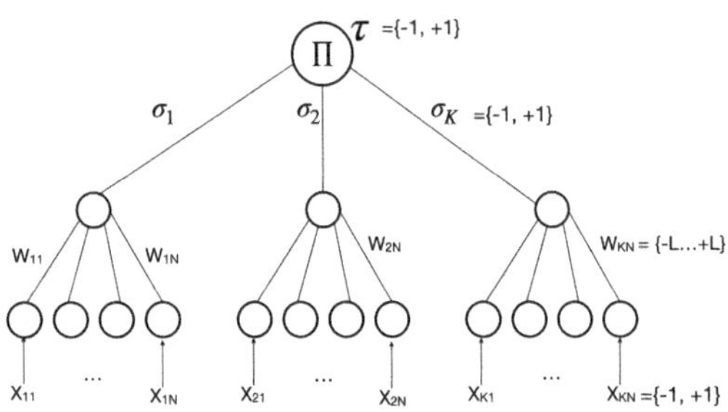

Fig. 1 Parity tree neural network structure

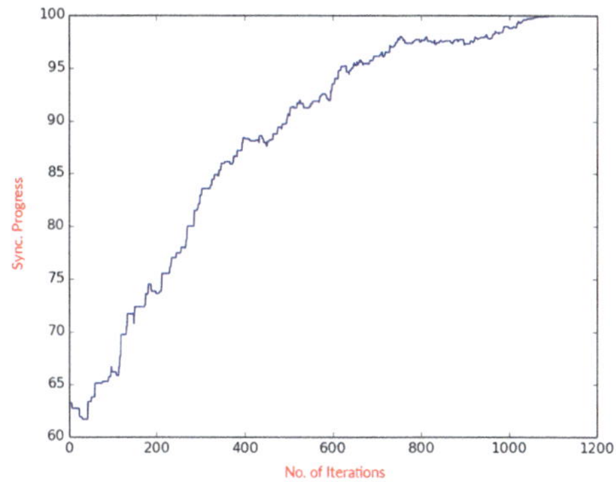

Fig. 2 The two neural networks took around 1200 iterations to fully synchronize

machines were able to synchronize in 13.52 seconds. Figure 2 shows the percent of synchronization of the two tree parity machines over time.

The tree parity machine key exchange [8] was very promising, especially because the authors mention mathematical proofs that their method is secure. Unfortunately, Klimov et al. [10] did look outside of the box in their work and found three quite unusual but working methods to break the system.

3 Analysis of Neural Cryptography

Klimov et al. [10] analyzed and proved the work done in [8] insecure and impracticable as it is vulnerable against different attacks that are explained below.

The Genetic Attack: This attack looked at the two neural networks willing to establish a connection from the biological point of view to make a biological attack using genetic algorithms. The general idea is to simulate a big population of neural networks that have the same structure of the two neural networks and train them with the same inputs. The neural networks from the population whose output is similar to the target two neural networks are now synchronized with the target neural networks and can read their communication. Ruttor and his collaborators including Kanter [18] affirmed the efficiency of this attack against neural networks, however their study shows that this attack can be worked around by increasing the value L that the weights can take which will increase the number of possibilities and therefore render the attack way more difficult to complete.

The Geometric Attack: In this attack, the authors simulate each input of the two target neural networks as a K random hyperplanes H_1 ... H_K corresponding to K perceptrons and the weights of each neural network as K points $W_1 ... W_K$ in the N-discrete space $U = \{-L, ..., L\}^N$ where $W_i = (w_{i1} ... w_{ik})$. Concretely, an attacker constructs a neural network with random weights but with the same neural network structure as the target and at each step of training, the weights are updated according to these rules:

- If the two target neural networks have different outputs, the attacker does not update his weights.
- If the two target neural networks have the same outputs and the attacker also has the same output, the neural network's weights will be updated in the normal way.
- If the two target neural networks have the same outputs but not the attacker, then the attacker should find an i_0 that minimizes this formula $\left|\sum_{j=0}^{N} w_{ij}^{C} \cdot x_{ij}\right|$ and updates the weights assuming the hidden bits and the target's outputs.

The probabilistic Attack: In the probabilistic attack, the attacking tree parity machine is actually a probabilistic tree parity machine and this means that the weights are actually probabilistic weights $p_{i,j}(l) = l \in [-L, +L]$ where each probabilistic weight is a probabilistic distribution that represents the probability of the Tree Parity Machine A taking l as a parameter. Then, by passively eavesdropping the inputs $x_{i,j}$ the attacker can use either the Hebbian learning rule or the Monte Carlo method to update $p_{i,j}(l)$ and end up with the identical weights to the parties communicating. This is mainly due to the limited possible values in $[-L, +L]$ which makes them easy to simulate.

4 Deep-Learning-Based Cryptography

When neural-network-based cryptography models were judged inefficient, research in that area has been a little neglected during the last decade. However, Google Brain researchers Abadi and Andersen proposed a model [1] that is based on Generative Adversarial Networks (GANs) [5].

GANs are a special type of neural networks that have been proposed in 2014 by Goodfellow et al. The general idea is that one neural network tries to learn to spot irregularities in images or other data sent by an adversary that are very similar to real data.

The model [1] proposed by Abadi et al. works in a similar way. There are two neural networks called for reference as Alice and Bob that are learning to protect a communication in the presence of an eavesdropper Eve. Alice and Bob challenge Eve to learn to decrypt their exchanged ciphertexts while they train to make them as unreadable for him as possible. The results show that Alice and Bob can indeed learn encryption and hide their communication from Eve.

The ciphertexts generated by GAN-based encryption models [1] have been analyzed by Zhou et al. [24] and it appears that the output from Alice and Bob is vulnerable against probabilistic attacks. The ciphertexts have failed multiple probabilistic attacks including the χ^2 approach, the Kolmogorov–Smirnov (KS) approach as well as the NIST statistical test.

After that the model [1] has been improved by Coutinho et al. [3]. In their model, they propose a more advanced structure for Alice, Bob, and Eve that allows Alice to generate more complicated ciphertexts. The neural networks Alice and Bob end up learning the One Time Pad which can be combined with a pseudorandom generator to generate perfectly secure ciphertexts.

In the mean time, Shi et al. [21] propose an improved model based on continuous variable quantum neural network. Other authors [25] build another variant of the model that is based on the topology of evolving neural networks [22] which work in a different way from GANs. They show that it possible to learn encryption based on the topology of evolving neural networks [22]. The performance of the model between multiple parties has also been investigated [11].

Other authors [6, 9, 23] also transformed the model into a model that learns to hide data into an image (i.e., steganography). Hayes et al. [6] evoked a problem where the neural networks cannot learn the same model when there are more than two parties training with Alice but it has been solved in [12].

Deep learning and GAN-based cryptography is the current research flow that is seeing more interest than the tree parity machine (TPM) and it was briefly introduced in this section but our focus will be on the TPM in the next sections.

5 Security Improvements to the Tree Parity Machine

This section shows and briefly describes the works that followed up the tree parity machine.

5.1 Security Improvement Using Private Inputs

The key exchange protocol using the TPM [8] has been revised and improved using private inputs by some researchers as in [14, 15].

Private inputs (or queries) are inputs generated by Alice and Bob in turn at every iteration. The inputs are correlated to their respective weights and allow to have a mutual influence between Alice and Bob which highly reduced the possibility of passive synchronization by attackers.

The researchers in [14, 15] implemented the model using queries but each of them used a different function to generate the private inputs.

Noh et al. [13] implemented two tree parity machines in a blockchain system and used the blockchain headers in every transaction as a secret. As the eavesdropping attacker cannot know the destination of the message, he cannot synchronize in secret with the two parties communicating. However, as this method relies on blockchain which headers are expanding, this results in a slower synchronization between the two parties.

5.2 Periodically Send Erroneous Outputs

Allam et al. [2] put the DTMP (Do not Trust My Partner) theory into practice in order to fool any attackers performing passive synchronizations as in [10]. They modified the synchronization protocol by sending probabilistic controlled wrong outputs from time to time. This puts the attacker into a situation where he cannot tell which output is correct and which is not.

5.3 Development of an Optimal Structure of Tree Parity Machine

Salguero et al. [19] did a study on the original TPM model and proposed an optimal structure that generates a 512-bit key. The authors did more than 10 million simulations with different parameters and number of neurons in each layer. The authors demonstrate that by using the values $K = 8$, $N = 16$, and $L = 2^3$ as parameters, Alice and Bob can synchronize in less than 6 seconds with a 0% success rate for attackers. The authors validate the results by showing that a small change in the parameters leads to a polynomial increase of the synchronization time.

Ruttor et al. [17] also demonstrate that increasing the number of hidden units exponentially increases the complexity of the geometric attack.

5.4 Permutation Parity Machines for Neural Cryptography

Reyes et al. [16] extended the structure of the tree parity machine to another variant that they called a Permutation Parity Machine (PPM). A permutation parity machine has the same overall neural network structure as a tree parity machine however the values of the parameters and their number is different.

A permutation parity machine is defined as a neural network with K hidden units just like the tree parity machine. These units are simple perceptrons (neurons) each with its own input. There are N units with N inputs that take binary inputs (either 0 or 1). As for the weights W, they are drawn from a state vector $S \in \{0, 1\}^G$ where G must be greater than $K \cdot N$. The ith hidden units are calculated using an exclusive or between the weight w and the input x. The final output is either 1 or 0.

Reyes et al. [16] conduct comparative attacks on the tree parity machine and the permutation parity machine. The results show that the permutation parity machine has a lower rate of successful attacks rather than the tree parity machine by conducting the attacks performed in [10]. The authors then demonstrate that the probability order of a successful attack on the permutation parity machine can be as low as 10^{-20} when the value of N is equal to 16 and the value of G is equal to 128.

The probability of a successful attack is demonstrated to be dependent on the value of G by the following formula: $P_E = \frac{1}{2^{G-1}}$. We can see that with $G = 128$ we have $P_E = 10^{-20}$.

The result is therefore lighter than the method proposed in [17] as they use a value of 1000 for N which will significantly increase the synchronization time compared to this method.

However, Seoane et al. [20] demonstrate a successful attack on the permutation parity machine which therefore renders the PPM discussed in [16] not 100% secure.

5.5 Summary Table

Table 1 summarizes the most important techniques that improved the tree parity machine model [8] or shown its weaknesses.

Table 1 Summary of works that improved or provided a security analysis to the TPM

Year	Work	Summary of contribution	Technique	Summary of vulnerabilities
2002	Analysis of neural cryptography [10]	Demonstrates that the tree parity machine is vulnerable to three different attacks. which makes the model unusable	1- Genetic attack: an attacker with knowledge of the NN structure can mimic their behavior and end up with the same key 2- Geometric attack: this attack simulates all the possible NN structures and uses probabilities to find the correct one to end up with the same key. 3- Probabilistic attack: this attack uses probabilities to predict the weights vector	Not applicable
2010	permutation parity machine for neural cryptography [16]	The authors improve the TPM by transforming it into a permutation parity machine (PPM) and show that it is resistant against probabilistic attacks	The authors introduce the PPM. A more robust version of the TPM that is resistant to probabilistic attacks. The authors keep the same neural network structure, however the values that the weights take differ	Broken in [20]. (see the row below)
2012	Successful attack on permutation parity machine-based neural cryptography [20]	The authors prove that the PPM is still vulnerable against probabilistic attacks	The authors make an attacker that simulates the space of possible weight vectors using the Monte Carlo probabilistic method and show that the attacker can predict the final weight vector before the two parties synchronizing and therefore obtain the secret key	Not applicable

(continued)

Table 1 (continued)

Year	Work	Summary of contribution	Technique	Summary of vulnerabilities
2019	On the development of an optimal structure of tree parity machine for the establishment of a cryptographic key [19]	Using the same model in to improve the security by changing the depth of the weights of the neural networks to generate a 512-bit key	The authors expand the range [-L, +L] that the weights take their values from. By doing so, the model operates very well against geometric and genetic attacks	The model does not show any security against the probabilistic attack

6 Discussion and Conclusion

The tree parity machine [8] is a promising work as it enables two users to quickly agree on a common secret key without relying on number theory which is known to be weak against quantum computers. As the tree parity machine does not rely on number theory, it has seen many positive feedback and many researchers tried to improve it after it have been broken in [10].

The most prominent follow-up work was in [16] where the authors developed a permutation parity machine similar to the tree parity machine but works with different values for the weights. The authors demonstrated that it can be secure for a large value of G with $N = 16$ and therefore removing the necessity of using a big value for N as it was proposed in [17] with a value of 1000 for N.

Another promising work by Salguero et al. [19] where the authors tried creating an optimal tree parity machine that generates a key of 512 bits in size. Their experiments show that the tree parity machine can indeed generate a 512-bit, key, however this is done by fixing a value of $L = 8$ which limits the space of the values that the weights of the neural network can take. As the space is limited, brute force attacks might be possible in a long but relatively short time compared to existing schemes that might take hundreds of years to decode with a brute force attack.

Therefore, as an answer to the title of this paper, we assume that the tree parity machine is not secure enough to be used in real-life scenarios and a more secure variant with probably higher number of hidden layers needs to be investigated.

As for deep learning and GAN-based cryptography, the models are more promising and perform better than the tree parity machine mainly because these models are more complicated and are trained in a more complicated way than the tree parity machines.

7 Notice

This research paper was originally presented at the SCIS2021 (The 38th Symposium on Cryptography and Information Security) symposium in Japan and was entitled “A recent progress of Neural Key Exchange : Are Tree Party Machines still alive ?”.

The current research paper is an extension of the previous work toward publication through an international conference.

Acknowledgements Ishak Meraouche is financially supported by the Ministry of Education, Culture, Sports, Science and Technology (MEXT), Japan for his studies at Kyushu University.

References

1. Abadi M, Andersen DG (2016) Learning to protect communications with adversarial neural cryptography. CoRR abs/1610.06918
2. Allam A, Abbas H (2010, 10) On the improvement of neural cryptography using erroneous transmitted information with error prediction. IEEE transactions on neural networks / a publication of the IEEE Neural Networks Council 21:1915–1924
3. Coutinho M, de Oliveira Albuquerque R, Borges F, García-Villalba LJ, Kim T (2018) Learning perfectly secure cryptography to protect communications with adversarial neural cryptography. Sensors 18(5):1306
4. Diffie W, Hellman M (2006, Sep) New directions in cryptography. IEEE Trans Inf Theor 22(6):644–654. https://doi.org/10.1109/TIT.1976.1055638
5. Goodfellow IJ, Pouget-Abadie J, Mirza M, Xu B, Warde-Farley D, Ozair S, Courville A, Bengio Y (2014) Generative adversarial networks
6. Hayes J, Danezis G (2017) Generating steganographic images via adversarial training. In: Guyon I, Luxburg UV, Bengio S, Wallach H, Fergus R, Vishwanathan S, Garnett R (eds.) Advances in Neural Information Processing Systems 30, pp. 1954–1963. Curran Associates, Inc. (2017), http://papers.nips.cc/paper/6791-generating-steganographic-images-via-adversarial-training.pdf
7. Jamil T (2004, April) The rijndael algorithm. IEEE Potentials 23(2):36–38
8. Kanter I, Kinzel W, Kanter E (2002, 02) Secure exchange of information by synchronization of neural networks. EPL (Europhysics Letters) 57
9. Ke Y, Zhang M, Liu J, Su T (2017) Generative steganography with kerckhoffs' principle based on generative adversarial networks. CoRR abs/1711.04916
10. Klimov A, Mityagin A, Shamir A (2002) Analysis of neural cryptography. In: Advances in Cryptology - ASIACRYPT 2002, 8th International Conference on the Theory and Application of Cryptology and Information Security, Queenstown, New Zealand, December 1-5, 2002, Proceedings. pp. 288–298
11. Meraouche I, Dutta S, Sakurai K (2020) 3-party adversarial cryptography. In: International Conference on Emerging Internetworking, Data & Web Technologies. pp. 247–258. Springer
12. Meraouche I, Dutta S, Sakurai K (2020) 3-party adversarial steganography. In: You I (ed) Information Security Applications. Springer International Publishing, Cham, pp 89–100
13. Noh S, Kyung-Hyune R (2020) Implicit Authentication in Neural Key Exchange Based on the Randomization of the Public Blockchain. AIChain Workshop
14. Prabakaran N, Vivekanandan P (2008) A new security on neural cryptography with queries
15. Revankar P, Dr Z, Rathod D (2010, 01) Private inputs to tree parity machine. Int J Comput Theory Eng 665–669
16. Reyes OM, Zimmermann KH (2010) Permutation parity machines for neural cryptography. Physical Review E 81(6):066117
17. Ruttor A, Kinzel W, Kanter I (2007) Dynamics of neural cryptography. Phys Rev E 75(5):056104
18. Ruttor A, Kinzel W, Naeh R, Kanter I (2006) Genetic attack on neural cryptography. Phys Rev E 73(3):036121

19. Salguero E, Fuertes W, Lascano J (2019, 03) On the development of an optimal structure of tree parity machine for the establishment of a cryptographic key. Security and Communication Networks, 1–10
20. Seoane LF, Ruttor A (2012) Successful attack on permutation-parity-machine-based neural cryptography. Phys Rev E 85(2):025101
21. Shi J, Chen S, Lu Y, Feng Y, Shi R, Yang Y, Li J (2020) An approach to cryptography based on continuous-variable quantum neural network. Sci Rep 10(1):2107. https://doi.org/10.1038/s41598-020-58928-1
22. Vargas DV, Murata J (2019) Spectrum-diverse neuroevolution with unified neural models. CoRR abs/1902.06703. http://arxiv.org/abs/1902.06703
23. Yedroudj M, Comby F, Chaumont M (2019) Steganography using a 3 player game. CoRR abs/1907.06956
24. Zhou L, Chen J, Zhang Y, Su C, James MA (2019) Security analysis and new models on the intelligent symmetric key encryption. Comput Sec 80:14–24
25. Zhu Y, Vargas DV, Sakurai K (2018) Neural cryptography based on the topology evolving neural networks. In: 2018 Sixth International Symposium on Computing and Networking Workshops (CANDARW), pp. 472–478

Chapter 59
A Divisive Hierarchical Clustering Algorithm to Find Clusters with Smaller Diameter to Cardinality Ratio

Sadman Sadeed Omee and Md. Saidur Rahman

1 Introduction

Clustering is arguably the most important section of unsupervised learning where data are partitioned into sensible groups. The goals of clustering are to divide an unlabelled data set into separate groups according to their similarity, to find underlying structure in data and to organize and summarize it through cluster prototypes [18].

Clustering algorithms can be generally divided into two classes—partitional and hierarchical [18]. Hierarchical clustering algorithms can be further divided into two categories—agglomerative and divisive. An agglomerative hierarchical clustering starts by assigning each data point into their own separate cluster and merges two clusters at each step analyzing their similarity. On the other hand, a divisive hierarchical clustering algorithm is the opposite of this process, i.e., starts with a single cluster composed of all data points initially and splits one of the available clusters into two at each iteration following some optimization criterion (top-down approach). Divisive approaches have several advantages over agglomerative approaches which are described in [13]. Divisive approaches usually have a tendency to generate more balanced dendrograms than agglomerative approaches. Also in the case when a few clusters are necessary to obtain, divisive algorithms are more efficient than agglomerative ones as we have to wait for the last iterations of an agglomerative approach to obtain this. Moreover, agglomerative methods do not have an idea about the global

S. S. Omee (✉) · Md. S. Rahman
Graph Drawing and Information Visualization Laboratory Department of Computer Science and Engineering, Bangladesh University of Engineering and Technology (BUET), Dhaka-1205, Bangladesh
e-mail: 1405065.sso@ugrad.cse.buet.ac.bd

Md. S. Rahman
e-mail: saidurrahman@cse.buet.ac.bd

G. Mathur et al. (eds.), *Proceedings of 2nd International Conference on Artificial Intelligence: Advances and Applications*, Algorithms for Intelligent Systems,
https://doi.org/10.1007/978-981-16-6332-1_59

distribution, so it proceeds by finding patterns locally. Whereas, divisive methods consider the global distribution when splitting a cluster into two.

Massive volumes of data are being produced every second nowadays and so the necessity of efficient data clustering algorithms has increased rapidly for the purpose of mining these data. The variety of the data and the constant accumulation of colossal amount of data have made it even more challenging. Traditional clustering algorithms are often not adequate to deal with data of high volume and variety [18, 29]. That is why new clustering techniques and criteria are needed to be adopted to analyze and investigate the data.

1.1 Problem Definition

Given a point set S consisting of n points lying on a d-dimensional space and a positive integer k, i.e., $S = \{p_1, p_2, p_3, \ldots, p_n\}$, where $p_i \in \mathbb{R}^d$ for $i = 1, 2, \ldots, n$ and $k \in \mathbb{Z}$, we consider the problem of partitioning the points of S into k clusters $C_1, C_2, \ldots, C_k$, where $\bigcup_{c=1}^{k} C_c = S$ and $C_i \cap C_j = \phi$ for $i, j = 1, 2, \ldots, k$ and $i \neq j$, so as to minimize the maximum diameter to cardinality ratio among all clusters. Our algorithm starts by putting all the points in a single cluster and then recursively picks the cluster with the largest diameter to cardinality ratio and divides it into two clusters so that this optimization criterion is satisfied.

1.2 Related Works

A bipartition is needed to perform at each step of a divisive hierarchical clustering algorithm of one of the currently available clusters. So naturally one of the possible approaches is to check all the possible bipartitions and choose one of them according to some criterion. Partitioning a set of n objects into two non-empty subsets can be done in $2^{n-1} - 1$ ways and Edwards and Cavalli-Sforza [6] adopted this approach and they chose the one having the smallest intra-cluster sum of squares among all the possible bipartitions. But checking all possible bipartitions is computationally very expensive.

Macnaughton-Smith et al. [21] and Kaufman and Rousseeuw [19] both constructed iterative divisive clustering procedures with a restricted number of bipartitions. Although these algorithms perform well, they are often sensitive to outliers.

Guénoche et al. [13] studied two divisive hierarchical clustering algorithms with the diameter criterion—an implementation of Hubert's method [16] and a refinement of Rao's method [26]. Divisive hierarchical clustering algorithms with diameter criterion recursively pick the cluster with the maximum diameter among the currently existing clusters at that moment and split that cluster into two clusters. The problem with this criterion is that sometimes it results in a *dissection effect* [5, 14]. When objects very similar to one another are assigned to separate clusters by the clustering algorithm instead of assigning them to the same cluster, it is called the dissection effect.

Xiong et al. [33] presented a divisive hierarchical clustering algorithm based on multiple correspondence analysis (MCA). But it is designed focusing only on categorical data.

1.3 Our Contributions

In this paper, we devise a divisive hierarchical clustering algorithm for partitioning a point set S consisting of n points on a d-dimensional space into k clusters such that the maximum diameter to cardinality ratio among all clusters is minimized. The reason to go for a divisive approach is described in Sect. 1. We compare the performance of our algorithm with that of other widely used algorithms. They are k-means clustering [20], DIvisive ANAlysis (DIANA) [19], Affinity Propagation [11] and Ward Agglomerative Clustering (WAC) [32]. We experiment our algorithm's performance with three different distance metrics against that of these clustering algorithms on some benchmark datasets and compare the results using the Adjusted Rand Index [25] and Silhouette Coefficient [27] metrics that validate our algorithm's efficiency. Finally we experiment the efficiency of both our algorithms, k-means clustering and Ward Agglomerative Clustering on increasing value of k on another benchmark dataset and show that our algorithm performs better than these algorithms.

The rest of the paper is organized accordingly. Section 2 consists of some preliminary terms. Section 3 describes the whole algorithm. Section 4 deals with the experiments, results, and analysis of our algorithm and comparison with other clustering algorithms. Finally we conclude our paper in Sect. 5 mentioning some future works.

2 Preliminaries

Given a cluster C consisting of n points $p_1, p_2, \ldots, p_n$, where $p_i \in \mathbb{R}^d$ for $i = 1, 2, \ldots, n$, the *centroid* c of C is a d-dimensional vector which is the arithmetic mean position of all the points of C. Centroid c can be expressed as follows –

$$c = \frac{1}{n} \sum_{i=1}^{n} p_i$$

The *diameter* of a cluster is the maximum distance between any two points of the cluster. Let, C be a cluster of n points, where $C = \{p_1, p_2, p_3, \ldots, p_n\}$. Suppose,

$dist(p_i, p_j)$ be the distance between points p_i and p_j of C. Then the diameter $d(C)$ of C is as follows –

$$d(C) = \max_{p_i, p_j \in C} dist(p_i, p_j)$$

The *cardinality* of a set or cluster is the number of elements it contains. We denote the cardinality of a cluster C by $|C|$.

k-*means* is an iterative clustering method where k predefined centroids are used to create k clusters. Each data is assigned to the cluster with the nearest mean. The objective function of k-means algorithm is to minimize intra-cluster variance i.e., sum of squared error (SSE). For every data point p and centroid c, the sum squared error can be defined as follows where p_j^i is the jth data point assigned to i^{th} cluster.

$$SSE = \sum_{i=1}^{k} \sum_{j=1}^{n} \left\| p_j^i - c_i \right\|^2$$

Given a set of points, the *convex hull* of the point set is defined as the smallest convex polygon that circumscribes all the points of the set. Convex hull is used in our algorithm for the farthest point pair query of a point set.

kd-*tree* is a binary tree data structure where each leaf node represents a point in a k-dimensional space. kd-tree is used in our algorithm for nearest neighbor search.

For two d-dimensional points $X = \{x_1, x_2, \ldots, x_d\}$ and $Y = \{y_1, y_2, \ldots, y_d\}$, the Minkowski distance of order p between two points is defined as,

$$Minkowski(X, Y) = \left(\sum_{i=1}^{d} \left\| x_i - y_i \right\|^p \right)^{\frac{1}{p}}$$

$p = 1, 2$ and ∞ is equivalent to Manhattan distance, Euclidean distance, and Chebyshev distance, respectively.

We define a term *ratio value* in this paper. This value is the ratio of the diameter to cardinality of a cluster. This ratio value is used as the criterion for splitting clusters in our divisive hierarchical clustering algorithm. Our objective is to minimize the maximum ratio value among all clusters.

William M. Rand [25] proposed Rand Index(RI) to define a measure of similarity between two different clusterings of a specific point set. One of them is usually the ground truth clustering so that one can measure how accurate another clustering algorithm's result is to that of the ground truth clustering. Suppose a set of n objects $S = \{\delta_1, \delta_2, \delta_3, \ldots, \delta_n\}$ is given. Let, $G = \{g_1, g_2, g_3, \ldots, g_M\}$ and $C = \{c_1, c_2, c_3, \ldots, c_N\}$ be two different clusterings of S where C is the clustering algorithm's result and G is the "ground truth" classification. We construct the contingency table in Table 1 to show the overlap between G and C. Each entry p_{ij} in Table 1 indicates total number of objects both g_i and c_j contains. Let a be the total pairs of objects those are in the same cluster in both G and C, b be the total pairs of objects those are in the same cluster in G but not in the same cluster in C, c be the

Table 1 Contingency table for comparing two different clustering output G and C of set S

Clustering G \ Clustering C	c_1	c_2	c_3	…	c_N	Total
g_1	p_{11}	p_{12}	p_{13}	…	p_{1N}	$p_{1.}$
g_2	p_{21}	p_{22}	p_{23}	…	p_{2N}	$p_{2.}$
g_3	p_{31}	p_{32}	p_{33}	…	p_{3N}	$p_{3.}$
⋮	⋮	⋮	⋮	⋱	⋮	⋮
g_M	p_{M1}	p_{M2}	p_{M3}	…	p_{MN}	$p_{M.}$
Total	$p_{.1}$	$p_{.2}$	$p_{.3}$	…	$p_{.N}$	$p_{..} = n$

total pairs of objects those are in the same cluster in C but not in the same cluster in G, and d be the total pairs of objects those are in different clusters in both G and C. These variables can be expressed by the terms of Table 1 by the following equations.

$$a = \sum_{i=1}^{M} \sum_{j=1}^{N} \binom{p_{ij}}{2}$$

$$b = \sum_{i=1}^{M} \binom{p_{i.}}{2} - a$$

$$c = \sum_{j=1}^{N} \binom{p_{.j}}{2} - a$$

$$d = \binom{n}{2} - a - b - c$$

Then RI can be calculated by the following equation.

$$RI = \frac{a + d}{a + b + c + d}$$

The range of RI is between 0 to 1 where higher values indicate good clustering and lower values indicate bad clustering.

However, there are some problems with this RI measure, such as the expected value of RI of two random clusterings never happens to be a constant value, it shows high variability [8] and it is very sensitive to the total clusters computed in each clustering [23]. Hubert and Arabie [17] proposed the Adjusted Rand Index(ARI) for removing these problems by assuming the generalized hypergeometrical distribution as a model of randomness. ARI can be expressed by the following equation.

$$ARI = \frac{\binom{n}{2}(a+d) - [(a+b)(a+c) + (b+d)(c+d)]}{\binom{n}{2}^2 - [(a+b)(a+c) + (b+d)(c+d)]}$$

When the "ground truth" labels are not known, Silhouette [27] is an efficient internal evaluation metric for measuring validity of clustering. If a point p is assigned to cluster C, then let $a(p)$ be the average distance of p to all other points of C and $d(p, C')$ be the average distance of p to all the objects of another cluster C'. Now let $b(p) = \min_{X \neq C}(d(p, X))$. The Silhouette Coefficient (SC) of the data point p can be expressed by the following equation.

$$s(p) = \frac{a(p) - b(p)}{max\{a(p), b(p)\}}$$

We use the mean Silhouette Coefficient of all samples as the validation metric. The range of SC is between -1 to 1 where higher values indicate better clustering.

3 The Algorithm

In this section, we describe our algorithm. Let $S = \{p_1, p_2, p_3, \ldots, p_n\}$ be a point set of n points on a d-dimensional space and k be a positive integer. The goal of any clustering algorithm is to divide the point set into meaningful clusters, i.e., similar points into the same cluster and dissimilar points into separate clusters. There are many ways to measure the similarity between two points. In this paper we select the distance between two points as the similarity measure. We propose a new optimization criterion for our divisive hierarchical clustering algorithm, that is to partition S into k clusters such that the maximum diameter to cardinality ratio among all clusters is minimized. We divide our algorithm into four major steps. They are briefly described below.

1. Initially dividing the point set S into two clusters C_1 and C_2 in such a way that the number of points in both the clusters is as close as possible. We name it the *initial divide* step.
2. Finding points those are actually nearer to the centroid of S than the centroid of the cluster they are assigned to. We remove those points from their corresponding cluster and store them in a temporary cluster C_{temp}. We name it the *temporary cluster creation* step.
3. Applying a greedy heuristic and merging the temporary cluster (C_{temp}) found in step 2 with one of the clusters between C_1 and C_2. We use the greedy heuristic to determine which cluster between C_1 and C_2 we need to merge C_{temp} with. We call it the *merge by greedy heuristic* step.

4. Finally we remove all the points of C_1 that are nearer to the centroid of C_2 than that of C_1 and assign them to C_2. Then we do the same for C_2, that is, we remove all the points of C_2 that are nearer to the centroid of C_1 than that of C_2 and assign them to C_1. We call it the *filtering* step.

These four steps create two clusters C_1 and C_2 from point set S. Our objective is to get k clusters. According to the rule of a divisive hierarchical clustering algorithm, we need to repeat the above-mentioned four steps in either C_1 or C_2 and so on to divide into more clusters. We need to use these steps $k - 1$ times for getting k clusters. Our algorithm selects the cluster that has the maximum diameter to cardinality ratio among the existing clusters for partitioning in each iteration as our objective is to minimize the maximum ratio value among all clusters. In Subsection 3.1–3.4, we discuss the four major steps in detail.

3.1 *Initial Divide*

Our algorithm first divides S into two clusters C_1 and C_2. Initially it considers the whole point set S as a single cluster.

Algorithm 1: $InitialDivide(C)$

```
Input: A cluster C
Output: Two clusters C1 and C2
C1, C2 ← φ
f1, f2 ← farthest point pair of C
C ← C \{f1, f2}
C1 ← C1 ∪ {f1}
C2 ← C2 ∪ {f2}
q1 ← f1
q2 ← f2
while C is not empty do
    x1 ← nearest point of q1 of C
    C ← C \{x1}
    C1 ← C1 ∪ {x1}
    q1 ← x1
    if C is not empty then
        x2 ← nearest point of q2 of C
        C ← C \{x2}
        C2 ← C2 ∪ {x2}
        q2 ← x2
    end
end
return C1, C2
```

Let, that cluster be C. The algorithm starts by finding the two farthest points f_1 and f_2 of C and assigning f_1 to C_1 and f_2 to C_2 and removing those points from C.

Then the algorithm computes nearest points of f_1 and f_2 in C and assigns them to C_1 and C_2, respectively and removes those points from C. The process is repeated until C is empty. A formal description of the algorithm is given in Algorithm 1.

For a 2-dimensional point set the farthest point pair can be computed in a total of $O(n \log n)$ time by rotating calipers method [28]. This method first requires computing the convex hull of the point set which takes $O(n \log n)$ time. Then it searches for the two farthest points in $O(n)$ time with the rotating calipers. So in total it takes $O(n \log n)$ time. The nearest point query of a specific point within a point set takes $O(\log n)$ time in the expected case by using a kd-tree [12].

3.2 Temporary Cluster Creation

In this step the algorithm creates a temporary cluster C_{temp}. Some points may lie closer to the centroid of the whole point set rather than the centroid of the cluster it is assigned to in the previous step. Those points are temporarily assigned to C_{temp} and removed from their respective cluster. A formal description of the algorithm is presented in Algorithm 2.

Algorithm 2: $CreateTemporaryCluster(C_1, C_2)$

Input: Two clusters C_1 and C_2
Output: Updated Clusters C_1, C_2 and a temporary cluster C_{temp}

$c_1 \leftarrow$ centroid of C_1
$c_2 \leftarrow$ centroid of C_2
$c \leftarrow$ centroid of C_1 and C_2 combined
$C_{temp} \leftarrow \phi$
for $i \leftarrow C_1.begin$ *to* $C_1.end$ **do**
    $x \leftarrow i^{th}$ point of C_1
    $dist(x, c_1) \leftarrow$ distance between x and c_1
    $dist(x, c) \leftarrow$ distance between x and c
    if $dist(x, c) < dist(x, c_1)$ **then**
        $C_{temp} \leftarrow C_{temp} \cup \{x\}$
        $C_1 \leftarrow C_1 \setminus \{x\}$
    end
end
for $i \leftarrow C_2.begin$ *to* $C_2.end$ **do**
    $x \leftarrow i^{th}$ point of C_2
    $dist(x, c_2) \leftarrow$ distance between x and c_2
    $dist(x, c) \leftarrow$ distance between x and c
    if $dist(x, c) < dist(x, c_2)$ **then**
        $C_{temp} \leftarrow C_{temp} \cup \{x\}$
        $C_2 \leftarrow C_2 \setminus \{x\}$
    end
end
return C_1, C_2, C_{temp}

We denote the distance between two point x_1 and x_2 by x_1x_2. Let, C be a cluster of a point set S consisting of k clusters and c be its centroid. Assume that f is the farthest from c within all the points of C. Then we define that any point p is within the *range* of cluster C if $pf \leq cf$. The intuition of this step comes from Lemma 1 where the meaning of C_1 and C_2 is stated in Algorithm 2.

Lemma 1 *Let, $p_1 \in C_1$ and $p_2 \in C_2$ and c_1 and c_2 be their centroid, respectively. Suppose, they are assigned to C_{temp} by Algorithm 2. Then for some point f_1 such that $p_1c_1 + c_1f_1 = p_1f_1$ and some point f_2 such that $p_2c_2 + c_2f_2 = p_2f_2$, it can be stated that, p_1p_2 is at least smaller than either p_1f_1 or p_2f_2.*

Proof Let, c be the centroid of the points of C_1 and C_2 combined. We choose $f_1(f_2)$ in such a way that $p_1c_1(p_2c_2) + c_1f_1(c_2f_2) = p_1f_1(p_2f_2)$. The reason is described below.

If p_1 is the farthest point of C_1 from c_1, then for any other point $p' \in C_1$, $p'c_1 \leq p_1c_1$. Else, at least for one point $p'' \in C_1$, $p''c_1 > p_1c_1$. So in either case, f_1 is in the range of C_1. So we can consider a point at f_1 and if that point belonged the actual point set, it would have been assigned to C_1 as it is well inside the range of C_1. Same goes for f_2 being in the range of C_2.

Now as p_1 and p_2 are assigned to C_{temp}, so $p_1c < p_1c_1$ and $p_2c < p_2c_2$. Combining these,

$$p_1c + p_2c < p_1c_1 + p_2c_2 \tag{1}$$

We consider 3 cases.
Case 1: $p_1c_1 > p_2c_2$
Equation (1) becomes,

$$\begin{aligned} p_1c + p_2c &< p_1c_1 + p_1c_1 \\ &= p_1c_1 + f_1c_1 \\ &= p_1f_1 \\ \therefore p_1c + p_2c &< p_1f_1 \end{aligned} \tag{2}$$

Now $p_1p_2 < p_1c + p_2c$, because of triangle inequality. Then (2) becomes,

$$\begin{aligned} p_1p_2 < p_1c + p_2c &< p_1f_1 \\ \therefore p_1p_2 &< p_1f_1 \end{aligned} \tag{3}$$

Case 2: $p_1c_1 < p_2c_2$
Similar to case 1 we prove it for f_2 this time. Finally we get,

$$p_1p_2 < p_2f_2 \tag{4}$$

Case 3: $p_1c_1 = p_2c_2$
Both (3) and (4) can be proved true in this case. So we get,

$$p_1p_2 < p_1f_1 \ and \ p_1p_2 < p_2f_2 \tag{5}$$

Combining (3), (4), and (5), we can say that, p_1p_2 is at least smaller than either p_1f_1 or p_2f_2.

Lemma 1 indicates that despite having a larger distance, if (p_1, f_1) pair or (p_2, f_2) pair deserve to be in the same cluster, then it can be argued that (p_1, p_2) pair also deserves to be in the same cluster. That is why all the points assigned to C_{temp} are merged with either C_1 or C_2 in the next step so that they remain in the same cluster.

3.3 Merge by Greedy Heuristic

In this step we give a greedy heuristic to merge cluster C_1 or C_2 with cluster C_{temp} obtained from the previous step. So final cluster pairs after this step would be either $(C_1 + C_{temp}, C_2)$ or $(C_1, C_2 + C_{temp})$. We denote $(C_1 + C_{temp})$ as C_1^* and $(C_2 + C_{temp})$ as C_2^*.

Algorithm 3: $CalculateRatio(C)$

Input: A cluster C
Output: Diameter to cardinality ratio of C
1 $d \leftarrow$ diameter of C
2 $ratio \leftarrow \frac{d}{|C|}$
3 **return** $ratio$

First we describe the greedy heuristic. We calculate the ratio value of C_1^* and C_2^*. If the ratio value of C_1^* is less than the ratio value of C_2^* we merge C_{temp} with C_1. Similarly, we merge C_{temp} with C_2 if the ratio value of C_2^* is less than the ratio value of C_1^*. As we are trying to minimize the maximum ratio value, in every iteration we perform the greedy operation of selecting between C_1 and C_2 for merging with C_{temp} in such a way that the maximum ratio value of the two clusters at the end of this step is minimized. We present a helper function for calculating the ratio value in Algorithm 3 and a formal description of this step is given in Algorithm 4.

Algorithm 4: $MergeByGreedyHeuristic(C_1, C_2, C_{temp})$

Input: Three clusters C_1, C_2 and C_{temp}
Output: Updated Clusters C_1 and C_2

$C_1^* \leftarrow C_1 \cup C_{temp}$
$C_2^* \leftarrow C_2 \cup C_{temp}$
$ratio_1 \leftarrow CalculateRatio(C_1^*)$
$ratio_2 \leftarrow CalculateRatio(C_2^*)$
if $ratio_1 \leq ratio_2$ **then**
    $C_1 \leftarrow C_1^*$
else
    $C_2 \leftarrow C_2^*$
end
delete C_1^*, C_2^* and C_{temp}
return C_1, C_2

3.4 Filtering

We got the updated C_1 and C_2 in the last step. The filtering step is plain and simple. It takes all the points of C_1 that are nearer to the centroid of C_2 than that of C_1 and assigns them to C_2. Similarly, it takes all the points of C_2 that are nearer to the centroid of C_1 than that of C_2 and assigns them to C_1. A formal description of the algorithm is given in Algorithm 5.

Algorithm 5: $Filtering(C_1, C_2)$

Input: Two clusters C_1 and C_2
Output: Updated Clusters C_1 and C_2

$c_1 \leftarrow$ centroid of C_1
$c_2 \leftarrow$ centroid of C_2
for $i \leftarrow C_1.begin\ to\ C_1.end$ **do**
    $x \leftarrow i^{th}$ point of C_1
    $dist(x, c_1) \leftarrow$ distance between x and c_1
    $dist(x, c_2) \leftarrow$ distance between x and c_2
    if $dist(x, c_1) > dist(x, c_2)$ **then**
        $C_1 \leftarrow C_1 \setminus \{x\}$
        $C_2 \leftarrow C_2 \cup \{x\}$
    end
end
for $i \leftarrow C_2.begin\ to\ C_2.end$ **do**
    $x \leftarrow i^{th}$ point of C_2
    $dist(x, c_1) \leftarrow$ distance between x and c_1
    $dist(x, c_2) \leftarrow$ distance between x and c_2
    if $dist(x, c_1) < dist(x, c_2)$ **then**
        $C_2 \leftarrow C_2 \setminus \{x\}$
        $C_1 \leftarrow C_1 \cup \{x\}$
    end
end
return C_1, C_2

3.5 Putting it all Together

Here we combine all the steps and complete our divisive hierarchical algorithm. As we said earlier, Algorithms 1–2 and 4–5 divide a point set into two clusters. Algorithm 6 is a combined version of all these steps which is a part of every iteration of our algorithm.

Algorithm 6: *CreateTwoClusters(C)*

```
Input: A cluster C
Output: Clustering of C to split it into two new clusters C1 and C2
C1, C2 ← InititalDivide(C)
C1, C2, Ctemp ← CreateTemporaryCluster(C1, C2)
C1, C2 ← MergeByGreedyHeuristic(C1, C2, Ctemp)
C1, C2 ← Filtering(C1, C2)
return C1, C2
```

Algorithm 7: *DHClustering(S)*

```
Input: The point set S and an integer k
Output: k clusters of S such that the maximum ratio value among all clusters is minimized
C ← φ
create a cluster CS with all the points of S
ratio ← CalculateRatio(CS)
create a state c with CS and ratio
insert c into C
for i ← 1 to k − 1 do
    state ← extract the state with the maximum ratio value from C
    cluster ← extract the cluster from state
    C1, C2 ← CreateTwoClusters(cluster)
    ratio1 ← CalculateRatio(C1)
    ratio2 ← CalculateRatio(C2)
    create a state c1 with C1 and ratio1
    insert c1 into C
    create a state c2 with C2 and ratio2
    insert c2 into C
end
clusters ← extract the clusters from C
delete C
return clusters
```

We discussed earlier that we pick the cluster that has the maximum ratio value among the existing clusters and select it for splitting using Algorithm 6 in each iteration as our optimization criterion is to minimize this ratio value. This indicates another greedy heuristic as we are always choosing the cluster that has the maximum ratio value for splitting. For this we calculate the ratio value of each cluster. We then create a state with the cluster along with its ratio value and store it in a set. Then we extract the cluster in that set that has the maximum ratio value because our objective is to minimize the maximum ratio value among all clusters. After $k - 1$ iterations of our algorithm, we get k clusters that satisfy our optimization criterion. A formal definition of the complete divisive hierarchical clustering algorithm is presented in Algorithm 7.

4 Results and Analysis

The experiments of this paper have been done on a computer that has an Intel(R) Core(TM) i7-6700 CPU with a processor base frequency of 3.40 gigahertz and a maximum turbo frequency of 4.00 gigahertz. The installed memory of the computer is 16.00 gigabytes. We have performed all necessary data analysis using Python libraries. Traditional data analysis tools are not feasible for larger datasets [7, 34] and so we performed our own additional analysis along with the tools for these cases. The program for our algorithm has been written in C++ and Python programming language (https://github.com/SadmanOmee/dhclustering_diameter_to_cardinality_ratio). For generating k-means, Ward Agglomerative Clustering, and Affinity Propagation output we have used the popular machine learning library Scikit-learn [24] for the Python programming language. Scikit-learn has an implementation of the k-means clustering that uses an efficient initialization method proposed by Arthur and Vassilvitskii [3]. In this paper we have run k-means clustering 30 times with different centroid seeds and then we have found the run with the minimum sum squared error and have chosen it as the final output. We have implemented the DIANA algorithm ourselves. All the figures have been drawn using libraries of the Python language.

We have compared our algorithm with the other mentioned clustering algorithms. Berkhin [4] stated in his survey that "The k-means algorithm is by far the most popular clustering tool used in scientific and industrial applications." This is why we have chosen k-means clustering for comparing. The Scikit-learn version uses Euclidean distance for measuring distance between points. DIANA [19] is one of the most popular divisive hierarchical algorithms available. We have also chosen Ward Agglomerative Clustering (WAC) [32] which uses an efficient linkage criteria called the "Ward Linkage." We have used Euclidean distance metric for WAC as it is the default setting in the Scikit-learn version. Finally we have also chosen Affinity Propagation [11] because it can find clusters very fast and so very suitable for cases where a large number of clusters are needed to be generated [31].

The convergence of k-means clustering has been highly researched and it is known to have a superpolynomial, i.e., exponential running time in worst case [2, 30]. The reason is that k-means can have exponentially many iterations in the worst case. As our algorithm have specific number of iterations, it can never have an exponential running time like k-means clustering. k-means finds clusters of convex shapes [1], so it is not suitable for finding clusters with non-convex shapes [15]. Also the accuracy of k-means clustering highly depends on the initialization because the iterations of k-means clustering can get stuck in a local minimum for bad initialization [22].

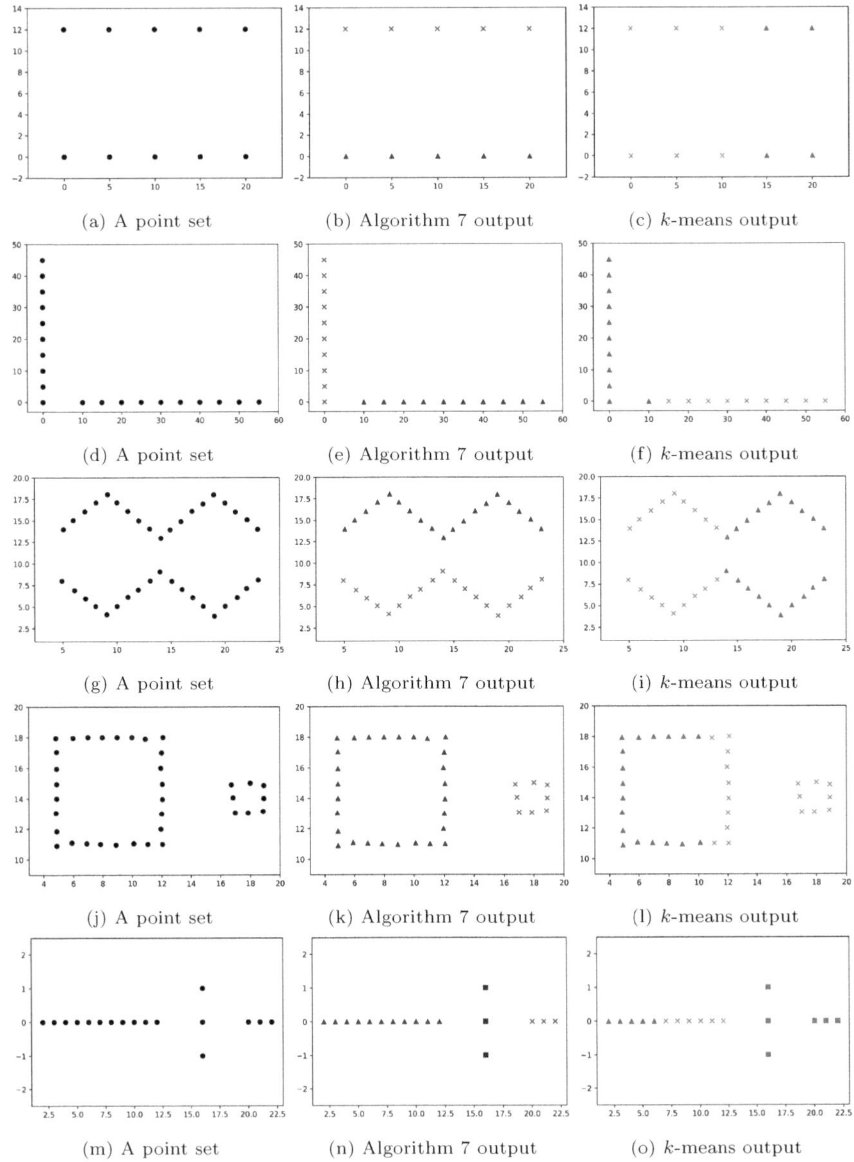

(a) A point set (b) Algorithm 7 output (c) k-means output

(d) A point set (e) Algorithm 7 output (f) k-means output

(g) A point set (h) Algorithm 7 output (i) k-means output

(j) A point set (k) Algorithm 7 output (l) k-means output

(m) A point set (n) Algorithm 7 output (o) k-means output

Fig. 1 Some cases where k-means clustering fails to generate proper clusters but our Algorithm excels

Figure 1 shows some cases where k-means clustering fails to generate good clusters but our algorithm succeeds to find proper clusters. k-means does not generate good clusters with these cases because it tries to minimize the intra-cluster variance.

Table 2 Comparison of Algorithm 7 and other clustering algorithms for different large Benchmark datasets

(a) Data for Alg 7

Dataset	Total Points	k	Alg. 7 (euclidean)		Alg 7 (manhattan)		Alg 7 (chebyshev)	
			ARI	SC	ARI	SC	ARI	SC
A1	3000	20	0.8724	0.6441	0.8314	0.5823	0.8179	0.542
A2	5250	35	0.849	0.6247	0.8158	0.5672	0.7874	0.5034
A3	7500	50	0.8573	0.6498	0.7835	0.5455	0.7623	0.5127
S1	5000	15	0.8341	0.6534	0.8214	0.6192	0.7466	0.5282
S2	5000	15	0.8414	0.5892	0.8253	0.5748	0.8045	0.5536
t4.8k	8000	6	–	0.4713	–	0.4188	–	0.3995
Unbalance	6500	8	0.892	0.8386	0.8798	0.8157	0.8486	0.7961

(b) Data for k-means and Ward Agglomerative Clustering

Dataset	Total Points	k	k-means		WAC	
			ARI	SC	ARI	SC
A1	3000	20	0.826	0.595	0.7926	0.5762
A2	5250	35	0.8277	0.5975	0.7885	0.5797
A3	7500	50	0.8213	0.5894	0.8091	0.5875
S1	5000	15	0.8523	0.7112	0.8332	0.7084
S2	5000	15	0.8672	0.6261	0.8527	0.6076
t4.8k	8000	6	–	0.4125	–	0.3667
Unbalance	6500	8	0.6478	0.4342	0.7884	0.7225

We have compared our algorithm with other clustering algorithms for some benchmark datasets. We have selected 8 datasets for this purpose – A-Sets (A1, A2, A3), S-Sets (S1, S2), t4.8k, Unbalance, and Birch2. Details about these datasets can be found in [9, 10]. We have also selected 4 benchmark shape sets [9] for comparing.

We have run our algorithm using Euclidean, Manhattan, and Chebyshev distance metrics and other clustering algorithms on these datasets. The results and comparison of our algorithm with k-means clustering and WAC with the larger benchmark datasets using the ARI and SC metrics have been shown in Table 2. As Affinity Propagation and DIANA are not suitable for handling larger datasets, we have excluded these two algorithms in this comparison. It has been shown from [10] that k-means performs poorly when cluster sizes have a strong unbalance. We have found from the table that the difference of accuracy between our algorithm and k-means clustering is huge with the Unbalance dataset as our algorithm has performed very well with this dataset but k-means performed poorly. On A-Sets, our algorithm with the Euclidean distance metric has the best accuracy among all.

There are no ground truth labels available for the t4.8k dataset, so we have only compared the SC value for this dataset. We have found that our algorithm with the Euclidean distance metric has the best accuracy with this dataset. The only datasets

Table 3 Comparison of algorithm 7 and other clustering algorithms for different small Benchmark datasets

(a) Data for Alg 7

Dataset	Total Points	k	Alg. 7 (euclidean)		Alg 7 (manhattan)		Alg 7 (chebyshev)	
			ARI	SC	ARI	SC	ARI	SC
Aggregation	788	7	0.8133	0.5023	0.8561	0.5325	0.8039	0.4881
Flame	240	2	0.762	0.4615	0.7475	0.4379	0.682	0.3942
Compound	399	6	0.5972	0.4916	0.5413	0.4692	0.5189	0.4251
R15	600	15	0.7553	0.4752	0.7266	0.4482	0.6945	0.4055

(b) Data for k-means, DIANA, Affinity Propagation and WAC

Dataset	Total Points	k	k-means		DIANA		Affinity Prop.		WAC	
			ARI	SC	ARI	SC	ARI	SC	ARI	SC
Aggregation	788	7	0.7952	0.4801	0.7745	0.4737	0.3456	0.4052	0.769	0.4716
Flame	240	2	0.4901	0.378	0.4825	0.359	0.4423	0.3418	0.711	0.4159
Compound	399	6	0.5364	0.4312	0.4923	0.487	0.3557	0.4151	0.5506	0.4175
R15	600	15	0.7512	0.472	0.714	0.4368	0.7034	0.4688	0.739	0.4622

with which both k-means and WAC have slightly better accuracy than our algorithm is with the S-Sets. On average, our algorithm with the Euclidean distance metric has the best accuracy among all.

We have also run our algorithm and other algorithms on different smaller size benchmark datasets. The results and comparison of our algorithm with k-means clustering, DIANA, Affinity Propagation, and WAC with these datasets have been shown in Table 3. In the Aggregation dataset the clusters are well separated from one another and our algorithm with the Manhattan distance metric have outperformed all with this dataset as Manhattan distance usually yields results that are robust to the influence of outliers but Euclidean distance is less robust to it [35]. We have also found from the table that our algorithm with the Euclidean distance metric has outperformed all others with the rest of the datasets.

The comparison plot of our algorithm with three different distance metrics with other clustering algorithms using ARI and SC separately with the benchmark datasets have been shown in Figs. 2 to 5.

The quality of k-means clustering also depends on the value of k. It has been shown from [10] that the success of k-means clustering has an inverse linear dependency on k. We have experimented the performance of our algorithm with Euclidean distance metric, k-means clustering, and WAC with increasing value of k. We have chosen the Birch2 dataset for this purpose which contains 100000 data points. We have run our algorithm, k-means, and WAC with $k = 10$ to 100 for an interval of 10 clusters on this benchmark dataset (Figs. 3 and 4).

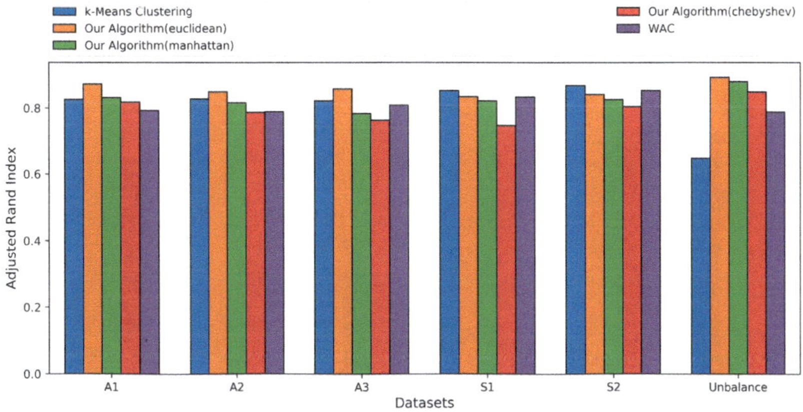

Fig. 2 Performance comparison of Algorithm 7 and other clustering algorithms for different large benchmark datasets using ARI

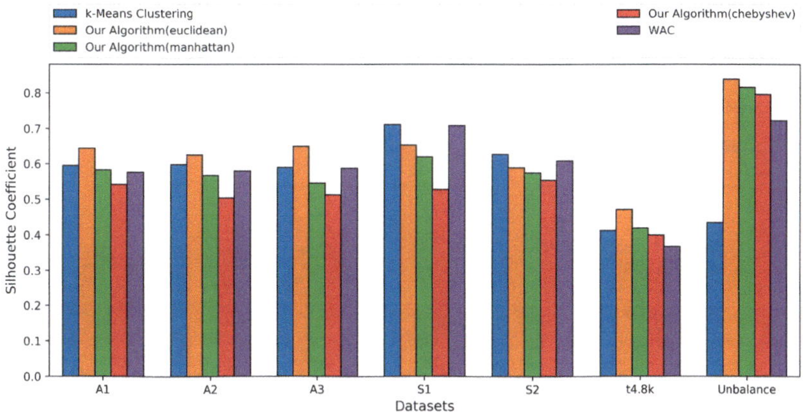

Fig. 3 Performance comparison of Algorithm 7 and other clustering algorithms for different large benchmark datasets using SC

We have plotted the results in ARI and it has been shown in Fig. 6. We have seen that the accuracy of k-means gradually decreases with increasing value of k which has also been described in [10]. The performance of our algorithm and WAC also decreases with increasing k but our algorithm still has better accuracy compared to that of k-means and WAC for all values of k. Also the accuracy of WAC is worse than k-means initially but its accuracy has surpassed that of k-means after $k = 80$ (Fig. 5).

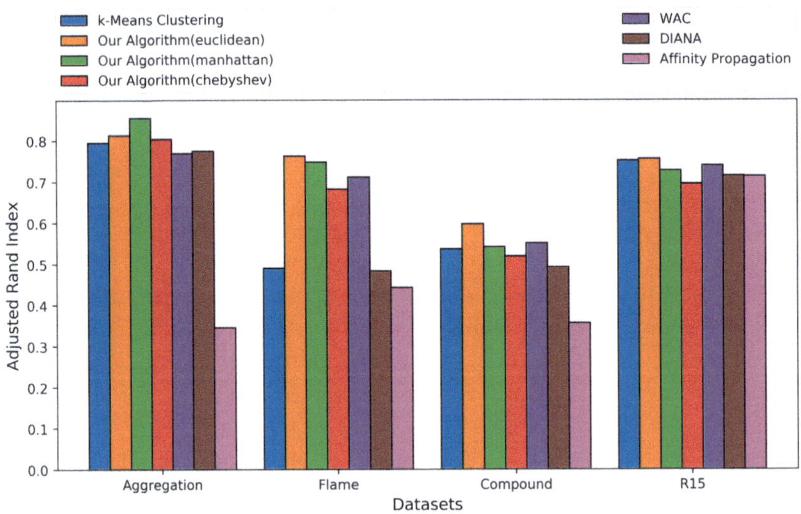

Fig. 4 Performance comparison of Algorithm 7 and other clustering algorithms for different small benchmark datasets using ARI

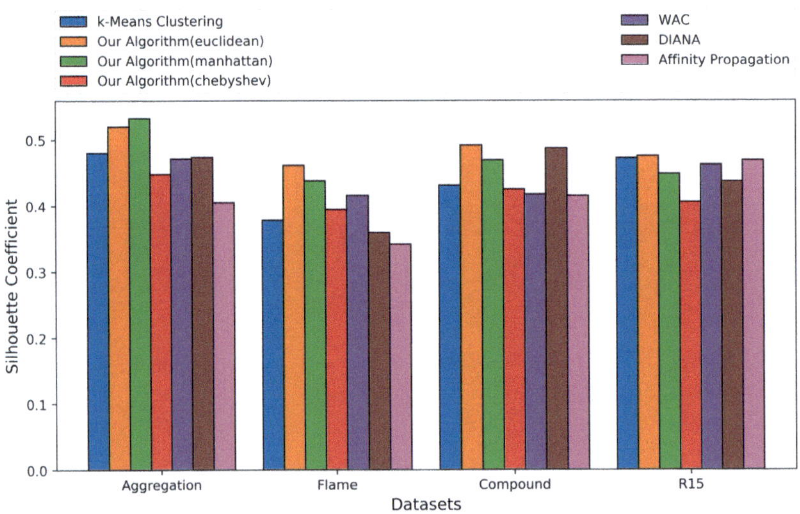

Fig. 5 Performance comparison of Algorithm 7 and other clustering algorithms for different small benchmark datasets using SC

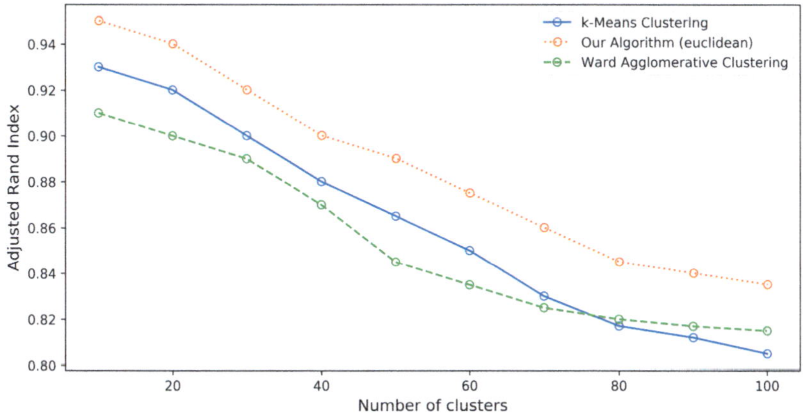

Fig. 6 Performance comparison of Algorithm 7 with k-means clustering and WAC for increasing number of clusters on Birch2 dataset using ARI

5 Conclusions

In this paper, we have presented a new and efficient divisive hierarchical clustering algorithm for a set of points such that the maximum diameter to cardinality ratio among all clusters is minimized. Our algorithm has basically four steps those repeat in every iteration. It follows a greedy heuristic to complete one of these steps. After each iteration, it also follows the optimization criterion to select a cluster from the remaining clusters to split it and create two new clusters.

Experimental results validate the efficiency of our algorithm. We have compared our algorithm with some widely popular clustering algorithms with some benchmark datasets using three different similarity metrics and found that our algorithm with the Euclidean distance metric has the best efficiency among all these. Then we have tested our algorithms' accuracy with other algorithms with increasing value of k and found that our algorithm performs better than those algorithms.

Our algorithm can also be tested on other similarity metrics such as cosine similarity and mahalanobis distance. Also finding a better way to overcome the problem of decreasing accuracy with increasing number of clusters would be a great job. As the real-world data is huge and can be really complex in structure, higher number of clusters are often needed to be produced.

Finally testing accuracy on different datasets can find more clues about how to increase the quality of our algorithm. Similarity based on distance measures really does not work on some datasets such as some specific image data and audio data etc. In these cases, we need to test our algorithm with completely new kinds of similarity metrics and see whether these metrics actually work with the flow of our algorithm.

References

1. Anderberg MR (1973) Cluster analysis for applications. Academic press
2. Arthur D, Vassilvitskii S (2006) How slow is the k-means method? In: Proceedings of the Twenty-second Annual Symposium on Computational Geometry. pp. 144–153
3. Arthur D, Vassilvitskii S (2007) k-means++: the advantages of careful seeding. In: Proceedings of the Eighteenth Annual ACM-SIAM Symposium on Discrete Algorithms (SODA). pp. 1027–1035
4. Berkhin P (2006) A survey of clustering data mining techniques. In: Grouping Multidimensional Data, pp. 25–71. Springer
5. Cormack RM (1971) A review of classification. J Royal Stat Soc Seri A (General) 134(3):321–353
6. Edwards AW, Cavalli-Sforza LL (1965) A method for cluster analysis. Biometrics pp. 362–375
7. Fan J, Han F, Liu H (2014) Challenges of big data analysis. Nat Sci Rev 1(2):293–314
8. Fowlkes EB, Mallows CL (1983) A method for comparing two hierarchical clusterings. J Am Stat Assoc 78(383):553–569
9. Fränti P (2018) Clustering basic benchmark. Online, (Sep 2018). https://cs.joensuu.fi/sipu/datasets/
10. Fränti P, Sieranoja S (2018) K-means properties on six clustering benchmark datasets. Appl Intell 48(12):4743–4759
11. Frey BJ, Dueck D (2007) Clustering by passing messages between data points. Science 315(5814):972–976
12. Friedman JH, Bentley JL, Finkel RA (1977) An algorithm for finding best matches in logarithmic expected time. ACM Trans Mathem Software (TOMS) 3(3):209–226
13. Guénoche A, Hansen P, Jaumard B (1991) Efficient algorithms for divisive hierarchical clustering with the diameter criterion. J Classif 8(1):5–30
14. Hansen P, Jaumard B (1997) Cluster analysis and mathematical programming. Mathem Program 79(1–3):191–215
15. Huang Z (1997) A fast clustering algorithm to cluster very large categorical data sets in data mining. In: Proceedings of the SIGMOD Workshop on Research Issues on Data Mining and Knowledge Discovery **3**(8), 34–39
16. Hubert L (1973) Monotone invariant clustering procedures. Psychometrika 38(1):47–62
17. Hubert L, Arabie P (1985) Comparing partitions. J Class 2(1):193–218
18. Jain AK (2010) Data clustering: 50 years beyond k-means. Pattern Recogn Lett 31(8):651–666
19. Kaufman L, Rousseeuw PJ (1990) Finding groups in data: an introduction to cluster analysis. John Wiley & Sons
20. Lloyd S (1982) Least squares quantization in PCM. IEEE Trans Inform Theory 28(2):129–137
21. Macnaughton-Smith P, Williams W, Dale M, Mockett L (1964) Dissimilarity analysis: a new technique of hierarchical sub-division. Nature 202(4936):1034–1035
22. MacQueen J et al (1967) Some methods for classification and analysis of multivariate observations. In: Proceedings of the Fifth Berkeley Symposium on Mathematical Statistics and Probability. vol. 1, pp. 281–297. Oakland, CA, USA
23. Morey LC, Agresti A (1984) The measurement of classification agreement: an adjustment to the rand statistic for chance agreement. Educ Psychol Measur 44(1):33–37
24. Pedregosa F, Varoquaux G, Gramfort A, Michel V, Thirion B, Grisel O, Blondel M, Prettenhofer P, Weiss R, Dubourg V, Vanderplas J, Passos A, Cournapeau D, Brucher M, Perrot M, Duchesnay E (2011) Scikit-learn: machine learning in Python. J Mach Learn Res 12:2825–2830
25. Rand WM (1971) Objective criteria for the evaluation of clustering methods. J Am Stat Assoc 66(336):846–850
26. Rao M (1971) Cluster analysis and mathematical programming. J Am Stat Assoc 66(335):622–626
27. Rousseeuw PJ (1987) Silhouettes: a graphical aid to the interpretation and validation of cluster analysis. J Comput Appl Math 20:53–65

28. Toussaint G (1983) Solving geometric problems with the rotating calipers. IEEE MELECON 1983:1–8
29. Tsai CW, Lai CF, Chao HC, Vasilakos AV (2015) Big data analytics: a survey. J Big data 2(1):1–32
30. Vattani A (2011) K-means requires exponentially many iterations even in the plane. Discrete Comput Geom 45(4):596–616
31. Wang K, Zhang J, Li D, Zhang X, Guo T (2008) Adaptive affinity propagation clustering. arXiv preprint arXiv:0805.1096
32. Ward JH Jr (1963) Hierarchical grouping to optimize an objective function. J Am Stat Assoc 58(301):236–244
33. Xiong T, Wang S, Mayers A, Monga E (2012) DHCC: divisive hierarchical clustering of categorical data. Data Mining Knowl Disc 24(1):103–135
34. Zaslavsky A, Perera C, Georgakopoulos D (2013) Sensing as a service and big data. arXiv preprint arXiv:1301.0159
35. Zhong Y, Deng Y, Jain AK (2012) Keystroke dynamics for user authentication. In: 2012 IEEE Computer Society Conference on Computer Vision and Pattern Recognition Workshops. pp. 117–123. IEEE

Chapter 60
Building Damage Detection Using Discrete Wavelet Transforms and Convolutional Neural Networks

Piyush Ranjan Biswal, Banhi Sanyal, and Ramesh Kumar Mohapatra

1 Introduction

Natural disasters can have a devastating impact on humanity [1]. A large-scale disaster can often cause significant damage to buildings and properties [2]. Damaged buildings can be used as a representative metric for affected population localisation. An accurate assessment of the damages caused can be of great help for emergency response and decision-making. However, to do this job manually, a sizeable workforce and other resources are needed along with a substantial amount of supervision to detect faulty assessments. This task can prove to be somewhat arduous, particularly in the aftermath of a deadly disaster. Time is also a crucial factor in the process. Hence a drawn-out manual assessment period proves to be futile [3]. However, using satellite imagery and deep learning techniques, this job can be done in a much more efficient way [4]. It takes significantly less effort and time to accurately predict the damaged and undamaged buildings. It can also work on a much larger scale efficiently than the manual method [5, 6].

The main challenge of using satellite imagery and deep learning techniques is that it needs a considerably large labelled dataset of damaged and undamaged building images. There is a limited number of datasets that provide such large number of high-resolution images useful for training deep learning neural network architectures. This can be overcome by creating datasets of aerial imagery of buildings affected by various disasters and making them available publicly for training various models. This will encourage development of many accurate and efficient model architectures which will lead to better disaster response strategies.

P. R. Biswal (✉) · B. Sanyal · R. K. Mohapatra
National Institute of Technology, Rourkela, Rourkela, Odisha, India
e-mail: 517CS1009@nitrkl.ac.in

R. K. Mohapatra
e-mail: mohapatrark@nitrkl.ac.in

G. Mathur et al. (eds.), *Proceedings of 2nd International Conference on Artificial Intelligence: Advances and Applications*, Algorithms for Intelligent Systems,
https://doi.org/10.1007/978-981-16-6332-1_60

The previous approaches on building damage detection use a variety of machine learning algorithms like Decision Trees, Random Forests, K-Nearest Neighbour, Naive Bayes, etc. to solve the classification problem. However these techniques are quite outdated and are not as accurate as the one used in the proposed architecture. Currently deep learning offers a wide variety of techniques to solve classification problems like this. The developments in the field of deep learning can be attributed to the improvements in using GPUs to speed up the learning process significantly. It can be noted there is no separate feature extraction module in the previous architectures. This is a suspected flaw in these systems. As a result, such architectures have considerably low accuracy, in the range of 70–80%. Therefore they are not feasible enough to be used in real-world use cases. In this paper we try to design a real-world architecture which mitigates the mentioned challenges and also achieves a higher accuracy. Despite the presence of various feature extraction methods, the Discrete Wavelet Transform (DWT) [7] is found to be very useful for the problem we are dealing with. Hence, DWT is used to identify useful features from the raw data and overcome the previous limitations. Conducted experiments indicate very high accuracy by DWT, and moreover, drastic reduction in the time taken for classification. It results in a scalable architecture that can be extended to other critical classification tasks in future. In classification module, a convolutional neural network is created and trained on images. The CNN consists of multiple convolution, pooling and fully connected layers that contribute significantly to classification accuracy for images. All this translates to a highly efficient and robust classification model. Image augmentation is also carried out in data preprocessing phase to expand the dataset. The model is trained with a small learning rate for many epochs to prevent over-fitting. Finally, the results are visualised in plots. The performance of this architecture is compared with that of other proposed architectures studied in the literature.

The paper is organised into the following sections. Various schemes proposed by researchers in this regard are discussed in Sect. 2 with their significant characteristics. Section 3 presents the method and other technical details of proposed architecture. Section 4 depicts a detailed analysis of the proposed work and comparison with other architectures. Section 5 delineates the concluding remark.

2 Literature Survey

There have been many notable developments in the domain of building damage detection using deep learning that have led to such efficient and robust architectures. The work by Lecun et al. [8] introduced convolutional neural networks for image classification which is highly efficient. This laid the foundation for the development of many architectures for image classification with high accuracy.

For feature extraction, there are a wide variety of methods available. The work by Ghazali et al. [7] uses 2D Discrete wavelet transforms for feature extraction from the images. This method is very effective for dimensionality reduction. Designing an efficient architecture and training it on datasets is computationally intensive. Transfer

Learning helps in transfer of features learned by an architecture in one classification task to another. The work by He et al. [9] introduced Residual Networks or ResNets that achieve significantly less classification error. ResNets are commonly used to identify features from images in the absence of a large training set for learning. There are many previous works in this field by different authors. The work by Bai et al. [10] uses dual polarisation to distinguish features of damaged and undamaged areas. The best accuracy achieved is 80.5% when using support vector machine for classification. The work by Wang et al. [11] utilises Oversegmentation using SLIC superpixel, Color-HED feature extraction and Maximal similarity merging process for superior feature extraction results. The best accuracy achieved is 82.35% on ResNet-50. The work by Ci et al. [12] uses Ordinal Regression for classification with very high accuracy. The best accuracy obtained is 77.39% for combination of VGG-16 and Ordinal Regression. The work by Xu et al. [13] uses four different CNN architectures similar to AlexNet for better classification results. The best accuracy recorded is 72%. The work by Zhang et al. [14] considers statistical modelling approach as well to decide the structural safety of damaged buildings. The best accuracy obtained is 91%. The work by Mangalathu et al. [15] uses machine learning techniques to rapidly predict building damage. The best accuracy achieved is 66% for random forest classifier.

3 Methodology

Before going to the detail of the proposed architecture, the dataset used in our experiment is discussed below followed by the system requirements and the experimental setup.

3.1 Dataset

The project requires a properly labelled dataset of images of damaged and undamaged buildings. The Kaggle dataset Satellite Images of Hurricane Damage by K. Scott Mader [16] consists of satellite images of buildings from Texas after Hurricane Harvey. Using exploratory data analysis, the following information was obtained. The data is divided into two classes, *damage* and *no_damage*. The dataset consists of 23,000 images of size 128 × 128 divided into:

- Training set: 5,000 images in each class.
- Validation set: 1,000 images in each class.
- Test set 1: 1,000 images in each class.
- Test set 2: 8,000 images in *damage* and 1,000 images in *no_damage*.

3.2 Experimental Setup

Deep learning is a field with intense computational requirements. Deep learning models running on GPUs take exponentially lesser time to train than CPUs [17]. There is also a need for enough RAM to load the model and dataset. For proposed model is conducted on system with following specifications:

- CPU: 2-core Intel(R) Xeon(R) CPU @ 2.30 GHz.
- Disk: 70 GB.
- RAM: 16 GB.
- GPU: NVIDIA Tesla T4.
- VRAM: 16 GB.

3.2.1 Data Preprocessing

Owing to the rather small size of the dataset, image data augmentation is used to expand the training dataset size by 100%. It also significantly increases the diversity of available training data [18]. The techniques applied for augmentation are random horizontal flip and random vertical flip with a probability of 0.5 and random rotation in a range of $\pm 60°$. Now, the training dataset is doubled to 20,000 images in total, i.e. 10,000 images for each class.

3.2.2 Feature Extraction

Discrete wavelet transform is very popular method of feature extraction when it comes to analysis, de-noising and compression of signals and images. Biorthogonal wavelets are commonly used in image processing to detect and filter white Gaussian noise, due to their high contrast of neighbouring pixel intensity values [19]. By experimentation, it was found that *biorthogonal 3.7* obtained best results in detecting features from images of damaged buildings. PyWavelets [20] package (*pywt*) is used to apply the wavelet transform. The function *dwt2* is used for 2D Discrete Wavelet Transform, with *bior3.7* wavelet and *periodization* mode. The transform needs to be applied on grayscale image. After converting original RGB image to grayscale, the wavelet transform is applied and image is saved.

3.2.3 Architecture

The images are loaded in batches of size 64. Then a convolutional neural network architecture is created and trained with a learning rate of 0.0001, momentum of 0.75 for 250 epochs, using cross entropy loss as error metric and stochastic gradient descent as optimiser. The CNN architecture is detailed in Table 1. For each epoch,

Table 1 CNN architecture

Layer	#Params
2D convolution	80
2D batch normalisation	16
ReLU activation	0
2D max pooling	0
2D convolution	1168
2D batch normalisation	32
ReLU activation	0
2D max pooling	0
2D convolution	4640
2D batch normalisation	64
ReLU activation	0
2D max pooling	0
2D convolution	18496
2D batch normalisation	128
ReLU activation	0
2D max pooling	0
Linear	262400
ReLU activation	0
Dropout regularisation	0
Linear	16448
ReLU activation	0
Dropout regularisation	0
Linear	520
ReLU activation	0
Dropout regularisation	0
Linear	18

the loss and accuracy obtained are displayed, both for training and validation set. Then, accuracy is calculated for training, validation and test sets.

3.3 Results

For any given neural net architecture, loss is calculated as the negative log-likelihood and residual sum of squares for classification. Naturally, the main objective in a learning architecture is to reduce (minimise) the loss function's value with respect to the architecture's parameters by changing the weight vector values through different optimization methods, such as backpropagation in neural networks. Accuracy of a

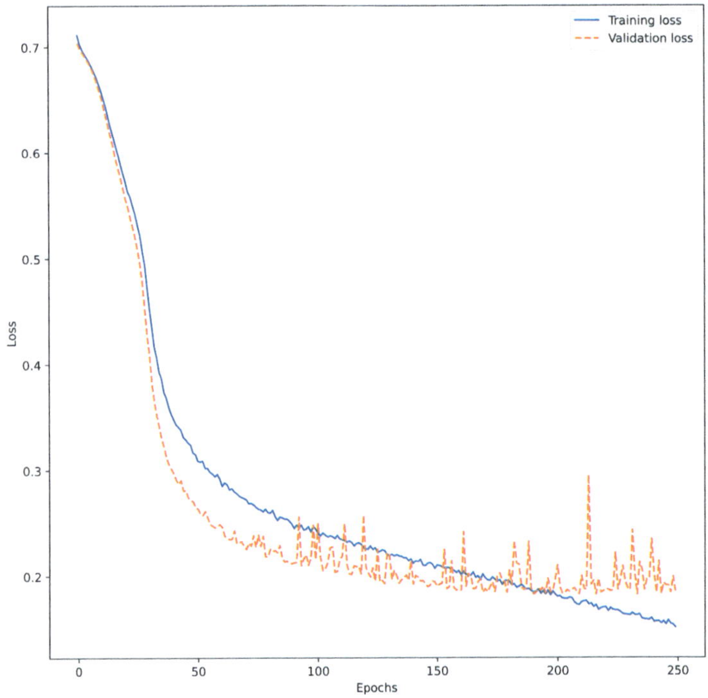

Fig. 1 Loss graph

architecture is usually determined after its parameters are learned and fixed and no learning is taking place. Then the test samples are fed to the architecture and the number of mistakes (zero-one loss) it makes is recorded, after comparison to the true targets. Then the percentage of misclassification is calculated. For the proposed architecture, the loss and accuracy are calculated for each epoch and plotted in Figs. 1 and 2.

It can be observed that the loss graph drops sharply at first till 0.3 when the architecture is still learning abstract features. However, as it starts learning the finer features that give it an edge in predicting the correct class, the decline in loss is marginal. In case of accuracy, the graph increases steeply at first till 80–85%, after which there is only a slight rise to approximately 93%. These trends are more or less similar for both training and validation data. At the end of the training phase, the final results are calculated. The final training loss is found to be 0.1309 and final validation loss is valued at 0.1712, both rounded to 4 places after decimal. Similarly, the final training accuracy turns out to be 95.24% and final validation accuracy results to 92.85%. Finally, the accuracy of test set 1 (balanced) is found to be 94.8% and accuracy of test set 2 (unbalanced) is found to be 92.44%. The confusion matrices for both test sets are also calculated and plotted in Fig. 3a, b. The model predicts 920 instances correctly in the *damage* class and 963 instances correctly in the *no_damage*

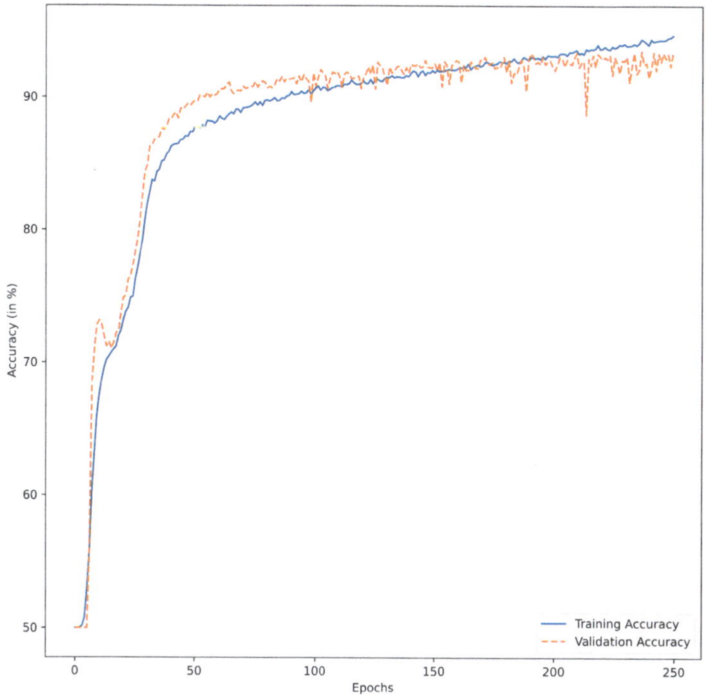

Fig. 2 Accuracy graph

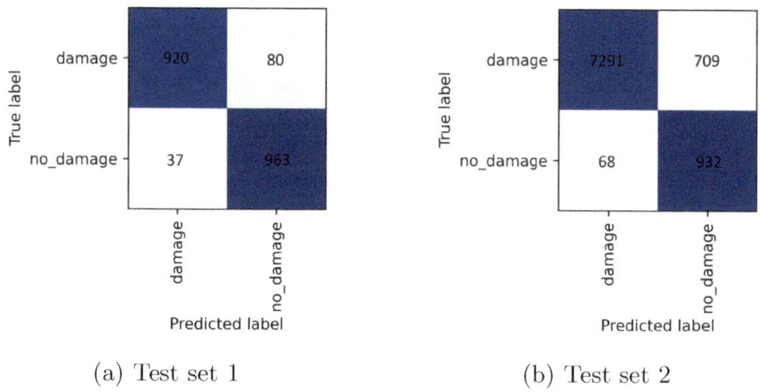

Fig. 3 Confusion matrices

class of Test set 1 consisting of 1000 instances in each class. For the unbalanced Test set 2, 7291 instances are classified accurately out of 8000 instances in *damage* class and 932 instances are classified accurately out of 1000 instances in *no_damage* class.

4 Comparative Analysis

Table 2 shows a comparison of several previous works in the field on the basis of database, architecture and final accuracy.

The work by Bai et al. [10] utilises a Polarimetric SAR scattering mechanism in the data preprocessing stage. Afterwards support vector machine and k-nearest

Table 2 Comparison of results of similar works

Authors	Database	Architecture	Accuracy (%)
Bai et al. [10]	PALSAR-2, Sentinel-1A and GTD by UNOSAT	Polarimetric SAR Scattering Mechanism, Support Vector Machine and K-Nearest Neighbour	80.5
Wang et al. [11]	High-Resolution Imagery by EagleView Technologies	Simple Linear Iterative Clustering, Colour-Holistically nested Edge Detection and ResNet	82.35
Ci et al. [12]	Aerial Imagery of 2010 Yushu and 2014 Ludian earthquakes	VGG-16, ResNet-50 and Ordinal Regression	75.39
Xu et al. [13]	Satellite Imagery of 2010 Haiti, 2017 Mexico city and 2018 Indonesia earthquakes	Concatenated Channel Model, Post-image Only Model, Twin-tower Concatenate Model and Twin-tower Subtract Model	72
Zhang et al. [14]	Aerial Imagery of tall buildings	Decision Tree and Random Forest	91
Mangalathu et al. [15]	EERI dataset	Spatial Interpolation, Discriminant Analysis, K-Nearest Neighbours, Decision Tree, Random Forest	66
Proposed work	Kaggle dataset Satellite Images of Hurricane Damage	Discrete Wavelet Transform and Convolutional neural network	94.8

neighbour classifier are used for classification, with higher accuracy reported on SVM classifier. This shows that SVM is a better classifier for this case.

In the work by Wang et al. [11], simple linear iterative clustering and colour-holistically nested edge detection are used in data preprocessing stage. Support vector machine and ResNet-50 are utilised for classification. ResNet-50 is found to achieve better accuracy than SVM, possibly due to the advantage of transfer learning.

In the work by Ci et al. [12], VGG-16 and ResNet-50 are used for feature extraction, followed by Ordinal Regression and Softmax for classification. As expected, the accuracy obtained in the case of VGG-16 is higher than that of ResNet-50, because of the higher number of parameters and features learnt.

The work by Xu et al. [13] employs four different cnn architectures—concatenated channel model, post-image only model, twin-tower concatenate model and twin-tower subtract model for classification. The performance of the architectures is evaluated using fivefold cross validation. A fairly accurate classification result is obtained. However, it is lower than that of other architectures.

The work by Zhang et al. [14] integrates statistical modelling, machine learning, stochastic simulation and optimization to assess the damage. Decision tree and random forest are used for classification. This unique approach results in a significantly accurate classification.

The work presented by Mangalathu et al. [15] first selects relevant predictor variables in preprocessing stage. Then it employs Discriminant Analysis, K-Nearest Neighbour, Decision Tree and Random Forest algorithms to classify images. The highest accuracy is achieved on random forests. However, this is significantly lower than the results obtained for other architectures.

Considering these previous works, it can be observed that applying transfer learning can lead to more accurate models. However, there is a limit up to which architecture can be improved with transfer learning. This might be due to the fact that these architectures learn a large number of features, not all of which are useful in this classification task [21].

The proposed architecture utilises an additional feature extraction module which helps in classification accuracy greatly. The feature extraction module is missing in many previous works and is probably the reason for the less accuracy in these architectures. This might indicate that direct application of classifiers (k-NN, ResNet, etc.) on images does not extract the optimal feature set. Using various transfer learning architectures for classification module, no significant increase in accuracy was observed. However by designing a custom CNN classification module with multiple convolution, padding and fully connected layers with dropout regularisation and batch normalisation applied, a remarkably accurate architecture is built. The competence of the proposed scheme on both balanced and unbalanced set are more promising than previous schemes.

5 Conclusion

Wavelet-based feature extraction combined with convolutional neural networks achieves a significant accuracy of 94.8% on a balanced test set and 92.44% on an unbalanced test set. This method proves to be very efficient and accurate compared to previous works in the field. Thus, this precise classification of damaged and undamaged buildings contributes to accurate damage assessments and swift risk mitigation. The main issue now is accurately assessing building damage for disasters that happen anywhere in the world and not just those that look similar to the ones the architecture has been trained on. This is challenging because the variety of training data available is limited to a handful of events that occurred in a few geographic locations. Generalising to future disasters that will likely occur in new locations is therefore still a challenge for the model [13]. However, this can be overcome by training model on geographically diverse datasets and fine-tuning the model to adjust accordingly. There is a lot of opportunity for future improvements on the model architecture. It can be developed further to help in other time critical classification tasks with significant accuracy.

References

1. Skidmore M, Toya H (2002) Do natural disasters promote long-run growth? Econ Inq 40(4):664–687
2. McDonald R (2003) Introduction to natural and man-made disasters and their effects on buildings, vol 23. Routledge
3. Leitmann J (2018) Five ways to do better post-disaster assessments. https://blogs.worldbank.org/sustainablecities/five-ways-do-better-post-disaster-assessments. Last updated 24 Oct 2018
4. Gueguen L, Hamid R (2015) Large-scale damage detection using satellite imagery. In: Proceedings of the IEEE conference on computer vision and pattern recognition, pp 1321–1328
5. Thomas J, Bowyer KW, Kareem A (2011) Towards a robust automated hurricane damage assessment from high resolution images. In: 13th international conference on wind engineering, pp 10–15
6. Khan A, Sohail A, Zahoora U, Qureshi AS (2020) A survey of the recent architectures of deep convolutional neural networks. Artif Intell Rev 53(8):5455–5516
7. Ghazali KH, Mansor MF, Mustafa MM, Hussain A (2007) Feature extraction technique using discrete wavelet transform for image classification. In: 5th student conference on research and development. IEEE, pp 1–4
8. LeCun Y, Bottou L, Bengio Y, Haffner P (1998) Gradient-based learning applied to document recognition. Proc IEEE 86(11):2278–2324
9. He K, Zhang X, Ren S, Sun J (2016) Deep residual learning for image recognition. In: Proceedings of the IEEE conference on computer vision and pattern recognition, pp 770–778
10. Bai Y, Adriano B, Mas E, Koshimura S (2017) Building damage assessment in the 2015 Gorkha, Nepal, earthquake using only post-event dual polarization synthetic aperture radar imagery. Earthq Spectra 33(1_suppl):185–195
11. Wang F, Kerekes JP, Xu Z, Wang Y (2018) Residential roof condition assessment system using deep learning. J Appl Remote Sens 12(1):016040

12. Ci T, Liu Z, Wang Y (2019) Assessment of the degree of building damage caused by disaster using convolutional neural networks in combination with ordinal regression. Remote Sens 11(23):2858
13. Xu JZ, Lu W, Li Z, Khaitan P, Zaytseva V (2019) Building damage detection in satellite imagery using convolutional neural networks. arXiv:1910.06444
14. Zhang Y (2019) Post-earthquake performance assessment and decision-making for tall buildings: integrating statistical modeling, machine learning, stochastic simulation and optimization. Doctoral dissertation, UCLA
15. Mangalathu S, Sun H, Nweke CC, Yi Z, Burton HV (2020) Classifying earthquake damage to buildings using machine learning. Earthq Spectra 36(1):183–208
16. Mader KS (2019) Satellite images of hurricane damage. https://www.kaggle.com/kmader/satellite-images-of-hurricane-damage. Last updated 16 Oct 2019
17. Raina R, Madhavan A, Ng AY (2009) Large-scale deep unsupervised learning using graphics processors. In: Proceedings of the 26th annual international conference on machine learning, pp 873–880
18. Shorten C, Khoshgoftaar TM (2019) A survey on image data augmentation for deep learning. J Big Data 6(1):60
19. Pragada S, Sivaswamy J (2008) Image denoising using matched biorthogonal wavelets. In: 6th Indian conference on computer vision, graphics & image processing. IEEE, pp 25–32
20. Lee GR, Gommers R, Waselewski F, Wohlfahrt K, O'Leary A (2019) PyWavelets: a Python package for wavelet analysis. J Open Source Softw 4(36):1237. Last updated 19 Oct 2019
21. Williams J, Tadesse A, Sam T, Sun H, Montanez GD (2020) Limits of transfer learning. In: International conference on machine learning, optimization, and data science. Springer, Cham, pp 382–393

Chapter 61
Forward and Backward Modelling of Wire and Arc Additive Manufacturing Process using Multiple Adaptive Neuro-Fuzzy Inference System

Dhrubajyoti Gupta, Ananda Rabi Dhar, Shibendu Shekhar Roy, and Nilrudra Mandal

1 Introduction

Wire and arc additive manufacturing (WAAM) has shown its potential in producing medium to large components. This advanced technology has high deposition rate because of higher efficacy of arc and can be potentially utilized for unlimited build size. The wire-feed system makes the process cleaner and environment-friendly. The electric arc either continuously or in pulsated mode melts the filler wire to form droplets, which are transferred into the molten pool to get deposited on solidification in the form of additive layers. The continuous heat input leading to instability in thermal and mechanical behaviour of the deposited layers is mitigated by the Fronius developed cold metal transfer (CMT) process. In this technology, a reciprocating wire-feed mechanism is adopted to forcibly short-circuit the wire tip contact with the substrate material [1]. CMT process offers improved arc stability, high-quality material deposition, controlled transfer of material, less distortion, low thermal input, almost zero spatter, and high process tolerance. Nevertheless, like other additive manufacturing (AM) technologies, an optimized process planning that provides uniform, reliable, and defect-free deposition is instrumental for the successful application of this technique in industry.

Any change in process parameters and working conditions in this complex process may have great impact on the dimensional tolerances and material properties of the parts produced. The most crucial factors that require optimal settings and ideally

D. Gupta · A. R. Dhar · S. S. Roy (✉)
Department of Mechanical Engineering, National Institute of Technology Durgapur, West Bengal, India
e-mail: shibendy.roy@me.nitdgp.ac.in

N. Mandal
Materials Processing and Microsystems Laboratory, Central Mechanical Engineering Research Institute, Durgapur, West Bengal, India

G. Mathur et al. (eds.), *Proceedings of 2nd International Conference on Artificial Intelligence: Advances and Applications*, Algorithms for Intelligent Systems,
https://doi.org/10.1007/978-981-16-6332-1_61

real-time monitoring along with enhanced decision support capabilities are heat dissipation, interlayer temperature, and quality of deposited previous layers. The geometry of the parts produced solely depends on the process parameters in a non-linear fashion. Various research activities have already contributed in this direction to model the advanced welding, especially the CMT process so as to understand the complex relationship between inputs and outputs [2].

A predictive system which can estimate the deposited layer geometry and other dependent process responses in function of the CMT input process parameters and the environmental conditions should be clearly recommended for improved efficiency. This kind of system can potentially reduce the time in selecting the process parameter values and also the trial phase wastes of costly materials often used in additive mode of manufacturing, thereby reducing the overall production cost. Many mathematical models and artificial neural networks (ANN)-based models were used [3–14] in the past addressing such typical advanced welding problems. The finite element method (FEM) was also adopted to predict the width and penetration of weld bead [15, 16]. In the FEM models, the temperature profile of the moving heat input as well as the material properties of filler and substrate were considered. The modelling work of complex welding process so far has been restricted to only simple geometries and required oversimplifying assumptions to reduce computation demands conforming to the added complexity. It may be rewarding to future researchers to apply machine learning techniques to aid the development of predictive models even involving numerical simulation.

The literature review particularly reveals that regardless of the approach adopted, the studies and simulations were strictly concentrated on the forward mapping of the input-output parameters, which involves determination of the output parameters from the known set of inputs. The responses of practical importance in WAAM, especially CMT process, namely the width and height of weld bead often ask for nominal values unlike maximum or minimum ones, which necessitates the development of a predictive system which can estimate the process parameters to obtain a desired quality of responses. To the best of the authors' knowledge no significant study has been reported on backward mapping of the input-output parameters in this case. It is reported that backward mapping model cannot be accurately obtained by using statistical regression because of the presence of multi-collinearity among the responses. This has necessitated the usage of computational intelligent-based approach like adaptive neuro-fuzzy inference system (ANFIS), which is essentially a hybridized approach of fuzzy reasoning and neural computing. The use of fuzzy reasoning takes care of the issues with imprecise and incomplete data in the real-world application while the neural network takes care of the function approximation by a supervised learning schema. This work tries to bridge the gap of unexplored fuzzy-based systems in the realm of WAAM especially CMT.

The present study has attempted to carry out both forward and backward mapping of an exemplary single deposition CMT process reported in past literature [2]. The model has been developed using MATLAB fuzzy toolbox functions. Here, wire diameter, wire-feed speed and ratio of wire-feed speed to travel speed have been considered as the input process parameters and 12 responses starting from wire diameter (WD) to heat source length (HSL) as depicted in Table 1 have been regarded as the dependent process outputs.

A forward mapping model has been developed using a multi-adaptive neuro-fuzzy inference system (MANFIS), which predicts twelve responses for a given set of values of three input parameters. The data generated from regression equations for the three responses plus the actual experimental data have been used as input data set for the model, out of which a definite portion has been used to validate and test the trained model. The training of the model has been carried out using traditional gradient descent algorithm. Similar procedure has been adopted for developing the backward mapping model to estimate three process parameters from the given twelve responses and testing has been carried out for validation purposes. The developed model has been further compared with an ANN-based model using gradient descent algorithm to get both qualitative and quantitative perspective of the advantages of using this method over the methodologies adopted by others [3, 4].

The remaining part of the paper is arranged as follows: The methodology has been described in Sect. 2: Materials and fabrication have been briefed in sub-section 2.1, while statistical modelling and data generation has been dealt in sub-section 2.2 and development of computational intelligence (CI)-based modelling with MANFIS is explained in the subsequent sub-sections. Results for both forward and backward mappings are presented and discussed in Sect. 3. Finally, some conclusions are reported in Sect. 4.

2 Methodology

2.1 Materials and Fabrication

A modified GMA-welding variant, which is an advanced technology utilizing controlled dip transfer mode, developed and coined as cold metal transfer (CMT) by Fronius was envisaged in this study. The CMT control panel allowed the selection of the pure CMT process, material type (G3Si1), shielding gas mixture, and wire diameter to be capitalized for obtaining desired welding characteristics and additional parameter settings. It is worth mentioning that an optimal shielding gas mixture containing Ar/CO2 (18%) from the group M21 was used to get the best welding characteristics. A comprehensive investigation was carried out [2] to reveal the impact of three important factors such as wire diameter (WD), wire feed speed (WFS), and the ratio of WFS to travel speed (TS) on process responses related to bead geometry and plate fusion characteristics as tabulated in Table 1.

Experimental investigations concerned with single deposition were conducted on as-rolled S355J2+N grade structural steel used as substrates (chemically C-Mn steel). A solid mild steel wire as a consumable electrode (G3Si1) was utilized throughout the study applied in three variants of diameters—0.8, 1.0, and 1.2 mm. The base material was selected as per the minimum requirements set from mechanical and chemical

Table 1 Input parameters and ranges along with responses

Process parameters and responses	Input/Response	Symbols	Units	Ranges considered	
				Max	Min
Wire diameter	Input	WD	m/min	1.2	0.8
Wire feed speed	Input	WFS	m/min	1.68	13.35
Ratio of wire-feed speed to travel speed	Input	(WFS/TS)		6.62	19.34
Bead width	Response	W	mm	8.83	2.52
Bead height	Response	H	mm	3.62	1.61
Remelting ratio	Response	RR		0.48	0.02
Penetration	Response	P	mm	2.13	0.27
Dilution	Response	D		32.62	2.4
Deposition rate	Response	DR	Kg/hr	3.1	0.6
Contact angle	Response	CA	°	132.4	42.8
Reinforcement area	Response	A_2	mm^2	18.35	4.16
Penetration area	Response	A_1	mm^2	7.31	0.26
Aspect ratio	Response	AR		1.147	0.289
Heat input	Response	HI	J/mm	635.35	115.8
Heat source length	Response	HSL	mm	21.7	4.9

aspects of the filler wire. Other reasons for selecting this were because of its low cost and prevalent application in almost every elementary structural fabrication. The dimensions of the base plates used in this study are $250 \times 200 \times 6mm$. A shielding gas flow rate of 15 l/min was regularized throughout this study for all experimental trials.

Weld bead geometry and fusion characteristics on substrate were measured using *AxioVision*—an image analysis software utilizing several built-in measurement functions. Interested readers may please refer to the original article [2] to get acquainted with a detailed description of the experimental methodology.

2.2 Statistical Modelling and Data Generation

For conducting experimental investigations on the CMT process, a Q-optimal (also known as IV-optimal) design has been adopted. A statistical regression model has been developed using Minitab software by feeding the experimental data [2] as metadata for defining a custom design for the response surface methodology (RSM). The input variables were codified using the respective maxima as +1 and respective minima as −1. Regression equations of coded type have been produced by applying RSM on the 70 experimental data set. They are shown in Table 2 using intercepts and coefficients. 95% confidence interval has been adopted and stepwise backward elimination method of all available terms (full quadratic) for selection has been applied. For the development of any computationally intelligent model especially involving evolutionary algorithms, a large number of training samples is often placed as a compulsory prerequisite. In order to cater to this need, 9 hypothetical levels have been created keeping the lower and upper bounds of input parameters intact. 729 numbers of data have been generated using all the possible combinations of input parameter hypothetical levels. This has been done essentially to get maximum coverage within the range of experimental design space. This is worth mentioning that the goodness of fit for each response has been found close and above 90%—an indicator favouring adequate models.

2.3 ANFIS Modelling for Forward and Backward Mapping

As the name suggests, ANFIS typically amalgamates the benefits of neural network capable of learning with substantial accuracy with the fuzzy reasoning which works pretty well for incomplete and imprecise data of real-world problems [17]. For handling multiple responses there are two variants of ANFIS—first, multiple ANFIS (MANFIS) which has as many as separate multiple ANFIS modules as the responses and second, C-ANFIS which houses all the responses as output nodes in one single ANFIS module. In this study, MANFIS has been adopted for developing the models. The three process parameters have been used as inputs whereas the twelve dependent process responses have been provided as the outputs for developing the forward mapping model. The MANFIS architecture for forward mapping model is shown in Fig. 1. The backward mapping of the process parameters has been achieved by swapping the inputs and outputs of the physical process model [18, 19]. Twelve dependent responses have been provided as the input parameters to the MANFIS model, whereas the three process parameters have been kept as outputs of the model (Fig. 2). The MANFIS architecture can be understood by following the layout of the layers and nodes contained therein and their contributions for producing the final outputs from a set of inputs, and thereafter, applying the learning algorithms for the parameters of the membership functions for both inputs and outputs. The layers are

Table 2 Intercepts and coefficients in regression equations for all responses

Responses	Inputs and their combinations									
	Intercept	WD	WFS	WFS/TS	WD * WFS	WD * (WFS/TS)	WFS * (WFS/TS)	WD * WD	WFS * WFS	(WFS/TS)* (WFS/TS)
		(A)	(B)	(C)	(AB)	(AC)	(BC)	(A^2)	(B^2)	(C^2)
W	7.056	1.589	1.287	1.46	–0.075	0.391	0.7	–0.06	–1.455	–0.003
H	2.388	0.479	–0.086	0.605	0.192	0.076	–0.184	0.074	0.391	–0.007
RR	0.4	0.148	0.311	–0.011	0.107	–0.02	0.047	–0.015	–0.034	0.003
P	1.64	0.842	0.909	0.39	0.443	0.187	0.28	0.014	–0.178	0.015
D	29.149	8.457	18.45	–1.046	4.636	–1.17	3.859	–1.531	–4.096	–0.547
DR	2.772	1.078	2.178	–0.023	0.834	–0.014	–0.017	0.06	0.023	–0.005
CA	48.92	1.672	–15.806	–2.589	9.749	0.774	–24.656	2.342	40.251	–2.903
A2	11.726	4.922	0.433	5.264	0.489	2.035	0.206	0.515	–0.386	–0.551
A1	4.567	3.146	3.568	1.528	2.163	0.687	0.973	0.222	–0.317	0.153
AR	0.312	0.012	–0.063	0.011	0.199	–0.04	–0.132	0.034	0.352	0.032
HI	331.275	116.458	–44.528	172.642	–48.108	70.651	18.328	–0.85	–27.516	–9.59
HSL	18.028	6.667	10.314	1.568	3.689	0.153	1.724	0.504	–2.186	0.111

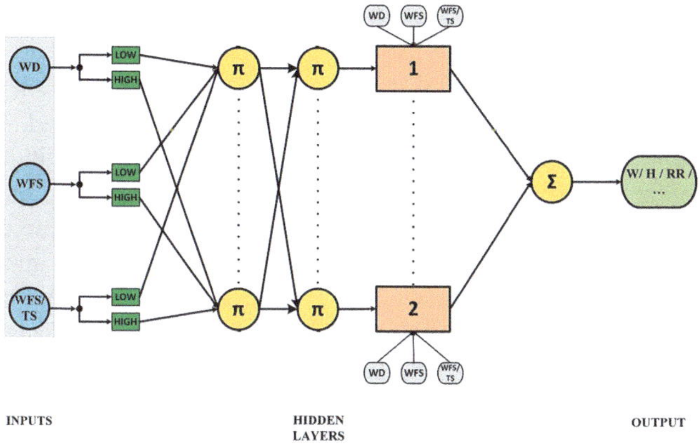

Fig. 1 MANFIS architecture of the forward mapping model

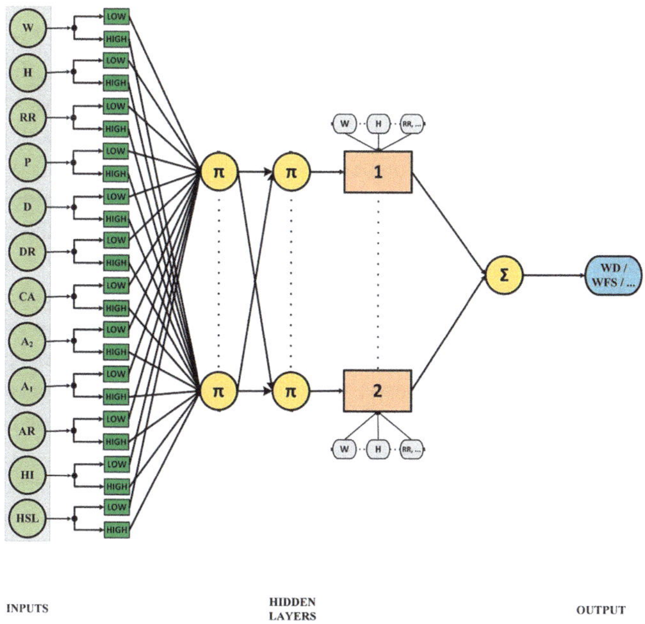

Fig. 2 MANFIS architecture of the backward mapping model

described here in this sub-section while the training algorithms are discussed in the following sub-section.

Layer 0: This layer consists of the input variables of the problem.

Layer 1: This layer consists of the membership functions for all the inputs. The membership function, μ, selected for this work is a gaussian function as supported by MATLAB toolbox functions, as shown in Eq. 1, which has two parameters viz.

m, the mean and s, the standard deviation. These parameters are known as premise parameters.

These parameters are usually initialized using a clustering mechanism. For this model, fuzzy-C-means (FCM) clustering algorithm has been applied on the training data set for the initialization purpose. The experimental input training data after normalization are fed to FCM for initializing the parameters based on clusters or centroids and number of membership functions. These parameters are adjusted via a supervised learning process through back propagation. The type and number of membership functions of each input variable is same, with only the premise parameters set at different initial values. Hence, the number of nodes in this layer is given by Eq. 2,

$$\mu = \frac{1}{1 + e^{-\frac{1}{2}\times(\frac{x-m}{s})^2}} \tag{1}$$

$$n = ni \times mf \tag{2}$$

where, n is the number of nodes, ni is number of input variables and mf is the number of membership functions for each input. The number of membership functions has to be selected on the basis of the performance of the model during training. For simplicity, in Fig. 1, two membership functions are shown. The actual model of forward mapping has a higher value than this.

Layer 2: This layer can be thought of as the product layer, symbolized by π, where each membership function value of one input is multiplied by the membership function values of all the other inputs cumulatively. The respective membership functions of all the inputs are connected as constituents of each inference rule for the model. There can be several ways to combine each input membership function with another. The full factorial combination would generate mf^{ni} number of rules. This would make the system less responsive during training as it would require enormous computational power and memory. In this study, a membership function pertaining to a particular order or level (high, medium or low) for one input variable is connected to the same order or level of membership functions for other input variables, which makes the number of nodes in this layer equal to the number of membership functions, mf. The value of a node representing each rule contains the combined firing strength μ_{rule} of the rule obtained by multiplying all the membership function values of this rule, as shown in Eq. 3,

$$\mu_{rule} = \mu_1 \times \mu_2 \times \mu_3 ... \times \mu_{ni} \tag{3}$$

where, μ_i is membership function value of ith input variable.

Layer 3: This is a normalization layer, symbolized with N. This layer contains the same number of nodes as in the previous layer. This essentially normalizes the firing strength of each rule by simply dividing the respective firing strength by the summation of all the firing strengths. All the nodes in this layer are connected to all the nodes in the previous layer by the virtue of the mathematical summation operator in the denominator, as shown in Eq. 4,

$$\mu_{normal(i)} = \frac{\mu_{rule(i)}}{\mu_{rule(1)} + \mu_{rule(2)} + \mu_{rule(3)} \cdots + \mu_{rule(mf)}} \tag{4}$$

where, $\mu_{normal(i)}$ is the normalized firing strength for ith rule and $\mu_{rule(i)}$ is the firing strength of the ith rule, and similarly, $\mu_{rule(mf)}$ is firing strength of the last i.e. mfth rule.

Layer 4: This layer also consists of same number of nodes as in the previous layer. These nodes are connected with previous layer by one to one mapping. Additionally, each node has connections with all the input variables. “Takagi and Sugeno” type of fuzzy inference mechanism, which is mostly followed for majority of ANFIS models, has been adopted here. In this case, three inputs along with one normalized firing strength value produce the total value for each node in this layer. This value is obtained using some linear coefficient terms known as consequent parameters as shown in Eq. 5. The number of consequent parameters is one more than the number of input variables used in the model.

$$f_{rule(i)} = \mu_{rule(i)}(p_{i1}a + p_{i2}b + p_{i3}c + p_{i4}) \tag{5}$$

where, $f_{rule(i)}$ is the value of node in layer 4 representing output or consequence for ith rule and $p_{i1}, p_{i2}, \ldots, p_{i4}$ are the consequent parameters for ith rule and $a, b, andc$ are the values of the input variables considering total three input variables.

Layer 5: This layer comprises of output nodes. The number of nodes is same as the number of responses used in the model i.e. three. In this layer, the consequent outputs from the previous layer are summed up and the final output value is produced, which is shown in Eq. 6. Each output node is connected with all the nodes in the previous layer, where, f_{final} is the final output value.

$$f_{final} = f_{rule(1)} + f_{rule(2)} + \ldots + f_{rule(mf)} \tag{6}$$

In the MANFIS model, learning has two passes—forward and backward. Here, batch mode of learning has been implemented which means, one forward and one backward pass for the entire training set is considered as an epoch. Two algorithms have been used for training the traditional MANFIS model viz. back propagation (BP) and recursive least square (RLS), which has made this a hybrid training process. BP has been used for training the premise parameters by computing the ordered derivatives of the errors obtained in the previous iteration propagated from layer 5 to layer 2. RLS method has been used to train the consequent parameters of layer 5 on each forward pass.

BP algorithm works on gradient-based steepest descent method. The adjustment of the independent variable is done in the direction of slope defined by tangent on the current position. MATLAB toolbox functions “GENFIS3” and “ANFIS” [20] have been envisaged by custom coding in order to achieve the desired MANFIS architecture. Parametric study has been performed by running several trials to determine the optimal number of membership functions and step length.

3 Results and Discussions

3.1 Simulation Results of Forward Modelling

The optimal configuration for the MANFIS layout in terms of number of membership functions for each predictor and optimal learning parameters namely, step length and maximum number of epochs have been found by running the validation trials after the training with a specific topology. The best setting has been saved as a pre-trained model for evaluating the test samples. The R-squared values for the training samples and mean absolute percentage error for the test data have been illustrated graphically in Fig. 3. Larger prediction errors in case of HSL (41.60%), HI (35.09%) have been witnessed, which has been mutually agreed by lower R-squared values (88.98% and 82.30%, respectively) observed for these responses in the trained model. The prediction accuracy has been encouraging in case of basic weld geometry parameters like W and H. The prediction accuracy and model fitness for D and DR have been also satisfactory. Though the penetration and reinforcement area has been very adequately modelled by this approach (more than 98% goodness of fit), the prediction accuracies for the test cases pertaining to these have been not so high. The forward model has been compared with the ANN model and the test results are shown in Table 3. For majority of the responses, this clearly demonstrates the superiority of the MANFIS model.

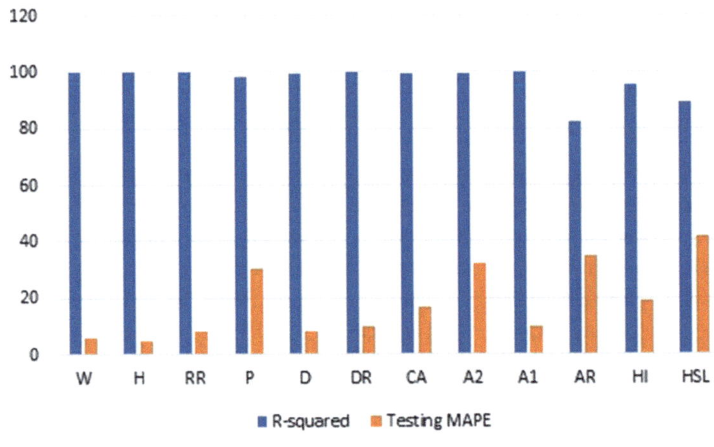

Fig. 3 Summary of performance in forward mapping model

Table 3 Comparison of test results with ANN model in forward mapping

Models	W	D	P	H	D	DR	CA	A_2	A_1	AR	HI	HSL
Testing MAPE MANFIS	6.22	4.90	8.39	30.51	8.39	10.00	16.72	31.85	10.16	35.09	18.73	41.6
Testing MAPE ANN	13.71	12.27	9.54	31.77	7.49	15.89	14.56	38.31	14.37	39.15	16.95	42.38

3.2 Simulation Results of Backward Modelling

Learning the optimal MANFIS configuration for the backward mapping model has been carried out along similar lines as that of the forward mapping model. The optimal step length has been found as 0.01, whereas the number of membership functions has been optimally set as 12. The maximum number of epochs conducted for each output has been also found as 100 by running several trials with the other optimal settings kept intact. The goodness of fit of the backwardly trained network model has been found considerably high and close to 99.9% for each output. The prediction accuracies of process parameters have been found adequate and comparatively better than the forward scenario. This may be attributed for the presence of so many predictors in backward model. The performance of the backward mapping is depicted in Fig. 4. The backward model has been further compared with the ANN model and the test results are shown in Table 4. This also reveals the developed MANFIS model can predict the backward mapping scenarios more accurately.

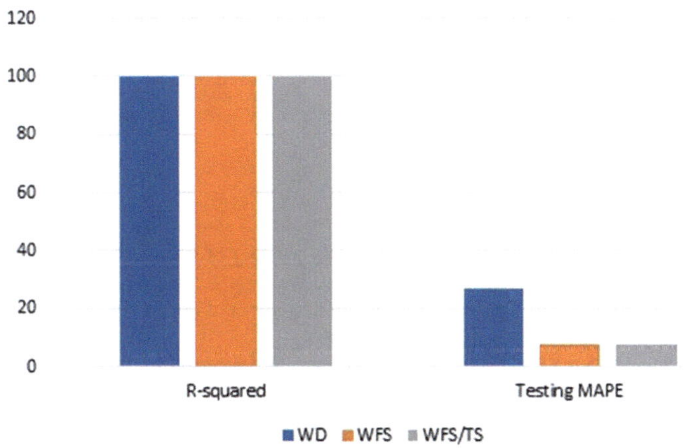

Fig. 4 Summary of performance in backward mapping model

Table 4 Comparison of test results with ANN model in backward mapping

Models	WD	WFS	WFS/TS
Testing MAPE MANFIS	26.79	7.91	7.73
Testing MAPE ANN	26.72	8.66	17.10

4 Conclusions

1. A CMT process has been studied using three input variables and 12 responses conforming geometry and plate fusion characteristics representing process stability.
2. The statistical regression-based model has been developed to generate a large number of data often required for training a neural network.
3. The regression model has been used to replace the missing/erroneous values in the experimental data.
4. The forward and backward models of input and output parameters using MANFIS paradigm have been successfully built.
5. The MANFIS architecture has been optimized by parametric study for number of membership functions and step length for learning using the training set of data.
6. The R-squared values of the developed models are more than 98% in the bi-directional mappings suggesting the adequacy of the same.
7. The MANFIS model when compared with equivalent ANN model on test data has shown the superiority of the model in both forward and backward mappings.
8. The average error using the test data has been found satisfactorily low for each response indicating adequate predictive systems capturing bi-directional mappings.

References

1. Furukawa K (2006) New CMT arc welding process-welding of steel to aluminium dissimilar metals and welding of super-thin aluminium sheets. Weld Int 20:440–445
2. Sequeira Almeida PM (2012) Process control and development in wire and arc additive manufacturing, Doctoral dissertation, Cranfield University
3. Meco S, Pardal G, Eder A, Quintino L (2013) Software development for prediction of the weld bead in CMT and pulsed-MAG processes. Int J Adv Manufac Technol 64(1–4):171–178
4. Thao DT, Jeong JW, Kim IS, Kim JWHJ (2008) Predicting lapjoint bead geometry in GMA welding process. Arch Mater Sci Eng 32:121–124
5. Palani PK, Murugan N (2006) Development of mathematical models for prediction of weld bead geometry in cladding by flux cored arc welding. Int J Adv Manuf Technol 30:669–676
6. Murugan N, Gunaraj V (2000) Prediction and optimization of weld bead volume for the submerged arc process-part 1. Weld J 79:286–294

7. Dey V, Pratihar DK, Datta GL, Jha MN, Saha TK, Bapat AV (2009) Optimization and prediction of weldment profile in beadon-plate welding of Al-1100 plates using electron beam. Int J Adv Manuf Technol 48:513–528
8. Karray FO, Silva CWD (2004) Soft computing and intelligent systems design: theory. Tools and Applications, Addison Wesley, New York
9. Andersen K, Cook GE, Karsai G, Ramaswamy K (1990) Artificial neural networks applied to arc welding process modeling and control. IEEE Trans Ind Appl 26:824–830
10. Pal S, Samantaray A (2008) Artificial neural network modeling of weld joint strength prediction of a pulsed metal inert gas welding process using arc signals. J Mater Process Technol 202:464–474
11. Kim I, Son J, Park C, Lee C, Prasad YKD (2002) A study on prediction of bead height in robotic arc welding using a neural network. J Mater Process Technol 131:229–234
12. Jeng J-Y, Mau T-F, Leu S-M (2000) Prediction of laser butt joint welding parameters using back propagation and learning vector quantization networks. J Mater Process Technol 99:207
13. Manikya Kanti K, Srinivasa Rao P (2008) Prediction of bead geometry in pulsed GMA welding using back propagation neural network. J Mater Process Technol 200:300–305
14. Nagesh DS, Datta GL (2002) Prediction of weld bead geometry and penetration in shielded metal-arc welding using artificial neural networks. J Mater Process Technol 123:303–312
15. Sathiya P, Jinu GR, Singh N (2009) Simulation of weld bead geometry in GTA welded duplex stainless steel (DSS). Sch Res Exch, pp. 1–9
16. Shanmugam NS, Buvanashekaran G, Sankaranarayanasamy K, Ramesh Kumar S (2010) A transient finite element simulation of the temperature and bead profiles of T-joint laser welds. Mater Des 31:4528–4542
17. Jang JS (1993) Anfis: adaptive-network-based fuzzy inference system IEEE transactions on systems, man, and cybernetics 23(3):665–685
18. Datta S, Pratihar DK (2010) A hybrid computing scheme for forward and reverse mappings of metal inert gas welding process. Int J Comput Intell Stud 1:256
19. Maji K, Pratihar DK (2010) Forward and reverse mappings of electrical discharge machining process using adaptive network-based fuzzy inference system. Expert Syst Appl 37:8566–8574
20. Chapman SJ (2003) Programação em MatLab para engenheiros. Cengage Learning, São Paulo

Chapter 62
Wireless Sensor Networks Localization by Improved Whale Optimization Algorithm

Nebojsa Bacanin, Milos Antonijevic, Timea Bezdan, Miodrag Zivkovic, and Tarik A. Rashid

1 Introduction

Wireless sensor network (WSN) is built of a finite number of autonomous sensors that are distributed spatially. The sensors are tiny and inexpensive devices that are equipped with processing capability, memory, sensing capabilities, wireless communication capabilities, battery (source of power). Their deployment is very popular in different smart applications, sensors enable sensing different physical characteristics, such as light, temperature, humidity, pressure, they are widely used in monitoring patients, tracking surveillance system, agriculture, monitoring the quality of water, etc. The sensor senses, processes, and sends the collected data from the environment either to the dedicated collector node or to the fixed radio station through a wireless ad hoc network. The deployed sensors in WSN need to know their spatial coordinates in order to function efficiently. In WSN, there are two main kinds of sensors, one is the anchor nodes and another type is the unknown nodes. The positions of the anchors are known, while the position of the unknown node should be located.

In practice, one of the major issues in WSNs is their localization (approximation of location) capability. The localization refers to determining the unknown sensor node position (or target node) by making use of the known sensor nodes positions (anchor nodes), based on different measurements, such as the time and the angle of arrival, the time difference of arrival, etc.

Global positioning system (GPS) is the standard technique for the sensor localization, where the sensors are supplied with GPS receivers and the devices are able to locate themselves based on the provided position from the GPS. The drawbacks of the GPS localization method are the high energy consumption, high cost of instal-

N. Bacanin (✉) · M. Antonijevic · T. Bezdan · M. Zivkovic
Singidunum University, Danijelova 32, 11000 Belgrade, Serbia

T. A. Rashid
Computer Science and Engineering Department, University of Kurdistan Hewler, Erbil, KRG, Iraq

G. Mathur et al. (eds.), *Proceedings of 2nd International Conference on Artificial Intelligence: Advances and Applications*, Algorithms for Intelligent Systems,
https://doi.org/10.1007/978-981-16-6332-1_62

lations, complicated deployment in different environments, such as indoor, urban environment, underground, etc. These problems motivated researchers to develop alternative methods for localization sensors, where instead of having each sensor a GPS, the number of GPS is reduced and the rest of the sensors without GPS (unknown nodes) would receive the localization information from the sensors equipped with GPS (from the anchor nodes).

The localization methods can be separated into two broad groups of algorithms, the range-based and range-free localization methods. The range-free algorithms approximate the sensor location by utilizing the connectivity between adjacent sensors and using the information of topology. On the other hand, range-based algorithms employ a ranging technique that measures the angle or point-to-point distance between the sensors and calculates the location based on the range information.

The range-based algorithms require the usage of hardware which results in higher costs, the range-free algorithms are less expensive, however, their error rate is higher. The most commonly used distance measuring techniques are the Time and the Angle of Arrival (ToA and AoA), Received Signal Strength (RSS), and Time Difference of Arrival (TDoA).

The sensor node localization problem is considered as an unconstrained optimization problem and it is one of the NP-hard optimization tasks. Deterministic approaches fail in solving NP-hard problems due to the time complexity. However, stochastic approximation approaches, such as nature-inspired metaheuristics are proven to be very efficient in finding optimal or close to optimal solutions in a reasonable time frame. Opposite to the classical optimization approaches, the implementation of metaheuristic-based optimization techniques is easier, they require less computing resources, and they have faster convergence.

Swarm intelligence (SI) algorithms are characterized by randomization, and in an iterative procedure, the algorithm during execution converges to the optimum or close-to-optimum solution based on local search technique (exploitation or intensification) and global search technique (exploration or diversification). It is crucial in SI algorithms to keep a proper trade-off among exploitation and exploration. Implementation of additional operators in existing algorithms can result in improvement, making better exploration-exploitation trade-off and having a better convergence rate. Although many algorithms are proposed in the literature for WSN localization, there is still a need to develop a more robust and more efficient algorithm that will reduce the localization error and speed up the convergence rate. In this work, we propose an improved nature-inspired SI approach, the whale optimization algorithm (WOA) [18] for WSN localization.

The manuscript is structured as follows: Section 2 provides the survey of nature-inspired metaheuristics and outlines the practical applications in the domain of WSN optimization, the proposed WOA approach is discussed in Sect. 3. After that, Sect. 4 shows the outcomes of the conducted practical experiments, while Sect. 5 concludes the research and proposes the future work.

2 Background and Related Work

The sensor nodes localization problem in WSNs is an NP-hard problem by nature, therefore it is not possible to solve it in acceptable time frame by utilizing the traditional deterministic approach [13, 28]. On the opposite, the stochastic approaches, such as SI metaheuristics algorithms, are able to find solutions that are good enough (not guaranteed to be the best) within the acceptable time frame.

SI metaheuristics algorithms were inspired by the various processes which can commonly be found in nature. Through billions of years of evolution, nature established itself as a very successful problem solver. Maximizing the feeding, hunting, and mating processes are just some examples which all species require for survival. Recent literature survey shows that a significant number of algorithms in the past twenty years were based on the above-mentioned fundamental behaviors expressed by different species. These approaches have been implemented for various NP-hard challenges in practice, and they have demonstrated to be very powerful in solving complex tasks in reasonable time frame.

Nature-inspired algorithms can be separated into two classes: algorithms belonging to the first group are called evolutionary algorithms (EA), and the second group is referred to as swarm intelligence (SI) algorithms. EA algorithms are derived from the concepts of evolution and natural selection, and the most important example of this group is the genetic algorithm (GA), which has been utilized in solving different practical NP-hard problems in the recent years, including machine learning [11], task and workflow scheduling and load balancing in the cloud computing domain [42], Internet of Things in health services [9]. It was also successfully utilized in the domain of ad hoc and WSN for maximum coverage deployment of WSN [41] and localization [21].

SI algorithms, on the other hand, draw inspiration from the collective intelligence and social behavior exhibited by ants, bees, butterflies, fish, wolves, etc. These relatively simple creatures are able to show very intelligent behavior when they form large groups called swarms. This particular property has inspired large number of SI algorithms. Some of the most famous algorithms belonging to the SI include artificial bee colony (ABC) [6, 14], bat algorithm (BA) [34, 40], cuckoo search algorithm (CS) [2], and firefly algorithm (FA) [39]. The ABC metaheuristics, for example, is one of the oldest SI metaheuristics, that imitates the honey-foraging behavior of the bee swarm. Other metaheuristics also mimic some characteristic behavior, such as the flashing behavior and attractiveness of the fireflies (FA), or echolocation-based hunting by the small bats (BA). Many other SI metaheuristics followed, including the monarch butterfly optimization (MBO) [32, 36], moth search algorithm (MSA) [25, 35], elephant herding optimization (EHO) [27, 30, 37], tree growth algorithm (TGA) [31], grey wolf optimization (GWO) [4, 44], and whale optimization algorithm (WOA) [5].

SI metaheuristics have recently proven to be very powerful optimizers, not only for the global optimization task [31], but also for solving different challenges from various practical engineering and machine learning domains, including the problem of task and workflow scheduling in the cloud computing domain [4, 7], time series prediction for the number of COVID-19 cases [43], IoT big data management [9], artificial neural networks (ANN) training and hyperparameter optimization [3, 20], and so on. SI metaheuristics have also been widely used in solving different WSN NP-hard challenges. Clustering task in WSN has been frequently addressed by SI, with the focus on the selection of the cluster heads [24, 38], with the main goal to prevent node depletion by utilizing a PSO approach. WSN lifetime was addressed in [45], by utilizing the enhanced dragonfly algorithm. The same task with the application of the enhanced grey wolf algorithm was addressed in [45].

Several SI approaches were used to tackle the localization problem in WSN. Performances of elephant herding algorithm and tree growth algorithm were discussed in [29] and [30], with promising results. MBO approach was utilized in [26], while chicken swarm optimization (CSO) approach was discussed in [1]. GA-based algorithm for localization was proposed in [21], where DV-Hop algorithm (implemented on the hop distance estimation) was improved with GA, with promising results. Firefly-based two-stage WSN localization was proposed in [33]. Finally, ABC-based localization was suggested in [15] and [8]. All SI-based approaches obtained better results when compared to traditional WSN localization algorithms.

3 Original and Improved Whale Optimization Algorithm

WOA method belongs to the group of novel SI approaches, and it was introduced in 2016 [19]. The algorithm draws inspiration from the humpback whales' hunting tactics called the bubble-net strategy. In the recent years, WOA established itself as a very promising optimizer, and it was applied in solving various problems from different domains, both in the original and in modified state [16, 17].

In the rest of this section, a quick introduction to the original WOA approach is given first. Afterward, the deficiencies and flaws of the original WOA are presented, which have been observed during conducted experiments. At the end, the enhanced WOA algorithm is presented, which is able to surpass some of the observed flaws of the basic WOA approach.

3.1 Original WOA

The humpback whales use a cooperative hunting technique to catch their prey, which is known in the written works as the bubble-net hunting maneuver. During this maneuver, whales dive below the school of fish, blow bubbles and move in circles toward the surface. Consequently, the prey is trapped by the circle of bubbles which forces it to swim upward [12]. Interested reader can find more information about this hunting strategy in [19].

WOA implements the search process by going through the global and local searches at the same time, and those two phases are known as exploration and exploitation, respectively. The exploitation phase is modeled by the whales' performing circles around prey and utilizing the bubble-net attack approach. On the other hand, the exploration phase can be modeled as the pseudo-random search for the possible prey. WOA is a population-based algorithm, where the current best candidate solution (the global best solution determined by the greatest fitness) corresponds to the prey. All other solutions in the population denote individual whales.

3.1.1 Exploitation Process

While performing the exploitation, every candidate conducts the search in the neighborhood, headed toward the currently best solution. Then, the fitness of every solution is obtained, and all solutions will update their respective positions in respect to the current best solution. This process is mathematically represented by the equations Eqs. (1) and (2): [19]:

$$\vec{D} = |\vec{C} \cdot \vec{X}^*(t) - \vec{X}(t)| \tag{1}$$

$$\vec{X}(t+1) = \vec{X}^*(t) - \vec{A} \cdot \vec{D}, \tag{2}$$

where $\vec{X}(t)$ and $\vec{X}^*(t)$ represent the candidates and the current best solution in round t, $\vec{A}$ and $\vec{C}$ are the coefficient vectors, and the operator $\cdot$ represents the element-wise multiplicator.

The above-mentioned vectors $\vec{A}$ and $\vec{C}$ can be computed with the application of Eq. [19]:

$$\vec{A} = 2\vec{a} \cdot \vec{r} - \vec{a} \tag{3}$$

$$\vec{C} = 2 \cdot \vec{r}. \tag{4}$$

where $\vec{r}$ denotes a pseudo-random vector from uniform distribution within the boundaries $[0, 1]$, $\vec{a}$ represents a vector which is linearly decreasing from the value 2 to 0 over the iterations.

Each candidate $\vec{X}$ is being directed in every round of execution toward a neighborhood of the current best solution $\vec{X}^*$ through the adjustment of the coefficients $\vec{A}$ and $\vec{C}$ by applying the pseudo-random vector $\vec{r}$.

The bubble-net hunting approach is mathematically modeled by decreasing the circles and by utilizing the spiral path. Circles are reduced by linear decrease of α from 2 to 0 in each round of algorithm's execution, as modeled by Eq. (5) of the original algorithm [19]:

$$\vec{a} = 2 - t\frac{2}{maxIter}, \tag{5}$$

where parameters t and $maxIter$ represent the current round and the maximum number of rounds.

The spiral path is conducted in two phases. In the beginning, the distance from the current solution ($\vec{X}(t)$) and the global best ($\vec{X}^*(t)$) is computed. Afterward, the new position of candidate ($\vec{X}(t+1)$) is obtained with the application of the spiral Eq. (6):

$$\vec{X}(t+1) = \vec{D}' \cdot e^{bl} \cdot \cos(2\pi l) + \vec{X}^*(t), \tag{6}$$

where $\vec{D}'$ represents the distance from the i-th candidate solution to the global best, which can be computed as $\vec{D}' = |\vec{X}^*(t) - \vec{X}(t)|$, and b denotes a fixed value utilized to control the shape of the logarithmic spiral. Finally, parameter l represents the pseudo-random value from the interval $(-1, 1)$.

The whales' behavior where they circle around the target by moving in the spiral paths and decreasing the circle radius at the same time can be modeled by switching between two described mechanisms in every round with equal probabilities p (determined in each iteration by a random value from 0 to 1), which is given by the Eq. (7):

$$\vec{X}(t+1) = \begin{cases} \vec{X}^*(t) - \vec{A} \cdot \vec{D}, \text{ if } p < 0.5 \\ \vec{D}' \cdot e^{bl} \cdot \cos(2\pi l) + \vec{X}^*(t), \text{ if } p \geq 0.5 \end{cases}. \tag{7}$$

3.1.2 Exploration Process

During the exploration phase, every candidate will update its position based on the location of a random solution instead of the global best. Vector $\vec{A}$ is utilized in such a way that if the generated random numbers are greater or equal to one ($|A| \geq 1$), the new position of the candidate is pointed in the direction of a random solution, thus performing the global search. This behavior can be modeled with the Eq. (8), as described in [19]:

$$\vec{X}(t+1) = \vec{X}_{rnd}(t) - \vec{A} \cdot \vec{D}, \tag{8}$$

where $\vec{D}$, which denotes the distance from the i-th candidate to the random solution rnd in round t of the algorithm's execution, is calculated as $\vec{D} = |\vec{C} \cdot \vec{X}_{rnd}(t) - \vec{X}(t)|$.

3.2 *Observed Deficiencies of Basic WOA*

After conducting practical experiments with the basic WOA approach, we were able to conclude that there is a space for the improvement of the original algorithm. Two drawbacks were observed, namely inadequate balance among the exploitation and the exploration processes, and converging too early.

The first problem, inadequate balance affects the performance of the algorithm, which can be clearly seen in the earlier rounds of execution of the metaheuristics.

In some cases, the exploration is not powerful enough, and WOA cannot locate the correct portion of the search space, where optimal solution resides, causing the algorithm to generate poor mean values. More precisely, the exploration of the basic WOA is set toward already existing random solutions, while new solutions are not being created.

The second observed problem in conducted experiments is caused in case that the trade-off among those two processes has been settled in favor of the latter. It can lead the algorithm to converge too early, and get trapped in suboptimal part of the search space. This early convergence can happen when control parameters of the algorithm $\vec{A}$ and $\vec{C}$ are not able to create better solutions in successive rounds, consequently lsing the diversity of the population. In such runs, speed of converging is also not acceptable.

3.3 Improved WOA

According to previous research [10, 22, 23], by incorporating quasi-reflection-based learning (QRL) procedure into metaheuristics, intensification-diversification trade-off can be improved. By improving the balance between these two main search processes, early convergence can also be mitigated.

Proposed enhanced WOA approach, which generates quasi-reflected population at the end of each iteration, is named WAO quasi-reflected learning (WOAQRL).

Quasi-reflected parameter j of solution x, x_j^{qr}, is created by using the following expression [22]:

$$x_j^{qr} = \text{rnd}\left(\frac{lb_j + ub_j}{2}, x_j\right) \tag{9}$$

where $\frac{lb_j + ub_j}{2}$ denotes the mean of j-th component's lower bound (lb_j) and upper bound (ub_j) and expression $\text{rnd}\left(\frac{lb_j + ub_j}{2}, x_j\right)$ generates random number from uniform distribution in range $\left[\frac{lb_j + ub_j}{2}, x_j\right]$

For every solution i from the population P at the end of each iteration quasi-reflected individual x_i^{qr} is generated. Afterward, newly created quasi-reflected population P^{qr} is merged with original population P, solutions in such combined population are sorted in descending order according to fitness and best N solutions are chosen for the next iteration.

Inner workings of proposed WOAQRL are summarized in the form of the pseudo-code, which is given in Algorithm 1.

Algorithm 1 The proposed WOAQRL algorithm

Initialization. Generation of the random starting population P with N solution $X_i (i = 1, 2, 3, ..., N)$ and initialization ot the control parameters' values.
Fitness calculating. Compute the fitness of every candidate and identify the global best X^*
while $(t < maxIter)$ **do**
  for each candidate **do**
    Update A, C, a, l and p
    if $p < 0.5$ **then**
      if |A|<1 **then**
        Update the current candidate solution X by applying Eq. (2)
      else
        Select random solution rnd from the population
        Update the current candidate solution X by applying Eq. (8)
      end if
    else
      Update the current candidate solution X by applying Eq. (6)
    end if
  end for
  In case any of the solutions goes outside the acceptable region of the search space, it must be modified
  All the solutions from the population must be evaluated by computing the fitness
  The global best X^* should be updated if necessary
  Generated quasi-reflected population P^{qr} by using Eq. (9)
  Merge P and P^{qr} $(P \cup P^{qr})$ and sort individuals in descending order according to its' fitness
  Choose best N solutions from $P \cup P^{qr}$
  $t = t + 1$
end while
return The best solution (X^*) from the population

4 Simulations and Discussion

This section provides the results of the conducted experiments. The utilized experimental environment and parameters are the same as in [1], in order to present comparative analysis of the proposed approach against other advanced metaheuristics that were also applied on the same localization problem. Finally, this section provides discussion of the obtained results.

The WSN nodes are placed in a random fashion across the area 100 x 100. Total number of nodes is 50, with 10 anchors and 40 nodes with the unknown position, where each node has a transmission range of 25m, as stated in [1]. In the conducted experiments, maximum number of rounds during execution of the algorithm was set to 150, which is the same setup as in [1]. The positions of the unknown nodes were calculated by utilizing the original WOA and the proposed WOAQRL method. Localization error of WOA method is shown in Fig. 1.

In the conducted experiments, localization simulations have been performed to compare the performances of the basic WOA and proposed WOAQRL approach.

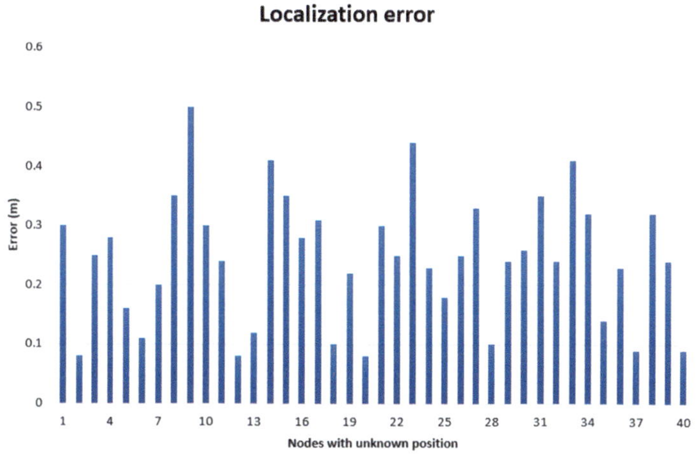

Fig. 1 WSN Localization error obtained by WOA

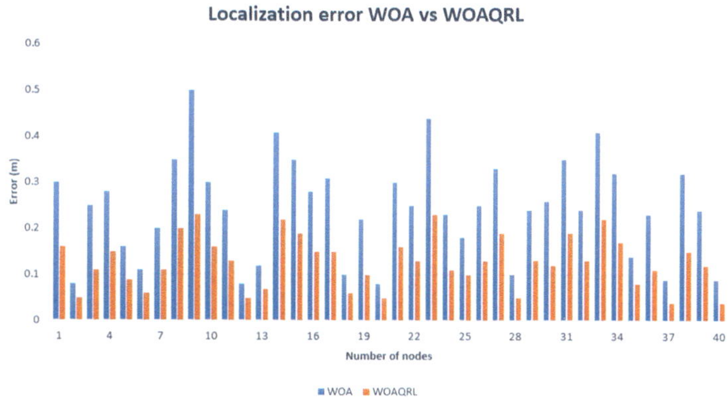

Fig. 2 Localization error—WAO versus WOAQRL

The simulation results are shown in Fig. 2, which suggests that the performances of the WOAQRL are better when compared to the basic WOA.

The range of transmission is considered to be one of the most valuable parameters in WSN localization problem. Increasing the range of transmission of sensor nodes leads to decreasing of the localization error. Simulation results with various values of the transmission range parameter for WOA and WOAQRL metaheuristics are given in Table 1, along with the results for CSO, PeSOA, PSO, and BPSO for comparative analysis, which were derived from [1]. Comparative results are also depicted in Fig. 3. The simulation outcomes indicate that the basic WOA has worse performances than the CSO metaheuristics, while the proposed WOAQRL method obtains slightly better results than the CSO.

Table 1 Comparative transmission range analysis

	Mean localization error(m)					
Range (m)	CSO	PeSOA	PSO	BPSO	WOA	WOAQRL
25	0.11	0.15	0.38	0.72	0.13	0.10
30	0.10	0.12	0.32	0.65	0.12	0.09
35	0.09	0.11	0.26	0.57	0.11	0.08
40	0.07	0.10	0.20	0.50	0.09	0.06

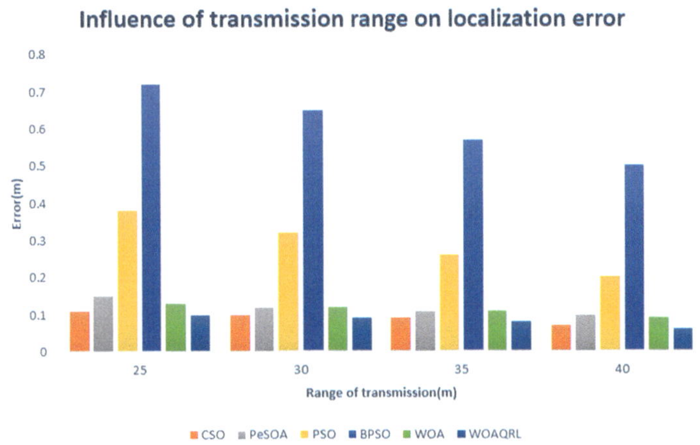

Fig. 3 WSN Localization error for different transmission range values

Another important parameter is the number of the anchors (sensors with known location). Increasing the number of anchor nodes in the WSN leads to easier, more precise localization of the unknown sensor nodes. Simulation results with various number of anchors for WOA and WOAQRL metaheuristics are given in Table 2, along with the results for CSO, PeSOA, PSO, and BPSO for comparative analysis. The CSO, PeSOA, PSO, and BPSO metaheuristics results for different values of anchor nodes were derived from [1]. Comparative results are also depicted in Fig. 4. From the conducted simulations, it is possible to conclude that the proposed WOAQRL approach achieved best performances, while the original WOA method performed slightly worse compared to the CSO approach.

Table 2 Comparison of the algorithms concerning the number of anchors

	Mean error(meters)					
Anchors	CSO	PeSOA	PSO	BPSO	WOA	WOAQRL
10	0.11	0.15	0.45	0.72	0.13	0.10
11	0.09	0.12	0.37	0.64	0.12	0.09
12	0.08	0.11	0.32	0.53	0.10	0.07
13	0.05	0.10	0.29	0.50	0.09	0.04

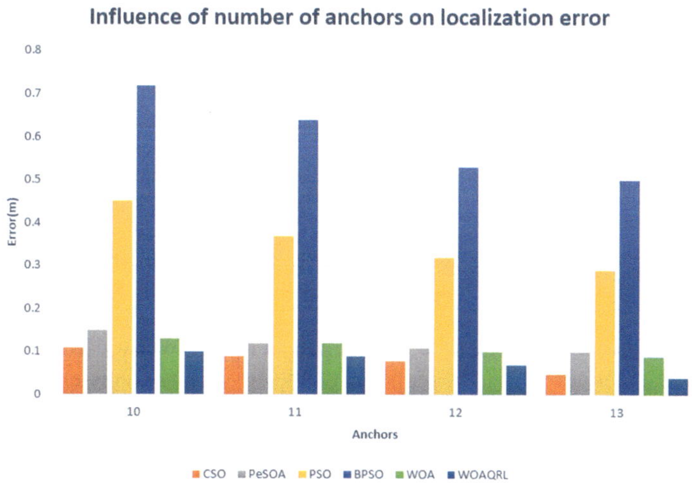

Fig. 4 WSN Localization error for different amount of anchor nodes

Generally speaking, from the practical observations, the number of located sensors after each round is considered to be an important factor for evaluation of the performances of the proposed approach. Comparative analysis of the amount of located nodes in respect to the iteration number is shown in Fig. 5. The CSO metaheuristics results were derived from [1]. From the conducted simulations, it can be concluded that the proposed WOAQRL method clearly outperformed all other metaheuristics, while the original WOA method performed slightly worse compared to the CSO approach. Figure 5 also shows that the WOAQRL algorithm can localize unknown sensor nodes in less iterations than other metaheuristics included in the research.

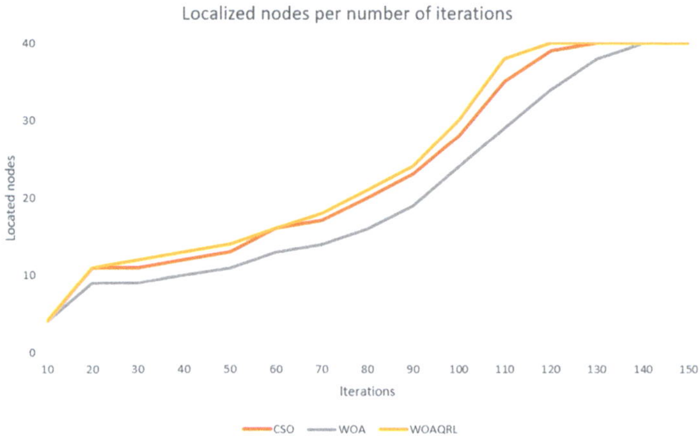

Fig. 5 WSN Localization error for different amount of anchor nodes

5 Conclusion

This research proposes an enhanced WOA method, named WOAQRL for addressing the localization task in WSN. The proposed method overcomes deficiencies observed in the original WOA, which can be seen from the outcomes of the conducted experiments.

The proposed approach has been evaluated and compared to other popular algorithms that were previously used in resolving the localization task in WSN. The approaches included in the research were CSO, PeSOA, PSO, BPSO, and the basic implementation of the WOA. The proposed WOAQRL method clearly outperforms other metaheuristics that were covered in the comparative analysis and practical experiments.

In future work, we aim to perform additional experiments with the proposed WOAQRL method, and to apply it in solving other problems from WSN and other application domains.

Acknowledgements The paper is supported by the Ministry of Education, Science and Technological Development of Republic of Serbia, Grant No. III-44006.

References

1. Al Shayokh M, Shin SY (2017) Bio inspired distributed wsn localization based on chicken swarm optimization. Wirel Pers Commun 97(4):5691–5706
2. Bacanin N (2010) Implementation and performance of an object-oriented software system for cuckoo search algorithm. Int J Mathem Comput Simul 6(1):185–193 December
3. Bacanin N, Bezdan T, Tuba E, Strumberger I, Tuba M (2020) Monarch butterfly optimization based convolutional neural network design. Mathematics 8(6):936

4. Bacanin N, Bezdan T, Tuba E, Strumberger I, Tuba M, Zivkovic M (2019) Task scheduling in cloud computing environment by grey wolf optimizer. In: 2019 27th Telecommunications Forum (TELFOR), pp. 1–4. IEEE
5. Bacanin N, Tuba E, Zivkovic M, Strumberger I, Tuba M (2019) Whale optimization algorithm with exploratory move for wireless sensor networks localization. In: International Conference on Hybrid Intelligent Systems, pp. 328–338. Springer
6. Bacanin N, Tuba M (2012) Artificial bee colony (ABC) algorithm for constrained optimization improved with genetic operators. Stud Inform Control 21(2):137–146 June
7. Bezdan T, Zivkovic M, Tuba E, Strumberger I, Bacanin N, Tuba M (2020) Multi-objective task scheduling in cloud computing environment by hybridized bat algorithm. In: International Conference on Intelligent and Fuzzy Systems, pp. 718–725. Springer
8. Cheng L, Qu L, Xu Y (2017, Sep) Artificial bee colony algorithm-based multiple-source localization method for wireless sensor network. In: 2017 2nd IEEE International Conference on Computational Intelligence and Applications (ICCIA), pp. 445–448. 10.1109/CIAPP.2017.8167257
9. Elhoseny M, Abdelaziz A, Salama AS, Riad A, Muhammad K, Sangaiah AK (2018) A hybrid model of internet of things and cloud computing to manage big data in health services applications. Fut Generat Comput Syst 86:1383–1394. https://doi.org/10.1016/j.future.2018.03.005 http://www.sciencedirect.com/science/article/pii/S0167739X17322021
10. Ewees AA, Abd Elaziz M, Houssein EH (2018) Improved grasshopper optimization algorithm using opposition-based learning. Expert Syst Appl 112:156–172. https://doi.org/10.1016/j.eswa.2018.06.023 http://www.sciencedirect.com/science/article/pii/S0957417418303701
11. Goldberg DE (1989) Genetic Algor Search, 1st edn. Optimization and Machine Learning. Addison-Wesley Longman Publishing Co., Inc, Boston, MA, USA
12. Goldbogen JA, Friedlaender AS, Calambokidis J, McKenna MF, Simon M, Nowacek DP (2013, 02) Integrative Approaches to the Study of Baleen Whale Diving Behavior, Feeding Performance, and Foraging Ecology. BioScience 63(2):90–100. 10.1525/bio.2013.63.2.5,
13. Goyal S, Patterh MS (2014) Wireless sensor network localization based on cuckoo search algorithm. Wirel Pers Commun 79:223–234. https://doi.org/10.1007/s11277-014-1850-8 May
14. Karaboga D, Basturk B (2008) On the performance of artificial bee colony (abc) algorithm. Appl soft comp 8(1):687–697
15. Kulkarni VR, Desai V, Kulkarni RV (2016, Dec) Multistage localization in wireless sensor networks using artificial bee colony algorithm. In: 2016 IEEE Symposium Series on Computational Intelligence (SSCI). pp. 1–8. 10.1109/SSCI.2016.7850273
16. Ling Y, Zhou Y, Luo Q (2017) Lévy flight trajectory-based whale optimization algorithm for global optimization. IEEE Access 5:6168–6186. https://doi.org/10.1109/ACCESS.2017.2695498
17. Mafarja MM, Mirjalili S (2017) Hybrid whale optimization algorithm with simulated annealing for feature selection. Neurocomputing 260:302–312. https://doi.org/10.1016/j.neucom.2017.04.053 http://www.sciencedirect.com/science/article/pii/S092523121730807X
18. Mirjalili S, Lewis A (2016) The whale optimization algorithm. Adv Eng Software 95:51–67
19. Mirjalili S, Lewis A (2016) The whale optimization algorithm. Advances in Engineering Software 95:51–67. https://doi.org/10.1016/j.advengsoft.2016.01.008 http://www.sciencedirect.com/science/article/pii/S0965997816300163
20. Panda N, Majhi SK (2019) Improved salp swarm algorithm with space transformation search for training neural network. Arab J Sci Eng, pp. 1–19
21. Peng B, Li L (2015) An improved localization algorithm based on genetic algorithm in wireless sensor networks. Cogn Neurodyn 9(2):249–256. https://doi.org/10.1007/s11571-014-9324-y Apr
22. Q, F, Z, C, Z, X (2020) A novel quasi-reflected harris hawks optimization algorithm for global optimization problems. Soft Computing. https://doi.org/10.1007/s00500-020-04834-7
23. Rahnamayan S, Tizhoosh HR, Salama MMA (2007) Quasi-oppositional differential evolution. In: 2007 IEEE Congress on Evolutionary Computation, pp. 2229–2236. 10.1109/CEC.2007.4424748

24. Sambo DW, Yenke BO, Förster A, Dayang P (2019) Optimized clustering algorithms for large wireless sensor networks: a review. Sensors **19**(2). 10.3390/s19020322
25. Strumberger I, Tuba E, Bacanin N, Beko M, Tuba M (2018, May) Hybridized moth search algorithm for constrained optimization problems. In: 2018 International Young Engineers Forum (YEF-ECE), pp. 1–5. 10.1109/YEF-ECE.2018.8368930
26. Strumberger I, Tuba E, Bacanin N, Beko M, Tuba M (2018, April) Monarch butterfly optimization algorithm for localization in wireless sensor networks. In: 2018 28th International Conference Radioelektronika (RADIOELEKTRONIKA), pp. 1–6. 10.1109/RADIOELEK.2018.8376387
27. Strumberger I, Bacanin N, Beko M, Tomic S, Tuba M (2017, November) Static drone placement by elephant herding optimization algorithm. In: Proceedings of the 24th Telecommunications Forum (TELFOR). 10.1109/TELFOR.2017.8249469
28. Strumberger I, Bacanin N, Tuba M (2017, June) Enhanced firefly algorithm for constrained numerical optimization, ieee congress on evolutionary computation. In: Proceedings of the IEEE International Congress on Evolutionary Computation (CEC 2017), pp. 2120–2127
29. Strumberger I, Beko M, Tuba M, Minovic M, Bacanin N (2018) Elephant herding optimization algorithm for wireless sensor network localization problem. In: Camarinha-Matos LM, Adu-Kankam KO, Julashokri M (eds) Technol Innov Resil Syst. Springer International Publishing, Cham, pp 175–184
30. Strumberger I, Minovic M, Tuba M, Bacanin N (2019) Performance of elephant herding optimization and tree growth algorithm adapted for node localization in wireless sensor networks. Sensors 19(11):2515. https://doi.org/10.3390/s19112515
31. Strumberger I, Tuba E, Zivkovic M, Bacanin N, Beko M, Tuba M (2019) Dynamic search tree growth algorithm for global optimization. In: Camarinha-Matos LM, Almeida R, Oliveira J (eds) Technological Innovation for Industry and Service Systems. Springer International Publishing, Cham, pp 143–153
32. Strumberger I, Tuba M, Bacanin N, Tuba E (2019) Cloudlet scheduling by hybridized monarch butterfly optimization algorithm. J Sens Act Netw 8(3):44. https://doi.org/10.3390/jsan8030044
33. Tuba E, Tuba M, Beko M (2018) Two stage wireless sensor node localization using firefly algorithm. In: Yang XS, Nagar AK, Joshi A (eds) Smart Trends in Systems, Security and Sustainability. Springer Singapore, Singapore, pp 113–120
34. Tuba M, Alihodzic A, Bacanin N (2015) Cuckoo Search and Bat Algorithm Applied to Training Feed-Forward Neural Networks, pp. 139–162. Springer International Publishing, Cham. 10.1007/978-3-319-13826-8_8
35. Wang GG (2016) Moth search algorithm: a bio-inspired metaheuristic algorithm for global optimization problems. Meme Comput. https://doi.org/10.1007/s12293-016-0212-3 Sep
36. Wang GG, Deb S, Cui Z (2015, MAy) Monarch butterfly optimization. Neural Computing and Applications, pp. 1–20. 10.1007/s00521-015-1923-y
37. Wang GG, Deb S, dos S. Coelho L (2015, December) Elephant herding optimization. In: Proceedings of the 2015 3rd International Symposium on Computational and Business Intelligence (ISCBI), pp. 1–5
38. Wang J, Gao Y, Liu W, Sangaiah AK, Kim HJ (2019) An improved routing schema with special clustering using pso algorithm for heterogeneous wireless sensor network. Sensors 19(3). 10.3390/s19030671
39. Yang XS (2009) Firefly algorithms for multimodal optimization. In: Watanabe O, Zeugmann T (eds) Stochastic Algorithms: Foundations and Applications. Springer, Berlin Heidelberg, Berlin, Heidelberg, pp 169–178
40. Yang XS, Hossein Gandomi A (2012) Bat algorithm: a novel approach for global engineering optimization. Eng Comput 29(5):464–483
41. Yourim Y, Yong-Hyuk K (2013) An efficient genetic algorithm for maximum coverage deployment in wireless sensor networks. IEEE Trans Cybern 43(5):1473–1483
42. Zhan ZH, Zhang GY, Ying-Lin, Gong YJ, Zhang J (2014) Load balance aware genetic algorithm for task scheduling in cloud computing. In: Dick G, Browne WN, Whigham P, Zhang M, Bui

LT, Ishibuchi H, Jin Y, Li X, Shi Y, Singh P, Tan KC, Tang K (eds.) Simulated Evolution and Learning, pp. 644–655. Springer International Publishing, Cham

43. Zivkovic M, Bacanin N, Venkatachalam K, Nayyar A, Djordjevic A, Strumberger I, Al-Turjman F (2021) Covid-19 cases prediction by using hybrid machine learning and beetle antennae search approach. Sustainable Cities and Society 66. https://doi.org/10.1016/j.scs.2020.102669 http://www.sciencedirect.com/science/article/pii/S2210670720308842
44. Zivkovic M, Bacanin N, Zivkovic T, Strumberger I, Tuba E, Tuba M (2020) Enhanced grey wolf algorithm for energy efficient wireless sensor networks. In: 2020 Zooming Innovation in Consumer Technologies Conference (ZINC), pp. 87–92. IEEE
45. Zivkovic M, Zivkovic T, Venkatachalam K, Bacanin N (2021) Enhanced dragonfly algorithm adapted for wireless sensor network lifetime optimization. In: Data Intelligence and Cognitive Informatics, pp. 803–817. Springer

Chapter 63
An Application of OMFO for Optimal Bidding Strategy in Pay-as-Bid Auction Environment

Pooja Jain and Akash Saxena

Nomenclature	
MUT_p	Generating Company-GC's minimum up-time of pth block
MDT_p	Generating Company-GC's minimum down-time of pth block
$\alpha_p^{on}(t)$	Generating Company-GC's 'ON' time for particular completion hour
$\alpha_p^{off}(t)$	Generating Company-GC's 'OFF' time for particular completion hour
M	Total no. of Competitors
T_{off}	Total no. of 'OFF' hours at the starting time
C_h	Hot-Startup Price
C_c	Cold-Startup Price
T_c	Steady State to assess cooling time
$C_p(t)$	Generating Company-GC's working expense for p^{th} block
C_p^u	Startup cost function as far as nonlinear remarkable capacity
C_p^d	Constant shut down price
q_p^k	K^{th} Rival's p^{th} block bid price
$C_p^{pc}(t)$	Production cost in terms of non-differentiable, non-convex function
$\overline{P_p}$	Generating Company-GC's maximum restriction of yield of pth block
$\underline{P_p}$	Generating Company-GC's minimum restriction of yield of pth block
$\chi_p(t)$	Paired Flag which is equivalent to 1 when unit is dispatching power or, in all likelihood it is zero
a,b & c	Coefficients for cost assessment
d,e	Constants for incorporation of Valve Point stacking impact

P. Jain (✉) · A. Saxena
SKIT M& G, Jaipur, Rajasthan, India

G. Mathur et al. (eds.), *Proceedings of 2nd International Conference on Artificial Intelligence: Advances and Applications*, Algorithms for Intelligent Systems,
https://doi.org/10.1007/978-981-16-6332-1_63

1 Introduction

The new era of restructuring has created a new opportunity for electrical engineers in modeling and management of electricity market. Although competition presents in both the supply side and demand side of the market but mainly competition in the supply side has emerged as the driving force for generating companies to become more efficient and use latest generation technologies. The various auction environments are used in electricity markets as practical approaches. Any sale climate is totally founded on three basic laws like, offering, clearing, and estimating laws. The offering law decides the quality and measure of given offers by different market members. The clearing law determines the way toward settling the market, picks the victors, and tells the measure of exchanged force. The evaluating law gives us the insight concerning at what value the produced force will be sold. The power market sell off conditions are fixed offered closeout climate, in which bidders present their offers to the power market with the goal that they don't know about one another offers cost. The main disadvantage with the sealed-bid auction is known as winner's curse" [1], i.e., winners do not have any idea about rival participant's revenue.

There are major two types of auction environments used in power markets, (i) UPA(Uniform price Auction), (ii) PABA (Pay-as-Bid auction) [2]. In UPA, all of the winners would be paid at an MCP, i.e., market clearing price, irrespective of what they had bid but in a PABA auction, each customer will be rewarded its bid price, and each supplier will be paid its offered price [3].

In this paper, PABA environment is used as a Pricing mechanism which is used to calculate profit and revenue obtained by a generating company. In a restructured electricity market, every generating company wishes to maximize its own profit. This issue of amplifying benefit is an enhancement issue which is known as Strategic Bidding(SB) problem. This SB issue is tackled through different strategies in writing. A literature review of these methods are presented in [4]. As this problem is stochastic in behavior, some stochastic approaches already exist in literature [5–13].

To maximize generating company's profit, pool market structure is used. All generating companies submit their bids which consist amount of power dispatch and their bid price. These bids are arranged in ascending order and intersection point of this supply and demand curve is obtained and this point is known as MCP. In PABA environment all the bids which are lower than MCP are chosen bids and all suppliers get paid their offered bids.

To simulate this problem, an OMFO [11], a recently proposed variant of MFO [14] algorithm which is based on opposition concept is used to solve this optimal bidding problem in PABA environment. The rest of the paper is coordinated as follows: Section 2 comprises Problem formulation. The proposed approach introduced in Sect. 3. Section 4 portrays every one of the mathematical and graphical outcomes got through improvement routine. The last section, i.e., Sect. 5 summarizes the paper.

2 Problem Formation

Let us expect that the offered costs are now known from the recorded information. Offer costs of all adversaries are assessed by a typical pdf function which is composed according to Eq. (1),

$$pdf({\beta_p}^k) = \frac{1}{\sqrt{2\pi}{\sigma_p}^k} \exp\left(-\frac{({\beta_p}^k - {\mu_p}^k)^2}{2({\sigma_p}^k)^2}\right) \tag{1}$$

Here ${\beta_p}^k$ is contender offer cost of k_{th} contender for p_{th}block. ${\sigma_p}^k$ and ${\mu_p}^k$ Std.Dev and normal of offer cost of k_{th}rival for p_{th}block. The working expense capacity of Generating Company-GC, $C_p(t)$ is indicated as follows,

$$C_p(t) = {C_p}^{pc}(t) + {C_p}^u(\chi_p(t)(1 - \chi_p(t-1)) + {C_p}^d\{(1 - \chi_p(t))\chi_p(t-1)\} \tag{2}$$

where

$${C_p}^{pc}(t) = a(P_p(t))^2 + bP_k(c) + d + e\left|\sin(w(\overline{P}_p - P_p(t)))\right| \tag{3}$$

$${C_p}^u = C_h + C_c\left(1 - \exp\left(-\frac{T_{off}}{T_c}\right)\right) \tag{4}$$

The target function for ideal offering issue is outlined according to given condition :

$$MaximizeRev = \sum_{t=1}^{H}\sum_{p=1}^{K} q_p * P_p(t) \tag{5}$$

Subject to limitations::

(1) Upper and Lower Power Limits :

$$\underline{P}_p(t)\chi_p(t) \le P_p(t) \le \overline{P}_p(t)\chi_p(t) \qquad \forall t \in H \tag{6}$$

(2) MUT and MDT :

$$(1 - \chi_p(t+1))MUT_p \le {\alpha_p}^{on}(t) \qquad if\,\chi_p(t) = 1 \tag{7}$$

$$\chi_p(t+1)MDT_p \le {\alpha_p}^{off}(t) \qquad if\,\chi_p(t) = 0 \tag{8}$$

(3) Bid Price Limitation

$$C_p(t) \le q_p(t) \le q_{\max} \qquad \forall t \in H \tag{9}$$

The issue explanation written in Eq. (5) is streamlined by utilizing proposed technique in next sections by considering every one of the limitations given in eq. nos.(6-9) to get the ideal answer for income procured by Generating Company-GC.

3 Proposed Methodology

To solve this strategic bidding problem, an improved version of recently proposed algorithm MFO [14] is used. This improved version is based on opposition concept and named as OMFO [11]. A few disadvantages are available in parent MFO like getting caught in nearby optima, moderate intermingling, and high computational time. Therefore, the OMFO is utilized to stay away from these intricacies by adding oppositional highlight. All the details about MFO and OMFO are taken from [11].

4 Simulation Results and Discussion

As OMFO [11] have fantastic convergence, exploitation, and investigation capability, the appropriateness of OMFO is tried on two distinctive test framework, i.e., IEEE-14 bus test framework for three diverse load levels, i.e., 1200 MW, 1700 MW, and 2200 MW and on 7-Generator framework for three distinctive cost models and three distinctive load levels, i.e., for 3000 MW, 4500 MW, and 6500 MW which depend on problem framed in Sect. 2. The enhancement routine is reenacted in PABA climate for singular hour exchanging. The complete data of generator cost characteristics and rivals' behviour data are taken from [11].

Every one of the reenactments are completed on MATLAB 2017, i5 processor CPU, 2.00GHz, and with 8 GB RAM. To demonstrate adequacy of OMFO in PABA sell off environment, the reasonable examinations of results is finished for certain standard calculations to be specific Moth Flame Optimizer (MFO [14]), Ant lion Optimizer (ALO [15]), Sine-Cosine Algorithm (SCA [16]), and Particle Swarm Optimization (PSO [17]). Tables 1 and 2 shows the similar examination of benefit got through OMFO and other standard stochastic improvement approaches referenced

Table 1 Profit analysis for IEEE-14 bus test system

System demand (S.D)	Profit obtained ($)				
	OB-MFO	MFO	ALO	SCA	PSO
S.D=1200 MW	$849.79	$763.78	$701.36	$659.45	$612.74
S.D=1700 MW	$2,978.12	$2,541.65	$2174.96	$2017.26	$1978..23
S.D=2200 MW	$4,089.74	$3,845.59	$3752.31	$3412.78	$3129.89

Table 2 Profit analysis for 7-generator test system

System demand	Cost models	Profit obtained				
		OB-MFO	MFO	ALO	SCA	PSO
S.D=3000 MW	Cost Model-I	$1,907.56	$1,893.78	$1703.78	$1653.78	$1547.96
	Cost Model-II	$2,246.64	$1,945.67	$1896.75	$1812.19	$1763.14
	Cost Model-III	$2,452.78	$2,165.94	$2014.65	$1963.78	$1889.49
S.D=4500 MW	Cost Model-I	$8,432.17	$8,295.74	$8157.69	$8017.42	$7945.19
	Cost Model-II	$6,145.34	$6,035.45	$5947.58	$5846.28	$5649.78
	Cost Model-III	$8,945.13	$8,697.42	$8547.69	$8469.74	$8317.45
S.D=6500 MW	Cost Model-I	$17,456.48	$16,632.43	$16023.98	$15945.12	$14456.49
	Cost Model-II	$15,473.59	$14,954.37	$14063.95	$13253.47	$12369.17
	Cost Model-III	$16,554.96	$15,971.28	$14965.25	$14127.92	$12978.98

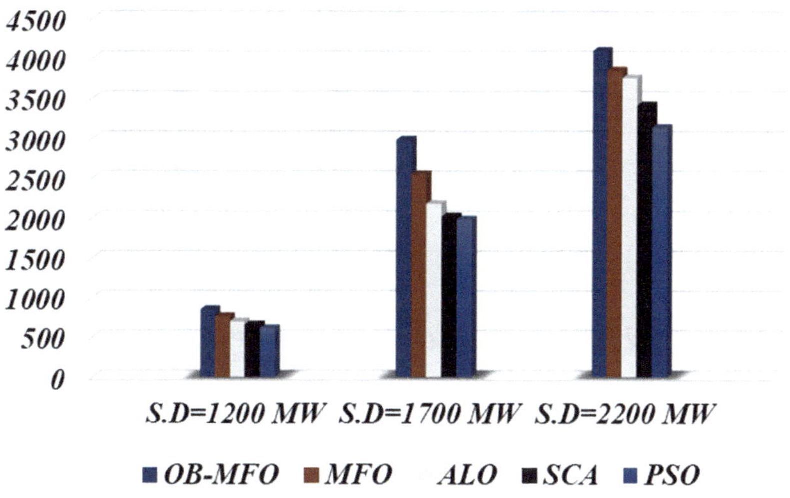

Fig. 1 Profit visualization for IEEE-14 test system for various demand levels

above. Also the structured presentations appeared in Fig. 1 address the graphical investigation of benefit acquired for 1200 MW, 1700 MW, and 2200 MW load levels of IEEE 14 bus test framework, respectively. Figures 2, 3 and 4 addresses the percep-

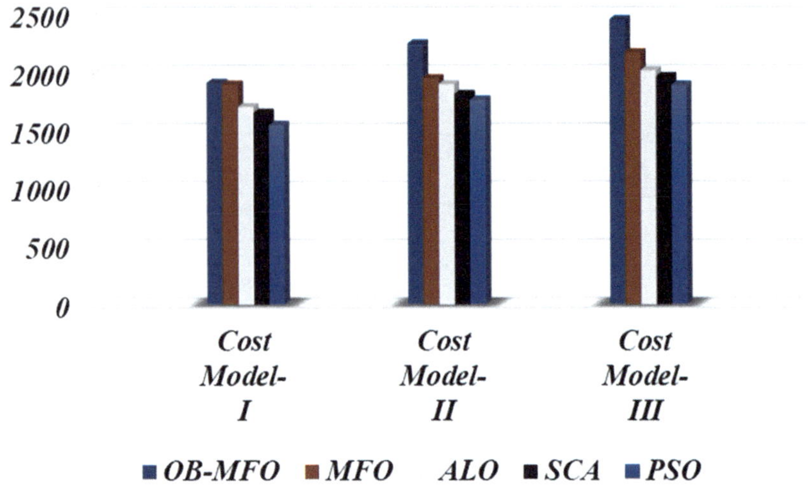

Fig. 2 Profit visualization for 7-generator system for different cost models for demand level of 3000 MW

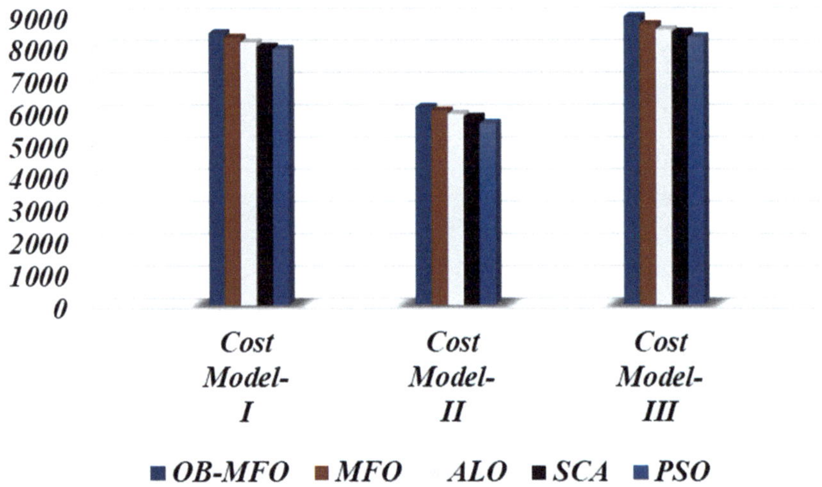

Fig. 3 Profit visualization for 7-generator system for different cost models for demand level of 4500 MW

tion of benefit acquired by Generating Company-GC for three distinctive cost models for various load levels for example 3000 MW, 4500 MW, and 6500 MW individually. The outcomes affirm the adequacy of OMFO over other stochastic advancement approaches.

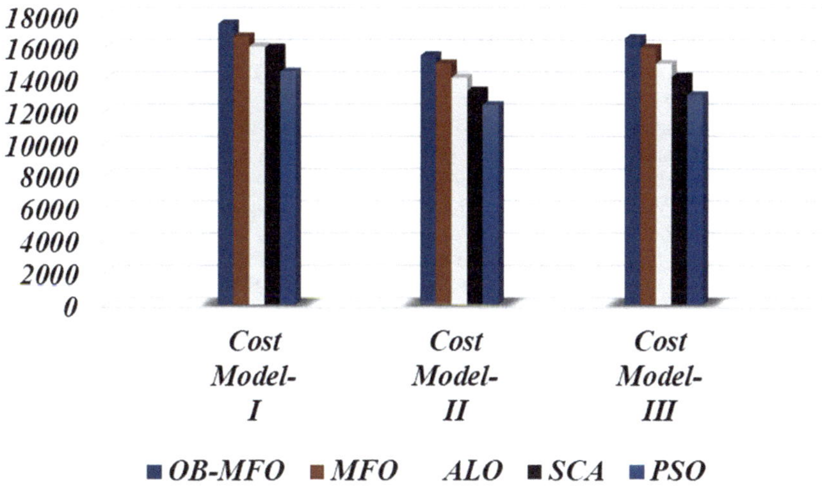

Fig. 4 Profit visualization for 7-generator system for different cost models for demand level of 6500 MW

5 Conclusion

To acquire most extreme benefit for generating company, this paper has introduced an ideal offering technique in PABA climate which is depends on OMFO for Generating Company-GC. The significant end drawn from above reenactment results are as per the following:-

1. By rival's conduct in the power market, most elevated benefit making offer costs of Generating Company-GC are acquired with the OMFO.
2. Different test frameworks and different load levels for singular exchanging hour have been considered to demonstrate the adequacy of the OMFO.

By examining the above expressed mathematical and graphical results, it can be reasoned that the OMFO's relevance improves the benefit got by Generating organization in PABA climate.

References

1. Maurer L, Barroso L (2011) Electricity auctions: an overview of efficient practices. The World Bank
2. Sugianto LF, Liao KZ (2014) Comparison of different auction pricing rules in the electricity market. Mod Appl Sci 8(1):147
3. Morey MJ et al (2001) Power market auction design: Rules and lessons in market based control for the new electricity industry. Prepared for Edison Electric Institute

4. Wen F, David AK (2001) Optimal bidding strategies and modeling of imperfect information among competitive generators. IEEE Transac Power Syst 16(1):15–21
5. Wen F, David A (2001) Strategic bidding for electricity supply in a day-ahead energy market. Electric Power Syst Res 59(3):197–206
6. Azadeh A, Ghaderi SF, Nokhandan BP, Sheikhalishahi M (2012) A new genetic algorithm approach for optimizing bidding strategy viewpoint of profit maximization of a generation company. Expert Syst Appl 39(1):1565–1574
7. Yucekaya AD, Valenzuela J, Dozier G (2009) Strategic bidding in electricity markets using particle swarm optimization. Electric Power Syst Res 79(2):335–345
8. Bajpai P, Singh S (2007) Fuzzy adaptive particle swarm optimization for bidding strategy in uniform price spot market. IEEE Trans Power Syst 22(4):2152–2160
9. Kumar JV, Kumar DV, Edukondalu K (2013) Strategic bidding using fuzzy adaptive gravitational search algorithm in a pool based electricity market. Appl Soft Comput 13(5):2445–2455
10. Shivaie M, Ameli MT (2015) An environmental/techno-economic approach for bidding strategy in security-constrained electricity markets by a bi-level harmony search algorithm. Renew Energy 83:881–896
11. Jain P, Saxena A. An opposition theory enabled moth flame optimizer for strategic bidding in uniform spot energy market. Eng Sci Technol Intern J 22:1047–1067
12. Saxena A, Soni BP, Kumar R, Gupta V (2018) Intelligent grey wolf optimizer-development and application for strategic bidding in uniform price spot energy market. Applied Soft Comput
13. Jain P, Saxena A (2020) A new redefined model of firefly algorithm with application to strategic bidding problem in power sector. Int Transac Elect Energy Syst 30(4):e12279
14. Mirjalili S (2015) Moth-flame optimization algorithm: A novel nature-inspired heuristic paradigm. Knowl Based Syst 89:228–249
15. Mirjalili S (2015) The ant lion optimizer. Adv Eng Soft 83:80–98
16. Mirjalili S (2016) Sca: a sine cosine algorithm for solving optimization problems. Knowl Based Syst 96:120–133
17. Eberhart R, Kennedy J (1995) A new optimizer using particle swarm theory. In: Micro Machine and Human Science, 1995. MHS'95., Proceedings of the Sixth International Symposium on. pp. 39–43. IEEE

Chapter 64
On-Device ML: An Efficient Approach to Classify Large Number of Images Using Multi-threading in Android Devices

Saurabh Kothari, Rayan Crasta, Alen Biju, Trupti Lotlikar, and Harshit Rai

1 Introduction

The capability of running machine learning models on Mobile devices has numerous benefits. The cloud-based approach was the industry standard until recently where the data was aggregated from many users, processed, and then sent to the cloud to train the machine learning model. The problem with cloud-based training is that it is slow because additional time is needed to transmit data to and fro the cloud. It also costs time and money while risking data privacy. Ensuring the privacy and integrity of the application's data is a crucial part [1]. In cloud-based inference methods, once the data leaves the user device, they no longer have control over the sensitive data which can be re-purposed without the user's consent. On-device machine learning can bring about many revolutionary changes, which can take advantage of powerful functions, such as real-time diagnosis and detection of plant diseases [2].

On-device machine learning is much more reliable as it does not depend on factors like internet availability and transfer speeds. With on-device machine learning, the neural networks reside in the mobile devices themselves, ensuring that they are always available and reliable regardless of connectivity. This offers a personalized experience for the users as the data stays on the device [3].

As mobile CPU is getting better and more powerful and with the introduction of GPU's in mobile devices, on-device machine learning is the future for performing complex tasks on mobile devices. When dealing with thousands of images, it is difficult to transfer such a huge amount of data to the cloud. Thus, machine learning proves to be very cost-efficient as well.

When we consider applying the machine learning algorithms on thousands of images, there is a high chance of the android terminating the app's process, and hence crashing the app [4]. By default, all components of the same application run

S. Kothari (✉) · R. Crasta · A. Biju · T. Lotlikar · H. Rai
Department of Information Technology, Fr. C. Rodrigues Institute of Technology, Navi Mumbai 400703, India

G. Mathur et al. (eds.), *Proceedings of 2nd International Conference on Artificial Intelligence: Advances and Applications*, Algorithms for Intelligent Systems,
https://doi.org/10.1007/978-981-16-6332-1_64

in the same process called main thread, thus the app cannot perform any other tasks simultaneously if it is already doing some task such as performing inference of images. The multi-threaded approach solves this problem. Using a multi-core CPU on a mobile device, we can utilize different threads to perform inference as well as other operations simultaneously.

2 Requirements

The solution is tested on an Android device. The android device should have a multi-core CPU to support multi-threading. The presence of a GPU makes the process much faster and is an added advantage [5]. The TFLite library is built against NDK API 19. It may work for Android API levels below 19 but is not guaranteed. TensorFlow lite libraries should be included in Gradle dependencies.

3 Proposed Approach

3.1 Need For Multi-threading

Reference [4] When an application launches, android creates a new Linux process along with an execution thread known as the main thread. This main thread is responsible for rendering UI and handling events like callbacks associated with the life-cycle. If any other block of work needs to be run, it is first pushed into a work queue and then serviced by the main thread [6]. If the main thread cannot finish executing blocks of work within 16 ms, the user may observe hitching, lagging, or a lack of UI responsiveness to input. If the main thread is continuously consumed by larger work, the system generates an 'Application Not Responding' or an 'ANR' dialog indicating the app is unresponsive [7]. Generally, the system displays an ANR dialog if an application cannot respond to user input. For example, if an application blocks some I/O operation on the main thread so the system can't process incoming user input events. Or perhaps the app spends too much time computing something (converting an image URI to bitmap format) on the main thread.

This paper aims to propose an efficient way to classify large numbers of images in one go. Sequential image classification is a very CPU intensive task and mobile CPUs have limited computing power. If this inference is done on the main thread, after a few seconds the app will likely crash and raise Application Not responding dialog because the application is doing a lot of work on the main thread such as fetching the image path via content providers, converting an image to bitmap format, preprocessing this bitmap image, and finally running inference to classify the image. Hence, as the images are processed one by one, the main thread is getting consumed continuously leading to unresponsive UI and android terminates the app by generating

ANR dialog. The size of the image is also one of the factors to take into consideration. If the image size is large, then it will take more time to inference, while a small-sized image will take less time. As the main thread does so much work, it's better to create new threads for longer work.

3.2 Producer–Consumer Approach

To solve the problem of classifying large numbers of images in one go, we propose a multi-threaded producer–consumer approach with blocking queue implementation. Two threads, namely, producer and consumer are created to run separate from the main thread and also distribute the overall workload required to classify an image [8]. The producer–consumer design separates the identification of work from the execution of work. In this proposed producer–consumer design, a shared blocking queue is used to control the flow [9]. Blocking Queue is part of the Java concurrent util package and is best suited for this application since it's thread-safe and the producer and consumer thread can work at different speeds. The Blocking Queue works as follows: If the shared queue is full and the producer thread tries to enqueue an element into the shared queue, Blocking Queue will invoke and block the producer thread until the consumer thread dequeues an element from the shared queue. Similarly, if the shared queue is empty and the consumer thread tries to dequeue an element, Blocking Queue will block the consumer thread until the producer thread enqueues an element into the shared queue.

The work of the producer thread is to fetch image paths via content providers (image acquisition) and insert its output into a shared queue. The work of the consumer thread is to consume those images from the shared queue and convert them into bitmap format and feed it into the classifier for inference. Multiple consumer threads can be implemented to further increase the throughput and reduce inference time for large numbers of images. Both the threads are implemented via the Runnable interface along with a Handler. Since the producer and consumer are running on separate threads, it reduces the load on the main thread of the app and allows parallel processing. This leaves the main thread to freely render UI, hence users can interact with the screen even while the classification of large numbers of images is going on. This prevents the app from crashing and distributes the load on the CPU (Fig. 1).

3.3 Image Acquisition

To get all the images from the android file system, content providers are used. A content provider manages access to the central data repository and provides the basic "CRUD" (create, retrieve, update, and delete) functions [10]. To access data in a content provider, we use the ContentResolver object in our application's Context to communicate with the provider as a client.

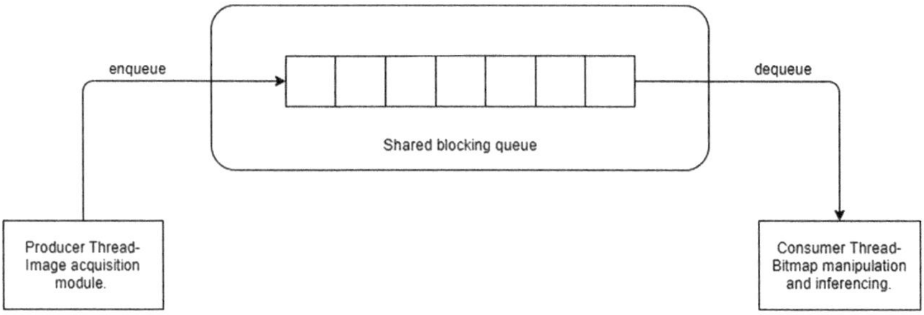

Fig. 1 Producer–consumer approach with Blocking Queue implementation

ContentResolver.query() method is used to get a list of the words and their locales from the User Dictionary Provider. The query() method then calls the ContentProvider.query() method which queries the given URI, returning a cursor over the result set [11]. Cursor provides random read-write access to the result set returned by a database query. This cursor object is used to parse through each entry of the database table (containing file paths) and save only those file paths that are images.

3.4 Running Inference on Thread

Reference [12] TensorFlow Lite is used as the primary tool to run machine learning models on Android devices. TensorFlow models can be converted into mobile compatible TensorFlow lite models using the TensorFlow lite converter provided by Google [13]. The TensorFlow lite interpreter is used to perform inference to make predictions based on the input image. The first step is to load the .tflite model into the app and create an instance of the tflite interpreter along with the GPU delegate. The image acquisition module is used to fetch the input image path and is run on the producer thread. The fetched images are enqueued into the shared blocking queue. The images are then dequeued by the consumer thread and are converted into bitmap format one by one. The bitmap image is then preprocessed and reshaped according to the model's requirements. This reshaped bitmap is then given to the interpreter. Bitmap conversion and preprocessing along with inference are done on the consumer thread. After the inference is done, the result is sent to the UI thread or the main thread for display purposes.

4 Experimental Results

To test the proposed approach, we built two android applications. App 'B' implements multi-threading using a blocking queue while the second app 'A' runs on the main thread without multi-threading. A CNN model was trained using TensorFlow and Keras. The model is a binary classifier. This TensorFlow model was converted into TensorFlow Lite (tflite) model and was integrated with both applications. A folder picker was implemented to select a particular folder containing images. The images were scanned from the selected folders and the results were noted. The average size of an image was between 50 KB and 10 MB.

4.1 Model Description

For the purpose of experimentation, a simple CNN model with two classes is implemented. The accuracy or the execution time of the model are not taken into consideration as our aim is to test the multi-threaded approach to sequentially classify a large number of images without throttling the mobile CPU. We measure the performance/success of our approach based on android app B's ability to classify all the images given to it as input while maintaining same (or less) execution time per image as app 'A'. The model's input shape is [1 160 160 3]. To save training time and maintain healthy accuracy, transfer learning technique is used. The MobileNet V2 is selected as the base model. It is pre-trained on the ImageNet dataset. The general idea is to use the previously learned knowledge of MobileNet V2 for our new task. After adding the base model, a Global Average Pooling 2D layer is added to convert the features to a single 1280-element vector per image. Next, a tf.keras.layers.Dense layer with 128 ReLU activation units is added. The last layer is tf.keras.layers.Dense layer with sigmoid activation function to convert these features into a single prediction per image. The model is compiled and trained using Adam optimizer. The output shape is [1 1]. After the model is trained, it is saved (.pb) and converted into a TensorFlow Lite model (.tflite) using a TensorFlow lite converter.

4.2 Dataset

The dataset consists of a total of 5,000 images divided between two classes. The data set is split into three parts: training dataset, testing dataset and validation dataset. The training set is used to train the model and contains 1,700 images per class. The Validation set is used to validate the model after training and it contains 500 images per class. The images are preprocessed and compressed to [1 160 160 3] before feeding to the model for training.

Table 1 Multi-threaded approach versus non multi-threaded approach

Sr no.	No. of images	App 'A'—Inference time without multi-threading (s)	App 'B'—Inference time with multi-threading (s)
1	250	23.1	22.24
2	500	37.8	37.3
3	1000	App crashed after 64 s	82.37
4	2000	App crashed	196.12

4.3 Results

The results of a comparative test between the app performing binary classification with multi-threading (app 'B') and without multi-threading (app 'A') are as follows. Four folders each containing 250, 500, 1000, and 2000 images, respectively, are prepared for the test. These are inputs to our classifier. Both the apps are given same input. The app 'A' built using our approach should classify any number of images without throttling the system (crashing the app). The mobile hardware used had the following specifications:
Chipset: Mediatek Helio P70 (12 nm)
CPU: Octa-core (4 $\times$ 2.1 GHz Cortex-A73 and 4 $\times$ 2.0 GHz Cortex-A53)
GPU: Mali-G72 MP3

The app 'A' crashes when given a large number of images as input, while app 'B' completes inference of all the images given to it. The above results prove that app 'B' built using our approach performs better(without throttling the hardware) than app 'A' built traditionally without multi-threading (Table 1).

5 Conclusion

We proposed a new approach based on multi-threading and producer-consumer paradigm to handle a large number of images for on-device classification. We described the thread structure, its need, and demonstrated how to run the inference on threads to achieve a seamless user experience without the app crashing. We built an android application using the above approach and were able to verify the claims made in this paper. This approach can be used to build applications like explicit content moderation that can search the entire Android device containing thousands of images and if detected, delete any explicit content. From the above research and studies, we can conclude that using on-device machine learning along with an approach that takes advantage of the Android operating System's multi-threading capabilities can be very useful for developing applications that require a large continuous computational task such as classifying a large number of images in one go.

References

1. Mather T, Kumaraswamy S, Latif S (2009) Cloud security and privacy: an enterprise perspective on risks and compliance. Published by O'Reilly Media, Inc.
2. Li L-H, Chu Y-S, Chu J-Y, Guo S-H (2019) A machine learning approach for detection plant disease: taking orchid as example. In: Proceedings of the 3rd international conference on vision, image and signal processing. Association for computing machinery, New York, NY, USA, Article 43, pp 1–6. https://doi.org/10.1145/3387168.3387238
3. Example on-device model personalization with TensorFlow Lite. https://blog.tensorflow.org/2019/12/example-on-device-model-personalization.html. Accessed December 2020
4. Processes and threads overview. https://developer.android.com/guide/components/processes-and-threads. Accessed December 2020
5. Interpreter TensorFlow Lite. https://www.tensorflow.org/lite/api_docs/java/org/tensorflow/lite/Interpreter. Accessed December 2020
6. Better performance through threading Android Developers. https://developer.android.com/topic/performance/threads. Accessed 06 December 2020
7. Keeping your app responsive. https://developer.android.com/training/articles/perf-anr. Accessed December 2020
8. Producer Consumer Design Pattern with Blocking Queue Example in Java (nd). https://javarevisited.blogspot.com/2012/02/producer-consumer-design-pattern-with.html. Accessed 04 December 2020
9. Interface BlockingQueue. https://docs.oracle.com/en/java/javase/11/docs/api/java.base/java/util/concurrent/BlockingQueue.html. Accessed December 2020
10. Content provider basics Android Developers . https://developer.android.com/guide/topics/providers/content-provider-basics. Accessed December 2020
11. Cursor Android Developers. https://developer.android.com/reference/android/database/Cursor. Accessed December 2020
12. TensorFlow Lite guide. https://www.tensorflow.org/lite/guide. Accessed December 2020
13. TensorFlow Lite inference. https://www.tensorflow.org/lite/guide/inference. Accessed December 2020

Chapter 65
A Conceptual Framework Based on Conversational Agents for the Early Detection of Cognitive Impairment

Moisés R. Pacheco-Lorenzo, Sonia Valladares-Rodríguez, Luis Anido-Rifón, and Manuel J. Fernández-Iglesias

1 Introduction

The majority of First-World countries are experimenting a process of population aging that all indicators suggest will last during the near future. This situation poses new social and economical challenges, which essentially revolve around facilitating an active and healthy lifestyle of senior citizens.

There are many diseases that are more common in the elderly, and their frequency increases with higher life expectancy. These include disorders that affect the nervous systems—for example, prevalence of dementia increases with age, affecting nearly 50% of the population over 85 years of age. Alzheimer's Disease (AD) is the most common form of dementia in older adults [22]. At the same time, there are other medical conditions that are less frequent among older adults, but have devastating consequences when they take place later in life, which raises the issue in terms of public health importance. One of such disorders is depression, as it is associated with increased risk of morbidity, increased risk of suicide, decreased physical, cognitive, and social functioning, and greater self-neglect [12].

In this context, information technologies became facilitators of a numerous range of services, devices, and applications aimed at highly improving people's lives. Among the latest advances in this field, the appearance of the so-called intelligent or smart solutions, paved the way for the provision of personalized services adapted to the needs and expectations of each individual. Moreover, these smart solutions have the capability of exchanging information with other devices or applications,

M. R. Pacheco-Lorenzo (✉) · S. Valladares-Rodríguez · L. Anido-Rifón ·
M. J. Fernández-Iglesias
AtlanTTic, Universidade de Vigo, Escola de Enxeñría de Telecomunicación, Campus Lagoas-Marcosende, Vigo, Spain
e-mail: moises.pacheco@gist.uvigo.es

G. Mathur et al. (eds.), *Proceedings of 2nd International Conference on Artificial Intelligence: Advances and Applications*, Algorithms for Intelligent Systems,
https://doi.org/10.1007/978-981-16-6332-1_65

interacting with each other and the environment through communication networks in an autonomous way to generate new knowledge or to adapt to evolving conditions.

The link between healthcare and technology dates back to 1999, when the concept of eHealth was introduced as technologically assisted medicine [8]. After more than two decades of this initial proposal, applications of Information and Communication Technologies (ICT) in health care are wide and diverse, including the gathering and storage of medical and clinical data [35]; the remote distribution of health-related services [1]; or the assistance to decision-making and diagnosis in clinical settings [38].

Regarding the third application field just mentioned, Artificial Intelligence (AI) can be identified as the key enabling technology to facilitate decision-making in eHealth. AI was defined in 1956 by McCarthy [26] as the "*science and engineering of making intelligent machines, especially intelligent computer programs*".

Thus, one of the basic objectives of AI is to imitate the human way of thinking. For this, Artificial Neural Network (ANN) [14] were developed. ANNs are inspired by the decision process in networks of biological nerve cells, where there is an input being transformed along the links of the network in order to produce an output. As their biological counterparts, ANNs are able to "learn" by themselves by tuning certain parameters, and have numerous applications such as computer vision and speech recognition.

The second key objective of AI is to achieve ways to behave in a rational way by means of intelligent agents. These are capable of processing stimuli from their environment and act consistently in a correct manner [5]. Virtual assistants are capable to behave in a human-like manner and interact with them using natural language, and they represent a valid example of this proposal.

According to a systematic review from 2018 [24], the introduction of conversational agents for health-related purposes was an emerging field of research, where the few published studies were mainly quasi-experimental, and rarely their efficacy or safety were evaluated.

The most popular implementation of smart virtual assistants is based on smart speakers, which became a technology trend in 2019[1] following the trail of chatbots [32], and even replacing them as the preferred solution for human–machine interaction. According to Value Market Research,[2] the global market value by revenue for this type of technology was USD 4.5 k million in 2017, and is expected to reach USD 30 k million by 2024 with a compound annual growth rate of 30%. Therefore, the number of homes with a smart speaker is already high, and is rapidly growing.

On the other hand, the diagnosis of neuropsychiatric disorders is currently performed by medical professionals in a safe and controlled clinical environment. The main diagnosis tools are a collection of standardized neuropsychiatric scales, or "classical tests", such as the Mini-mental State Examination (MMSE) for dementia [13] or the Patient Health Questionnaire - 9 (PHQ-9) for depression [23]. These tests

[1] https://disruptionhub.com/disruption-trends-9-for-2019/.

[2] https://www.valuemarketresearch.com/report/smart-speaker-market.

offer as a result an evaluation of the cognitive or mood-related state of the patient, and require an interpretation by a trained medical practitioner.

The main disadvantage of the present-day diagnosing process is its late detection characteristics [28] both for dementia [16] and depression [34], since in most cases it is when the symptoms become evident that the patient visits the doctor. Therefore, this hinders or even fully prevents the ability to intervene and treat such diseases, as the disease is typically in an advanced state.

In addition, another main inconvenience of the traditional detection mechanism is that tests are perceived as intrusive and unnatural by patients, which in turn makes them feel uncomfortable when taking them [6, 45].

This work is aimed at proposing a conceptual framework for the detection of cognitive impairment in senior adults using smart conversational agents. Taking as a reference standard procedures for the diagnosis of neurophysiological diseases, and by applying Machine Learning (ML) and Deep Learning (DL) techniques to the oral interactions between the agent and a senior individual, the proposed framework will be able to support the identification of signs of cognitive deterioration in a nonintrusive way.

Thus, using AI-based software components, the proposed framework will support the development of applications that could provide additional evidence about the cognitive status of elder adults, help to discriminate between cognitive impairment and other confounding conditions such as depression, and even to adapt the interactions of other applications in the smart home ecosystem promoting an active lifestyle to the changing communication abilities of target users.

Finally, the proposed framework will facilitate the scalability and long-term sustainability of these novel applications, by supporting cloud-based systems' design, off-the-shelf smart devices, and open and interoperable AI-based software components.

The structure of this paper is as follows: Sect. 2 introduces the state of the art in this field. In Sect. 3, we propose a methodology to follow consisting on a set of tasks or milestones to complete. Section 4 presents the high-level design for the conceptual framework devised. Finally, Sect. 5, we discuss the future implications of the proposed solution.

2 Background

In this section, we will explain the previous contributions within this subject found in the literature, and we will summarize the available tools that we have for implementing our solution.

2.1 Review of the Literature

In order to identify existing research proposals for the use of conversational systems for detection of neuropsychiatric disorders, the authors have conducted a systematic review of the literature [31]. The main outcomes from this process are summarized below.

When it comes to the main objective of the works analyzed, the most common scenario is to perform an analysis based on both verbal and audio-visual features, with a conversational agent conducting the dialogue. This is the case with [18, 27, 39, 42, 43]. Another common approach is to employ only audio-visual features, that is how the user sounds and/or looks, but without taking into account what the user says, and therefore without the need for a conversational agent. In this category, we can find [11, 15, 19, 20, 37]. An additional objective found within the reviewed works is studying existing previously developed conversational systems and evaluate their performance in different ways, particularly from their detection capabilities, for instance, in [9, 17, 33].

As for the targeted disorders, many of the articles address depressive disorders [4, 9, 17, 18, 25, 33, 37]. On the other hand, another substantial amount are targeted at neurocognitive disorders. In this broad category, there is a big concern with dementia in general [30, 39, 43], but some of its representations are addressed too. The common Alzheimer's Disease is targeted in [19], and its previous state known as Mild Cognitive Impairment (MCI) is targeted in [27, 40, 42].

We classified the articles by their hardware and software technologies used as well. The most common hardware platform used are mobile devices [11, 17, 20, 42], since they are affordable and easy to use. Software technologies, on the other hand, vary considerably and are not commonly used more than once. Furthermore, there is also a considerable amount of works that either do not use or do not specify which hardware [4, 9, 15, 18, 25, 37, 40] or software [17, 19, 20, 30, 37, 40] technologies they have employed.

As for the data acquisition process, the vast majority of works use automatic feature extraction, both for verbal and audio-visual inputs. Then, they normally apply one or more Machine Learning algorithms to detect disorders on the user's mental state. Support Vector Machine (SVM) is a particularly popular algorithm, used in [15, 19, 30, 37, 39, 40, 43], however more advanced Deep Learning-based models such as Multilayer Perceptron (MLP) are increasingly popular [4, 15, 25, 37, 40, 42].

Finally, when it comes to perform a validation process on the detection system, most of the reviewed works do not perform a rigorous reliability and validity measurement, as recommended by the methodology proposed by the American Psychiatric Association [3]. They generally only carry out a criterion validity measurement, relying on a gold standard test, such as MMSE [20, 39, 42, 43] or PHQ [4, 17, 18, 25] for MCI and depression, respectively. However, only two works [4, 33] perform a reliability analysis, and only two [9, 33] compute any sort of user experience metrics. Furthermore, the majority of the experiments that rely on real users trying their systems only involve 54 participants or less [11, 18–20, 30, 39, 40, 43].

2.2 *Smart Speakers*

Out of the currently available smart speaker products, there are two that stand out as being the most widespread among people's homes: Google Nest and Amazon Echo [21]. Particularly, their versions Google Nest Mini[3] and Amazon Echo Dot[4] are both quite affordable (around $50 each) and include their virtual assistant software: Google Assistant and Amazon Alexa respectively.

In addition, these products also come integrated with their respective cloud services and SDKs integrated with them. This means that any developer is able to create their own applications for each of the platforms, interacting with the provided APIs, and then use them directly from the speaker itself.

For instance, for Amazon's speaker, there are "Alexa Skills" which are voice-activated applications of the smart virtual assistant that provide a general functionality. These can be fully customizable and any of the user's needs that the developer may want to give support to. The launch of a skill is quite straightforward: after "waking up" Alexa, the user can use the skill's invocation name to run it. Furthermore, a particular functionality may be asked for, and generic or custom variables (or "slots") may be defined and provided to the skill. Alexa will also take care of obtaining and validating all the required variables if the developer establishes so, and there can even be a two-step confirmation by the user.

Additionally, there is session-based persistency during the use of the skill, and it can also store information on the S3 AWS platform if the developer defines so. The skill may as well access third party APIs available online in order to obtain any additional information or perform some sort of task.

For Google's smart speaker, the provided functionalities are called "Actions" instead of Skills, but the *modus operandi* is very similar. Google Assistant and Google's cloud services also provide equivalent tools to develop these actions.

One important option that both speakers offer is the possibility of setting reminders for the user, which can be repeated periodically and contain some sort of message. This means that not only the user may start the action or skill but also the application itself can start running at a certain point, initiating the interaction with the user. We believe this to be a key element for our use case, since it can help motivate the daily use of these devices by elderly people—users who are typically reluctant to do so.

---

[3] https://store.google.com/product/google_nest_mini.

[4] https://www.amazon.com/Echo-Dot/dp/B07FZ8S74R.

2.3 Gold Standard Tests

As we mentioned in the introduction, the main diagnosing tool currently available are classical tests, also known as the "gold standard". In the systematic review carried out, we have identified one of them, the Mini-mental State Examination [13]. This is the most widely used classical test for evaluating and detecting cognitive impairment. It examines orientation, word recall, language abilities, attention and calculation, and visuospatial ability. It takes about 8 min to administer which makes it more suitable for initial screening.

Another commonly used classical test is the Montreal Cognitive Assessment (MoCA) [29]. This test takes about 11 min and, like the MMSE, it is also a routine cognitive screening test rated on a 30-point scale. However, it doesn't have such a strong language component as the MMSE does.[5] Consequently, the MoCA has been found to be a more "sensitive" test, meaning that it can better discern between the early stages of dementia such as Mild Cognitive Impairment [41]. Whereas the MMSE is likely a better test for more severe conditions, the MoCA is more sensitive to MCI, and this difference on sensitivity is likely to be the biggest factor for determining which one is used.

At the same time, the California Verbal Learning Test (CVLT) [7] is a commonly used verbal learning and memory test that measures key cognitive constructs such as repetition learning, serial position effects, semantic organization, intrusions, and proactive interference. The current version contains recalling and recognition of two lists of 16 words over immediate and delayed trials. Administration time is about 45 min with an additional 20-min delay.

There is a revised version of this test, the CVLT-II, that adds an alternate form with different word categories (thus preventing artificially inflated scores when re-testing) and a short form with nine word lists [46]. Alternatively, there exists the previous Rey Auditory Verbal Learning Test (RAVLT) which is non-proprietary (unlike the CVLT) and provides a brief, valid, and cost-effective alternative to it for screening verbal learning impairments [36].

Finally, it is worth mentioning the "Test de Aprendizaje Verbal España-Complutense (TAVEC)" [2], which is a Spanish verbal learning test based on both the RAVLT and CVLT-II. It uses three element lists, each of them aimed at learning, inference, and recognition. The first two lists contain 16 words, but the recognition list contains 44 words. It takes up to 60 min to administer and it allows to evaluate 33 different cognitive variables.

---

[5] https://www.todaysgeriatricmedicine.com/news/ex_012511_01.shtml.

3 Methodology

After considering the state of the art, and systematically reviewing the literature on the subject, we were able to identify a research gap. This then becomes our working hypothesis and general objective:

Early detection of cognitive impairment with the use of Conversational Agents and Machine Learning, in a psychometrically valid and ecological way.

In order to demonstrate this hypothesis, we propose the following specific milestones, which are explained in a generalized way:

(i) Select and adapt the classical test or tests to the conversational channel. This can be done in a relatively direct fashion, similarly to what was implemented in [4]. Thus, the idea is to migrate the test's tasks into a voice conversation driven by a virtual assistant, typically embedded in a commercial product such as a smart speaker. Therefore, the test's rating scale output can be directly obtainable from the user's responses.

(ii) Transform the classical test so that it is more ecological to the user. After adapting the test to the conversational platform, we need to overcome one of the biggest drawbacks of classical tests, which is their intrusiveness. This means the test should feel not forced on the user—ideally, such that the user is not even aware that he/she is being subjected to a cognitive test. At the same time, the conversational data obtained should still be carrying the different features addressed by the classical test.

(iii) Design and implement a prediction model to detect cognitive impairment, that takes features from the conversational data in order to make an inference of the user's mental state based on previous interactions.

(iv) Design and implement voice-enabled applications for active ageing support. Considering that one of the main challenges for elderly population is to keep a socially and physically active lifestyle, a set of functionalities to facilitate this would definitely come in handy in the person's everyday life. Thus, we would be providing help to the user's physical independence and integration in social life, reducing dependency, fragility, and loneliness. Therefore, the idea is to try to take a step further than traditional tools (e.g., paper schedule, TV, post mail, etc.), improving their services such that they are easily available to the user from a single device in a unified and integrated way. Hence, this device should provide voice assistance to elderly, including functionalities such as establishing reminders, send and receive messages from acquaintances, suggest social activities nearby, etc.

(v) Carry out a psychometric validation of the implemented system, in a systematic and rigorous way. For this, it is necessary to perform a pilot study with a statistically significant amount of users and, taking the result of a gold standard test as a baseline truth, obtain a comprehensive reliability (i.e., test–retest, inter-observer, parallel and internal) as well as validity (i.e., face, content, construct, criterion, internal and external) assessment by analyzing the predictions of the inference model.

4 Proposed Solution

In Fig. 1, we can see a high-level overview of the solution's architecture that we propose in this framework. In it, we can see the different architecture components that integrate the system. The usual flow of events that will happen is as follows:

(1) The elderly person at home starts using a smart speaker, either by using the "wake up" word or because a reminder has been set periodically.
(2) The smart speaker includes Alexa or Google Assistant and it is connected wirelessly to a WiFi router, which has Internet access.
(3) The virtual assistant connects with the appropriate cloud platform, where the developed functionality (i.e., Alexa Skill or Google Action) is launched.
(4) The cloud function has their own logic programmed so that it can manage the conversation. It also can connect to a platform-specific database, such as S3 for Amazon or the Cloud Storage for Google. This allows to perform a direct adaptation of the classical test.
(5) Moreover, these functions can access a different cloud service where the trained Machine Learning prediction model would be hosted. Whenever a user completes a cognitive test (either traditional or ecological), the data will input the model and it will inference the cognitive state based on previous training data.
(6) Finally, the test data and result will be stored in an additional database. Undoubtedly, the user will be able to access and remove this data at any moment.

In order to use the Machine Learning model for prediction, we need to train it first. This process is depicted in Fig. 2, where both the training and assessment phases are shown.

In the training phase, a group of elderly users is needed to provide training data. First, a medical professional administers them one or several classical tests, such as the ones mentioned in Sect. 2.3: MMSE, CVLT, MoCA or TAVEC. From this, and from previous analysis of these tests, we are able to extract both the cognitive evaluation as well as the cognitive areas addressed (i.e., recall, language abilities, semantic organization, proactive inference, etc.) for using later on.

Then, the elderly users take the non-intrusive conversational test designed. Through a voiced conversation by a smart speaker, the virtual agent is able to extract the same cognitive areas, represented by audio and text features. As stated in [15], blending audio-visual (i.e., how the user talks) and verbal (i.e., what the user says) features will generally improve detection performance. With this data, together with the cognitive evaluation established by the gold standard test, we are able to train a Machine Learning model in order to fit the audio and text input data to the desired output result. For this, there are several Machine Learning algorithms available, including the ones analyzed in [10, 31, 44]: Logistic Regression, Support Vector Machine (SVM) or Decision Trees. Once the algorithm is trained, a new user may take the ecological conversational test and obtain a cognitive evaluation inferenced by the Machine Learning model, as we can see in the assessment phase of Fig. 2.

Moreover, by extracting and comparing the features and results from both the classical and ecological tests, we are able to perform a psychometric validation, as stated in the last point of Sect. 3.

In order to implement the ecological test, the deployment of an additional cloud server with a smart conversational manager is proposed. This will allow us to implement the concept of *personal health*, which means the adaptation of the test to the user profile and previous answers. The idea is to drive the conversation introducing a component of Deep Learning, and especially Reinforcement Learning. Something very similar has already been proposed and implemented in a recent study [40], where an adaptive dialogue algorithm was modelled as a Markov chain in order to identify MCI.

Finally, we would like to mention the limitations of the proposed solution. While the number of users in the pilot study will be statistically significant, there is always a limitation on the validation process if we consider a sub-sample of the population of interest. At the same time, the concept of an ecological conversational test by means of an adaptive dialogue algorithm is still state of the art, which potential setbacks in a real environment's application are still unknown.

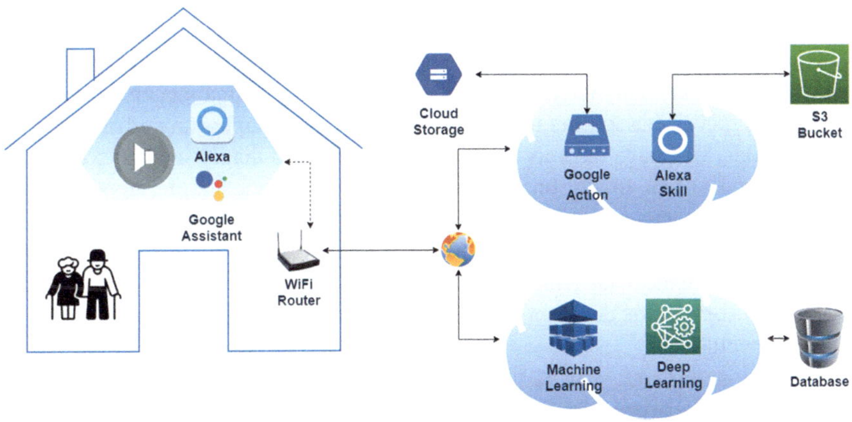

Fig. 1 High-level architecture overview

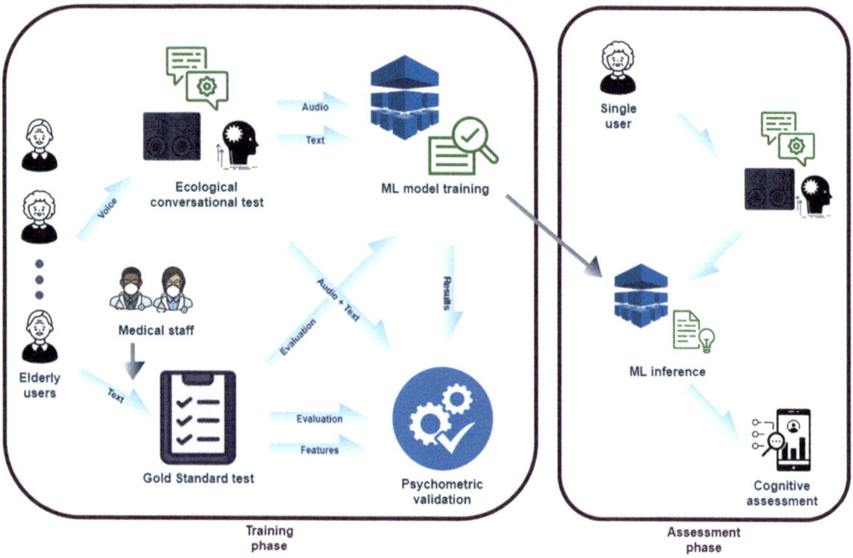

Fig. 2 Training and assessment phases

5 Conclusion

Under a scientific perspective, this framework is intended to serve as a technological reference framework to investigate new instruments for the diagnosis of neurophysiological conditions, beyond the classical paper-and-pen tests that serve today as gold standards. The characteristics of intelligent conversational agents, and specifically their speech-based interaction capabilities and the supporting advanced analysis methods, will pave the way for novel approaches to diagnosis that expected to overcome most of the limitations the classical methods. For example, advanced classification and discrimination algorithms would support to reduce the impact of confounding factors such as the existence of concurrent diseases, language barriers, limiting physical impairments consequence of advanced age, or low education. Besides, they would support the development of more ecological solutions not depending on cultural specificities.

In addition, this framework could serve as a reference platform to extend existing knowledge in Machine Learning, especially its application to pattern recognition and classification, and eventually discover new algorithms to infer the cognitive state of a person from their conversational interaction. Moreover, new research initiatives in this context may also contribute to other application fields within Machine Learning and Deep Learning, such as smart conversational management.

To sum up, the expected application scope of the proposed framework goes beyond cognitive impairment detection, but is designed to support new contributions in the broader scope Artificial Intelligence applications.

From the perspective of the target user base, this proposal is expected to have a positive impact in several end-user groups.

- *Primary users*. These would be the elderly people who would directly use the virtual assistant to support their activities of daily living. Smart conversational agents would facilitate activating their social and physical activity, as well as providing a non-intrusive way for the early detection of cognitive impairment.
- *Secondary users*. Family members and close friends of primary users would also benefit from the proposed framework, both in terms of facilitating communication by means of existing messaging features and, most importantly, it could contribute to relief part of the pressure experienced by caregivers, as they would have additional tools to contribute to the well-being of primary users, since they would obtain evaluations and alerts in case the cognitive assessment tool detects some anomaly, that can be complemented with guidelines and counseling.
- *Tertiary users*. Healthcare institutions and social services will also benefit from this approach, as they would have available a non-intrusive, scalable and effective mechanism for patients' cognitive assessment, which incorporates the benefits discussed above.

Finally, from an economic standpoint, this framework will contribute to make a more effective use of existing public resources, as an early detection of cognitive impairment can have a dramatic contribution to facilitate early interventions. Note that these diseases, when the initial intervention occurs in their more advanced stages, will require a considerable amount of resources.

Once this framework has been developed, our work is aimed to field-test it with the development of a proof-of-concept implementation. Firstly, this will allow us to check the suitability of our solution, and secondly, we will generate more knowledge within the utility of smart conversational agents for the early detection of cognitive impairment.

Acknowledgements This research was partially funded by Ministerio de Ciencia, Innovación y Universidades under the grant reference FPU19/01981 (Formación de Profesorado Universitario). The Article Processing Charge (APC) was funded by the University of Vigo.

References

1. Agarwal S, Tong Lau C (2010) Remote health monitoring using mobile phones and web services. Telemed E Health 603–607
2. Alejandre M, Benedet M (1998) Test de aprendizaje verbal españa-complutense. Manual
3. APA (1966) Standards for educational and psychological tests and manuals. American Psychological Association
4. Arrabales R (2020) Perla: a conversational agent for depression screening in digital ecosystems. Design, implementation and validation
5. Brenner W, Wittig H, Zarnekow R (1998) Intelligent software agents: foundations and applications, 1st ed. Springer-Verlag

6. Chaytor N, Schmitter-Edgecombe M (2004) The ecological validity of neuropsychological tests: a review of the literature on everyday cognitive skills. Neuropsychol Rev 13:181–197
7. Delis D, Kramer J, Kaplan E, Thompkins B, Psychological Corporation (Firm), Harcourt Brace Jovanovich, Inc. (Firm) (1987) CVLT: California verbal learning test-adult version: manual. Psychological Corporation
8. Della Mea V (2001) What is e-health (2): the death of telemedicine? J Med Internet Res 3(2)
9. DeVault D, Artstein R, Benn G, Dey T, Fast E, Gainer A, Georgila K, Gratch J, Hartholt A, Lhommet M, Lucas G, Marsella S, Morbini F, Nazarian A, Scherer S, Stratou G, Suri A, Traum D, Wood R, Xu Y, Rizzo A, Morency L-P (2014) SimSensei kiosk: a virtual human interviewer for healthcare decision support. In: 13th international conference on autonomous agents and multiagent systems, AAMAS 2014, vol 2, pp 1061–1068
10. Facal D, Valladares S, Lojo-Seoane C, Pereiro AX, Anido-Rifón L, Juncos-Rabadán O (2019) Machine learning approaches to studying the role of cognitive reserve in conversion from mild cognitive impairment to dementia. Int J Geriatr Psychiatry 34
11. Faurholt-Jepsen M, Busk J, Frost M, Vinberg M, Christensen EM, Winther O, Bardram JE, Kessing LV (2016) Voice analysis as an objective state marker in bipolar disorder. Transl Psychiatry 6
12. Fiske A, Wetherell JL, Gatz M (2009) Depression in older adults. Annu Rev Clin Psychol 5(1):363–389
13. Folstein MF, Folstein SE, McHugh PR (1975) "Mini-mental state": a practical method for grading the cognitive state of patients for the clinician. J Psychiatr Res 12(3):189–198
14. Graupe D (2007) Principles of artificial neural networks, 2nd ed. World Scientific Publishing Co., Inc.
15. Griol D, Molina JM, Callejas Z (2019) Combining speech-based and linguistic classifiers to recognize emotion in user spoken utterances. Neurocomputing 326–327:132–140
16. Holtzman DM, Morris JC, Goate AM (2011) Alzheimer's disease: the challenge of the second century. Sci Transl Med 3(77):77sr1
17. Inkster B, Sarda S, Subramanian V (2018) An empathy-driven, conversational artificial intelligence agent (Wysa) for digital mental well-being: real-world data evaluation mixed-methods study. JMIR Mhealth Uhealth6(11)
18. Jaiswal S, Valstar M, Kusumam K, Greenhalgh C (2019) Virtual human questionnaire for analysis of depression, anxiety and personality, pp 81–87. In: Proceedings of the 19th ACM international conference on intelligent virtual agents, IVA 2019
19. Khodabakhsh A, Kuscuoglu S, Demiroglu C (2014) Detection of Alzheimer's disease using prosodic cues in conversational speech. In: 2014 22nd signal processing and communications applications conference (SIU), pp 1003–1006
20. Kobayashi M, Kosugi A, Takagi H, Nemoto M, Nemoto K, Arai T, Yamada Y (2019) Effects of age-related cognitive decline on elderly user interactions with voice-based dialogue systems
21. Kowalczuk P (2018) Consumer acceptance of smart speakers: a mixed methods approach. J Res Interact Mark 12(4):418–431
22. Kowalska M, Owecki M, Prendecki M, Wize K, Nowakowska J, Kozubski W, Lianeri M, Dorszewska J (2017) Aging and neurological diseases. In: Senescence - physiology or pathology
23. Kroenke K, Spitzer R, Williams J (2001) The PHQ-9 validity of a brief depression severity measure. J Gen Intern Med 16
24. Laranjo L, Dunn A, Tong HL, Kocaballi AB, Chen J, Bashir R, Surian D, Gallego B, Magrabi F, Lau A, Coiera E (2018) Conversational agents in healthcare: a systematic review. J Am Med Inform Assoc
25. Mallol-Ragolta A, Zhao Z, Stappen L, Cummins N, Schuller B (2019) A hierarchical attention network-based approach for depression detection from transcribed clinical interviews
26. Mccarthy J, Minsky M, Rochester N, Shannon C (2006) A proposal for the dartmouth summer research project on artificial intelligence. AI Mag 27
27. Mirheidari B, Blackburn D, O'Malley R, Walker T, Venneri A, Reuber M, Christensen H (2019) Computational cognitive assessment: investigating the use of an intelligent virtual agent for the detection of early signs of dementia

28. Muñoz RF, Mrazek PJ, Haggerty RJ (1996) Institute of medicine report on prevention of mental disorders: summary and commentary. Am Psychol 51(11):1116–1122
29. Nasreddine ZS, Phillips NA, Bédirian V, Charbonneau S, Whitehead V, Collin I, Cummings JL, Chertkow H (2005) The montreal cognitive assessment, MoCA: a brief screening tool for mild cognitive impairment. J Am Geriatr Soc 53(4):695–699
30. Okada S, Inoue K, Imai T, Noguchi M, Kuwamura K (2019) Dementia scale classification based on ubiquitous daily activity and interaction sensing
31. Pacheco-Lorenzo MR, Valladares-Rodríguez SM, Anido-Rifón LE, Fernández-Iglesias MJ (2021) Smart conversational agents for the detection of neuropsychiatric disorders: a systematic review. J Biomed Inform 113:103632
32. Peitzker T (2016) Why chatbots are so disruptive. VentureBeat
33. Philip P, Micoulaud Franchi J-A, Sagaspe P, de Sevin E, Olive J, Bioulac S, Sauteraud A (2017) Virtual human as a new diagnostic tool, a proof of concept study in the field of major depressive disorders. Sci Rep 7:42656
34. Picardi A, Lega I, Tarsitani L, Caredda M, Matteucci G, Zerella MP, Miglio R, Gigantesco A, Cerbo M, Gaddini A, Spandonaro F, Biondi M, Group T (2016) A randomised controlled trial of the effectiveness of a program for early detection and treatment of depression in primary care. J Affect Disord 198
35. Rahmani A, Thanigaivelan NK, Gia TN, Granados J, Negash B, Liljeberg P, Tenhunen H (2015) Smart e-health gateway: bringing intelligence to internet-of-things based ubiquitous healthcare systems. In: 2015 12th annual IEEE consumer communications and networking conference (CCNC)
36. Schmidt M et al (1996) Rey auditory verbal learning test: a handbook. Western Psychological Services Los Angeles, CA
37. Sidorov M, Minker W (2014) Emotion recognition and depression diagnosis by acoustic and visual features: a multimodal approach. Association for Computing Machinery, Inc., pp 81–86
38. Sloot PMA, Tirado-Ramos A, Altintas I, Bubak M, Boucher C (2006) From molecule to man: decision support in individualized e-health. Computer 39(11):40–46
39. Tanaka H, Adachi H, Ukita N, Ikeda M, Kazui H, Kudo T, Nakamura S (2017) Detecting dementia through interactive computer avatars. IEEE J Transl Eng Health Med 5:1–11
40. Tang F, Uchendu I, Wang F, Dodge H, Zhou J (2020) Scalable diagnostic screening of mild cognitive impairment using AI dialogue agent. Sci Rep 10
41. Trzepacz P, Hochstetler H, Wang S, Walker B, Saykin A (2015) Relationship between the montreal cognitive assessment and mini-mental state examination for assessment of mild cognitive impairment in older adults. BMC Geriatr 15:107
42. Tsai Y-T, Lin W-A (2019) Design of an intelligent cognition assistant for people with cognitive impairment. In: Proceedings - 20th international conference on high performance computing and communications, 16th international conference on smart city and 4th international conference on data science and systems, HPCC/SmartCity/DSS 2018, pp 1207–1212
43. Ujiro T, Tanaka H, Adachi H, Kazui H, Ikeda M, Kudo T, Nakamura S (2018) Detection of dementia from responses to atypical questions asked by embodied conversational agents. In: Interspeech. 19TH annual conference of the international speech communication association (Interspeech 2018). Speech research for emerging markets in multilingual societies, vols 1–6, pp 1691–1695
44. Valladares-Rodriguez S, Perez-Rodriguez R, Facal D, Fernández-Iglesias MJ, Anido-Rifón L, Mouriño-Garcia M (2017) Design process and preliminary psychometric study of a video game to detect cognitive impairment in senior adults. Peer J 5:e3508
45. Van der Elst W, Van Boxtel MP, Van Breukelen GJ, Jolles J (2008) A large-scale cross-sectional and longitudinal study into the ecological validity of neuropsychological test measures in neurologically intact people. Arch Clin Neuropsychol 23(7–8):787–800
46. Woods SP, Delis DC, Scott JC, Kramer JH, Holdnack JA (2006) The California verbal learning test - second edition: test-retest reliability, practice effects, and reliable change indices for the standard and alternate forms. Arch Clin Neuropsychol 21(5):413–420

Chapter 66
Context-Based Quote Generation from Images

Aditi Sharma, Divya Gupta, and Mukesh Kumar

1 Introduction

Our Project is built upon Deep Learning-based Paradigm. It is based on the concept of transfer learning used in extracting features from the image using a pre-trained Convolutional Neural Network (CNN) model, and computing on those features, caption describing the image using Long Short Term Memory (LSTM) is generated. Text analysis of the generated caption is used to suggest quotes similar to the image. The caption generating algorithm is based on the implementation of the "merge-model" described by [1] in What is the Role of Recurrent Neural Networks (RNNs) in an Image Caption Generator? 2017. The pipeline of the proposed algorithm is given as in Fig. 1.

2 Related Work

Kiros et al. [2], inspired by machine translation, introduced an approach based on an encoder-decoder pipeline that learns a multi-modal joint embedding space with images and text and decodes the distributed representations from the space. Kiros et al. [3] modified the above approach and introduced neural networks to generate natural language descriptions of images. Mitchell et al. [4] proposed the concept of syntactic trees constructed from Flickr images and text descriptions by using syntactically informed word co-occurrence statistics, generator filters and constraints. Mnih et al. [5] takes advantage of human perception and introduces a Recurrent Neural Network model with an attention module to enhance the computation of images.

A. Sharma · D. Gupta · M. Kumar (✉)
UIET, Panjab University, Chandigarh, India
e-mail: mukesh_rai9@pu.ac.in

G. Mathur et al. (eds.), *Proceedings of 2nd International Conference on Artificial Intelligence: Advances and Applications*, Algorithms for Intelligent Systems,
https://doi.org/10.1007/978-981-16-6332-1_66

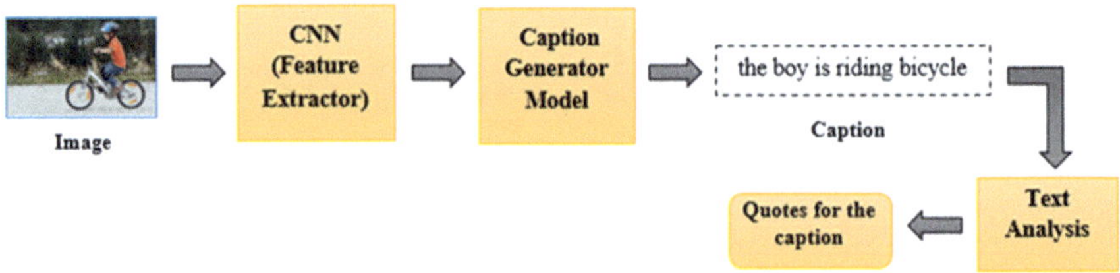

Fig. 1 The proposed algorithm

Goel et al. [6] introduce an approach that forms the foundation of our work. In the original work, a description of an image in natural language is generated which is matched with a list of candidate quotes and relevant quotes are shown as output. We modify the approach in the following way:

1. In generating caption for the image, we experiment with various CNN models like VGG16, MobileNetV2, InceptionV3 and ResNet50 and choose the one which gives the best performance in our case. Previous works included caption generation considering only VGG16.
2. Instead of matching the generated image description to all the quotes in the dataset, we take the categorized quotes and find categorywise common words which are matched with the image descriptions using the Jaccard Similarity [7] and the Gestalt Pattern Matching [8]. This saves the matching time and makes the computation faster.

3 Proposed Work

The proposed system caters to the need of associating inspirational quotes with the image. Some of the existing systems only generate the English captions and others generate quotes by it with the quotes present in the database but the proposed system first measures the similarity of the English description of the image (caption) with the categories (e.g., philosophy, travel, adventure, etc.) and then top ten quotes are fetched from the most similar category. Those quotes are again matched with the caption to generate the most similar quotes. The similarity measures used are the same as mentioned above. The top three quotes along with their authors are then displayed. The proposed work involves the following phases: (i) Data Collection, (ii) Vocabulary Creation, (iii) Building Feature Sets, (iv) Defining Caption Generating Model, (v) Training Model, (vi) Filtering out the best performing CNN and (vii) Outputting Quotes for the generated caption.

(i) Data Collection:
Data-set used for training and testing the model architecture is Flickr8K caption data-set [9]. Using each of pre-trained CNN (e.g., VGG16, MobileNetV2,

X1,	X2 (text sequence),	y (word)
image	startseq,	dog
image	startseq, dog,	is
image	startseq, dog, is,	sitting
image	startseq, dog, is, sitting,	on
image	startseq, dog, is, sitting, on,	grass
image	startseq, dog, is, sitting, on, grass,	endseq

Fig. 2 Feature sets

ResNet50 and InceptionV3), the feature vector sets of all images in the data-set are created and stored in distinct pickle files for each of the four CNNs. Preprocessing of text data is also done in this phase which includes creating a description named dictionary of the list with a key having image file's name and value is the list of captions corresponding to each image, lower-casing all the descriptions, removing punctuation and inserting <startseq> at the beginning and <endseq> at the ending of each description for letting deep learning network that from where to start and where to stop. The maximum length of the generated caption is considered equal to the maximum length of the caption in captions' data-set (Observed maximum length is 34).

(ii) Vocabulary Creation:
Vocabulary set is created using the description dictionary by making set of all the words from the descriptions. Captions will be generated using this vocabulary set.

(iii) Building Feature Sets:
Feature sets to be used in the training and testing will be like: [image_feature_vector, input_string, next_word], e.g., the input sequence "dog is sitting on grass" will be split in the following pairs (Fig. 2):
Here X1= image_feature_vector, X2= input_string, y= next_word. The Feature set sequences are created by encoding the descriptions into numerical sequences. Separate 3 arrays are created for X1, X2 and y where X1 is holding image_feature_vector, X2 will be holding list of numerical sequence split in the way as shown in the example and y will be holding list of next words corresponding to each X1, X2.

(iv) Defining Caption Generating Model:
The model is divided into three parts:

(a) Image Processor Model:
It considers the image feature vector of n length (n = length of feature vector generated by CNNs) as input which is then further processed by a Dense layer to convert it into a feature vector of length 256. Dropout Layer with 50% dropout is used to provide regularization

(b) Sequence Processor Model:
This model considers a feature vector length of 34 (maximum caption length) as an input which is then further processed by Embedding layer followed by the LSTM layer of 256 memory units.

(c) Decoder Model:
As Image Processor Model and Sequence Processor Model outputs fixed-length vector. The decoder Model merges the two outputs using add operation. This is then provided to a Dense 256 neuron layer followed by a final output Dense layer which makes a softmax prediction over the entire output vocabulary for the next word in the sequence.
Pictorial representation of the model is shown in Fig. 3. In the figure, n = feature vector length of CNN used for feature extraction and vocabulary length is observed to be 7579.

(v) Training Model:
Four models each with four CNN is trained for 20 epochs on Google Colab platform and the epoch at which model with less validation loss is saved and later testing of the model is evaluated using BLEU score to choose the best performing CNN for feature extraction process.

(vi) Filtering out the best performing CNN:
On the basis of larger BLEU score (BLEU-1: 0.578895, BLEU-2: 0.322668, BLEU-3: 0.224482, BLEU-4: 0.108088) and less validation loss (3.7576) than other three CNNs, we chose MobileNetV2 as our image feature extractor. The plot of caption generation model is shown in Fig. 3.

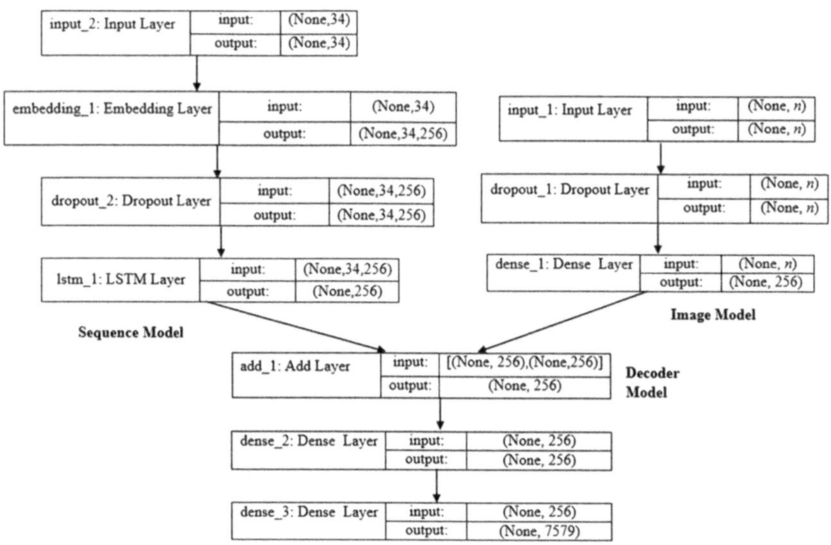

Fig. 3 Caption generation model

(vii) Outputting Quotes for generated caption:
Text analysis is then done on the generated caption to produce an appropriate quote. For text analysis, a data-set of quotes (quotes_all data-set [10]) was taken with each quote belonging to a certain category. The quotes were tokenized, removed of stop words and lemmatized. A list of categorywise common words in quotes was made. Out of those common words, the top 50 nouns and adjectives belonging to each category were retained. The list is used to find similarities with the generated captions and to produce appropriate quotes.

In text analysis, these steps are followed:

1. The generated caption is tokenized, removed of stop words and lemmatized.
2. The caption is then compared with each category on the basis of common words found categorywise. The similarity is found using the following:
   The Jaccard Similarity: To compare two objects, it looks at the elements they have in common (the intersection) and divides it by the number of elements the two objects have in total (the union), i.e., the Jaccard Similarity = $(X\cap Y)/(X\cup Y)$.
   Gestalt pattern matching: The idea is to find the longest contiguous matching sub-sequence that contains no "junk" elements. The same idea is then applied recursively to the pieces of the sequences to the left and to the right of the matching sub-sequence. (It is nearly 0.5 times the Jaccard similarity)
3. The similarity scores are then used to find the most similar category to which caption matches. It is given by the category having the maximum value of the similarity score.
4. The top 10 quotes are fetched from the most similar category.
5. Those quotes are again matched with the caption to generate the most similar quotes. The similarity measures used are the same as mentioned above.
6. The top 3 quotes are then displayed.

The detailed design of the Proposed Method is shown in Fig. 4.

Experimentation and Results

The proposed quotes association system has been implemented using Python programming language, Keras [11] and NLTK (Natural Language Tool Kit) [12] functionalities. Using Python and Keras, the caption generation model is trained and tested. The text analysis system is built using Python and NLTK functionalities. Training and testing of the caption generating model are done on Google Colab Platform [13]. As the quotes are mapped using generated captions so estimating the quality of suggested quotes is a user-dependent criteria and is different for every human being. In the testing part, we have evaluated the performance of four models each using different pre-trained CNN for feature extraction. We used the following measures for testing of various pre-trained Convolutional Neural Network models with the LSTM networks:

Bilingual Evaluation Understudy Score (BLEU Score): The 1-gram, 2-gram, 3-gram and n-gram BLEU Scores were calculated for various training models using the ground truth provided in the data-set, and the model with the best BLEU Score was

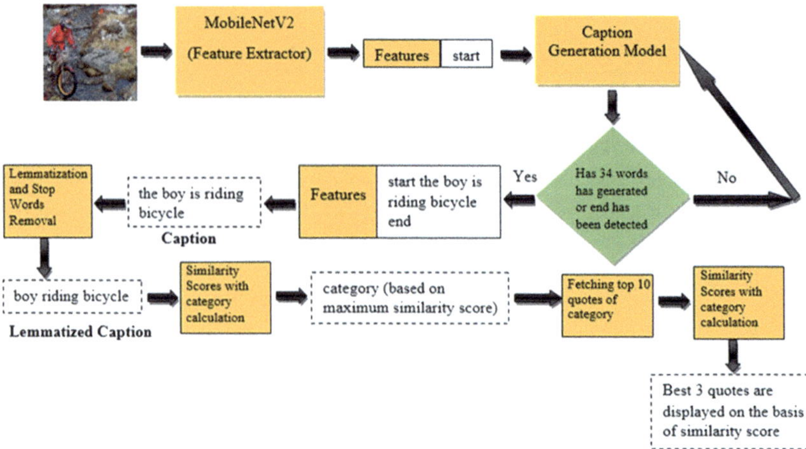

Fig. 4 Detailed design

CNN	Validation Loss	BLEU-1 Score	BLEU-2 Score	BLEU-3 Score	BLEU-4 Score
VGG16	3.8569	0.569074	0.312886	0.211893	0.096428
MobileNetV2	3.7576	0.578895	0.322668	0.224482	0.108088
InceptionV3	3.7659	0.566663	0.299221	0.204520	0.100033
ResNet50	3.8054	0.559542	0.309317	0.214915	0.103256

Fig. 5 Results

Fig. 6 Plots of loss values

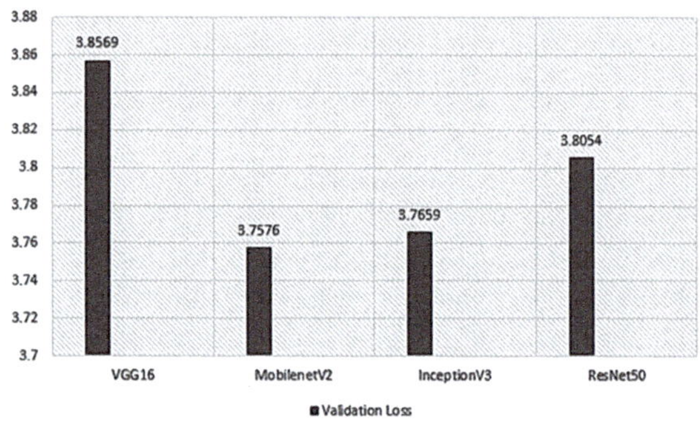

finally used. Validation loss: The validation loss for various models was calculated and the model with minimum validation loss was finally used. Results and graphical plot of results are shown in Figs. 5 and 6.

As can be seen, the feature extractor model with the best performance measures in comparison to other models experimented is MobileNetV2. Figure 7 below

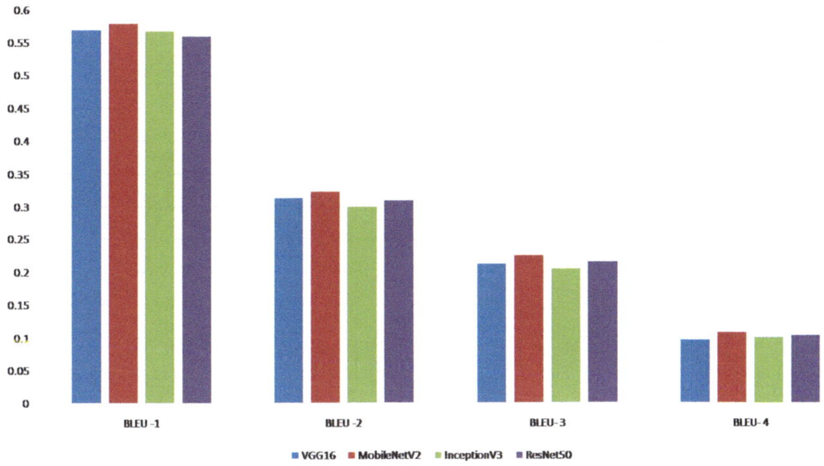

Fig. 7 Comparison statistics

displays the comparison of BLEU score and validation loss for the above pre-trained convolutional neural network models present in Keras library of Python language.

Sample quotes generated for the image input by the user

4 Funding Details

The work reported in this paper is funded by the Capacity Building and Research Entrepreneurship Centre sanctioned to Panjab University, Chandigarh by the Ministry of Electronics and Information Technology (MeitY), Government of India.

5 Conclusion

The feature extractor model with the best performance measures in comparison to other models used is MobileNetV2. Moreover, in our web application, speed matters and we require results in less time which MobileNetV2 is capable to deliver. So we use MobileNetV2 in our application. To calculate better results, we use an ensemble of similarity measures (the Jaccard Similarity and the Gestalt Pattern Matching). The application is provided in the form of a web application to facilitate ease of use. Sample of User Interface of the web application is shown in Fig. 8.

	True Strength is delicate.
	I have my own strength.
	Strength does not come from physical capacity. It comes from an indomitable will.
	Yes, forget your weakness, whatever that weakness may be. It is egotism; it is selfishness after all, for it is a dwelling on self. Forget your weakness and remember your strength.
	The failure of academic feminists to recognize difference as a crucial strength is a failure to reach beyond the first patriarchal lesson. In a world, divide and conquer must become define and empower.
	People think of Latina women as being fiery and fierce, which is usually true. But I think the quality that many Latinas possess is strength. I'm very proud to have Latin blood.

Fig. 8 Sample quotes generated

6 Future Scope

The web application can be ported to mobile devices in the form of mobile applications to provide a dedicated platform for quote generation. Moreover, we can perform similarity matching on the basis of context for better results. We can also generate quotes using the LSTM model by giving as input a suitable data-set to train the model to write in poetic style.

References

1. Tanti M, Gatt A, Camilleri KP (2017) What is the role of recurrent neural networks (RNNs) in an image caption generator? arXiv:1708.02043
2. Kiros R, Salakhutdinov R, Zemel RS (2014) Unifying visual-semantic embeddings with multimodal neural language models. arXiv:1411.2539
3. Kiros R, Salakhutdinov R, Zemel R (2014) Multimodal neural language models. In: International conference on machine learning. PMLR, pp 595–603
4. Mitchell M, Dodge J, Goyal A, Yamaguchi K, Stratos K, Han X, Mensch A, Berg A, Berg T, Daumé H III (2012) Midge: generating image descriptions from computer vision detections. In: Proceedings of the 13th conference of the European chapter of the Association for Computational Linguistics, pp 747–756
5. Mnih V, Heess N, Graves A, Kavukcuoglu K (2014) Recurrent models of visual attention. arXiv:1406.6247
6. Goel S, Madhok R, Garg S (2018) Proposing contextually relevant quotes for images. In: European conference on information retrieval. Springer, pp 591–597
7. Chaudhuri S, Ganti V, Kaushik R (2006) A primitive operator for similarity joins in data cleaning. In: 22nd international conference on data engineering (ICDE'06). IEEE, p 5
8. Ratcliff JW, Metzener DE (1988) Pattern-matching-the gestalt approach. Dr Dobbs J 13(7):46

9. Hodosh M, Young P, Hockenmaier J (2013) Framing image description as a ranking task: data, models and evaluation metrics. J Artif Intell Res 47:853–899
10. Famous quotes database. https://thewebminer.com/buy-famous-quotes-database. Accessed 15 Dec 2018
11. Keras documentation. https://github.com/keras-team/keras. Accessed 10 Oct 2018
12. NLTK documentation. https://www.nltk.org. Accessed 12 Dec 2018
13. Google colab platform. https://colab.research.google.co. Accessed 15 Dec 2018

Chapter 67
Flood Hazard Mapping of Kuttanaad Region, Kerala

Jayati Vijaywargiya and Rama Rao Nidamanuri

1 Introduction

Flood is a major natural disaster that affects millions of people and is prevalent every year at many places around the globe. It needs to be monitored at regional scale. As a preventive measure at regional scale it is necessary to identify the flood prone areas so as to facilitate quick response, provide early alert and decrease the count of tragedies due to flood. Flood impacts not only the socio-economic stability of the region but also has environmental consequences. The vulnerability due to floods can even result in the exodus from the flood affected region.

1.1 Flood History of Kuttanaad Region

In 2018, Kuttanaad region experienced worst flood in the last twenty years. According to the reports of Agricultural Department, these floods resulted in the ravaging of 7316 ha of paddy fields of this region. Kuttanaad is amongst the lowest altitude regions of India. Major part of this region lies 1.2–3m below the sea level. Moreover, it lies in the foothills of Western Ghats and is drained by four rivers namely, Pamba, Meenachil, Achankovil and Manimala. The spatial location of this region brings flood with almost every monsoon in this region. According to the satellite-based assessment of the August 2018 flood in parts of Kerala, India, the rise in the water level of Kuttanaad region during the floods of 2018 was approximately 5m.

J. Vijaywargiya (✉) · R. R. Nidamanuri
Department of Earth and Space Sciences, Indian Institute of Space Science and Technology, Thiruvananthapuram, Kerala, India
e-mail: jayativijaywargiya@ieee.org

G. Mathur et al. (eds.), *Proceedings of 2nd International Conference on Artificial Intelligence: Advances and Applications*, Algorithms for Intelligent Systems,
https://doi.org/10.1007/978-981-16-6332-1_67

1.2 Flood Hazard Mapping

Flooding has many hazards, primarily the loss of life and damage of property, secondarily, as the after effects of floods, famine and diseases can outburst over large area, also over a long time it can cause damage to the environment of the region, leaching and erosion of soil or change in the position of river channel. Flood Hazard Mapping is helpful to give the first hand precautionary measures to reduce the causality due to flood. It tells about the regions which are more vulnerable to flood due to heavy rainfall. These expected flood regions can be determined using the topographic data and historical rainfall data of the region. Due to advancement of technology, the satellite data along with the ground information can be used to derive the flood hazard map of the region. In this work Landsat-8 land use land cover data has been used, the Digital Elevation Model of the region was taken from the SRTM DEM, Rainfall data was provided by Irrigation Design and Research Board, Thiruvananthapuram, Kerala. These data together have been used to derive the flood hazard map of the Kuttanaad region.

2 Literature Review

Flood risk assessment using multiple criterion started after year 2000. There have been many approaches to map the hazardous areas. One of them being index-based analytical hierarchical approach. This was also implemented by Nerantzis Kazakis, Ioannis Kougias, and Thomas Patsialis (2015) over the Evros region in Greece [1].

Multi-parameter-based approaches have been previously used to determine the flood hazard map. Few such works are weighted approach over ten parameters to study the flood risk on the Malaysian region by Tahrany et al. in the year 2014. This work also took into consideration the parameter 'Distance from River/Drainage' and it is also considered a physical parameter in this work. Other general parameters considered in this work were rainfall, digital elevation model, slope and soil drainage and few parameters that were considered were, curvature, geology, green farmland, timber age, timber density, etc. [2].

Flood risk mapping using GIS was also implemented by Daniela Rincon Romero et al. to study the flood hazardous area over the Toronto region. They also used multiple criteria like census data, Land use pattern, DEM, etc. [3].

Flash Flood Hazard Mapping Using Satellite Images and GIS Tools over Najran City, Kingdom of Saudi Arabia (KSA) was done by Ismail Elkhrachy using index-based approach and analytical hierarchical process [4].

In recent days, due to the devastation by flood there is a growing trend to develop flood atlas over large area. In India flood atlas is made for the state of Assam and now soon will be ready for Odisha. Similar work was done by Md. Yousuf Gazi and Md. Ashraful Islam for making the regional scale flood hazard atlas of Bangladesh. In this work also multi criteria evaluation technique was used and the weight values were based on expert knowledge using Analytical hierarchical process. Elevation, slope, distance from river, LULC were among the few parameters used for mapping [5].

In this work, the two major things were selection of parameters and selection of algorithm to derive the weightage of each parameter. On the basis of previous work done in this field and recent algorithmic development, seven parameters were considered for this work and index-based approach using analytical hierarchical process is used to map the flood hazard regions of Kuttanaad region.

2.1 Study Area and Data

Kuttanaad region, which lies in the west coast of India was chosen as the study area. It lies in the Alappuzha and Kottayam Districts of Kerala, India. This region is also known as the 'Rice bowl of Kerala'. The primary occupation of people of this region is agriculture, specifically rice. It is extended over the latitudinal range of 9.248869° N to 9.791042° N and longitudinal range of 76.328700° E to 76.604953° E. This region has an approximate temperature range of 20–35 °C. It is a coastal region and has average elevation of −2.2 m. The average annual rainfall over this region is more than 300 cm. The Kuttanaad region under the Alappuzha district of Kerala is well known for backwaters and canals. This region majorly has stretches of paddy fields. Some of the villages of this region are, Champakkulam, Payippad, Karichal, Cheruthana, Karuvaatta, Narakathara, Mamkompu, etc.

In 2018 monsoon, this area faced severe floods. As the consequence of flood the 9,907 hectares of paddy fields were wiped off. Roadways were water logged and rowboats were used for transportation. Unavailability of the basic amenities forced the people to take shelter in relief camps. After flood a few reasons were determined for such a socio-economic loss. Thus, future construction and relief should be planned based on such flood maps so as to delineate socio-economic loss (Fig. 1).

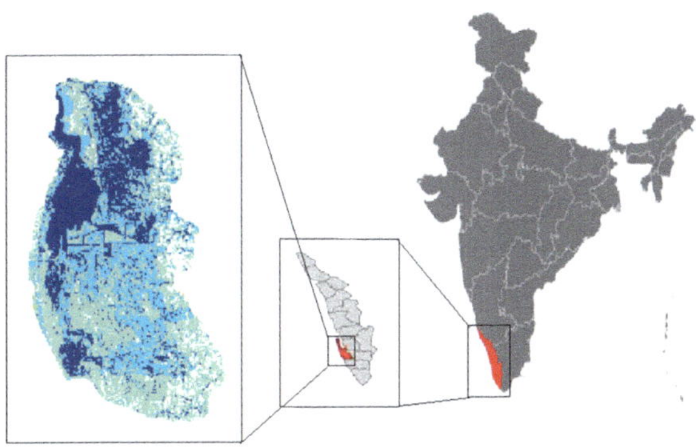

Fig. 1 Kuttanaad Region, Kerala, India

3 Methodology of Mapping

To retrieve the flood hazardous areas, we have developed multi criteria Flood Hazard Index-based methodology. The methodology is represented in the flow diagram in Fig. 2.

Flood hazard Index gives the relative risk of the region to the flood. In this work we have used AHP, Analytical Hierarchical Process, method to allot the relative weight to the parameters. This method reduces the complexity and allows pairwise comparison of the attributes. The parameters are given ranking according to their relative relevance.

In this work, maximum weightage is given to flow accumulation. Distance from drainage and Elevation are given equal importance. Here, soil and land use pattern are also given equal importance. Then, followed by Rainfall intensity and slope. Thus, flow accumulation accounts the most in flood hazard whereas slope accounts the least.

3.1 Analytical Hierarchical Process

Analytical Hierarchical Process is a structural technique to analyse the problems where interrelated attributes are involved. In this process first the weights are ranked as per their relative relevance. This sorts the attributes in hierarchical manner. Then a pairwise comparison amongst all the attributes is done and is tabulated. Each row attributes to its importance with respect to all other attributes. These values are normalised and their mean corresponds to weight of each factor.

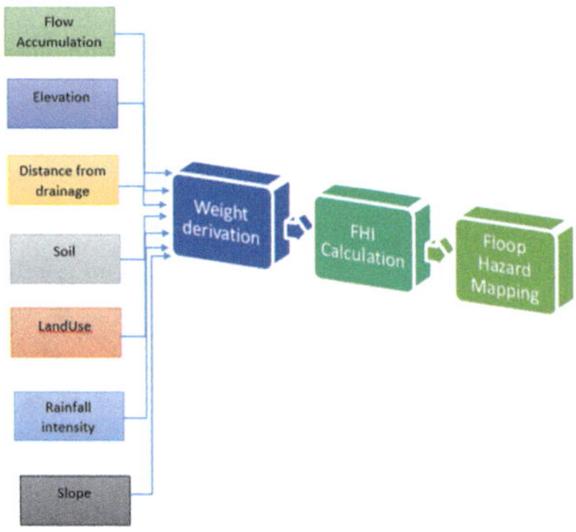

Fig. 2 Flow chart for FHM of Kuttanaad Region, Kerala, India

3.2 *Physical Parameters*

In this work, each grid point is evaluated on the basis of every parameter. Then each cell is allotted a value, that is ranked based on summation of product of weight of the physical parameter and parameter value at that cell. Qualitative as well as quantitative parameters are used. Each value of parameter is ranked and the parameter value is referred to the rank.

For, rainfall intensity calculation, we received annual data from four rain gauge stations in that region. Average annual rainfall was calculated using IDW, Inverse Distance Weighted, method and this data was interpolated over the entire region. The ranking is given according to the average annual rainfall with the highest average annual rainfall given the maximum ranking. Land use is one of the influential factor in flood mapping. In urban terrain there is complete runoff whereas in vegetation or soil there is partial runoff. Thus the ranking is given to various land use parameters according to their infiltration capability.

The areas near the water bodies, streams or any other drainage network are more susceptible to flood. Overflow from the water bodies makes the areas near them more crucial during floods. The influence of drainage decreases as the distance from drainage increases. Thus, distance from drainage is considered a relevant parameter in FHM. A multi ring buffer is created around the drainage network at the distance of 200m, 500m, 1000m, 15000m and 20000m and are ranked from 5 to 1 according to their susceptibility for flood. The soil type greatly influences the run off during the flood. The water bodies give complete runoff and thus are ranked as 5, followed by clay, gravelly, loamy and sandy are ranked from 4 to 1, respectively. Sand gives maximum infiltration whereas clay gives the least. Flow accumulation is the most relevant feature in Flood Hazard Mapping. It is a derived parameter from DEM. The higher the flow accumulation the higher is the threat of flood. In this work the Flow Accumulation rank ranges from 5 to 1. Higher values of Flow Accumulation are ranked greater than lower values of FA.

Higher elevation are less susceptible to be influenced by flood, whereas the low lying region is more vulnerable to flood. In this region most of the area is low lying. Only some parts in north east of Kuttanaad region are at relatively higher elevation. Thus, overall the whole area is very much vulnerable to flood. Slope attributes to the amount of surface runoff and the water in filtered. Steep slopes will result in runoff to low lying areas whereas the lower slope regions have high chances of water accumulation. As per our classification, the region is close to flat terrain. Here, low slope values are given higher ranking then higher slope region.

4 Results and Discussions

The proposed methodology combines the seven parameters and can be thought of superimposing various maps according to their weights resulting in the Flood Hazard Map of Kuttanaad region. It has 4 classes based on the value of Flood Hazard Index

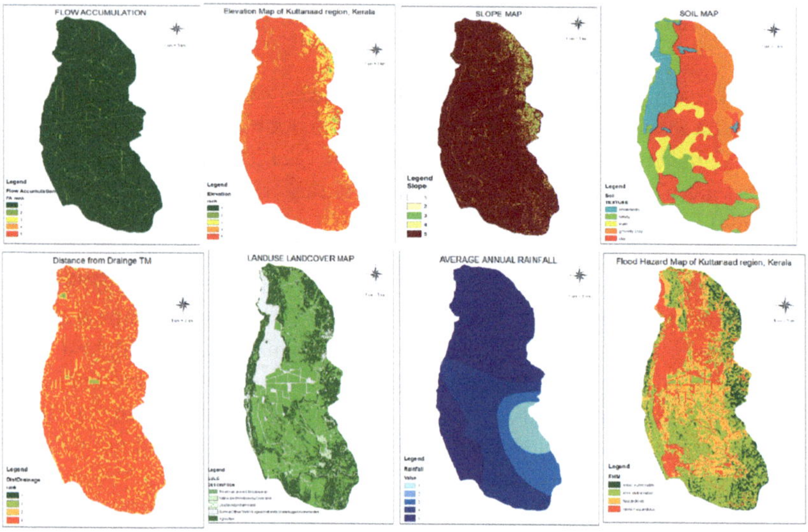

Fig. 3 Physical parameter maps with flood hazard map of Kuttanaad region

they are most vulnerable indicated by red colour, vulnerable indicated by orange colour, less vulnerable given by light green colour and the safe zone is indicated by dark green colour. This technique can be applied on other regions as well and parameter set chosen can differ according to the terrain of region selected. There have been some limitations in this work. The technique used for making this flood map has also been used by Kazakis et al. [6] and he has validated this technique over his region of study. These kinds of Flood Hazard Maps play a prominent role in decision-making and implementing mitigating measures during floods. As this region majorly lies below sea level and is also well drained with water bodies, it is very much venerable to floods (Fig. 3).

References

1. Patsialis T, Kazakis N, Kougias I (2015) Assessment of flood hazard areas at a regional scale using an index-based approach and analytical hierarchy process: application in Rhodope-Evros region, Greece. Sci Total Environ 538(2015):555–563
2. Tehrany M, Lee M-J, Pradhan B, Jebur M, Lee S (2014) Flood susceptibility mapping using integrated bivariate and multivariate statistical models. Environ Earth Sci 72:1–15. https://doi.org/10.1007/s12665-014-3289-3
3. Romero DR, Usman T, Armenakis C (2018) Flood risk mapping using GIS and multi-criteria analysis: a greater Toronto area case study. Geosciences 8:275. https://doi.org/10.3390/geosciences8080275
4. Elkhrachy I (2015) Flash flood hazard mapping using satellite images and GIS tools: a case study of Najran City, Kingdom of Saudi Arabia (KSA). Egypt J Remote Sens Space Sci. https://doi.org/10.1016/j.ejrs.2015.06.007

5. Gazi MY, Islam MA, Hossain S (2019) Flood-hazard mapping in a regional scale—way forward to the future hazard atlas in Bangladesh. Malays J Geosci 3:1–11. https://doi.org/10.26480/mjg.01.2019.01.11
6. Kuttanad Region. https://en.wikipedia.org/wiki/Kuttanad
7. Menon VA (2018) Kuttanad after the flood. Econ Polit Wkly 53
8. Cherqui F et al (2015) Assessing urban potential flooding risk and identifying effective risk-reduction measures. Sci Total Environ 514C. https://doi.org/10.1016/j.scitotenv.2015.02.027
9. Napolitano P, Fabbri A (1996) Single-parameter sensitivity analysis for aquifer vulnerability assessment using DRASTIC and SINTACS. IAHS-AISH Publication 235
10. Ouma Y, Tateishi R (2014) Urban flood vulnerability and risk mapping using integrated multi-parametric AHP and GIS: methodological overview and case study assessment. Water 6. https://doi.org/10.3390/w6061515
11. Tehrany M, Pradhan B, Jebur M (2014) Flood susceptibility mapping using a novel ensemble weights-of-evidence and support vector machine models in GIS. J Hydrol 512:332–343. https://doi.org/10.1016/j.jhydrol.2014.03.008
12. Understanding floods. https://www.chiefscientist.qld.gov.au/publications/understanding-floods

Chapter 68
Forest Fire Damage and Recovery Assessment of Bandipur Forest, India

Jayati Vijaywargiya and Rama Rao Nidamanuri

1 Introduction

Almost all ecosystems face disturbances due to natural or anthropogenic reasons on spatial or temporal scale. Few of the forest fires are natural but at many instances they have been used by man to clear the forest region for agricultural, industrial, or any other economic activity. Forest fires have the potential to destroy physical, biological, ecological, and environmental stability of the region. Natural forest fires sometimes help in maintaining the health and stability of ecosystem but misuse of fire to clear the vegetated region for anthropogenic activity has led to inculcation of protected biodiversity under the threat of destruction.

Forest fires not only cause damage to the fauna but also result in smoke, which reduces the activity of photosynthesis and can cause the health damage to the residents nearby. Topography, temperature, moisture, forest type, and growing season are some of the prevalent factors that decide the intensity and the spread of the fires. Rehabilitation or regrowth in the burnt forest depends on the intensity of burn, natural factors thereafter and local and government initiatives undertaken. It is essential to map the forest fire to understand the extent of burnt area, temporal regrowth, and the undesired damage.

Remote sensing has been prominent in the areas of change detection, damage assessment, phenological change assessment, and other related fields. Several methodologies and datasets have been proposed to unsheathe the extent of burnt area. Contributions have been made in past to study the extent of forest fires using multispectral remote sensing technologies. Various methodologies have been implemented and are described in previous literatures. Majorly, classification of burnt area

J. Vijaywargiya (✉) · R. R. Nidamanuri
Department of Earth and Space Sciences, Indian Institute of Space Science and Technology, Thiruvananthapuram, Kerala, India
e-mail: jayativijaywargiya@ieee.org

G. Mathur et al. (eds.), *Proceedings of 2nd International Conference on Artificial Intelligence: Advances and Applications*, Algorithms for Intelligent Systems,
https://doi.org/10.1007/978-981-16-6332-1_68

using visual interpretation and ground truth reference and use of indices such as Normalized Burn Ratio (NBR) and delta Normalized Burn Ratio (dNBR) were prevalent [1, 2].

A number of studies have used multispectral remotely sensed data from Sentinel-2 (S2) earth observation mission due to its relevance for terrestrial mapping applications [3, 4]. Traditional methods of the burnt area mapping had limitations in terms of computational resources, data availability, and analysis. Data used in previous studies are the pre-fire and post-fire single-day imagery for which cloud masking becomes a major limitation. In addition, forest fires is a continuous phenomenon consisting of small and widespread fires which needs to be traced on continuous temporal scale to extract the total affected area. To map the area affected by forest fire high-resolution and dense time series data is required. The limitations of data and its preprocessing challenges can be resolved using big geospatial datacube. Google Earth Engine (GEE) which is a cloud-based computing platform gives advantage by providing the users with huge catalog of geospatial datacube from various satellite sources. Thus, GEE can be used as an efficient tool to overcome the above challenges of data availability and data preprocessing.

Recently, in 2019 Bandipur National Park, which is one of the world biodiversity ecosystems, observed widespread forest fire. The spatial– temporal burnt and regrowth area assessment of this region is critical to determine the spatial extent and in due course damage and recovery response over Bandipur National Park region. This research study presents methodology to identify burnt and regrowth area using multitemporal series of S2 data on Google Earth Engine Platform.

2 Study Area

Bandipur National park is spread over an area of 937 square kilometers and was established under Project tiger as a Tiger Reserve in 1974. This National Park is located in Karnataka and encompasses some areas of Nagarhole National Park, Mudumalai National Park and Wayanad Wildlife Sanctuary. It is located in the extent of 75° 12′ 17″; E to 76° 51′ 32″; E and 11° 35′ 34″; N to 11° 57′ 02″; N and between the altitude ranges of 680 and 1454 m. It forms the region of merger of the Deccan Plateau and the Western Ghats. The National Park predominantly has dry and moist deciduous forest and shrub lands, which make it more vulnerable to spread and intensify the fire. This region mainly has tropical climate which consist of two major seasons dry season and wet season. The region remains relatively warm and experiences intense sunlight throughout the year.

This National Park is the habitat of the endangered and the vulnerable species of fauna, viz. Tiger, Indian elephant, Sloth beer, Sambar, and others. The major flora species found here are, Teak, Rosewood, Sandalwood and Bamboo. Due to its geographical location, Bandipur National Park is not much susceptible to natural forest fires. The fires aroused in February 2019 in this region was reported as an act of vandalism, as stated by C Jayaram, Principal Chief Conservator of Forests at the

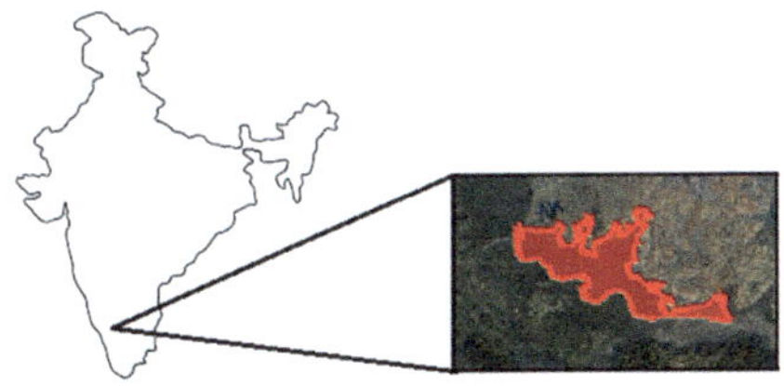

Fig. 1 Study area, Bandipur national forest

Forest Department of Karnataka. Accidental or intentional, these fires have caused damage to the vegetation cover in the region. Replenishment of the natural forest is a prolong process. Many factors like temperature, precipitation, soil, topographic location, etc., contribute to the regrowth of burnt forest regions. Bandipur National Park, being one of the world hotspot regions has a profound significance due to wide variety of indigenous and few endangered species present here (Fig. 1).

3 Methodology

Forest fires are characterized as rapidly spreading fires due to availability of prodigious woodland. These fires immensely influence of the region as they partially or completely demolish the vegetation cover of the region, which may or may not be replenished with time. Thus, having an impact on global ecosystem and also, contributing to the climate change. The regrowth of these fires can be studied at both spatial and temporal scales. The structural outline of the methodology for this work is shown in Fig. 2.

3.1 Selection of Temporal Range

This work presents the burnt and regrowth spatial temporal analysis of forest fire over Bandipur National Park. For this purpose, imagery over different time slots was extracted, viz. from 1–15 January 2019 to 1–15 March 2019 are used for burnt area analysis and 1–15 March 2019 to 15–30 April 2019 are taken into account for regrowth assessment.

3.2 Selection of Satellite Data

GEE (Google Earth Engine) combines petabytes of data from multiple sensors at global scale on a cloud-based platform. It facilitates us with cloud-based big geospatial data cube from various satellite sensors. The data provided by GEE is in Analy-

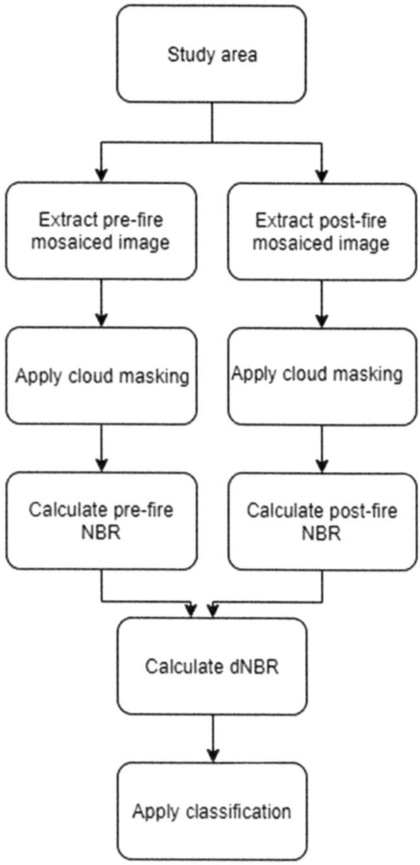

Fig. 2 Structural outline of methodology

sis Ready Data (ARD) format. Numerous datasets from Sentinel, Landsat, MODIS, climate data, temperature data, etc. are available in this platform. The burnt and regrowth area mapping requires high spatial and temporal resolution data. In this work, Sentinel-2 series multispectral dataset was used, these datasets were available in the Sentinel collections of Google Earth Engine datasets. It has temporal resolution of 5 days and the spatial resolution of 30 m. The sentinel-2 dataset has an ascendency due to its wide swath and high resolution, making it apt for vegetation, soil, and water monitoring. Thus, in this work spatial temporal analysis of burn and regrowth is done using Sentinel-2 imagery.

3.3 Data Preprocessing

In this work, the data is taken as an Image collection and then is filtered by time slots, i.e., date and the spatial boundaries. The boundary of the Bandipur National Park is taken as an asset, shape file in GEE. This mask the region of interest and all the analysis and computation is performed on the selected region. The cloud mask function is then applied on the images. GEE facilitates with the sensor-based and Pixel-based cloud-masking algorithm, making the clouds and their shadows efface from the image collection. In the mosaicking operation, the best possible pixel value in the date range selected is extracted from the composite of mosaicked image.

3.4 Calculation of the NBR and dNBR

The NBR (Normalized Burn Ratio) is calculated for both pre- and post-fire images.

$$\text{NBR} = (\text{NIR} - \text{SWIR2})/(\text{NIR} + \text{SWIR2})$$
$$\text{dNBR} = (\text{NBR prefire} - \text{NBR postfire}) * 1000$$

In Sentinel-2 the Band 8 and Band 12 represent NIR and SWIR2 regions, respectively. Then the difference of NBR between the pre- and post-happening is calculated and is scaled by 1000. This variable is termed as dNBR (delta Normalized Burn Ratio) (Table 1).

3.5 Classification

Here, thresholding based classification is implemented. Each pixel is assigned one of the eight classes, high severity burn, high-moderate severity burn, moderate-low severity burn, low severity burn, unburned, NA, low regrowth, and high regrowth, based on the value of dNBR. The thresholding limits used in this work for burnt area mapping is proposed by USGS. Post classification the statistical methods are used to extract the area under each class (Table 2).

Table 1 Sentinel-2 MSI bands used for the study

Band	Resolution	Wavelength	Description
B8	10 m	835.1 nm(S2A)/833 nm(S2B)	NIR
B12	20 m	2202.4 nm (S2A) / 2185.7 nm (S2B)	SWIR 2

Table 2 Burn severity classes and threshold as proposed by USGS

Class	Scaled dNBR
High severity burn	660 to 1300
High-moderate severity burn	440 to 659
Moderate-low severity burn	270 to 439
Low severity burn, unburned	100 to 269
Unburned	−100 to 99
Low regrowth	−250 to −101
High regrowth	−500 to −251

4 Results

The area under each category is calculated, burnt area due to forest fire of February 2019 and the regrowth of forest till April 2019 was analyzed. It was seen that approximately 87.52% of area under Bandipur National Park was affected by local and widespread forest fire in a span of 1 month. Overtime the severity of the burn abridged and recovery of forest by anthropogenic efforts and natural climatic conditions led 97.59% of the total area to come under the category of unburnt, low, or high regrowth. Table 3 shows the calculated area under each class of burn severity and regrowth and Fig. 3 burnt area classified map of Bandipur National Forest affected by forest fires between January and March 2019.

Table 3 Area under each class of over temporal scale

Class	Area (hectares)	
	Period I	Period II
	January 2019 to March 2019	March 2019 to April 2019
NA	0.09	0
High severity burn	644.76	0.45
High-moderate severity burn	2900.88	28.08
Moderate-low severity burn	30606.48	196.92
Low severity burn, unburned	47867.12	2032.47
Unburned	1499.12	16342.56
Low regrowth	180.54	22868.82
High regrowth	11.88	52241.57

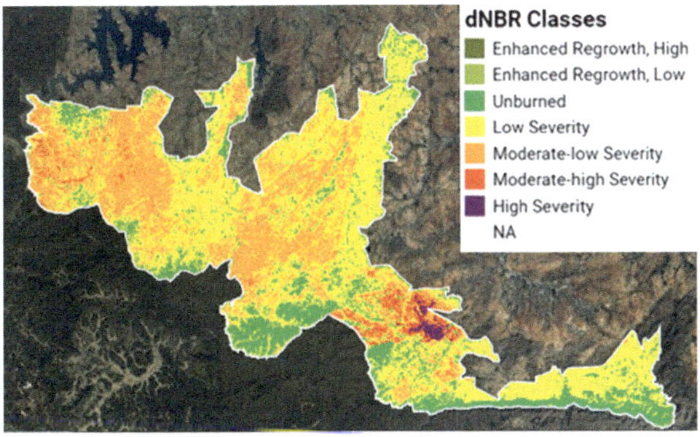

Fig. 3 Burnt area classified map of Bandipur National Forest

5 Discussion

In this work, data over the spatial region was accumulated over a period of 15 days, thus avoiding the limitations due to cloud cover. For spatial–temporal analysis over a large area, GEE has remarkable advantage in terms of datasets, which are available in Analysis Ready Data (ARD) form which can be analyzed and visualized on cloud platform, refraining the user from data download and mosaicking task. The approach of Big Geospatial Data cube allows seamless accumulation of datasets over temporal scale, making the signature of the burnt area to be traced on continuous scale. In this work, long periods for burnt area mapping and regrowth are considered. Thus, the damage caused by small and widespread fires over a period of time, more than a month was mapped. Also, the regrowth occurred over a long period of time is extracted.

6 Conclusion

Our results demonstrated that approximately 87.52% of area under Bandipur National Park was affected by forest fire. Overtime the severity of the burn and recovery of forest by anthropogenic efforts and natural climatic conditions led 97.59% of the total area to come under the category of unburnt, low, or high regrowth. The burnt area classified Map is shown in Fig. 3. In this work, the burnt and regrowth after the forest fire in Bandipur National, February 2019, was interpreted using remotely sensed image collection on a free and open-source platform, Google Earth Engine. The limitations of the present studies include natural growth of tress to be classified as one of the classes, no ground truth verification was conducted and the results are purely based on remotely sensed collection of Sentinel-2 imagery. This study comes up with a new approach which overcomes the limitations of previous approaches, easy to implement, and is extensible to map the burnt forest areas. These studies are particularly important when investigating the forest fire and regrowth on an approximating scale. It can be deduced from his work that the forest fire burned a large part of forest but the regrowth and the effect of burn depleted over time.

References

1. Reddy S (2008) Forest burnt area mapping using remote sensing and GIS. In: Proceedings of Andhra Pradesh Academy of Sciences, Hyderabad, 14–16 November 2008
2. Babu S, Roy A, Aggarwal R (2018) Mapping of forest fire burned severity using the sentinel datasets. In: ISPRS: international archives of the photogrammetry, remote sensing and spatial information sciences, XLII-5
3. Long T, Zhang Z, He G, Jiao W, Tang C, Wu B, Zhang X, Wang G, Yin R (2019) 30 m resolution global annual burned area mapping based on landsat images and Google earth engine. Remote Sens 11(5):489
4. Verhegghen A, Eva H, Ceccherini G, Frédéric A, Gond V, Gourlet-Fleury S, Cerutti P (2016) The potential of sentinel satellites for burnt area mapping and monitoring in the Congo Basin forests. Remote Sens 8(12):986
5. Nautiyal S, Rao NR (2010) Conserving biodiversity in protected area of biodiversity hotspot in India: a case study. Int J Ecol Environ Sci 36(2):195–200
6. Joseph S, Anitha K, Murthy MSR (2009) Forest fire in India: a review of the knowledge base. J Forest Res 14(3):127–134
7. Parks S, Holsinger L, Voss M, Loehman R, Robinson N (2018) Mean composite fire severity metrics computed with Google earth engine offer improved accuracy and expanded mapping potential. Remote Sens 10(6):879
8. Ganjam M, Reddy S (2015) Geospatial monitoring and prioritization of forest fire incidences in Andhra Pradesh, India. Environ Monitor Assess 187(10):616
9. Shimizu K, Ota T, Mizoue N (2019) Detecting forest changes using dense landsat 8 and sentinel-1 time series data in tropical seasonal forests. Remote Sens 11(16):1899

Chapter 69
Wearable Fall-Detection Using Deep Embedded Clustering Algorithm

R. Jothi

1 Introduction

Falls are one of the leading causes of injury in elderly persons. Falls may lead to severe health complications and sometimes even to death. Immediate medical attention is necessary to reduce negative consequences of falls. In this context, assistive devices such as fall detectors are suggested by healthcare professionals. Main objective of a fall-detection system is to alert when a fall event has occurred.

Currently, there are several fall-detection systems most of which are based on supervised machine learning approaches [2–4, 12]. These algorithms regard fall-detection as activity classification problem, whose objective is to classify a set of input features into one of the activities, namely sitting, running, falling, walking, etc. Source of these input features may be videos or wearable sensors. In this paper, we consider only wearable fall-detection.

Özdemir et al. [8] have performed a comprehensive analysis of fall-detection using different classification algorithms, namely, support vector machines, k-nearest neighbor (k-NN), least squares method (LSM), Bayesian classifier, dynamic time warping and artificial neural networks. As reported in [8], K-NN and LSM methods have shown higher fall recognition rate than other algorithms. Some researchers have focused on fall subtype classification along with fall-detection. Albert et al. [2] have demonstrated fall-detection as well as fall-type classification using five different machine learning algorithms: k-NN, Decision trees, support vector machines (SVM), sparse multinomial logistic regression (SMLR), and Naive Bayes. They have

R. Jothi (✉)
Vellore Institute of Technoloy, Chennai, Tamilnadu, India
e-mail: jothi.r@vit.ac.in

G. Mathur et al. (eds.), *Proceedings of 2nd International Conference on Artificial Intelligence: Advances and Applications*, Algorithms for Intelligent Systems,
https://doi.org/10.1007/978-981-16-6332-1_69

used mobile phone's accelerometers for generating simulated falls. Among the five algorithms, SVM has shown highest fall-recognition accuracy and SMLR has shown good performance in identifying the type of falls.

Artificial neural network (ANN) algorithms have been widely used for various classification problems including fall-detection systems [3]. Recently, deep learning algorithms have also been employed for fall-detection [9]. Santos et al. [9] presented a convolution neural network (CNN)-based deep learning model for predicting falls. Experimental analysis on three bench marking fall datasets has reported that CNN model has obtained 93.8% accuracy.

As can be seen from the discussion, most of existing fall-detection systems employ supervised learning algorithms. Performance of these algorithms rely on huge training data and generating such data is tedious process. This paper presents fall-detection using cluster analysis. Being an unsupervised learning approach, cluster analysis extracts hidden patterns from the given dataset by grouping the similar data instances into a set of clusters. Clustering is rich in literature and different clustering techniques such as K-means [6], spectral clustering [5] are used for various exploratory analysis.

Inspired from massive success of deep learning algorithms, we employ deep embedded clustering (DEC) for fall-detection. DEC is one of the unsupervised learning models from deep learning paradigm [13]. DEC simultaneously learns feature representations and cluster assignments using deep neural networks. Traditional clustering approaches focus on grouping the data instances using distance functions, whereas DEC maps the given data into a lower dimensional feature space and simultaneously optimizes a clustering objective in this lower dimensional space. Principal component analysis (PCA) is also widely used for lower dimensional data embedding, but it becomes inefficient for non-linear data mapping. Ability to perform non-linear embedding, DEC has gained lots of attention in image segmentation and other pattern recognition problems [7]. Deep convolutional autoencoder (CAE)-based fall-detection system has been proposed in [10], where they use autoencoder for classifying human activities from radar signals.

To the best of our knowledge, application of DEC for wearable fall-detection is limited. This motivated us to carry out fall-detection using DEC algorithm. For experimental purpose, we consider Fall dataset from Kaggle repository [1]. Fall-detection performance of DEC is compared against three different clustering algorithms namely K-means, hierarchical agglomerative, and spectral clustering. Results indicate that DEC achieves higher fall-detection accuracy as compared to other clustering algorithms considered in the study.

Rest of the paper is organized as follows. Section 2 briefly describes DEC algorithm. Experimental analysis is presented in Sect. 3. Conclusion and future directions are given in Sect. 4.

2 Methodology

2.1 Deep Embedded Clustering

DEC is one of the recent advancements in unsupervised learning and it has gained lots of attention into this field [7]. Objective of DEC is to map a set of data points from one domain X to another domain Z and simultaneously learn cluster assignment during mapping. The mapping parameter θ is learned using stacked encoder and decoder units, both of them are implemented using fully connected deep neural network. The encoder unit consists of an input layer, three dense layers and one output layer for embedding. The decoder unit has the configuration that is reverse to the encoder as shown in Fig. 1.

Given input features x_i, the encoder obtains latent features z_i, generally in lower dimensions; the decoder function reconstructs the data x_i from the latent features z_i. The encoder is trained for extracting the features by minimizing the reconstruction loss objective function L_{rec}. Let f_θ be the encoder function. Let g_θ be the decoder function. Then, reconstruction loss L_{rec} is computed as per the Eq. (1) [7].

$$L_{\text{rec}} = \| x_i - g_\theta(f_\theta(z_i)) \|^2 \tag{1}$$

Once the features are extracted, the decoder unit is dropped. Extracted features act as input for clustering process. Then, the network is adjusted so as to keep reconstruction loss L_{rec} minimum. In the meantime, the points in embedded space are clustered using K-means algorithm. Beginning with randomly chosen k cluster centroids, each point z_i is assigned to a nearest cluster centroid. This soft assignment is based on Student t-distribution kernel as defined in Eq. (2):

$$q_{ij} = \frac{(1+ \| z_i - \mu_j \|^2 / \alpha)^{-\frac{\alpha+1}{2}}}{\sum_l (1+ \| z_i - \mu_l \|^2 / \alpha)^{-\frac{\alpha+1}{2}}}, 1 \leq l \leq k, \tag{2}$$

where $1 \leq i \leq n$ and $1 \leq j \leq k$. Here, n denotes the number of input samples in the dataset and k is the number of clusters (activity classes). The degrees of freedom α

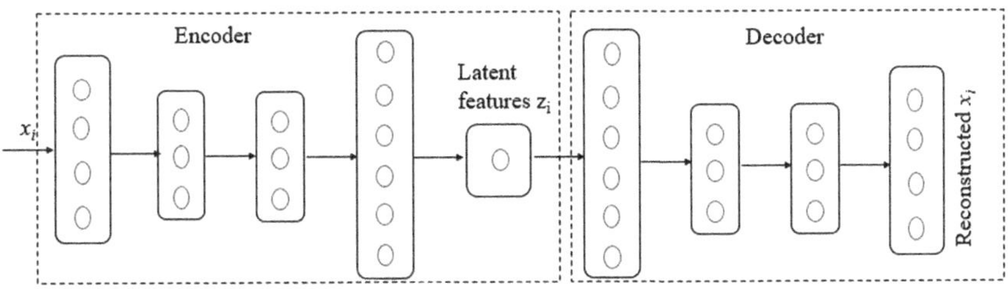

Fig. 1 Architecture of deep embedded clustering [13]

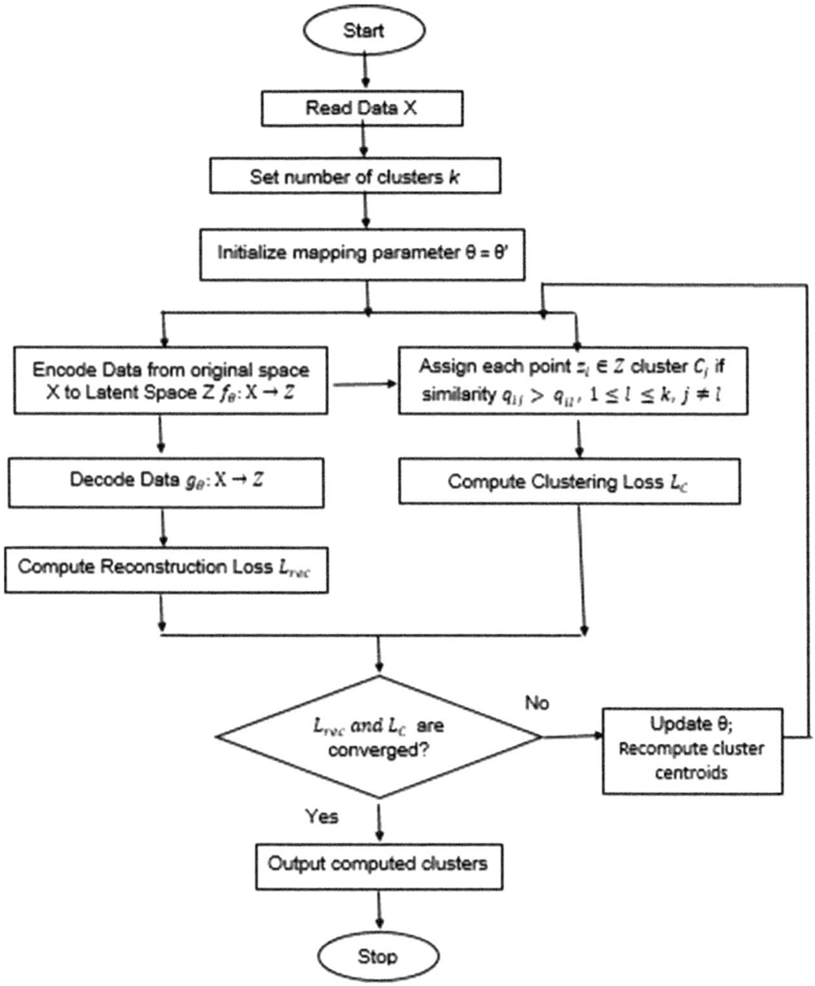

Fig. 2 Flowchart of deep embedded clustering algorithm

is set to 1. The q_{ij} value measures the similarity between an embedded point z_i to a cluster centroid μ_j, If a point has more similarity to a cluster, then it will be assigned to that particular cluster.

The clusters are iteratively fine-tuned to minimize the clustering loss function L_c defined using Kullback–Leibler (KL) divergence measure. The KL function actually measures the deviation between the distribution of soft cluster assignment and the auxiliary target distribution [13]. Based on L_{rec} and L_c, the mapping function θ as well as cluster centroids are updated. The process of encoding and clustering continues until convergence is met. The complete steps in DEC algorithm is illustrated in the flowchart shown in Fig. 2.

3 Experimental Analysis

3.1 Datatset

Dataset considered for the present study is taken from Kaggle repository [1]. The dataset represents six different physical activities performed by the old-age patients from Chinese hospital. Different activities considered are: Standing, Walking, Sitting, Falling, Cramps, and Running. Each of the 16,382 samples in the dataset represents features collected from wearable motion sensors. Various features are Sugar level (SL), EEG monitoring rate (EEG), Blood pressure (BP), Heart beat rate (HR), and Blood circulation. Figure 3 illustrates the distribution of samples in each of the activities. It is noted that number of samples in falling activity is much greater than that of cramps. So if a cramping activity in an elderly person is carefully observed, the falling situations could be avoided. The histogram of different features are shown in Fig. 4. As seen from the figure, most of the patients had sugar level within 0.25 units. EEG level is also same for most of the patients. However, it is noted that blood pressure, sugar level, and blood circulation levels seem to be varying across different participants. It is observed that the given feature space is non-linear and applying a DEC algorithm helps us to predict the probability of fall effectively.

3.2 Experimental Setup

As per the guidelines given in [13], the network dimensions of DEC are set as f–500–500–2000–3, where f is the number of features in the dataset. We set f = 5, considering the features SL, BP, HR, EEG, and circulation. All layers are fully connected. Rectified linear units (ReLUs) is used as activation function in all encoder–decoder pairs. Batch size is set to 128 and learning rate is set 1. DEC obtains clusters through iterative refinement procedure, and the maximum number of iterations is set to 1000.

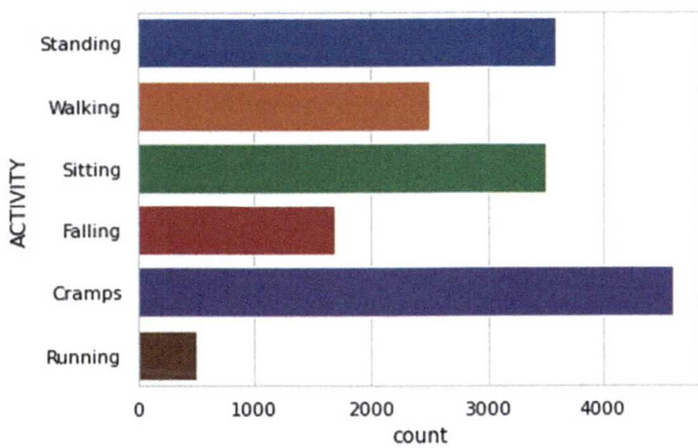

Fig. 3 Class distribution: the number of samples pertaining to different activity category

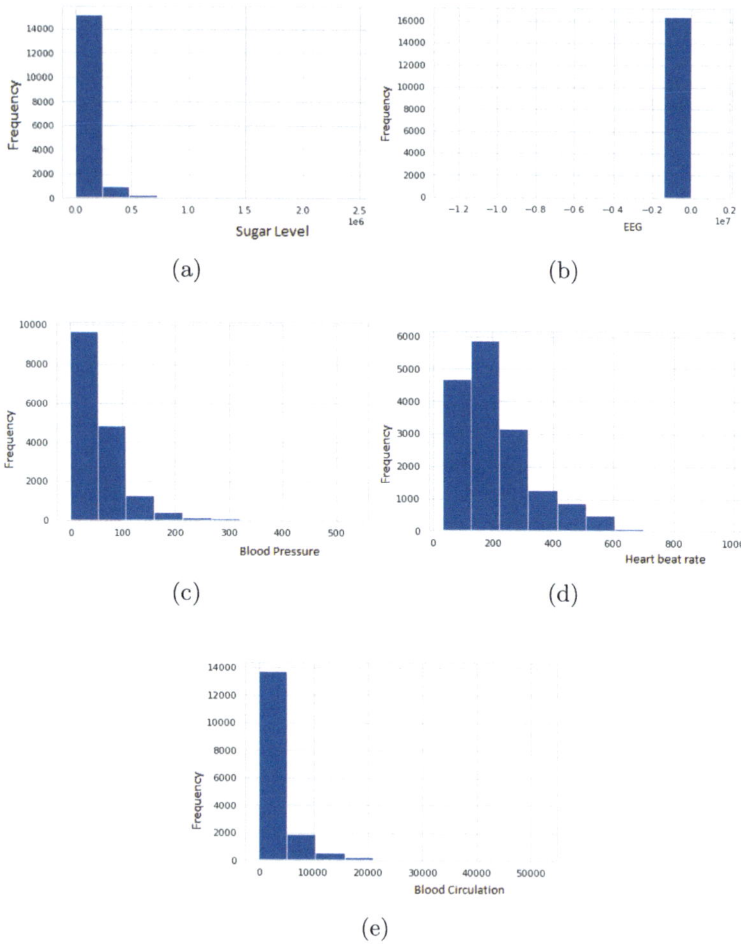

Fig. 4 Histogram of various features of the dataset: sugar level (SL), EEG, blood pressure (BP), heart rate (HR), and blood circulation

DEC algorithm is compared with three other clustering algorithms, namely, K-means, hierarchical agglomerative clustering (HAC), and spectral clustering. Euclidean distance is used as dissimilarity measure in all four algorithms. Number of clusters is set to the number of activities, i.e., 6. For HAC algorithm, average linkage metric is used as cluster-merging criteria. The core of spectral clustering is the similarity graph which may be fully connected graph, k-nearest neighbor (k-NN) graph, or ϵ-neighborhood graph [5]. We use k-NN graph as the similarity graph in the present study and the number of nearest neighbors k is chosen as $\log(n) + 1$ as suggested in [11].

Algorithms are compared with respect to both internal and external cluster quality indices. Silhouette index (SI) is used as internal cluster quality measure which denotes the degree of cohesion of data instances in a cluster while maintaining inter-cluster separation. SI values range from -1 to $+1$, where higher the SI value, more the intra-cluster cohesion and inter-cluster separation. Adjusted rand index (ARI) and Jaccard Index are used as external cluster quality measures, which are defined as follows. Let $P1$ be the ground truth partitions of the given dataset. Let $P2$ be the partitions obtained by an algorithm. Then, ARI represents the all-pairwise agreement of cluster membership with respect to $P1$ and $P2$. ARI also ranges from -1 to $+1$. Higher value of ARI indicates that the partitioning results of the algorithm mostly agrees with ground truth results. While ARI considers all-pairwise comparisons, Jaccard index disregards the pairs of data instances that disagree in $P1$ and $P2$.

3.3 Results and Discussion

Table 1 presents the fall-detection results of different algorithms with respect to Silhouette Index, Adjusted Rand Index and Jaccard Index. The DEC algorithm has attained highest fall recognition performance with respect to both internal as well as external cluster validation measures as shown in the table. We have also observed the network loss (reconstruction loss) and clustering loss after every epoch. As reported in Fig. 5, the reconstruction loss reaches minimum as the number of epochs increases. Similar pattern observed with clustering loss also. As the umber of epochs is also one of the crucial parameter in training any deep learning model, performance analysis of the DEC algorithm with respect to varying number of epochs is carried out. Figure 6 shows the cluster quality measures for different epochs. It is observed that SI and ARI measures are maximum at epochs = 100.

The algorithms are also evaluated using clustering accuracy measure. TK-means algorithm has obtained 51.75% accuracy. HAC algorithm has obtained 52.20%, which is closer to the performance of K-means. Spectral clustering 60.5% accuracy, which is quite better than K-means and HAC. DEC algorithm outperforms other algorithms with accuracy of 71.75%.

Table 1 Fall-detection results of different algorithms with respect to Silhouette index, adjusted rand index, and Jaccard index

Algorithm	Silhouette index	Adjusted rand index	Jaccard index
K-means	0.3047	0.4861	0.3944
HAC	0.2966	0.4711	0.2963
Spectral clustering	0.3105	0.6435	0.5881
DEC	**0.4584**	**0.7841**	**0.6602**

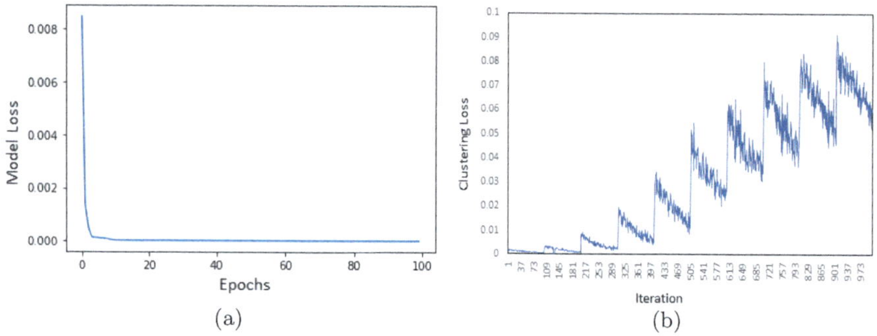

Fig. 5 Network loss and clustering loss: **a** Network loss with respect to number of epochs. **b** Clustering loss observed during iterations of one particular epoch

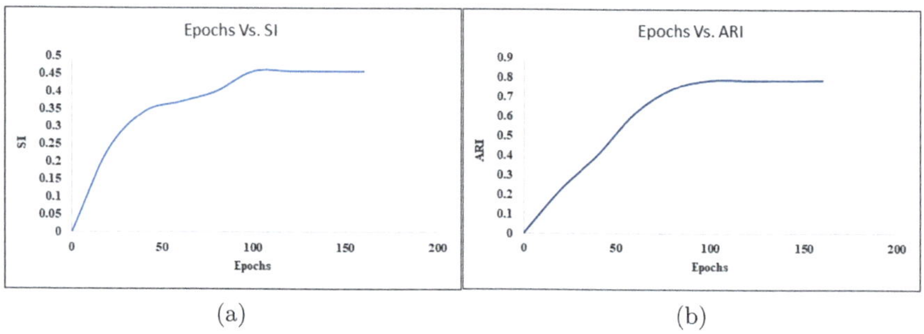

Fig. 6 Cluster quality measures for varying number of epochs

4 Conclusion

Fall-detection from wearable devices is gaining lots of attention in the recent years. Most of the existing work on fall-detection are based on supervised learning which demand tedious training data generation process. This paper presented a study on wearable fall-detection using deep embedded clustering (DEC), which is one of the recent unsupervised algorithms using deep learning model. Experimental analysis is carried out Kaggle fall dataset. Results have shown that DEC algorithm has performed better than traditional clustering algorithms such as K-means, HAC, and spectral clustering. As a future work, extensive analysis on various fall datasets will be carried using different DECs.

References

1. http://www.kaggle.com/alexanderguentert/falldetection/
2. Albert MV, Kording K, Herrmann M, Jayaraman A (2012) Fall classification by machine learning using mobile phones. PLoS ONE 7(5):e36556

3. Casilari-Pérez E, García-Lagos F (2019) A comprehensive study on the use of artificial neural networks in wearable fall detection systems. Expert Syst Appl 138:112811
4. Igual R, Medrano C, Plaza I (2013) Challenges, issues and trends in fall detection systems. Biomed Eng Online 12(1):66
5. Jothi R, Mohanty SK, Ojha A (2016) Functional grouping of similar genes using eigenanalysis on minimum spanning tree based neighborhood graph. Comput Biol Med 71:135–148
6. Jothi R, Mohanty SK, Ojha A (2016) On careful selection of initial centers for k-means algorithm. In: Proceedings of 3rd international conference on advanced computing, networking and informatics. Springer, pp 435–445
7. Min E, Guo X, Liu Q, Zhang G, Cui J, Long J (2018) A survey of clustering with deep learning: from the perspective of network architecture. IEEE Access 6:39501–39514
8. Özdemir AT, Barshan B (2014) Detecting falls with wearable sensors using machine learning techniques. Sensors 14(6):10691–10708
9. Santos GL, Endo PT, Monteiro KHdC, Rocha EdS, Silva I, Lynn T (2019) Accelerometer-based human fall detection using convolutional neural networks. Sensors 19(7):1644
10. Seyfioğlu MS, Özbayoğlu AM, Gürbüz SZ (2018) Deep convolutional autoencoder for radar-based classification of similar aided and unaided human activities. IEEE Trans Aerosp Electron Syst 54(4):1709–1723
11. Von Luxburg U (2007) A tutorial on spectral clustering. Stat Comput 17(4):395–416
12. Wang X, Ellul J, Azzopardi G (2020) Elderly fall detection systems: a literature survey. Front Robot AI 7:71
13. Xie J, Girshick R, Farhadi A (2016) Unsupervised deep embedding for clustering analysis. In: International conference on machine learning, pp 478–487

Chapter 70
Assamese Character Recognition Using Convolutional Neural Networks

Mihir Yadav, Divyansh Mangal, Srinivasan Natesan, Marcin Paprzycki, and Maria Ganzha

1 Introduction

Optical character recognition (OCR) is the oldest subdomain of pattern recognition. It refers to the conversion of machine-printed, or handwritten, symbols into a machine-encoded text.

A lot of work has been devoted to OCR in Western scripts, and languages like Chinese, Japanese, etc. (see, [11]). Over the past two decades, significant progress has been made also in Indian scripts, like Devanagari [2], Gurumukhi [7], Bengali [1], Tamil, Telugu, or Malayalam [9]. However, character recognition for Assamese, the major Northeastern language of India, spoken by over 14 million people, as their native language, is still lacking. Specifically, only a few attempts have been made towards handwritten Assamese Character recognition. Sarma et al. [10] used template matching and reached up to 80% accuracy. Bania et al. [3] applied a feedforward back propagation Neural Network (NN). The maximum reported accuracy was 90.34%. In the context of their work, the major limitation of NNs is their inability to capture spatial features in an image. This limitation can be overcome by using the CNN, which eliminated the need for manual feature extraction, as they learn features during training, directly from the images. However, to the best of our knowledge, CNNs haven't been applied on Assamese characters yet. Doing this became the goal of our work.

M. Yadav · D. Mangal · S. Natesan
Indian Institute of Technology, Guwahati, Assam, India
e-mail: natesan@iitg.ac.in

M. Paprzycki (✉)
Systems Research Institute Polish Academy of Sciences, Warsaw, Poland
e-mail: paprzyck@ibspan.waw.pl

M. Ganzha
Warsaw University of Technology, Warsaw, Poland

G. Mathur et al. (eds.), *Proceedings of 2nd International Conference on Artificial Intelligence: Advances and Applications*, Algorithms for Intelligent Systems,
https://doi.org/10.1007/978-981-16-6332-1_70

Since, there is no standard Assamese Character image dataset available on the Internet, we have collected handwritten character samples and generated a dataset containing more than 12 k images, which can also facilitate further research in this domain.[1]

The rest of the contribution is organized as follows. Section 2 describes dataset generation, preprocessing steps applied, architecture of the developed CNN models, and the relevant deep learning concepts. Section 3 summarizes the obtained experimental results. Section 4 outlines the future work directions.

2 Proposed Methodology

In this section, we outline how CNNs have been applied on the handwritten Assamese characters, to assess their potential in this context.

2.1 Datasets Used in Experiments

According to the best of knowledge, there is no "standard" image dataset of Assamese language handwritten characters, available on the Internet. Hence, we decided to generate our own. Specifically, we have created two datasets (DATASET_1 and DATASET_2), which, when combined, contain total of 12,863 images. The combined dataset is almost balanced, as each character appears in almost equal number of images.

DATASET_1 has been generated by collecting handwritten samples, each containing the 52 Assamese characters (11 vowels and 41 consonants), provided by a group of over 200 students, faculty, and staffs members of IIT Guwahati. This group comprised of persons representing several age groups, education levels, and both genders. Specifically, participants were given two sheets, divided into boxes, and were asked to write the characters, in the order in which they appear in the Assamese alphabet. Next, these sheets were scanned (see, Fig. 1) and individual characters' images cropped out, preserving the order in which they were written. As a result, a total of 10,994 images have been generated.

DATASET_2 was formed using the Tezpur University—Online Handwritten Assamese Character (TU-OHAC) dataset,[2] generated by Baruah and Hazarika [4]. This dataset, collected from over 45 writers, has characters represented as writer's pen movement, on a digitizing tablet. A stroke is defined as a collection of (x, y) coordinates recorded between a single PEN_DOWN and a PEN_UP. To generate the image, all strokes have been plotted. Prior to plotting, preprocessing is performed as described in Sect. 2.2. In addition to the 52 characters, the TU-OHAC dataset

[1] This dataset is available from: Srinivasan Natesan at natesan@iitg.ac.in.

[2] Available from the UCI Machine Learning repository [5].

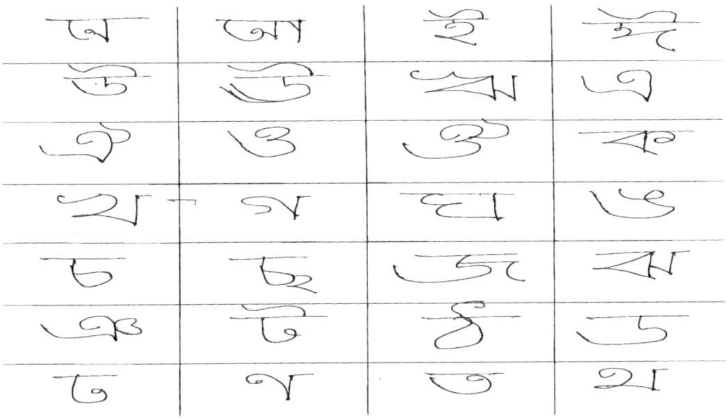

Fig. 1 Sample of a scanned page, from DATASET_1

contains extra conjunct consonants, which we ignored to unify both datasets. Hence, DATASET_2 contains 1869 images of 52 characters.

2.2 Dataset Preprocessing

Otsu Binarization. Binarization is a very important noise filtering technique applied during the preprocessing step in OCR-related problems. Here, all pixels, classified as background are colored white and the remaining ones are colored black. Since images in DATASET_2 are obtained by plotting graphs (not photographed or scanned) binarization is not needed there. Binarization methods can be divided into three categories: threshold-based, clustering-based, and hybrid [13]. Threshold methods are further divided into global threshold and locally adaptive. Otsu Binarization is a global threshold method. Here, instead of arbitrarily deciding the value of the threshold for the entire image, an optimal value of threshold is calculated. Before performing Otsu Binarization, image is converted to grayscale. Given the histogram $P(i)$ of the input image, this technique evaluates an optimal threshold t, to be used to segmented pixels into two clusters. Process iterates over possible threshold values and finds the one that minimizes the weighted sum of the intra-class variance of the two clusters $(\sigma_w^2(t))$, which is calculated as follows:

$$\sigma_w^2(t) = q_1(t)\sigma_1^2(t) + q_2(t)\sigma_2^2(t), \tag{1}$$

where

$$q_1(t) = \sum_{i=1}^{t} P(i), \qquad q_2(t) = \sum_{i=t+1}^{255} P(i),$$

$$\mu_1(t) = \sum_{i=1}^{t} \frac{iP(i)}{q_1(t)}, \qquad \mu_2(t) = \sum_{i=t+1}^{255} \frac{iP(i)}{q_2(t)}$$

$$\sigma_1^2(t) = \sum_{i=1}^{t} [i - \mu_1(t)]^2 \frac{P(i)}{q_1(t)}, \qquad \sigma_2^2(t) = \sum_{i=t+1}^{255} [i - \mu_2(t)]^2 \frac{P(i)}{q_2(t)}.$$

Here, μ_1 and μ_2 denote the means, while σ_1^2 and σ_2^2 are the variances within the clusters. To perform Otsu Binarization, we have used the OpenCV library function cv.threshold and passed cv.THRESH_OTSU as the flag.

Outlier Elimination. While writing/drawing the characters, writers mistakenly scribbled some random dots near the corners of the boxes. To remove them, we employed a simple technique, taking motivation from the statistical interquartile range method. We evaluate the median x and y coordinates and remove the farthest 2% of the points, measured from this median. Then, we crop out the desired area, which varies from the minimum to the maximum x and y coordinates, of the leftover 98% of the points.

Normalization. Writers were asked to write the characters inside a square box. Since different people write in different orientations, the relative positions of the characters were different. Furthermore, font-size of the characters written by different persons also varied significantly. Hence, the values of x and y were normalized, to ensure that the character's plot occupies the complete area, and is not skewed in a particular direction. The latter point will be addressed in the future research, to bring more flexibility to the trained models.

For DATASET_2, the (x, y) coordinates are immediately available. For images in DATASET_1, coordinates were extracted using thresholding. Next, the maximum and minimum values of x and y coordinates were calculated and used to re-scale all coordinate values to the interval $[0, M]$. The following expression describes this process:

$$x(i) = \left(\frac{x(i) - x_{\min}}{x_{\max} - x_{\min}}\right) * M, \quad y(i) = \left(\frac{y(i) - y_{\min}}{y_{\max} - y_{\min}}\right) * M, \quad \forall\ 1 \le i \le N. \quad (2)$$

Here, N, M are the number of points and size of final image, respectively.

Image smoothing. DATASET_1 is developed by scanning the characters written on paper. However, each image in DATASET_2 results from plotting a finite number of points (300 to 500). Here, the curve obtained by joining these points may fail to depict the actual shape, resulting in sharp edges and irregularities in the plotted curve. Therefore, we applied smoothing to images in DATASET_2. Specifically, a 3-point moving average filter method, was used to smooth individual strokes, for $1 < i < N$:

$$x(i) = \frac{x(i-1) + x(i) + x(i+1)}{3}, \quad y(i) = \frac{y(i-1) + y(i) + y(i+1)}{3} \tag{3}$$

where N is the number of points and we iterate over i in an increasing order.

2.3 Generating Final Image Dataset

Preprocessing is done directly on the images of DATASET_1 and on the (x, y) coordinates of the text files in DATASET_2. With the help of a Python script, we shall plot these coordinates (using Matplotlib) to obtain the character images for DATASET_2. The axes and labels are removed and the figure is resized to 96×96 pixels. The process is summarized in Fig. 2.

2.4 CNN Model Architecture

Here, architecture refers to the number of layers, size of the filters, and types of activation and pooling function used. In this work, we have experimented with four CNN models: LeNet-5, ResNet-50, InceptionV3, and DenseNet-201, and analyzed their performance on the, above described, dataset. The underlying learning rule was the same in all cases, i.e., backpropagation using gradient descent. The activation function was ReLu (Rectified Linear Unit) and the pooling strategy was MaxPooling. Let us now describe in more details each CNN architecture.

LeNet-5 Default LeNet 5 contains two sets of convolutional layers, followed by pooling layers [8]. We increased this count to four. Number of filters, used in each layer, was also increased. Finally, we added a dropout layer, and a fully connected layer, which uses Softmax as the activation function. It generates output between 0 and 1, the class with maximum value of the output is selected as the predicted class.

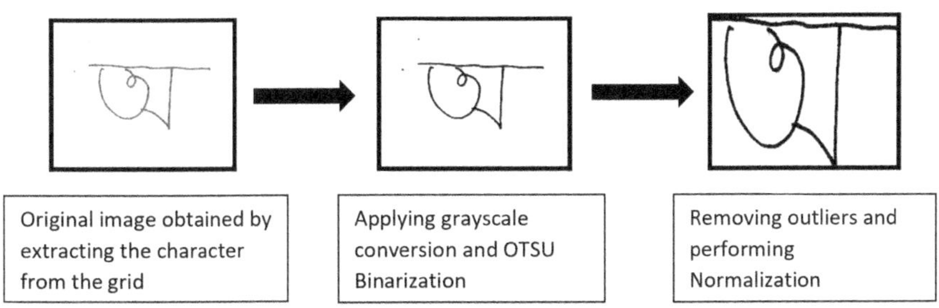

Fig. 2 Preprocessing applied step-by-step on the character label 'A' (DATASET_1)

ResNet-50 The main hindrance of training very deep neural networks was the vanishing gradient problem. This may be conceptualized as the rapid diminishing of the gradient of the loss function, during back propagation, due to successive multiplication. To overcome this issue, ResNet employs skip connection, which adds the output from the earlier layers to later layers, thus allowing a shortcut path for the gradient to follow. ResNet-50 comprises of 5 stages, each with a convolution and identity block. Convolution and identity blocks have three convolutional layers each.

InceptionV3 Original incarnation (InceptionV1) of this architecture was called GoogleNet. The central idea behind GoogleNet was the inception module, which performs convolution on an input with multiple sizes of filters, concatenating the resulting outputs, and sending them to the next layer. This network consists of such modules "stacked upon each other", with occasional max-pooling layers with stride 2, to halve the resolution of the grid. InceptionV3 architecture has several innovations over the initial GoogleNet, like convolution factorization, auxiliary classifiers, and grid size reduction (see [12] for further details).

DenseNet-201 This is a very recent architecture, introduced in 2018. It connects each layer to every other layer, in a feedforward fashion. For each layer, feature maps of all preceding layers are used as inputs. It concatenates them, instead of summing them up. They alleviate the vanishing gradient problem, and strengthen feature propagation. Through feature reuse, they are easy to train and highly parameter efficient. Their overall structure contains "dense blocks" connected by convolutional and pooling layers (see [6]).

The total number of trainable parameters of ResNet, DenseNet, and InceptionV3 models is of the order of a million. The complexity of their architecture makes it difficult to make changes to the network. Scaling up may cause large parts of computational gains to be lost. Therefore, we have decided to not to tweak their internal structure, as we did for Lenet-5 and use the default implementation that Keras provides. However, this is also one of interesting research directions that we plan to explore in the future.

2.5 *Transfer Learning*

Training a Deep CNN, like ResNet, with random initialization of model weights, is highly computationally expensive. It may even be practically infeasible, due to the limited size of our dataset. So, in such cases, it is suggested that the concept of transfer learning may be applied. Here, one uses a pre-developed model as a starting point, for another task, to improve the performance of the second task.

Typically, transfer learning is used by two approaches, develop model approach and pre-trained approach [2]. Pre-trained approach has two main methods, namely, fine-tuning and fixed feature extractor. In the latter, all the layers of the pre-trained model are frozen except the last fully connected layer. In this context, we have decided to use fine-tuning pre-trained approach, wherein instead of random initialization, weights are initialized with a network pre-trained on the ImageNet dataset, comprising of 1.2 million images belonging to over 1000 different categories. Thus, some fundamental features are already learned and weights are adjusted. We now feed our dataset to the entire model (no layer is freezed here) and start training it further. Note, however, that this is also one of points worthy further exploration.

3 Experimental Results and Analysis

The models were implemented using Python frameworks Tensorflow and Keras. They were trained on 70% of data, validated on 20%, and then tested on the remaining 10% of the dataset. The split was done in a balanced way, ensuring that the number of instances of each character image is proportionally the same in each part. The number of epochs was set to 100. The checkpoint callback has been used while training, which after each epoch, checks the validation accuracy of the model, and saves the weights corresponding to the model that gives the highest validation accuracy, into an external hdf5 file. This highest validation accuracy is displayed in Table 1. After the training is complete, the model is saved in the external hdf5 file, and is run on the test data and the test accuracy is evaluated by applying it to the selected 10% of the original dataset.

All models except LeNet used transfer learning for weight initialization. The models have been run on several combinations of hyperparameters, namely, learning rate and batch size and, the best-performing combination was selected for each model. Observed in all cases stability of results, with the increasing number of epochs shows that models have reached saturation and can be reasonably expected that they are at their "peak performance".

Table 1 summarizes the results obtained by various architectures. As noted, these results are the best results obtained when running experiments. Figure 3 shows the obtained validation accuracy curves. Overall, DenseNet shows the best performance in terms of accuracy. This was expected, as DenseNet is the most recent architecture that incorporates all the beneficial aspects of ResNet and GoogleNet, while introducing some additional useful concepts, like feature reuse. Based on analysis of literature, we can state that the 94.62% accuracy can be seen as state of the art in Assamese character recognition.

Table 1 Validation and test accuracies

Architecture	Learning rate	Batch size	Val acc (%)	Test acc (%)
LeNet 5	0.0008	48	88.82	86.25
ResNet 50	0.0008	128	94.88	93.55
InceptionV3	0.001	48	94.73	94.09
DenseNet 201	0.001	48	95.37	**94.62**

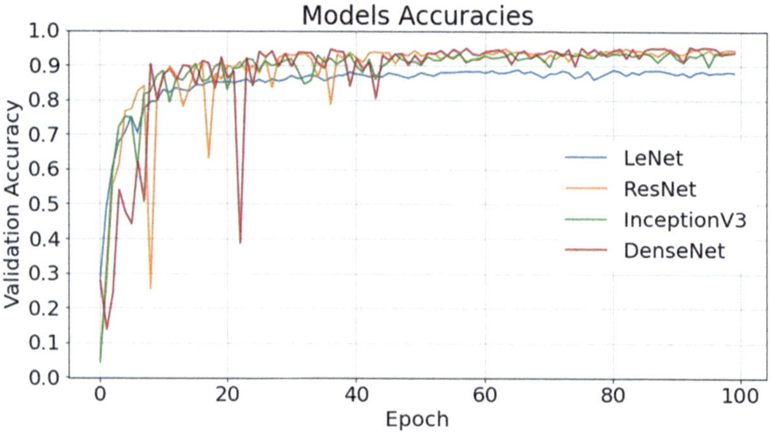

Fig. 3 Validation accuracy plot

4 Concluding Remarks

The aim of this contribution was to present preliminary results of our research into recognition of handwritten Assamese characters. Specifically, based on literature survey, we have applied four different CNNs to the problem. Before doing so, we had to create a dataset to train our models, as no Assamese character dataset could be found online. The best obtained result, ~94% accuracy, can be seen as the benchmark result for the future improvements. We have also indicated research areas that we plan to pursue in the future.

References

1. Al Rabbani Alif M, Ahmed S, Hasan MA (2017) Isolated Bangla handwritten character recognition with convolutional neural network. In: International conference on computer and information technology, Dhaka, Bangladesh, pp 1–6. https://doi.org/10.1109/ICCITECHN.2017.8281823
2. Aneja N, Aneja S (2019) Transfer learning using CNN for handwritten Devanagari character recognition. In: International conference on advances in information technology, Chikmagalur, India, pp 293–296
3. Bania RK, Khan R (2018) Handwritten Assamese character recognition using texture and diagonal orientation features with artificial neural network. Int J Appl Eng Res 13. https://doi.org/10.1007/978-981-15-0339-9
4. Baruah U, Hazarika SM (2015) A dataset of online handwritten Assamese characters. J Inf Process Syst 11:325–341. https://doi.org/10.3745/JIPS.02.0008.
5. Dua D, Graff C (2019) UCI machine learning repository Irvine. University of California, School of Information and Computer Science, CA
6. Huang G, Liu Z, Van Der Maaten L, Weinberger KQ (2017) Densely connected convolutional networks. CVPR. Honolulu, HI, USA, pp 2261–2269

7. Jindal U, Gupta S, Jain V, Paprzycki M (2020) Offline handwritten Gurumukhi character recognition system using deep learning. Advances in intelligent systems and computing, vol 1064
8. Lecun Y, Bottou L, Bengio Y, Haffner P (1998) Gradient-based learning applied to document recognition. Proc IEEE 86(11):2278–2324. https://doi.org/10.1109/5.726791
9. Nair PP, James A, Saravanan C (2017) Malayalam handwritten character recognition using convolutional neural network. In: International conference on inventive communication and computational technologies, Coimbatore
10. Sarma P, Chourasia CK, Barman M (2019) Handwritten Assamese character recognition. In: IEEE conference for intelligent technologies, Bombay. India. https://doi.org/10.1109/ICAIT47043.2019.8987286
11. Singh H, Sharma RK, Singh VP (2021) Online handwriting recognition systems for Indic and non-Indic scripts: a review. Artif Intell Rev 54:1525–1579
12. Szegedy C, Vanhoucke V, Ioffe S, Shlens J, Wojna Z (2016) Rethinking the inception architecture for computer vision. In: Conference on computer vision and pattern recognition (CVPR)
13. Wen J, Li S, Sun J (2013) A new binarization method for non-uniform illuminated document images. Pattern Recogn 46(6):1670–1690. https://doi.org/10.1016/j.patcog.2012.11.027

Chapter 71
Modelling and Optimizing City Traffic Using an Agent Platform

Robert Pniewski, Katsiaryna Stankevich, Maria Ganzha, and Marcin Paprzycki

1 Introduction

The concept of a smart city aims, among others, at the development of ubiquitous transport ecosystems, in which participants move to their destinations with minimal disruptions. Here, minimal disruption may mean that the public transport schedules are appropriately synchronized to minimize transport time between most popular destinations. It can also mean that in case of disruptions, the system can provide alternative ways of reaching the destination in such a way that the total "time loss" will be minimized. In this context, Raphael Gindrat points out (see, [1]) that more than 50% of the world's population lives in metropolitan areas, which directly contributes to the increasing need for intelligent traffic management.

In our previous work (see, [2] for all details), an agent-based city traffic simulator has been introduced. The developed system involved two main capabilities: (1) intelligent traffic light management, based on vehicle-light communication and (2) smart bus transport enhancements, based on traveller-to-bus-stop communication. Initial experiments indicated that the proposed approach may be able to reduce the travel time, encouraging further exploratory research.

The aim of this contribution is to discuss how the initial simulator has been extended, to manage commuting obstacles, such as road accidents, public transportation failures, or unexpected roadblocks. Specifically, the following scenarios have been considered: (1) Let us assume that cars could communicate with an "obstacle

R. Pniewski · K. Stankevich · M. Ganzha
Warsaw University of Technology, Warsaw, Poland
e-mail: r.pniewski@student.mini.pw.edu.pl

M. Ganzha
e-mail: m.ganzha@mini.pw.edu.pl

M. Paprzycki (✉)
Systems Research Institute Polish Academy of Sciences, Warsaw, Poland
e-mail: marcin.paprzycki@ibspan.waw.pl

G. Mathur et al. (eds.), *Proceedings of 2nd International Conference on Artificial Intelligence: Advances and Applications*, Algorithms for Intelligent Systems,
https://doi.org/10.1007/978-981-16-6332-1_71

management entity" (OME). The OME may inform the vehicles about road accidents, and route obstructions, so that they can react accordingly. This, in turn, may prevent unnecessary increase of travel time. (2) Assume that travellers and buses can communicate with a "public transport management entity" (PTME). In the case of a bus failure, its passengers may be provided (by the PTME) with information about best solutions to reach their travel destination. Again, both message-based communication, and strategic decision-making, are the core concepts of the proposed agent-based system.

Before proceeding further, let us note that, due to the space restrictions and the fact that no further pertinent literature has been found, we have decided to direct interested readers to the state-of-the-art summary presented in [2].

The remaining parts of this paper are organized as follows. In Sect. 2, we discuss the design of the extended version of traffic simulator. Next, in Sect. 3, the way that the simulator was implemented is briefly outlined. Finally, in Sect. 4, description of experiments has been presented.

2 Agent-Based Simulator Design

The main assumptions of the traffic simulator, outlined in [2], remained unchanged. However, roles of selected agents have been adjusted, and few new agents have been introduced. Note that, in the system, all modelled entities are represented as agents. One should keep this in mind to understand that, for instance, *Bus* agent is an agent that represents a bus moving on the streets of the city.

- *SmartCity*: single agent responsible for overseeing transport movement. Its role, was considerably reduced, in comparison with the original simulator (it remains in the new system mainly for technical reasons—as an entity that simplifies the implementation of the simulator).
- *TroubleManager*: single agent that, during the simulation, monitors traffic jams, construction sites, and accidents. Using information from *Car* agents, *TroubleManager* marks "trouble points" and broadcasts them to travellers.
- *LightManager*: agents managing traffic lights (one per crossroad). The role of the original *LightManager* was extended by the capability of detecting traffic jams. Information about such situations is forwarded to the *TroubleManager*.
- *StationManager*: bus stop agent, responsible for communication between buses and travellers, to allow making the departure times more flexible.
- *Bike*: configurable number of agents that facilitate change of mode of communication. They represent persons travelling by bikes and inform the *SmartCity* when the destination is reached. they communicate with *LightManager*s, providing specific crossroads with ETA of reaching them.
- *Car*: configurable number of agents. Representing situation when drivers notice an obstacle, during the simulation *Car*s "generate" trouble spots, (construc-

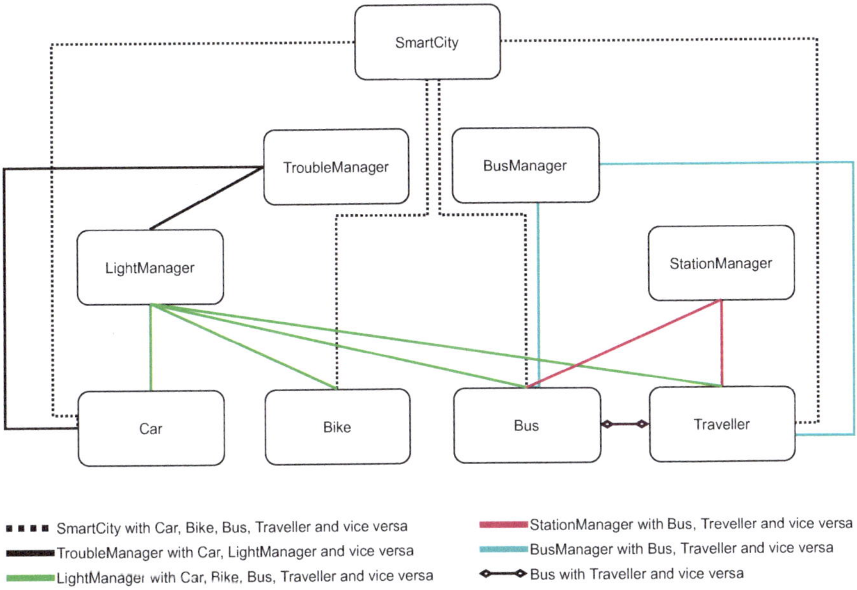

Fig. 1 Agent knowledge and information exchange map

tion/accident) and inform *TroubleManager* about them (type and location). *Car*s adjust routes, based on information about obstacles.

- *Traveller*: configurable number of agents, representing travellers. The new feature is dealing with bus crashes. Here, the *Traveller* can choose from two options: wait for the next bus, or travel by a bike. The decision is based on the length of the remaining journey and bike availability.
- *Bus*: number of *Bus* agents depends on the number of bus lines passing through the chosen simulation area (simulations used information about actual bus lines in Warsaw). After a bus crash, all travellers leave the bus.
- *BusManager*: delivers information about bus departure times, (e.g. by responding to *Traveller*s requests). Moreover, it is informed (by *Bus* agents) about bus crashes. This info is used to update the bus schedule.

Overall, the simulator was extended by the following features: trouble places (locations which cause traffic disruption, e.g. accidents or construction sites), traffic jam management, and bus crashes. Therefore, to facilitate needed functions, new communication between concerned entities was introduced. The agent knowledge map is shown in Fig. 1. It overlaps with the information exchange map, and represents the interactions described above. Connected agents send messages both ways.

Let us now describe the key communication patterns (represented in Fig. 1) in more detail. As noted, *Car*s are responsible for detection of construction/accident locations. When an obstacle is spotted, *Car* agent sends an INFORM message to the *TroubleManager*, with information about the position and type of obstruction. The

TroubleManager checks if the problem is already registered. If not, then *TroubleManager* saves the received data (updates the internal representation of the traffic situation in the city). The traffic-update-broadcast, in the form of a PROPOSE message, is sent to all road users, allowing change of the route, to evade the construction/accident location.

When a traffic jam occurs, the *LightManager* gathers the necessary information: position, length of the jam (measured by the number of cars), and sends an INFORM message to the *TroubleManager*. The *TroubleManager* processes this information and broadcasts a PROPOSE message to all *Car*s. Hence, *Car*s can make decisions about their route. Next, when the traffic jam disappears, the *LightManager* sends an INFORM message to the *TroubleManager*, reporting that the traffic jam is over. *TroubleManager* broadcasts a PROPOSE message to all *Car*s, allowing them to, possibly, adjust travel routes, again.

In case of a crash, the *Bus* sends an INFORM messages to all concerned *Traveller*s, *Station*s, and the *BusManager*. This message includes all key data: crash coordinates, bus line number, and accident time. When the *BusManager* receives this information, it updates the schedule. *Traveller*s react according to their strategy: switch to the bicycle, or wait for the next bus. When *StationManager*s get the message, they send INFORM messages to the upcoming *Traveller*s, and those already waiting at bus-stops, so that they may adjust travel plans. Since the *BusManager* agent was created, communication pattern, between *Traveller*s and *StationManager*, has been modified accordingly.

As far as communication *Traveller* $\leftrightarrow$ *StationManager, BusManager* is concerned, the most suitable bus line is obtained by *Traveller*s using the REQUEST message sent to *BusManager*, specifying the destination. Based on the content of the response, *Traveller* sends an INFORM message to the pertinent *StationManager*, containing the ETA, and the desired bus line. When the *Traveller* arrives at the bus stop, REQUEST-WHEN message is sent to the *StationManager*, with the request to enter the bus straightaway. When the *Bus* arrives, the *StationManager* sends a REQUEST message, asking *Traveller* to enter the bus. En route, at each station, *Bus* informs concerned *Traveller*s about their arrival.

The *Bike* agent inherited the communication with the *LightManager* from the *Car*. The remaining aspects of communication remained untouched.

3 Simulator Implementation

The new release of the simulator has been implemented. Here, the backend did not change (other than updates and bug fixes). It is implemented as a Maven project, chosen for convenient dependency management. The code is written in Java. Moreover, JADE framework is used in order to implement agents and agent communication. However, the frontend is now implemented as a web page. It is written in JavaScript, and is using React framework. This dramatically increased the responsiveness of the system, allowing for larger simulations. Communication between frontend and backend is handled via WebSockets.

Moreover, a closer look had to be taken at the simulator design. The system had been heavily dependent on the quality of the internet connection, due to frequent calls to external APIs [2]. The current simulator, is no longer so sensitive to the connection issues, due to the introduction of a new, complex caching functionality. Specifically, the data retrieved from the external APIs is cached for future reuse. Hence, the simulator can generate cars and other objects significantly faster, allowing captured traffic to be denser. It can be claimed that, deployment of a system based on the principles outlined above, could be possible in a real city, due to the increased tool reliability.

4 Experimental Validation

Series of simulations have been performed, with an increasing number of cars 5, 10, …, 40, while maintaining the same size of the simulation area. For instance—5 cars means, that the journey time has been measured for the last (5th) car, which joined the simulation. The following questions have been posed: (1) Does the proposed traffic jam response improve cars' travel times? (2) Does the agent communication help in case of obstacles?, and (3) Does the implemented communication improve travellers' commuting experience in case of bus failures?

4.1 Traffic Jam Strategy

To answer the first question, the average time of car to travel from point A to point B, with and without smart traffic jam management was measured. (1) Without the strategy, cars remain on the jammed route. When the traffic jam management is applied, affected cars may be able to change their route accordingly. The main factor of the strategy is the anticipated journey time on the jammed route vs. the predicted trip duration through the route, which evades the jam. Based on this, the traffic jam management system decides whether the car shall choose the alternative route. The comparison is depicted in Fig. 2.

It is obvious that the traffic jam strategy vastly improves the cars' journey times. In the largest experiment, an improvement of $\approx$**63%** was observed.

4.2 Traffic Obstacle Management Strategy

To answer the second question, the average trip time has been measured for cars travelling from point A to point B, with and without obstacle management. In "normal

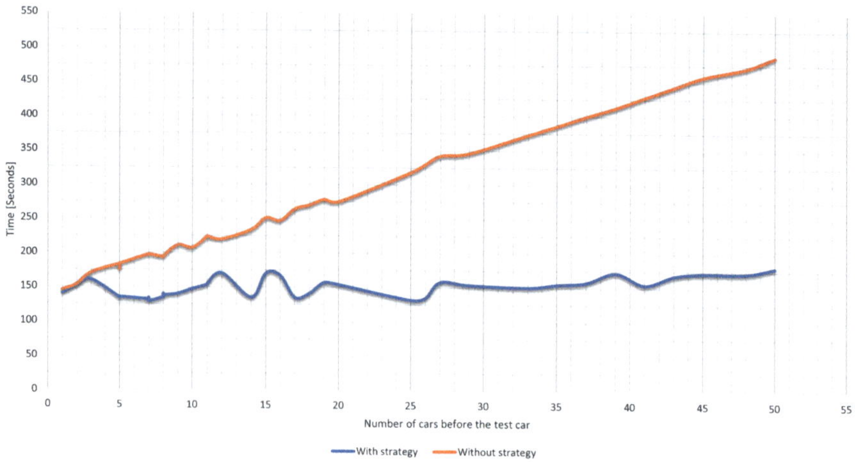

Fig. 2 Traffic jam strategy results

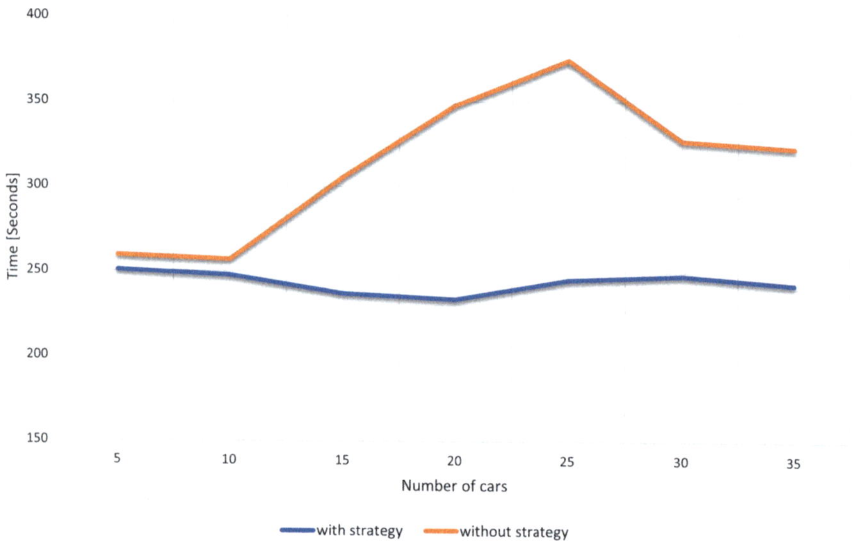

Fig. 3 Trouble point strategy results

conditions", if the cars' routes involve an obstacle, cars do not change routes, until they stumble upon the trouble point (generated over a fixed period). On the other hand, if the traffic obstacle management is active, affected cars adjust their routes as soon as they are informed about it by the system (even if it is located "far away"). The results are presented in Fig. 3.

It is clearly visible that the trouble point strategy helps in case of road construction sites and unexpected accidents. Again in the largest experiment, an enhancement of

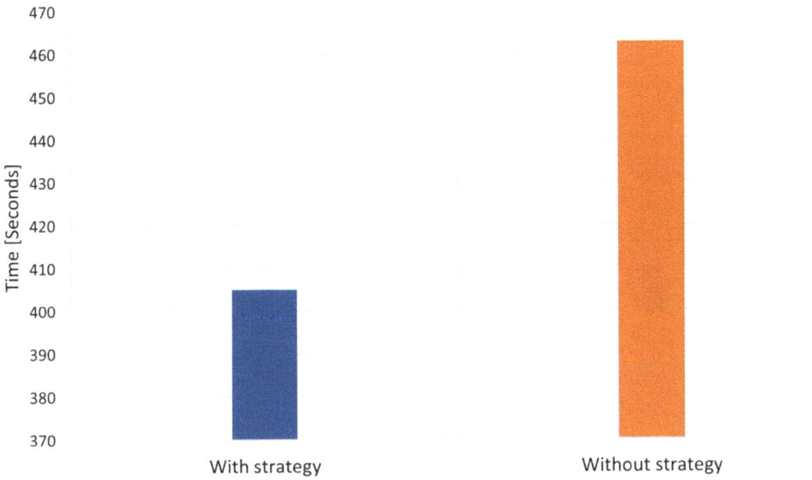

Fig. 4 Transport change strategy results

≈**34**% could be observed. Moreover, it was established that the average gain was ≈**21**%.

4.3 Transport Mode Switching Strategy

For the last question, the average trip time was measured for those travelling, with and without bus failure management. The test scenario involved a bus, which crashed, letting the passengers disembark at the failure point. Normally, the traveller, upon leaving the damaged bus, goes to the nearest station and awaits the next bus. When the bus failure management was applied, travellers, upon bus failure, could choose to continue their journey either by the next bus or by a nearby city bike. The decision was made on the basis of bike availability and resulting travel time (to establish if switching to a bike makes sense or not). The results are shown in Fig. 4.

As can be seen, the transport mode change strategy has improved the average travel time by up to ≈**12.5**%.

Apart from the described, newly-introduced traffic management strategies, the system has also been improved in terms of its core functionalities, namely, in the field of smart traffic lights. The average journey time has been measured for cars travelling from point A to point B, with and without smart traffic light management. Results of experiments for the old simulator design can be found in [2], while the results for experiments performed with the current, improved simulation tool, have been depicted in Fig. 5.

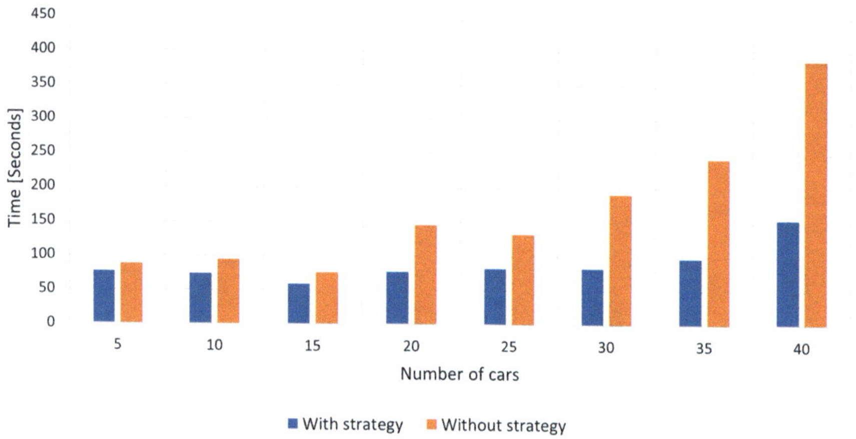

Fig. 5 Traffic light strategy results achieved with current simulator iteration

Overall, cars' average journey times' improvement has risen up to ≈**48%** in the largest case, **as opposed to** the ≈**6%** enhancement achievable in the previous version of the simulator.

5 Concluding Remarks

This work concerned modelling and optimization of traffic in a smart city. Strategies, responding to traffic congestion, accidents and public transport failures have been proposed and experimented with. It was observed that the proposed optimization strategies result in travel time improvements reaching 60% for the largest simulations.

References

1. Gildrat R (2019) In a world of autonomous vehicles, this is why we'll need more public transport than ever. In: BigThink, July 2019. https://bigthink.com/technology-innovation/autonomous-vehicles
2. Pniewski R, Sellin D, Stankevich K, Ganzha M, Paprzycki M (2021) Applying software agents to make city traffic management smarter. In: Goyal D, Gupta AK, Piuri V, Ganzha M, Paprzycki M (eds) Proceedings of the second international conference on information management and machine intelligence. Lecture notes in networks and systems, vol 166. Springer, Singapore. https://doi.org/10.1007/978-981-15-9689-6_72

Author Index

G. Mathur et al. (eds.), *Proceedings of 2nd International Conference on Artificial Intelligence: Advances and Applications*, Algorithms for Intelligent Systems,
https://doi.org/10.1007/978-981-16-6332-1

Printed by Printforce, the Netherlands